Theory and Applications of Computability

In cooperation with the association Computability in Europe

More information about this series at http://www.springer.com/series/8819

Books published in this series will be of interest to the research community and graduate students, with a unique focus on issues of computability. The perspective of the series is multidisciplinary, recapturing the spirit of Turing by linking theoretical and real-world concerns from computer science, mathematics, biology, physics, and the philosophy of science.

The series includes research monographs, advanced and graduate texts, and books that offer an original and informative view of computability and computational paradigms.

John Longley • Dag Normann

Higher-Order Computability

Springer

John Longley
School of Informatics
The University of Edinburgh
Edinburgh, UK

Dag Normann
Department of Mathematics
The University of Oslo
Oslo, Norway

ISSN 2190-619X ISSN 2190-6203 (electronic)
Theory and Applications of Computability
ISBN 978-3-662-51711-6 ISBN 978-3-662-47992-6 (eBook)
DOI 10.1007/978-3-662-47992-6

Springer Heidelberg New York Dordrecht London

Softcover re-print of the Hardcover 1st edition 2015

Printed on acid-free paper

Springer-Verlag GmbH Berlin Heidelberg is part of Springer Science+Business Media (www.springer.com)

To Caroline and Svanhild

Preface

This book serves as an introduction to an area of computability theory that originated in the 1950s, and since then has fanned out in many different directions under the influence of workers from both mathematical logic and theoretical computer science. Whereas the work of Church and Turing from the 1930s provided a definitive concept of computability for natural numbers and similar discrete data, our present inquiry begins by asking what 'computability' might mean for data of more complex kinds. In particular, can one develop a good theory of computability in settings where 'computable operations' may themselves be passed as inputs to other computable operations? What does it mean to 'compute with' (for example) a function whose arguments are themselves functions?

Concepts of computability in higher-order settings are of interest for a variety of reasons within both logic and computer science. For example, from a metamathematical point of view, questions of computability arise naturally in the attempt to elucidate and study notions of *constructivity* in mathematics. In order to give a constructive interpretation for some mathematical theory involving functions, real numbers, operators on spaces of real-valued functions, or whatever, one needs, first of all, a clear idea of what is meant by a 'constructive presentation' of these mathematical objects, and secondly, an understanding of what ways of manipulating these presentations are accepted as legitimate constructions—in other words, a notion of *computability* for the objects in question. From a computer science point of view, concepts of higher-order computability bear naturally on the question of what can and cannot be computed (in principle) within various kinds of programming languages—particularly languages that manipulate *higher-order* data which cannot be reduced to natural numbers in the requisite sense. There are also applications in which the logic and computer science strands intertwine closely, for instance in the extraction of computer programs from mathematical proofs.

A particular emphasis of the present book will be the way in which both logic and computer science have had a deep and formative influence on the theory of higher-order computability. This dual heritage is to some extent reflected in the collaboration that has given rise to this book, with the second author representing the older mathematical logic tradition, and the first author the more modern computer science one. As we shall see, the theory in its present form consists of ideas developed sometimes jointly, though often independently, by a variety of research communities each with its own distinctive motivations—however, the thesis of the book is that these can all be seen, with hindsight, as contributions to a single coherent subject.

In contrast to the situation for the natural numbers, where the *Church–Turing thesis* encapsulates the idea that there is just a single reasonable notion of effective computability, it turns out that at higher types there are a variety of possible notions competing for our attention. Indeed, in the early decades of the subject, a profusion of different concepts of computability were explored, each broadly giving rise to its own strand of research: Kleene's 'S1–S9' computability in a set-theoretic setting, computability for the total continuous functionals of Kleene and Kreisel, the partial computable functionals of Scott and Ershov, the sequentially computable functionals as represented by Plotkin's programming language PCF, the sequential algorithms of Berry and Curien, and others besides. All of this has given rise to a large and bewildering literature, with which both authors have previously attempted to grapple in their respective survey papers [178, 215]. More recently, however, many of these strands have started to re-converge, owing largely to increased interaction among the research communities involved, and a unifying perspective has emerged which allows all of them to be fitted into a coherent and conceptually satisfying picture, revealing not only the range of possible computability notions but the relationships between them. The purpose of this book is to present this integrated view of the subject in a systematic and self-contained way, accessible (for instance) to a beginning graduate student with a reasonable grounding in logic and the theory of computation.

Our emphasis will be on the 'pure theory' of computability in higher-order settings: roughly speaking, for each notion of computability that we consider, we shall develop a body of definitions, theorems and examples broadly analogous to that encountered in a typical introductory course on classical (first-order) computability. Rather little space will be devoted to applications of these ideas in other areas of logic or computer science, although we shall sometimes allude to such applications as motivations for our study.

The book is divided into two parts. In Part I (Chapters 1–4) we outline our vision of the subject as a whole, introducing the main ideas and reviewing their history, then developing the general mathematical framework within which we conduct our investigation. In Part II (Chapters 5–13) we consider, in turn, a succession of particular computability notions (approximately one per chapter) that have emerged as conceptually natural and mathematically interesting. Our advice to the reader is to begin with Chapter 1 for an informal introduction to the main concepts and the intuitions and motivations behind them, then proceed to Chapter 3 for the general theory along with some key examples that will play a pervasive role later on. Thereafter, the remaining chapters may be tackled in more or less any order with only a moderate amount of cross-referral. Further details of the book's structure and the logical dependencies between chapters may be found in Section 1.3.

The idea of writing this book matured during the second author's visit to the University of Edinburgh in the spring of 2009. Whilst the second author's earlier book [207] is often cited as a reference work in the area, it only covers one part of the story, and that part only up to 1980, and we both felt that the time was ripe for a more comprehensive work mapping out the current state of the field for the benefit of future researchers. In 2009 we agreed on an outline for the contents of such a book, but our writing began in earnest in 2011, with each of us producing initial drafts for designated parts of the volume. In order to ensure coherence at the level of linguistic style and use of terminology, the first author then prepared a full version of the manuscript, which was then revised in the light of the second author's comments. The authors have been in steady contact with each other throughout the whole process, and they accept equal responsibility for the final product.[1]

It has been our privilege over many years to have developed a close acquaintance with this part of the mathematical landscape through our various traversals thereof, and our journeys have been immeasurably enriched by the company of the numerous fellow travellers from whom we have learned along the way and with whom we have shared ideas. Our thanks go to Samson Abramsky, Andrej Bauer, Ulrich Berger, Jan Bergstra, Chantal Berline, Antonio Bucciarelli, Pierre-Louis Curien, Thomas Ehrhard, Yuri Ershov, Martín Escardó, Solomon Feferman, Mike Fourman, Pieter Hofstra, Martin Hyland, Achim Jung, Jim Laird, Guy McCusker, Paul-André Melliès, Johan Moldestad, Yiannis Moschovakis, Hanno Nickau, Luke Ong, Jaap van Oosten, Gordon Plotkin, Jim Royer, Vladimir Sazonov, Matthias Schröder, Helmut Schwichtenberg, Dana Scott, Alex Simpson, Viggo Stoltenberg-Hansen, Thomas Streicher and Stan Wainer, among many others. We also wish to pay particular tribute to the memory of Robin Gandy, whose influence cannot be overestimated.

We are very grateful to Barry Cooper and the Computability in Europe editorial board for the opportunity to publish this work under their auspices, and for their support and encouragement throughout the project. We also thank Ronan Nugent of Springer-Verlag for his guidance and assistance with the publication process, and the copy editor for detecting numerous minor errors.

Our academic host organizations, the School of Informatics at the University of Edinburgh and the Department of Mathematics at the University of Oslo, have for many years provided congenial and stimulating environments for our work, and we have each benefited from the hospitality of the other's institution. Finally, our heartfelt thanks go to our respective wives, Caroline and Svanhild, for their patience and love throughout the writing of the book, and for their cheerful tolerance of the levels of mental preoccupation that such undertakings seem invariably to engender.

Edinburgh, Oslo
October 2014

John Longley
Dag Normann

[1] A list of known errata and other updates relating to the content of the book will be maintained online at `http://homepages.inf.ed.ac.uk/jrl/HOC_updates.pdf`.

Contents

Part I
Background and General Theory

Whereas for ordinary first-order functions on the natural numbers (i.e. functions of type $\mathtt{N} \to \mathtt{N}$) there is only one reasonable concept of computability, this is not the case for operations of more complex types such as $((\mathtt{N} \to \mathtt{N}) \to \mathtt{N}) \to \mathtt{N}$. The main purpose of this book is to study a range of possible concepts of 'higher-order' computability—that is, computability in settings where any operation or 'program' can itself serve as the input to another operation or program. In Part I of the book, we discuss the conceptual and historical background to our study, and present the mathematical framework within which we will conduct our investigation. This will provide some general theory which will support our investigation of particular computability concepts in Part II.

In Chapter 1, we introduce somewhat informally the main intuitions and motivations behind our work, drawing largely on familiar examples from computing. The viewpoint we adopt is that notions of higher-order computability can be seen as forming part of a much broader landscape of computability notions in general. Such notions are embodied by *computability models*, which may be typed or untyped, total or partial, extensional or intensional, 'effective' or 'continuous', to mention just some of the axes of variation. We also survey some applications of higher-order computability in mathematical logic and in computer science, and delineate more precisely the scope and contents of the book.

In Chapter 2, we complement this conceptually oriented overview with a survey of the field from a more historical perspective. We begin with Church's introduction of the λ-calculus, and discuss the important contributions of Gödel, Kleene, Kreisel, Platek, Scott, Ershov, Plotkin and many others. We conclude with a brief account of some of the main themes that characterize current work in the area. Throughout this chapter, our emphasis is on the interplay of ideas from both mathematical logic and theoretical computer science.

Our formal mathematical exposition begins in Chapter 3. We present a general theory of computability models (including higher-order models) and simulations between them, drawing largely on ideas from combinatory logic. On the basis of some relatively simple definitions, we are able to develop a useful body of general theory which is applicable to all the models we wish to study. By way of examples, we introduce some models of particular importance that will play key roles later in the book. In this way, the chapter serves as a gateway to the technical subject matter of the book as a whole.

Many, though not all, of the computability models we will study are (typed or untyped) *λ-algebras*: that is, they satisfy the usual equations of the λ-calculus. In Chapter 4 we develop some further general theory that is applicable to all λ-algebras. This includes a range of different mathematical characterizations of λ-algebras, a theory of *retractions* explaining which types can be conveniently embedded in which other types, a comparison of different ways of representing *partial* operations, and the general proof technique of *logical relations*. Any λ-algebra with a 'type of numerals' admits an interpretation of Gödel's *System T*, which offers a higher-order generalization of the concept of primitive recursion; we conclude the chapter by establishing some important properties of this system.

Chapter 1
Introduction and Motivations

The mathematical study of the concept of computability originated with the fundamental work of Church and Turing in the 1930s. These workers focussed on computations whose inputs and outputs consisted of *discrete* data, such as natural numbers or finite strings of symbols over a finite alphabet, but in this book we shall mainly be concerned with what 'computability' might mean for data of a more complex, dynamic kind. Typical questions include:

- What concepts of computability are possible for data of some given kind? Is there just a single reasonable concept of computability, or a range of different concepts?
- Which definitions of computability turn out to give rise to equivalent concepts, and which do not?
- What mathematical properties do computable operations possess? What structural properties does the *class* of computable operations possess?
- Are there any particularly interesting, surprising or useful examples of computable operations? Or of non-computable operations?
- How are the various concepts of computability related to one another? Can some be said to be 'stronger' than others?

More specifically, this book will be about the possible meanings of computability in a *higher-order* setting. This means that we will be working with models of computation in which any operation or 'program' can itself serve as the input to another operation or program: for instance, we will often be considering not just ordinary first-order functions, but second-order functions acting on such functions, third-order functions acting on these, and so on. However, many of the key ideas we shall employ make equally good sense in settings that do not naturally support such higher-order computation. We shall therefore start by explaining the basics of our approach in a very general context, before narrowing the focus to the higher-order models that will form the main subject matter of the book.

As one might expect, much of the interest in higher-order computability stems from its relevance to computer science—in particular, the light it sheds on the expressive power of different kinds of programming languages or other frameworks for computation. Historically, however, the development of the subject was closely intertwined with mathematical logic, and in particular with the problem of clarifying the constructive content of various kinds of logical systems. In this introductory chapter we shall touch on both these aspects. We begin in Section 1.1 with an informal introduction to some of the key concepts that will pervade the book, drawing on examples from computing that are likely to be familiar to most readers. In Section 1.2, by way of motivation, we discuss how the same fundamental ideas find applications in both mathematical logic and theoretical computer science. Finally, in Section 1.3, we explain in more detail what the book contains, how it is structured, and what the reader should expect from it.

1.1 Computability Models

By 'computation', we shall mean the manipulation of data of various kinds, typically with the aim of generating other data. Of course, one could understand this to include the manipulation of data by human agents or by other kinds of natural processes, but to keep the discussion focussed, we will concentrate here on computations performed by mechanical computers in the usual sense of the term.

Central to our approach will be a general notion of what it is to be a mathematical *model of computability*. This notion comprises two ingredients: the concept of *data*, and the concept of a *computable operation* on such data. We begin by explaining the informal understanding of these two concepts that underpins our work.

1.1.1 Data and Datatypes

Basic kinds of data that commonly feature in computing include integers, booleans, floating-point reals, and characters drawn from some finite alphabet such as the ASCII character set. From these, many other kinds of data can be built up, such as strings, arrays and labelled records. More complex types of data include:

- list, tree and graph structures (e.g. binary trees with nodes labelled by integers),
- abstract data values (e.g. finite *sets* of integers, perhaps implemented using lists or trees of some kind),
- lazy data structures (e.g. potentially infinite lists whose elements are computed on demand),
- objects, in the sense of object-oriented languages such as C++ and Java,
- functions, procedures and processes, considered as data items that can themselves be stored and passed as arguments.

Computations might also manipulate 'data' that is never physically stored in its entirety on the computer. For instance, in an interactive computer game, the potentially endless sequence of mouse clicks supplied by the user (along with associated time and position information) can be regarded as the 'input' to the program, whilst the resulting behaviour of the game, extended over time, can be regarded as its 'output'.

As the above examples suggest, it is often useful to classify items of data according to their *type*—for example, whether they are integers, strings, arrays of real numbers, mouse input streams, functions from lists of integers to booleans, or whatever. Many programming languages help the programmer to avoid basic errors by enforcing some kind of segregation between different types: for instance, it would typically be illegal to store a string in a variable declared to be of integer type. However, programming languages vary widely in the extent to which they enforce such typing constraints, and in the type systems that they use to do so. As an introduction to many of the powerful type systems in use in modern languages, we recommend Pierce [232]. In the present book, we shall adopt a broad approach, embracing both 'typed' and 'untyped' views of computation.

The starting point for our mathematical treatment of data and types will be very simple indeed: underlying every *computability model* $\mathbf{C}$ that we will consider, there will be a set T of *type names*, and an indexed family $|\mathbf{C}| = \{\mathbf{C}(\tau) \mid \tau \in \mathsf{T}\}$ of sets which we think of as *datatypes*. The elements of these sets will be particular *values* of these datatypes. For instance, T might contain just the type names `int` and `bool`, with $|\mathbf{C}|$ consisting of the sets

$$\mathbf{C}(\mathtt{int}) = \mathbb{Z} = \{\ldots,-2,-1,0,1,2,\ldots\}\,, \qquad \mathbf{C}(\mathtt{bool}) = \mathbb{B} = \{\mathit{tt},\mathit{ff}\}\,.$$

An *untyped* model will be one for which T consists of a single element $*$; here all data values are treated as having the same type $\mathbf{C}(*)$.

Even at this early stage, there is a crucial point to be noted. In choosing to represent a certain kind of data using a certain set of values, we are inevitably selecting some particular *level of abstraction* at which to model our data—a level appropriate to the kind of computation we want to consider. For example, suppose that in some high-level programming language, we have defined an abstract datatype `S` for finite sets of integers, equipped with the following primitive constants and operations:

$$\begin{array}{rlrl} \mathtt{empty} : \mathtt{S}\,, & & \mathtt{isEmpty} : \mathtt{S} \rightarrow \mathtt{bool}\,, \\ \mathtt{add} : \mathtt{S} \times \mathtt{int} \rightarrow \mathtt{S}\,, & & \mathtt{isMember} : \mathtt{int} \times \mathtt{S} \rightarrow \mathtt{bool}\,. \end{array}$$

The point of this datatype is that for programming purposes, elements of `S` may be thought of simply as finite sets. Suppose, for instance, that a programmer defines

$$\begin{array}{rcl} \mathtt{X} & = & \mathtt{add}\,(\mathtt{add}\,(\mathtt{empty},3),\ 5)\,, \\ \mathtt{Y} & = & \mathtt{add}\,(\mathtt{add}\,(\mathtt{empty},5),\ 3)\,. \end{array}$$

It may well be that the concrete representations of `X` and `Y` inside the computer are quite different, but the whole point of the abstract type is to hide these differences from the programmer: since there is no way for any high-level program to distinguish between `X` and `Y`, both can be modelled by the unordered set $\{3,5\}$. So if our intention is to model computations that can be undertaken in the high-level language, it will be appropriate to represent S in a computability model by $\mathscr{P}_{\mathrm{fin}}(\mathbb{Z})$, the set of finite sets of integers.

On the other hand, it might be that we specifically want to model more closely the way values of type S are represented. In one implementation of S, for instance, they might be represented by (unsorted) lists of integers; it may then happen that the values of X and Y above are represented concretely by, say, the lists $[5,3]$ and $[3,5]$ respectively. To capture this, we might choose to represent our data mathematically using $\mathbb{Z}^*$, the set of finite sequences of integers. Of course, if a different implementation of S were being modelled (say, one using balanced trees), then the mathematical representation of data would be different again, but the same principle would apply.

Even this is not the full story. At a still more concrete level, what we call lists (or trees) are typically represented in memory by *linked* data structures, consisting of cells which may contain links to other cells in the form of the memory addresses at which they are stored. From this point of view, the 'same' list or tree might be represented by any number of different memory configurations depending on exactly where in memory its cells reside. This is a level of detail which programmers usually try to avoid considering. Nonetheless, to give a concrete account of computations at this machine level, it would be appropriate to work with some suitable mathematical representation of these memory configurations, for example by functions from addresses to byte values. One can even imagine continuing this process yet further, considering for instance the multitude of ways in which such a memory configuration might itself be embodied as a concrete physical system.

Examples of this kind abound. For instance, a high-level language might provide (explicitly or implicitly) a type of *functions* from integers to integers. For the purpose of modelling computation within the high-level language itself, it might be appropriate to represent this by some set of (possibly partial) functions from $\mathbb{Z}$ to $\mathbb{Z}$. Such a function, however, is typically an infinite mathematical object, so cannot be represented directly within a finite computer. Instead, a value of function type will be represented concretely by some *program* that computes particular values of the function on demand. Of course, there will typically be very many programs that compute the same function. Moreover, the 'programs' we have in mind here might be those written in the high-level language itself, or the executable machine code to which these are compiled, or perhaps something in between. The point is simply that there are many different (though related) levels of abstraction at which a 'function' might be considered. This observation will prove to be of crucial importance in higher-order settings, where the data in question will typically consist of functions or operations of some kind.

The moral of all this, then, is that a computability model will not only represent some world of data and computable operations, but will embody a particular *view* of this world at some chosen level of description or abstraction. Indeed, the idea of varying the level at which we consider some given computational scenario will turn out to be a major theme of this book. The question of how to relate different levels of description to one another will be explored in Section 1.1.4.

1.1.2 Computable Operations

Typically, a computation will consist of applying some kind of program, algorithm or 'computable operation' to an item of data called the *input*; the computation, if successful, will then yield another item of data called the *output*. As we have seen, the 'computation' might be some process that completes and returns its output within a finite time, or it might be an ongoing process whose input and output must be regarded as potentially infinite streams of events. Whilst one can certainly point to important aspects of computation that this simple input–output view does not account for, it is clearly sufficiently general to embrace a very wide range of computational phenomena, and will be the perspective from which we study computation in this book.

In a typed setting, any given computable operation f will accept inputs of some fixed type A and will produce outputs of some fixed type B. If we think of a datatype in the naive way simply as the *set* of its possible values (e.g. the set of integers, or the set of all finite lists of strings), then the effect of applying such an operation f may be modelled simply as a *partial function* $f : A \rightharpoonup B$.

By working with partial rather than total functions, we allow for the possibility that for certain inputs $a \in A$ the computation might be unsuccessful: for example, it might get stuck in an infinite loop and so never return a result, or perhaps be aborted owing to some error signal that we do not wish to regard as a genuine output value. As already noted, many different operations or 'programs' will typically give rise to the same partial function: for instance, many different algorithms for sorting a list are known. For the time being, however, it will suffice to think of a computable operation purely in terms of the partial function it gives rise to.

By a 'model of computability', then, we shall broadly mean a collection of datatypes of interest together with a collection of operations that we consider to be 'computable'. This suggests a definition along the following lines:

A *computability model* $\mathbf{C}$ over a set T of type names consists of

- an indexed family $|\mathbf{C}| = \{\mathbf{C}(\tau) \mid \tau \in \mathsf{T}\}$ of sets which we think of as *datatypes*,
- for each $\sigma, \tau \in \mathsf{T}$, a set $\mathbf{C}[\sigma, \tau]$ of partial functions $f : \mathbf{C}(\sigma) \rightharpoonup \mathbf{C}(\tau)$, which we think of as 'computable operations' from σ to τ.

Of course, in order to develop a reasonable mathematical theory, we would expect to require that our models satisfy certain axioms. In fact, the axioms we need in order to get started are very modest:

- For each $\tau \in \mathsf{T}$, the identity function $id : \mathbf{C}(\tau) \rightarrow \mathbf{C}(\tau)$ is present in $\mathbf{C}[\tau, \tau]$.
- For any $f \in \mathbf{C}[\rho, \sigma]$ and $g \in \mathbf{C}[\sigma, \tau]$, the composition $g \circ f$ is present in $\mathbf{C}[\rho, \tau]$.

Readers familiar with category theory will note that a computability model is nothing other than a subcategory of $\mathcal{S}et_p$, the category of sets and partial functions.

Even weaker definitions of computability model are possible: for example, in Subsection 3.1.6 we shall introduce the notion of a *lax* model, and it is shown in Longley [183] how our theory can also be generalized to cover *non-deterministic* computation. For the present, however, the simple definition above will serve our purpose.

The key point is that in constructing a computability model, we are free to choose our sets $\mathbf{C}[\sigma, \tau]$ according to the kind of computation we are trying to account for. Very often, this will be closely tied to the level of abstraction at which we are representing our data. To return to our abstract datatype S for finite sets, suppose we wish to construct a model $\mathbf{C}$ that represents what can be computed in the high-level language. In this model, we may have $\mathbf{C}(\mathtt{S}) = \mathscr{P}_{\mathrm{fin}}(\mathbb{Z})$, but the set $\mathbf{C}[\mathtt{S}, \mathtt{int}]$ will lack a function that returns the *size* of a given set, even though this is a mathematically well-defined function, because the primitive operations we chose for S do not allow this operation to be computed. On the other hand, in a model $\mathbf{D}$ that is designed to capture what can be computed if one is granted access to the underlying list representations, the size operation will naturally be computable.[1]

The spirit of our work, then, is that we do not take any single concept of computability as 'given' to us—rather, we are free to define as many such concepts as we like, and our objective is to discover which are the most fruitful concepts to consider. Indeed, for the purposes of theoretical study, we may sometimes be led to consider computability models whose operations are not 'effectively computable' in any realistic sense at all, since our intention may be merely to abstract out some important mathematical property that computable operations can be expected to possess. We will see an illustration of this in Example 1.1.4 below.

1.1.3 Examples of Computability Models

We now consider some typical examples of computability models, in order to reinforce the above intuitions and to provide an initial idea of the scope of the concept. (We will not require complete, precise definitions at this point.) These examples will also be used later to illustrate the concept of *simulation*, which we will introduce in Section 1.1.4.

The particular examples discussed here are admittedly somewhat unrepresentative of the models to which the bulk of the book will be devoted. This is because they are chosen so as to make contact with familiar computing intuitions and to require only a minimum of mathematical setting up; as a consequence, they are relatively concrete and to some degree ad hoc. By contrast, the models we shall mostly be studying are typically more abstract, and their definitions are shaped by more purely mathematical considerations. Nonetheless, the present examples will suffice to illustrate many of the general concepts and problems that we shall encounter.

Example 1.1.1 (Turing machines). Turing's original analysis of computability was based on a mathematical idealization of the behaviour of a human computing agent

[1] Notions of 'abstract' versus 'concrete' computation for algebraic datatypes of this kind have been extensively studied by Tucker and Zucker [290]. We will in fact say rather little about such datatypes in this book—our purpose here is simply to introduce a fundamental distinction that will play a major role in our study of higher-order computation.

now known as the *Turing machine* model. In summary, a Turing machine is a device with a finite number of possible internal states (including a designated *start state*), and with access to a memory tape consisting of a sequence of cells, infinite in both directions, where each cell may at any point in time contain a single symbol drawn from some finite alphabet Γ. At any given time, the machine is able to see a single cell of the tape at the current *read position*. Depending on the symbol currently appearing in this cell, and also on the machine's current internal state, the machine may then jump to a new internal state and optionally perform one of the following actions:

- Replace the contents of the current cell by some other specified symbol.
- Move the current read position one step to the left or right.

This process is then repeated with the new read position and internal state, until the machine enters a designated 'halting state', at which point it stops. Full details of the mathematical formalization of this model are given in Turing's original paper [292] and in most basic texts on computability theory, e.g. [245, 65, 274].

There are several ways in which the Turing machine model can be turned into a computability model in our sense; we give here a selection. We assume Γ is some fixed finite alphabet containing at least two symbols.

1. At the most concrete level, we can consider a computability model with a single datatype M of *memory states*. A memory state is simply a function $m : \mathbb{Z} \to \Gamma$ specifying the contents of the tape, with integers representing cells on the tape and 0 representing the current read position. Any Turing machine T can now be regarded as 'computing' a certain partial function $f_T : M \rightharpoonup M$ in the following way: let $f_T(m) = m'$ if, starting from the memory state m and the internal start state of T, the execution of T eventually halts yielding the final memory state m'. By definition, the computable operations of our model are the partial functions arising as f_T for some machine T. We shall refer to this model as $\mathbf{T}_1$; it is easy to show that it satisfies the identity and composition axioms.
   An important point about this model is that although the operations are 'effectively computable' in a clear sense, the data itself need not be: it makes perfectly good mathematical sense to ask how a given Turing machine would behave on a certain 'non-computable' initial memory state $m : \mathbb{Z} \to \Gamma$.
2. Somewhat more abstractly, we may define a computability model $\mathbf{T}_2$ consisting of those partial functions $\mathbb{N} \rightharpoonup \mathbb{N}$ computable by Turing machines. For this, one should fix on some convention for representing natural numbers via memory states. For example, we might say that a memory state m *represents* a number n if the decimal representation of n appears written in memory cells $0, \ldots, k-1$, with the most significant digit appearing in cell 0, and the end of the input marked by a special 'blank' symbol in cell k. We may then say (for example) that a machine T *computes* a partial function $f : \mathbb{N} \rightharpoonup \mathbb{N}$ if whenever m represents n, we have $f(n) = n'$ iff the execution of T starting from m eventually halts in some memory state m' representing n'. We now define $\mathbf{T}_2$ to consist of the single datatype $\mathbb{N}$ together with all Turing-computable partial functions $f : \mathbb{N} \rightharpoonup \mathbb{N}$.

It is well known that this class of *partial computable functions* $\mathbb{N} \rightharpoonup \mathbb{N}$ is extremely robust: for example, it is not sensitive to the choice of representation of natural numbers or to other points of detail in the above definitions. Indeed, the so-called *Church–Turing thesis* amounts to the claim that this coincides with the class of all partial functions computable (in principle) by mechanical means. The model $\mathbf{T}_2$ will play a major role in this book—cf. Example 1.1.11 below.

3. The Turing machine model can be modified to allow for the presence of three tapes: a read-only *input tape*, a write-only *output tape*, and a *working tape* that permits both reading and writing. We may suppose the read and write tapes are unidirectional and that the current position can be moved only rightwards, whilst the working tape is bidirectional as usual. Assuming for simplicity that all three tapes employ the same alphabet Γ, we can now represent the contents of an input or output tape by a function $d : \mathbb{N} \to \Gamma$. Let D be the set of all such functions.
A machine T may be said to *compute* a partial function $f : D \rightharpoonup D$ defined as follows: $f(d) = d'$ iff, when T is run on the input tape d and a blank working tape, the resulting execution will write the infinite sequence d' on the output tape. We then obtain a computability model $\mathbf{T}_3$ consisting of the single datatype D and all machine-computable partial functions $f : D \rightharpoonup D$.[2] This illustrates the possibility of modelling computations that run for infinite time.

Example 1.1.2 (Programming language models). For another class of examples, let $\mathscr{L}$ be some deterministic typed programming language, such as Haskell or a suitable deterministic fragment of Java. Somewhat informally, we can construct a computability model $\mathbf{C} = \mathbf{C}(\mathscr{L})$ from $\mathscr{L}$ as follows:

- For each type σ of $\mathscr{L}$, we have a datatype $\mathbf{C}(\sigma)$ consisting of the $\mathscr{L}$-definable values of type σ.
- For each $\sigma, \tau \in \mathbf{T}$, we let $\mathbf{C}[\sigma, \tau]$ consist of those partial functions $f : \mathbf{C}(\sigma) \rightharpoonup \mathbf{C}(\tau)$ that are computable in $\mathscr{L}$. Here we assume $\mathscr{L}$ admits function declarations for any given argument type σ and return type τ (whether or not such functions are themselves treated as first-class values in $\mathscr{L}$).

To make this more precise, we would need to clarify what we mean by a *value* of a given type σ, and in particular when two programs of $\mathscr{L}$ define the *same* value. In fact, there are several possible ways of responding to these questions (see e.g. Subsections 3.2.3 and 3.6.1), but we shall not enter into details here.

We think of $\mathbf{C}(\mathscr{L})$ as embodying the notion of 'computability in $\mathscr{L}$'. Many computability models of interest arise from programming languages in this way—and as we shall shortly see, different languages often give rise to radically different models. For the purpose of theoretical study, we shall work mainly with small-scale proto-languages, but it is worth noting that (deterministic fragments of) realistic programming languages also give rise to computability models in the present sense. If one adopts the more general framework presented in [183], one can likewise obtain computability models even from non-deterministic languages (including process calculi

[2] It is not quite immediate (although true) that these computable partial functions are closed under composition—the reader may enjoy thinking this through. What is more immediate is that they form a *lax* computability model in the sense of Subsection 3.1.6.

such as Milner's π-calculus), although we shall not venture into such territory in this book.

Example 1.1.3 (The λ-calculus). Another model of computation frequently discussed in introductory courses on computability is the untyped *λ-calculus*, which provided the setting for Church's original investigations in computability theory (see Subsection 2.1.1). This is arguably the simplest programming language of interest, and will provide a useful example in what follows.

Terms M of the λ-calculus are generated from variables x by means of the following grammar:

$$M ::= x \mid MM \mid \lambda x.M .$$

The idea behind the untyped λ-calculus is that every term can play the role of a function: a term MM' should be understood as 'M applied to M'', whilst $\lambda x.M$ may be informally read as 'the function that maps a given x to M' (where M will typically be some term involving x). Thus, the 'function' $\lambda x.xx$ will map N to NN.

In a certain sense, then, we wish to say that the terms $(\lambda x.xx)N$ and NN have the same 'meaning'. This can be done by endowing the set Λ of λ-terms with an equivalence relation $=_\beta$ known as *β-equality*. (For the definition of $=_\beta$, see Example 3.1.6.) Writing L for the quotient set $\Lambda/{=_\beta}$, and $[M]$ for the $=_\beta$-equivalence class of M, any term $P \in \Lambda$ will induce a well-defined mapping $[M] \mapsto [PM] : \mathsf{L} \to \mathsf{L}$. This enables us to view L as a computability model: its sole datatype is L itself, and the operations on it are exactly the mappings induced by terms P as above. We also obtain a submodel L^0 by restricting to the set Λ^0 of *closed* terms: the elements of L^0 are the equivalence classes $[M]$ where M is closed, and the computable operations are those induced by closed terms P.

Among other things, natural numbers can be encoded within the λ-calculus: the number n can be represented by the *Church numeral* $\lambda f.\lambda x.f^n x$, where $f^n x$ denotes the n-fold application $f(f(\cdots(fx)\cdots))$. One of the early achievements of computability theory was to show that the functions $\mathbb{N} \to \mathbb{N}$ computable within the λ-calculus coincide with those computable by Turing machines, and likewise within many other models of computation.

Example 1.1.4 (Event-style models). Our final example is very loosely inspired by the *event structure* model of Winskel [303], although we include here only the bare minimum of structure needed for the purpose of illustration. Whilst still involving data of a relatively concrete kind, this illustrates the possibility of a more abstract view of computability that isolates certain mathematical properties of computable operations.

We have in mind a framework for modelling the behaviour of systems that react to a stream of *input events* such as keystrokes or mouse clicks (perhaps tagged with the time at which they occur), and produces a stream of *output events* such as sounds or changes to the visual display (these too may be timestamped if desired). All that really matters for our purpose is that each individual event must be bounded in time, and that within any finite time period, only finitely many events may occur.

Suppose that A is the set of possible input events, and that S is some associated set of subsets of A. The idea is that S specifies which sets of input events are to be regarded as 'consistent' or compatible with one another. For instance, if the event e represents a mouse click in the top left of the screen at time t, and e' represents a click in the bottom right at the same time t, then we would expect S to reflect the incompatibility of these events by not including any set that contains both e and e'.

It seems reasonable to expect that S will satisfy certain conditions. First, if $Y \subseteq X \in S$ then $Y \in S$: any subset of a 'compatible' set of events should itself be compatible. Secondly, if every finite subset of X is present in S, then $X \in S$: informally, if there were some 'clash' within the event set X, we would expect this to manifest itself within some finite time, so that the problem would already be present within some finite set of events.

Now suppose B is a set of possible output events, with $T \subseteq \mathcal{P}(B)$ a family of 'consistent' event sets also satisfying the above conditions. The behaviour of a particular system can now be modelled by a function $f : S \to T$: intuitively, if $X \in S$, then $f(X)$ consists of all output events that are guaranteed to occur given that all the input events in X have occurred.[3] Even from this simple idea, it already follows that f must satisfy certain properties:

1. If $X, Y \in S$ and $X \subseteq Y$, then $f(X) \subseteq f(Y)$. That is, the output events guaranteed by the input events in Y will include those already guaranteed by the 'smaller' event set X. We express this by saying f is *monotone* with respect to subset inclusion.
2. If $X_0 \subseteq X_1 \subseteq \cdots$ are event sets in S, then $f(\bigcup_j X_j) = \bigcup_j f(X_j)$. Informally, every output event $y \in f(\bigcup_j X_j)$ must occur within some finite time, so the computation will only have been able to register the occurrence of finitely many input events $x_0, \ldots, x_{r-1}$ by the time y occurs. By choosing j large enough so that $x_0, \ldots, x_{r-1} \in X_j$, we see that y is already present in $f(X_j)$. We express this idea by saying that f is *(chain) continuous* with respect to inclusion.[4]

Thus, monotonicity and continuity are necessary properties of a 'behaviour function' f that depend on next to nothing about how the function is implemented: for instance, they would still hold even if our system had access to some magical technology for computing classically non-computable functions $\mathbb{N} \to \mathbb{N}$. We may therefore construct an *abstract* model of computability that builds in only these two aspects of computable operations, in order to investigate just how much can be derived from these properties alone. Specifically, we can build a computability model E along the following lines:

- Datatypes are sets $S \subseteq \mathcal{P}(A)$ (for A some event set) such that if $Y \subseteq X \in S$ then $Y \in S$, and if S contains every finite subset of X then $X \in S$.
- Computable operations from $S \subseteq \mathcal{P}(A)$ to $T \subseteq \mathcal{P}(B)$ are functions $f : S \to T$ that are monotone and continuous with respect to subset inclusion.

The interplay between the concepts of computability and continuity will emerge as one of the major themes of this book.

---

[3] Of course, the occurrence of some output events might in some cases depend on the *absence* of a positive event such as a mouse click within a certain time period. We can account for this simply by including such detectable 'negative events' in our set of input events.

[4] The alert reader may notice that condition 2 actually implies condition 1, but it is conceptually useful to consider monotonicity as a condition in its own right.

1.1.4 Simulating One Model in Another

We indicated earlier that the same computational situation can often be represented by many different computability models corresponding to different levels of abstraction or description, and this raises the question of how such different views should be related to one another. One natural way to do this is provided by the concept of a *simulation*. Broadly speaking, if **C** and **D** are computability models, a simulation $\gamma : \mathbf{C} \longrightarrow\!\!\!\!\triangleright\, \mathbf{D}$ will be a way of simulating or representing the model **C** using **D**; in this context, we typically think of **C** as 'more abstract' and **D** as 'more concrete'.

The general concept of a simulation (informally understood) is ubiquitous in computing. For example, an *interpreter* for a language $\mathscr{L}$ written in a language $\mathscr{L}'$ is in effect a way of simulating $\mathscr{L}$ in $\mathscr{L}'$; likewise, a *compiler* which translates from $\mathscr{L}$ to $\mathscr{L}'$ provides a way of simulating any individual $\mathscr{L}$-program by an $\mathscr{L}'$-program. Indeed, any high-level programming language, whether interpreted or compiled, can ultimately be seen as implemented in terms of (hence 'simulated by') a low-level machine language. Conversely, a programmer may sometimes use a simulator for a low-level language written in a high-level language, for the purpose of developing and debugging low-level code.

Our definition of simulation abstracts out certain essential features of these and many other situations. To say that we can simulate **C** using **D**, we shall require the following:

1. Every data value in **C** (that is, every element of every datatype $\mathbf{C}(\tau)$) should be 'represented' by at least one data value in **D**.
2. Every computable operation in **C** should be simulated or mimicked by a computable operation in **D** with respect to this representation of data.

In order to make sense of condition 2, it is natural to assume that elements of the same datatype in **C** are represented by elements of the same type in **D**. A crucial point, however, is that a single data *value* in **C** may in general be represented by several different values in **D**. Intuitively, this is to be expected in the light of our earlier discussion: what appears as a single data value at one level of abstraction (e.g. the set $\{3,5\}$) might correspond to several different values at some more concrete level of representation (e.g. the lists $[3,5]$ and $[5,3]$). We will shortly see that many naturally arising examples of simulations are indeed multi-valued in this sense.

We can collect these ideas into a definition as follows. Suppose **C** and **D** are computability models with types indexed by T, U respectively. A *simulation* γ of **C** in **D** (written $\gamma : \mathbf{C} \longrightarrow\!\!\!\!\triangleright\, \mathbf{D}$) consists of:

- a mapping associating to each type $\tau \in \mathsf{T}$ a 'representing type' $\gamma\tau \in \mathsf{U}$,
- for each $\tau \in \mathsf{T}$, a relation $\Vdash^{\gamma}_{\tau}$ between elements of $\mathbf{D}(\gamma\tau)$ and those of $\mathbf{C}(\tau)$,

subject to the following conditions:

- For each $\tau \in \mathsf{T}$ and each $a \in \mathbf{C}(\tau)$, there is some $a' \in \mathbf{D}(\gamma\tau)$ such that $a' \Vdash^{\gamma}_{\tau} a$.
- Every operation $f \in \mathbf{C}[\sigma, \tau]$ is *tracked* by some $f' \in \mathbf{D}[\gamma\sigma, \gamma\tau]$; that is to say, if $f(a)$ is defined and $a' \Vdash^{\gamma}_{\sigma} a$, then $f'(a')$ is defined and $f'(a') \Vdash^{\gamma}_{\tau} f(a)$.

We may read $a' \Vdash^{\gamma}_{\sigma} a$ as 'a' represents a' (or more traditionally, 'a' *realizes* a') with respect to γ. Note that if f' tracks f and both a', a'' realize a where $f(a)$ is defined, then both $f'(a'), f'(a'')$ will realize $f(a)$, though they need not be equal. Note also that $f'(a')$ may be defined even if $f(a)$ is undefined.

Let us now revisit our examples of computability models from Subsection 1.1.3, and give some examples of simulations that can be constructed between them. Again, our emphasis here is on the intuitions rather than the formal details.

Example 1.1.5. As regards the Turing machine models of Example 1.1.1, it is immediate from the definition of $\mathbf{T}_2$ that it comes equipped with a simulation $\mathbf{T}_2 \longrightarrow\!\!\!\!\rhd \mathbf{T}_1$: if $n \in \mathbb{N}$ and m is a memory state, we take $m \Vdash n$ iff m represents n in the sense we have defined. It is also immediate from the definition that every computable operation in $\mathbf{T}_2$ is tracked by one in $\mathbf{T}_1$.

In the other direction, we cannot have a non-trivial simulation $\mathbf{T}_1 \to \mathbf{T}_2$: since there are uncountably many memory states, these cannot be faithfully encoded as natural numbers. However, there is an evident variation $\mathbf{T}_1^{\mathrm{fin}}$ on $\mathbf{T}_1$ in which we restrict our memory states to those with a designated blank symbol in all but finitely many cells. Such memory states may be coded as natural numbers, and it is clear that the action of any Turing machine may then be emulated by a partial computable function $\mathbb{N} \rightharpoonup \mathbb{N}$. We thus obtain a simulation $\mathbf{T}_1^{\mathrm{fin}} \longrightarrow\!\!\!\!\rhd \mathbf{T}_2$.

We leave it as an exercise for the reader to investigate what simulations exist to and from $\mathbf{T}_3$ and obvious variants thereof.

Example 1.1.6. Let $\mathscr{L}$ be a programming language as in Example 1.1.2. Assuming programs of $\mathscr{L}$ are finite combinatorial objects, any program may be effectively encoded as a natural number via a system of *Gödel numbering*; this will in turn lead to an encoding of values of $\mathscr{L}$ as natural numbers. (This encoding will in general be multi-valued if several programs are deemed to define the 'same' value.) Furthermore, assuming that evaluation of programs is an effective operation (as must be the case if the language is to be implementable on a conventional computer), the behaviour of any $\mathscr{L}$-definable function will be emulated by a partial computable function $\mathbb{N} \rightharpoonup \mathbb{N}$. In this way we obtain a simulation $\mathbf{C}(\mathscr{L}) \longrightarrow\!\!\!\!\rhd \mathbf{T}_2$. Alternatively, one can simulate $\mathbf{C}(\mathscr{L})$ directly in $\mathbf{T}_1$—this is perhaps closer in spirit to the way programming languages are actually implemented via sequences of binary digits.

As a concrete example, let $\ulcorner - \urcorner$ denote some effective Gödel numbering for closed λ-terms as defined in Example 1.1.3. We may say a natural number n realizes an equivalence class $[M] \in \mathsf{L}^0 = \Lambda^0/{=_\beta}$ if $n = \ulcorner M' \urcorner$ for some $M' \in [M]$. For any term P, the mapping $M \mapsto PM$ is clearly an effective operation at the level of Gödel numbers, so we obtain a simulation $\mathsf{L}^0 \longrightarrow\!\!\!\!\rhd \mathbf{T}_2$.

Example 1.1.7. Going in the other direction, if our language $\mathscr{L}$ contains a type for natural numbers and is Turing complete (that is, any partial computable function $\mathbb{N} \rightharpoonup \mathbb{N}$ can be implemented in $\mathscr{L}$), we clearly obtain a simulation $\mathbf{T}_2 \longrightarrow\!\!\!\!\rhd \mathbf{C}(\mathscr{L})$.

Note that this may be the case even if the natural numbers are only indirectly present in $\mathscr{L}$. Consider, for instance, the encoding of natural numbers by Church numerals in untyped λ-calculus. Since the untyped λ-calculus is Turing complete, this yields a simulation $\mathbf{T}_2 \longrightarrow\!\!\!\!\rhd \mathsf{L}^0$.

Example 1.1.8. Our final example relates the Turing machine model $\mathbf{T}_3$ to the event model E of Example 1.1.4. Recall that in $\mathbf{T}_3$, our single datatype D consists of all total functions $\mathbb{N} \to \Gamma$; these serve to represent both the content of the input tape (read one cell at a time) and that of the output tape (written one cell at a time). Our idea here is to represent computable partial functions $D \rightharpoonup D$ by means of processes that can generate certain (finite or infinite) portions of output in response to certain portions of input.

Formally, we can take our set A of events to consist of pairs (n,a) where $n \in \mathbb{N}$ and $a \in \Gamma$: such a pair represents the appearance of the symbol a in cell n of the input or output tape. Next, we take the set S to consist of all sets $X \subseteq A$ such that $(n,a),(n,a') \in X$ implies $a = a'$; thus, in effect, S is the set of all *partial* functions $\mathbb{N} \rightharpoonup \Gamma$. This means that we have a natural inclusion $\iota : D \hookrightarrow S$; we may therefore set $s \Vdash d$ iff $s = \iota(d)$.

Suppose now that a machine T computes a partial function $f : D \rightharpoonup D$. We can see that T will actually yield some function $f' : S \to S$ which restricts to f: given any portion of input $X \in S$, take $f'(X)$ to consist of the output produced by T without inspecting any more of the input than X. By the general considerations discussed in Example 1.1.4, the function f' is necessarily monotone and continuous. Thus, every operation $D \rightharpoonup D$ computable in $\mathbf{T}_3$ is tracked by an operation $S \to S$ computable in E, so we obtain a simulation $\mathbf{T}_3 \to \mathsf{E}$.

Note that different machines T that compute the same $f : D \rightharpoonup D$ may give rise to different functions $S \to S$: for example, they may differ in how much input they read before generating five symbols of output. Functions $S \to S$ thus embody a finer level of detail about computational processes than the bare functions $D \rightharpoonup D$ that they represent. This is very typical of what happens when we represent operations on 'total' data by operations on 'partial' data—another idea that will recur frequently in this book.

We also note that if one works with the wider class of computability models studied in [183], the concept of simulation becomes correspondingly broader in scope. For instance, it is shown in [183] how a well-known translation from the λ-calculus to Milner's π-calculus can be presented as a simulation in this sense.

Sometimes there will be more than one way to simulate a given model $\mathbf{C}$ in a given $\mathbf{D}$. For instance, suppose $\mathscr{L}$ is a programming language along the lines of Example 1.1.2. On the one hand, we might directly define a simulation $\gamma : \mathbf{C}(\mathscr{L}) \longrightarrow\!\triangleright \mathbf{T}_1$ based on some bit-level representation of $\mathscr{L}$-programs. On the other hand, we might define a translation or compilation K from $\mathscr{L}$ to some other language $\mathscr{L}'$ which itself admits a direct simulation $\gamma' : \mathbf{C}(\mathscr{L}') \longrightarrow\!\triangleright \mathbf{T}_1$. At least in good cases, the compilation yields a simulation $\mathbf{C}(\mathscr{L}) \longrightarrow\!\triangleright \mathbf{C}(\mathscr{L}')$, and by composing this with γ' in the obvious way we obtain a second simulation $\delta : \mathbf{C}(\mathscr{L}) \longrightarrow\!\triangleright \mathbf{T}_1$.

Although distinct, the simulations γ, δ will typically be related. If the translation $K : \mathscr{L} \to \mathscr{L}'$ is itself effectively computable (as will certainly be the case for any practical compilation process), then for any $\mathscr{L}$-program M, a bit representation of M in $\mathbf{T}_1$ can be transformed *via a computation in* $\mathbf{T}_1$ into a bit representation of $K(M)$. This typically amounts to saying that γ-realizers for values in $\mathbf{C}(\mathscr{L})$ can be uniformly transformed to δ-realizers for the same values via an operation of $\mathbf{T}_1$.

More precisely, for each type σ of $\mathbf{C}(\mathscr{L})$, there is an operation $t_\sigma \in \mathbf{T}_1[\gamma\sigma, \delta\sigma]$ such that whenever $m \Vdash^{\gamma}_{\sigma} v$ we have $t_\sigma(m) \Vdash^{\delta}_{\sigma} v$. In this situation, we say γ is *transformable* to δ, and write $\gamma \preceq \delta$. Note that it may or may not be the case that $\delta \preceq \gamma$: this will depend on whether the translated term $K(M)$ retains sufficient information to allow an effective reconstruction of the source term M up to value equivalence. We write $\gamma \sim \delta$ if both $\gamma \preceq \delta$ and $\delta \preceq \gamma$.

The framework of computability models and simulations (presented in Section 3.4 as a category preorder-enriched by $\preceq$) will provide the general setting for much of our investigation of notions of computability. It also suggests an interesting notion of *equivalence* for computability models, to which we turn our attention next.

1.1.5 Equivalences of Computability Models

Traditional presentations of computability theory often emphasize that Turing machines, the untyped λ-calculus and many other models of computation are *equivalent* in their computational power, insofar as they give rise to the same class of partial computable functions $\mathbb{N} \rightharpoonup \mathbb{N}$. Whilst this is certainly true, our focus here will be on a more subtle concept of 'equivalence'—one that will allow us to draw non-trivial distinctions between different models of interest.

What definition of equivalence is suggested by our framework of computability models and simulations? For models $\mathbf{C}$ and $\mathbf{D}$ to be considered equivalent, we would at least expect that each should be capable of simulating the other: that is, there should be simulations $\gamma : \mathbf{C} \multimap \mathbf{D}$ and $\delta : \mathbf{D} \multimap \mathbf{C}$. But is this all? What we propose here is that one should also require γ, δ to be 'mutually inverse' in some sense—this, after all, is the usual mathematical requirement for a pair of mappings to constitute an equivalence. We will show that this leads to a mathematically interesting and fruitful notion of equivalence which highlights some essential differences between models—differences that are not visible if we merely consider the computable functions $\mathbb{N} \rightharpoonup \mathbb{N}$ that the models can represent. Even within the realm of practical programming languages, there are many interesting things that can be done in some languages but not others, in a sense that is not undermined by the coarse-grained observation that every such language can simulate every other.[5]

In the case of simulations, literally requiring $\delta \circ \gamma = id_{\mathbf{C}}$ and $\gamma \circ \delta = id_{\mathbf{D}}$ would be too restrictive to be useful. However, we obtain an interesting concept of equivalence if we require γ and δ to be *mutually inverse up to intertransformability*: $\delta \circ \gamma \sim id_{\mathbf{C}}$ and $\gamma \circ \delta \sim id_{\mathbf{D}}$. Intuitively, if we view $\delta \circ \gamma$ as a way of simulating or 'encoding' $\mathbf{C}$ within itself, this says that suitable encoding and decoding operations are computable within $\mathbf{C}$ (and likewise for $\mathbf{D}$).

[5] The informal intuition that there are interesting expressivity distinctions even among Turing complete programming languages is also the motivation for the work of Felleisen [95]. However, Felleisen is chiefly concerned with distinctions of a more fine-grained nature than those we consider here.

We can get an intuition for what this equivalence means by looking at two particular scenarios:

Example 1.1.9. Recall the Turing machine models $\mathbf{T}_1$ and $\mathbf{T}_2$ of Example 1.1.1, and the variant $\mathbf{T}_1^{\mathrm{fin}}$ from Example 1.1.5. We have seen that there are simulations $\gamma : \mathbf{T}_1^{\mathrm{fin}} \longrightarrow\!\!\!\!\triangleright\, \mathbf{T}_2$ and $\delta : \mathbf{T}_2 \longrightarrow\!\!\!\!\triangleright\, \mathbf{T}_1^{\mathrm{fin}}$, and it is easy to see that $\gamma \circ \delta \sim id_{\mathbf{T}_2}$. However, as things stand, these simulations do not form an equivalence. Although $\delta \circ \gamma \preceq id_{\mathbf{T}_1^{\mathrm{fin}}}$, the reverse is not true: in the absence of any effective way to recognize the boundaries of the non-blank portion of the input, we cannot (effectively within $\mathbf{T}_1^{\mathrm{fin}}$) transform a general finite memory state into a (coded) numeral for it.

The problem may be remedied if we restrict the memory states of $\mathbf{T}_1^{\mathrm{fin}}$ still further by requiring, for example, that the potentially non-blank portion in input and output states must be delimited by certain special 'boundary markers' (we leave the details to the reader). The resulting *bounded input* model $\mathbf{T}_1^{\mathrm{bdd}}$ then turns out to be equivalent to $\mathbf{T}_2$.

Example 1.1.10. More interestingly, let us now consider the relationship between $\mathbf{T}_2$ and L^0. We have seen in Examples 1.1.6 and 1.1.7 that there are simulations $\gamma : \mathsf{L}^0 \longrightarrow\!\!\!\!\triangleright\, \mathbf{T}_2$ (based on Gödel numbering) and $\delta : \mathbf{T}_2 \longrightarrow\!\!\!\!\triangleright\, \mathsf{L}^0$ (based on Church numerals). Furthermore, given a little knowledge of untyped λ-calculus, it is not hard to see that $\gamma \circ \delta \sim id_{\mathbf{T}_2}$. On the one hand, from n we may readily compute a Gödel number for its Church numeral $\widehat{n} = \lambda f.\lambda x.f^n x$; conversely, from a Gödel number for any $M =_\beta \widehat{n}$, it is easy to extract n itself by an effective computation.

Less trivially, a theorem of Kleene [133] says that there is a closed λ-term E (called an *enumerator*) such that $E\,\widehat{\ulcorner M \urcorner} =_\beta M$ for any closed M; this amounts to saying that $\delta \circ \gamma \preceq id_{\mathsf{L}^0}$. However, the reverse is not true: we cannot *within the λ-calculus itself* pass from a term M to (a Church numeral for) the Gödel number for M, or even for some $M' =_\beta M$. Intuitively, the issue here is that the λ-calculus cannot 'inspect its own syntax': if some term P is applied to M, then P is able to apply M to other terms and utilize the results, but it cannot look inside M to observe its syntactic structure.

In fact, it is quite easy to prove that *no* equivalence exists between $\mathbf{T}_2$ and L^0 (see Example 3.5.3). This, allied with the above intuition for why $id \not\preceq \delta \circ \gamma$, suggests that there is indeed a deep difference in character between the two models—a difference not visible at the level of the representable functions $\mathbb{N} \rightharpoonup \mathbb{N}$.

This last example illustrates two important aspects of our framework of models and simulations, both of which will be recurring features of this book. Firstly, the existence of certain simulations or transformations between them can sometimes offer a convenient way of encapsulating certain theorems or constructions of interest. In this way, the framework offers a uniform setting that helps us to see how a wide range of results and constructions fit together. Secondly, *inequivalences* between models can sometimes highlight interesting differences in their 'computational power', even when both models are effective and Turing complete.

What evidence is there that our notion of equivalence is a mathematically natural one? One answer to this question is given by a *correspondence theorem* which we shall outline in Subsection 3.4.2. This involves a construction whereby each computability model $\mathbf{C}$ gives rise to a certain category $\mathcal{A}sm(\mathbf{C})$, the *category of assemblies* over $\mathbf{C}$. We will define this category formally in Subsection 3.3.6, but broadly speaking, one can think of $\mathcal{A}sm(\mathbf{C})$ as consisting of all datatypes implementable in $\mathbf{C}$, and all $\mathbf{C}$-computable functions between them. In typical cases, the category $\mathcal{A}sm(\mathbf{C})$ is very rich in structure, and is of interest for many reasons quite apart from its connection with simulations and equivalences (again see Subsection 3.3.6).

What the correspondence theorem tells us is that simulations $\mathbf{C} \longrightarrow\!\!\triangleright\, \mathbf{D}$ correspond precisely to functors $\mathcal{A}sm(\mathbf{C}) \longrightarrow\!\!\triangleright\, \mathcal{A}sm(\mathbf{D})$ of a certain specified kind, and (hence) that $\mathbf{C}$ and $\mathbf{D}$ are equivalent if and only if $\mathcal{A}sm(\mathbf{C})$ and $\mathcal{A}sm(\mathbf{D})$ are equivalent as categories. Such results provide an initial indicator that the notions of simulation and equivalence have good mathematical credentials.[6]

For the most part, $\mathcal{A}sm(\mathbf{C})$ will play an implicit background role in this book as a 'universe of computable functions' arising from $\mathbf{C}$. One reason for its importance is that it helps us to see just what equivalences imply: if $\mathbf{C}$ and $\mathbf{D}$ are equivalent, which properties of $\mathbf{C}$ will automatically be shared by $\mathbf{D}$ as well? From the correspondence theorem, we see that anything that can be cast as a purely categorical property of $\mathcal{A}sm(\mathbf{C})$ will be shared by $\mathbf{D}$, since it will be equally true of $\mathcal{A}sm(\mathbf{D})$.

1.1.6 Higher-Order Models

Having arrived at the idea that there may, in some sense, be many interestingly different notions of 'computability' even within the realm of effective and Turing complete models, one naturally wishes to explore the extent of this phenomenon. How many such notions are there, and how widely do they vary? Which of them are the 'important' ones? Can we classify them in some way?

Whilst these questions are certainly tantalizing, it would appear at the present time that the range of reasonable computability models on offer is simply too wide to allow us to form any overall map of the landscape (see [183] for further discussion). Our purpose in this book, however, is to provide at least a partial map for a more restricted tract of territory: that consisting of *higher-order* models.

In our basic definition of computability models in Section 1.1, data values were sharply distinguished from the computable operations that act upon them. The essential characteristic of higher-order models, however, is that computable operations can themselves be represented by data values, and thus can serve as inputs to other computable operations. That is to say, a computability model $\mathbf{C}$ over T is higher-order if for every pair of types $\sigma, \tau \in \mathsf{T}$ there is some type $\sigma \to \tau \in \mathsf{T}$ such that computable operations in $\mathbf{C}[\sigma, \tau]$ can be represented by elements of $\mathbf{C}(\sigma \to \tau)$.

To nail this down just a little further, let us assume for convenience that our computability model has 'product types' in some mild sense. We can then require that for each σ, τ there is a type $\sigma \to \tau$ equipped with an 'application' operation

[6] We should note, however, that certain other variations on our definitions can also claim somewhat analogous credentials. See Longley [183, Section 2] for a more thorough discussion.

$$app_{\sigma\tau} \in \mathbf{C}[(\sigma \to \tau) \times \sigma, \tau],$$

with respect to which every operation in $\mathbf{C}[\sigma, \tau]$ is represented within $\mathbf{C}(\sigma \to \tau)$. In fact, to obtain a well-behaved theory, we shall require that this is the case uniformly in a parameter of any type ρ: specifically, the partial functions $f : \mathbf{C}(\rho) \times \mathbf{C}(\sigma) \rightharpoonup \mathbf{C}(\tau)$ representable by operations in $\mathbf{C}[\rho \times \sigma, \tau]$ must correspond (via $app_{\sigma\tau}$) to those representable by operations in $\mathbf{C}[\rho, \sigma \to \tau]$. The precise formulation of the concept of higher-order model will be treated in several complementary ways in Section 3.1.

Our notion of higher-order model will be quite liberal in several respects. First, there is no requirement that the element of T that we have called $\sigma \to \tau$ is actually a syntactic type expression of the form $\sigma \to \tau$: for instance, if T is the one-element type world $\{*\}$, we can have $* \to * = *$. Secondly, we do not require that *every* element of $\mathbf{C}(\sigma \to \tau)$ represents a computable operation in $\mathbf{C}[\sigma, \tau]$—only that every computable operation is represented by an element. Thirdly, we allow that there may be many elements of $\mathbf{C}(\sigma \to \tau)$ representing the same computable operation. Thus, in contrast to the operations themselves, elements of $\mathbf{C}(\sigma \to \tau)$ may be *intensional* objects: programs, algorithms or computation strategies that give rise to partial functions $\mathbf{C}(\sigma) \rightharpoonup \mathbf{C}(\tau)$ in a many-to-one way (cf. the discussion in Subsection 1.1.2). Finally, our basic axioms for higher-order models will be relatively weak ones—although we shall also be interested in stronger notions such as that of a *λ-algebra* (see Chapter 4).

We have refrained from introducing higher-order models up till now, because we wished to emphasize that the crucial concepts of simulation, transformation and equivalence make sense in a purely 'first-order' setting. Having reached this point, however, we can now see that several of the models we mentioned earlier are indeed higher-order models.

Example 1.1.11. Consider for instance the Turing model $\mathbf{T}_2$ from Example 1.1.1. If a function $f : \mathbb{N} \rightharpoonup \mathbb{N}$ is computed by a Turing machine T, the details of the construction of T can themselves be encoded by a number $t \in \mathbb{N}$, which can be viewed as representing f within $\mathbb{N}$ itself. Furthermore, there is an 'application' operation $app : \mathbb{N} \times \mathbb{N} \rightharpoonup \mathbb{N}$ (computable by a *universal Turing machine*) which takes a code t for a machine T and an input n, and returns the result (if any) of applying T to n. Endowed with this higher-order structure, the model $\mathbf{T}_2$ is known as *Kleene's first model*, or K_1; it will play a crucial role as our standard reference model for effective computability on finitary data. For instance, we are able to define an *effective* computability model $\mathbf{C}$ to be one that admits a (certain kind of) simulation $\mathbf{C} \longrightarrow\!\!\triangleright K_1$ (see Subsection 3.5.1).

It turns out that $\mathbf{T}_1$ is not a higher-order model (the reader may enjoy discovering why not with reference to Section 3.1), but $\mathbf{T}_3$ is one. In fact, $\mathbf{T}_3$ is equivalent to a relativized version of *Kleene's second model*, which we shall introduce in Section 3.2.1.

Example 1.1.12. Let $\mathscr{L}$ be a programming language as in Example 1.1.2, and suppose now that $\mathscr{L}$ is a *higher-order* language: that is, $\mathscr{L}$-definable functions from σ

to τ are themselves represented by values of some type $\sigma \to \tau$. Under mild conditions, $\mathbf{C}(\mathscr{L})$ will then be a higher-order computability model. Note that a higher-order language $\mathscr{L}$ might have an explicit type constructor '$\to$' (as in Haskell or Standard ML), or it might simply allow functions from σ to τ to be *represented* using a third type. For instance, in Java, for any σ, τ we can form a type of objects possessing a method `m` with argument type σ and return type τ.

The untyped λ-calculus is a paradigmatic example of a higher-order language: any computable operation on L^0 is of the form $[M] \mapsto [PM]$, and so is representable within L^0 itself by the element $[P]$.

Our event model of Example 1.1.4 can also be shown to be a higher-order model, close in spirit to the model PC which we shall introduce in Subsection 3.2.2.

For much of this book, we shall be considering models with an explicit type constructor $\to$: for instance, they might have a type $\mathtt{N}$ for natural numbers, a type $\mathtt{N} \to \mathtt{N}$ for (total or partial) *first-order* functions on $\mathbb{N}$, a type $(\mathtt{N} \to \mathtt{N}) \to \mathtt{N}$ for *second-order* functions, and so on. As we shall see, such models can differ from one another in quite blatant ways: even if they share the same computable operations at type $\mathtt{N} \to \mathtt{N}$, they might differ at type $(\mathtt{N} \to \mathtt{N}) \to \mathtt{N}$ or some higher type. To give just one example, consider the partial function $P : [\mathbb{N} \rightharpoonup \mathbb{N}] \rightharpoonup \mathbb{N}$ (where $[\mathbb{N} \rightharpoonup \mathbb{N}]$ is the set of all partial functions $\mathbb{N} \rightharpoonup \mathbb{N}$) defined by

$$P(f) = \begin{cases} 0 & \text{if } f(0) = 0 \text{ or } f(1) = 0, \\ \text{undefined} & \text{otherwise.} \end{cases}$$

The idea is that to compute $P(f)$, we would need to evaluate $f(0)$ and $f(1)$ 'in parallel' and see whether either yields 0; it will not do to choose one of these to evaluate first and then proceed to the other one. Thus, the function P will be deemed 'computable' if our concept of computation supports parallel function calls, but not if we are working within a purely 'sequential' model. This shows that even the answer to the question 'Which functions of higher type are computable?' may depend on the particular concept of computation we have in mind. (There is a tie-up here with Example 1.1.10: it turns out that, in a suitable sense, $\mathbf{T}_2$ supports parallel operations since it allows computations to be 'interleaved', whereas L^0 is purely sequential.) In fact, there are several reasonable candidates for a notion of higher type computable function, each yielding its own brand of 'computability theory'—we shall be exploring these in considerable detail as part of our general programme of mapping out the higher-order computability landscape.

Given that a survey of the whole spectrum of computability models seems out of reach, there are several reasons why the higher-order models are a coherent and natural subclass to consider. Firstly, as the above examples might suggest, many prominent examples of computability models naturally turn out to be higher-order anyway. Secondly, higher-order models have received a good deal of attention to date (we will survey the history in Chapter 2), and substantial parts of this territory are now reasonably well mapped out. Indeed, by virtue of its relatively mature development, it seems reasonable to hope that the study of higher-order models might serve as a paradigm case for wider explorations in the future. Thirdly, there are mathematical reasons why higher-order models form an interesting subclass—in particular, there is an appealing alternative account of simulations that works well in the higher-order setting (see Subsection 3.4.1).[7] Fourthly, certain recurring ideas and methods are characteristic of the study of higher-order models (e.g. use of λ-calculi, proofs by induction on type levels, various standard ways of constructing one higher-order model from another). This means that results and proofs in the area often share a 'common flavour', even when they concern quite different models of computation. Finally, there are some well-established applications of computability models in mathematical logic that depend characteristically on their higher-order structure—we shall review some of these in the next section.

The purpose of this book, then, is to present a detailed, though by no means exhaustive, study of the landscape of higher-order computability notions, covering their general theory along with a selection of particularly prominent models and the respective 'computability theories' they give rise to. Fuller details of the precise scope of the book will be given in Section 1.3 below.

1.2 Motivations and Applications

We see the task of mapping out the space of reasonable computability notions as having its own intrinsic interest. However, there are also other motivations stemming from areas where these ideas have found application. We now look briefly at some of these, giving pointers to the literature for those topics that are not covered in detail in this book. Other related areas will be reviewed in Chapter 13.

1.2.1 Interpretations of Logical Systems

Many of the early investigations of higher-order notions of computability were focussed on the problem of giving illuminating interpretations for various systems of first-, second- or higher-order arithmetic—particularly systems of a 'constructive' or 'semi-constructive' character. We briefly mention a few broad classes of such interpretations, in order to explain how it is that higher types enter into the picture.

First, there is a large class of what can be broadly called *realizability interpretations* for constructive systems. In constructive approaches to mathematics from Brouwer [40] onwards, a formula of the kind $\forall x.\exists y.\phi(x,y)$ is typically understood to mean that there is some sort of operation or 'construction' which, given a value for x, produces a corresponding value for y, along with suitable evidence that $\phi(x,y)$ holds. From a metatheoretical standpoint, this immediately raises a question: what does it mean for us to be 'given' a value for x, and how are we allowed to manipulate this value in order to construct or compute a corresponding value for y? If x and

[7] This should, however, be counterbalanced by the observation that the class of higher-order models is rather lacking in good closure properties; see Longley [183].

y here range over entities of some complex mathematical type, such as functions over the real numbers, this already calls for a suitable computability notion for such entities. Thus, even to elucidate the fundamental notion of a 'constructive presentation of a mathematical object', we are naturally led into the study of non-trivial (and perhaps higher-order) computability notions.

Higher types also enter the story on account of the complexity of logical formulae. For instance, an implication formula $\phi \Rightarrow \psi$ is typically understood to mean that there is a construction which, given evidence for ϕ, transforms it into evidence for ψ. Since logical formulae often involve nested implications, as in $(\phi \Rightarrow \psi) \Rightarrow \theta$, we are led to consider computable operations that can act on other operations.

The idea of realizability interpretations, then, is to try to clarify, or at least approximate, some notion of constructive meaning by replacing the informally understood notion of a 'construction' by some explicitly defined notion of a 'computable operation', which will typically be higher-order in character. For instance, we might choose to use a computability notion embodied by an *untyped* higher-order model (as in Kleene's original approach to realizability [135], which in effect uses the model we have called K_1), or we might use a *typed* one, in which the 'type' of an operation will vary according to the shape of the formula it is witnessing (as in Kreisel's *modified realizability* [153] and its successors). Much of the interest here lies in the way in which different computability notions give rise to different realizability interpretations, reflecting subtly different kinds of constructive meaning.

There are also other interpretations that work in a somewhat different way (such as Gödel's *Dialectica interpretation* and its extension by Spector to a system of analysis), but which still share the basic feature that the meaning of the logical connectives is understood in terms of 'computable operations', which must necessarily be of a higher-order character in order to deal with complex logical formulae. In some cases, interpretations in this vein can be applied even to *classical* systems, either with the help of a proof-theoretic translation from a classical to a constructive system (such as the Friedman–Dragalin *A-translation*), or by giving a direct computational interpretation of certain classical proof principles. Indeed, the search for ways to realize certain classical rules (e.g. choice principles) has sometimes stimulated the discovery of new kinds of computable higher type operations (e.g. new forms of bar recursion: see Subsection 7.3.3).

In another broad class of interpretations (represented by the work surveyed in Feferman [94]), one makes no attempt to 'interpret' the logical connectives as such—they inherit their meanings from the ambient meta-level—rather, we use higher-order models simply as a way of constraining the class of mathematical *objects* that are deemed to exist. The idea is that in certain definitionist or predicativist views of mathematics, entities such as real numbers, or functions on reals, are only considered to 'exist' if they are explicitly definable by certain means. Since reals, functions on them, and many other mathematical entities can readily be coded up as first-, second- or third-order functions over $\mathbb{N}$, we can choose some higher-order model consisting of 'definable' functions, and regard it as our mathematical universe for the purpose of interpreting logical formulae. For instance, taking 'definable' to mean 'effectively computable' in some way would typically lead to a very restricted universe; allowing quantification over $\mathbb{N}$ as an acceptable means of definition would lead to a more generous one. In any case, computability models of the kind studied in this book can furnish a rich source of examples and inspiration.

The general programme of interpreting logical systems using higher-order operations in some way has at least three broad motivations. The first is conceptual: they can offer at least an approximate 'picture' of the mathematical universe as it might appear to a constructivist or predicativist of a certain school—a picture that is accessible to mathematicians who do not necessarily share the philosophical presuppositions of that school. Such pictures or interpretations can help to clarify the fundamental concepts underpinning a given constructive approach, and shed light on the relationships between various foundational positions.

The second motivation is metatheoretical: such interpretations can lead to *consistency* and *independence* results for logical systems. The idea is simple: if we can give an interpretation of (say) second-order arithmetic validating some combination A of axioms (some of which might be classically false), we at least know that A is consistent. Likewise, if we can give an interpretation satisfying the combination $A \cup \{\phi\}$ and another satisfying $A \cup \{\neg\phi\}$, we know that the axiom ϕ is *independent* of the axioms in A.

This story can be given a further twist if we pay attention to the meta-level system within which the definition of the interpretation is conducted. Suppose, for instance, that we are able to construct an interpretation for some logical system $\mathbf{L}'$, and prove its soundness, by working entirely within some meta-level system $\mathbf{L}$ that is weaker than $\mathbf{L}'$. In this case, what will appear 'from inside $\mathbf{L}$' simply as a semantics for $\mathbf{L}'$ will appear 'from outside' as a syntactic translation $\widehat{-}$ from statements and proofs in $\mathbf{L}'$ to those in $\mathbf{L}$. Now suppose (for instance) that the interpretation preserves the meaning of all $\mathbf{L}$-formulae in some class C, in the sense that any $\phi \in C$ is provably equivalent (within $\mathbf{L}$) to its translation $\widehat{\phi}$. In this case, it follows that if $\phi \in C$ is provable in $\mathbf{L}'$ then $\widehat{\phi}$ is provable in $\mathbf{L}$, whence ϕ itself is provable already in $\mathbf{L}$. That is, $\mathbf{L}'$ is *conservative* over $\mathbf{L}$ for statements in C. Arguments in a similar spirit can also be used to establish other metatheoretical results, such as the *disjunction* and *existence* properties for a variety of constructive systems.

In view of the potential for such proof-theoretic applications, it is natural when developing theories of higher-order computability to have at least half an eye on the strength of the meta-level machinery required to develop them. Although in this book we shall not give details of the metasystems needed to formalize our results, we will often flag up uses of reasoning principles that are not available within a 'pure' constructive framework such as that of Bishop [37];[8] however, we make no claim to have pinpointed all such uses.

The seminal article for metamathematical applications of these kinds is Kreisel's 1959 paper [152]. For work in the area up to 1973, the indispensable reference is Troelstra's book [288]. Other useful sources include the survey articles of Feferman [94] and Troelstra [289], and more recent works by Kohlenbach [149, 148].

---

[8] Typical examples of such principles include: the law of excluded middle, the limited principle of omniscience, Markov's principle, König's lemma, and various forms of the axiom of choice.

The third motivation for computational interpretations of logical systems has to do with clarifying and extracting the computational content implicit in mathematical proofs. We discuss this important area of application under a new heading.

1.2.2 Extracting Computational Content from Proofs

The starting point for this class of applications was a question posed by Kreisel: 'What more do we know if we have proved a theorem by restricted means than if we merely know that it is true?' The idea is that a proof in some restricted (e.g. constructive or semi-constructive) system is in principle likely to be 'more informative' than an unrestricted one, and this raises the question of what useful information we can extract from it.

As a prototypical example, suppose we have a constructive proof of some statement $\forall x.\exists y.\phi(x,y)$ (the types of x and y do not matter for now). As already noted, such a proof will provide a 'construction' which, given any value of x, yields a value of y such that $\phi(x,y)$ holds (along with a proof of $\phi(x,y)$). Moreover, if a realizability interpretation of our proof system can be given, then as we have seen, the notion of 'construction' here can be replaced by some concrete notion of 'computable operation'. The upshot is that a proof of $\forall x.\exists y.\phi(x,y)$ can automatically be translated into a realizer for this formula, from which we can readily extract a computable function f mapping each x to a suitable y. Higher types play an important role here, because even if x and y (and hence f) are of 'low type' and $\phi(x,y)$ is structurally simple, the *proof* of $\forall x.\exists y.\phi(x,y)$ might involve logically complex formulae whose realizers are of high type, and these intermediate realizers play a role in the extraction of f.

Information of this kind can sometimes be extracted even from *classical* proofs. As mentioned above, a realizability interpretation (or similar) can sometimes be combined with a proof-theoretic translation to yield an interpretation of a classical system, and in good cases, the translation of a formula $\forall x.\exists y.\phi(x,y)$ will still be of the form $\forall x.\exists y.\overline{\phi}(x,y)$, so that a witnessing function f can still be extracted. In other cases, we may not get a function f that gives a precise value of y for each x, but rather a function g that yields a *bound* on the smallest suitable y (supposing y to be of natural number type).

Such 'proof mining' techniques have proved fruitful in both mathematics and computer science. In mathematics, information extracted from known proofs has led to new methods and results in approximation theory, fixed point theory and other areas (see Kohlenbach's book [149]). In computer science, the idea has suggested a possible programming methodology based on *extracting programs from proofs*. For instance, to create a program f meeting the specification that some input–output relation $\phi(x,f(x))$ should hold for all x, we can first prove the theorem $\forall x.\exists y.\phi(x,y)$, then extract from this both a program f and a proof that it is correct. Work of Berger and others (e.g. [28, 24]) has shown that, at least within certain application domains, this methodology leads to new and unexpected programs that sometimes surpass the 'obvious' implementations in terms of efficiency.

A broadly related application of higher-order computation is the technique of *normalization by evaluation* due to Berger and Schwichtenberg [27]. The idea here is that an interpretation of λ-terms as higher type operators can be used as a way of actually evaluating the λ-terms: roughly speaking, by computing the semantic interpretation of a term M and then converting it back into syntax, we can obtain a normal form for M. Somewhat surprisingly, this can in some cases be more efficient than the usual method of applying a sequence of reduction steps. There is a connection here with the ideas of proof extraction: if one takes the usual proof (due to Tait) of normalization for typed λ-calculus and extracts its computational content, one ends up with a version of the normalization by evaluation algorithm [29].

For further reading in this area, see Schwichtenberg and Wainer [268].

1.2.3 Theory of Programming Languages

Another area in which higher-order models have found application is in the theory and design of programming languages. A guiding idea here is that one ought to be able to design a programming language in a clean and principled way by starting with a mathematically pleasant model embodying some suitable concept of computation, then using our structural understanding of this model to arrive at a tidy set of language primitives for defining computable operations. An advantage of this principled approach is that it ensures that there will be a clear conceptual model that suffices for the description of program behaviour—something that may not be the case if a language is developed simply by adding in new features at will. Likewise, the pleasing mathematical structure of the model is likely to lead to clean logical principles for reasoning about programs—again, such principles are less likely to be forthcoming for a language designed in a more ad hoc way.

A process of this kind can be seen in the early history of functional programming languages such as Standard ML, Haskell and Miranda. All these have their ancestry in Scott's Logic for Computable Functions [256], a system designed as a notation for computable elements of the mathematical model of partial continuous functions, as well as a framework for reasoning about such elements using principles such as *fixed point induction*. (For more on the history, see Section 2.3.) The transparency and ease of reasoning much extolled by advocates of pure functional programming are thus due in large measure to the mathematical roots of these languages in higher-order computability models. More recently (and more speculatively), models arising in *game semantics* have been examined by Longley [182] with a view to extracting new language primitives for object encapsulation, coroutining and backtracking.

Of course, one can also proceed the other way round: start from a language embodying some commonly arising programming concept (e.g. exceptions, local state), and try to find the simplest and cleanest mathematical model for computable operations in which it can be interpreted. This, indeed, is the goal of much work in *denotational semantics*, a subject that has given rise to an extensive body of mathematical research, for instance in domain theory [2] and in game semantics [3]. The hope is that such models will give us a mathematical handle on program behaviour which leads to simpler ways of reasoning than those based directly on syntactic descriptions of program execution.

In whichever direction we proceed, the idea is to end up with a programming language $\mathscr{L}$ together with a mathematical model $\mathscr{M}$ in which it can be interpreted, in such a way that the latter provides ways to reason about the former. A typical use for such models is to show that two given programs P and P' of $\mathscr{L}$ are *observationally equivalent* ($P \sim_{\mathrm{obs}} P'$): that is, that P and P' may be freely interchanged in any larger program of which they form a part without affecting the program's observable behaviour.[9] If $[\![-]\!]$ is an interpretation of $\mathscr{L}$ in a model $\mathscr{M}$ satisfying some standard conditions, it is generally easy to show that $[\![P]\!] = [\![P']\!]$ implies $P \sim_{\mathrm{obs}} P'$. The point is that to establish $P \sim_{\mathrm{obs}} P'$ by purely syntactic means would typically be cumbersome, whereas showing $[\![P]\!] = [\![P']\!]$ is often just a matter of simple calculation.

Naturally, the better the 'fit' between $\mathscr{L}$ and $\mathscr{M}$, the more use we can make of $\mathscr{M}$ for reasoning about $\mathscr{L}$. In particular, the following two properties of interpretations are often seen as desirable:

- *Full abstraction*: $P \sim_{\mathrm{obs}} P'$ implies $[\![P]\!] = [\![P']\!]$. In other words, all observational equivalences of the language are captured by the model.
- *Definability*, also called *universality*: for every element x of $\mathscr{M}$ (of suitable type) there exists a closed program P with $[\![P]\!] = x$. In other words, the model contains no 'junk' that is not relevant to modelling the language.

For certain kinds of languages, it is shown by Longley and Plotkin [186] that if both these conditions hold, we can create a logic of programs using our understanding of the structure of $\mathscr{M}$, but which also has a direct operational interpretation in terms of $\mathscr{L}$ that could be grasped by programmers with no knowledge of $\mathscr{M}$. For one preliminary outworking of this idea, see Longley and Pollack [187].

Note too that if an interpretation satisfies both full abstraction and definability, then the relevant part of $\mathscr{M}$ is in essence nothing other than the set of closed programs of $\mathscr{L}$ modulo observational equivalence. We thus have both a syntactic and a semantic way of characterizing the same mathematical structure up to isomorphism, and these will often give different handles on its structure and properties. The search for such alternative and complementary presentations of the same computability notion will occupy a good deal of our attention in this book.

[9] Traditionally, observational equivalence of programs is often understood to mean *contextual* equivalence: e.g. we might say $P \sim_{\mathrm{obs}} P'$ if for every $\mathscr{L}$-context $C[-]$ and every observable value v, we have that $C[P]$ evaluates to v iff $C[P']$ does. However, more semantically defined notions of 'observation' can also be considered, taking account of all the ways in which a program might interact with its external environment. As argued by Ghica and Tzevelekos [103], the latter may be better suited to discussing program behaviour in real-world settings.

1.2.4 Higher-Order Programs and Program Construction

One may also ask whether the theory of higher-order models yields particular examples of higher-order programs that are useful in computing practice. To date, such examples have been rather scarce. This is largely because most of the theoretically interesting examples of higher-order operations live at *third order* or above, and whilst second-order operations are ubiquitous in modern programming, practical uses for third- or higher-order operations are something of a rarity.[10] Notable exceptions occur within the specialized field of *exact real number computation*, where non-trivial ideas from higher-order computability have been successfully applied by Simpson [272], Escardó [82] and others. There are also prospects for applications of wider significance, e.g. in the work of Escardó on the surprising possibility of exhaustive searches over certain infinite sets [81], or that of Escardó and Oliva on optimal strategies for a very general class of sequential games [86].

Aside from such possibilities, one can also consider general approaches to the *construction* of programs (or to the analysis of existing programs) which view them as composed by plugging together certain building blocks which may be of higher type. For example, much programming 'in the large' (i.e. at the level of program modules) is implicitly higher-order in character: e.g. a functor which takes an implementation of a module A and returns an implementation of B is typically a higher-order operation (and there are even languages that support higher-order functors). In another vein, preliminary work by Longley [180] suggests how a simple treatment of phenomena such as *fresh name generation* can be given by treating the relevant language constructs as higher-order operators. It may be that such use of higher-order machinery as 'scaffolding' for the development or analysis of low-order programs proves more influential in the long run than any direct practical uses of specific higher-order operations.

1.3 Scope and Outline of the Book

The present book offers an in-depth study of higher-order computability models, encompassing:

- the general theory of such models,
- the construction of particular models of interest, including equivalences between seemingly diverse constructions,
- the computability theories these models give rise to,

[10] In the standard Haskell libraries, for instance, the only third-order operations we are aware of are those associated with continuations. In the standard Java API library for reflection, a typical use of the `Method.invoke` method may be conceptually of third order or above. Readers of this book will also enjoy the proposal by Okasaki [224] regarding a practical use for a sixth-order function in the domain of combinator parsing—although the function in question is actually an instantiation of a conceptually third-order polymorphic function.

- examples of interesting computable operations within these models,
- relationships between different models.

We shall give a certain priority to material that has not previously been covered in book form—in places where the material is well covered in other textbooks, we have allowed ourselves to be somewhat more sketchy in our treatment. Some results and proofs (e.g. in Chapter 6) appear here for the first time. Furthermore, our synthesis of the material is new: whilst many of the models we consider have been studied elsewhere, we are here presenting them as instances of a common phenomenon, using the general concepts of simulation and transformation to describe the rich network of interrelationships between them.[11]

We shall approach our subject essentially as a branch of pure mathematics, with an emphasis on clarifying and consolidating the foundations of the theory, and with rather little space given to applications. The spirit of our endeavour is to try to identify and study those computability notions that have some claim to intrinsic significance, both conceptually and mathematically.

1.3.1 Scope of the Book

To give a more specific idea of the scope of the book, we mention here some of the broad 'axes of variation' among the kinds of models we shall consider.

Effective, continuous and finitary models. Our primary interest will be in 'effective' computability notions that might reasonably be regarded as higher-order generalizations of Turing computability. However, to give a rounded and coherent account of such models, it is convenient to widen our scope to take in other kinds of structures. On the one hand, we shall cover more abstract models of computability such as those based on *continuity* rather than effectivity (in the spirit of Example 1.1.4). Often this will be tantamount to considering what is effectively computable in the presence of an arbitrary 'oracle' of type $\mathbb{N} \to \mathbb{N}$. On the other hand, we shall also pay some attention to computability over *finitary* base types such as the booleans—as we shall see, continuous models can sometimes be usefully regarded as 'completions' of their finitary parts.

Domain of operations. Even if the operations in our model are effective, there remains the question of whether they act only on *effective* data (as will generally be the case for models of the form $\mathbf{C}(\mathscr{L})$), or on some wider class of data values (as with the model $\mathbf{T}_1$ of Example 1.1.1). In many situations, it makes perfectly good sense to speak of computable operations acting on potentially non-computable data.

Total versus partial models. We shall consider models in which all computable operations are total functions, as well as others in which operations may be par-

[11] Among other things, the present book fulfils the role of the projected sequels (Parts II and III) to the first author's survey paper [178].

tial. Indeed, in a higher-order context, we are naturally led to consider *hereditarily total* operations (e.g. total functions acting on total functions acting on total functions), as well as *hereditarily partial* ones. Partial functions will often be represented mathematically by total ones using a special element $\perp$ to signify an undefined result.

Data representation. Our models of computation will vary as regards the level of representation of data that computations have access to, as in the discussion of Section 1.1. For example, a data value of function type might be given to us solely as an 'oracle' or 'black box' to which inputs can be fed, or as a more *intensional* object such as a program or algorithm that can also be manipulated in other ways. Indeed, much of the richness of the subject stems from the possibility of many different levels of intensionality, typically related to one another via simulations.

Style of computation. Intertwined with the question of data representation is that of the kinds of computations we are allowed to perform. For instance, we have already noted the difference between 'sequential' and 'parallel' flavours of computation, and the impact this has on the computability of functions. Also in this category are questions such as whether a computation can jump out of a function call without completing it, or what kinds of 'internal state' a computation is allowed to maintain.

Typed versus untyped models. Many of the models we consider will have an explicit type structure. For instance, our type names might be syntactically built up from some base type name (often $\mathtt{N}$) by means of the constructor '$\rightarrow$'; we refer to these as the *simple types* (or sometimes the *finite types*) over the given base type. However, we have already seen that there are important examples with just a single type for data (e.g. K_1 or L^0).

Extensional versus intensional models. A model $\mathbf{C}$ may be called *extensional* if elements of $\mathbf{C}(\sigma \rightarrow \tau)$ can be treated simply as functions from $\mathbf{C}(\sigma)$ to $\mathbf{C}(\tau)$, regardless of the level of intensionality at which the computation of these functions takes place. In a non-extensional model, elements of $\mathbf{C}(\sigma \rightarrow \tau)$ will necessarily be presented as finer-grained objects such as algorithms or strategies.

Whilst these axes in principle indicate the extent of the territory of interest, we should note that our coverage of this territory will be far from uniform in its depth. In particular, the lion's share of the book (Chapters 5 to 11) will be devoted to *simply typed, extensional* models—that is, to classes of computable functionals of finite type, typically over a ground type of natural numbers. This is the part of the territory that has been the most comprehensively explored, and here we believe we are in a position to offer a fairly definitive map of the landscape. Other kinds of models, both 'untyped' or 'intensional', will be surveyed in a much briefer, more fragmentary way in Chapters 3 and 12, in some cases with pointers to more detailed treatments elsewhere. We emphasize, however, that our treatment of such models is integral to the book as a whole. Computational processes or algorithms are by nature intensional objects, and many of the leading constructions of extensional models of computability proceed via intensional ones; it would therefore not be intellectually coherent to study extensional models without some study of the intensional ones as well. Likewise, many important constructions of typed models proceed via untyped ones, so we are naturally led to pay attention to the latter as well as the former.

Other related classes of models could also be comprehended by further extending the above axes of variation, or by adding new ones. Some such extensions will be briefly touched on in Chapter 13.

1.3.2 Chapter Contents and Interdependences

We now proceed to a more detailed outline of the structure of the book and the contents of each chapter. The book is in two parts: Part I (Chapters 1–4) covers background and general theory, while Part II (Chapters 5–13) is devoted to the study of particular models and their associated concepts of computability.

In the present introductory chapter, we have outlined our vision of a plethora of computability notions related via simulations. However, this picture is based on relatively recent ideas, and is not representative of how the study of higher-order computability actually evolved. In Chapter 2 we supplement this picture with a more historically oriented account of the subject's development.

Chapters 3 and 4 are devoted to the general theory of higher-order models, considered at two different levels of structure. In Chapter 3 we develop more systematically the theory of (higher-order) models and simulations outlined in Section 1.1. Although the definitions in question involve only relatively weak structure, they suffice for the development of a useful body of general theory. We shall also illustrate the general concepts with some key examples that will play a pervasive role in the rest of the book. In Chapter 4 we consider the somewhat stronger notion of a *λ-algebra*—essentially a higher-order model in which the usual equations of λ-calculus hold. For such models, we are able to develop some more powerful general machinery, including a theory of embeddings or *retractions* between types, and the proof technique of *(pre)logical relations*. Most, though not all, of our models of primary interest will turn out to be λ-algebras.

From Chapter 5 onwards, we consider particular notions of higher type computability. Roughly speaking, we shall work our way from 'weaker' to 'stronger' computability notions—although we cannot follow such a scheme exactly, since the notions we wish to consider do not form a neat linear order. This will also be correlated with a general progression from purely 'extensional' flavours of computation (in which functions are treated simply as oracles) to increasingly 'intensional' ones (where more representation information is available).

We begin in Chapter 5 with the notion of higher type computable functional due to Kleene [138], originally defined in the context of *total* functionals (over $\mathbb{N}$) using his schemas S1–S9. Broadly speaking, this is the notion we are naturally led to if we insist on treating functions of higher type as oracles or black boxes in a very strict way. In Chapter 5 we cover the 'classical' theory of Kleene computability, retaining his emphasis on total functionals. In Chapter 6 we consider a more recent model based on what we call *nested sequential procedures* (NSPs), also known in the literature as *PCF Böhm trees*. An NSP can be seen as an abstract representation of the underlying 'algorithm' that a Kleene program embodies, and an approach to Kleene computability via NSPs yields some technically new proof methods and results. We also use NSPs to present various possible interpretations of Kleene computability in the setting of *partial* functionals. One such interpretation turns out to coincide in expressive power with Plotkin's PCF, a programming language much studied in the theoretical computer science literature. We devote Chapter 7 to studying PCF and its models in further detail. Our point of view throughout Chapters 6 and 7 is that nested sequential procedures capture the implicit notion of algorithm that is common to both PCF and classical Kleene computability.

For the next two chapters, we return to total functionals over $\mathbb{N}$. In Chapter 8 we study the *total continuous functionals*, defined independently and in different ways by Kleene [140] and Kreisel [152]. This turns out to be the definitive class of total functionals based on the paradigm of 'continuous functions acting on continuous data'; it has an interesting subclass of 'effectively computable functions' which turns out to be richer than the class of Kleene computable ones. In Chapter 9 we turn our attention to the *hereditarily effective operations*, the definitive class based on the paradigm of 'effective functions acting on effective data'. Interestingly, this turns out to be incomparable with the class of (effective) total continuous functions; moreover, there can be no reasonable class of total computable functionals over $\mathbb{N}$ that embraces them both (Theorem 9.1.13).

To find a single computability notion that subsumes both these classes, one must turn again to partial functionals. In Chapter 10 we consider Scott's model of *partial continuous functionals*, one of the most mathematically pleasant of all higher-order models. This (or rather its effective submodel) can also be characterized in terms of computability in PCF extended with certain 'parallel' operations. In Chapter 11, we consider a radically different way of extending PCF computability, represented by the class of *sequentially realizable* functionals. Essentially, these are functionals computable in a purely sequential manner, but taking advantage of a finer level of intensional representation than is needed for the PCF-computable functionals alone. Again, we obtain an incompatibility result: no reasonable class of partial computable functionals over $\mathbb{N}$ subsumes both the partial continuous and the sequentially realizable functionals (Theorem 11.1.4).

In Chapter 12, we turn briefly to *intensional* models—those in which objects of higher type must be presented as something more fine-grained than just functions. As mentioned earlier, there is a wide field of such models to be explored, and a comprehensive treatment is out of the question here, but we investigate just a small selection of intensional models that play significant roles elsewhere in the book: the *sequential algorithms* model of Berry and Curien, and Kleene's first and second models (see Example 1.1.11). In particular, Kleene's first model can be regarded as the limit point of our subject, in that higher-order computability here collapses to first-order computability. We conclude in Chapter 13 with a short discussion of related areas and possible directions for future research.

The diagram on page 33 serves as an approximate roadmap showing the principal computability notions considered in this book and their interrelationships; the lines typically indicate the existence of some canonical type-respecting simulation from the 'lower' notion to the 'higher' one. The diagram glosses over many details—for instance, we do not differentiate between 'effective' and 'full continuous' variants—but it should nonetheless suffice to provide some initial orientation.

One consequence of our plan of proceeding from weaker to stronger notions is that results and examples from earlier chapters often have some applicability in later ones too. Indeed, results concerning a computability model **C** often have a kind of twofold interest: on the one hand, they may give information about the mathematical structure of **C** itself, and on the other, they may have something to say about all models that support the notion of computability embodied by **C** (for instance, all models **D** that admit a simulation $\mathbf{C} \longrightarrow\!\!\triangleright \mathbf{D}$ of a certain kind). Thus, for example, much of the material on Kleene computability in Chapters 5 and 6 can be viewed as general theory applicable to many of the other models in the book. In this way, these chapters can be viewed as continuing the trajectory begun in Chapters 3 and 4, in which we develop general theories of increasing strength and increasing specificity.

However, this expository strategy also comes with a cost: many of the key constructions of more 'extensional' models depend on more 'intensional' ones discussed later in the book. We have addressed this problem by selecting the examples described in Chapter 3 so as to give the reader enough information on these intensional models to read the rest of the book in sequence.

As regards the logical interdependence of chapters, the broad principle is that Chapters 3 and 4 provide the foundation on which everything else rests, but thereafter Chapters 5 to 12 are mostly independent of one another, insofar as they deal with distinct computability notions. In practice, of course, we do not expect all readers to work through Chapters 3 and 4 in their entirety. Whilst we would advise the reader to acquire a reasonable familiarity with Chapter 3 at the outset, it will probably suffice to give Chapter 4 just a brief perusal and then refer back to it as necessary. Certain interdependences among the later chapters should be noted: Sections 6.2 and 6.4 rely on Section 5.1, while some parts of Chapter 7 rely on Section 6.1. There are also numerous other interconnections between chapters (e.g. between Chapters 8 and 9), but these are not such as to constrain the order of reading.

1.3.3 Style and Prerequisites

This book is intended to serve as an essentially self-contained exposition of higher-order computability theory, accessible for instance to beginning graduate students in mathematical logic or theoretical computer science.

Our emphasis is on conveying the broad sweep of a large tract of territory rather than doing everything in meticulous technical detail. For example, we shall sometimes give an informal description of some algorithm, then assert that it can be formalized in some particular programming language without giving full details of how to do so. Even where our treatment is somewhat sketchy, however, we have sought

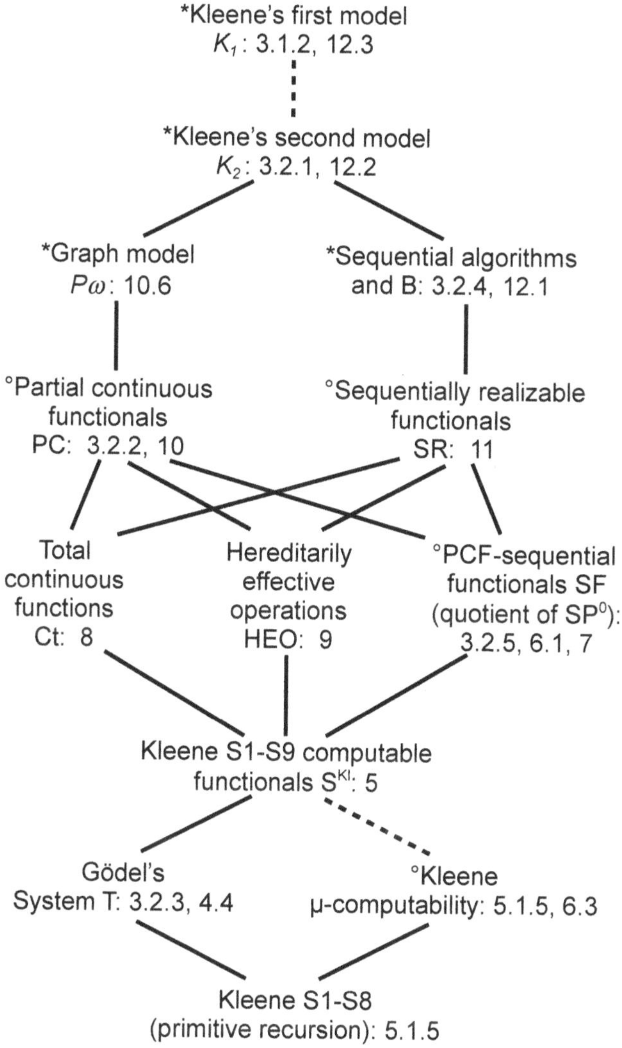

Fig. 1.1 Approximate roadmap for the main computability notions studied in this book. Numbers are chapter and section references. Intensional models are marked *; partial extensional models are marked ○; total extensional models are unmarked. Note that S1–S9 is more powerful than μ-computability, but we do not have a true inclusion here since S^{Kl} consists only of the *total* S1–S9 computable functionals. A simulation in K_1 exists only for the *effective* version of K_2.

to give enough detail to allow interested readers to fill in the rest for themselves. Results whose proofs are straightforward are often explicitly labelled as exercises.

On the whole, we have aimed at a concrete and intuitive style of presentation rather than an abstract and formal one, with a generous supply of examples and supporting intuitions (although rather little, as mentioned earlier, by way of applications). Although the abstract language of category theory plays an important role in the book, we have in general tried to economize on our use of it, taking the view that abstraction is 'a good servant but a bad master'. We have also tried not to pursue 'maximum generality' as an end in itself.

The mathematical prerequisites for reading this book are broadly as follows:

- *Computability theory*, also traditionally known as *recursion theory*. The reader should be familiar with the concepts of Turing machine (or some equivalent model of computation), total and partial computable functions on $\mathbb{N}$, Church's thesis, decidable and semidecidable (= computably enumerable) sets, the undecidability of the halting problem, Kleene's T-predicate, the s-m-n theorem, and preferably Kleene's first and second recursion theorems. The reader should also be aware of the concept of Turing machine computations relative to an *oracle* $f : \mathbb{N} \rightarrow \mathbb{N}$ (or $f : \mathbb{N} \rightharpoonup \mathbb{N}$)—this provides an important starting point for the ideas of higher-order computability. All of this material is covered in introductory texts on computability theory, for example [245, 65, 274].
- *Category theory*. We assume familiarity with the basic notions of category, functor, natural transformation, limit, colimit, adjunction, and cartesian closed category. (Knowledge of monads is not required.) The notions of 2-category and 2-functor play an occasional role in Chapters 3 and 4, but here only the basic definitions are assumed. Again, all of this is covered in basic texts, e.g. [189, 231].
- *Formal languages*. A little experience of formal languages as defined by context-free grammars is assumed, along with the notions of variable binding and substitution, and the idea of *type systems* defined via formal rules (e.g. for λ-calculi or simple programming languages). Experience of some kind of *operational semantics* (either small-step or big-step) would also be advantageous.
- *Logic*. We presuppose basic acquaintance with the language of first-order predicate logic and its classical model-theoretic semantics. We also assume basic familiarity with the concept of ordinals. In a few places we refer to the ideas of *logical complexity*—in particular the notions of Π^0_n, Σ^0_n, Π^1_n and Σ^1_n predicates—but the necessary background is summarized in Subsection 5.2.1.
- *Topology*. In Chapters 8 and 10 especially, we assume knowledge of the point-set definitions of topological space, open and closed sets, continuous map, subspace, compact space and Hausdorff space, along with some fluency in the use of these concepts.

As other books that enjoy some overlap with ours and may provide further detail in certain areas, we recommend the texts by Amadio and Curien [6], Barendregt [9], Beeson [16], Streicher [283], and Stoltenberg-Hansen, Lindström and Griffor [279].

A few notational conventions used pervasively in the book will be introduced at the start of Chapter 3.

Chapter 2
Historical Survey

As explained in Chapter 1, this book is concerned with higher-order notions of computability, and particularly with computable operations of finite type over the natural numbers. In this chapter, we briefly survey the subject's development from a historical perspective, as a complement to the more conceptually oriented overview of Chapter 1. Readers interested in a more comprehensive account of the subject's history up to around 2000 should consult the surveys by Longley [178] and Normann [215].

Our account here will give special emphasis to two themes in particular. The first (also emphasized in [178, 215]) is that the theory in its present form is the result of a sometimes joint, but often independent, effort from many different research communities. The second is that many of the key ideas originated not from the study of computability itself, but from logic and foundational investigations, and it was only later that their significance for computability theory was fully realized.

In Section 2.1 we review some of the ideas and ingredients that pre-dated the study of higher-order computability, and out of which the subject initially emerged. In Sections 2.2 and 2.3 we survey the main lines of research on computability at higher types from around 1959 to 2000, while in Section 2.4 we mention some trends in more recent and current research.

2.1 Formative Ingredients

2.1.1 Combinatory Logic and λ-Calculus

Clearly, it is impossible to draw clear boundaries to the history of a subject, but we choose to start in the 1920s and 1930s with the introduction of *combinatory logic* by Schönfinkel [262] and Curry [64], and of the *λ-calculus* by Church [51].[1] Both of these systems were syntactic calculi encapsulating very general properties

[1] Church actually worked with a variant known nowadays as the *λI-calculus*.

of functions, applications and substitution, based on a conception of functions *as rules*, in contrast to the set-theoretic view of a function as a set of pairs. Furthermore, both systems were *untyped*, and in particular admitted *self-application* (terms of the form MM)—something that made no obvious sense from a classical set-theoretic standpoint. These can be seen as the first examples of 'higher-order' systems in the sense of this book, insofar as any 'function' could also play the role of 'data'.

In the λ-calculus, the fundamental notion was that of *variable abstraction* (as in a term $\lambda x.M$), whereas the idea of combinatory logic was to eliminate bound variables entirely in favour of constants, principally those known as k and s (cf. Definition 3.1.16). Despite this difference, there are well-behaved translations between the two systems, and in many respects they can be seen as capturing the same phenomena. In this book, we shall use combinatory logic as the basis for our general theory in Chapter 3, and λ-calculus for the stronger theory of Chapter 4.

Both combinatory logic and the λ-calculus were originally motivated by logical and foundational concerns: for instance, what we now call the λ-calculus was originally part of a larger system which attempted (albeit unsuccessfully) to provide an alternative foundation for mathematics. However, Church soon realized that the λ-calculus was in practice powerful enough to express all known algorithmically computable functions on the natural numbers (represented by *Church numerals* as in Example 1.1.3). This led him to propose λ-definability as a *definition* of the notion of 'effective calculability' for such functions [52], and in this way, computability theory was born. The mathematical study of the class of λ-definable functions was vigorously pursued by Church's student Kleene [133].

In the decades following these ground-breaking developments, however, the λ-calculus had little obvious impact on computability theory, at least as studied by mathematicians and logicians. Kleene himself came to prefer the alternative (equivalent) definitions of computability via Turing machines or Herbrand–Gödel equations, and it is nowadays generally accepted that Turing's analysis [292] provides a sounder conceptual foundation for computability theory than Church's.

In contrast to this, the computational potential of the λ-calculus was enthusiastically taken up by early workers in computer science. The basic constructs of the λ-calculus (although with a somewhat different meaning) were incorporated by McCarthy into the LISP programming language [190], and the λ-calculus was provided with a machine-oriented semantics by Landin [169], who used it to explain the mechanism of procedure abstraction and invocation in ALGOL 60. The language CUCH, developed by Böhm and others [39], was a further step in the realization that the ideas of Curry and Church could form the basis of a usable programming language (as opposed to a purely theoretical analysis of effective computability). In this way, the link between λ-calculus and computation was once again reinforced.

An important difference in attitude between mathematicians and computer scientists manifested itself early on. Mathematicians tended to view formal languages as ways of referring to mathematical structures that exist quite independently of the languages, whilst computer scientists generally considered languages as purely symbolic systems which need not 'refer' to anything beyond themselves (though mathematical models might help us to understand them better). This partly explains why the *untyped* λ-calculus, in particular, was willingly accepted by computer scientists as a tool, but treated with some suspicion by mathematicians in view of the difficulty of ascribing any clear meaning to its concept of self-application (see for example Scott [256, 260]). This cultural divide, and the efforts to bridge it, have had a significant shaping effect on the development of our subject.

No such reservations applied, however, to the *simply typed* version of the λ-calculus, as introduced in 1940 by Church [53]. Here, types are built up inductively from one or more basic types by means of the function type constructor $\rightarrow$, and the associated typing rules imposed on terms mean that self-application is excluded. The existence of a simple set-theoretic interpretation of this system was clear from the outset, and Church's student Henkin [115] also formulated a more general notion of 'model' for it. Once again, Church's motivation was to provide an alternative logical foundation for a significant portion of mathematics, not to introduce some concept of computability: indeed, in its pure form, the simply typed λ-calculus can only express a rather limited class of functions $\mathbb{N} \rightarrow \mathbb{N}$. However, the work of Church and Henkin bequeathed the general concept of a *type structure* (see Section 3.6.4), which later became one of the foundation stones of higher type computability, and plays a key role in the present book.

For more on the history of λ-calculus and combinatory logic, we warmly recommend the detailed survey by Cardone and Hindley [47].

2.1.2 Relative Computability and Second-Order Functionals

Another formative idea that developed early on—this time coming from within computability theory itself—was that of *relative computability*. Already in [293], Turing had formulated the idea of computing one function $\mathbb{N} \rightarrow \mathbb{N}$ using another as an *oracle*. This is a basic classical notion, and provides the foundation for degree theory. Although Turing's paper only considered computation relative to a *fixed* oracle, it is a small step from here to regard the choice of oracle as itself an input to the algorithm; in this way, the notion of a Turing machine working with an oracle of type $\mathbb{N} \rightarrow \mathbb{N}$ yields a definition of a class of *second-order* computable functionals. This step was made explicit by Kleene [136].

At the same time, it was recognized that there was another possible way to 'compute with' functions $f : \mathbb{N} \rightarrow \mathbb{N}$ if the latter were themselves computable: one could simply perform an ordinary first-order computation with a Kleene index e for f as its input. However, a remarkable theorem from 1957 by Kreisel, Lacombe and Shoenfield [155] showed that these two quite different approaches give rise to exactly the same class of total computable second-order functionals. A somewhat analogous result for *partial* functionals acting on partial functions had been obtained in 1955 by Myhill and Shepherdson [202], although this theorem did not refer to computation with oracles as such. These two landmark theorems set the pattern for many of the equivalence results that were to follow—precise statements will be given as Theorems 9.2.1 and 10.4.4.

2.1.3 Proof Theory and System T

A further source of inspiration came from proof theory. As outlined in Subsection 1.2.1, the idea is to consider higher-order operations as witnesses for the truth of logical statements, and use these to obtain information about logical systems.

The first person to do this was Kleene [135], who used the untyped model we call K_1 (see Examples 1.1.11 and 3.1.9) to give a *realizability interpretation* in this vein for first-order Heyting Arithmetic. Another early example was provided by Gödel's System T [108], a simply typed λ-calculus over the natural numbers that extends the classical notion of primitive recursion to all finite types (see Subsection 3.2.3). Gödel used this to give his so-called *Dialectica interpretation* of first-order Heyting Arithmetic in terms of *quantifier-free* formulae involving objects of higher type.

Once again, the motivation was metamathematical, and the goal of defining a class of 'computable operations' was rather far from Gödel's intentions (his presentation was in terms of equational axioms rather than reduction rules). Nonetheless, his system provided a spur to the development of higher type computability notions. Later workers such as Tait [285] did treat System T more as a 'programming language', giving reduction rules and proving normalization results. One measure of the complexity of System T is that such normalization proofs demand more than Peano Arithmetic (PA), since the consistency of PA is an elementary consequence of normalization. An excellent survey of later research related to System T can be found in Troelstra's introductory note to the reprint in [109] of Gödel's [108].

2.2 Full Set-Theoretic Models

Whilst the above ideas were all to play important roles, none of them by themselves offered anything like a full-blown generalization of Church–Turing computability to all finite type levels. The subject came of age around 1959, when several such generalizations were proposed. In this section and the next, we survey these proposals and the respective strands of research they gave rise to.

As already hinted, there are broadly two ways to introduce a class of computable functionals: we can either give a standalone syntactic formalism or 'programming language' that gives rise to them (as with System T), or start with some mathematically given type structure of higher-order functionals and provide some way to single out the computable ones. In this section, we will consider three examples of the latter approach: Kleene's definition via the schemes S1–S9, Platek's alternative approach using partial functionals, and finally Kleene's later, more algorithmic approach. All of these are in some way concerned with the *full* set-theoretic type structure S, in which $\mathsf{S}(\mathtt{N}) = \mathbb{N}$ and $\mathsf{S}(\sigma \to \tau)$ consists of all total functionals from $\mathsf{S}(\sigma)$ to $\mathsf{S}(\tau)$. We draw attention to these examples both because they are of interest in themselves, and because they represent stages of development displaying a progression from logical to computational concerns, and paving the way for the computer science investigations that we shall meet in Section 2.3.

2.2.1 Kleene

In 1959, Kleene published a major paper [138] proposing a generalization of his concept of 'computation with oracles' to all finite types. Yet again, the initial motivation came from logic. Kleene had successfully developed a theory of *logical complexity* for first- and second-order predicate formulae (see Subsection 5.2.1), and now wanted to generalize this to higher-order logic. Since Kleene's approach to logical complexity was grounded in a notion of 'primitive recursive' predicate, and some of the basic theorems also referred to a general notion of 'semidecidable' predicate, a generalization of these concepts to all finite types was called for.

A further motivation stemmed from Kleene's own analysis of one particular logical complexity class: the Δ^1_1 *sets*. Kleene had shown that these coincide exactly with the *hyperarithmetical* sets—the class obtained by iterating the *Turing jump* operator through the computable ordinals. (In degree theory, if a degree d is represented by a set $D \subseteq \mathbb{N}$, the *jump* of d consists of all sets Turing equivalent to the halting problem relative to D.) Kleene's idea was to capture the power of the jump operator using the second-order functional ${}^2\exists$ embodying existential quantification over $\mathbb{N}$, and then to consider the process of testing membership of a hyperarithmetical set as an infinitary 'computation' using ${}^2\exists$ as an oracle. Once again, all of this entailed the development of an appropriate notion of computability at higher types.

Kleene's main definition in [138] consisted of nine computation schemes, known as S1–S9, which together constituted an inductive definition of a relation

$$\{e\}(\Phi_0,\ldots,\Phi_{n-1}) = a\,,$$

which we may read as 'algorithm number e with inputs $\Phi_0,\ldots,\Phi_{n-1}$ terminates yielding the result a'. Here $\Phi_0,\ldots,\Phi_{n-1}$ are elements of S which may be of arbitrarily high (pure) type. This was the first serious attempt to define a general concept of computation for such functionals; we shall study it in detail in Chapter 5.

Once Kleene's definition of computability was in place, his approach to the hyperarithmetical hierarchy could be readily generalized to higher quantifiers ${}^k\exists$ (embodying quantification over type $k-2$) and other functionals of interest. Computing relative to ${}^3\exists$, for instance, yields a *hyperanalytical* hierarchy, a transfinite extension of the hierarchy of second-order formulae arising in analysis. For much of the 1960s and 1970s, a significant body of research concentrated on computability relative to the so-called *normal* functionals (those functionals F of type $k \geq 2$ such that ${}^k\exists$ is computable relative to F), with key contributions by Moschovakis, Gandy, Grilliot, Sacks, MacQueen, Harrington, Kechris and others. Normann's *set recursion* represented a further extension of this theory; contributions to set recursion were also made by Sacks and Slaman.

Insofar as they concern infinitary 'computations' relative to highly non-computable objects, normal and set recursion are best seen as contributions to *definability theory* rather than computability theory proper (although we shall glance at these topics in Section 5.4 and Subsection 13.2.3). We suggest Sacks [248] for a thorough treatment of these areas, and Normann [218] for references to related work.

2.2.2 Platek

Although Kleene's definition represented a new way of thinking about computing using oracles, and provided logicians with a powerful tool in definability theory, there were some difficulties with his definition.

One problem is that Kleene essentially *axiomatizes* the existence of 'universal' algorithms, rather than deriving this from some more primitive definition. Specifically, to each Kleene algorithm there is attached an index e, a natural number that codes the algorithm. Then one of his schemes, the *enumeration scheme* S9, simply states that we can execute an algorithm uniformly in the index, and it is possible here that the algorithm may itself involve further invocations of S9. Although the construction is mathematically sound, there is a flavour of self-reference in Kleene's concept. One might consider replacing the enumeration scheme with one for *minimization* (that is, searching for the smallest natural number that passes a given test), but this considerably diminishes the computational power of the system, and the connection with definability theory is lost.

Another problem is the restriction of Kleene's concept to the realm of *total* functionals. The consequence is that composing two computable operations is problematic: the first may yield a partial functional as its output, whilst the second expects a total one as its input. This problem and its ramifications are discussed in Chapter 5.

In his thesis, Platek [233] gave an alternative definition of computable functionals that avoids both these problems. First of all, he extended the Kleene computable functionals, not only to all finite types, but to what he called *hereditarily consistent partial functionals* as well. At the base level, there are just the integers, but at higher levels Platek considered those partial functionals F that are 'consistent' in the sense that if $F(x) = n$ and the graph of y extends that of x, then $F(y) = n$. (In modern parlance, such functionals are usually called *monotone*.) In particular, if a functional F is defined on two compatible inputs x and x', then $F(x) = F(x')$.

Each Kleene functional corresponds to a certain equivalence class of *hereditarily total* Platek functionals, so that Platek's world is in a sense an extension of Kleene's world. However, the thoroughgoing use of partial functionals obviates the problems with composition mentioned above. This can be seen as a significant step towards a more natural treatment of computable operations, insofar as any notion of general computation will inevitably give rise to partial functions.

Secondly, Platek used a system of (typed) *combinators* to define his computable operations. Most significantly, his primary definition of computations is devoid of Kleene indices, but instead uses typed *least fixed point* operators. An ordinal-indexed increasing chain of hereditarily consistent functionals of a fixed type will have a least upper bound that is hereditarily consistent as well. As a consequence, each hereditarily consistent functional F of type $\sigma \to \sigma$ will have a least fixed point of type σ, obtained by iterating F transfinitely on the empty input, taking least upper bounds at limit stages. Whilst Platek's construction involves 'computations' of transfinite length relative to functionals of immense complexity, the idea of fixed points was later to have a major impact in computer science (see Section 2.3).

Platek compared his combinatory definition using least fixed point operators with Kleene's approach using indices and schemes. He proved that as long as the latter was interpreted in the Platek model, the two definitions are equivalent. Platek's thesis is not generally available, but a published account of the material can be found in Moldestad [196]; see also Section 6.4.

There were also various attempts, beginning with Moschovakis [199], to find axioms capturing the properties of current generalizations of computability theory (see Fenstad [96] for a survey). In retrospect, we can see that Platek's computation theory for hereditarily consistent functionals fails to satisfy some of the axioms then considered to capture the soul of computability. In particular, in a Platek computation relative to a functional, there is not in general a *unique* minimal computation tree representing the exact information leading from the input functional to the output. (In the terminology of [199], Platek's notion is only a *pre-computation theory* rather than a *computation theory* in the full sense.) We may note, however, that many subsequent computer science notions likewise fail to satisfy these axioms, suggesting that the fault may lie more with the axioms than with Platek's definition.

2.2.3 Kleene Again

Kleene realized that there were problems with his approach from [138, 139]. In a series of papers [141, 142, 143, 144, 145], he worked out an alternative approach in which functionals are defined via intensional objects representing *algorithms* (cf. Subsection 1.3.1). This time, Kleene's stated motivation was overtly computability theoretic: he was looking for a generalization of Church's thesis to all finite types.

Kleene's construction is complex, and we shall not include a full account of it in this book, largely because the more modern frameworks of *sequential algorithms* and *nested sequential procedures* capture most of the essential ideas more crisply (see Sections 2.3 and 2.4). The basic idea, however, is transparent. The intensional description of a functional will be a decorated tree in which paths may be of transfinite length, embodying a strategy or protocol to be followed by an idealized agent computing the functional. Each node will be labelled either with the final value of the computation (in case of a leaf node) or with an oracle call to be performed next; the branches emanating from such a node correspond to different responses from the oracle. Each such tree T will then define a functional Φ_T where we compute $\Phi_T(F)$ by following the (possibly transfinite) branch of T determined by the responses of F to the oracle calls until we either reach a leaf node or find no relevant node in T. Application of one such tree to another of the appropriate type will then be a *dialogue* of ordinal length.

Kleene used this setup to give an 'intensional' semantics for a system S1–S8 + S11, where S11 embodies the first recursion theorem and does duty for Platek's fixed point operator. Historically, this was the first substantial attempt to give a mathematical model for *algorithms themselves* in a higher-type setting—a project that was later to receive a good deal of attention in computer science (see Subsection 2.3.4).

2.3 Continuous and Effective Models

The two approaches due to Kleene, and the one due to Platek, represent *generalized* notions of computability in a strong sense. Computations run in 'time' measured by transfinite ordinals, or are represented as well-founded 'computation trees' of infinite rank in which the set of branches emanating from a given node may have a cardinality well above the continuum. Thus, these theories share some features of classical computability theory, but are far removed from what could be implemented on a computer. There was, however, a parallel development of concepts closer to genuinely 'effective' computability theory, and we now turn to discussions of some of these. Here we still find motivations from logic and the foundations of mathematics, but we will also see that the concepts in question are of interest in computer science. The key difference from the highly non-constructive approaches of Kleene and Platek was based on one simple idea: we restrict our type structures by imposing a *continuity* requirement, and insist that computations must terminate within a finite number of steps if they terminate at all.

2.3.1 Kleene–Kreisel

The year 1959, in which Kleene introduced his system S1–S9, also saw the publication of two other landmark papers, by Kleene [140] and Kreisel [152].

Kleene isolated a subclass of functionals within the full type structure S: the *countable functionals*. His idea was that numbers $n \in \mathbb{N}$ contain a finite amount of information, and functions $f : \mathbb{N} \to \mathbb{N}$ contain a countable amount of information. Since there are uncountably many functions at level 1, an arbitrary functional G at level 2 could only be fully described by uncountably much information. However, if we are told that G is *continuous*, we can give a complete description in a countable way, and this can be coded by a type 1 function $g : \mathbb{N} \to \mathbb{N}$, known as an *associate* for G.

At higher type levels, Kleene considered a functional to be *countable* if its action on the set of countable inputs can likewise be coded by a type 1 associate. Kleene showed that the class of countable functionals is closed under relative S1–S9 computability, and that every S1–S9 computable functional has a computable associate.

Historically, the key idea here was the introduction of type 1 associates, and in particular of the operation for applying one associate to another. The structure defined by this is known as *Kleene's second model*; it will be one of our primary models of interest (Subsection 3.2.1 and Section 12.2), and also has important metamathematical applications (see Subsection 13.3.1).

Kreisel's paper, which appeared in the same volume as Kleene's, introduced the *continuous functionals*. He first defined a typed class of *formal neighbourhoods*, representing finite pieces of information about the behaviour of functionals. The formal neighbourhoods are ordered in a natural way by their information content, and form a lower semilattice. Kreisel's continuous functionals are then represented by equivalence classes of *filters* of formal neighbourhoods—that is, as idealized 'limit points' determined by consistent ensembles of information about them.

Kreisel's prime motivation was to provide a tool for analysing the constructive content of statements of analysis (i.e. second-order number theory). Using Gödel's Dialectica interpretation [108], he transcribed each formula A in analysis to one of the form

$$\exists \Psi . \forall \phi . A^{\circ} ,$$

where A° is quantifier free, and Ψ and ϕ range over continuous functionals whose types depend on the structure of A. He then defined A to be *constructively true* if

$$\exists \Psi . (\Psi \text{ is computable} \wedge \forall \phi . A^{\circ}) .$$

Viewed as an attempt to explain the constructive meaning of statements of analysis in terms of 'simpler' statements, this was not a success. Although the above formula superficially appears to be of low quantifier complexity, the definition of the ranges of Ψ and ϕ may in principle be just as complex as the formula we are trying to interpret. Indeed, there are formulae A in analysis such that it is consistent with Zermelo–Fraenkel (ZF) set theory that A is constructively true according to Kreisel's definition, and consistent that it is not. This seems to fly in the face of what constructive truth ought to mean: if A does admit a 'constructive' demonstration, all models of ZF ought to agree on this. Nevertheless, the paper was historically important for many reasons. The metamathematical research it inspired was described in Section 1.2; we mention here two aspects of its influence on computability theory itself.

- Although Kreisel was interested in hereditarily total functionals, it is only a short step from his formal neighbourhoods to the compact elements used by Scott to define his *partial* continuous functionals. This latter construction provided the cornerstone for the development of denotational semantics, as we shall see below.
- The *Kreisel representation theorem* reduces the quantifier $\forall \phi$ in the transcribed form of a formula to a quantifier $\forall n$ over $\mathbb{N}$. This turned out to be very useful in developing the computability theory of Kreisel's functionals: see Section 8.4.

Despite some superficial differences between the constructions of Kleene and Kreisel,[2] it was soon realized that they were in essence defining the same extensional type structure of *total continuous functionals* over $\mathbb{N}$. In this book we will denote this model by Ct, and will study it in Chapter 8. The equivalence was first proved in [116]; we shall give a modern proof in Section 8.2.

An important point of contact between Kleene's and Kreisel's work is that both authors proved a *density theorem*, stating (loosely) that in Ct, every 'neighbourhood' determined by a consistent finite piece of information is indeed inhabited by

[2] Specifically, Kreisel's functionals were defined for all finite types and Kleene's only for the pure ones. Also, Kleene's countable functionals in principle act on both countable and non-countable arguments, although when studying their computability theory one soon realizes it is better to exclude non-countable functionals from the definition altogether.

a functional. Kreisel proved that every formal neighbourhood contains a continuous functional (or more precisely, is an element in some filter representing one), and Kleene showed that at each type level, the set of finite sequences that can be extended to an associate is primitive recursive. The proofs of these theorems are basically the same, and provide an effectively enumerated dense set at each type. The density theorem is essential in the proof of the Kreisel representation theorem.

Kreisel also defined a continuous functional to be *recursive* if it is represented by a *computably enumerable* ideal of neighborhoods. Kleene had two concepts of computability: one (agreeing with Kreisel's) induced by the computability of associates, and one induced by his S1–S9 definition. In the terminology of Section 1.3, these respectively correspond to computing with intensional representations and with the extensional functions themselves. Tait [284] showed that these two computability notions are not equivalent: the famous *fan functional* (dating back to ideas of Brouwer) is computable at the intensional level but not the extensional one.

In the 1960s there was little research activity on Ct. This changed in the 1970s and early 1980s with contributions from Gandy, Hyland, Bergstra, Wainer, Normann and others; we shall describe these developments in Chapter 8.

Kreisel [151, 152] also considered *effective* type structures, in which all functionals are specifiable with only a *finite* amount of information; such structures were rediscovered independently by Troelstra around 1970. One approach is to modify the definition of Ct so as to involve only *computable* associates, or equivalently computably enumerable filters. Alternatively, we can replace associates by Kleene-style indices $e \in \mathbb{N}$ (embodying only finitely much information); a type $k+1$ function is thus representable by a computable function acting on indices for type k functions, and thus by a Kleene index for such a function. The underlying applicative structure here is Kleene's first model K_1.

A very pleasing result is that these two very different approaches to building an effective type structure coincide at all type levels. This can be seen as a higher-order generalization of the Kreisel–Lacombe–Shoenfield theorem; it was noted by Kreisel in [152], and a complete published proof appeared in Troelstra [288]. Interestingly, the resulting type structure, which we call HEO, is neither 'larger' nor 'smaller' than the effective submodel of Ct itself (cf. the diagram on page 33). These and other results on HEO will be covered in Chapter 9.

2.3.2 Scott–Ershov

In 1969, Scott wrote a very influential note [256], published many years later as [261]. At the time, as we have seen, the untyped λ-calculus was being used as a tool in computer science, notably in Böhm's language CUCH. The problem, as observed by Scott, was that no natural models for this calculus were known—only term models constructed from the calculus itself. Indeed, by its nature, the self-application admitted by the λ-calculus appears to contradict the *axiom of foundation* in set theory: no function (considered as a graph) can include itself in its domain.

This led Scott to claim that the untyped λ-calculus was unsatisfactory and dubious as a foundation for program semantics. He also claimed that most programs deal with *typed* data. Thus he suggested that the untyped λ-calculus should be replaced by a typed system as a framework for programming and reasoning about programs.

Inspired by Platek's thesis [233], Scott constructed a logic, LCF, based on typed combinatory logic with a least fixed point operator. As a typed language containing terms for functionals and expressions for relations between them, LCF was suitable for mechanization, and had a considerable influence on the development of machine-assisted proofs. For our purposes, however, it is more significant that Scott provided LCF with a mathematical model. If we represent undefinedness by a special element $\perp$, and add to Platek's definition of the hereditarily consistent functionals a continuity condition saying that functionals must preserve least upper bounds of increasing chains, we obtain a type structure PC of *partial continuous functionals*. In this way, Scott combined Platek's use of partial functionals with the Kleene–Kreisel restriction to continuous functionals to obtain a remarkably simple and mathematically pleasing structure. The model PC will be one of our prime examples: see Subsection 3.2.2 and Chapter 10.

The historical importance of Scott's note is clear. Already by the end of 1969, he himself had used these ideas to produce mathematical models of the untyped λ-calculus [257] (thus confounding one aspect of his own objections to such a system), and had initiated the study of domain theory and the use of complete lattices as a setting for the denotational semantics of programming languages.

Other perspectives on the partial and total continuous functionals were developed in the 1970s by Ershov, using the topological framework of *complete f_0-spaces* (closely related to Scott domains), along with his theory of *enumerated sets* (see Ershov [75]). In the terminology of Subsection 1.1.5, an enumerated set is a certain kind of assembly over Kleene's first model K_1—that is, a set in which each element is equipped with some set of *indices* $e \in \mathbb{N}$. Ershov showed that enumerated sets and complete f_0-spaces both gave rise to the same type structure $\mathsf{PC}^{\mathrm{eff}}$, the natural 'effective submodel' of PC. This can be viewed as a higher type generalization of the Myhill–Shepherdson theorem. Ershov also characterized the Kleene–Kreisel model Ct via the hereditarily total elements of PC [72, 74]; this can be seen as a cleaned-up version of Kreisel's construction via formal neighbourhoods, and is the construction of Ct favoured by most modern authors. We shall follow this approach in Chapter 8.

The structure PC also admits an interpretation of Kleene's schemes S1–S9. Indeed, following an argument due to Platek [233], S1–S9 has exactly the same expressive power as LCF when interpreted over PC. There are, however, some subtleties regarding the interpretation of S1–S9 here: see Section 6.4.

In fact, even for the purpose of computing hereditarily *total* functionals, S1–S9 turns out to be stronger when interpreted over PC than over Ct directly—intuitively because the totality of intermediately computed objects is no longer required (see Section 6.4.2). Indeed, Berger [18] proved that there is an S1–S9 computable representative of the fan functional. (Apparently, this was also known to Gandy, though he never published the result.) Later, Normann [212] showed that in fact *every* effective element of Ct is represented by an S1–S9 computable element of PC.

2.3.3 LCF *as a Programming Language*

The language LCF was further investigated by Milner [193] and Plotkin [236]. In a sense, the Scott model for LCF is too rich: there are effective and even finite elements of PC that are not interpretations of LCF terms. One example is the 'parallel' operator P mentioned in Subsection 1.1.6 (see also Subsection 7.3.1). Because of such elements, there will also be pairs of LCF terms that have different interpretations in the Scott model but nonetheless are *observationally equivalent* (cf. Subsection 1.2.3). We therefore say that the Scott model is not *fully abstract* for LCF. Milner showed that, within the framework of domain theory, there is exactly one fully abstract model for LCF.

Plotkin [236] reformulated LCF as PCF, replacing combinators with traditional λ-expressions, and providing the language with a sound and complete *evaluation* strategy, so that given any closed term M of type N, M will symbolically evaluate to a numeral $\widehat{n}$ iff M is interpreted as n in the Scott model. In this way, Plotkin gave a standalone syntactic definition of PCF as an implementable programming language without reference to any denotational model. The language PCF will be studied in Chapter 7; we shall henceforth use the name PCF even when referring to papers originally written in the context of LCF.

Plotkin [236] also introduced a 'parallel conditional' operator $:\supset$ and a continuous existential quantifier $\exists$ over $\mathbb{N}$, and proved that

- all finitary elements in PC are definable relative to $:\supset$ (or equivalently relative to a closely related 'parallel or' operation),
- all effective elements of PC (and only these) are definable relative to $:\supset$ and $\exists$.

Essentially the same results were independently obtained by Sazonov [251], who noted that neither $:\supset$ nor $\exists$ can be defined from the other in PCF. Sazonov also introduced the idea of *degrees of parallelism*: see Subsection 10.3.3.

Milner's fully abstract model of PCF [193] is essentially a term model: its building blocks are terms in the finitary fragment of PCF, and the model is obtained as the *ideal completion* of the observational ordering on such terms. The problem with term models is that they hardly give any new information about the calculus that they are models for, since they are defined via the calculus itself. Thus a natural question arose: Is there a more 'mathematical' characterization of a fully abstract model for PCF? Whilst this so-called *full abstraction problem* was not itself mathematically precise, it led to much interesting research on the nature of higher-order computation, some of which we now survey.

2.3.4 The Quest for Sequentiality

Perhaps the most significant aspect of the full abstraction problem is that it called for an independent mathematical understanding of the notion of *sequentiality* implicit in Plotkin's evaluation strategy for PCF—informally, the idea of 'doing only one subcomputation at a time', in contrast to the 'parallel' strategies needed to compute $:\supset$ or $\exists$. This turned out to be a non-trivial problem, and generated a good deal of research in the 1980s and 1990s. (For a more detailed survey, see Ong [225].)

It is not obvious how to define the 'sequentially computable' functionals mathematically. Milner [193] and Vuillemin [297] provided equivalent definitions for first-order functions of several arguments. Even this is not as trivial as it may seem, and at higher types (or for functions between more general datatypes), things are even less obvious. We mention here two attempts which, whilst they both failed to capture the essence of PCF computability, highlighted alternative notions of sequential computability that turned out to be of interest in their own right.

The first of these is due to Berry and Curien [33], who defined a concept of *sequential algorithm*, and hence a notion of *sequential functional*, on a class of *concrete domains* previously introduced by Kahn and Plotkin [128]. Whilst this captured the idea of sequentiality in a compelling way, there was a mismatch with PCF computability in that it yielded an *intensional* type structure of 'computable operations' rather than an extensional one of computable functionals. Nonetheless, this structure turned out to be an interesting and very natural one; we will consider it in Section 12.1. (In fact, it is close to the model of Kleene mentioned in Subsection 2.2.3, as shown by Bucciarelli [43].)

Another line of attack led to Berry's *stable* functions [32], and thence to the *strongly stable* functions of Bucciarelli and Ehrhard [45]. These were both extensional type structures, but failed to exclude all non-PCF-definable finite elements. Nonetheless, it was shown by Ehrhard [69] that every strongly stable functional was in a sense computable via a Berry–Curien sequential algorithm. Later work of van Oosten [294] and Longley [176] brought this type structure further into focus as an important class of computable functionals in its own right; we will study it in Chapter 11.

As regards the original notion of PCF computability, in 1993, three closely related solutions to the full abstraction problem were obtained simultaneously and essentially independently: by Abramsky, Jagadeesan and Malacaria [5], by Hyland and Ong [124], and by Nickau [203]. All these approaches used a concept of *games* and *strategies* to model PCF-style processes at an intensional level; the resulting structure can then be collapsed to yield the desired extensional model. The version of Hyland and Ong, in particular, was influenced by Kleene's later approach as described in Subsection 2.2.3. The abovementioned papers led to an extensive body of work in *game semantics* for various extensions of PCF and other programming languages, most of which we will be unable to cover in this book for reasons of space (although see Sections 7.4 and 12.1).

The work of Sazonov should also be mentioned here. As early as the 1970s, he had characterized the class of PCF-definable elements in the Scott model (relative to functions $f : \mathbb{N} \rightarrow \mathbb{N}$) via what he called *strategies*: purified forms of PCF-style procedures that operate on functional *oracles* [251, 252]. (One can in fact obtain from this a construction of the Milner model, although this was not done at the time.) Thirty years later, Sazonov revisited these ideas to give a standalone construction of the PCF-computable functionals [253, 254]—in contrast to the game models, this uses only the *extensions* of such functionals at level k in order to define the functionals at level $k+1$.

There is a sense, however, in which none of these constructions of a fully abstract model achieved what had been hoped for, since even for programs in *finitary* PCF (say, PCF over the booleans), they did not yield an evident effective way to decide observational equivalence. In 1996, a major negative result was proved by Loader [172]: observational equivalence even in finitary PCF is actually undecidable. (We shall discuss this result in Subsection 7.5.3, although we will not give the proof.) This implies that one cannot hope for a fully abstract model constructed in, say, the 'extensional' spirit of traditional domain theory—the kinds of intensional models described above may therefore be the best we can do.

2.4 The Third Millennium

2.4.1 A Unifying Framework

As the above account makes clear, research on higher-order computability prior to 2000 had a tendency to fan out in many different directions, with each new definition of computability giving rise to a distinct strand of research. Since the start of the millennium, however, the trend has been for some of the strands to re-converge, or at least for the relationships between them to be made explicit, and a unifying perspective on the entire subject has begun to emerge. This has been due partly to a more serious engagement between the logic and computer science traditions, and partly to the use of the general framework of (higher-order) computability models and simulations between them, as outlined in Chapter 1. This framework arose out of ideas originally developed by Longley [173] in an untyped setting, and later generalized to encompass a wider range of models [174, 183]. Here, the ideas of combinatory logic provide some structure common to all higher-order models of interest, and interesting relationships between different models are frequently expressible as simulations. This 'stepping back' from the study of individual models to consider the space of all of them (and their interrelationships) is one of the distinguishing features of the approach taken in this book; we will present the mathematical framework for this in Chapter 3.

This general framework has also enabled us to articulate and prove new kinds of results. For instance, it has allowed us to give general definitions of the concepts of *continuous* and *effective* higher-order model, each covering a wide range of examples, and to prove non-trivial theorems that apply to *all* such models subject to some mild conditions. The most substantial example to date is given by Longley's *ubiquity theorems* [179], which identify the (total) *extensional collapse* of a wide range of type structures. Extensional collapse is a simple construction for turning a possibly partial, possibly intensional type structure into a total extensional one. There were already several known theorems (some of which we have mentioned above) stating that the collapse (relative to $\mathbb{N}$) of some particular continuous or effective type structure yielded Ct or HEO respectively. The ubiquity theorems, on the other hand, state that the collapse of *any* full continuous model satisfying certain mild conditions is Ct, while the collapse of any effective model satisfying some mild conditions is HEO. This shows that Ct and HEO are highly robust mathematical structures and in a sense represent 'canonical' notions of intensionally computable total functional. We will discuss the ubiquity theorems in Sections 8.6 and 9.2, though space will not permit us to include full proofs.

Besides such new results, the framework of simulations has in some cases led us to more direct proofs of existing results (see e.g. Theorem 8.2.1), as well as clarifying how the different parts of the subject fit together.

2.4.2 Nested Sequential Procedures

Another noteworthy trend in recent research has been the prominence given to what we shall call *nested sequential procedures* (NSPs) as a fruitful way of analysing both PCF-style and S1–S9-style computability. An NSP is in essence a tree representation of a Sazonov-style algorithm for nested sequential computation with oracles, and the NSP framework is in a sense simply a recasting of Sazonov's ideas. At the same time, NSPs capture many of the essential intuitions behind Kleene's work, and have much in common with the spirit of his later approach. In Chapters 6 and 7, we shall present NSPs as a way of articulating the implicit notion of algorithm that lies behind both PCF and Kleene computability, consolidating and extending previous work. We here briefly mention some specific results and problems that have been the subject of attention.

At the turn of the millennium, the following important question was left open: Is every element of Milner's model computable by an NSP? (The question can equivalently be posed in terms of other intensional models such as game models.) Since both Milner's model and the NSP model can be defined as the ideal completions of their finite elements, the problem can be rephrased as:

> Does the ideal completion of the observational quotient of the finite NSP model coincide with the observational quotient of its ideal completion?

Normann [214] proved that the answer is no: there is an increasing sequence (in the observational ordering) of finite sequential functionals of pure type 3 that is not bounded by any sequential functional. This in turn led to two further kinds of questions:

1. In the observational quotient of the NSP model, which chains have least upper bounds, and can we restrict our concept of continuity to the preservation of least upper bounds of such chains?
2. How far are the sequential functionals from being chain complete, i.e. how common are sequences like the one constructed in [214]?

Such questions were addressed by Escardó and Ho [84], by Sazonov [253, 254] and by Normann and Sazonov [221]; see also Section 7.6. However, the full picture is not yet clear, and there is room for further research.

Finally, nested sequential procedures have proved valuable in giving a combinatorial handle on notions weaker than full PCF (or S1–S9) computability. As we shall see in Chapter 6, structural conditions on NSPs such as *left-well-foundedness* can be used to pick out interesting submodels; these have led to some new technical results, such as the non-computability of the well-known *bar recursion* operator in System T plus minimization (see Subsection 6.3.4).

2.5 Summary

In this chapter we have surveyed the main landmarks in the history of higher-order computability, or at least those that are of significance for the approach we adopt in this book. The subject has its origin in mathematical logic and computability theory, but since Scott's seminal note [256], the computer science aspects have come to the fore, and the influence on the subject from theoretical computer science has been profound.

Researchers who worked on this subject from the perspective of mathematical logic and generalized computability theory have (with the exception of Platek and of Kleene in his later work) focussed mainly on total functionals, and primarily on extensional type structures. This was natural, because traditional logical formalizations of mathematics tend to use total objects (for example, the possibility of non-denoting expressions can introduce complications into a two-valued classical logic), and because objects such as sets and functions in traditional mathematics are extensionally conceived. Researchers from computer science, on the other hand, focussed more on partial operations, and on intensional objects. Again, this was natural, since a program is essentially an intensional object, and in order to analyse the behaviour of programs we have to take account of the fact that they do not always terminate.

However, independently of the primary interests of the researchers involved, the most important achievements have arguably been of a mathematical nature rather than a practical one. Ideas from topology, proof theory, computability theory and category theory have combined to yield a subject that in a sense is interdisciplinary with respect to methods, motivations and applications, but which nevertheless enjoys its own coherence and autonomy as a branch of pure mathematics. It is this view of the subject that we shall seek to present in this book.

Chapter 3
Theory of Computability Models

We now begin the formal development of our material. Our purpose in this chapter is to set up the general framework within which we shall work, consolidating on the notions of computability model, simulation and equivalence as outlined in Section 1.1, and introducing several important examples.

In Section 3.1 we discuss our basic notion of 'higher-order computability model' in some detail, presenting three essentially equivalent approaches giving different perspectives on this notion, and offering some preliminary examples. In Section 3.2 we present a selection of more elaborate examples, many of which will play pervasive roles in the rest of the book. In Section 3.3 we start to explore what can be done within our general framework, developing some computational machinery common to all the models we shall consider.

Section 3.4 introduces the concept of a *simulation* and the related notion of *equivalence* between models. The scope of these concepts is illustrated in Section 3.5 with a wide range of examples. In Section 3.6 we consider some general constructions for obtaining new models from old ones.

Throughout the book, we will make frequent use of the following notational conventions. If e and e' are mathematical expressions whose value is potentially undefined (for example, because they involve applications of partial functions), we write

- $e\downarrow$ to mean 'the value of e is defined',
- $e\uparrow$ to mean 'the value of e is undefined',
- $e = e'$ to mean 'the values of both e and e' are defined and they are equal',
- $e \simeq e'$ to mean 'if either e or e' is defined then so is the other and they are equal',
- $e \succeq e'$ to mean 'if e' is defined then so is e and they are equal'.

If e is a mathematical expression possibly involving the variable x, we write $\Lambda x.e$ to mean the ordinary (possibly partial) function f defined by $f(x) \simeq e$, where x ranges over some specified domain.[1] We also allow some simple 'pattern matching' in Λ-expressions: for instance: $\Lambda(x,y).x+y$ maps a pair (x,y) to $x+y$.

[1] Warning: our use of the uppercase symbol Λ here has little to do with its use as a type abstraction operator in System F and similar calculi. We will reserve the lowercase λ for use as part of the syntax of formal languages (see e.g. Section 4.1).

We write $\mathbb{N}$ for the mathematical set of natural numbers (starting at 0), and $\mathtt{N}$ for the type of natural numbers within various formal systems. For finite sequences of length n, we shall adopt the computer science convention of indexing from 0 and write for example $x_0, \ldots, x_{n-1}$, allowing this to denote the empty sequence when $n = 0$. We sometimes abbreviate such sequences using vector notation $\vec{x}$. The set of finite sequences over a set X is denoted by X^*.

If $P(n)$ is a predicate or a boolean expression with a natural number variable n, we write $\mu n.P(n)$ for the unique n (if one exists) such that $P(0), \ldots, P(n-1)$ are false and $P(n)$ is true.

3.1 Higher-Order Computability Models

We begin by setting up the basic notion of model with which we shall work. Our exposition builds on ideas developed by the first author in [173, 179, 183], and is also influenced by the work of Cockett and Hofstra [54, 55]. The reader may find it helpful to read this section in conjunction with the discussion of Section 1.1.

3.1.1 Computability Models

Definition 3.1.1. A *computability model* $\mathbf{C}$ over a set T of *type names* consists of

- an indexed family $|\mathbf{C}| = \{\mathbf{C}(\tau) \mid \tau \in \mathsf{T}\}$ of sets, called the *datatypes* of $\mathbf{C}$,
- for each $\sigma, \tau \in \mathsf{T}$, a set $\mathbf{C}[\sigma, \tau]$ of partial functions $f : \mathbf{C}(\sigma) \rightharpoonup \mathbf{C}(\tau)$, called *operations* of $\mathbf{C}$,

such that

1. for each $\tau \in \mathsf{T}$, the identity function $id : \mathbf{C}(\tau) \to \mathbf{C}(\tau)$ is in $\mathbf{C}[\tau, \tau]$,
2. for any $f \in \mathbf{C}[\rho, \sigma]$ and $g \in \mathbf{C}[\sigma, \tau]$ we have $g \circ f \in \mathbf{C}[\rho, \tau]$, where $\circ$ denotes ordinary composition of partial functions.

Thus, a computability model is nothing other than a category of sets and partial functions. A somewhat more general notion of model, in which composites are only required to exist in a 'lax' sense, will be discussed in Section 3.1.6.

Note that $\mathbf{C}(\sigma)$, $\mathbf{C}(\tau)$ may be the same set even when σ, τ are distinct. We shall use uppercase letters $A, B, C, \ldots$ to denote *occurrences* of sets within $|\mathbf{C}|$: that is, sets $\mathbf{C}(\tau)$ implicitly tagged with a type name τ. We shall write $\mathbf{C}[A, B]$ for $\mathbf{C}[\sigma, \tau]$ if $A = \mathbf{C}(\sigma)$ and $B = \mathbf{C}(\tau)$. These conventions will often allow us to avoid explicit reference to the type names $\tau \in \mathsf{T}$.

In typical cases of interest, the operations of $\mathbf{C}$ will be 'computable' maps of some kind between datatypes. However, the data values themselves will not necessarily be 'computable' in nature: indeed, it is often of interest to consider computable operations acting on potentially non-computable data (see Example 3.1.7 below).

Note that computable operations are purely *extensional* objects: that is, they are simply partial functions from inputs to outputs. This accords with the informal idea that we are primarily trying to capture 'what' is computable rather than 'how' it is computed. However, it is as well to bear in mind that often there will be *intensional* objects (for example, programs or algorithms of some sort) lurking behind these operations. The modelling of such intensional objects will be made explicit when we introduce higher-order models in the next subsection.

Definition 3.1.2. A computability model $\mathbf{C}$ is *total* if every operation $f \in \mathbf{C}[A,B]$ is a total function $f : A \to B$.

The terms 'total' and 'partial' require a little caution. A notion of computable operation that we naturally think of as 'partial' (in the sense that non-terminating computations are admitted) might be represented either by a genuinely non-total computability model, or by a total model which builds in a special value $\bot$ to represent non-termination. For example, the model that we shall call the 'partial continuous functionals' (see Section 3.2.2) is technically a total model in the above sense. However, little confusion should arise, as the intention will generally be clear from the context.

Typically, we shall be interested in models that admit some kind of *pairing* of operations. The following is inspired by the categorical notion of products:

Definition 3.1.3. A computability model $\mathbf{C}$ has *weak (binary cartesian) products* if there is an operation assigning to each $A,B \in |\mathbf{C}|$ a datatype $A \bowtie B \in |\mathbf{C}|$ along with operations $\pi_A \in \mathbf{C}[A \bowtie B, A]$ and $\pi_B \in \mathbf{C}[A \bowtie B, B]$ (known as *projections*), such that for any $f \in \mathbf{C}[C,A]$ and $g \in \mathbf{C}[C,B]$, there exists $\langle f,g \rangle \in \mathbf{C}[C, A \bowtie B]$ satisfying the following for all $c \in C$:

- $\langle f,g \rangle(c) \downarrow$ iff $f(c) \downarrow$ and $g(c) \downarrow$.
- In this case, $\pi_A(\langle f,g \rangle(c)) = f(c)$ and $\pi_B(\langle f,g \rangle(c)) = g(c)$.

We say that $d \in A \bowtie B$ *represents* the pair (a,b) if $\pi_A(d) = a$ and $\pi_B(d) = b$.

In contrast to the usual definition of categorical products, the operation $\langle f,g \rangle$ need not be unique, since many elements of $A \bowtie B$ may represent the same pair (a,b). We do not formally require that every (a,b) is represented in $A \bowtie B$, though in all cases of interest this will be so. The reader is also warned that $\pi_A \circ \langle f,g \rangle$ will not in general coincide with f.

As a mild convenience, we assume our models come endowed with a specific choice of a datatype $A \bowtie B$ for each A and B. However, for many purposes the choice of $A \bowtie B$ is immaterial, since given another datatype $A \bowtie' B$ with the same properties, one may easily construct an operation in $\mathbf{C}[A \bowtie B, A \bowtie' B]$ that preserves representations of pairs.

Virtually all the models we shall consider can be endowed with weak products. Often, though not always, they will actually have *standard products*: that is, for each A and B we can choose $A \bowtie B$ to be a set isomorphic to $A \times B$ (with the expected projections). We shall see in Section 3.4 that every computability model with weak products is 'equivalent' to one with standard products.

The above is not the only reasonable notion of a 'model with products' that one might consider. The weaker notions of *affine* and *monoidal* products have also proved to be of considerable interest in computer science, especially in connection with linear logic (see Subsection 13.1.3). These weaker kinds of products are accounted for by the more general framework of Longley [183], but adopting either of them as our basic notion of product would open up a field too wide for a book of this length.

The following complements Definition 3.1.3 by providing a notion of nullary weak product:

Definition 3.1.4. A *weak terminal* in a computability model $\mathbf{C}$ consists of a datatype $I \in |\mathbf{C}|$ and an element $i \in I$ such that for any $A \in |\mathbf{C}|$, the constant function $\Lambda a.i$ is in $\mathbf{C}[A,I]$.

If $\mathbf{C}$ has weak products and a weak terminal (I,i), it is easy to see that for any $A \in |\mathbf{C}|$ there is an operation $t_A \in \mathbf{C}[A, I \bowtie A]$ such that $\pi_A \circ t_A = id_A$.

3.1.2 Examples of Computability Models

To provide some initial orientation, we give here three simple but important examples of computability models—we shall use these as running examples throughout this section and the next. Several more examples will be given in Section 3.2.

Example 3.1.5. Perhaps the most fundamental of all computability models is the model with the single datatype $\mathbb{N}$ (where we think of natural numbers as representative of all kinds of discrete finitary data), and whose operations $\mathbb{N} \rightharpoonup \mathbb{N}$ are precisely the Turing-computable partial functions (cf. Example 1.1.1). This model has standard products, since the well-known computable pairing operation

$$\langle m,n \rangle \;=\; (m+n)(m+n+1)/2 + m$$

defines a bijection $\mathbb{N} \times \mathbb{N} \to \mathbb{N}$ (and the remainder of Definition 3.1.3 is clearly satisfied). Any element $i \in \mathbb{N}$ may serve as a weak terminal, since $\Lambda n.i$ is computable.

Once this model has been endowed with some additional structure in the next subsection, it will be referred to as *Kleene's first model* or K_1.

Example 3.1.6. Recall from Example 1.1.3 that the terms of the *untyped* λ*-calculus* are generated from a set of variable symbols x by means of the following grammar:

$$M \;::=\; x \mid MM \mid \lambda x.M \,.$$

We shall assume familiarity with the notions of free and bound variable and that of α*-conversion* (renaming of bound variables). Strictly speaking, we shall define Λ to be the set of untyped λ-terms modulo α-equivalence, though for convenience we will usually refer to elements of Λ simply as terms. We write $M[x \mapsto N]$ for the result of substituting N for all free occurrences of x within M, renaming bound variables in M as necessary to avoid the capture of free variables within N.

Let $\sim$ be any equivalence relation on Λ with the following properties:

$$(\lambda x.M)N \sim M[x \mapsto N] , \qquad M \sim N \Rightarrow PM \sim PN .$$

(The first of these is known as the *β-rule*.) From these two properties, we can derive $M \sim N \Rightarrow MP \sim NP$. As a key example, we may define $=_\beta$ to be the smallest equivalence relation $\sim$ satisfying the above properties and also

$$M \sim N \Rightarrow \lambda x.M \sim \lambda x.N .$$

Writing $[M]$ for the $\sim$-equivalence class of M, any term $P \in \Lambda$ induces a well-defined mapping $[M] \mapsto [PM]$ on $\Lambda/\sim$. The mappings induced by some P in this way are called *λ-definable*.

We may regard $\Lambda/\sim$ as a total computability model: the sole datatype is $\Lambda/\sim$ itself, and the operations on it are exactly the λ-definable mappings. It also has weak products: a pair (M,N) may be represented by the term *pair* M N where *pair* $= \lambda xyz.zxy$, and suitable projections are induced by the terms *fst* $= \lambda p.p(\lambda xy.x)$ and *snd* $= \lambda p.p(\lambda xy.y)$. (It is easy to check that *fst* (*pair* M N) $\sim M$ and *snd* (*pair* M N) $\sim N$.) However, it can be shown that this model does not possess a *standard* product structure (the reader familiar with Böhm trees and their partial ordering may find this an interesting exercise). Once again, any element at all can play the role of a weak terminal.

We also obtain a submodel $\Lambda^0/\sim$ consisting of the equivalence classes of *closed* terms M, with operations induced by closed terms P.

Example 3.1.7. Let B be any family of *base sets*, and let $<B>$ denote the family of sets generated from B by adding the singleton set $1 = \{()\}$ and closing under binary products $X \times Y$ and set-theoretic function spaces Y^X. We shall consider some computability models whose family of datatypes is $<B>$.

First, we may define a computability model $\mathsf{S}(B)$ with $|\mathsf{S}(B)| = <B>$ (often called the *full set-theoretic model over* B) by letting $\mathsf{S}(B)[X,Y]$ consist of all set-theoretic functions $X \to Y$ for $X,Y \in <B>$: that is, we consider all functions to be 'computable'. However, this model is of limited interest since it does not represent an interesting concept of computability.

To do better, we may start by noting that whatever the 'computable' functions between these sets are supposed to be, it is reasonable to expect that they will enjoy the following closure properties:

1. For any $X \in <B>$, the unique function $X \to 1$ is computable.
2. For any $X,Y \in <B>$, the projections $X \times Y \to X$, $X \times Y \to Y$ are computable.
3. For any $X,Y \in <B>$, the application function $Y^X \times X \to Y$ is computable.
4. If $f : Z \to X$ and $g : Z \to Y$ are computable, so is their pairing $\langle f,g \rangle : Z \to X \times Y$.
5. If $f : X \to Y$ and $g : Y \to Z$ are computable, so is their composition $g \circ f : X \to Z$.
6. If $f : Z \times X \to Y$ is computable, so is its transpose $\widehat{f} : Z \to Y^X$.

One possible (albeit rather conservative) approach is therefore to start by specifying some set C of functions between our datatypes that we wish to regard as 'basic computable operations', and define a computability model $\mathsf{K}(B;C)$ over $<B>$ whose operations are exactly the functions generated from C under the above closure conditions. Needless to say, all such models $\mathsf{K}(B;C)$ are total and have standard products. In view of condition 1, the element $()$ serves as a weak terminal.

As a typical example, take $B = \{\mathbb{N}\}$; we shall often denote $\mathsf{S}(\{\mathbb{N}\})$ simply by S. Let C consist of the following basic operations: the zero function $\Lambda().0 : 1 \to \mathbb{N}$; the successor function $suc : \mathbb{N} \to \mathbb{N}$; and for each $X \in <B>$, the primitive recursion operator $rec_X : (X \times X^{X\times\mathbb{N}} \times \mathbb{N}) \to X$ defined by

$$rec_X(x,f,0) = x\,, \qquad rec_X(x,f,n+1) = f(rec_X(x,f,n),n)\,.$$

As we shall see in Section 3.2.3, the resulting model $\mathsf{K}(B,C)$ consists of exactly those operations of S definable in Gödel's *System T*.

3.1.3 Weakly Cartesian Closed Models

Our main objects of study will be 'higher-order' computability models—those in which operations (or entities representing them) can themselves play the role of data. We now give some definitions that capture this idea.

It turns out that there is something of a gap between the definitions that are easiest to motivate conceptually within the general framework of computability models, and those that are technically most convenient to work with in practice. To span this gap, we shall actually give a sequence of three essentially equivalent definitions (here and in the next two subsections) showing how the conceptually and practically motivated approaches are related.

For the most conceptually transparent definition, it is convenient to assume the presence of weak products. Readers familiar with category theory will notice the resemblance to the definition of a *cartesian closed category*.

Definition 3.1.8. Suppose $\mathbf{C}$ has weak products and a weak terminal. We say $\mathbf{C}$ is *weakly cartesian closed* if it is endowed with the following for each $A,B \in |\mathbf{C}|$:

- a choice of datatype $A \Rightarrow B \in |\mathbf{C}|$,
- a partial function $\cdot_{AB} : (A \Rightarrow B) \times A \rightharpoonup B$, external to the structure of $\mathbf{C}$,

such that for any partial function $f : C \times A \rightharpoonup B$ the following are equivalent:

1. f is represented by some $\overline{f} \in \mathbf{C}[C \bowtie A, B]$, in the sense that if d represents (c,a) then $\overline{f}(d) \simeq f(c,a)$.
2. f is represented by some total operation $\widehat{f} \in \mathbf{C}[C, A \Rightarrow B]$, in the sense that

$$\forall c \in C, a \in A.\ \widehat{f}(c) \cdot_{AB} a \simeq f(c,a)\,.$$

As explained in Chapter 1, the idea here is that operations $A \rightharpoonup B$ can be represented by data of type $A \Rightarrow B$; moreover, the passage between operations and data values may be effected uniformly in a parameter of type C. It follows easily from the above definition that $\cdot_{AB}$ is itself represented by an operation $app_{AB} \in \mathbf{C}[(A \Rightarrow B) \bowtie A, B]$.

Crucially, and in contrast to the definition of cartesian closed category, there is no requirement that $\widehat{f}$ is unique. This highlights an important feature of our framework: in many models of interest, elements of $(A \Rightarrow B)$ will be *intensional* objects (programs or algorithms), and there may be many intensional objects giving rise to the same partial function $A \rightharpoonup B$.

We may now verify that each of our three examples from Section 3.1.2 is a weakly cartesian closed model.

Example 3.1.9. Consider again the model of Example 3.1.5, comprising the partial Turing-computable functions $\mathbb{N} \rightharpoonup \mathbb{N}$. Here $\mathbb{N} \Rightarrow \mathbb{N}$ can only be $\mathbb{N}$, so we must provide a suitable operation $\cdot : \mathbb{N} \times \mathbb{N} \to \mathbb{N}$. This is done using the concept of a *universal Turing machine*. Let $T_0, T_1, \ldots$ be some sensibly chosen enumeration of all Turing machines for computing partial functions $\mathbb{N} \rightharpoonup \mathbb{N}$. Then there is a Turing machine that accepts two inputs e, a and returns the result (if any) of applying the machine T_e to the single input a. We may therefore take $\cdot$ to be the partial function computed by U.

Clearly, the partial functions $f : \mathbb{N} \times \mathbb{N} \rightharpoonup \mathbb{N}$ representable within the model via the pairing operation from Example 3.1.5 are just the partial computable ones. We may also see that these coincide exactly with those represented by some total computable $\tilde{f} : \mathbb{N} \to \mathbb{N}$, in the sense that $f(c,a) \simeq \tilde{f}(c) \cdot a$ for all $c, a \in \mathbb{N}$. One half of this is immediate: given a computable $\tilde{f}$ the operation $\Lambda(c,a).\tilde{f}(c) \cdot a$ is clearly computable. The other half is precisely the content of Kleene's *s-m-n theorem* from basic computability theory (in the case $m = n = 1$): for any Turing machine T accepting two arguments, there is a machine T' accepting one argument such that for each c, $T'(c)$ is an index for a machine computing $T(c,a)$ from a.

When endowed with this weakly cartesian closed structure, this computability model is known as *Kleene's first model* or K_1.

Example 3.1.10. Now consider the model $\Lambda/{\sim}$; we shall write L for the set $\Lambda/{\sim}$ considered as the sole datatype in this model. Recall the terms *pair*, *fst*, *snd* from Example 3.1.6, and set $\mathsf{L} \Rightarrow \mathsf{L} = \mathsf{L} \bowtie \mathsf{L} = \mathsf{L}$. We may obtain a weakly cartesian closed structure by letting $\cdot$ be given by application. If $M \in \Lambda$ induces an operation in $[\mathsf{L} \bowtie \mathsf{L}, \mathsf{L}]$ representing some $f : \mathsf{L} \times \mathsf{L} \to \mathsf{L}$ then $\lambda x.\lambda y.M(pair\,x\,y)$ induces the corresponding operation in $[\mathsf{L}, \mathsf{L} \Rightarrow \mathsf{L}]$; conversely, if N induces an operation in $[\mathsf{L}, \mathsf{L} \Rightarrow \mathsf{L}]$ then $\lambda z.N(fst\,z)(snd\,z)$ induces the corresponding one in $[\mathsf{L} \bowtie \mathsf{L}, \mathsf{L}]$.

Example 3.1.11. For models of the form $\mathsf{K}(B;C)$, we naturally define $X \Rightarrow Y = Y^X$ and take $\cdot_{XY}$ to be ordinary function application. These models are endowed with binary products, and it is immediate from closure condition 6 in Example 3.1.7 that they are weakly cartesian closed (and even cartesian closed in the usual sense).

Such models show that not every element of $X \Rightarrow Y$ need represent an operation in $\mathbf{C}[X,Y]$, or equivalently one in $\mathbf{C}[1, X \Rightarrow Y]$. This accords with the idea that our models consist of 'computable' operations acting on potentially 'non-computable' data: operations in $\mathbf{C}[X,Y]$ are computable, whereas elements of X need not be.

3.1.4 Higher-Order Models

As mentioned above, virtually every model we shall wish to consider in practice has weak products and a weak terminal. Nevertheless, our main interest is in the type operator $\Rightarrow$ rather than $\bowtie$, so for the sake of conceptual economy it is natural to ask whether the essential properties of $\Rightarrow$ may be axiomatized without reference to $\bowtie$. We show here how this may be achieved via the concept of *higher-order model*. We present the definition in two stages.

Definition 3.1.12. A *higher-order structure* is a computability model $\mathbf{C}$ possessing a weak terminal (I,i), and endowed with the following for each $A,B \in |\mathbf{C}|$:

- a choice of datatype $A \Rightarrow B \in |\mathbf{C}|$,
- a partial function $\cdot_{AB} : (A \Rightarrow B) \times A \rightharpoonup B$ (external to the structure of $\mathbf{C}$).

Note that the choice of weak terminal is not deemed part of the structure here. We write $\cdot_{AB}$ as $\cdot$ when A,B are clear from the context. We treat $\Rightarrow$ as right-associative and $\cdot$ as left-associative so that for example $f \cdot x \cdot y \cdot z$ means $((f \cdot x) \cdot y) \cdot z$.

The significance of the weak terminal (I,i) here is that it allows us to pick out a subset $A^\sharp$ of each $A \in |\mathbf{C}|$, namely the set of elements of the form $f(i)$ where $f \in \mathbf{C}[I,A]$ and $f(i)\downarrow$. Clearly, this is independent of the choice of (I,i): if $a = f(i)$ and (J,j) is another weak terminal, then composing f with $\Lambda x.i \in \mathbf{C}[J,I]$ gives $f' \in \mathbf{C}[J,A]$ with $f'(j) = a$. Intuitively, we think of $A^\sharp$ as playing the role of the 'computable' elements of A, and i as some generic computable element. On the one hand, if $a \in A$ were computable, we would expect each $\Lambda x.a$ to be computable so that $a \in A^\sharp$; on the other hand, the image of a computable element under a computable operation should be computable, so that every element of $A^\sharp$ is computable.

Any weakly cartesian closed model $\mathbf{C}$ is a fortiori a higher-order structure. The following definition, though perhaps mysterious at first sight, identifies those higher-order structures that correspond to weakly cartesian closed models (this idea will be made precise by Theorem 3.1.15).

Definition 3.1.13. A *higher-order (computability) model* is a higher-order structure $\mathbf{C}$ satisfying the following conditions for some (or equivalently any) weak terminal (I,i):

1. A partial function $f : A \rightharpoonup B$ is present in $\mathbf{C}[A,B]$ iff there exists $\widehat{f} \in \mathbf{C}[I, A \Rightarrow B]$ such that
$$\widehat{f}(i)\downarrow\,, \qquad \forall a \in A.\ \widehat{f}(i) \cdot a \simeq f(a)\,.$$
2. For any $A,B \in |\mathbf{C}|$, there exists $k_{AB} \in (A \Rightarrow B \Rightarrow A)^\sharp$ such that
$$\forall a.\ k_{AB} \cdot a\downarrow\,, \qquad \forall a,b.\ k_{AB} \cdot a \cdot b = a\,.$$

3. For any $A,B,C \in |\mathbf{C}|$, there exists

$$s_{ABC} \in ((A \Rightarrow B \Rightarrow C) \Rightarrow (A \Rightarrow B) \Rightarrow (A \Rightarrow C))^\sharp$$

such that

$$\forall f,g.\ s_{ABC} \cdot f \cdot g \downarrow, \qquad \forall f,g,a.\ s_{ABC} \cdot f \cdot g \cdot a \simeq (f \cdot a) \cdot (g \cdot a)\,.$$

The elements k and s correspond to well-known combinators from combinatory logic (we shall omit the subscripts on k,s when these can be easily inferred). The significance of k is relatively easy to understand: it allows us to construct *constant* maps in a computable way. (The condition $\forall a.\, k \cdot a \downarrow$ is needed in case B is empty.) A possible intuition for s is that it somehow does duty for an application operation $(B \Rightarrow C) \times B \rightharpoonup C$ within $\mathbf{C}$ itself, where the application may be performed uniformly in a parameter of type A.

The particular choice of elements k,s is not officially part of the structure of a higher-order model. Nonetheless, for convenience we shall sometimes tacitly assume that a higher-order model comes equipped with a choice of suitable k_{AB} and s_{ABC} for each A,B,C.

The following proposition illustrates some consequences of Definition 3.1.13. We leave the proofs as exercises; they may be undertaken either at this point, or more conveniently using machinery to be introduced in Section 3.3.1.

Proposition 3.1.14. *Suppose* $\mathbf{C}$ *is a higher-order model.*

(i) For any $j < m$*, there exists* $\pi_j^m \in (A_0 \Rightarrow \cdots \Rightarrow A_{m-1} \Rightarrow A_j)^\sharp$ *such that*

$$\forall a_0,\ldots,a_{m-1}.\ \pi_j^m \cdot a_0 \cdot \ldots \cdot a_{m-1} \;=\; a_j\,.$$

(ii) Suppose $m,n > 0$*. Given*

$$f_j \in (A_0 \Rightarrow \cdots \Rightarrow A_{m-1} \Rightarrow B_j)^\sharp \quad (j = 0,\ldots,n-1)\,,$$
$$g \in (B_0 \Rightarrow \cdots \Rightarrow B_{n-1} \Rightarrow C)^\sharp\,,$$

there exists $h \in (A_0 \Rightarrow \cdots \Rightarrow A_{m-1} \Rightarrow C)^\sharp$ *such that*

$$\forall a_0,\ldots,a_{m-1}.\ h \cdot a_0 \cdot \ldots \cdot a_{m-1} \;\simeq\; g \cdot (f_0 \cdot a_0 \cdot \ldots \cdot a_{m-1}) \cdot \ldots \cdot (f_{n-1} \cdot a_0 \cdot \ldots \cdot a_{m-1})\,.$$

(iii) Suppose $m > 0$*. For any element* $f \in (A_0 \Rightarrow \cdots \Rightarrow A_{m-1} \Rightarrow B)^\sharp$*, there exists* $f^\dagger \in (A_0 \Rightarrow \cdots \Rightarrow A_{m-1} \Rightarrow B)^\sharp$ *such that*

$$\forall a_0,\ldots,a_{m-1}.\ f^\dagger \cdot a_0 \cdot \ldots \cdot a_{m-1} \;\simeq\; f \cdot a_0 \cdot \ldots \cdot a_{m-1}\,,$$
$$\forall k < m.\ \forall a_0,\ldots,a_{k-1}.\ f^\dagger \cdot a_0 \cdot \ldots \cdot a_{k-1} \downarrow\,.$$

To see why Definition 3.1.13 is an appropriate generalization of Definition 3.1.8, we shall employ the following concept. If $\mathbf{C}$, $\mathbf{D}$ are higher-order structures, we say $\mathbf{C}$ is a *full substructure* of $\mathbf{D}$ if

- $|\mathbf{C}| \subseteq |\mathbf{D}|$,

- $\mathbf{C}[A,B] = \mathbf{D}[A,B]$ for all $A,B \in |\mathbf{C}|$,
- some (or equivalently any) weak terminal in $\mathbf{C}$ is also a weak terminal in $\mathbf{D}$,
- the meanings of $A \Rightarrow B$ and $\cdot_{AB}$ in $\mathbf{C}$ and $\mathbf{D}$ coincide.

In relation to the third condition, note that if (I,i) and (J,j) are weak terminals in $\mathbf{C}$ then $\Lambda x.j \in \mathbf{C}[I,J] = \mathbf{D}[I,J]$, so if (I,i) is a weak terminal in $\mathbf{D}$ then so is (J,j).

The following theorem now provides the justification for Definition 3.1.13:

Theorem 3.1.15. *A higher-order structure is a higher-order model iff it is a full substructure of a weakly cartesian closed model.*

Proof. Let $\mathbf{C}$ be a higher-order structure. For the right-to-left implication, suppose $\mathbf{D}$ is weakly cartesian closed with a weak terminal (I,i), and suppose $\mathbf{C}$ is a full substructure of $\mathbf{D}$. We shall show that $\mathbf{C}$ satisfies the conditions of Definition 3.1.13.

For condition 1, first note that for any $f \in \mathbf{C}[A,B] = \mathbf{D}[A,B]$ we have that $f \circ \pi_A \in \mathbf{D}[I \bowtie A, B]$ represents $\Lambda(x,a).f(a)$, which by Definition 3.1.8 is in turn represented by some total $\widehat{f} \in \mathbf{D}[I, A \Rightarrow B]$ and this clearly has the property required. Conversely, any $g \in \mathbf{D}[I, A \Rightarrow B]$ with $g(i)\downarrow$ represents $\Lambda(x,a).g(x)\cdot a$ which is in turn represented by some $\overline{g} \in \mathbf{D}[I \bowtie A, B]$, and $g \circ \langle \Lambda a.i, id_A\rangle \in \mathbf{D}[A,B]$ then has the required property.

For condition 2, suppose $A,B \in |\mathbf{C}|$. Let $k' \in \mathbf{D}[A, B \Rightarrow A]$ correspond to $\pi_A \in \mathbf{D}[A \bowtie B, A]$ as in Definition 3.1.8, let $\widehat{k'} \in \mathbf{D}[I, A \Rightarrow (B \Rightarrow A)]$ correspond to $k' \circ \langle \Lambda a.i, id_A\rangle$, and take $k = \widehat{k'}(i)$; it is easy to show that k has the required property. The verification of condition 3 is recommended as an exercise.

Conversely, suppose $\mathbf{C}$ is a higher-order model, with (I,i) a weak terminal. We build a weakly cartesian closed model $\mathbf{C}^{\times}$ into which $\mathbf{C}$ embeds fully as follows:

- Datatypes of $\mathbf{C}^{\times}$ are sets $A_0 \times \cdots \times A_{m-1}$, where $m > 0$ and $A_0, \ldots, A_{m-1} \in |\mathbf{C}|$.
- If $D = A_0 \times \cdots \times A_{m-1}$ and $E = B_0 \times \cdots \times B_{n-1}$ where $m,n > 0$, the operations in $\mathbf{C}^{\times}[D,E]$ are those partial functions $f : D \rightharpoonup E$ of the form

$$f = \Lambda(a_0,\ldots,a_{m-1}).\,(f_0 \cdot a_0 \cdot \ldots \cdot a_{m-1}, \ldots, \ f_{n-1} \cdot a_0 \cdot \ldots \cdot a_{m-1}),$$

  where $f_j \in (A_0 \Rightarrow \cdots \Rightarrow A_{m-1} \Rightarrow B_j)^{\sharp}$ for each j; we say that $f_0, \ldots, f_{n-1}$ *witness* the operation f. Note that for $(f_0 \cdot a_0 \cdot \ldots \cdot a_{m-1}, \ldots, f_{n-1} \cdot a_0 \cdot \ldots \cdot a_{m-1})$ to be defined, it is necessary that all its components be defined.

It remains to check the relevant properties of $\mathbf{C}^{\times}$. That $\mathbf{C}^{\times}$ is a computability model is straightforward: the existence of identities follows from part (i) of Proposition 3.1.14, and that of composites from part (ii). It is also easy to see that $\mathbf{C}^{\times}$ has standard products (again, the existence of suitable projections follows from Proposition 3.1.14(i)), and that (I,i) is a weak terminal in $\mathbf{C}^{\times}$.

Let us now show that $\mathbf{C}^{\times}$ is weakly cartesian closed. Given $D = A_0 \times \cdots \times A_{m-1}$ and $E = B_0 \times \cdots \times B_{n-1}$ with $m,n > 0$, take $C_j = A_0 \Rightarrow \cdots \Rightarrow A_{m-1} \Rightarrow B_j$ for each j, and let $D \Rightarrow E$ be the set of tuples $(f_0, \ldots, f_{n-1}) \in C_0 \times \cdots \times C_{n-1}$ witnessing operations in $\mathbf{C}^{\times}[D,E]$. The application $\cdot_{DE}$ is then given by

$$(f_0, \ldots, f_{n-1}) \cdot_{DE} (a_0, \ldots, a_{m-1}) \simeq (f_0 \cdot a_0 \cdot \ldots \cdot a_{m-1}, \ldots, f_{n-1} \cdot a_0 \cdot \ldots \cdot a_{m-1}).$$

Next, given an operation $g \in \mathbf{C}^\times[G \times D, E]$ witnessed by operations $g_0, \ldots, g_{n-1}$ in $\mathbf{C}$, take $g_0^\dagger, \ldots, g_{n-1}^\dagger$ as in Proposition 3.1.14(iii); then $g_0^\dagger, \ldots, g_{n-1}^\dagger$ witness the corresponding total operation $\widehat{g} \in \mathbf{C}^\times[G, D \Rightarrow E]$. Conversely, the witnesses for any such total $\widehat{g}$ also witness the corresponding g.

With all this structure in place, it is now immediate that $\mathbf{C}$ is a full substructure of $\mathbf{C}^\times$. This completes the proof. □

In principle, there are more higher-order models than weakly cartesian closed ones, since the former need not possess weak products. Even if a higher-order model $\mathbf{C}$ has weak products and a weak terminal, it need not be weakly cartesian closed, since it may lack internal *pairing* operations (see Section 3.3.2). We will see, however, that the existence of such operations is automatic under mild assumptions on $\mathbf{C}$ (Exercise 3.3.8).

Informally, the force of Theorem 3.1.15 is that the theory of weakly cartesian closed models is 'conservative' over that of higher-order models: any property of higher-order structures implied by the definition of weakly cartesian closed model is already implied by the definition of higher-order model. Indeed, one can formulate a precise theorem to this effect, but it would take us too far astray to do so here.

3.1.5 Typed Partial Combinatory Algebras

Our definitions so far have been chosen to show how higher-order models fit into the wider framework of computability models. However, Definitions 3.1.8 and 3.1.13 are somewhat cumbersome to work with in practice, owing in part to the formal distinction between *data values* and *operations*. Whilst this is an inevitable distinction in the setting of general computability models, in the higher-order case there is evidently some redundancy here, since operations can themselves be represented by data values. This suggests the possibility of an equivalent formulation of higher-order models purely in terms of data values, in which we single out a special class of values intended to play the role of operations.

The following definition captures roughly what is left of a higher-order model once the operations are discarded.

Definition 3.1.16. (i) A *partial applicative structure* $\mathbf{A}$ consists of

- an inhabited family $|\mathbf{A}|$ of datatypes $A, B, \ldots$ (indexed by some set T),
- a (right-associative) binary operation $\Rightarrow$ on $|\mathbf{A}|$,
- for each $A, B \in |\mathbf{A}|$, a partial function $\cdot_{AB} : (A \Rightarrow B) \times A \rightharpoonup B$.

As before, we often omit the subscripts from $\cdot_{AB}$ and treat $\cdot$ as left-associative.

(ii) A *typed partial combinatory algebra* (TPCA) is a partial applicative structure $\mathbf{A}$ satisfying the following conditions:

1. For any $A, B \in |\mathbf{A}|$, there exists $k_{AB} \in A \Rightarrow B \Rightarrow A$ such that

$$\forall a.\ k \cdot a \downarrow, \qquad \forall a, b.\ k \cdot a \cdot b = a .$$

2. For any $A,B,C \in |\mathbf{A}|$, there exists $s_{ABC} \in (A \Rightarrow B \Rightarrow C) \Rightarrow (A \Rightarrow B) \Rightarrow (A \Rightarrow C)$ such that

$$\forall f,g.\ s \cdot f \cdot g \downarrow , \qquad \forall f,g,a.\ s \cdot f \cdot g \cdot a \simeq (f \cdot a) \cdot (g \cdot a) .$$

A TPCA is *total* if all the application operations $\cdot_{AB}$ are total.

Clearly, any higher-order model in the sense of Definition 3.1.13 yields an underlying TPCA. However, in passing to this TPCA we lose the information that says which elements of $A \Rightarrow B$ are supposed to represent operations. We may remedy this by considering TPCAs equipped with a substructure of notionally 'computable' elements.

Definition 3.1.17. (i) If $\mathbf{A}^\circ$ denotes a partial applicative structure, a *partial applicative substructure* $\mathbf{A}^\sharp$ of $\mathbf{A}^\circ$ consists of a subset $A^\sharp \subseteq A$ for each $A \in |\mathbf{A}^\circ|$ such that

- if $f \in (A \Rightarrow B)^\sharp$, $a \in A^\sharp$ and $f \cdot a \downarrow$ in $\mathbf{A}^\circ$, then $f \cdot a \in B^\sharp$.

Such a pair $(\mathbf{A}^\circ;\mathbf{A}^\sharp)$ is called a *relative partial applicative structure.*

(ii) A *relative TPCA* is a relative partial applicative structure $(\mathbf{A}^\circ;\mathbf{A}^\sharp)$ such that there exist elements k_{AB}, s_{ABC} in $\mathbf{A}^\sharp$ witnessing that $\mathbf{A}^\circ$ is a TPCA.

The term 'relative TPCA' comes from the role played by such structures in *relative realizability*: see for example Birkedal and van Oosten [36].

We may say a relative TPCA $(\mathbf{A}^\circ;\mathbf{A}^\sharp)$ is *full* if $\mathbf{A}^\sharp = \mathbf{A}^\circ$. We will use $\mathbf{A}$ to range over both ordinary TPCAs and relative ones (writing $\mathbf{A}^\circ, \mathbf{A}^\sharp$ for the two components of $\mathbf{A}$ in the latter case), so that in effect we identify an ordinary TPCA $\mathbf{A}$ with the relative TPCA $(\mathbf{A};\mathbf{A})$. Indeed, we may sometimes refer to ordinary TPCAs as 'full TPCAs'. Clearly, the models K_1 and $\Lambda/\!\sim$ (considered as relative TPCAs) are full, while in general $\mathsf{K}(B;C)$ is not: rather, it is a relative TPCA $\mathbf{A}$ in which $\mathbf{A}^\circ$ is a full set-theoretic type structure whilst $\mathbf{A}^\sharp$ consists of only the C-computable elements.

Theorem 3.1.18. *There is a canonical bijection between higher-order models (in the sense of Definition 3.1.13) and relative TPCAs.*

Proof. First suppose $\mathbf{C}$ is a higher-order model, and let $\mathbf{A}^\circ$ be its underlying partial applicative structure. Take (I,i) a weak terminal in $\mathbf{C}$, and for any $A \in |\mathbf{C}|$, define $A^\sharp = \{g(i) \mid g \in \mathbf{C}[I,A],\ g(i)\downarrow\}$ as in Section 3.1.4. As noted there, this is independent of the choice of (I,i); in fact, it is easy to see that $a \in A^\sharp$ iff (A,a) is a weak terminal. To see that the $A^\sharp$ form an applicative substructure, suppose $f \in (A \Rightarrow B)^\sharp$ is witnessed by $f' \in \mathbf{C}[I,A \Rightarrow B]$, and $a \in A^\sharp$ is witnessed by $a' \in \mathbf{C}[I,A]$, and suppose further that $f \cdot a = b$. Take $\check{f}' \in \mathbf{C}[A \Rightarrow B]$ corresponding to f'; then $\check{f}'(a) = b$ and so $\check{f}' \circ a'$ witnesses $b \in B^\sharp$.

Let $\mathbf{A}^\sharp$ denote the substructure formed by the sets $A^\sharp$. It is directly built into Definition 3.1.13 that there are elements k_{AB}, s_{ABC} in $\mathbf{A}^\sharp$ with the properties required by Definition 3.1.16; thus $(\mathbf{A}^\circ;\mathbf{A}^\sharp)$ is a relative TPCA.

For the converse, suppose $\mathbf{A}$ is a relative TPCA. Take $|\mathbf{C}| = |\mathbf{A}^\circ|$, and for $A,B \in |\mathbf{C}|$, let $\mathbf{C}[A,B]$ consist of all partial functions $\Lambda a.\, f \cdot a$ for $f \in (A \Rightarrow B)^\sharp$. To see that $\mathbf{C}$ has identities, for any $A \in |\mathbf{C}|$ we have

$$i_A = s_{A(A\Rightarrow A)A} \cdot k_{A(A\Rightarrow A)} \cdot k_{AA} \in (A \Rightarrow A)^\sharp$$

and clearly i_A induces $id_A \in \mathbf{C}[A,A]$. For composition, given operations $f \in \mathbf{C}[A,B]$, $g \in \mathbf{C}[B,C]$ induced by $f' \in (A \Rightarrow B)^\sharp$, $g' \in (B \Rightarrow C)^\sharp$, we have that $g \circ f \in \mathbf{C}[A,C]$ is induced by $s_{ABC} \cdot (k_{(B\Rightarrow C)A} \cdot g) \cdot f$. Thus $\mathbf{C}$ is a computability model.

For a weak terminal, take any $U \in |\mathbf{C}|$ and let $I = U \Rightarrow U$ and $i = i_U$ as defined above. Then for any A we have that $k_{IA} \cdot i \in (A \Rightarrow U \Rightarrow U)^\sharp$ induces $\Lambda a.i \in \mathbf{C}[I,A]$.

To turn $\mathbf{C}$ into a higher-order structure, we of course take $\Rightarrow$ and $\cdot$ as in $\mathbf{A}^\circ$. We may now verify that for any A we have

$$A^\sharp = \{g(i) \mid g \in \mathbf{C}[I,A],\ g(i)\downarrow\}\,,$$

so that the present meaning of $A^\sharp$ coincides with its meaning in Section 3.1.4. For given $a \in A^\sharp$ we have $k_{IA} \cdot a \in (I \Rightarrow A)^\sharp$ inducing an operation g with $g(i) = a$. Conversely, given $g \in \mathbf{C}[I,A]$ with $g(i)\downarrow$ we have that $g(i) = g' \cdot i$ for some $g' \in (I \Rightarrow A)^\sharp$; but $i \in I^\sharp$ so $g(i) \in A^\sharp$.

By applying the above equation to the type $A \Rightarrow B$, we see that conditions 1 and 2 of Definition 3.1.13 are satisfied, and conditions 3 and 4 are immediate from the k,s conditions in Definition 3.1.16. Thus $\mathbf{C}$ is a higher-order model.

Finally, it is clear from the above discussion that the two constructions we have given are mutually inverse. □

It is also worth remarking that in the setting of a relative TPCA $\mathbf{A}$, we have a natural *degree structure* on the elements of $\mathbf{A}^\circ$. Specifically, if $a \in A$ and $b \in B$ where $A, B \in |\mathbf{A}^\circ|$, let us write $a \gg b$ if there exists $f \in \mathbf{A}^\sharp(A \Rightarrow B)$ with $f \cdot a = b$. Then $\gg$ is a preorder on the elements of $\mathbf{A}$ whose associated equivalence classes may be considered as 'degrees'. Clearly, $\gg$ induces a partial order on these degrees, and there is a lowest degree, consisting of exactly the elements of $\mathbf{A}^\sharp$.

The case where $|\mathbf{A}|$ consists of just a single datatype deserves special mention. In the full setting, such a TPCA is essentially just a single set A equipped with a partial 'application' operation $\cdot : A \times A \rightharpoonup A$, such that for some $k, s \in A$ we have

$$\forall x,y.\ k \cdot x \cdot y = x\,, \qquad \forall x,y.\ s \cdot x \cdot y\downarrow\,, \qquad \forall x,y,z.\ s \cdot x \cdot y \cdot z \simeq (x \cdot z) \cdot (y \cdot z)\,.$$

(The condition $k \cdot x\downarrow$ may be omitted here since A is necessarily inhabited.) We call such a structure an (untyped) *partial combinatory algebra* (or PCA). The concept of a PCA has been studied extensively in the literature, starting with Staples [277] and Feferman [92]; our models K_1 and $\Lambda/\!\sim$ are typical examples. The notion of a relative PCA may be defined similarly.

Of our three definitions of models for higher-order computability, the relative TPCA formulation is the one we shall most often favour, although we shall pass freely between formulations when it is convenient to do so. For both TPCAs and higher-order models, we shall often omit the symbol $\cdot$ and denote application simply by juxtaposition (again left-associatively, so that $fxyz$ means $((fx)y)z$). We shall also sometimes use other symbols for application, particularly where there is a need to distinguish between different models. For instance, the partial application operation in Kleene's first model K_1 will be written as $\bullet$, so that $e \bullet n$ is the result (if defined) of applying the eth Turing machine to n.

Interesting alternative notions of higher-order model and TPCA may be obtained by using affine or monoidal products (as mentioned at the end of Subsection 3.1.1) instead of cartesian ones in the definition of weakly cartesian closed model. These give us the notions of *affine* and *linear* higher-order model respectively; in contradistinction to these, the models we have defined may be called *intuitionistic* higher-order models, because the type operator '$\Rightarrow$' behaves like implication in intuitionistic logic. For some brief remarks on affine and linear models, see Subsection 13.1.3.

3.1.6 Lax Models

For simplicity, we have worked so far with a simple definition of computability model in which operations are required to be closed under ordinary composition of partial functions. It turns out, however, that with a few refinements, practically all the general theory presented in this chapter goes through under a somewhat milder assumption. We define the notion of a *lax computability model* by replacing condition 2 in Definition 3.1.1 with the following:

- For any $f \in \mathbf{C}[\rho, \sigma]$ and $g \in \mathbf{C}[\sigma, \tau]$, there exists $h \in \mathbf{C}[\rho, \tau]$ with $h(a) \succeq g(f(a))$ for all $a \in \mathbf{C}(\rho)$.

We may refer to h here as a *supercomposition* of f and g. However, we do not require $\mathbf{C}$ to be equipped with an operation specifying a choice of supercomposition for any given f and g, so issues such as associativity do not arise here.

We sometimes refer to our standard computability models as *strict* when we wish to emphasize the contrast with lax models. Of course, for total computability models, the distinction evaporates completely.

One possible motivation for the concept of lax model is that it is often natural to think of an application $f(a)$ in terms of some computational agent F representing f being placed 'alongside' a representation A of a to yield a composite system $F \mid A$, which may then evolve in certain ways via interactions between F and A. If an agent G representing g is then placed alongside this to yield a system $G \mid F \mid A$, there is the possibility that G may interact 'directly' with F rather than just with the result obtained from $F \mid A$; thus, $G \mid F$ might admit other behaviours not accounted for by $g \circ f$. (For a precise example of this in process algebra, see Longley [183].)

Once a few subtle points have been addressed (as explained below), it turns out that the corresponding relaxation in the setting of TPCAs is as follows. The notion of a *(relative) lax TPCA* is given by replacing the axioms for s_{ABC} in Definition 3.1.16 with:

$$\forall f,g.\ s \cdot f \cdot g \downarrow, \qquad \forall f,g,a.\ s \cdot f \cdot g \cdot a \succeq (f \cdot a) \cdot (g \cdot a)\,.$$

This is the notion of 'lax model' that we shall most often work with in the sequel. As a rule of thumb, theorems about ordinary (relative) TPCAs may be turned into theorems about lax ones by suitably replacing occurrences of '$\simeq$' with '$\succeq$'.

The additional freedom offered by lax models is sometimes useful. For instance, *Kleene's second model* K_2 is most naturally seen as a lax PCA; it can also be shown to be a strict PCA, but only by a slightly artificial construction (see Section 12.2). Moreover, Kleene's definition of partial computable functions over the full set-theoretic type structure yields only a lax model (see Section 5.1.4).

Insofar as lax models are more general and almost all our theory works for them, the lax framework would seem to be the preferred setting for the theory. However, working with lax models is sometimes more delicate, and the material of Sections 3.1.1–3.1.5 provides a case in point. We now briefly sketch the refinements needed to obtain a lax analogue of the foregoing theory (the details will not be required in the sequel).

The definitions of weak products and weak terminal may be carried over unchanged to the setting of lax computability models; note that $\langle f,g \rangle$ is still required to be a pairing in the 'strict' sense that its domain coincides precisely with $\mathrm{dom}\, f \cap \mathrm{dom}\, g$. The definition of weakly cartesian closed model is likewise unchanged, although one should note that in the lax setting, whether a given model is weakly cartesian closed may be sensitive to the choice of the type operator $\bowtie$.

For the definition of a lax higher-order model, we simply replace '$\simeq$' by '$\succeq$' in condition 4 of Definition 3.1.13. It can be checked that if $\mathbf{C}$ is a lax higher-order model and satisfies the '$\simeq$' version of condition 4, then $\mathbf{C}$ has strict composition.

As already noted, the definition of lax (relative) TPCA is obtained simply by changing '$\simeq$' to '$\succeq$' in the axiom for s. Theorem 3.1.18 and its proof then go through essentially unchanged: there is a canonical bijection between lax higher-order models and lax relative TPCAs.

However, a slight technical complication arises in trying to obtain an analogue of Theorem 3.1.15. In any model with weak products, let us say a weak terminal (I,i) is a *weak unit* if for any $A,B \in |\mathbf{C}|$, the operations in $f \in \mathbf{C}[A,B]$ are exactly the partial functions $f : A \rightharpoonup B$ representable by some $\overline{f} \in \mathbf{C}[I \bowtie A, B]$, in the sense that if $\pi_I(d) = i$ and $\pi_A(d) = a$ then $\overline{f}(d) \simeq f(a)$. (This is of course automatic if $\mathbf{C}$ has strict composition.) The following now gives most of the content of Theorem 3.1.15 in the lax setting:

Theorem 3.1.19. *(i) Any lax higher-order model is a full substructure of a lax weakly cartesian closed model.*

(ii) If $\mathbf{D}$ *is a lax weakly cartesian closed model in which some weak terminal* (I,i) *is a weak unit, any full substructure of* $\mathbf{D}$ *containing* I *is a lax higher-order model.*

In the proof of part (i), we require a small modification of the construction of $\mathbf{C}^{\times}$ as given in the proof of Theorem 3.1.15: the operations f_j that witness an operation $f \in \mathbf{C}^{\times}[D,E]$ must be explicitly required to satisfy

$$\forall k < m.\ \forall a_0,\ldots,a_{k-1}.\ f_j \cdot a_0 \cdot \ldots \cdot a_{k-1} \downarrow .$$

3.1.7 Type Worlds

We now turn our attention to an ingredient of our definitions which we have kept in the background so far: the set T used to index the datatypes in a computability model (Definition 3.1.1) or relative TPCA (Definition 3.1.16). We may think of elements $\sigma \in \mathsf{T}$ as 'names' for datatypes: they may, for example, be syntactic expressions in some formal language for types. To facilitate comparisons between models, it is natural to classify them according to the repertoire of 'types' they are trying to represent—that is, to group together models that share the same index set T. The following ad hoc definition provides some simple machinery to support this.

Definition 3.1.20. (i) A *type world* is simply a set T of *type names* σ, optionally endowed with any or all of the following:

1. a *fixing map*, assigning a set $\mathsf{T}[\sigma]$ to certain type names $\sigma \in \mathsf{T}$,
2. a *product structure*, consisting of a total binary operation $(\sigma, \tau) \mapsto \sigma \times \tau$,
3. an *arrow structure*, consisting of a total binary operation $(\sigma, \tau) \mapsto \sigma \to \tau$.

(ii) A *computability model over* a type world T is a computability model $\mathbf{C}$ with index set T (so that $|\mathbf{C}| = \{\mathbf{C}(\sigma) \mid \sigma \in \mathsf{T}\}$) subject to the following conventions:

1. If T has a fixing map, then $\mathbf{C}(\sigma) = \mathsf{T}[\sigma]$ whenever $\mathsf{T}[\sigma]$ is defined.
2. If T has a product structure, then $\mathbf{C}$ has weak products and for any $\sigma, \tau \in \mathsf{T}$ we have $\mathbf{C}(\sigma \times \tau) = \mathbf{C}(\sigma) \bowtie \mathbf{C}(\tau)$.
3. If T has an arrow structure, then $\mathbf{C}$ is a higher-order model and for any $\sigma, \tau \in \mathsf{T}$ we have $\mathbf{C}(\sigma \to \tau) = \mathbf{C}(\sigma) \Rightarrow \mathbf{C}(\tau)$.
4. If T has both a product and an arrow structure, then $\mathbf{C}$ is weakly cartesian closed.

Thus, a computability model over T possesses the appropriate kind of structure for the elements present in T. We have already noted that being weakly cartesian closed is stronger than simply being a higher-order model with products; the difference will be spelt out in Subsection 3.3.2.

By a *(lax) relative TPCA over* a type world T with arrow structure, we shall mean a (lax) relative TPCA $\mathbf{A}$ such that the corresponding higher-order model $\mathbf{C}$ (as in Theorem 3.1.18) is a computability model over T.

The type worlds we shall most frequently consider are the following.

Example 3.1.21. The one-element type world $\mathsf{O} = \{*\}$ with just the arrow structure $* \to * = *$. TPCAs over this type world are precisely (untyped) PCAs as discussed in Section 3.1.5; both K_1 and $\Lambda/\!\sim$ are examples.

Example 3.1.22. If $\beta_0, \ldots, \beta_{n-1}$ are distinct *basic type names* and $B_0, \ldots, B_{n-1}$ are sets, we may define the type world $\mathsf{T}^{\to}(\beta_0 = B_0, \ldots, \beta_{n-1} = B_{n-1})$ to consist of formal type expressions freely constructed from $\beta_0, \ldots, \beta_{n-1}$ via $\to$, fixing the interpretation of each β_i at B_i. This type world has a fixing map and an arrow structure, but no product structure. We may write just $\mathsf{T}^{\to}(\beta_0, \ldots, \beta_{n-1})$ if we do not wish to constrain the interpretation of the β_i.

A typical example is the type world $\mathsf{T}^{\rightarrow}(\mathtt{N} = \mathbb{N})$. Models over this type world correspond to *finite type structures* over $\mathbb{N}$; the models $\mathsf{K}(B;C)$ are examples.

A related example is the type world $\mathsf{T}^{\rightarrow}(\mathtt{N} = \mathbb{N}_{\perp})$, where $\mathbb{N}_{\perp}$ is the set of natural numbers together with an additional element $\perp$ representing 'non-termination'. Whereas $\mathbb{N}$ may be used to model actual *results* of computations (the values they return), we may think of $\mathbb{N}_{\perp}$ as representing some computational *process* which may or may not return a natural number. Models over $\mathsf{T}^{\rightarrow}(\mathtt{N} = \mathbb{N}_{\perp})$ and similar type worlds can be used to represent intuitively 'partial' notions of computability.

Example 3.1.23. Similarly, we define $\mathsf{T}^{\rightarrow\times}(\beta_0 = B_0, \dots, \beta_{n-1} = B_{n-1})$ to be the type world consisting of type expressions freely constructed from $\beta_0, \dots, \beta_{n-1}$ via $\rightarrow$ and $\times$, fixing the interpretation of each β_i at B_i. Again, if no fixing map is required, we write just $\mathsf{T}^{\rightarrow\times}(\beta_0, \dots, \beta_{n-1})$.

Type worlds featuring a *unit type* (denoted by $\mathtt{1}$) are also useful. We shall write $\mathsf{T}^{\rightarrow\times\mathtt{1}}(\beta_0 = B_0, \dots, \beta_{n-1} = B_{n-1})$ for the type world

$$\mathsf{T}^{\rightarrow\times}(\mathtt{1} = \{()\},\ \beta_0 = B_0,\ \dots,\ \beta_{n-1} = B_{n-1})\ .$$

We will often refer to the type names in a type world simply as *types*, and use ρ, σ, τ to range over them. When dealing with formal type expressions, we adopt the usual convention that $\rightarrow$ is right-associative, so that $\rho \rightarrow \sigma \rightarrow \tau$ means $\rho \rightarrow (\sigma \rightarrow \tau)$. For definiteness, we may also declare that $\times$ is right-associative, although in practice we shall not always bother to distinguish clearly between $(\rho \times \sigma) \times \tau$ and $\rho \times (\sigma \times \tau)$. We consider $\times$ as binding more tightly than $\rightarrow$, so that $\rho \times \sigma \rightarrow \tau$ means $(\rho \times \sigma) \rightarrow \tau$, and $\rho \rightarrow \sigma \times \tau$ means $\rho \rightarrow (\sigma \times \tau)$.

We shall also use the notation $\sigma_0, \dots, \sigma_{r-1} \rightarrow \tau$ as an abbreviation for $\sigma_0 \rightarrow \sigma_1 \rightarrow \dots \sigma_{r-1} \rightarrow \tau$ (allowing this to mean τ in the case $r = 0$). This allows us to express our intention regarding which objects are to be thought of as 'arguments' to a given operation: for instance, the types $\mathtt{N},\mathtt{N},\mathtt{N} \rightarrow \mathtt{N}$ and $\mathtt{N},\mathtt{N} \rightarrow (\mathtt{N} \rightarrow \mathtt{N})$ are formally the same, but in the first case we are thinking of a three-argument operation returning a natural number, while in the second we are thinking of a two-argument operation returning (say) a function $\mathbb{N} \rightarrow \mathbb{N}$. We also write $\sigma^{(r)} \rightarrow \tau$ for the type $\sigma, \dots, \sigma \rightarrow \tau$ with r arguments. The notation σ^r is reserved for the r-fold product type $\sigma \times \cdots \times \sigma$.

The following observation, although trivial, will often play an important role:

Proposition 3.1.24. *Any type $\sigma \in \mathsf{T}^{\rightarrow}(\beta_0, \dots, \beta_{n-1})$ may be uniquely written in the form $\sigma_0, \dots, \sigma_{r-1} \rightarrow \beta_i$.* $\square$

We shall call this the *argument form* of σ. The importance of this is that it provides a useful induction principle for types: if a property holds for $\sigma_0, \dots, \sigma_{r-1} \rightarrow \beta_i$ whenever it holds for each of $\sigma_0, \dots, \sigma_{r-1}$, then it holds for all $\sigma \in \mathsf{T}^{\rightarrow}(\beta_0, \dots, \beta_{n-1})$. We shall refer to this as *argument induction*; it is often preferable as an alternative to the usual *structural induction* on types.

Closely associated with argument form is the notion of the *level* of a type σ: informally, the stage at which σ appears in the generation of $\mathsf{T}^{\rightarrow}(\beta_0, \dots, \beta_{n-1})$ via argument induction.

$$\begin{aligned} \mathrm{lv}(\beta_i) &= 0\,, \\ \mathrm{lv}(\sigma_0,\ldots,\sigma_{r-1}\to\beta_i) &= 1+\max_{i<r}\mathrm{lv}(\sigma_i) \quad (r\geq 1)\,. \end{aligned}$$

When working with $\mathsf{T}^{\to\times}(\beta_0,\ldots,\beta_{n-1})$, it is natural to augment this definition with

$$\mathrm{lv}(\sigma\times\tau) = \max(\mathrm{lv}(\sigma),\mathrm{lv}(\tau))\,.$$

We may also define the *pure type of level k over* σ, written $\overline{k}[\sigma]$:

$$\overline{0}[\sigma] \;=\; \sigma\,, \qquad \overline{k+1}[\sigma] \;=\; \overline{k}[\sigma]\to\sigma\,.$$

For type worlds generated by a single base type β, we may write simply $\overline{k}$ for $\overline{k}[\beta]$. For instance, in the type world $\mathsf{T}^{\to}(\mathtt{N})$ we write $\overline{2}$ for the type $(\mathtt{N}\to\mathtt{N})\to\mathtt{N}$. We often write $\overline{k}$ just as k in contexts where there is no danger of confusion, e.g. $\mathbf{C}(3)$ for $\mathbf{C}(\overline{3})$, or $\mathbf{C}(1,2\to 0)$ for $\mathbf{C}(\overline{1},\overline{2}\to\overline{0})$.

In the context of total TPCAs over $\mathsf{T}^{\to}(\mathtt{N}=\mathbb{N})$, the level of a type is a good measure of its structural complexity: in a large family of models, every type will actually be computably isomorphic to the pure type of the same level (Theorem 4.2.9). This is not so for total TPCAs over $\mathsf{T}^{\to}(\mathtt{N}=\mathbb{N}_\perp)$, where the stratification of types is typically much more fine-grained. We shall return to these issues in Section 4.2.

3.2 Further Examples of Higher-Order Models

We now present some more elaborate examples of higher-order models, in order to give an impression of the scope of our general theory. Whilst some of these examples will receive more detailed treatment in a chapter or section of their own, they will also play a major role in other parts of the book, and our purpose here is to summarize the basic information about these models with which we shall assume familiarity throughout the book as a whole. We shall omit many of the proofs here, deferring these to the parts of the book where the models in question are studied in detail.

3.2.1 Kleene's Second Model

Whereas K_1 captures a notion of computability for finitary data such as natural numbers, our next example embodies a notion of computability in an *infinitary* setting, where a data item may consist of a countably infinite list of finitary pieces of data. We may prototypically represent such an item by a total function $g\in\mathbb{N}^{\mathbb{N}}$, which we may visualize as a list of natural numbers written on an infinite tape:

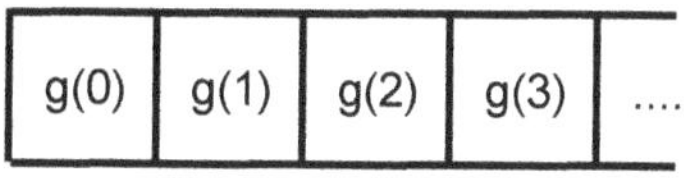

Many kinds of infinitary data arising in mathematics and computer science, such as exact real numbers, lazy lists or infinite trees, can readily be represented in this form.

Let us first consider informally the nature of a (total) 'computable' operation $F : \mathbb{N}^{\mathbb{N}} \to \mathbb{N}$. Whilst we allow our input $g \in \mathbb{N}^{\mathbb{N}}$ to involve an infinite amount of information, we have in mind that computations themselves should be 'finitary' in character. More precisely, when our computation returns a result $F(g) \in \mathbb{N}$, it must do so after only a finite number of 'computation steps', and this means that it will only be able to inspect finitely many values of g (say $g(i_0), g(i_1), \ldots, g(i_{r-1})$) before completing. It follows that if g' is any other data item within the set

$$\{h \in \mathbb{N}^{\mathbb{N}} \mid h \upharpoonright \{i_0, \ldots, i_{r-1}\} = g \upharpoonright \{i_0, \ldots, i_{r-1}\}\} ,$$

then we must also have $F(g') = F(g)$. This amounts exactly to saying that F is *continuous* with respect to the standard (Baire) topology on $\mathbb{N}^{\mathbb{N}}$ and the discrete topology on $\mathbb{N}$.

Of course, there will be many continuous functions F that are not 'effectively computable' in any reasonable sense. The point is merely that the concept of continuity isolates one key property of finitary computations, so that an *abstract* model of computation based solely on continuity will already capture some interesting aspects of computability. The connections between computability and continuity will be a major theme of this book.

Note also that in the above situation, since g is total, it is in principle harmless to assume that $i_j = j$ for each j: in effect, to insist that F must ask for values of g in the order $g(0), g(1), \ldots$. This idea will be incorporated into the way we define K_2.

It seems most natural to approach the definition via the application operation, rather than via the precise class of (partial) continuous endofunctions it gives rise to. The idea is that if $F : \mathbb{N}^{\mathbb{N}} \to \mathbb{N}$ is computable in the above sense, the function F can be completely described by recording, for each finite sequence $m_0, \ldots, m_{r-1}$, whether the information '$g(0) = m_0, \ldots, g(r-1) = m-1$' suffices to determine the value of $F(g)$, and if so, what that value is. This is itself just a countable collection of finite pieces of data, and can be represented by an element $f \in \mathbb{N}^{\mathbb{N}}$ as follows. Here and later, we write $\langle \cdots \rangle$ for some standard computable coding operation $\mathbb{N}^* \to \mathbb{N}$.

$$\begin{array}{ll} f(\langle m_0, \ldots, m_{r-1} \rangle) = m+1 & \text{if } F(g) = m \text{ whenever } g(0) = m_0, \ldots, g(r-1) = m_{r-1} , \\ f(\langle m_0, \ldots, m_{r-1} \rangle) = 0 & \text{if `} g(0) = m_0, \ldots, g(r-1) = m_{r-1}\text{' does not suffice} \\ & \text{to determine } F(g). \end{array}$$

We can then compute $F(g)$ from just f and g:

$$F(g) \;=\; (f \mid g) =_{\mathrm{def}} \; f(\langle g(0),\ldots,g(r-1)\rangle)-1\,,$$
$$\text{where } r=\mu r.f(\langle g(0),\ldots,g(r-1)\rangle)>0\,.$$

Even if $f \in \mathbb{N}^{\mathbb{N}}$ does not represent some F in this way, we can regard the above formula as defining a *partial* computable function $(f \mid -) : \mathbb{N}^{\mathbb{N}} \rightharpoonup \mathbb{N}$, which will be continuous on its domain (note that the search for a suitable r may fail to terminate).

A small tweak is now required to obtain an application operation with codomain $\mathbb{N}^{\mathbb{N}}$ rather than $\mathbb{N}$. In effect, the computation now accepts an additional argument $n \in N$, which can be assumed to be known before any queries to g are made. We may therefore define

$$(f \odot g)(n) =_{\mathrm{def}} \; f(\langle n, g(0),\ldots,g(r-1)\rangle)-1\,,$$
$$\text{where } r=\mu r.f(\langle n, g(0),\ldots,g(r-1)\rangle)>0\,.$$

In general, this will define a *partial* function $\mathbb{N} \rightharpoonup \mathbb{N}$. At this point, a somewhat brutal measure is needed: we shall henceforth consider $f \odot g$ to be 'defined' only if the above formula yields a *total* function $\mathbb{N} \to \mathbb{N}$. In this way, we obtain a partial application operation

$$\odot \;:\; \mathbb{N}^{\mathbb{N}} \times \mathbb{N}^{\mathbb{N}} \rightharpoonup \mathbb{N}^{\mathbb{N}}\,.$$

The structure $(\mathbb{N}^{\mathbb{N}}, \odot)$ is known as *Kleene's second model*, or K_2; it was studied by Kleene and Vesley in [146]. We will show in Section 12.2 that K_2 is a PCA.

We shall henceforth take K_2 to be our definitive model for continuous (finitary) computation over countably infinite data. The apparent artificiality in the choice of domain of $\odot$ will not prove to be a hindrance in practice, since it will be enough that the model behaves well *within* this domain.

We can also build effectivity into the model by restricting $\odot$ to the set of total *computable* functions $\mathbb{N} \to \mathbb{N}$ (it is easy to check that if f,g are computable then so is $f \odot g$ when defined). Since the natural candidates for k,s in K_2 are computable, this gives us both a PCA K_2^{eff} (for 'effective operations acting on effective data') and a relative PCA $(K_2;K_2^{\mathrm{eff}})$ (for 'effective operations acting on arbitrary data').

3.2.2 *The Partial Continuous Functionals*

Having introduced the idea that continuity offers a reasonable idealization of the notion of finitary computability, we now look at another important model based on this principle: the model PC consisting of *partial continuous functionals* over the natural numbers. This model is due independently to Ershov [71] and Scott [256]; we shall present it here as a model over the type world $\mathsf{T} = \mathsf{T}^{\to\times 1}(\mathtt{N} = \mathbb{N}_\perp)$ in the sense of Section 3.1.7.

The idea is that we shall construct sets $\mathsf{PC}(\sigma)$ for each $\sigma \in \mathsf{T}$, where each $\mathsf{PC}(\sigma \to \tau)$ consists of all functions $f : \mathsf{PC}(\sigma) \to \mathsf{PC}(\tau)$ that are 'continuous' in the sense that any finite piece of observable information about an output $f(x)$ can only rely on a finite amount of information about x. To make sense of this, we need to provide, along with each $\mathsf{PC}(\sigma)$, a reasonable notion of 'finite piece of information' for elements of $\mathsf{PC}(\sigma)$. It is natural to identify a 'piece of information' with the set of elements of which that information is true, so in the first instance, we shall specify a family $\mathscr{F}_\sigma$ consisting of those subsets of $\mathsf{PC}(\sigma)$ which we regard as representing finitary properties of elements. (In fact, $\mathscr{F}_\sigma$ will form a basis for what is known as the *Scott topology* on $\mathsf{PC}(\sigma)$.)

We construct these data by induction on types. At base type, we take $\mathsf{PC}(\mathtt{N}) = \mathbb{N}_\bot$ and $\mathscr{F}_\mathtt{N} = \{\mathbb{N}_\bot\} \cup \{\{n\} \mid n \in \mathbb{N}\}$; the idea is that each n corresponds to a finite piece of information, while the whole set $\mathbb{N}_\bot$ represents 'no information'. We also take $\mathsf{PC}(\mathtt{1}) = \{()\}_\bot$ and $\mathscr{F}_\mathtt{1} = \{\{()\}_\bot, \{()\}\}$. For products, we take $\mathsf{PC}(\sigma \times \tau) = \mathsf{PC}(\sigma) \times \mathsf{PC}(\tau)$ as expected, and let $\mathscr{F}_{\sigma\times\tau}$ consist of all sets $U \times V$ where $U \in \mathscr{F}_\sigma$, $V \in \mathscr{F}_\tau$. At function types, we take $\mathsf{PC}(\sigma \to \tau)$ to consist of all functions $f : \mathsf{PC}(\sigma) \to \mathsf{PC}(\tau)$ that are *continuous* in that for all $x \in \mathsf{PC}(\sigma)$ and all $V \in \mathscr{F}_\tau$ containing $f(x)$, there is some $U \in \mathscr{F}_\sigma$ containing x with $f(U) \subseteq V$.

The continuity of a function f may be manifested by its *graph*, namely the set of pairs $(U,V) \in \mathscr{F}_\sigma \times \mathscr{F}_\tau$ with $f(U) \subseteq V$. This suggests a possible notion of 'finite piece of information' for elements of $\mathsf{PC}(\sigma \to \tau)$. Since each pair (U,V) in the graph of f can be reasonably seen as giving finite information about f, it seems natural to identify finite pieces of information with finite subsets of this graph. We therefore decree that $\mathscr{F}_{\sigma\to\tau}$ should contain all finite intersections of sets of the form

$$(U \Rightarrow V) \ =_{\text{def}} \ \{f \in \mathsf{PC}(\sigma \to \tau) \mid \forall x \in U.\, f(x) \in V\}$$

for $U \in \mathscr{F}_\sigma$, $V \in \mathscr{F}_\tau$. (It makes no real difference whether we allow the empty set as a member of $\mathscr{F}_{\sigma\to\tau}$.)

This completes the construction of the sets $\mathsf{PC}(\sigma)$. For any σ, τ we have the application operation $(f,x) \mapsto f(x) : \mathsf{PC}(\sigma \to \tau) \times \mathsf{PC}(\sigma) \to \mathsf{PC}(\tau)$, and it is not hard to verify that PC is a TPCA (and indeed corresponds to a cartesian closed category).

The structure PC is mathematically very well-behaved. We list here some of its most important properties. The proofs are straightforward exercises; details can be found in standard domain theory texts such as [6, 279].

First, we may define a partial order $\sqsubseteq_\sigma$ on each $\mathsf{PC}(\sigma)$ by

$$x \sqsubseteq_\sigma y \ \Leftrightarrow \ (\forall U \in \mathscr{F}_\sigma.\, x \in U \Rightarrow y \in U)\ .$$

We sometimes omit the type subscript in $\sqsubseteq_\sigma$. We may informally read $x \sqsubseteq y$ as saying that y contains at least as much 'information' as x, in that it satisfies all finitary properties of x and maybe more. At base type, we of course have $x \sqsubseteq_\mathtt{N} y$ iff $x = \bot \vee x = y$, and $x \sqsubseteq_\mathtt{1} y$ iff $x = y$. The following proposition collects together some other useful properties of these orderings:

Proposition 3.2.1. *(i) For types $\sigma \times \tau$, we have $(x,y) \sqsubseteq (x',y')$ iff $x \sqsubseteq x'$ and $y \sqsubseteq y'$. For types $\sigma \to \tau$, the ordering is pointwise: $f \sqsubseteq_{\sigma\to\tau} g$ iff $\forall x.\, f(x) \sqsubseteq_\tau g(x)$.*

(ii) Any $x,y \in \mathsf{PC}(\sigma)$ *have a greatest lower bound* $x \sqcap y$*.*

(iii) If $x,y \in \mathsf{PC}(\sigma)$ *have an upper bound, they have a least upper bound* $x \sqcup y$*.*

(iv) A subset $D \subseteq \mathsf{PC}(\sigma)$ *has a least upper bound* $\bigsqcup D$ *if and only if any two elements* $x,y \in D$ *have a (least) upper bound.*

(v) Every inhabited set $U \in \mathscr{F}_\sigma$ *is upwards closed and has a least element* x_U*, so that* $U = \{y \mid x_U \sqsubseteq y\}$*. (We call such elements* x_U *the* finite *elements of* $\mathsf{PC}(\sigma)$*.)*

(vi) The finite elements x_U *coincide exactly with the* compact *elements of* $\mathsf{PC}(\sigma)$*: that is, those elements x such that if* $D \subseteq \mathsf{PC}(\sigma)$ *has a least upper bound* $\sqsupseteq x$*, then some finite subset of D also has least upper bound* $\sqsupseteq x$*.*

(vii) Every $x \in \mathsf{PC}(\sigma)$ *is the least upper bound of the finite elements* $y \sqsubseteq x$*.*

Since each $U \in \mathscr{F}_\sigma$ is completely determined (in the presence of $\sqsubseteq$) by the finite element x_U, and the notion of compactness gives an independent characterization of these elements, we may at this point dispense entirely with the families $\mathscr{F}_\sigma$, representing finitary information via compact elements rather than subsets of $\mathsf{PC}(\sigma)$.

It is useful to have a more explicit description of what the compact elements look like. If $a \in \mathsf{PC}(\sigma)$ and $b \in \mathsf{PC}(\tau)$ are compact elements, let us write $(a \Rightarrow b)$ for the element of $\mathsf{PC}(\sigma \to \tau)$ defined by

$$(a \Rightarrow b)(x) \;=\; \begin{cases} b & \text{if } a \sqsubseteq x\,, \\ \bot & \text{otherwise.} \end{cases}$$

Note that this agrees with our earlier notation $(U \Rightarrow V)$ for finitary properties in $\mathsf{PC}(\sigma \to \tau)$. We may now verify the following:

- In $\mathsf{PC}(\mathtt{N})$ and $\mathsf{PC}(\mathtt{1})$, every element is compact.
- In $\mathsf{PC}(\sigma \times \tau)$, (a,b) is compact iff a and b are compact.
- In $\mathsf{PC}(\sigma \to \tau)$, an element is compact iff it is a least upper bound of finitely many elements $(a \Rightarrow b)$ where $a \in \mathsf{PC}(\sigma)$ and $b \in \mathsf{PC}(\tau)$ are compact.

This suggests the possibility of a formal language for denoting compact elements, reinforcing the idea that compact elements correspond to finite information. Consider the set of syntactic expressions e generated by the following grammar (where n ranges over $\mathbb{N}$):

$$e \;::=\; \bot \mid n \mid () \mid (e_0 \Rightarrow e_0') \sqcup \cdots \sqcup (e_{r-1} \Rightarrow e_{r-1}')\,.$$

Not every such expression will denote a compact element, for example because the least upper bound in question might not exist in PC. Nevertheless, it is fairly easy to prove that at each type σ, it is decidable whether a given e denotes some compact element $[\![e]\!] \in \mathsf{PC}(\sigma)$, whether $[\![e]\!], [\![e']\!]$ have an upper bound, and whether $[\![e]\!] \sqsubseteq [\![e']\!]$. (See Section 10.1 for more detail.)

Since expressions e are themselves finitary data and may be effectively coded up as natural numbers, we may speak of a *computably enumerable (c.e.)* set of compact elements of $\mathsf{PC}(\sigma)$. We say an element $x \in \mathsf{PC}(\sigma)$ is *c.e.* if it is the least upper bound of some c.e. set of compact elements (or equivalently if the set of all compact elements below x is c.e.). It is easy to check that the c.e. elements of PC

are closed under application and in themselves constitute a cartesian closed model $\mathsf{PC}^{\mathrm{eff}}$. Moreover, there is a cartesian closed model whose data values are those of PC but whose operations are those representable by elements of $\mathsf{PC}^{\mathrm{eff}}$; this corresponds to a relative TPCA $(\mathsf{PC};\mathsf{PC}^{\mathrm{eff}})$.

The models PC and $\mathsf{PC}^{\mathrm{eff}}$ will be studied as computability models in their own right in Chapter 10; they will also be used to construct the model Ct of *total* continuous functionals studied in Chapter 8, and the analogous model HEO of hereditarily effective operations in Chapter 9.

3.2.3 Typed λ-Calculi

In Example 3.1.6 we saw that a higher-order model could be constructed out of terms of the untyped λ-calculus. We now consider how similar syntactic models arise from *typed* λ-calculi. Such calculi will often feature in this book as rudimentary 'programming languages' corresponding to various computability notions, and the relationships between syntactic models and semantic ones will be a frequently recurring theme in later chapters.

For simplicity, we shall work here with types drawn from $\mathsf{T} = \mathsf{T}^{\rightarrow}(\beta_0,\ldots,\beta_{r-1})$, where the β_i are some choice of base types. We assume given some set C of *constant symbols* c, each with an assigned type $\tau_c \in \mathsf{T}$; we also suppose we have an unlimited supply of *variable symbols* x^σ for each type σ. The set of well-typed *terms* over C is generated inductively by the following rules, where we write $M : \sigma$ to mean 'M is a well-typed term of type σ':

$$c : \tau_c \qquad x^\sigma : \sigma \qquad \frac{M : \tau}{\lambda x^\sigma.M : \sigma \to \tau} \qquad \frac{M : \sigma \to \tau \quad N : \sigma}{MN : \tau}$$

We shall often omit type superscripts, and shall abbreviate $\lambda x_0.\cdots.\lambda x_{n-1}.M$ to $\lambda x_0 \cdots x_{n-1}.M$. The notions of free and bound variable, closed term, and capture-avoiding substitution $M[x^\sigma \mapsto N]$ are defined as usual. It is easy to check that if $M : \tau$ and $N : \sigma$ then $M[x^\sigma \mapsto N] : \tau$.

We will write $\mathscr{L}(C)_\sigma$ for the set of terms over C of type σ, and $\mathscr{L}^0(C)_\sigma$ for the set of closed terms.

Since we wish to think of such systems as programming languages, we shall usually want to give some operational rules for evaluating closed terms. Typically this will be done by specifying some set Δ of formal δ-*rules* of the form

$$cM_0 \ldots M_{r-1} \rightsquigarrow N\,,$$

where the left and right sides are terms of the same type, and the free variables of N are among those of $cM_0 \ldots M_{r-1}$. The idea is that constants c may be thought of as 'primitive operators' of the language, and the δ-rules specify the computational behaviour of these operators. The free variables allow the δ-rules to be treated as 'templates' from which closed instances may be generated as required.

We may also specify some set Ξ of *basic evaluation contexts* $K[-]$. These are 'terms with holes' which we may use to indicate that the evaluation of a term $K[M]$ may proceed by first evaluating the subterm M.

Before giving the precise rules for evaluating closed terms, we illustrate the above concepts with two important examples. In both of them, we take $\mathsf{T} = \mathsf{T}^{\rightarrow}(\mathtt{N})$.

Example 3.2.2 (Gödel's T). Let C consist of the following constants:

$$\begin{aligned} \widehat{0} &: \mathtt{N}\,, \\ suc &: \mathtt{N} \to \mathtt{N}\,, \\ rec_\sigma &: \sigma \to (\sigma \to \mathtt{N} \to \sigma) \to \mathtt{N} \to \sigma \qquad \text{for each } \sigma \in \mathsf{T}\,. \end{aligned}$$

Let Δ consist of the δ-rules

$$\begin{aligned} rec_\sigma x f \widehat{0} &\rightsquigarrow x\,, \\ rec_\sigma x f (suc\; n) &\rightsquigarrow f (rec_\sigma x f n) n\,, \end{aligned}$$

and let Ξ consist of the contexts of the forms

$$[-]N\,, \qquad suc[-]\,, \qquad rec_\sigma PQ[-]\,.$$

(For instance, the first of these will mean that if $M \rightsquigarrow M'$ then $MN \rightsquigarrow M'N$.) This defines (a version of) Gödel's *System T*, a language embodying a natural higher type generalization of primitive recursion (cf. Subsection 5.1.5).

As an aside, it is easy to see how terms of this language may be interpreted as operations in the full set-theoretic model S over $\mathbb{N}$, and that the operations thus obtained are precisely the 'System T definable operations' as identified under Example 3.1.7. (See also Example 3.5.9.)

Example 3.2.3 (Plotkin's PCF). Let C consist of the following constants:

$$\begin{aligned} \widehat{n} &: \mathtt{N} \qquad \text{for each } n \in \mathbb{N}\,, \\ suc,\, pre &: \mathtt{N} \to \mathtt{N}\,, \\ ifzero &: \mathtt{N},\mathtt{N},\mathtt{N} \to \mathtt{N}\,, \\ Y_\sigma &: (\sigma \to \sigma) \to \sigma \qquad \text{for each } \sigma \in \mathsf{T}\,. \end{aligned}$$

Let Δ consist of rules of the forms

$$\begin{aligned} suc\; \widehat{n} &\rightsquigarrow \widehat{n+1}\,, \\ pre\; \widehat{n+1} &\rightsquigarrow \widehat{n}\,, \\ pre\; \widehat{0} &\rightsquigarrow \widehat{0}\,, \\ ifzero\; \widehat{0} &\rightsquigarrow \lambda xy.x\,, \\ ifzero\; \widehat{n+1} &\rightsquigarrow \lambda xy.y\,, \\ Y_\sigma f &\rightsquigarrow f(Y_\sigma f)\,. \end{aligned}$$

Again we let Ξ consist of the contexts of the forms

$$[-]N\,, \qquad suc\,[-]\,, \qquad pre\,[-]\,, \qquad ifzero\,[-]\,.$$

The resulting language is known as Plotkin's PCF (*Programming* language for *Computable Functions*).

The constants Y_σ are *general recursion* operators, whilst the others provide a necessary minimum of arithmetic. One could of course make do with just the numeral $\widehat{0}$ as in the previous example, but the above definition meshes more easily with the model to be described in Section 3.2.5. Many other inessential variations in the definition of PCF may be found in the literature—for instance, Plotkin's definition in [236] includes a basic type of booleans, along with constants for equality testing and general conditionals.

PCF will be important in its own right as a prototypical language for a natural notion of nested sequential computation, in which there is intuitively just a single 'thread of control'. We will study this notion in detail in Chapters 6 and 7. Furthermore, PCF will serve as a basis for numerous extensions corresponding to other computability notions, as discussed from Chapter 10 onwards.

We now return to the general rules for the evaluation of programs. Given Δ and Ξ as above, let us define a *one-step reduction* relation $\rightsquigarrow$ between closed terms of the same type by means of the following inductive definition (in which all terms are assumed to be well-typed):

1. $(\lambda x^\sigma.M)N \rightsquigarrow M[x^\sigma \mapsto N]$ (β*-reduction*).
2. If $(c\,M_0 \ldots M_{r-1} \rightsquigarrow N) \in \Delta$, and $-^*$ denotes some type-respecting substitution of closed terms for the free variables of $c\,M_0 \ldots M_{r-1}$, then $(c\,M_0^* \ldots M_{r-1}^*) \rightsquigarrow N^*$.
3. If $M \rightsquigarrow M'$ and $K[-] \in \Xi$, then $K[M] \rightsquigarrow K[M']$.

We call the entire system the *(call-by-name) typed* λ*-calculus* specified by $\vec{\beta}, C, \Delta$ and Ξ.

Depending on the choice of Δ and Ξ, the relation $\rightsquigarrow$ may or may not be *deterministic* in the sense that $M \rightsquigarrow M' \wedge M \rightsquigarrow M'' \Rightarrow M' = M''$. Note that both System T and PCF have deterministic reduction relations. In any case, we may now define the *many-step reduction* relation $\rightsquigarrow^*$ to be the reflexive-transitive closure of $\rightsquigarrow$.

Typically we are interested in the possible *values* of a closed term M, that is, those terms M' such that $M \rightsquigarrow^* M'$ and M' can be reduced no further. (If $\rightsquigarrow$ is deterministic, there is at most one value for a given M.) We shall see in Section 4.4 that in System T, every closed $M : \mathtt{N}$ reduces to a unique numeral $suc^k\,\widehat{0}$. In PCF, by contrast, there are terms such as $Y_{\mathtt{N}}(\lambda x^{\mathtt{N}}.x)$ that can never be reduced to a value. Thus System T is in essence a language for *total* functionals, whilst PCF is a language for *partial* ones.

To fashion a TPCA from a typed λ-calculus of this kind, let $=_{\mathrm{op}}$ be the *operational equivalence* relation on closed terms generated by

$$M \rightsquigarrow M' \;\Rightarrow\; M =_{\mathrm{op}} M'\,, \qquad M =_{\mathrm{op}} M' \;\Rightarrow\; MN =_{\mathrm{op}} M'N \wedge PM =_{\mathrm{op}} PM'\,.$$

For each $\sigma \in \mathsf{T}$, we therefore have the set $\mathscr{L}^0(C)_\sigma/{=_{\mathrm{op}}}$ of closed terms of type σ modulo $=_{\mathrm{op}}$. It is easy to see that with the evident application operations these constitute a TPCA $\mathscr{L}^0(C)/{=_{\mathrm{op}}}$: as in Example 3.1.6, the elements k, s are given by terms $\lambda xy.x$ and $\lambda xyz.xz(yz)$.

In the case of the examples we have mentioned, we shall refer to the closed term models as $\mathrm{T}^0/{=_{\mathrm{op}}}$ and $\mathrm{PCF}^0/{=_{\mathrm{op}}}$. Of course, we would like to be reassured that these quotients do not collapse the structures completely. In the former case, this can be seen from the standard interpretation $[\![-]\!]$ of System T in the set-theoretic model S (see Example 3.5.9): by induction on the generation of $=_{\mathrm{op}}$, it is easy to see that if $M =_{\mathrm{op}} M'$ then $[\![M]\!] = [\![M']\!]$. In the latter case, a similar argument applies using the interpretation of PCF in the model PC (see Example 3.5.6 below).

3.2.4 The van Oosten Model

Our next model can be seen as an analogue of K_2, but one capturing a specifically 'sequential' flavour of computation related to PCF. In place of $\mathbb{N}^{\mathbb{N}}$, we take our single set of data values to be the set $\mathbb{N}_\bot^{\mathbb{N}}$ of functions $\mathbb{N} \to \mathbb{N}_\bot$ (essentially partial functions $\mathbb{N} \rightharpoonup \mathbb{N}$). By analogy with K_2, a 'sequentially computable' operation $F : \mathbb{N}_\bot^{\mathbb{N}} \to \mathbb{N}_\bot$ may proceed by interrogating its argument $g \in \mathbb{N}_\bot^{\mathbb{N}}$ at various values $n \in \mathbb{N}$ before (possibly) returning a result. In the present case, however, it is *not* harmless to assume that g is queried in the order $g(0), g(1), \ldots$: for instance, if we wish to compute $F(g) = g(1)$, we should not begin by asking for $g(0)$ first, since if $g(0) = \bot$ then the computation of $F(g)$ will hang up. Instead, a procedure for computing a function $F : \mathbb{N}_\bot^{\mathbb{N}} \to \mathbb{N}_\bot$ should be represented by a *decision tree* as illustrated in the figure below: at each node $?n$, the procedure queries its argument g at n and branches on the result $g(n)$, whilst on reaching a leaf $!k$, the procedure returns k as the value of $F(g)$.

In general, there are three reasons why the result of $F(g)$ may be undefined:

- The computation may hit a node at which the label is undefined.
- The computation may query g at an argument n at which its value is undefined.
- The computation may keep putting queries to g forever, tracing out an infinite path in the decision tree (this is not illustrated by the above example).

As with K_2, a small refinement is needed to represent an operation that returns an element of $\mathbb{N}_\bot^{\mathbb{N}}$ rather than just $\mathbb{N}_\bot$. In effect, we require an infinite *forest* consisting of a tree T_m for each value m of the additional argument in $\mathbb{N}$. We may visualize such a forest as a tree with no root node, but with a top-level branch labelled by m leading to the tree T_m for each m.

We now consider how sequentially computable operations of this kind might themselves be represented as elements of $\mathbb{N}_\bot^{\mathbb{N}}$. The idea is that the position of a node α in such a forest may be specified by a non-empty sequence $m_0, \ldots, m_{i-1}$ of natural numbers, namely the sequence of labels along the path leading to α, and this may itself be coded by the number $\langle m_0, \ldots, m_{i-1} \rangle$. Likewise, labels on nodes may be

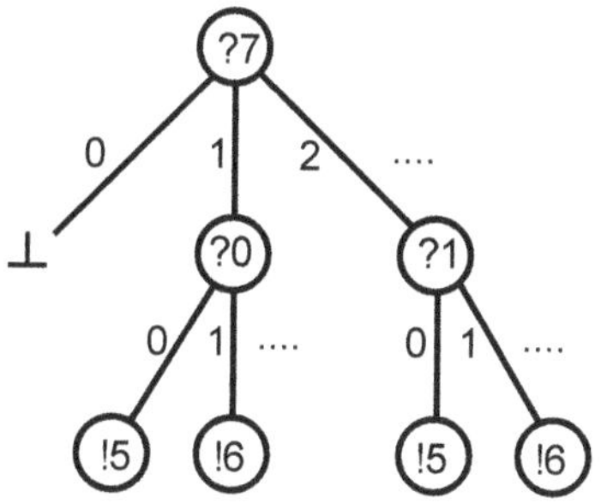

Fig. 3.1 Decision tree for $F_0 : g \mapsto g(g(7)-1)+5$ (taking $0-1$ to be undefined).

coded by natural numbers: we can code the label $?n$ by $2n$, and $!k$ by $2k+1$. Since a decision forest can be viewed as a partial function from positions to labels, it may be coded by an element $f \in \mathbb{N}_\perp^\mathbb{N}$.

These ideas are formalized by the following definition, which specifies how an element $f \in \mathbb{N}_\perp^\mathbb{N}$ (construed as coding a decision tree) may be 'applied' to an element $g \in \mathbb{N}_\perp^\mathbb{N}$ (construed as a plain function) to yield a new element $f \cdot g \in \mathbb{N}_\perp^\mathbb{N}$.

$$
\begin{aligned}
(f \cdot g)(m) \; = \; k \text{ iff } & \exists m_0, n_1, m_1, \ldots n_r, m_r. \\
& m_0 = m \;\wedge \\
& \forall i < r.\, f(\langle m_0, \ldots, m_i \rangle) = 2n_{i+1} \;\wedge \\
& \forall i < r.\, g(n_{i+1}) = m_{i+1} \;\wedge \\
& f(\langle m_0, \ldots, m_r \rangle) = 2k+1 \; .
\end{aligned}
$$

In Chapter 12 we will show that this defines a (total) combinatory algebra, which we denote by B. This model was discovered by van Oosten [294], and was further studied in Longley [176]. We shall see in Chapter 12 that it is intimately related to the earlier *sequential algorithms* model of Berry and Curien [33].

As with K_2, we may, if we wish, restrict ourselves to the *effective submodel* $\mathsf{B}^{\mathrm{eff}}$ consisting of just the computable partial functions on $\mathbb{N}$ (it is easy to check that the set of such functions is closed under application). We also have the relative PCA $(\mathsf{B};\mathsf{B}^{\mathrm{eff}})$, in which effective sequential operations are allowed to act on possibly non-computable data.

3.2.5 Nested Sequential Procedures

We conclude this section with a typed model of sequential computation which corresponds very closely to PCF, and which will also play an important role in our study of Kleene computability (Chapters 5 and 6). Whereas the model B described above captures the basic idea of sequentiality in a pleasing way, the present model elaborates on this by building in the notion of *nesting* of subcomputations: when a computation procedure Φ of type level k applies its formal parameter G of level $k-1$ to some argument, this argument may itself be a complex computation procedure h of level $k-2$ which G in its turn may invoke if it so chooses. Furthermore, these function invocations are required to conform to an appropriate nesting discipline: e.g. if Φ invokes G on the argument h and G in turn invokes h on some x, then the result of $h(x)$ must be returned before G can return a value for $G(h)$, and this itself must happen before the rest of the computation of Φ is able to proceed.

We shall call the model in question the *nested sequential procedure (NSP)* model. We shall not give a complete technical definition here, but rather concentrate on conveying the intuitions informally by illustrative examples. The model will be formally defined and studied in Chapter 6.

Intuitively, the NSP model occupies a kind of middle ground between syntax and semantics: NSPs are significantly more abstract and semantic than the language PCF (which can be interpreted in terms of NSPs), but much more intensional and syntactic than, say, the model PC of Section 3.2.2 (within which NSPs themselves may be interpreted). The construction we present here has the advantage that no significant theorems are needed in order to show that the structure is well-defined—although showing that the structure is indeed a higher-order model is non-trivial.

We start with some very simple examples of type $\mathtt{N} \to \mathtt{N}$. As in the partial continuous model PC, the idea is that these should represent functions $\mathbb{N}_\perp \to \mathbb{N}_\perp$, where we view $\mathbb{N}_\perp$ as the type of computational processes which, when activated, may evaluate to a natural number or may diverge. Consider for example a predecessor operation $\mathit{pred} : \mathbb{N}_\perp \to \mathbb{N}_\perp$, where $\mathit{pred}(0) = \perp$ by convention (and $\mathit{pred}(\perp) = \perp$ by necessity). We represent this as follows:

$$\mathit{pred} \;=\; \lambda x^{\mathtt{N}}.\,\mathtt{case}\,x\,\mathtt{of}\;(0 \Rightarrow \perp \mid 1 \Rightarrow 0 \mid \cdots)\,.$$

This will be a simple example of an NSP, and we will use this example to explain the semantics of the `case` construct. We think of x as an object of type $\mathbb{N}_\perp$; such objects may be passed around as parameters without attempting to evaluate them. However, the semantics of `case` will demand that if we wish to evaluate $\mathtt{case}\,x\,\mathtt{of}\;(0 \Rightarrow \perp \mid 1 \Rightarrow 0 \mid \cdots)$ then we are forced to start by evaluating x: if this yields a natural number n, we select the appropriate conditional branch, but if the evaluation of x diverges, the entire computation will diverge. (We may express this by saying that `case` is *strict* in its first argument.)

Let us see how the above NSP gets applied to NSPs of type $\mathtt{N}$. As two specimen cases, we consider the elements $\perp, 5 \in \mathbb{N}_\perp$; these will be represented by the NSPs $\lambda.\perp$ and $\lambda.5$ respectively. (The zero-argument λ-abstractions here are needed to make our general syntax smooth and uniform; see the formal definition below.) According to a system of rules to be spelt out in Section 6.1, we will have the following sequences of reductions:

$$\begin{array}{rll} pred\,(\lambda.5) & \rightsquigarrow \texttt{case}\,(\lambda.5)\,\texttt{of}\,(0 \Rightarrow \bot \mid 1 \Rightarrow 0 \mid \cdots) & \\ & \rightsquigarrow \texttt{case}\,5\,\texttt{of}\,(0 \Rightarrow \bot \mid 1 \Rightarrow 0 \mid \cdots) & \rightsquigarrow 4\,, \\ pred\,(\lambda.\bot) & \rightsquigarrow \texttt{case}\,(\lambda.\bot)\,\texttt{of}\,(0 \Rightarrow \bot \mid 1 \Rightarrow 0 \mid \cdots) & \\ & \rightsquigarrow \texttt{case}\,\bot\,\texttt{of}\,(0 \Rightarrow \bot \mid 1 \Rightarrow 0 \mid \cdots) & \rightsquigarrow \bot\,. \end{array}$$

There will also be NSPs of type $\mathtt{N} \to \mathtt{N}$ that may return a value even on the argument $(\lambda.\bot)$. Consider for instance

$$g_0 = \lambda x.\,5 \qquad \text{and} \qquad g_1 = \lambda x.\,\texttt{case}\,x\,\texttt{of}\,(0 \Rightarrow 5 \mid 1 \Rightarrow 5 \mid \cdots)\,.$$

The first of these will be an NSP that returns the result 5 without even bothering to evaluate its argument x; the second evaluates its argument and returns 5 if this succeeds. These behaviours will be exemplified by the following reduction sequences (again, see the formal definition below):

$$\begin{array}{rl} g_0\,(\lambda.\bot) & \rightsquigarrow 5\,, \\ g_1\,(\lambda.\bot) & \rightsquigarrow \texttt{case}\,(\lambda.\bot)\,\texttt{of}\,(0 \Rightarrow 5 \mid 1 \Rightarrow 5 \mid \cdots) \\ & \rightsquigarrow \texttt{case}\,\bot\,\texttt{of}\,(0 \Rightarrow 5 \mid 1 \Rightarrow 5 \mid \cdots) \rightsquigarrow \bot\,. \end{array}$$

As a simple exercise at this point, the reader is invited to construct two different NSPs representing the everywhere undefined function $\Lambda x.\bot : \mathbb{N}_\bot \to \mathbb{N}_\bot$. This already shows that NSPs are more 'intensional' objects than the plain mathematical functions that constitute the model PC.

There will be NSPs that accept several arguments. For instance, the following procedure of type $\mathtt{N} \to \mathtt{N} \to \mathtt{N}$ will be an NSP representing the addition operator:

$$\begin{array}{rl} plus = \lambda x^{\mathtt{N}} y^{\mathtt{N}}.\,\texttt{case}\,y\,\texttt{of}\,(& \\ & 0 \Rightarrow \texttt{case}\,x\,\texttt{of}\,(0 \Rightarrow 0 \mid 1 \Rightarrow 1 \mid \cdots) \\ & \mid 1 \Rightarrow \texttt{case}\,x\,\texttt{of}\,(0 \Rightarrow 1 \mid 1 \Rightarrow 2 \mid \cdots) \\ & \mid \cdots)\,. \end{array}$$

Here we have chosen an NSP that evaluates its second argument y before evaluating its first argument x. Of course, there is also an NSP that represents the same mathematical function $+ : \mathbb{N}_\bot \times \mathbb{N}_\bot \to \mathbb{N}_\bot$ but evaluates x first. This again illustrates the intensional nature of NSPs, a feature that they share with the van Oosten model.

Let us now turn to some higher-order NSPs. First, any second-order 'sequential' computation procedure of the kind considered in the previous subsection may readily be translated into an NSP. For instance, the decision tree depicted in Section 3.2.4 would yield the following:

$$
\begin{aligned}
F_0 \;=\; \lambda g^{\mathtt{N}\to\mathtt{N}}.\ & \mathtt{case}\ g(\lambda.7)\ \mathtt{of}\ (\\
& \quad 0 \Rightarrow \bot \\
& \mid 1 \Rightarrow \mathtt{case}\ g(\lambda.0)\ \mathtt{of}\ (0 \Rightarrow 5 \mid 1 \Rightarrow 6 \mid \cdots) \\
& \mid 2 \Rightarrow \mathtt{case}\ g(\lambda.1)\ \mathtt{of}\ (0 \Rightarrow 5 \mid 1 \Rightarrow 6 \mid \cdots) \\
& \mid \cdots) .
\end{aligned}
$$

There will also be NSPs of type $(\mathtt{N} \to \mathtt{N}) \to \mathtt{N}$ that do not arise from decision trees in this way: for instance, those of the form $\lambda g.\, \mathtt{case}\ g(\lambda.\bot)\ \mathtt{of}\ (\cdots)$. Intuitively, such NSPs correspond to operations of type $(\mathbb{N}_\bot \to \mathbb{N}_\bot) \to \mathbb{N}_\bot$ (for a suitable meaning of '$\to$') rather than $(\mathbb{N} \to \mathbb{N}_\bot) \to \mathbb{N}_\bot$. Nested function calls are also permitted: for instance, the operation $F_1 : g^{\mathtt{N}\to\mathtt{N}} \mapsto g(g(2)+1)$ may be represented by

$$
\lambda g^{\mathtt{N}\to\mathtt{N}}.\ \mathtt{case}\ \ g\ (\lambda.\mathtt{case}\ g(\lambda.2)\ \mathtt{of}\ (0 \Rightarrow 1 \mid 1 \Rightarrow 2 \mid \cdots))\ \mathtt{of}\ (0 \Rightarrow 0 \mid \cdots) .
$$

Next, consider the third-order operation Φ_0 of type $((\mathtt{N} \to \mathtt{N}) \to \mathtt{N}) \to \mathtt{N}$ represented by the program $\lambda F^{(\mathtt{N}\to\mathtt{N})\to\mathtt{N}}.\ F(\lambda x.x-1)+5$ (we again interpret $0-1$ as undefined). Here the argument passed to the 'oracle' F is a first-order operation which must itself be specified by an NSP, in this case *pred*:

$$
\Phi_0 \;=\; \lambda F.\ \mathtt{case}\ F\ \mathit{pred}\ \mathtt{of}\ (0 \Rightarrow 5 \mid 1 \Rightarrow 6 \mid \cdots) .
$$

As a further example of how the model is supposed to work, let us try applying Φ_0 to the procedure

$$
F_2 \;=\; \lambda g.\ \mathtt{case}\ g(\lambda.7)\ \mathtt{of}\ (0 \Rightarrow 3 \mid 1 \Rightarrow 4 \mid \cdots) ,
$$

which represents $\lambda g.g \mapsto g(7)+3$. The key idea here is that we can 'compute' directly with the NSPs themselves rather than with the syntactic definitions (essentially PCF terms) from which we obtained them. According to the rules that we shall specify, we will have the following sequence of reductions:

$$
\begin{aligned}
\Phi_1(F_2) \;&\rightsquigarrow\; \mathtt{case}\ F_2\ \mathit{pred}\ \mathtt{of}\ (0 \Rightarrow 5 \mid \cdots) \\
&\rightsquigarrow\; \mathtt{case}\ (\mathtt{case}\ \mathit{pred}(\lambda.7)\ \mathtt{of}\ (0 \Rightarrow 3 \mid \cdots))\ \mathtt{of}\ (0 \Rightarrow 5 \mid \cdots) \\
&\rightsquigarrow\; \mathtt{case}\ \mathit{pred}(\lambda.7)\ \mathtt{of}\ (0 \Rightarrow \mathtt{case}\ 3\ \mathtt{of}\ (0 \Rightarrow 5 \mid \cdots) \mid \cdots) \\
&\rightsquigarrow\; \mathtt{case}\ (\mathtt{case}\ 7\ \mathtt{of}\ (0 \Rightarrow \bot \mid 1 \Rightarrow 0 \mid \cdots)) \\
&\qquad\quad \mathtt{of}\ (0 \Rightarrow \mathtt{case}\ 3\ \mathtt{of}\ (0 \Rightarrow 5 \mid \cdots) \mid \cdots) \\
&\rightsquigarrow\; \mathtt{case}\ 7\ \mathtt{of}\ (0 \Rightarrow \bot \mid 1 \Rightarrow \mathtt{case}\ 0\ \mathtt{of}\ (\mathtt{case}\ 3\ \mathtt{of}\ (0 \Rightarrow 5 \mid \cdots) \mid \cdots)) \\
&\rightsquigarrow\; \mathtt{case}\ 6\ \mathtt{of}\ (0 \Rightarrow \mathtt{case}\ 3\ \mathtt{of}\ (0 \Rightarrow 5 \mid \cdots) \mid \cdots) \\
&\rightsquigarrow\; \mathtt{case}\ 9\ \mathtt{of}\ (0 \Rightarrow 5 \mid \cdots) \\
&\rightsquigarrow\; 14 .
\end{aligned}
$$

At third order, it is also possible for the argument passed to an oracle F to be an NSP that is itself constructed using F. For example, the *Kierstead functional* represented by

$$
\lambda F.\ F(\lambda x.F(\lambda y.x))
$$

corresponds to an NSP which is best displayed diagrammatically:

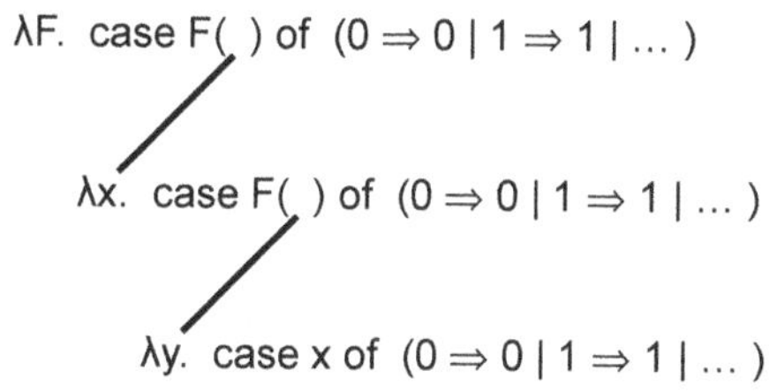

It is even possible for calls to an oracle F to be nested to infinite depth in this way. We shall meet some important examples of this in Chapters 6 and 7.

Having presented some motivating examples, we now give a very brief sketch of how the NSP model is formally constructed; full details will be given in Section 6.1. Assume that we have an infinite supply of typed variables x^σ for each $\sigma \in \mathsf{T}^{\rightarrow}(\mathtt{N})$. We consider nested sequential procedures to be syntax trees in the infinitary language defined by the following grammar:

$$\begin{array}{lrl} \textit{Procedures:} & p,q ::= & \lambda x_0 \ldots x_{r-1}.e \\ \textit{Expressions:} & d,e ::= & \bot \mid n \mid \mathtt{case}\ a\ \mathtt{of}\ (i \Rightarrow e_i \mid i \in \mathbb{N}) \\ \textit{Applications:} & a ::= & x q_0 \ldots q_{r-1} \end{array}$$

Here $(i \Rightarrow e_i \mid i \in \mathtt{N})$, sometimes abbreviated to $(i \Rightarrow e_i)$, is a piece of infinitary syntax involving a subterm e_i for each natural number i. We allow $r = 0$ in the clauses for procedures and applications. We shall use t as a metavariable that may range over terms of any of these three kinds.

We here interpret the above grammar *coinductively* rather than inductively, so as to allow terms that are infinitely deep as well as infinitely wide. We shall implicitly identify terms that differ only in the names of bound variables: that is, we work with terms only up to (infinitary) α-equivalence.

We also impose some *typing rules* on the raw terms given by the above grammar in a fairly straightforward way (see Section 6.1 for details). We write $\vdash e$ and $\vdash a$ to mean that e and a are well-typed, and $\vdash p : \sigma$ to mean p is a well-typed procedure of type σ. We now define $\mathsf{SP}^0(\sigma) = \{p \text{ closed} \mid \vdash p : \sigma\}$ for each σ. The superscript 0 indicates that we are dealing with closed procedures.

To define the application operations in SP^0, we introduce a reduction system to describe the computational dynamics of sequential procedures. Note, however, that many of the intermediate forms appearing in the examples of reductions above fail to be procedure terms according to the above syntax. To accommodate these intermediate forms, we therefore introduce a wider class of *meta-terms* and endow them with a reduction relation $\rightsquigarrow$; the terms proper will then play the role of 'normal forms'. Again, the details will be given in Section 6.1.

From this, it is easy to define a notion of (infinitary) *evaluation* for meta-terms. Specifically, if $\rightsquigarrow^*$ is the reflexive-transitive closure of $\rightsquigarrow$, and $\sqsubseteq$ is the evident syntactic ordering on meta-terms, we may define for any meta-term T the ordinary term $\ll T \gg$ to which it evaluates:

$$\ll T \gg \;=\; \bigsqcup \{t \mid \exists T'.\, T \rightsquigarrow^* T' \wedge t \sqsubseteq T'\}\,.$$

Thus, evaluation is in principle an infinitary process since T and $\ll T \gg$ are in general infinitely deep trees; however, any finite portion of $\ll T \gg$ may be obtained by a finite computation. This is closely analogous to the situation for *Böhm trees* in the context of untyped λ-calculus: see Barendregt [9].

We are now in a position to define the (total) application operations in SP^0. If $p = (\lambda x_0 \cdots x_r.e) \in \mathsf{SP}^0(\sigma)$ and $q \in \mathsf{SP}^0(\sigma_0)$ where $\sigma = \sigma_0, \cdots, \sigma_r \to \mathtt{N}$, we set

$$p \cdot q = \lambda x_1 \ldots x_r.\; \ll e[x_0 \mapsto q] \gg \,.$$

This completes the construction. We will show in Section 6.1 that SP^0 is a TPCA, and indeed a $\lambda\eta$*-algebra*. The proof involves a detailed but essentially elementary analysis of the reduction behaviour of meta-terms.

3.3 Computational Structure in Higher-Order Models

Next, we review some basic 'computability theory' that can be developed uniformly for all or some large class of higher-order models. We shall consider a variety of pieces of computational structure that a model may possess, approximately in order of 'increasing strength'. Most of this structure will be common to all of the main models of interest in this book. We shall also see that in the setting of untyped PCAs, all of this structure is available for free.

The material here is mostly a routine adaptation of well-established ideas from combinatory logic (see for example Barendregt [9]). We shall here work within the framework of lax TPCAs, and the term 'model' used without qualification will mean 'lax relative TPCA'. Strict analogues of the notions and results in this section may be obtained by replacing '$\succeq$' everywhere by '$\simeq$'.

3.3.1 Combinatory Completeness

We begin with a property, known as *combinatory completeness*, that can be seen as a syntactic counterpart to the notion of weakly cartesian closed model. In essence, combinatory completeness asserts that any operation definable by means of a formal expression over $\mathbf{A}$ (constructed using application) is representable by an element of $\mathbf{A}$ itself.

The notion of 'formal expression' is made precise as follows. In contrast to the languages of well-typed terms introduced in Section 3.2.3 for the purpose of constructing particular TPCAs, we are here regarding formal expressions as constituting a *metalanguage* for denoting elements of an arbitrary partial applicative structure.

Definition 3.3.1. Suppose $\mathbf{A}$ is a relative partial applicative structure over T.

(i) The set of well-typed *applicative expressions* $e : \sigma$ over $\mathbf{A}$ is defined inductively as follows:

- For each $\sigma \in \mathsf{T}$, we have an unlimited supply of *variables* $x^\sigma : \sigma$.
- For each $\sigma \in \mathsf{T}$ and $a \in \mathbf{A}^\sharp(\sigma)$, we have a *constant* symbol $c_a : \sigma$ (we shall often write c_a simply as a).
- If $e : \sigma \to \tau$ and $e' : \sigma$ are applicative expressions, then ee' is an applicative expression of type τ. As usual, we take juxtaposition to be left-associative, so that $ee'e''$ means $(ee')e''$.

We write $V(e)$ for the set of variables appearing in e.

(ii) A *valuation* in $\mathbf{A}$ is a function ν assigning to certain variables x^σ an element $\nu(x^\sigma) \in \mathbf{A}^\circ(\sigma)$. Given an applicative expression e and a valuation ν covering $V(e)$, the *value* $[\![e]\!]_\nu$, when defined, is given inductively by

$$[\![x^\sigma]\!]_\nu = \nu(x)\,, \qquad [\![c_a]\!]_\nu = a\,, \qquad [\![ee']\!]_\nu \simeq [\![e]\!]_\nu \cdot [\![e']\!]_\nu\,.$$

Note that if $e : \tau$ and $[\![e]\!]_\nu$ is defined then $[\![e]\!]_\nu \in \mathbf{A}^\circ(\tau)$. The property of interest may now be formulated as follows.

Definition 3.3.2. Let $\mathbf{A}$ be a relative partial applicative structure. We say $\mathbf{A}$ is *lax combinatory complete* if for every applicative expression $e : \tau$ over $\mathbf{A}$ and every variable x^σ, there is an applicative expression $\lambda^* x^\sigma.e$ with $V(\lambda^* x^\sigma.e) = V(e) - \{x^\sigma\}$, such that for any valuation ν covering $V(\lambda^* x^\sigma.e)$ and any $a \in \mathbf{A}^\circ(\sigma)$, we have

$$[\![\lambda^* x^\sigma.e]\!]_\nu \downarrow\,, \qquad [\![\lambda^* x^\sigma.e]\!]_\nu \cdot a \succeq [\![e]\!]_{\nu, x \mapsto a}\,.$$

We say $\mathbf{A}$ is *strictly combinatory complete* if this holds with '$\simeq$' in place of '$\succeq$'.

Theorem 3.3.3. *A (relative) partial applicative structure* $\mathbf{A}$ *is a lax (relative) TPCA iff it is lax combinatory complete.*

Proof. If $\mathbf{A}$ is lax combinatory complete, then for any ρ, σ, τ we may define

$$\begin{aligned} k_{\sigma\tau} &= [\![\lambda^* x^\sigma.(\lambda^* y^\tau.x)]\!]_\emptyset\,, \\ s_{\rho\sigma\tau} &= [\![\lambda^* x^{\rho\to\sigma\to\tau}.(\lambda^* y^{\rho\to\sigma}.(\lambda^* z^\rho.xz(yz)))]\!]_\emptyset\,. \end{aligned}$$

Clearly these satisfy the usual axioms for k and s in their lax version.

Conversely, if $\mathbf{A}$ is a lax TPCA, then given any suitable choice of elements k and s for $\mathbf{A}$, we may define $\lambda^* x^\sigma.e$ by induction on the structure of e:

$$\begin{aligned} \lambda^* x^\sigma.x &= s_{\sigma(\sigma\to\sigma)\sigma}\, k_{\sigma(\sigma\to\sigma)}\, k_{\sigma\sigma}\,, & \\ \lambda^* x^\sigma.e &= k_{\tau\sigma} e & \text{if } e : \tau \text{ does not contain } x\,, \\ \lambda^* x^\sigma.ee' &= s_{\sigma\tau\tau'}(\lambda^* x^\sigma.e)(\lambda^* x^\sigma.e') & \text{if } e : \tau \to \tau',\ e' : \tau, \text{ and } ee' \text{ contains } x\,. \end{aligned}$$

(Variations on this definition are also possible; see Barendregt [9, Section 7.3].) It is an easy exercise to check that $\lambda^* x^\sigma .e$ has the required properties. □

The same argument shows that **A** is a strict TPCA iff it is strictly combinatory complete.

As mentioned in Section 3.1.5, we often tacitly suppose that a TPCA **A** comes equipped with some choice of k and s drawn from $\mathbf{A}^\sharp$, and in this case we shall use the notation $\lambda^* x.e$ for the applicative expression given by the above proof. Since all the constants appearing in e are drawn from $\mathbf{A}^\sharp$, the same will be true for $\lambda^* x.e$.

In TPCAs constructed as syntactic models for untyped or typed λ-calculi (as in Example 3.1.6 or Section 3.2.3), the value of $\lambda^* x.e$ coincides with $\lambda x.e$. However, the notational distinction is worth retaining, since the term $\lambda^* x.e$ as defined above is not syntactically identical to $\lambda x.e$.

If $x_0, \ldots, x_{n-1}$ are distinct variables, we may write $\lambda^* x_0 \ldots x_{n-1}.e$ as an abbreviation for

$$\lambda^* x_0.(\lambda^* x_1.(\cdots(\lambda^* x_{n-1}.e)\cdots))$$

(where we expand the λ^*-abstractions to applicative expressions from the inside out). If $n \geq 1$ and $V(e) \subseteq \{x_0, \ldots, x_{n-1}\}$, then $\lambda^* x_0 \ldots x_{n-1}.e$ contains no variables and its value is some fixed element of $\mathbf{A}^\sharp$. Where no confusion can result, we shall write simply $\lambda^* x_0 \ldots x_{n-1}.e$ in place of $[\![\lambda^* x_0 \ldots x_{n-1}.e]\!]_\emptyset$.

More generally, we may consider terms of the λ-calculus (possibly with constants c_a, and subject to the obvious typing constraints) as *meta-expressions* for applicative expressions. Specifically, any such λ-term M can be regarded as denoting an applicative expression $M^\dagger$ as follows:

$$x^\dagger = x\,, \qquad c_a^\dagger = c_a\,, \qquad (MN)^\dagger = M^\dagger N^\dagger\,, \qquad (\lambda x.M)^\dagger = \lambda^* x.(M^\dagger)\,.$$

Some caution is needed here, however, because β-equivalent meta-expressions do not always have the same meaning, as the following example shows.

Example 3.3.4. Consider the two meta-expressions $(\lambda x.(\lambda y.y)x)$ and $\lambda x.x$. Although these are β-equivalent, the first expands to $s(ki)i$ and the second to i, where $i \equiv skk$. In general, these may denote completely different elements in a TPCA, even a total one. Intuitively, the operation that feeds its input x to $\lambda^* y.y$ and returns the result might be distinguishable at a fine-grained intensional level from the operation that simply returns its input x.

The moral here is that β-reductions are not valid underneath λ^*-abstractions: in this case, the reduction $(\lambda^* y.y)x \rightsquigarrow x$ is not valid underneath $\lambda^* x$. However, it is useful to note that, at least for the definition of λ^* given above, β-reductions at top level are valid.

Proposition 3.3.5. *(i) If M and N are meta-expressions where all bound variables within M are distinct from the free variables of N and from x, then the applicative expressions $M^\dagger[x \mapsto N^\dagger]$ and $(M[x \mapsto N])^\dagger$ are syntactically identical.*

(ii) In this situation, $[\![((\lambda x.M)N)^\dagger]\!]_\nu \succeq [\![M[x \mapsto N]^\dagger]\!]_\nu$ for any valuation ν covering the relevant variables. □

Proof. Part (i) is an easy induction on the nesting depth of abstractions within M: note that if λy is such an abstraction then $(\lambda y.D[x])^\dagger \equiv k(D[x])^\dagger$ and $(\lambda y.D[N])^\dagger \equiv k(D[N])^\dagger$ for any context $D[-]$ not containing y free. Part (ii) is immediate from this and the proof of Theorem 3.3.3. □

For other definitions of λ^*, the validity of β-equality is sometimes even more restricted. We refer to Longley [179, Section 2] for a discussion of these subtleties, along with a more elaborate language of meta-expressions that allows for finer control of the 'evaluation time' of subexpressions underneath abstractions.

From now on, we will not need to distinguish formally between meta-expressions and the applicative expressions they denote. For the remainder of this chapter we shall use the λ^* notation for such (meta-)expressions, retaining the asterisk as a reminder that the usual rules of λ-calculus are not always valid.

3.3.2 Pairing

Let $\mathbf{A}$ be a relative TPCA over a type world T with arrow structure, and suppose also that $\mathbf{A}$ (considered as a higher-order model) has weak products, inducing a product structure $\times$ on T. This means that for any $\sigma, \tau \in \mathsf{T}$ there are elements

$$fst \in \mathbf{A}^\sharp((\sigma \times \tau) \to \sigma)\,, \qquad snd \in \mathbf{A}^\sharp((\sigma \times \tau) \to \tau)\,,$$

possessing a certain property as detailed in Definition 3.1.3. However, we have already noted that this is not enough to imply that $\mathbf{A}$ is weakly cartesian closed. To bridge the gap, we also require for each $\sigma, \tau \in \mathsf{T}$ a *pairing* operation

$$pair \in \mathbf{A}^\sharp(\sigma \to \tau \to (\sigma \times \tau))$$

such that

$$\forall a \in \mathbf{A}^\circ(\sigma), b \in \mathbf{A}^\circ(\tau).\ fst \cdot (pair \cdot a \cdot b) \;=\; a \,\wedge\, snd \cdot (pair \cdot a \cdot b) \;=\; b\,.$$

We leave it as an easy exercise to show that a higher-order model with weak products has pairing iff it is weakly cartesian closed—that is, the binary partial functions representable in $\mathbf{A}^\sharp((\rho \times \sigma) \to \tau)$ are exactly those representable in $\mathbf{A}^\sharp(\rho \to \sigma \to \tau)$. Henceforth we shall generally work with *pair* in preference to the 'external' pairing of operations given by Definition 3.1.3, and will write $pair \cdot a \cdot b$ as $\langle a, b \rangle$ when there is no danger of confusion.

All the models of interest will have pairing in the above sense. Moreover, in untyped models, pairing is automatic: generalizing the definitions from Example 3.1.6, we may take

$$pair \;=\; \lambda^* xyz.\, zxy\,, \qquad fst \;=\; \lambda^* p.\, p(\lambda^* xy.x)\,, \qquad snd \;=\; \lambda^* p.\, p(\lambda^* xy.y)\,.$$

The verification of the above properties is an exercise in the use of Proposition 3.3.5.

We regard the models with pairing as those in which the weak product and higher-order structures fit together well. As we shall see in the next few subsections, a recurring pattern is that some concept of interest can be defined either in terms of product structure or in terms of arrow structure, but pairing is needed to show that the two definitions coincide.

3.3.3 Booleans

A very natural condition to consider on models concerns the existence of elements playing the role of 'true' and 'false', intuitively allowing for conditional *branching* in the course of computations. In the TPCA setting, we may capture this notion as follows.

Definition 3.3.6. A model **A** *has booleans* if for some type $\mathtt{B}$ there exist elements

$$\begin{aligned} tt, ff &\in \mathbf{A}^\sharp(\mathtt{B})\,, \\ if_\sigma &\in \mathbf{A}^\sharp(\mathtt{B},\sigma,\sigma \to \sigma) \quad \text{for each } \sigma \end{aligned}$$

such that for all $x,y \in \mathbf{A}^\circ(\sigma)$ we have

$$if_\sigma \cdot tt \cdot x \cdot y \;=\; x\,, \qquad if_\sigma \cdot ff \cdot x \cdot y \;=\; y\,.$$

Note that tt, ff need not be the sole elements of $\mathbf{A}^\sharp(\mathtt{B})$.

Alternatively, we may define a notion of having booleans in the setting of a computability model **C** with weak products: just replace if_σ above with $if'_\sigma \in \mathbf{C}[\mathbb{B} \times \sigma \times \sigma, \sigma]$. It is easy to check that in a TPCA with products and pairing the two definitions coincide.

All the examples of TPCAs appearing in this book are easily seen to have booleans. Moreover, in untyped models, the existence of booleans is automatic: we may define $tt = \lambda^* xy.x$, $ff = \lambda^* xy.y$ and $if = \lambda^* zxy.zxy$.

Obviously, the value of an expression $if_\sigma \cdot b \cdot e \cdot e'$ cannot be defined unless the values of both e and e' are defined. However, there is a useful trick that allows us to build conditional expressions whose definedness requires only that the chosen branch of the conditional is defined. This trick is specific to the higher-order setting, and is known as *strong definition by cases*:

Proposition 3.3.7. *Suppose* **A** *has booleans as above. Given applicative expressions $e, e' : \sigma$, there is an applicative expression $(e \mid e') : \mathtt{B} \to \sigma$ such that for any valuation ν covering $V(e)$ and $V(e')$ we have*

$$[\![(e \mid e')]\!]_\nu \downarrow, \qquad [\![(e \mid e') \cdot tt]\!]_\nu \succeq [\![e]\!]_\nu\,, \qquad [\![(e \mid e') \cdot ff]\!]_\nu \succeq [\![e']\!]_\nu\,.$$

Proof. Let ρ be any type such that $\mathbf{A}^\circ(\rho)$ is inhabited by some element a, and define

$$(e \mid e') \;=\; \lambda^* z^{\mathtt{B}}.\,(if_\sigma z\,(\lambda^* r^\rho.e)\,(\lambda^* r^\rho.e')\,c_a)\,,$$

where z, r are fresh variables. It is easy to check that $(e \mid e')$ has the required properties. □

The expressions $\lambda^* r.e$, $\lambda^* r.e'$ in the above proof are known as *suspensions* or *thunks*: the idea is that $[\![\lambda^* r.e]\!]_\nu$ is guaranteed to be defined, but the actual evaluation of e_ν (which may be undefined) is 'suspended' until the argument c_a is supplied.

Exercise 3.3.8. Show that if **A** has booleans in the sense of Definition 3.3.6, the corresponding higher-order model has weak products, and each $\mathbf{A}^\sharp(\sigma)$ is inhabited, then **A** has pairing.

Exercise 3.3.9. Formulate a suitable definition of a higher-order model with *weak sums*. Show that any model with weak products and booleans also has weak sums.

3.3.4 Numerals

Another important condition on TPCAs concerns the existence of a reasonable representation of the natural numbers.

Definition 3.3.10. A model **A** *has numerals* if for some type $\mathtt{N}$ there exist

$$\widehat{0}, \widehat{1}, \widehat{2}, \ldots \in \mathbf{A}^\sharp(\mathtt{N}) ,$$
$$suc \in \mathbf{A}^\sharp(\mathtt{N} \to \mathtt{N}) ,$$

and for any $x \in \mathbf{A}^\sharp(\sigma)$ and $f \in \mathbf{A}^\sharp(\mathtt{N} \to \sigma \to \sigma)$ an element

$$Rec_\sigma(x, f) \in \mathbf{A}^\sharp(\mathtt{N} \to \sigma)$$

such that for all $x \in \mathbf{A}^\sharp(\sigma)$, $f \in \mathbf{A}^\sharp(\mathtt{N} \to \sigma \to \sigma)$ and $n \in \mathbb{N}$ we have

$$suc \cdot \widehat{n} = \widehat{n+1} ,$$
$$Rec_\sigma(x, f) \cdot \widehat{0} = x ,$$
$$Rec_\sigma(x, f) \cdot \widehat{n+1} \succeq f \cdot \widehat{n} \cdot (Rec_\sigma(x, f) \cdot \widehat{n}) .$$

The operators Rec_σ in effect allow for definitions by 'primitive recursion' (generalized to arbitrary types σ). The above definition has the advantage that it naturally adapts to the setting of a computability model **C** with products: just replace the types of f and $Rec_\sigma(x, f)$ above with $\mathbf{C}[\mathtt{N} \times \sigma, \sigma]$ and $\mathbf{C}[\mathtt{N}, \sigma]$ respectively. In the higher-order setting, however, the definition is equivalent to a superficially stronger one that requires the primitive recursion operators to exist 'internally' in **A**. This is the characterization we shall most often use in the sequel:

Proposition 3.3.11. *A model* **A** *has numerals iff it has elements* $\widehat{n}$, *suc as above and*

$$rec_\sigma \in \mathbf{A}^\sharp(\sigma \to (\mathtt{N} \to \sigma \to \sigma) \to \mathtt{N} \to \sigma) \quad \textit{for each } \sigma$$

such that for all $x \in \mathbf{A}^\circ(\sigma)$, $f \in \mathbf{A}^\circ(\mathtt{N} \to \sigma \to \sigma)$ *and* $n \in \mathbb{N}$ *we have*

$$\begin{aligned} suc \cdot \widehat{n} &= \widehat{n+1}\,, \\ rec_\sigma \cdot x \cdot f \cdot \widehat{0} &= x\,, \\ rec_\sigma \cdot x \cdot f \cdot \widehat{n+1} &\succeq f \cdot \widehat{n} \cdot (rec_\sigma \cdot x \cdot f \cdot \widehat{n})\,. \end{aligned}$$

Proof. The right-to-left implication is easy. For the other direction, define

$$rec_\sigma \;=\; Rec_{\sigma \to (\mathtt{N} \to \sigma \to \sigma) \to \sigma}\,(\lambda^* xf.x,\ \lambda^* nr.\lambda^* xf.\, f\, n\,(r\, x\, f))\,.$$

It is easy to check that rec_σ has the required properties. □

Exercise 3.3.12. Show that if **A** has numerals, then **A** has booleans.

Again, all the main TPCAs featuring in this book are easily seen to have numerals. In most of our leading models, the natural candidate for $\mathbf{A}(\mathtt{N})$ is the set $\mathbb{N}$ or $\mathbb{N}_\perp$.

Proposition 3.3.13. *Every untyped model has numerals.*

Proof. Using the encodings for pairings and booleans given above, we may define the *Curry numerals* $\widehat{n}$ in any untyped model as follows:

$$\widehat{0} \;=\; \langle tt, tt \rangle\,, \qquad \widehat{n+1} \;=\; \langle ff, \widehat{n} \rangle\,.$$

We may now take $suc = \lambda^* x.\langle ff, x \rangle$. We also have elements for the zero testing and predecessor operations: take $iszero = fst$ and $pre = \lambda^* x.\, if\,(iszero\; x)\,\widehat{0}\,(snd\; x)$. These will be used in the construction of a suitable element *rec*, which we defer until the next subsection. □

In any model with numerals, a rich class of functions $\mathbb{N}^r \to \mathbb{N}$ is representable. For example, the (first-order) *primitive recursive* functions on $\mathbb{N}$ are defined as the smallest class of functions $\mathbb{N}^r \to \mathbb{N}$ containing certain 'basic' functions (projections, constant functions, and successor) and closed under composition and primitive recursion. An easy induction over the construction of such functions shows:

Proposition 3.3.14. *For any primitive recursive* $f : \mathbb{N}^r \to \mathbb{N}$, *there is an applicative expression* $e_f : \mathtt{N}^{(r)} \to \mathtt{N}$ *(involving constants* $0, suc, rec_{\mathtt{N}}$*) such that in any model* $(\mathbf{A}^\circ; \mathbf{A}^\sharp)$ *with numerals we have* $[\![e_f]\!]_\nu \in \mathbf{A}^\sharp$ *(where* ν *is the obvious valuation of the constants) and*

$$\forall n_0, \ldots, n_{r-1}, m.\;\; f(n_0, \ldots, n_{r-1}) \;=\; m \;\Rightarrow\; [\![e_f]\!]_\nu \cdot \widehat{n_0} \cdot \ldots \cdot \widehat{n_{r-1}} \;=\; \widehat{m}\,.$$

In fact, this is only the beginning. We shall see later (Example 3.5.9 and Exercise 3.6.6) that a model **A** with numerals admits an interpretation of arbitrary closed terms of Gödel's System T, in such a way that all System T definable functions $\mathbb{N}^r \to \mathbb{N}$ are representable in $\mathbf{A}^\sharp$, and even System T definable functionals of higher type are in some sense computable within **A**.

3.3.5 Recursion and Minimization

We now turn our attention to two other kinds of recursion operators more powerful than the primitive recursors considered above. Once again, we will introduce these first as operations external to the model **A**, then show that this style of definition may be upgraded to a superficially stronger 'internal' one.

Definition 3.3.15. (i) A total model **A** *has general recursion*, or *has fixed points*, if for every element $f \in \mathbf{A}^\sharp(\rho \to \rho)$ there is an element $Fix_\rho(f) \in \mathbf{A}^\sharp(\rho)$ such that $Fix_\rho(f) = f \cdot Fix_\rho(f)$.

(ii) An arbitrary model **A** *has guarded recursion*, or *guarded fixed points*, if for every element $f \in \mathbf{A}^\sharp(\rho \to \rho)$ where $\rho = \sigma \to \tau$, there is an element $GFix_\rho(f) \in \mathbf{A}^\sharp(\rho)$ such that $GFix_\rho(f) \cdot x \succeq f \cdot GFix_\rho(f) \cdot x$ for all $x \in \mathbf{A}^\circ(\sigma)$.

Proposition 3.3.16. *(i) A total model* **A** *has general recursion iff for every type* ρ *there is an element* $Y_\rho \in \mathbf{A}^\sharp((\rho \to \rho) \to \rho)$ *such that for all* $f \in \mathbf{A}^\circ(\rho \to \rho)$ *we have*

$$Y_\rho \cdot f \;=\; f \cdot (Y_\rho \cdot f)\,.$$

(ii) **A** *has guarded recursion iff for every type* $\rho = \sigma \to \tau$ *there is an element* $Z_\rho \in \mathbf{A}^\sharp((\rho \to \rho) \to \rho)$ *such that for all* $f \in \mathbf{A}^\circ(\rho \to \rho)$ *and* $x \in \mathbf{A}^\circ(\sigma)$ *we have*

$$Z_\rho \cdot f \downarrow\,, \qquad Z_\rho \cdot f \cdot x \succeq f \cdot (Z_\rho \cdot f) \cdot x\,.$$

Proof. The right-to-left implications are easy. For the other direction, define

$$Y_\rho \;=\; Fix_{(\rho \to \rho) \to \rho}(\lambda^* y.\lambda^* f.f(yf))\,, \qquad Z_\rho \;=\; GFix_{(\rho \to \rho) \to \rho}(\lambda^* z.\lambda^* fx.f(zf)x)\,.$$

It is easy to check that these have the required properties. □

In a total model, a general recursor at a type $\sigma \to \tau$ is also a guarded recursor. We will see that even guarded recursion is sufficient for many purposes of interest.

Not all models of interest possess such recursion operators. Clearly, if **A** is a *total* model with $\mathbf{A}(\mathtt{N}) = \mathbb{N}$ a type of numerals as above, then **A** cannot have general or even guarded recursion: if $\rho = \mathtt{N} \to \mathtt{N}$ and $f = \lambda^* gx.suc(gx)$ then we would have $Z \cdot f \cdot \widehat{n} = suc \cdot Z \cdot f \cdot \widehat{n}$, which is impossible. However, many models with $\mathbf{A}(\mathtt{N}) = \mathbb{N}_\perp$ will have general recursion—roughly speaking, these are the models that are suitable for interpreting Plotkin's PCF (see Section 3.5).

Any *untyped* total model has general recursion, since we may take

$$W \;=\; \lambda^* wf.f(wwf)\,, \qquad Y \;=\; WW\,.$$

(This element Y is known as the *Turing fixed point combinator.*) Likewise, every untyped model, total or not, has guarded recursion, since we may take

$$V \;=\; \lambda^* vfx.f(vvf)x\,, \qquad Z \;=\; VV\,.$$

Note in passing that Kleene's *second recursion theorem* from classical computability theory is tantamount to the existence of a guarded recursion operator in K_1.

We may now fill the gap in the proof of Proposition 3.3.13. In any untyped model, let Z be a guarded recursion operator, define

$$R = \lambda^* rxfm.\ if(iszero\ m)(kx)(\lambda^* y.f(pre\ m)(rxf(pre\ m)\widehat{0}))\ ,$$

and take $rec = \lambda^* xfm.(ZR)xfmi$. Again, it is a good exercise in the use of Proposition 3.3.5 to check that this satisfies the axioms for *rec*. Note the use of strong definition by cases in the above meta-expression—this ensures, for instance, that $rec \cdot x \cdot f \cdot \widehat{0} \downarrow$ even if $f \cdot \widehat{n}$ is undefined for all n.

Exercise 3.3.17. Check that the above meta-expressions are 'well-typed', so that the above construction of primitive recursors works uniformly in any typed model with suitable elements $\widehat{n}$, *suc*, *pre*, *iszero* and Z.

Another related property, albeit much weaker than guarded recursion, concerns the existence of a *minimization* operator, also known as a *μ-operator*.

Definition 3.3.18. A model **A** with numerals *has minimization* if it contains an element $min \in \mathbf{A}^\sharp((\mathtt{N} \to \mathtt{N}) \to \mathtt{N})$ such that whenever $\widehat{g} \in \mathbf{A}^\circ(\mathtt{N} \to \mathtt{N})$ represents some total $g : \mathbb{N} \to \mathbb{N}$ and m is the least number such that $g(m) = 0$, we have $min \cdot \widehat{g} = \widehat{m}$.

Proposition 3.3.19. *There is an applicative expression Min involving constants* $\widehat{0}$*, suc, iszero, if and Z such that in any model with numerals and guarded recursion,* $[\![Min]\!]_\nu$ *(for the evident* ν*) is a minimization operator.*

Proof. Take $Min = Z(\lambda^* M.\lambda^* g.\ if\ (iszero\,(g\,\widehat{0}))\ \widehat{0}\ (M(\lambda^* n.g(suc\ n))))$. □

Recall that the first-order (Turing) computable partial functions $\mathbb{N}^r \rightharpoonup \mathbb{N}$ are generated from the basic functions via composition, primitive recursion and minimization. By analogy with Proposition 3.3.14, it is easy to see that all such functions are computable in any model with numerals and minimization:

Proposition 3.3.20. *For any partial computable* $f : \mathbb{N}^r \rightharpoonup \mathbb{N}$*, there is an applicative expression* $e_f : \mathtt{N}^{(r)} \to \mathtt{N}$ *(involving constants* $0, suc, rec_{\mathtt{N}}, min$*) such that in any model* **A** *with numerals and minimization we have* $[\![e_f]\!]_\nu \in \mathbf{A}^\sharp$ *(with the obvious valuation* ν*) and*

$$\forall n_0,\ldots,n_{r-1},m.\ f(n_0,\ldots,n_{r-1}) = m \Rightarrow [\![e_f]\!]_\nu \cdot \widehat{n_0} \cdot \ldots \cdot \widehat{n_{r-1}} = \widehat{m}\,.$$

Proof. Since our definition of minimization refers only to *total* functions $g : \mathbb{N} \to \mathbb{N}$, we appeal to the *Kleene normal form* theorem: there are primitive recursive functions $T : \mathbb{N}^{r+2} \to \mathbb{N}$ and $U : \mathbb{N} \to \mathbb{N}$ such that any partial computable f has an 'index' $e \in \mathbb{N}$ such that $f(\vec{n}) \simeq U(\mu y.T(e,\vec{n},y) = 0)$ for all $\vec{n}$. Using this, the result follows easily from Propositions 3.3.14 and 3.3.19. □

The above is sometimes expressed by saying that every partial computable function on $\mathbb{N}$ is *weakly representable* in any model with numerals and minimization. It appears that this is the best one can do in the general setting, even if the models are assumed to be strict. In all naturally arising models, however, it will be the case that every partial computable function is *strongly representable*—that is, the above formula may be strengthened to

$$\forall n_0,\ldots,n_{r-1},m.\ f(n_0,\ldots,n_{r-1}) = m \iff [\![e_f]\!]_V \cdot \widehat{n_0} \cdot \ldots \cdot \widehat{n_{r-1}} = \widehat{m}\,.$$

In Section 3.5 we shall extend this idea to higher types, showing that any model with numerals and guarded recursion admits an interpretation of PCF in some sense.

3.3.6 The Category of Assemblies

We have now seen several examples of computational structure often found within our models, many of them relating to the representation of various 'datatypes' such as booleans, natural numbers, products of existing datatypes, and (implicitly) certain spaces of total functions. From a conceptual point of view, it is illuminating to see all these as fitting into a broader framework, namely the *category of assemblies* over the given model **A**, which can loosely be seen as the world of 'all datatypes' that can be represented or 'implemented' in **A**. In fact, the basic definition makes sense for an arbitrary computability model **C**:

Definition 3.3.21. Let **C** be a lax computability model over T. The *category of assemblies over* **C**, written $\mathscr{A}sm(\mathbf{C})$, is defined as follows:

- Objects X are triples $(|X|, \rho_X, \Vdash_X)$, where $|X|$ is a set, $\rho_X \in T$ names some type, and $\Vdash_X \subseteq \mathbf{C}(\rho_X) \times |X|$ is a relation such that $\forall x \in |X|. \exists a \in \mathbf{C}(\rho_X).\ a \Vdash_X x$. (The formula $a \Vdash_X x$ may be read as '*a realizes x*'.)
- A morphism $f : X \to Y$ is a function $f : |X| \to |Y|$ that is *tracked* by some $\overline{f} \in \mathbf{C}[\rho_X, \rho_Y]$, in the sense that for any $x \in |X|$ and $a \in \mathbf{C}(\rho_X)$ we have

$$a \Vdash_X x \;\Rightarrow\; \overline{f}(a) \Vdash_Y f(x)\,.$$

An assembly X is called *modest* if $a \Vdash_X x \wedge a \Vdash_X x'$ implies $x = x'$. We write $\mathscr{M}od(\mathbf{C})$ for the full subcategory of $\mathscr{A}sm(\mathbf{C})$ consisting of modest assemblies.

It is routine to check that $\mathscr{A}sm(\mathbf{C})$ is indeed a category. In the case that **C** arises from a lax relative TPCA **A** as in Theorem 3.1.18, we may also denote the above categories by $\mathscr{A}sm(\mathbf{A})$, $\mathscr{M}od(\mathbf{A})$, or by $\mathscr{A}sm(\mathbf{A}^\circ;\mathbf{A}^\sharp)$, $\mathscr{M}od(\mathbf{A}^\circ;\mathbf{A}^\sharp)$. Note that realizers for elements $x \in |X|$ may be arbitrary elements of $\mathbf{A}^\circ(\rho_X)$, whereas a morphism $f : X \to Y$ must be tracked by an element of $\mathbf{A}^\sharp(\rho_X \to \rho_Y)$.

Intuitively, we regard an assembly X as an 'abstract datatype' for which we have a concrete implementation on the 'machine' **C**. The underlying set $|X|$ is the set of values of the abstract type, and for each $x \in |X|$, the elements $a \Vdash_X x$ are the possible machine representations of this abstract value. (Note that an abstract value x may have many possible machine representations a.) The morphisms $f : X \to Y$ may then be regarded as the 'computable mappings' between such datatypes—that is, those functions that can be implemented by some computable operation working on machine representations.

Viewed in this way, all the datatypes we shall typically wish to consider in fact live in the subcategory $\mathscr{M}od(\mathbf{C})$: an abstract data value is uniquely determined by any of its machine representations. Note also that if Y is modest, a morphism $f : X \to Y$ is completely determined by any $\overline{f}$ that tracks it.

The following shows that many of the concepts introduced so far have a natural significance in terms of the category of assemblies.

Theorem 3.3.22. *Let* $\mathbf{C}$ *be a lax computability model.*
(i) If $\mathbf{C}$ *has a weak terminal, then* $\mathscr{A}sm(\mathbf{C})$ *has a terminal object* 1.
(ii) If $\mathbf{C}$ *has weak products, then* $\mathscr{A}sm(\mathbf{C})$ *has binary cartesian products.*
(iii) If $\mathbf{C}$ *is weakly cartesian closed, then* $\mathscr{A}sm(\mathbf{C})$ *is cartesian closed.*
(iv) If $\mathbf{C}$ *has a weak terminal and booleans,* $\mathscr{A}sm(\mathbf{C})$ *has the coproduct* $1+1$.
(v) If $\mathbf{C}$ *has a weak terminal and numerals,* $\mathscr{A}sm(\mathbf{C})$ *has a natural number object.*

Proof. We give the key ingredients for (ii) and (iii), leaving the rest as exercises.

(ii) If X and Y are assemblies and ρ is a weak product of ρ_X and ρ_Y, define the assembly $X \times Y$ by

$$|X \times Y| = |X| \times |Y|, \quad \rho_{X\times Y} = \rho, \quad a \Vdash_{X\times Y} (x,y) \text{ iff } \pi_X(a) \Vdash_X x \wedge \pi_Y(a) \Vdash_Y y.$$

(iii) If X and Y are assemblies, let us say that an element $t \in \mathbf{C}(\rho_X \to \rho_Y)$ *tracks* a function $f : |X| \to |Y|$ if

$$\forall x \in |X|, a \in \mathbf{C}(\rho_X).\ a \Vdash_X x \Rightarrow t \cdot_{XY} a \Vdash_Y f(x).$$

Now define the assembly Y^X as follows:

$$\begin{aligned} |Y^X| &= \{f : |X| \to |Y| \mid f \text{ is tracked by some } t \in \mathbf{C}(\rho_X \to \rho_Y)\}, \\ \rho_{Y^X} &= \rho_X \to \rho_Y, \\ t \Vdash_{Y^X} f &\text{ iff } t \text{ tracks } f. \end{aligned}$$ □

It is easy to see that Theorem 3.3.22 also holds with $\mathscr{M}od(\mathbf{C})$ in place of $\mathscr{A}sm(\mathbf{C})$, and indeed that the inclusion $\mathscr{M}od(\mathbf{C}) \hookrightarrow \mathscr{A}sm(\mathbf{C})$ preserves all the relevant structure. Notice the pattern: 'weak' structure in $\mathbf{C}$ (defined without uniqueness conditions) gives rise to 'strong' structure in $\mathscr{A}sm(\mathbf{C})$ and $\mathscr{M}od(\mathbf{C})$.

If $\mathbf{C}$ is an untyped higher-order model, it turns out that much more is true: $\mathscr{A}sm(\mathbf{C})$ is a *quasitopos*, and also provides a model for powerful impredicative type theories. We refer to van Oosten [295] for more on these models from a categorical perspective, as well as an account of the *realizability toposes* to which they give rise.

Extrapolating from the examples of datatypes that we have seen, we can loosely regard $\mathscr{A}sm(\mathbf{C})$ as the world of all datatypes implementable within $\mathbf{C}$. Although we shall not always mention assemblies explicitly in the sequel, much of what we shall do with particular models can naturally be seen as working within $\mathscr{A}sm(\mathbf{C})$. We shall also see below that two models $\mathbf{C}, \mathbf{D}$ are 'equivalent' (in a sense to be defined) iff $\mathscr{A}sm(\mathbf{C})$, $\mathscr{A}sm(\mathbf{D})$ are equivalent as categories.

3.4 Simulations Between Computability Models

Having discussed computability models in some detail, we now seek an appropriate notion of *morphism* between models. Clearly one can formulate an evident notion of algebraic homomorphism, but a much more flexible and interesting notion is given by the concept of a *simulation* of one model in another. As indicated in Chapter 1, the question of which models may be simulated in which others, and in what ways, will be a core concern of this book.

We now formally develop the theory of simulations that we outlined in Chapter 1, starting with simulations between general computability models, and then specializing to higher-order ones. We also introduce the related notion of a *transformation* between simulations. We refer to Longley [183] for a somewhat more general treatment of this material and for fuller proofs of certain results. In the following section, we shall illustrate the scope of these concepts with numerous examples.

3.4.1 Simulations and Transformations

The basic notions of our theory are most simply presented in the general setting of (lax) computability models. The following definition is repeated from Section 1.1.4:

Definition 3.4.1. Let $\mathbf{C}$ and $\mathbf{D}$ be lax computability models over type worlds T, U respectively. A *simulation* γ of $\mathbf{C}$ in $\mathbf{D}$ (written $\gamma : \mathbf{C} \longrightarrow\!\!\!\triangleright\, \mathbf{D}$) consists of:

- a mapping $\sigma \mapsto \gamma\sigma : T \to U$,
- for each $\sigma \in T$, a relation $\Vdash^{\gamma}_{\sigma} \subseteq \mathbf{D}(\gamma\sigma) \times \mathbf{C}(\sigma)$

satisfying the following:

1. For all $a \in \mathbf{C}(\sigma)$ there exists $a' \in \mathbf{D}(\gamma\sigma)$ such that $a' \Vdash^{\gamma}_{\sigma} a$.
2. Every operation $f \in \mathbf{C}[\sigma, \tau]$ is *tracked* by some $f' \in \mathbf{D}[\gamma\sigma, \gamma\tau]$, in the sense that whenever $f(a)\downarrow$ and $a' \Vdash^{\gamma}_{\sigma} a$, we have $f'(a') \Vdash^{\gamma}_{\tau} f(a)$.

If A is the datatype occurrence associated with $\sigma \in T$, we may alternatively write γA for $\gamma\sigma$ and $\Vdash_A$ for $\Vdash_\sigma$.

Clearly, for any $\mathbf{C}$, we have the *identity* simulation $id_{\mathbf{C}} : \mathbf{C} \longrightarrow\!\!\!\triangleright\, \mathbf{C}$ given by $id_{\mathbf{C}}\sigma = \sigma$ and $a' \Vdash^{id_{\mathbf{C}}}_{\sigma} a$ iff $a' = a$. Moreover, given simulations $\gamma : \mathbf{C} \longrightarrow\!\!\!\triangleright\, \mathbf{D}$ and $\delta : \mathbf{D} \longrightarrow\!\!\!\triangleright\, \mathbf{E}$, we have the composite simulation $\delta \circ \gamma : \mathbf{C} \longrightarrow\!\!\!\triangleright\, \mathbf{E}$ defined by $(\delta \circ \gamma)\sigma = \delta(\gamma\sigma)$ and $a' \Vdash^{\delta\circ\gamma}_{\sigma} a$ iff there exists $a'' \in \mathbf{D}(\gamma\sigma)$ with $a'' \Vdash^{\gamma}_{\sigma} a$ and

$a' \Vdash^{\delta}_{\gamma\sigma} a''$. It is thus easy to see that lax computability models and simulations form a category, which we shall denote by $\mathfrak{CMod}$.

There is some further structure of interest here. If both γ and δ are simulations of $\mathbf{C}$ in $\mathbf{D}$, we may write $\gamma \preceq \delta$ if, informally, it is possible to translate γ-representations of values in $\mathbf{C}$ into δ-representations by means of a computable operation within $\mathbf{D}$. More formally:

Definition 3.4.2. Let $\mathbf{C}$, $\mathbf{D}$ be lax computability models and suppose $\gamma, \delta : \mathbf{C} \longrightarrow\!\!\!\!\triangleright\, \mathbf{D}$ are simulations. We say γ is *transformable* to δ, and write $\gamma \preceq \delta$, if for each $\sigma \in |\mathbf{C}|$ there is an operation $t \in \mathbf{D}[\gamma\sigma, \delta\sigma]$ such that

$$\forall a \in \mathbf{C}(\sigma), a' \in \mathbf{D}(\gamma\sigma).\ a' \Vdash^{\gamma}_{\sigma} a \Rightarrow t(a') \Vdash^{\delta}_{\sigma} a\,.$$

We write $\gamma \sim \delta$ if both $\gamma \preceq \delta$ and $\delta \preceq \gamma$.

From this and Definition 3.1.1, it is immediate that $\preceq$ is reflexive and transitive: that is, $\preceq$ is a *preorder* on the set of all simulations $\mathbf{C} \longrightarrow\!\!\!\!\triangleright\, \mathbf{D}$. Regarding this preorder as a category, if $\gamma \preceq \delta$ we may sometimes refer to the (unique) morphism $\xi : \gamma \rightarrow \delta$ as a *transformation* from γ to δ. It is also an easy exercise to check that composition of simulations in $\mathfrak{CMod}$ is monotone with respect to $\preceq$, so we have:

Theorem 3.4.3. $\mathfrak{CMod}$ *is a preorder-enriched category.*

The category $\mathfrak{CMod}$ naturally gives rise to the following notion:

Definition 3.4.4. We say that models $\mathbf{C}, \mathbf{D}$ are *equivalent* ($\mathbf{C} \simeq \mathbf{D}$) if there exist simulations $\gamma : \mathbf{C} \longrightarrow\!\!\!\!\triangleright\, \mathbf{D}$ and $\delta : \mathbf{D} \longrightarrow\!\!\!\!\triangleright\, \mathbf{C}$ such that $\delta \circ \gamma \sim id_{\mathbf{C}}$ and $\gamma \circ \delta \sim id_{\mathbf{D}}$.

Note that $\mathbf{C}$ and $\mathbf{D}$ may be equivalent even if their underlying type worlds T and U are quite different. Indeed, a question that will often be of interest is whether a given model is *essentially untyped*, that is, equivalent to a model over the singleton type world O. We shall see in Chapter 10, for example, that the model PC is essentially untyped in this sense.

Exercise 3.4.5. Show that a model $\mathbf{C}$ is essentially untyped iff it contains a *universal type*: that is, a datatype U such that for each $A \in |\mathbf{C}|$ there exist operations $e \in \mathbf{C}[A, U]$, $r \in \mathbf{C}[U, A]$ with $r(e(a)) = a$ for all $a \in A$.

In practice, we shall restrict our attention to simulations that satisfy a further property. We have here an example of the principle mentioned in Section 3.3.2: the property of interest can be defined in terms of either product or arrow types, and in the case of models with pairing, the two definitions can be shown to agree. Our first definition captures the idea of a simulation that 'respects products' in a mild sense:

Definition 3.4.6. Suppose $\mathbf{C}, \mathbf{D}$ are lax models with weak products and weak terminals $(I, i), (J, j)$ respectively. A simulation $\gamma : \mathbf{C} \longrightarrow\!\!\!\!\triangleright\, \mathbf{D}$ is *cartesian* if

1. for each $\sigma, \tau \in |\mathbf{C}|$ there exists $t \in \mathbf{D}[\gamma\sigma \bowtie \gamma\tau, \gamma(\sigma \bowtie \tau)]$ such that

$$\pi_{\gamma\sigma}(d) \Vdash^{\gamma}_{\sigma} a \wedge \pi_{\gamma\tau}(d) \Vdash^{\gamma}_{\tau} b \Rightarrow$$
$$\exists c \in \mathbf{C}(\sigma \bowtie \tau).\ \pi_{\sigma}(c) = a \wedge \pi_{\tau}(c) = b \wedge t(d) \Vdash^{\gamma}_{\sigma\bowtie\tau} c\,,$$

2. there exists $u \in \mathbf{D}[J, \gamma I]$ such that $u(j) \Vdash^{\gamma}_{I} i$.

The existence of suitable comparison operations in the other direction (that is, in $\mathbf{D}[\gamma(\sigma \bowtie \tau), \gamma\sigma \bowtie \gamma\tau]$ and $\mathbf{D}[\gamma I, J]$) is automatic. We write $\mathfrak{CMod}^{\times}$ for the category of such models with cartesian simulations between them. Cartesian simulations are referred to as *monoidal* in the more general setting of [183].

In the setting of higher-order models, a definition in a quite different spirit is possible: instead of requiring each operation in $\mathbf{C}$ to be simulable in $\mathbf{D}$, it is enough to require that the *application* operations are simulable. We work this out here in the setting of lax relative TPCAs.

Definition 3.4.7. Let $\mathbf{A}$ and $\mathbf{B}$ be lax relative TPCAs over type worlds T, U respectively. An *applicative simulation* $\gamma : \mathbf{A} \longrightarrow\!\!\!\triangleright\, \mathbf{B}$ consists of:

- a mapping $\sigma \mapsto \gamma\sigma : \mathsf{T} \to \mathsf{U}$,
- for each $\sigma \in \mathsf{T}$, a relation $\Vdash^{\gamma}_{\sigma} \subseteq \mathbf{B}^{\circ}(\gamma\sigma) \times \mathbf{A}^{\circ}(\sigma)$

satisfying the following:

1. For all $a \in \mathbf{A}^{\circ}(\sigma)$ there exists $b \in \mathbf{B}^{\circ}(\gamma\sigma)$ with $b \Vdash^{\gamma}_{\sigma} a$.
2. For all $a \in \mathbf{A}^{\sharp}(\sigma)$ there exists $b \in \mathbf{B}^{\sharp}(\gamma\sigma)$ with $b \Vdash^{\gamma}_{\sigma} a$.
3. 'Application in $\mathbf{A}$ is effective in $\mathbf{B}$': that is, for each $\sigma, \tau \in \mathsf{T}$ there exists some $r \in \mathbf{B}^{\sharp}(\gamma(\sigma \to \tau) \to \gamma\sigma \to \gamma\tau)$, called a *realizer for γ at σ, τ*, such that for all $f \in \mathbf{A}^{\circ}(\sigma \to \tau)$, $f' \in \mathbf{B}^{\circ}(\gamma(\sigma \to \tau))$, $a \in \mathbf{A}^{\circ}(\sigma)$ and $a' \in \mathbf{B}^{\circ}(\gamma\sigma)$ we have

$$f' \Vdash_{\sigma\to\tau} f \wedge a' \Vdash_{\sigma} a \wedge f \cdot a\downarrow \ \Rightarrow\ r \cdot f' \cdot a' \Vdash_{\tau} f \cdot a\,.$$

Theorem 3.4.8. *Suppose $\mathbf{C}$ and $\mathbf{D}$ are (lax) weakly cartesian closed models, and suppose $\mathbf{A}$ and $\mathbf{B}$ are the corresponding (lax) relative TPCAs with pairing via the correspondence of Theorem 3.1.18. Then cartesian simulations $\mathbf{C} \longrightarrow\!\!\!\triangleright\, \mathbf{D}$ correspond precisely to applicative simulations $\mathbf{A} \longrightarrow\!\!\!\triangleright\, \mathbf{B}$.*

Proof. Suppose first that $\gamma : \mathbf{C} \longrightarrow\!\!\!\triangleright\, \mathbf{D}$ is a cartesian simulation; we must check that the relations $\Vdash^{\gamma}_{\sigma}$ satisfy the three conditions of Definition 3.4.7. Condition 1 is already given by Definition 3.4.1. For condition 2, suppose $a \in \mathbf{A}^{\sharp}(\sigma)$ where $\mathbf{A}^{\circ}(\sigma) = A$. Then as in the proof of Theorem 3.1.18, we may find $g \in \mathbf{C}[I, A]$ with $g(i) = a$, where (I, i) is a weak terminal in $\mathbf{C}$. Take $g' \in \mathbf{D}[\gamma I, \gamma A]$ tracking g as in Definition 3.4.1, and supercompose with $u \in \mathbf{D}[J, \gamma I]$, where (J, j) is a weak terminal in $\mathbf{D}$, to obtain $g'' \in \mathbf{D}[J, \gamma A]$. Then $g''(j) \in \mathbf{B}^{\sharp}(\gamma\sigma)$, and it is easy to see that $g''(j) \Vdash^{\gamma}_{\sigma} a$.

For condition 3, let σ, τ be any types; then as noted after Definition 3.1.8, we have $app_{\sigma\tau} \in \mathbf{C}[(\sigma \to \tau) \times \sigma, \tau]$ tracked by some $app'_{\sigma\tau} \in \mathbf{D}[\gamma((\sigma \to \tau) \times \sigma), \gamma\tau]$. Supercomposing with the operation $t \in \mathbf{D}[\gamma(\sigma \to \tau) \times \gamma\sigma, \gamma((\sigma \to \tau) \times \sigma)]$ from

Definition 3.4.6, and transposing as in Definition 3.1.8, we obtain an operation in $\mathbf{D}[\gamma(\sigma \to \tau), \gamma\sigma \to \gamma\tau]$, and hence a realizer $r \in \mathbf{B}^\sharp(\gamma(\sigma \to \tau) \to \gamma\sigma \to \gamma\tau)$ with the required properties.

Conversely, suppose $\gamma : \mathbf{A} \longrightarrow\!\!\!\triangleright \mathbf{B}$ is an applicative simulation. To see that γ is a simulation $\mathbf{C} \longrightarrow\!\!\!\triangleright \mathbf{D}$, it suffices to show that every operation in $\mathbf{C}$ is tracked by one in $\mathbf{D}$. But given $f \in \mathbf{C}[\sigma, \tau]$, we may find a corresponding element $a \in \mathbf{A}^\sharp(\sigma \to \tau)$ as in Theorem 3.1.18, whence some $a' \in \mathbf{B}^\sharp(\gamma(\sigma \to \tau))$ with $a' \Vdash_{\sigma\to\tau} a$; hence (using a realizer $r \in \mathbf{B}^\sharp$ for γ at σ, τ) an element $a'' \in \mathbf{B}^\sharp(\gamma\sigma \to \gamma\tau)$ and so a corresponding operation $f' \in \mathbf{D}[\gamma\sigma, \gamma\tau]$. It is easy to check that f' tracks f.

It remains to show that γ is cartesian. For any types σ, τ, we have by assumption an element $pair_{\sigma\tau} \in \mathbf{A}^\sharp(\sigma \to \tau \to \sigma \times \tau)$, yielding some $p \in \mathbf{C}[\sigma, \tau \to \sigma \times \tau]$. Since γ is a simulation, this is tracked by some $p' \in \mathbf{D}[\gamma\sigma, \gamma(\tau \to \sigma \times \tau)]$. From the weak product structure in $\mathbf{D}$ we may thence obtain an operation

$$p'' \in \mathbf{D}[\gamma\sigma \times \gamma\tau,\ \gamma(\tau \to \sigma \times \tau) \times \gamma\tau]\,,$$

and together with a realizer for γ at τ and $\sigma \times \tau$, this yields an operation $t \in \mathbf{D}[\gamma\sigma \times \gamma\tau,\ \gamma(\sigma \times \tau)]$ with the required properties. The existence of a suitable $u \in \mathbf{D}[J, \gamma I]$ is left as an easy exercise. □

The notion of a transformation between simulations carries across immediately to the relative TPCA setting: an applicative simulation $\gamma : \mathbf{A} \longrightarrow\!\!\!\triangleright \mathbf{B}$ is transformable to δ if for each type σ there exists $t \in \mathbf{B}^\sharp(\gamma\sigma \to \delta\sigma)$ such that $a' \Vdash^{\gamma}_{\sigma} a$ implies $t \cdot a' \Vdash^{\delta}_{\sigma} a$. For much of this book, we shall implicitly be working within the 2-category $\mathfrak{RTpca}^\times$ of lax relative TPCAs with pairing, applicative simulations, and transformations between them. By Theorem 3.4.8 above, $\mathfrak{RTpca}^\times$ is a full sub-2-category of $\mathfrak{CMod}^\times$, where the inclusion 2-functor is an isomorphism of preorders on each homset.

A few other specialized kinds of simulations will also be of interest:

Definition 3.4.9. Suppose $\mathbf{A}$ and $\mathbf{B}$ are (lax) relative TPCAs over T and U respectively, and suppose $\gamma : \mathbf{A} \longrightarrow\!\!\!\triangleright \mathbf{B}$ is an applicative simulation.

(i) We say γ is *discrete* if $b \Vdash^\gamma a$ and $b \Vdash^\gamma a'$ imply $a = a'$.

(ii) We say γ is *single-valued* if for all $a \in \mathbf{A}$ there is exactly one $b \in \mathbf{B}$ with $b \Vdash^\gamma a$. We say γ is *projective* if $\gamma \sim \gamma'$ for some single-valued γ'.

(iii) If $\mathbf{A}$ and $\mathbf{B}$ have booleans $tt_\mathbf{A}, f\!f_\mathbf{A}$ and $tt_\mathbf{B}, f\!f_\mathbf{B}$ respectively, we say γ *respects booleans* if there exists $q \in \mathbf{B}^\sharp$ such that

$$b \Vdash^\gamma tt_\mathbf{A} \;\Rightarrow\; q \cdot b = tt_\mathbf{B}\,, \qquad b \Vdash^\gamma f\!f_\mathbf{A} \;\Rightarrow\; q \cdot b = f\!f_\mathbf{B}\,.$$

(iv) Likewise, if $\mathbf{A}$ and $\mathbf{B}$ have numerals $\widehat{n}$ and $\widetilde{n}$ respectively, we say γ *respects numerals* if there exists $q \in \mathbf{B}^\sharp$ such that for all $n \in \mathbb{N}$,

$$b \Vdash^\gamma \widehat{n} \;\Rightarrow\; q \cdot b = \widetilde{n}\,.$$

(v) If $\mathsf{T} = \mathsf{U}$, we say γ is *type-respecting* if γ is the identity on types, and moreover:

- $\Vdash^{\gamma}_{\sigma} = id_{\mathsf{T}[\sigma]}$ whenever T fixes the interpretation of σ.
- Application is itself a realizer for γ at each σ, τ: that is,

$$f' \Vdash^{\gamma}_{\sigma \to \tau} f \wedge a' \Vdash^{\gamma}_{\sigma} a \wedge f \cdot a \downarrow \;\Rightarrow\; f' \cdot a' \Vdash^{\gamma}_{\tau} f \cdot a .$$

- If T has product structure, then **A**, **B** have pairing and

$$d' \Vdash^{\gamma}_{\sigma} a \wedge b' \Vdash^{\gamma}_{\tau} b \Rightarrow pair \cdot a' \cdot b' \Vdash^{\gamma}_{\sigma \times \tau} pair \cdot a \cdot b ,$$
$$d' \Vdash^{\gamma}_{\sigma \times \tau} d \Rightarrow fst \cdot d' \Vdash^{\gamma}_{\sigma} fst \cdot d \wedge snd \cdot d' \Vdash^{\gamma}_{\sigma} snd \cdot d .$$

Clearly, each of these properties of simulations is preserved under composition and so gives rise to a subcategory of $\mathfrak{RTpca}^{\times}$.

Exercise 3.4.10. (i) Show that if γ respects numerals then γ respects booleans.

(ii) Show that if $\gamma : \mathbf{A} \longrightarrow\!\!\!\triangleright\, \mathbf{B}$ respects booleans where **A** has numerals and **B** has guarded recursion, then γ respects numerals. Thus, any simulation between untyped models that respects booleans also respects numerals.

3.4.2 Simulations and Assemblies

Some evidence for the mathematical credentials of $\mathfrak{CMod}$ comes from its connections with categories of assemblies. Although technically these results will not be required in the sequel, they form an important part of the conceptual justification for our framework, and so a brief account of them is in order. For more detail see Longley [183]; the ideas are also covered in van Oosten [295] in slightly less generality.

A simulation $\gamma : \mathbf{C} \longrightarrow\!\!\!\triangleright\, \mathbf{D}$ naturally induces a functor $\gamma_* : \mathscr{A}sm(\mathbf{C}) \to \mathscr{A}sm(\mathbf{D})$, capturing the evident idea that any datatypes implementable within **C** must also be implementable in **D**:

- On objects X, define $\gamma_*(X)$ by $|\gamma_*(X)| = |X|$, $\rho_{\gamma_*(X)} = \gamma\rho_X$, and $b \Vdash_{\gamma_*(X)} x$ iff $\exists a \in \mathbf{C}(\rho_X).\, a \Vdash_X x \wedge b \Vdash_{\gamma\rho_X} a$.
- On morphisms $f : X \to Y$, define $\gamma_*(f) = f$. Note that if $r \in \mathbf{C}[\rho_X, \rho_Y]$ tracks f as a morphism in $\mathscr{A}sm(\mathbf{C})$, and $r' \in \mathbf{D}[\gamma\rho_X, \gamma\rho_Y]$ tracks r with respect to γ, then r' tracks f as a morphism in $\mathscr{A}sm(\mathbf{D})$.

Moreover, a transformation $\xi : \gamma \to \delta$ yields a natural transformation $\xi_* : \gamma_* \to \delta_*$: just take $\xi_{*X} = id_{|X|} : \gamma_*(X) \to \delta_*(X)$, and note that if t witnesses $\gamma \preceq \delta$ at X, then t tracks ξ_{*X}. Furthermore, it is easy to see that the constructions $\gamma \mapsto \gamma_*$, $\xi \mapsto \xi_*$ are themselves functorial, so that the operation $\mathscr{A}sm$ may be viewed as a 2-functor from $\mathfrak{CMod}$ to $\mathfrak{Cat}$, the (large) 2-category of categories, functors and natural transformations.

It is convenient to regard each category of assemblies $\mathscr{A}sm(\mathbf{C})$ as equipped with the evident forgetful functor $\Gamma_{\mathbf{C}} : \mathscr{A}sm(\mathbf{C}) \to \mathscr{S}et$; note that if $\gamma : \mathbf{C} \longrightarrow\!\!\!\triangleright\, \mathbf{D}$ then $\Gamma_{\mathbf{D}} \circ \gamma_* = \Gamma_{\mathbf{C}}$. In fact, we can precisely characterize (up to natural isomorphism) those functors $\mathscr{A}sm(\mathbf{C}) \to \mathscr{A}sm(\mathbf{D})$ arising from simulations $\mathbf{C} \longrightarrow\!\!\triangleright \mathbf{D}$ via a categorical preservation property. For the benefit of the categorically minded reader we summarize the relevant structure here, referring to [183] for proof details.

Definition 3.4.11. Let $\mathscr{C}$ be any category, Γ a faithful functor $\mathscr{C} \to \mathscr{S}et$.

(i) $(\mathscr{C},\Gamma)$ *has subobjects* if for any object X in $\mathscr{C}$ and any mono $s : S \rightarrowtail \Gamma(X)$ in $\mathscr{S}et$, there exists a morphism $\bar{s} : Y \to X$ in $\mathscr{C}$ (necessarily mono) such that $\Gamma(\bar{s}) = s$, and for any morphism $f : Z \to X$ such that $\Gamma(f)$ factors through s, there is a unique $g : Z \to Y$ such that $f = \bar{s} \circ g$.

(ii) $(\mathscr{C},\Gamma)$ *has quotients* if for any object X in $\mathscr{C}$ and any epi $q : \Gamma(X) \twoheadrightarrow Q$ in $\mathscr{S}et$, there exists a morphism $\bar{q} : X \to V$ in $\mathscr{C}$ (necessarily epi) such that $\Gamma(\bar{q}) = q$, and for any morphism $f : X \to W$ such that $\Gamma(f)$ factors through q, there is a unique $g : V \to W$ such that $f = g \circ \bar{q}$.

(iii) $(\mathscr{C},\Gamma)$ *has copies* if for any $X \in \mathscr{C}$ and $S \in \mathscr{S}et$, there is an object $X \propto S$ in $\mathscr{C}$ equipped with morphisms

$$\pi : X \propto S \to X \,, \qquad \theta : \Gamma(X \propto S) \to S$$

such that for any $f : Z \to X$ and $\phi : \Gamma(Z) \to S$ there is a unique $g : Z \to X \propto S$ with $f = \pi \circ g$ and $\phi = \theta \circ \Gamma(g)$.

(iv) We say $(\mathscr{C},\Gamma)$ is a *quasi-regular category over* $\mathscr{S}et$ if it has subobjects, quotients and copies.

It is easy to check that $(\mathscr{A}sm(\mathbf{C}),\Gamma_{\mathbf{C}})$ is a quasi-regular category in this sense, and moreover that any functor $\gamma_* : \mathscr{A}sm(\mathbf{C}) \to \mathscr{A}sm(\mathbf{D})$ preserves the subobject, quotient and copy structure on the nose. In general, if $(\mathscr{C},\Gamma_{\mathscr{C}})$ and $(\mathscr{D},\Gamma_{\mathscr{D}})$ are quasi-regular categories, a *quasi-regular functor* between them will consist of a functor $F : \mathscr{C} \to \mathscr{D}$ equipped with a natural isomorphism $\iota : \Gamma_{\mathscr{C}} \cong \Gamma_{\mathscr{D}} \circ F$ such that F preserves subobjects, quotients and copies modulo ι in an evident sense. The main results (proved in [183, Section 3]) are as follows:

Theorem 3.4.12. *(i) If $(F,\iota) : (\mathscr{A}sm(\mathbf{C}),\Gamma_{\mathbf{C}}) \to (\mathscr{A}sm(\mathbf{D},\Gamma_{\mathbf{D}}))$ is a quasi-regular functor, then there is a simulation $\gamma : \mathbf{C} \longrightarrow\!\!\triangleright \mathbf{D}$ such that $\gamma_* \cong F$ (and the natural isomorphism is then uniquely determined).*

(ii) If γ, δ respectively correspond to (F,ι_F), (G,ι_G) in this way, then $\gamma \preceq \delta$ iff there exists a natural transformation $\alpha : F \to G$. (Any such natural transformation automatically satisfies $(\Gamma_{\mathbf{D}} * \alpha) \circ \iota_F = \iota_G$, where $*$ denotes horizontal composition.)

Cartesian simulations also fit well into this picture: if $\mathbf{C}$, $\mathbf{D}$ have weak products and weak terminals, it is easy to check that $\gamma : \mathbf{C} \longrightarrow\!\!\triangleright \mathbf{D}$ is cartesian iff $\gamma_* : \mathscr{A}sm(\mathbf{C}) \to \mathscr{A}sm(\mathbf{D})$ preserves finite products. In addition, many of the other properties of simulations mentioned in Section 3.4.1 have a natural significance in terms of the corresponding functors. The following facts may be proved as exercises:

Proposition 3.4.13. *Suppose* **A**, **B** *and γ are as in Definition 3.4.9.*

(i) γ is discrete iff $\gamma_* : \mathscr{A}sm(\mathbf{A}) \to \mathscr{A}sm(\mathbf{B})$ preserves modest assemblies.

(ii) γ *is projective iff* γ_* *preserves projective assemblies. (We call an assembly* projective *iff it is isomorphic to one with a single realizer for each element.)*

(iii) γ *respects booleans iff* γ_* *preserves booleans (that is,* $\gamma_*(1+1)$ *is canonically isomorphic to* $1+1$*).*

(iv) γ *respects numerals iff* γ_* *preserves the natural number object.*

Note, however, that the condition that γ is type-respecting is not mirrored by any categorical property of γ_*. This is consonant with the fact that $\mathscr{A}sm(\mathbf{A})$ does not in general contain traces of the underlying type world of $\mathbf{A}$. Indeed, as noted at the end of Section 3.4.1, 'typed' models are sometimes equivalent to 'untyped' ones.

Finally, we obtain a pleasing characterization of *equivalence* of models in terms of assemblies. Any equivalence (γ,δ) between models $\mathbf{C}$ and $\mathbf{D}$ clearly gives rise to an equivalence of categories $(\gamma_*,\delta_*) : \mathscr{A}sm(\mathbf{C}) \simeq \mathscr{A}sm(\mathbf{D})$ in which both γ_* and δ_* commute with the Γ functors. Conversely, if (F,G) is any equivalence of categories $\mathscr{A}sm(\mathbf{C}) \simeq \mathscr{A}sm(\mathbf{D})$ commuting with the Γ functors up to natural isomorphism, it is easily shown that F,G are quasi-regular, and so arise up to isomorphism from suitable γ,δ. We therefore have:

Theorem 3.4.14. $\mathbf{C} \simeq \mathbf{D}$ *iff there is an equivalence* $\mathscr{A}sm(\mathbf{C}) \simeq \mathscr{A}sm(\mathbf{D})$ *that commutes with the* Γ*-functors up to natural isomorphism.*

Note that the simulations γ,δ constituting an equivalence are necessarily cartesian.

Insofar as an equivalence between $\mathscr{A}sm(\mathbf{C})$ and $\mathscr{A}sm(\mathbf{D})$ says that $\mathbf{C}$ and $\mathbf{D}$ give rise to the same 'universe of computation', it appears natural to try to study and classify computability models up to equivalence: if $\mathbf{C} \simeq \mathbf{D}$, we may regard $\mathbf{C}$ and $\mathbf{D}$ simply as different presentations of the same underlying *notion of computability*. This is indeed the point of view we will adopt in much of this book.

For more recent progress in the theory of applicative simulations and their associated functors, see Faber and van Oosten [90].

3.5 Examples of Simulations and Transformations

We now introduce a range of examples to illustrate the scope of the concepts of simulation and transformation, drawing on the supply of models introduced in Sections 1.1 and 3.2. Many of these examples will feature later in the book—indeed, the selection of simulations described here will serve as an initial 'roadmap' for much of the territory we shall cover in detail later on. Generalizing from some of these examples, we shall also introduce the notions of *effective* and *continuous* computability models.

Example 3.5.1. Suppose $\mathbf{C}$ is any (strict or lax) computability model with weak products, and consider the following variation on the 'product completion' construction described in the proof of Theorem 3.1.15. Let $\mathbf{C}^\times$ be the computability model whose datatypes are sets $A_0 \times \cdots \times A_{m-1}$ where $A_i \in |\mathbf{C}|$, and whose operations $f \in \mathbf{C}^\times[A_0 \times \cdots \times A_{m-1}, B_0 \times \cdots \times B_{n-1}]$ are those partial functions represented (in an obvious way) by some operation in $\mathbf{C}[A_0 \bowtie \cdots \bowtie A_{m-1}, B_0 \bowtie \cdots \bowtie B_{n-1}]$. Clearly the inclusion $\mathbf{C} \hookrightarrow \mathbf{C}^\times$ is a simulation, as is the mapping $\mathbf{C}^\times \to \mathbf{C}$ sending $A_0 \times \cdots \times A_{m-1}$ to $A_0 \bowtie \cdots \bowtie A_{m-1}$. Moreover, it is easy to check that these simulations constitute an equivalence $\mathbf{C} \simeq \mathbf{C}^\times$. This shows that every strict [resp. lax] computability model with weak products is equivalent to one with standard products.

Example 3.5.2. Let $\mathbf{A}$ be any untyped (lax relative) PCA, or more generally any model with numerals $\widehat{0}, \widehat{1}, \ldots$ and minimization. We may then define a single-valued applicative simulation $\kappa : K_1 \longrightarrow\!\!\!\!\triangleright\, \mathbf{A}$ by taking $a \Vdash^\kappa n$ iff $a = \widehat{n}$. Condition 3 of Definition 3.4.7 is satisfied because the application operation $\bullet : \mathbb{N} \times \mathbb{N} \rightharpoonup \mathbb{N}$ of K_1 is representable by an element of $\mathbf{A}^\sharp$ (see Proposition 3.3.20).

In particular models, many choices of numerals may be available. For instance, if $\mathbf{A} = \mathbf{A}^\sharp = \Lambda/\!\sim$ as in Definition 3.1.6, then besides the *Curry numerals* available in any untyped model (see Section 3.3.4), we also have the *Church numerals* $\widetilde{n} = \lambda f.\lambda x.f^n x$. For this choice of numerals, it is somewhat delicate to construct a suitable element *rec*. The key ingredient is the following implementation of the predecessor function, due in essence to Kleene:

$$\lambda z.\, \mathit{fst}\,(z\ (\lambda x.\, \mathit{if}\,(\mathit{snd}\ x)\ \langle \widetilde{0}, \mathit{ff}\rangle\ \langle \mathit{suc}(\mathit{fst}\ x), \mathit{ff}\rangle)\ \langle \widetilde{0}, \mathit{tt}\rangle)\ .$$

An element *iszero* is easy to construct, so we may now use the definition of *rec* given in Section 3.3.5.

The Church numerals thus give rise to an alternative simulation $\kappa' : K_1 \longrightarrow\!\!\!\!\triangleright\, \Lambda/\!\sim$. However, it is easy to show that if a model $\mathbf{A}$ has numerals and recursion and $\kappa' : K_1 \longrightarrow\!\!\!\!\triangleright\, \mathbf{A}$ is any simulation whatever, then $\kappa' \sim \kappa$.

Example 3.5.3. We now revisit a situation discussed in Chapter 1. Recall from Example 1.1.3 the model $\Lambda^0/\!=_\beta$ consisting of closed λ-terms modulo β-equivalence. Let $\ulcorner - \urcorner$ be any effective *Gödel numbering* of λ-terms as natural numbers, and define $\gamma : \Lambda^0/\!=_\beta \longrightarrow\!\!\!\!\triangleright\, K_1$ by

$$m \Vdash^\gamma [M] \ \text{ iff } \ m = \ulcorner M' \urcorner \text{ for some } M' =_\beta M\ .$$

This is emphatically not a single-valued relation, since every representative of the equivalence class $[M]$ will give rise to a realizer for $[M]$. However, it clearly satisfies condition 3 of Definition 3.4.7, since juxtaposition of λ-terms is tracked by a computable function $(\ulcorner M \urcorner, \ulcorner N \urcorner) \mapsto \ulcorner MN \urcorner$ at the level of Gödel numbers. Thus γ is an applicative simulation.

As indicated in Chapter 1, we clearly have transformations $id_{K_1} \preceq \gamma \circ \kappa$ and $\gamma \circ \kappa \preceq id_{K_1}$; these correspond to the observation that the 'encoding' and 'decoding' mappings $n \mapsto \ulcorner \widehat{n} \urcorner$ and $\ulcorner \widehat{n} \urcorner$ are readily computable. Furthermore, a non-trivial *enumeration theorem* due to Kleene [133] tells us that $\kappa \circ \gamma \preceq id_{\Lambda^0/=_\beta}$: there is a term $P \in \Lambda^0$ such that $P(\ulcorner M \urcorner\!\widehat{\ }) =_\beta M$ for any $M \in \Lambda^0$.

However, the simulations κ, γ do not constitute an equivalence, since we do not have $id_{\Lambda^0/=_\beta} \preceq \kappa \circ \gamma$ (cf. the discussion in Example 1.1.3). Somewhat informally, we may say that K_1 is 'more intensional' than $\Lambda^0/=_\beta$.

In fact, it can be easily shown that K_1 is not equivalent to $\Lambda^0/=_\beta$, or indeed to any total untyped model (this result is due to Johnstone and Robinson [126]). Let us say a relative PCA **A** has *decidable equality* if there is an element $q \in \mathbf{A}^\sharp$ such that

$$q \cdot x \cdot y \;=\; \begin{cases} tt \text{ if } x = y \\ ff \text{ otherwise} \end{cases}$$

Clearly K_1 has decidable equality, and it is easy to see that if $\mathbf{A} \simeq \mathbf{B}$ then **A** has decidable equality iff **B** does. However, if a total relative PCA **A** were to contain such an element q, we could define $v = Y(q\,ff)$ so that $v = q\,ff\; v$, which would yield a contradiction.

Syntactic translations between formal languages provide a fruitful source of applicative simulations. We give two examples here.

Example 3.5.4. With reference to the languages introduced in Section 3.2.3, we may translate System T terms M to PCF terms M^θ simply by replacing all constants rec_σ by suitable implementations of these recursors in PCF (indeed, the implementation given in Section 3.3.5 adapts immediately to the typed setting of PCF). Such a translation clearly induces a type-respecting applicative simulation $\theta : \mathrm{T}^0/=_{\mathrm{op}} \longrightarrow\!\!\triangleright\, \mathrm{PCF}^0/=_{\mathrm{op}}$. This simulation is single-valued, since it is not hard to show that if $M =_{\mathrm{op}} M'$ then $M^\theta =_{\mathrm{op}} M'^\theta$. It is also easy to see that θ respects numerals.

Example 3.5.5. We may also translate from PCF into the untyped λ-calculus. One such translation ϕ may be defined on constants as follows (application and abstraction are translated by themselves). We borrow the terms *pair*, *fst*, *snd* from Example 3.1.6, and write $\langle M,N\rangle$ for *pair M N*. Following Section 3.3.4, we then define $\widehat{0} = \langle \lambda xy.x, \lambda xy.x\rangle$ and $\widehat{n+1} = \langle \lambda xy.y, \widehat{n}\rangle$.

$$\begin{aligned}
\widehat{n}^\phi &= \widehat{n}\,,\\
suc^\phi &= \lambda z.\langle \lambda xy.y, z\rangle\,,\\
pre^\phi &= \lambda z.\,(fst\; z)\,\widehat{0}\,(snd\; z)\,,\\
ifzero^\phi &= fst\,,\\
Y_\sigma^\phi &= (\lambda xy.y(xxy))(\lambda xy.y(xxy))\,.
\end{aligned}$$

This induces an applicative simulation $\phi : \mathrm{PCF}^0/=_{\mathrm{op}} \longrightarrow\!\!\triangleright\, \Lambda^0/=_\beta$ respecting numerals. Moreover, it is not hard to show that if $M =_{\mathrm{op}} M'$ then $M^\phi =_\beta M'^\phi$.

Both these examples are of a rather simple kind insofar as application in the source model is tracked directly by application in the target model. However, this is by no means a necessary property of syntactic translations. For example, Plotkin in [235] introduced the so-called *continuation passing style* translations in both directions between the standard 'call-by-name' λ-calculus (exemplified by our model $\Lambda^0/{=_\beta}$) and the *call-by-value* λ-calculus. As shown in Longley [173, Section 3.2], these can be viewed as applicative simulations between the corresponding term models; the realizers for application are non-trivial.

Another important family of applicative simulations arises from denotational interpretations of syntactic calculi in more semantic models. The following is perhaps the paradigmatic example.

Example 3.5.6. Recall the model PC of partial continuous functions from Section 3.2.2. There is a well-known interpretation of PCF in PC, assigning to each term $M:\tau$ with free variables drawn from $\Gamma = x_0^{\sigma_0},\dots,x_{r-1}^{\sigma_{r-1}}$ an element $[\![M]\!]_\Gamma$ of $\mathsf{PC}(\sigma_0\times\cdots\times\sigma_{r-1}\to\tau)$. (By convention we let $\sigma_0\times\cdots\times\sigma_{r-1}=1$ when $r=0$.) The definition of $[\![-]\!]$ may be outlined as follows; see Section 7.1 for more detail.

- The constants $\widehat{n}, suc, pre, ifzero$ are interpreted as elements of PC in the obvious way. (Note that for example $[\![ifzero]\!](0)(x)(\bot)=x$ and $[\![ifzero]\!](1)(\bot)(y)=y$.)
- Each constant Y_σ is interpreted by the element

$$\bigsqcup_n (\Lambda f.f^n(\bot)) \in \mathsf{PC}((\sigma\to\sigma)\to\sigma)\,.$$

  Note that for any $f\in\mathsf{PC}(\sigma\to\sigma)$, $[\![Y_\sigma]\!](f)$ is the *least fixed point* of f.
- Application and abstraction are interpreted as follows, where $\Gamma = x_0^{\sigma_0},\dots,x_{r-1}^{\sigma_{r-1}}$.

$$\begin{aligned}[\![MN]\!]_\Gamma &= app\circ\langle[\![M]\!]_\Gamma,[\![N]\!]_\Gamma\rangle\,,\\ [\![\lambda y^\sigma.M]\!]_\Gamma &= \Lambda(x_0,\dots,x_{r-1}).\,\Lambda y.[\![M]\!]_{\Gamma,y^\sigma}(x_0,\dots,x_{r-1},y)\,.\end{aligned}$$

It is easy to check that if $M\rightsquigarrow M'$ then $[\![M]\!]=[\![M']\!]$, and that $[\![M]\!]$ is always a c.e. element of PC. The interpretation $[\![-]\!]$ therefore gives rise to a single-valued, type-respecting applicative simulation $\mathrm{PCF}^0/{=_{\mathrm{op}}} \longrightarrow\!\!\triangleright\, \mathsf{PC}^{\mathrm{eff}}$.

Many other interpretations arising in denotational semantics also give rise to applicative simulations in this way: for instance, in Chapter 7 we shall see that there is a canonical applicative simulation $\mathrm{PCF}^0/{=_{\mathrm{op}}} \longrightarrow\!\!\triangleright\, \mathsf{SP}^0$.

Example 3.5.7. We noted in Section 3.2 that both K_2 and PC embody a concept of *continuous* operations on finitary data. The close relationship between them is reflected in the fact that each can be simulated in the other, although the simulations do not constitute an equivalence.

First, recall that $\mathsf{PC}(1)$ consists of all monotone functions $\mathbb{N}_\bot\to\mathbb{N}_\bot$. Thus, every $g\in K_2=\mathbb{N}^\mathbb{N}$ can be represented by $\delta(g)\in\mathsf{PC}(\mathtt{N}\to\mathtt{N})$, where $\delta(g)(n)=g(n)$ and $\delta(g)(\bot)=\bot$. It is evident from the definition of the K_2 application $\odot$ that it can be tracked by an operation in $\mathsf{PC}(1,1\to 1)$, so we obtain a single-valued applicative simulation $\delta: K_2 \longrightarrow\!\!\triangleright\, \mathsf{PC}$ (which clearly restricts to a simulation $K_2^{\mathrm{eff}} \longrightarrow\!\!\triangleright\, \mathsf{PC}^{\mathrm{eff}}$).

In the other direction, recall from Section 3.2.2 that any element x of PC can be specified by a countable set of finite pieces of information, namely the set of compact elements $y \sqsubseteq x$. For each type σ, take some effective coding of compact elements of $\mathsf{PC}(\sigma)$ as natural numbers, and write c_n^σ for the compact element coded by n (assume for convenience that every $n \in \mathbb{N}$ codes some compact element). We may then define an applicative simulation $\psi : \mathsf{PC} \longrightarrow\!\!\triangleright K_2$ by taking for each σ:

$$g \Vdash_\sigma^\psi x \text{ iff } x = \bigsqcup \{c_n^\sigma \mid n \in \text{ran } g\} .$$

To see that application in PC can be tracked in K_2, note that if $h \Vdash_{\sigma\to\tau}^\psi f$ and $g \Vdash_\sigma^\psi x$, we may readily define an element $h \diamond g \in K_2$ so that, for example,

$$\forall n.\ c_{(h \diamond g)(n)}^\tau = \Big(\bigsqcup_{m<n} c_{h(m)}^{\sigma\to\tau}\Big)\Big(\bigsqcup_{m<n} c_{g(m)}^\sigma\Big) .$$

Clearly $h \diamond g \Vdash_\tau^\psi f(x)$; moreover, the operation $\diamond$ is continuous and hence itself realizable in K_2. Again, it is easy to see that ψ restricts to a simulation $\mathsf{PC}^{\mathrm{eff}} \longrightarrow\!\!\triangleright K_2^{\mathrm{eff}}$.

It is easy to check that $\psi \circ \delta \sim id_{K_2}$. We also have $\delta \circ \psi \preceq id_{\mathsf{PC}}$, since within PC one may continuously pass from an element $g \in \mathsf{PC}(1)$ to the element of $\mathsf{PC}(\sigma)$ that it realizes. However, we clearly do not have $id_{\mathsf{PC}} \preceq \delta \circ \psi$, since there is no way within PC itself to pass from an element $x \in \mathsf{PC}(\sigma)$ to an enumeration of the compact elements below x.

Example 3.5.8. The nested sequential procedure model SP^0 enjoys a relationship to the van Oosten model B closely analogous to that between PC and K_2. One can readily simulate B using $\mathsf{SP}^0(1)$; conversely, a sequential procedure is a countably branching tree and so can be coded by a partial function $\mathbb{N} \rightharpoonup \mathbb{N}$ along lines similar to the representation of trees described in Section 3.2.4. In the latter case, the application operations of SP^0 turn out to be 'sequential' in the sense of B; see Section 6.1.5.

Examples 3.5.3, 3.5.7 and 3.5.8 display a common pattern. In each case, we have a pair of models $\mathbf{A}, \mathbf{B}$, where we think of $\mathbf{A}$ as 'more intensional' or 'more concrete' than $\mathbf{B}$, and a pair of simulations $\gamma : \mathbf{A} \longrightarrow\!\!\triangleright \mathbf{B}$, $\delta : \mathbf{B} \longrightarrow\!\!\triangleright \mathbf{A}$ such that $\delta \circ \gamma \sim id_{\mathbf{A}}$ and $\gamma \circ \delta \preceq id_{\mathbf{B}}$ but not $id_{\mathbf{B}} \preceq \gamma \circ \delta$. Informally, the absence of this latter transformation is attributable to the inability of $\mathbf{B}$ to 'see' its own elements at a sufficiently detailed level. In this situation, we may say (γ, δ) is an *applicative retraction* from $\mathbf{A}$ to $\mathbf{B}$. It is heuristically useful to think of the existence of such a retraction as yielding a kind of 'intensionality ordering' on models: although in principle there may be models $\mathbf{A} \not\simeq \mathbf{B}$ with applicative retractions in both directions, it seems that this does not arise in naturally occurring cases, so that in practice we have a partial order on the models of interest.

Our next two examples pick up threads from Sections 3.3.4 and 3.3.5.

Example 3.5.9 (Interpretations of System T). Let $\mathbf{A}$ be any total relative TPCA with numerals $\widehat{0}, \widehat{1}, \ldots$ of type $\mathtt{N}$, and associated operations suc, rec_σ. To any closed term $M : \sigma$ in Gödel's System T (see Section 3.2.3) we may associate an element $[\![M]\!] \in \mathbf{A}^\sharp(\sigma)$ as follows: replace each occurrence of λ by λ^* to obtain a meta-expression M^* as in Section 3.3.1, then expand M^* to an applicative expression $M^\dagger$ and evaluate $M^\dagger$ in $\mathbf{A}$, interpreting the constants $\widehat{0}$, suc, rec_σ in the obvious way.

Clearly $[\![MN]\!] = [\![M]\!] \cdot [\![N]\!]$, and it is easy to see that if $M \rightsquigarrow M'$ then $[\![M]\!] = [\![M']\!]$. We therefore obtain a type-respecting simulation $[\![-]\!] : \mathrm{T}^0/{=_{\mathrm{op}}} \longrightarrow\!\!\!\!\triangleright\, \mathbf{A}$. We will see in Exercise 3.6.6 that this construction can also be made to work for non-total models.

Now suppose $\gamma : \mathbf{A} \longrightarrow\!\!\!\!\triangleright\, \mathbf{B}$ is a type- and numeral-respecting applicative simulation between total models with numerals. By the above, we have applicative simulations $[\![-]\!]_{\mathbf{A}} : \mathrm{T}^0/{=_{\mathrm{op}}} \longrightarrow\!\!\!\!\triangleright\, \mathbf{A}$ and $[\![-]\!]_{\mathbf{B}} : \mathrm{T}^0/{=_{\mathrm{op}}} \longrightarrow\!\!\!\!\triangleright\, \mathbf{B}$, but it need not in general be the case that $\gamma \circ [\![-]\!]_{\mathbf{A}} \sim [\![-]\!]_{\mathbf{B}}$. For a simple counterexample, we may take $\mathbf{B}$ to be $\mathrm{T}^0/{=_{\mathrm{op}}}$ itself, and $\mathbf{A}$ to be its *observational quotient* with respect to $\mathbb{N}$ (see Section 3.6.1 below). This shows that a model $\mathbf{B}$ may in general admit several inequivalent applicative simulations of $\mathrm{T}^0/{=_{\mathrm{op}}}$.

Example 3.5.10 (Interpretations of PCF**).** Let $\mathbf{A}$ be any strict relative TPCA with numerals and general recursion. Since $\mathbf{A}$ contains elements playing the role of the PCF constants $\widehat{n}$, *suc*, *pre*, *ifzero* and Y_σ, we may translate closed PCF terms $M : \sigma$ into *expressions* $\widetilde{M} : \sigma$ over A by simply replacing λ with λ^* and expanding. Note that $M =_{\mathrm{op}} M'$ implies $[\![\widetilde{M}]\!] \simeq [\![\widetilde{M'}]\!]$ in $\mathbf{A}$. This does not itself give us an applicative simulation $\mathrm{PCF}^0/{=_{\mathrm{op}}} \longrightarrow\!\!\!\!\triangleright\, \mathbf{A}$, since $[\![\widetilde{M}]\!]$ may sometimes be undefined. However, we may obtain such a simulation $\theta_{\mathbf{A}}$ via a suspension trick: let $\theta_{\mathbf{A}}\sigma = \mathbb{N} \rightarrow \sigma$ for each σ, and take $a \Vdash_\sigma^{\theta_{\mathbf{A}}} [M]$ iff $a \cdot \widehat{0} \simeq [\![\widetilde{M}]\!]$. (For example, we have $[\![\lambda^* u.\widetilde{M}]\!] \Vdash [M]$ for any M.) It is easy to see that $\theta_{\mathbf{A}}$ is an applicative morphism, being realized at each σ, τ by $\lambda^* fxu.(fu)(xu)$.

Even if $\mathbf{A}$ has only guarded recursion, the same idea still works: we just need to translate the PCF constant Y_σ by the expression $\lambda^* f.Z(\lambda^* tv.f(tv))\widehat{0}$, checking that $\widetilde{Y_\sigma M} \simeq \widetilde{M(Y_\sigma M)}$, and we thence obtain a simulation $\theta_{\mathbf{A}} : \mathrm{PCF}^0/{=_{\mathrm{op}}} \longrightarrow\!\!\!\!\triangleright\, \mathbf{A}$ as before.

Our final example concerns a mild generalization of the notion of PCA that sometimes features in the literature (see e.g. van Oosten [295]).

Example 3.5.11. Define a (strict) *conditional PCA* to be a set A equipped with a partial application operation $\cdot$ such that for some $k, s \in A$ we have

$$\forall x,y.\ k \cdot x \cdot y = x\,, \qquad \forall x,y,z.\ s \cdot x \cdot y \cdot z \simeq (x \cdot z) \cdot (y \cdot z)\,.$$

Thus, an ordinary PCA is a conditional PCA in which $s \cdot x \cdot y \downarrow$ for all x, y. An example of a conditional PCA that is not an ordinary PCA is the set of *strongly normalizing* closed terms of untyped λ-calculus. The definitions of applicative simulation, transformation and equivalence carry over without changes to conditional PCAs.

If $(A, \cdot)$ is a conditional PCA, we may define a new application $*$ on A by $a * b \simeq a \cdot k \cdot b$. It is shown in [295, Chapter 1] by means of a suspension trick that $(A, *)$ is an ordinary PCA, and moreover is equivalent to $(A, \cdot)$, although the proof involves a non-constructive case split. Thus, from a classical point of view at least, the generalization to conditional PCAs does not admit any genuinely new computability notions.

In a similar vein, Faber and van Oosten [91] have recently shown by a more elaborate construction that any lax (untyped) PCA is equivalent to a strict one.

3.5.1 Effective and Continuous Models

It is natural to want to identify a class of computability models that are genuinely 'effective', in the sense that their data can be represented in some reasonable finitary way and their operations are ultimately implementable on a Turing machine. Abstracting from the case of $\Lambda^0/\!\sim$ above (Example 3.5.3), we see that applicative morphisms provide a way to capture this idea:

Definition 3.5.12. An *effective (relative) TPCA* is a (relative) TPCA $\mathbf{A}$ with numerals equipped with a numeral-respecting applicative simulation $\gamma : \mathbf{A} \longrightarrow\!\triangleright K_1$.

Notice that it need not be decidable whether a given number m realizes an element of $\mathbf{A}$. However, the requirement of respecting numerals does ensure a basic commensurability between the meanings of data in $\mathbf{A}$ and in K_1.

Syntactic models in general, such as $\Lambda^0/\!\sim$ and those described in Section 3.2.3, provide typical examples of effective TPCAs: the simulation γ is easily given by a Gödel numbering of syntactic terms. Other intuitively 'effective' models such as K_2^{eff} also provide examples: recall from Section 3.2.1 that the underlying set of K_2^{eff} consists of the total computable functions $\mathbb{N} \to \mathbb{N}$, so we may define $\gamma : K_2^{\mathrm{eff}} \longrightarrow\!\triangleright K_1$ by setting $m \Vdash^{\gamma} f$ iff $\forall n.\, m \bullet n = f(n)$. It is easy to see from the definition of application in K_2^{eff} that it is realizable within K_1. Similar considerations show that models such as $\mathsf{PC}^{\mathrm{eff}}$ and $\mathsf{B}^{\mathrm{eff}}$ are effective TPCAs.

Example 3.5.13. The following shows what can happen if the numeral-respecting condition in Definition 3.5.12 is omitted. Let $f : \mathbb{N} \to \mathbb{N}$ be some fixed non-computable function. The definition of K_1 may be relativized to f as follows: let $T_0', T_1', \ldots$ be a sensibly chosen enumeration of all Turing machines equipped with an *oracle* for f, so that a computation may ask for the value of some $f(m)$ as a single step. Now take $e \bullet^f n$ to be the result (if defined) of applying T_e' to the input n, and let $K_1^f = (\mathbb{N}, \bullet^f)$. The proof that K_1 is a PCA readily carries over to K_1^f.

Perhaps counterintuitively, there is an applicative simulation $\gamma : K_1^f \longrightarrow\!\triangleright K_1$: we may take $m \Vdash^{\gamma} n$ iff $m \bullet^f 0 = n$. That is, computations in K_1^f may after a fashion be simulated in K_1, except that all calls to f are 'postponed' until some supposed final $\bullet^f$-application to 0. Clearly, this simulation does not respect numerals, since the 'decoding' operation required by Definition 3.4.9(iv) depends crucially on the availability of f. Intuitively, the simulation γ is not a 'useful' one, since the results of computations remain encrypted in an unbreakable code. (See also Longley [173, Section 3.1].)

Many of our leading models that are not effective nonetheless conform to the paradigm of 'continuous computation' outlined in Section 3.2.1: data values can be represented by countably much information and computation is 'finitary' in the sense discussed in Section 3.2.1. This idea is enshrined in the following definition which closely parallels Definition 3.5.12:

Definition 3.5.14. A *continuous (relative) TPCA* is a (relative) TPCA $\mathbf{A}$ with numerals equipped with a numeral-respecting applicative simulation $\gamma : \mathbf{A} \longrightarrow\!\triangleright K_2$.

The choice of K_2 in this definition is admittedly somewhat arbitrary: it could equally well be replaced by any other model that both simulates and is simulated by K_2 in a numeral-respecting way. We choose K_2 as our point of reference simply because its construction embodies the intended notion of 'finitary computation on infinitary data' in a particularly direct way, as explained in Section 3.2.1.

A typical example of a continuous TPCA is PC, as shown by Example 3.5.7 above. Note also that the van Oosten model B may be readily simulated within $\mathsf{PC}(1)$; it follows that B too is a continuous PCA.

It is also convenient to single out those continuous models that contain *all* set-theoretic functions $\mathbb{N} \to \mathbb{N}$, in a manner compatible with their native representation in K_2:

Definition 3.5.15. A continuous TPCA $(\mathbf{A}, \gamma)$ is *full continuous* if the following hold:

1. For every $f : \mathbb{N} \to \mathbb{N}$, there is some $a \in \mathbf{A}(\mathtt{N} \to \mathtt{N})$ that represents f.
2. Moreover, a realizer for some such a may be computed from f within K_2—that is, there exists $h \in K_2$ such that for all $f \in \mathbb{N}^{\mathbb{N}}$ we have $h \odot f \Vdash^{\gamma}_{\mathtt{N}\to\mathtt{N}} a$ for some a representing f.

For example, the models PC and B are full continuous, whilst their effective submodels are not.

Another useful aspect of these definitions is that they allow us to discuss questions of *logical complexity* for effective and continuous TPCAs. For instance, we can say that a continuous TPCA (A, γ) is Π^1_1 if the corresponding partial equivalence relation on $\mathbb{N}^{\mathbb{N}}$ is definable by a Π^1_1 formula with two free variables of type $\mathbb{N}^{\mathbb{N}}$. The basic notions of logical complexity will be recalled in Section 5.2, where the complexity of the Kleene computable functionals will be analysed; analogous results for other models of interest will be noted later in the book.

3.6 General Constructions on Models

We conclude this section by describing some simple but important constructions for building new models from old ones, via *quotients*, *subsets* and *partial equivalence relations* (PERs) respectively. These will play significant roles in the rest of the book; we have delayed discussion of them until now in order to present them in the context of simulations. In connection with the PER construction, we also make some useful general observations concerning *extensional* higher-order models.

3.6.1 *Quotients*

The following simple notion is often useful, particularly in connection with syntactically constructed models.

Definition 3.6.1. Suppose $\mathbf{C},\mathbf{D}$ are computability models over the same type world T. We say $\mathbf{D}$ is a *quotient* of $\mathbf{C}$ if we have surjections $q_\sigma : \mathbf{C}(\sigma) \twoheadrightarrow \mathbf{D}(\sigma)$ for each $\sigma \in T$ satisfying the following:

1. For every operation $f \in \mathbf{C}[\sigma,\tau]$ there exists $\overline{f} \in \mathbf{D}[\sigma,\tau]$ such that $\overline{f}(q_\sigma(a)) \simeq q_\tau(f(a))$ for all $a \in \mathbf{C}(\sigma)$.
2. For every $\overline{f} \in \mathbf{D}[\sigma,\tau]$ there exists $f \in \mathbf{C}[\sigma,\tau]$ corresponding to $\overline{f}$ as above.

In this situation, the family of maps q_σ can be construed either as a single-valued type-respecting applicative simulation $\gamma : \mathbf{C} \longrightarrow\!\!\triangleright\, \mathbf{D}$, or as a discrete type-respecting applicative simulation $\delta : \mathbf{D} \longrightarrow\!\!\triangleright\, \mathbf{C}$. In this case we clearly have $\gamma \circ \delta = id_{\mathbf{D}}$ and $id_{\mathbf{C}} \preceq \delta \circ \gamma$, though not $\delta \circ \gamma \preceq id_{\mathbf{C}}$ in general.

An important source of quotients is the following construction inspired by ideas from synthetic domain theory (see for example Hyland [122]). Suppose $\mathbf{C}$ is a strict computability model with weak terminal (I,i), and $O \subseteq T \in |\mathbf{C}|$ is some set (called a *set of observations*) containing at least one 'computable' element $c \in T^\sharp$. The intention is to regard the elements of O as the only data values that can be observed 'directly' (for example, they might correspond to strings of characters printable on a computer screen); any other values of type $A \in |\mathbf{C}|$ may be observed only indirectly by first applying some operation $g \in \mathbf{C}[A,T]$. We now define the *observational quotient* $\mathbf{C}/{\sim_O}$ as follows. For $a,a' \in A$ where $A \in |\mathbf{C}|$, set $a \sim_O a'$ iff for all $g \in \mathbf{C}[A,T]$ and all $v \in O$ we have $g(a) = v \Leftrightarrow g(a') = v$. Let the datatypes of $\mathbf{C}/{\sim_O}$ be the sets $A/{\sim_O}$, and the operations of $\mathbf{C}/{\sim_O}$ be the partial functions $A/{\sim_O} \rightharpoonup B/{\sim_O}$ represented by some $f \in \mathbf{C}[A,B]$. It is easy to check that every $f \in \mathbf{C}[A,B]$ induces a well-defined operation $A/{\sim_O} \rightharpoonup B/{\sim_O}$, so that we indeed obtain a quotient $\mathbf{C}/{\sim_O}$ of $\mathbf{C}$ in the above sense.

Proposition 3.6.2. *If* $\mathbf{C}$ *is a higher-order model, so is* $\mathbf{C}/{\sim_O}$.

Proof. For notational simplicity, consider first the case $O = T$. If $f \sim_O f'$ in $A \Rightarrow B$ and $a \sim_O a'$ in A, then for any $g \in \mathbf{C}[B,T]$ we have

$$\begin{aligned} g(fa) &\simeq (\lambda^* x.g(xa))f \simeq (\lambda^* x.g(xa))f' \simeq g(f'a) \\ &\simeq (\lambda^* y.g(f'y))a \simeq (\lambda^* y.g(f'y))a' \simeq g(f'a')\,. \end{aligned}$$

Thus $fa \sim_O f'a'$ if both sides are defined; moreover, by specializing the above to $g = \lambda^* z.c$, we see that $fa\downarrow$ iff $f'a'\downarrow$. Hence the application operations in $\mathbf{C}$ induce well-defined application operations in $\mathbf{C}/{\sim_O}$, and the remaining ingredients of Definition 3.1.13 are easily verified. The argument adapts easily to the general case $O \subseteq T$. $\Box$

For many natural mathematical models $\mathbf{C}$, the quotient $\mathbf{C}/{\sim_O}$ typically coincides with $\mathbf{C}$ itself. For 'syntactic' models, however, such quotients are often of considerable interest. Let $\mathscr{L}^0(C)/{=_{\mathrm{op}}}$ be the closed term model for some typed λ-calculus $\mathscr{L}(C)$ as in Section 3.2.3. Choose some type τ of observations, and some set $O \subseteq \mathscr{L}^0(C)(\tau)$. (One often takes τ to be $\mathtt{N}$, and O to be the set of numerals $\widehat{n}$.) By identifying elements of O with their $=_{\mathrm{op}}$-classes, we may construct a quotient model

$(\mathscr{L}^0(C)/{=_{\mathrm{op}}})/{\sim_O}$, called the *closed term model for $\mathscr{L}(C)$ modulo observational equivalence*. This can be presented in terms of just a single quotienting operation by the relation $\sim_{\mathrm{obs}}$ on closed terms, where

$$M \sim_{\mathrm{obs}} M' : \sigma \quad \text{iff} \quad \forall P : \sigma \to \tau.\, \forall V \in O.\; PM =_{\mathrm{op}} V \iff PM' =_{\mathrm{op}} V\,.$$

In typical cases we will have $N =_{\mathrm{op}} V$ iff $N \rightsquigarrow^* V$ for $V \in O$, so that $=_{\mathrm{op}}$ may be replaced by $\rightsquigarrow^*$ in the above formula. Examples of such quotients will appear in Theorems 4.4.6 and 7.1.16.

3.6.2 The Subset Construction

Recall that a computability model is *total* if all the application operations are total functions. Given any model, we may define an associated total model:

Definition 3.6.3. For any computability model $\mathbf{C}$, let $\mathsf{Sub}(\mathbf{C})$ be the computability model defined as follows.

- Datatypes are subsets $S \subseteq A$ where $A \in |\mathbf{C}|$ (we assume each such S is implicitly tagged with a suitable choice of A).
- If $S \subseteq A$ and $T \subseteq B$, an operation in $\mathsf{Sub}(\mathbf{C})[S,T]$ is a total function $S \to T$ arising as the restriction of some $f \in \mathbf{C}[A,B]$.

Clearly $\mathsf{Sub}(\mathbf{C})$ is a total computability model; in fact, it is easily seen to be a full subcategory of $\mathscr{A}sm(\mathbf{C})$.

Of course, a simpler way to obtain a total model from $\mathbf{C}$ would be to retain the same family of datatypes as $\mathbf{C}$ and restrict attention to the total operations. However, the above construction has much better properties, as the following illustrates:

Proposition 3.6.4. *If $\mathbf{C}$ is a higher-order model, so is $\mathsf{Sub}(\mathbf{C})$.*

Proof. If (I,i) is a weak terminal in $\mathbf{C}$, clearly $(\{i\},i)$ is a weak terminal in $\mathsf{Sub}(\mathbf{C})$. Given datatypes $S \subseteq A$ and $T \subseteq B$ in $\mathsf{Sub}(\mathbf{C})$, define

$$S \Rightarrow_t T \;=\; \{f \in A \Rightarrow B \mid \forall a \in S.\, f \cdot a \in T\}\,,$$

and let $\cdot_{ST} : (S \Rightarrow_s T) \times S \to T$ be induced by application in $\mathbf{C}$. Conditions 1–3 of Definition 3.1.13 are now easily verified; note that if $S \subseteq A$, $T \subseteq B$ and $U \subseteq C$ then elements k_{AB}, s_{ABC} from $\mathbf{C}$ serve as k_{ST}, s_{STU} in $\mathsf{Sub}(\mathbf{C})$. □

It is also easy to see that $\mathsf{Sub}(\mathbf{C})$ inherits other properties of $\mathbf{C}$: if $\mathbf{C}$ has pairing, booleans or numerals then so does $\mathsf{Sub}(\mathbf{C})$. Note, however, that recursion is not inherited: indeed, if $\mathbf{C}$ has booleans then $\mathsf{Sub}(\mathbf{C})$ *cannot* have recursion.

The Sub construction transfers easily to the setting of relative TPCAs: take $\mathsf{Sub}(\mathbf{A})$ to be the relative TPCA $(\mathbf{S}^\circ;\mathbf{S}^\sharp)$ where $\mathbf{S}^\circ$ consists of all subsets of datatypes from $\mathbf{A}^\circ$ (with $\Rightarrow$ and $\cdot$ given as above), and if $(S \subseteq A) \in \mathbf{S}^\circ$ then $S^\sharp = S \cup A^\sharp$. Clearly this agrees with Definition 3.6.3 via the correspondence of Theorem 3.1.18.

In the general setting of computability models, there is an evident single-valued simulation $\mathsf{Sub}(\mathbf{C}) \longrightarrow\!\!\triangleright\, \mathbf{C}$. Moreover, Sub interacts well with simulations in the following sense:

Exercise 3.6.5. Show that the Sub construction Sub extends to a 2-functor Sub : $\mathfrak{CMod} \to \mathfrak{CMod}$.

In the case of higher-order models, often our main interest will not be in the whole of $\mathsf{Sub}(\mathbf{C})$ but in the simple types generated by some base type $B \in |\mathsf{Sub}(\mathbf{C})|$ (typically $B = \mathbb{N}$). Let us therefore write $\mathsf{Tot}(\mathbf{C};B)$ for the full substructure of $\mathsf{Sub}(\mathbf{C})$ consisting of types generated from B under $\Rightarrow$; we call this the model of *hereditarily total* elements of $\mathbf{C}$ relative to B, and may regard it as a model over $\mathsf{T}^{\rightarrow}(\beta = B)$.

Certain structures of this form have standard names in the literature. For instance, the structure $\mathsf{Tot}(K_1;\mathbb{N})$ is widely known as HRO, the type structure of *hereditarily recursive operations*. Likewise, $\mathsf{Tot}(K_2;\mathbb{N}^{\mathbb{N}})$ corresponds to (a mild variant of) ICF, the type structure of *intensional continuous functionals*. We shall consider both these structures briefly in Chapter 12.

Note crucially, however, that a simulation $\mathsf{Sub}(\gamma) : \mathsf{Sub}(\mathbf{C}) \longrightarrow\!\!\triangleright\, \mathsf{Sub}(\mathbf{D})$ will *not* in general respect the type structure: an element of $\mathsf{Sub}(\gamma)S \Rightarrow \mathsf{Sub}(\gamma)T$ does not in general yield one of $\mathsf{Sub}(\gamma)(S \Rightarrow T)$. Thus, $\mathsf{Sub}(\gamma)$ does not in general restrict to a simulation $\mathsf{Tot}(\mathbf{C};B) \longrightarrow\!\!\triangleright\, \mathsf{Tot}(\mathbf{D};B)$, even if γ itself is type-respecting over $\mathsf{T}^{\rightarrow}(\beta = B)$. This means that, in contrast to Exercise 3.6.5, the construction $\mathsf{Tot}(-;B)$ is *non-functorial* even on models over $\mathsf{T}^{\rightarrow}(\beta = B)$.

One may find models $\mathbf{C}$ and $\mathbf{D}$ that are equivalent over some $\mathsf{T}^{\rightarrow}(\beta = B)$ but such that $\mathsf{Tot}(\mathbf{C};B)$ and $\mathsf{Tot}(\mathbf{D};B)$ are inequivalent. An example may be obtained from an open term model for a typed combinatory logic with two function type constructors $\to$ and $\to'$, each with its own set of k and s combinators; we leave the details to the interested reader. This indicates that Tot is mathematically a much less well-behaved construction than Sub itself.

Exercise 3.6.6. Let $\mathbf{C}$ be any higher-order model with numerals (not necessarily total), with $\widehat{n}$, *suc* and rec_σ as in Section 3.3.4. Let $N = \{\widehat{n} \mid n \in \mathbb{N}\}$, and consider $\mathsf{Tot}(\mathbf{C};N)$. Show that this model contains the elements *suc* and rec_σ. Deduce from Example 3.5.9 that there is a numeral-respecting simulation $\mathrm{T}^0/{=_{\mathrm{op}}} \longrightarrow\!\!\triangleright\, \mathbf{C}$.

Exercise 3.6.7. A 'partial' analogue of the subset construction may be given. For any computability model $\mathbf{C}$, let $\mathsf{PSub}(\mathbf{C})$ have the same datatypes as $\mathsf{Sub}(\mathbf{C})$, and let the operations in $\mathsf{PSub}(\mathbf{C})[S,T]$ be those *partial* functions $S \rightharpoonup T$ arising as the restriction to S of some $f \in \mathbf{C}[A,B]$. Show that if $\mathbf{C}$ is a higher-order model, then so is $\mathsf{PSub}(\mathbf{C})$. Investigate which other properties of Sub hold also for PSub.

3.6.3 Partial Equivalence Relations

Our second construction somewhat resembles the subset construction, except that we here consider subsets subject to *equivalence relations*.

Definition 3.6.8. For any computability model $\mathbf{C}$, define the category $\mathscr{P}er(\mathbf{C})$ of *partial equivalence relations* over $\mathbf{C}$ as follows:

- Objects are datatypes $A \in |\mathbf{C}|$ together with a symmetric, transitive relation $\approx$ on A (or equivalently, subsets S of some $A \in |\mathbf{C}|$ together with an equivalence relation $\approx$ on S). We write $A/{\approx}$ for the set of equivalence classes, and $|\approx|$ for the set $\{a \in A \mid a \approx a\}$.
- Morphisms $f : (A, \approx) \to (B, \approx')$ are functions $f : A/{\approx} \to B/{\approx'}$ that are *tracked* by some $\overline{f} \in \mathbf{C}[A,B]$, in the sense that for all $a \approx a$ we have $[\overline{f}(a)] = f([a])$.

Exercise 3.6.9. Show that $\mathscr{P}er(\mathbf{C})$ is equivalent to the category $\mathscr{M}od(\mathbf{C})$ defined in Section 3.3.6.

We may regard $\mathscr{P}er(\mathbf{C})$ as itself a total computability model by associating to each object (A,R) the datatype A/R, and viewing the morphisms of $\mathscr{P}er(\mathbf{C})$ as operations. We shall denote this computability model by $\mathsf{PER}(\mathbf{C})$.

Proposition 3.6.10. *If* $\mathbf{C}$ *is a higher-order model, so is* $\mathsf{PER}(\mathbf{C})$.

Proof. This is closely related to part (iii) of Theorem 3.3.22, but for convenience we repeat the construction here. First, if (I,i) is a weak terminal in $\mathbf{C}$, the partial equivalence relation $\{(i,i)\}$ on I clearly serves as a weak terminal in $\mathsf{PER}(\mathbf{C})$. Next, given partial equivalence relations $(A,\approx)$ and $(B,\approx')$ over $\mathbf{C}$, define a partial equivalence relation $\approx''$ on $A \Rightarrow B$ by

$$f \approx'' g \ \text{ iff } \ \forall a,b \in A.\ a \approx b \Rightarrow f \cdot a \approx' g \cdot b\,,$$

and take $(A,\approx) \Rightarrow (B,\approx') = (A \Rightarrow B, \approx'')$, with the obvious application operation. The conditions of Definition 3.1.13 are then easily verified. □

It is also clear that if $\mathbf{C}$ has pairing, booleans or numerals then so does $\mathsf{PER}(\mathbf{C})$ (cf. Theorem 3.3.22).

It is easy to see how the PER construction transfers to TPCAs. In general, if $\mathbf{A}$ is a relative TPCA, we will write $\mathscr{P}er(\mathbf{A})$ or $\mathscr{P}er(\mathbf{A}^\circ;\mathbf{A}^\sharp)$ for the category whose objects are partial equivalence relations on some $\mathbf{A}^\circ(\sigma)$ and whose morphisms are realized by elements of $\mathbf{A}^\sharp$; note that $\mathscr{P}er(\mathbf{A}) \simeq \mathscr{M}od(\mathbf{A})$.

Note that there is an evident simulation $\mathsf{PER}(\mathbf{C}) \longrightarrow\!\!\triangleright\, \mathbf{C}$. It is also easy to check that the PER construction can be extended to a 2-functor on $\mathfrak{CMod}^d$, the 2-category of computability models, *discrete* simulations, and transformations between them. Indeed, if $\gamma : \mathbf{C} \longrightarrow\!\!\triangleright\, \mathbf{D}$ is any discrete simulation, the simulation $\mathsf{PER}(\gamma) : \mathsf{PER}(\mathbf{C}) \longrightarrow\!\!\triangleright\, \mathsf{PER}(\mathbf{D})$ is essentially the restriction to modest assemblies of γ_* as defined in Section 3.3.6.

The main property of $\mathsf{PER}(\mathbf{C})$ not shared by $\mathsf{Sub}(\mathbf{C})$ is *extensionality*. This is a notion specific to higher-order models:

Definition 3.6.11. A higher-order model $\mathbf{C}$ is *extensional* if for all $A,B \in |\mathbf{C}|$ we have

$$\forall f,f' \in A \Rightarrow B.\ (\forall a \in A.\ f \cdot a \simeq f' \cdot a) \ \Rightarrow\ f = f' \ .$$

From the proof of Proposition 3.6.10 it is immediate that if $\mathbf{C}$ is a higher-order model then $\mathsf{PER}(\mathbf{C})$ is extensional: if $(S \subseteq A, \approx)$ and $(T \subseteq B, \approx')$ are partial equivalence relations and $f,f' \in \mathbf{C}[A,B]$ induce the same mapping $A/\approx \longrightarrow B/\approx'$, then by definition $[f] = [f']$ in $(S,\approx) \Rightarrow (T,\approx')$.

As with $\mathsf{Sub}(\mathbf{C})$, we shall often be interested in the substructure of $\mathsf{PER}(\mathbf{C})$ generated by some base type. This structure is of sufficient importance to merit a name:

Definition 3.6.12. Let $\mathbf{C}$ be any higher-order structure, $\approx$ a partial equivalence relation on some $B \in |\mathbf{C}|$. The *extensional collapse* of $\mathbf{C}$ with respect to $\approx$, written $\mathsf{EC}(\mathbf{C};\approx)$, is the full substructure of $\mathsf{PER}(\mathbf{C})$ consisting of types generated from $\approx$ via $\Rightarrow$, regarded as a higher-order model over $\mathsf{T}^{\rightarrow}(\beta = B/\approx)$.

If $S \subseteq B$, we may write $\mathsf{EC}(\mathbf{C};S)$ for $\mathsf{EC}(\mathbf{C};\approx_S)$ where $a \approx_S b$ iff $a = b \in S$.

Typically the ground relation $\approx$ will be chosen so that its set of equivalence classes is isomorphic to $\mathbb{N}$ or $\mathbb{N}_\perp$. We may thus obtain a variety of total, extensional models over $\mathbb{N}$ or $\mathbb{N}_\perp$ using this construction; see Section 3.7 below.

Since this construction will play a major role, it is worth spelling out explicitly how it looks in the setting of TPCAs. If $\mathbf{A}$ is a full TPCA over $\mathsf{T}^{\rightarrow}(\beta)$ and $\approx$ a partial equivalence relation on $\mathbf{A}(\beta)$, we may define on each $\mathbf{A}(\sigma)$ a partial equivalence relation $\approx_\sigma$ by

$$\begin{aligned} a \approx_\beta b \ &\Leftrightarrow\ a \approx b\,, \\ f \approx_{\sigma\to\tau} g \ &\Leftrightarrow\ \forall a,b \in \mathbf{A}(\sigma).\ a \approx_\sigma b \Rightarrow f \cdot a \approx_\tau g \cdot b\,. \end{aligned}$$

We shall write $\mathsf{Ext}(\mathbf{A};\approx)$ for the substructure of $\mathbf{A}$ consisting of elements $a \in \mathbf{A}(\sigma)$ such that $a \approx_\sigma a$; we refer to these as the *hereditarily extensional* elements.[2] We may then define $\mathsf{EC}(\mathbf{A};\approx)$ as the quotient of $\mathsf{Ext}(\mathbf{A};\approx)$ modulo $\approx$.

Likewise, if $\mathbf{A}$ is a relative TPCA with $\approx$ a PER on $\mathbf{A}^\circ(\beta)$, we may define the relative TPCA

$$\mathsf{Ext}(\mathbf{A};\approx) \ = \ (\mathsf{Ext}(\mathbf{A}^\circ;\approx);\ \mathsf{Ext}(\mathbf{A}^\circ;\approx) \cap \mathbf{A}^\sharp)$$

and its evident quotient $\mathsf{EC}(\mathbf{A};\approx)$, where $\mathsf{EC}(\mathbf{A};\approx)^\sharp$ consists of the equivalence classes in $\mathsf{EC}(\mathbf{A}^\circ;\approx)$ that contain an element of $\mathbf{A}^\sharp$. Note that $\mathsf{EC}(\mathbf{A};\approx)^\sharp$ need not be an extensional TPCA in its own right, although in many important cases it will be: see e.g. Corollary 8.1.10.

Exercise 3.6.13. Suppose $\mathbf{A}$ is total, and $\approx$ is a *total* equivalence relation on $\mathbf{A}^\circ(\beta)$. We say $\mathbf{A}$ is *pre-extensional* relative to $\approx$ if for $f,g \in \mathbf{A}^\circ(\sigma_0,\ldots,\sigma_{r-1} \to \beta)$ we have

---

[2] The reader is warned that in the literature these are sometimes referred to as the *(hereditarily) total* elements, a term we are reserving for elements of $\mathsf{Tot}(\mathbf{A};B)$. In the case of primary interest, the two concepts coincide: thus $\mathsf{Tot}(\mathsf{PC};\mathbb{N}) = \mathsf{Ext}(\mathsf{PC};\mathbb{N})$ (see Corollary 8.1.2).

$$(\forall a_0,\ldots,a_{r-1}.\ f\cdot a_0\cdot\ldots\cdot a_{r-1}\approx g\cdot a_0\cdot\ldots\cdot a_{r-1})\ \Rightarrow\ (\forall h.\ h\cdot f\approx h\cdot g)\,.$$

Show that in this situation, the resulting relations $\approx_\sigma$ are all total equivalence relations, and that $\mathsf{EC}(\mathbf{A};\approx)$ is a quotient of $\mathbf{A}$ in the sense of Section 3.6.1.

In fact, both Tot and EC can be seen as instances of a more general construction framed in terms of *logical relations*. We shall take up this point of view in the following chapter in the context of λ-algebras.

Unlike the Tot construction from Section 3.6.2, the EC construction at least respects equivalences of models:

Proposition 3.6.14. *Suppose the higher-order models* $\mathbf{C}$, $\mathbf{D}$ *are equivalent via simulations* $\gamma:\mathbf{C}\longrightarrow\!\!\!\triangleright\,\mathbf{D}$, $\delta:\mathbf{D}\longrightarrow\!\!\!\triangleright\,\mathbf{C}$, *and suppose* $\approx,\approx'$ *are PERs on* $A\in|\mathbf{C}|$, $B\in|\mathbf{D}|$ *respectively such that* $\gamma_*(\approx)$ *and* $\approx'$ *are isomorphic in* $\mathscr{P}er(\mathbf{D})$ *(and so necessarily* $\delta_*(\approx')$ *and* $\approx$ *are isomorphic in* $\mathscr{P}er(\mathbf{C})$*). Then* $\mathsf{EC}(\mathbf{C};\approx)\cong\mathsf{EC}(\mathbf{D};\approx')$.

This follows immediately from the fact that γ_* and δ_* constitute an equivalence $\mathscr{P}er(\mathbf{C})\simeq\mathscr{P}er(\mathbf{D})$ (see Theorem 3.4.14); alternatively, it is a routine exercise to prove it directly.

Exercise 3.6.15. Show that if $\mathbf{D}$ is a quotient of $\mathbf{C}$ and the PER $\approx'$ on $\mathbf{D}(\sigma)$ induces $\approx$ on $\mathbf{C}(\sigma)$ in the obvious way, then $\mathsf{EC}(\mathbf{C};\approx)\cong\mathsf{EC}(\mathbf{D};\approx')$.

However, this is about all that can be said in general regarding the 'robustness' of EC. Like Tot, EC is non-functorial with respect to simulations: indeed, even when $\gamma:\mathbf{C}\longrightarrow\!\!\!\triangleright\,\mathbf{D}$ is a type-respecting *inclusion* of a smaller model into a larger one, and $\mathsf{EC}(\mathbf{C};\approx)$ and $\mathsf{EC}(\mathbf{D};\approx)$ agree at the base type, there need be no particular relationship between $\mathsf{EC}(\mathbf{C};\approx)$ and $\mathsf{EC}(\mathbf{D};\approx)$ at higher types. Informally, the issue is that if for some type σ there are 'more' elements of $\mathsf{EC}(\mathbf{D};\approx)(\sigma)$ than of $\mathsf{EC}(\mathbf{C};\approx)(\sigma)$—or even just more realizers for the same elements—then it may be 'harder' to realize an extensional operation of type $\sigma\to\tau$ in $\mathbf{D}$ than in $\mathbf{C}$, so there may be operations in $\mathsf{EC}(\mathbf{C};\approx)[\sigma,\tau]$ with no counterpart in $\mathbf{D}$. We shall meet some interesting and important examples of this phenomenon in later chapters (see for instance Subsection 9.1.2 and Example 12.2.7).

The EC construction also suffers from other anomalies. For instance, if $\mathbf{C}$ is a model over $\mathsf{T}^{\to}(\beta)$, $\approx$ is a PER on $\mathbf{C}(\beta)$ and $\approx'$ is a PER on $\mathbf{C}(\beta)/\!\approx$ inducing the PER $\approx''$ on $\mathbf{C}(\beta)$, it need not be the case that $\mathsf{EC}(\mathsf{EC}(\mathbf{C};\approx);\approx')$ coincides with $\mathsf{EC}(\mathbf{C};\approx')$. For further discussion and examples, see Honsell and Sannella [119].

In view of these peculiarities, the mathematical significance of EC should perhaps not be overplayed. Whilst it is tempting to regard $\mathsf{EC}(\mathbf{C};\approx)$ as comprising 'the extensional operations computable in $\mathbf{C}$', there may also be other total extensional models $\mathbf{E}$ simulable in $\mathbf{C}$ (via a simulation yielding $\approx$ at ground type) that are not comparable with $\mathsf{EC}(\mathbf{C};\approx)$ in any simple way. Among all such models $\mathbf{E}$, there may be nothing very distinguished about $\mathsf{EC}(\mathbf{C};\approx)$ beyond the fact that it *is* the extensional collapse. However, the EC construction is certainly useful as a means of *defining* many of the models of interest and giving a handle on their properties.

The Tot and EC constructions may sometimes be fruitfully combined. In particular, if $S \subseteq B \in |\mathbf{C}|$ we shall write $\mathsf{MEC}(\mathbf{C};S)$ for the model $\mathsf{EC}(\mathsf{Tot}(\mathbf{C};S);S)$, and refer to this as the *modified extensional collapse* of $\mathbf{C}$ with respect to S. The name arises from the fact that this structure arises naturally within the category $\mathscr{A}sm(\mathsf{Sub}(\mathbf{C}))$ which forms part of the *modified realizability model* over $\mathbf{C}$ (see van Oosten [295]). The MEC construction will play a role in Chapters 8 and 9.

3.6.4 Type Structures

We conclude the chapter with a few general observations regarding extensional models, of which the extensional collapses described above form prominent examples. By a *type structure over* some base set B, we shall mean a (partial or total) extensional higher-order model over the type world $\mathsf{T}^{\rightarrow}(\beta = B)$. (Type structures with products may be defined similarly.) In most cases that we shall consider, B will be either $\mathbb{N}$ or $\mathbb{N}_{\perp}$, though in Subsection 12.1.6 we will also encounter a type structure over $\mathbb{N}_{\perp}^{\top} = \mathbb{N} \cup \{\perp, \top\}$. Type structures enjoy some pleasing properties not shared by higher-order models in general.

First, an easy but important observation. Let us say a type structure $\mathbf{C}$ over B is *canonical* if every set $\mathbf{C}(\sigma \rightarrow \tau)$ is actually a set of partial functions from $\mathbf{C}(\sigma)$ to $\mathbf{C}(\tau)$ (and application in $\mathbf{C}$ is simply function application). By an easy induction on types we have:

Proposition 3.6.16. *Every type structure* $\mathbf{C}$ *over* B *is isomorphic to a unique canonical type structure over* B. □

A further pleasing aspect of *total* type structures over B is that they can be partially ordered according to their 'computational strength'—in contrast to arbitrary higher-order models for which we have a notion of simulation, but no natural partial ordering in general. Specifically, if both $\mathbf{C}$ and $\mathbf{D}$ are total type structures over B, let us write $\mathbf{C} \preceq \mathbf{D}$ if there exists a simulation $\mathbf{C} \longrightarrow\!\!\triangleright\, \mathbf{D}$, type-respecting over $\mathsf{T}^{\rightarrow}(\beta = B)$. Clearly $\preceq$ is reflexive and transitive. Furthermore:

Proposition 3.6.17. *If* $\mathbf{C} \preceq \mathbf{D}$ *and* $\mathbf{D} \preceq \mathbf{C}$ *then* $\mathbf{C} \cong \mathbf{D}$.

Proof. Suppose $\gamma : \mathbf{C} \longrightarrow\!\!\triangleright\, \mathbf{D}$ and $\delta : \mathbf{D} \longrightarrow\!\!\triangleright\, \mathbf{C}$ are type-respecting simulations over $\mathsf{T}^{\rightarrow}(\beta = B)$, and let $\Vdash$ abbreviate $\Vdash^{\delta \circ \gamma}$. We first show by induction on types that $\Vdash_{\sigma}$ is the identity on $\mathbf{C}(\sigma)$ for all $\sigma \in \mathsf{T}^{\rightarrow}(\beta)$. For $\sigma = \beta$ this is immediate. So assume $\sigma = \tau \rightarrow \upsilon$ where the hypothesis holds for τ and υ. Suppose $f, f' \in \mathbf{C}(\tau \rightarrow \upsilon)$ where $f \Vdash_{\tau \rightarrow \upsilon} f'$. Since $\delta \circ \gamma$ is type-respecting and $\mathbf{C}$ is total, we have for all $a \in \mathbf{C}(\tau)$ that $a \Vdash_{\tau} a$ so $f' \cdot a \Vdash_{\upsilon} f \cdot a$, whence $f' \cdot a = f \cdot a$. Thus $f' = f$ since $\mathbf{C}$ is extensional. On the other hand, for every f there is clearly at least one f' with $f' \Vdash_{\tau \rightarrow \upsilon} f$, so $\Vdash_{\tau \rightarrow \upsilon}$ is the identity. Thus $\delta \circ \gamma = id_{\mathbf{C}}$, and similarly we have $\gamma \circ \delta = id_{\mathbf{D}}$.

From this it follows easily that both γ and δ are discrete, and hence also that γ and δ are single-valued. Thus γ, δ constitute an isomorphism $\mathbf{C} \cong \mathbf{D}$. □

We write $\mathscr{T}(B)$ for the poset of total type structures over B modulo isomorphism.

Exercise 3.6.18. If $\mathbf{C}, \mathbf{D} \in \mathscr{T}(B)$, let us say a simulation $\gamma : \mathbf{C} \longrightarrow\!\!\!\triangleright\, \mathbf{D}$ is *B-respecting* if there are operations e, d in $\mathbf{D}$ such that for all $b \in B$ and all $a \in \mathbf{D}(\gamma\beta)$ we have

$$e \cdot b \Vdash^{\gamma} b\,, \qquad a \Vdash^{\gamma} b \Rightarrow d \cdot a = b\,.$$

(i) Suppose $\mathbf{C}$ and $\mathbf{D}$ are total type structures over B, and suppose $\mathbf{C}, \mathbf{D}$ are equivalent as higher-order models via B-respecting simulations γ, δ (which need not be type-respecting). Show that $\mathbf{C}$ and $\mathbf{D}$ are isomorphic.

(ii) In particular, suppose $\mathbf{C}$ and $\mathbf{D}$ are total type structures over $\mathbb{N}$, and that in both $\mathbf{C}$ and $\mathbf{D}$ the elements of $\mathbb{N}$ form a system of numerals as in Subsection 3.3.4. Show that if $\mathbf{C} \simeq \mathbf{D}$ as higher-order models then $\mathbf{C} \cong \mathbf{D}$.

Finally, we remark that total type structures are always λ-*algebras* (see Proposition 4.1.3), so the further theory to be developed in Chapter 4 is applicable to them.

3.7 Prospectus

In this chapter we have presented the general framework for our investigation of higher-order computability, and along the way have also introduced many of the specific models that we shall study in more detail in Part II: K_1, K_2, PC, B, SP^0, and the term models for System T and PCF. In addition, now that we have the extensional collapse construction available, it is worth pausing to give simple definitions of some of the other models appearing in the roadmap of Section 1.3, for the sake of filling out a little further our overview of the book as a whole. Specifically:

- The model SP^0 has an extensional quotient SF, the type structure of *sequential functionals* over $\mathbb{N}_\perp$. This coincides with the class of functionals computable in PCF relative to some first-order oracle; it will be studied in detail in Chapter 7.
- The model Ct of *total continuous functionals*, to be studied in Chapter 8, may be defined as $\mathsf{EC}(\mathsf{PC};\mathbb{N})$; this turns out to be isomorphic to $\mathsf{EC}(K_2;\mathbb{N})$.
- The model HEO of *hereditarily effective operations*, to be studied in Chapter 9, may be defined as $\mathsf{EC}(\mathsf{PC}^{\mathrm{eff}};\mathbb{N})$, or equivalently as $\mathsf{EC}(K_2^{\mathrm{eff}};\mathbb{N})$ or $\mathsf{EC}(K_1;\mathbb{N})$.
- The model SR of *sequentially realizable functionals*, studied in Chapter 11, may be defined as $\mathsf{EC}(\mathsf{B};\mathbb{N}_\perp)$, for an evident representation of $\mathbb{N}_\perp$ within B.

Other models, and further characterizations of the above models, will be introduced as the book proceeds.

Chapter 4
Theory of λ-Algebras

In the previous chapter, broadly speaking, we axiomatized and studied higher-order models as *combinatory algebras*—that is, as structures possessing elements k and s satisfying certain axioms. In the present chapter, we shall consider models from the related but somewhat different perspective of the *λ-calculus*. Not all the models mentioned in Chapter 3 satisfy the fundamental rules of the λ-calculus, but for those that do—the so-called *typed λ-algebras*—a wealth of further structure is available, and our purpose here is to study this structure in a general setting.

We have already met the (typed) λ-calculus in Section 3.2.3, and have seen in Section 3.3.1 how λ-terms may be 'interpreted' in an arbitrary TPCA. In the first instance, we may characterize typed λ-algebras simply as the TPCAs for which this interpretation validates β-equality. However, this syntactic style of definition does not by itself give much mathematical insight into the essential character of λ-algebras, so in Section 4.1 we supplement this with a series of other characterizations at increasing levels of abstraction, emphasizing in particular the categorical perspective and the relationship to cartesian closed categories. Taken together, these characterizations give a rounded understanding of what it means for a model to be a λ-algebra.

Along the way, we introduce the fundamental concept of a *retraction*, whereby one datatype may be embedded in or 'coded by' another, and the related *Karoubi envelope* construction, which extends a given λ-algebra by adding in all datatypes that are implicitly present as retracts of existing ones. The importance of retractions is that many properties of interest transfer automatically from a type to its retracts, so it is of interest to ask which types arise as retracts of which others. In Section 4.2, we show that a rich repertoire of datatypes can typically be obtained as retracts of the simple types, or indeed of just the *pure* types $\overline{k}$. This significantly extends the applicability of many theorems to be obtained later in the book.

In one respect, our definition of λ-algebra may appear arbitrarily constrained: it is limited to *total* models (i.e. those in which applications are always defined). We have chosen to focus on these because the theory of partial λ-algebras turns out to be much more subtle and intricate, and indeed some of the fundamental concepts here are still being worked out.[1] Nevertheless, one might reasonably ask whether we are thereby excluding some important and natural computability notions from consideration. In Section 4.3, we present some results that partially address this concern. In the *extensional* setting, for instance, there is an exact correspondence between non-total models over $\mathbb{N}$ and total ones over $\mathbb{N}_\perp$ satisfying a mild condition—this means that little is lost by concentrating on total structures. We also obtain some related results for non-extensional models; these will be useful in Chapter 12.

In Section 4.4 we switch topic somewhat, introducing the method of *(pre)logical relations*. This is a powerful proof technique for establishing properties of the elements of a model definable by λ-terms (perhaps relative to certain constants). We apply this method to show, for instance, that Gödel's System T is *strongly terminating* and has other agreeable properties. This is of significance because System T—essentially the simply typed λ-calculus extended with a well-behaved type of natural numbers—can be seen as defining a core class of computable total functionals that will be present in most of the typed models we wish to study (see Example 3.5.9).

4.1 Typed λ-Algebras

We begin by presenting several equivalent characterizations of (typed) λ-algebras and establishing their general structural theory. We start with some 'syntactic' definitions based on the interpretation of λ-terms (Subsections 4.1.1 and 4.1.2), then work our way towards progressively more abstract characterizations formulated in the language of category theory (Subsections 4.1.3 and 4.1.5). The categorical perspective provides the natural setting for much of the theory, in particular for the *Karoubi envelope* construction, which we introduce in Subsection 4.1.4, and which will provide the basis for the theory of retractions in Section 4.2.

Almost everything in this section is either well-established material or a mild adaptation thereof: see for example Barendregt [9] or Lambek and Scott [167].

4.1.1 Interpretations of λ-Calculus

We start with the syntax of the simply typed λ-calculus, already introduced briefly in Subsection 3.2.3. Let T be a type world with arrow structure as in Subsection 3.1.7, and let C be a set of constant symbols c each with an assigned type $\tau_c \in \mathsf{T}$. By a *(type) environment* we shall mean a finite list $\Gamma = x_0^{\sigma_0}, \ldots, x_{r-1}^{\sigma_{r-1}}$ of distinct typed variables. We shall write $\Gamma \vdash M : \sigma$ to mean that M is a term of type σ whose free variables are among those of Γ. Recall from Section 3.2.3 that the set of such well-typed terms over C is generated by the following rules (we give them here in a form that makes the relevant environments explicit):

[1] See for example Moggi [195] for several possible definitions, or Cockett and Hofstra [56] for recent progress in this area.

$$\Gamma \vdash c : \tau_c \qquad \Gamma \vdash x^\sigma : \sigma \quad (x^\sigma \in \Gamma) \qquad \frac{\Gamma, x^\sigma \vdash M : \tau}{\Gamma \vdash \lambda x^\sigma . M : \sigma \to \tau}$$

$$\frac{\Gamma \vdash M : \sigma \to \tau \qquad \Gamma \vdash N : \sigma}{\Gamma \vdash MN : \tau}$$

For most of this chapter we will take $C = \emptyset$ and consider the so-called *pure* λ-terms.

We shall consider λ-terms as subject to the usual relation $=_\beta$ of β-*equality*, namely the equivalence relation on terms of the same type generated by the clauses

$$\begin{array}{rlr} (\lambda x.M)N = M[x \mapsto N]\,, & & M = N \Rightarrow PM = PN\,, \\ M = N \Rightarrow \lambda x.M = \lambda x.N\,, & & M = N \Rightarrow MP = NP\,. \end{array}$$

We shall also write $=_{\beta\eta}$ for the equivalence relation generated by these clauses along with the η-*rule*:

$$\lambda x.Mx = M \quad (x \notin \mathrm{FV}(M))\,.$$

We have already seen in Section 3.3.1 how pure λ-terms may be interpreted in an arbitrary higher-order model. Recall that a well-typed λ-term $\Gamma \vdash M : \tau$ may be expanded to an applicative expression $\Gamma \vdash M^\dagger : \tau$ via

$$x^\dagger = x\,, \qquad (MN)^\dagger = M^\dagger N^\dagger\,, \qquad (\lambda x.M)^\dagger = \lambda^* x.(M^\dagger)\,,$$

where

$$\begin{array}{rll} \lambda^* x^\sigma . x = & s_{\sigma(\sigma\to\sigma)\sigma}\, k_{\sigma(\sigma\to\sigma)}\, k_{\sigma\sigma}\,, & \\ \lambda^* x^\sigma . e = & k_{\tau\sigma} e & \text{if } e : \tau \text{ does not contain } x\,, \\ \lambda^* x^\sigma . ee' = & s_{\sigma\tau\tau'}(\lambda^* x^\sigma . e)(\lambda^* x^\sigma . e') & \text{if } e : \tau \to \tau',\ e' : \tau, \text{ and } ee' \text{ contains } x\,. \end{array}$$

Note that $M^\dagger$ is a well-typed applicative expression built from k, s and variables; we refer to such entities as *combinatory expressions*.

At this point, we may note a pleasing property of our translation $-^\dagger$:

Proposition 4.1.1. *For any combinatory expression e, let e° be the typed λ-term obtained by replacing each occurrence of k and s by $\lambda xy.x$ and $\lambda xyz.xz(yz)$ respectively. Then for any λ-term M we have $(M^\dagger)^\circ =_\beta M$.*

Proof. A routine structural induction shows that $(\lambda^* x.e)^\circ =_\beta \lambda x.(e^\circ)$ for any e, and the proposition follows by an easy induction on M. □

Now suppose we are given a TPCA $\mathbf{A}$ over T equipped with a choice of elements $k_{\sigma\tau}$, $s_{\rho\sigma\tau}$ for each ρ, σ, τ satisfying the axioms of Definition 3.1.16. If Γ is an environment over T, a *valuation* of Γ in $\mathbf{A}$ is a function assigning to each variable x^σ in Γ an element $\nu(x) \in \mathbf{A}(\sigma)$. If $\Gamma \vdash e : \tau$ is a combinatory expression and ν any valuation of Γ in $\mathbf{A}$, we may attempt to define an interpretation $[\![e]\!]_\nu \in \mathbf{A}(\tau)$ by

$$[\![k]\!]_\nu = k\,, \qquad [\![s]\!]_\nu = s\,, \qquad [\![x]\!]_\nu = \nu(x)\,, \qquad [\![ee']\!]_\nu \simeq [\![e]\!]_\nu \cdot [\![e']\!]_\nu\,.$$

If **A** is total, the interpretation $[\![e]\!]_\nu$ will always be defined. If $\Gamma \vdash M : \tau$, we may then set $[\![M]\!]^*_\nu = [\![M^\dagger]\!]_\nu$. We write $[\![M]\!]^*$ for $[\![M]\!]^*_\nu$ when ν is empty. It is an easy exercise to check that $[\![\lambda x.M]\!]^*_\nu \cdot a = [\![M]\!]^*_{\nu(x \mapsto a)}$.

We have seen in Example 3.3.4 that such an interpretation will not necessarily validate β-equality in all instances. One way to introduce λ-algebras is simply as those total higher-order models that do satisfy β-equality. This highlights one basic reason for being interested in λ-algebras: they are the models in which certain natural kinds of equational reasoning work smoothly.

Definition 4.1.2. Let T be a type world with arrow structure.

(i) A *(total) λ-algebra* over T is a total TPCA **A** over T equipped with a choice of elements $k_{\sigma\tau}, s_{\rho\sigma\tau}$ (subject to the usual axioms) such that the induced interpretation $[\![-]\!]^*_-$ satisfies the following:

1. If $M =_\beta M'$ then $[\![M]\!]^*_\nu = [\![M']\!]^*_\nu$ for all ν.
2. $k_{\sigma\tau} \;=\; [\![\lambda xy.x]\!]^*$.
3. $s_{\rho\sigma\tau} \;=\; [\![\lambda xyz.xz(yz)]\!]^*$.

Such a structure is a $\lambda\eta$-algebra if condition 1 holds with $=_\beta$ replaced by $=_{\beta\eta}$.

(ii) A λ-algebra $\mathbf{A}^\circ$ equipped with a substructure $\mathbf{A}^\sharp$ containing the designated elements k, s is called a *relative λ-algebra*; similarly for relative $\lambda\eta$-algebras.

Condition 1 in the above definition is the important one; conditions 2 and 3 merely serve to eliminate spurious freedom in the choice of k, s. Indeed, if k, s are such that the induced interpretation $[\![-]\!]^*_-$ satisfies only condition 1, we may define $k' = [\![\lambda xy.x]\!]^*$ and $s' = [\![\lambda xyz.xz(yz)]\!]^*$ and from this obtain a second interpretation $[\![-]\!]'^*_-$ which satisfies all three conditions. This follows easily from the fact that

$$[\![M]\!]'^*_\nu \;=\; [\![M]\!]^*_\nu \text{ for all } M \text{ and } \nu\ ,$$

which in turn is a consequence of condition 1 for $[\![-]\!]^*_-$ together with Proposition 4.1.1.

In contrast to the definition of TPCA (Definition 3.1.16), the choice of elements k, s is deemed to be part of the structure of a λ-algebra. One can contrive syntactic examples of TPCAs that admit multiple λ-algebra structures via different choices of k and s, although it would seem that in most naturally arising semantic examples only one such choice is possible.

The following easy result ensures a plentiful supply of λ-algebras. Recall from Section 3.6.3 that a relative TPCA **A** is *extensional* if for all $A, B \in |\mathbf{A}|$ we have

$$\forall f, f' \in (A \Rightarrow B)^\circ.\ \ (\forall a \in A^\circ.\ f \cdot a \simeq f' \cdot a) \ \Rightarrow\ f = f'\ .$$

Proposition 4.1.3. *Suppose* **A** *is a total extensional relative TPCA. Then the choice of elements k, s is unique, and these elements make* **A** *into a relative $\lambda\eta$-algebra.*

Proof. The uniqueness of k and s, and likewise conditions 2 and 3 in Definition 4.1.2, are immediate from extensionality. For condition 1, we show by induction on the derivation of $M =_{\beta\eta} M'$ that

$$M =_{\beta\eta} M' \;\Rightarrow\; \forall \nu.\, [\![M]\!]^*_\nu \;=\; [\![M']\!]^*_\nu \,.$$

For the β-rule itself, the fact that $[\![(\lambda x.M)N]\!]^*_\nu = [\![M[x \mapsto N]]\!]^*_\nu$ follows easily from $(M[x \mapsto N])^\dagger = M^\dagger[x \mapsto N^\dagger]$, which is clear by inspection of the definition of λ^*, together with the fact that $[\![\lambda x.M]\!]^*_\nu \cdot a = [\![M]\!]^*_{\nu(x \mapsto a)}$. For the η-rule, it is easy to check that if x is not free in M then $[\![\lambda x.Mx]\!]_\nu$ and $[\![M]\!]_\nu$ have the same extension and hence are equal. Now suppose inductively that $[\![M]\!]_{\nu(x \mapsto a)} = [\![M']\!]_{\nu(x \mapsto a)}$ for all ν, a; then we have $[\![\lambda x.M]\!]_\nu = [\![\lambda x.M']\!]_\nu$ for all ν by extensionality. The remaining two induction cases are trivial. □

There are also many non-extensional λ-algebras. For example, the van Oosten model B and the sequential procedure model SP^0 from Section 3.2 will turn out to be, respectively, a λ-algebra and a $\lambda\eta$-algebra (see Sections 12.1 and 6.1). In SP^0, elements of a type $A \Rightarrow B$ are not functions from A to B, but intensional objects (nested sequential procedures) representing such functions. Likewise, in B, elements of $\mathsf{B} \Rightarrow \mathsf{B} = \mathsf{B}$ are not themselves functions $\mathsf{B} \to \mathsf{B}$, but representations of decision trees for such functions. Thus, in each of these models, the λ-terms $\lambda xy.x+y$ and $\lambda xy.y+x$ will receive different interpretations (given a suitable valuation for $+$), because both models are sensitive to *evaluation order*. Even so, the β-rule is still valid for these models, so that they qualify as λ-algebras.

Intuitively, a higher-order model will be a λ-algebra provided it is not 'too intensional': that is, provided its modelling of algorithms, processes or programs is not so fine-grained that the β-rule breaks down. Examples of total higher-order models that are *not* λ-algebras include the syntactic model for untyped combinatory logic (that is, the set of untyped applicative expressions in k, s modulo the congruence generated by $kxy = x$ and $sxyz = xz(yz)$), as well as various syntactic and semantic models for the *lazy λ-calculus* as described in [4]. Furthermore, the highly intensional models K_1 and K_2 fail even to be 'partial λ-algebras', in that they do not validate any reasonable version of β-equality (see Sections 12.2 and 12.3).

Exercise 4.1.4. Show that if $\mathbf{A}$ is a λ-algebra, then so are the models $\mathsf{Sub}(\mathbf{A})$ and $\mathsf{Tot}(\mathbf{A}; B)$ for any base type B (see Subsection 3.6.2).

4.1.2 The Internal Interpretation

The above definition of a λ-algebra is a natural one if we are starting from the theory of TPCAs and the interpretation of λ^*-expressions. In other respects, however, this definition is rather unsatisfactory. Firstly, our definition of the λ^* operation involves some seemingly arbitrary ingredients, and one may worry that the notion of λ-algebra might be unduly sensitive to the details of this definition. Secondly, the interpretation $[\![M]\!]^*$ of an open term $\Gamma \vdash M : \sigma$ is given in a *pointwise* style—that is, separately for each valuation ν of Γ—with no account of how the meaning of M might be 'computed' from the values of the free variables x_i within the model itself.

Related to this, the interpretation is decidedly *non-compositional* in general: knowing $[\![M]\!]_{\nu(x^\sigma \mapsto a)}$ for all values of $a \in \mathbf{A}(\sigma)$ is not in general enough to determine the value of $[\![\lambda x^\sigma.M]\!]_\nu$.[2]

These deficiencies may be overcome by resorting to a more 'internal' style of interpretation: if $\Gamma = x_0^{\sigma_0}, \ldots, x_{r-1}^{\sigma_{r-1}}$, then rather than interpreting a term $\Gamma \vdash M : \sigma$ as (in effect) an extensional function $[\![M]\!]^*_\Gamma : \mathbf{A}(\sigma_0) \times \cdots \times \mathbf{A}(\sigma_{r-1}) \to \mathbf{A}(\sigma)$, we interpret it as a value $[\![M]\!]^i_\Gamma \in \mathbf{A}(\Gamma \to \sigma)$, where $\Gamma \to \sigma$ denotes $\sigma_0, \ldots, \sigma_{r-1} \to \sigma$. (We may as well work here in a way that does not require products.) Indeed, we may define such an interpretation simply by setting

$$[\![M]\!]^i_\Gamma \;=\; [\![\lambda x_0 \ldots x_{r-1}.M]\!]^* \,.$$

In general, the interpretation $[\![M]\!]^i_\Gamma$ may be more 'informative' than $[\![M]\!]^*_\Gamma$, since elements of $\mathbf{A}(\Gamma \to \sigma)$ may contain intensional information (such as evaluation order); it is this that enables us to recover compositionality.

The essential properties of such an internal interpretation may be captured as follows. We here write $\nu \wr \Gamma$ to mean that ν is a valuation for Γ in $\mathbf{A}$.

Definition 4.1.5. Suppose $\mathbf{A}$ is a relative total applicative structure as in Definition 3.1.17. An *internal interpretation* of λ-terms in $\mathbf{A}$ associates to each $\Gamma \vdash M : \sigma$ an element $[\![M]\!]^i_\Gamma \in \mathbf{A}^\sharp(\Gamma \to \sigma)$ in a compositional way (so that $[\![MN]\!]^i_\Gamma$ is determined by $[\![M]\!]^i_\Gamma$ and $[\![N]\!]^i_\Gamma$, and $[\![\lambda x^\sigma.M]\!]^i_\Gamma$ is determined by $[\![M]\!]^i_{\Gamma,x^\sigma}$), and such that the following conditions are satisfied. We here suppose $\Gamma = x_0, \ldots, x_{r-1}$, and if $\nu \wr \Gamma$ we write $[\![M]\!]_\nu$ for $[\![M]\!]^i_\Gamma \cdot \nu(x_0) \cdot \ldots \cdot \nu(x_{r-1})$.

1. $[\![x_i]\!]_\nu = \nu(x_i)$ for any $\nu \wr \Gamma$.
2. If $\Gamma \vdash M : \sigma \to \tau$ and $\Gamma \vdash N : \sigma$, then $[\![MN]\!]_\nu = [\![M]\!]_\nu \cdot [\![N]\!]_\nu$ for any $\nu \wr \Gamma$.
3. If $\Gamma, x \vdash M : \tau$, then $[\![\lambda x.M]\!]^i_\Gamma \;=\; [\![M]\!]^i_{\Gamma,x}$.
4. If $\Gamma \vdash M : \sigma$ and $\nu' \wr \Gamma'$ extends $\nu \wr \Gamma$ (possibly with some reordering of variables), then $[\![M]\!]_{\nu'} = [\![M]\!]_\nu$.
5. *β-rule*: $[\![(\lambda x.M)N]\!]^i_\Gamma \;=\; [\![M[x \mapsto N]]\!]^i_\Gamma$.

We write $[\![M]\!]^i$ for $[\![M]\!]^i_\Gamma$ when Γ is empty.

Theorem 4.1.6. *Let* $\mathbf{A}$ *be a relative total applicative structure.*

(i) If $\mathbf{A}$ *carries a relative λ-algebra structure as in Definition 4.1.2, the interpretation defined by* $[\![M]\!]^i_{x_0,\ldots,x_{r-1}} \;=\; [\![\lambda x_0 \ldots x_{r-1}.M]\!]^*$ *satisfies the conditions of Definition 4.1.5.*

(ii) Conversely, if $\mathbf{A}$ *admits an internal interpretation as above, then setting* $k = [\![\lambda xy.x]\!]^i$ *and* $s = [\![\lambda xyz.xz(yz)]\!]^i$ *makes* $\mathbf{A}$ *into a relative λ-algebra as in Definition 4.1.2.*

(iii) There is a canonical bijection between relative λ-algebra structures on $\mathbf{A}$ *and internal interpretations of λ-terms in* $\mathbf{A}$*. This restricts to a bijection between*

[2] In fact, the structures for which $[\![-]\!]_-$ is compositional in this sense are exactly the (typed) λ-*models*. Being a λ-model is stronger than being a λ-algebra, but weaker than being an extensional higher-order model. For more information on λ-models, see Barendregt [9].

relative $\lambda\eta$-algebra structures and internal interpretations validating the η-rule: $[\![\lambda x.Mx]\!]^i_\Gamma = [\![M]\!]^i_\Gamma$.

Proof. (i) Suppose **A** is a relative λ-algebra, and define $[\![-]\!]^i_-$ as above. To see that $[\![-]\!]^i_-$ is compositional, note that

$$\begin{aligned}[\![MN]\!]^i_\Gamma &= [\![\lambda\vec{x}.MN]\!]^* \\ &= [\![(\lambda fg.\lambda\vec{x}.(f\vec{x})(g\vec{x}))(\lambda\vec{x}.M)(\lambda\vec{x}.N)]\!]^* \\ &= [\![\lambda fg.\lambda\vec{x}.(f\vec{x})(g\vec{x})]\!]^* \cdot [\![\lambda\vec{x}.M]\!]^* \cdot [\![\lambda\vec{x}.N]\!]^* \\ &= [\![\lambda fg.\lambda\vec{x}.(f\vec{x})(g\vec{x})]\!]^* \cdot [\![M]\!]^i_\Gamma \cdot [\![N]\!]^i_\Gamma\end{aligned}$$

and that

$$[\![\lambda y^\sigma.M]\!]^i_\Gamma \;=\; [\![\lambda\vec{x}y.M]\!]^* \;=\; [\![M]\!]^i_{\Gamma,y^\sigma} \,.$$

This also establishes condition 3 of Definition 4.1.5. Conditions 1, 2 and 4 follow from the evident fact that

$$[\![M]\!]^i_\Gamma \cdot a_0 \cdot \ldots \cdot a_{r-1} \;=\; [\![M]\!]^*_{x_0\mapsto a_0,\ldots,x_{r-1}\mapsto a_{r-1}} \,,$$

while condition 5 holds because $M =_\beta M'$ implies $\lambda\vec{x}.M =_\beta \lambda\vec{x}.M'$.

(ii) Suppose $[\![-]\!]^i_-$ is an internal interpretation. Define k,s as above, and thence obtain $[\![-]\!]^*_-$ as in Subsection 4.1.1. We first check that $k = [\![\lambda xy.x]\!]^*$ and $s = [\![\lambda xyz.xz(yz)]\!]^*$. For the former, note that $(\lambda xy.x)^\dagger = s(kk)(skk)$, and both $[\![-]\!]^*$ and $[\![-]\!]^i$ satisfy $[\![MN]\!] = [\![M]\!] \cdot [\![N]\!]$ for closed M,N; hence $[\![\lambda xy.x]\!]^* = [\![s(kk)(skk)]\!] = [\![(s(kk)(skk))^\circ]\!]^i = [\![(\lambda xy.x)^{\dagger\circ}]\!]^i$, with $-^\circ$ as in Proposition 4.1.1. But $(\lambda xy.x)^{\dagger\circ} =_\beta \lambda xy.x$ as an instance of Proposition 4.1.1, whence $[\![(\lambda xy.x)^{\dagger\circ}]\!]^i = [\![\lambda xy.x]\!]^i$ by condition 5 and compositionality. Thus $k = [\![\lambda xy.x]\!]^*$. The argument for s is similar.

It follows that $[\![M]\!]^i = [\![M]\!]^*$ for all *closed* M, since by Proposition 4.1.1, any such M is β-equivalent to one built from the terms for k,s via application. So if M,M' are closed and β-equivalent then $[\![M]\!]^* = [\![M']\!]^*$. Condition 1 of Definition 4.1.2 now follows easily, since in general $[\![M]\!]^*_\nu = [\![\lambda\vec{x}.M]\!]^* \cdot \nu(x_0) \cdot \ldots \cdot \nu(x_{r-1})$.

(iii) The above argument already makes clear that the constructions of (i) and (ii) are mutually inverse. The clause regarding the η-rule is left as an easy exercise. □

In practice, Definition 4.1.5 sometimes offers the most convenient way of showing that a particular (non-extensional) structure is a λ-algebra: see for instance Theorem 6.1.18.

4.1.3 Categorical λ-Algebras

Both of the characterizations of λ-algebras considered so far have been 'syntactic' in that they have involved interpretations of a calculus of λ-expressions. We now turn to a more mathematical presentation of a λ-algebra as a *category* of a certain kind. Although the set-theoretic framework of Chapter 3 will be the dominant one in this book, a strong case can be made that for λ-algebras, it is really the associated category that is the essential mathematical object of interest, and this category will indeed provide the appropriate setting for much of the theory that follows.

Given a relative λ-algebra $\mathbf{A}$, let $\mathscr{C}(\mathbf{A})$ be the category whose objects are datatypes of $\mathbf{A}^{\circ}$, and whose morphisms $A \to B$ are elements $f \in (A \Rightarrow B)^{\sharp}$ with the property that $f = \lambda x.fx$. (From now on, we shall freely use the λ-calculus as an internal language, omitting semantic brackets and treating given elements of $\mathbf{A}$ as 'constants': for example, $\lambda x.fx$ refers, strictly speaking, to the element $[\![\lambda x.fx]\!]_{f \mapsto f}$.) Identity morphisms are given by terms $\lambda x.x$, while the composition of $f : A \to B$ and $g : B \to C$ is given by $\lambda y.g(fy)$. The unit and associativity laws for $\mathscr{C}(\mathbf{A})$ emerge as instances of β-equality:

- $\lambda x.fx \;=\; \lambda y.(\lambda x.x)(fy) \;=\; \lambda y.f((\lambda x.x)y)$.
- $\lambda y.h((\lambda z.g(fz))y) \;=\; \lambda y.h(g(fy)) \;=\; \lambda y.(\lambda z.h(gz))(fy)$.

The fact that we do obtain a category in this way is one of the pleasant features of λ-algebras. Moreover, the results of the previous subsection already suggest that elements of $(A \Rightarrow B)^{\sharp}$ are likely to yield a better theory than the corresponding functions $A \to B$. The condition $f = \lambda x.fx$ on morphisms is needed for the unit laws to work; it can also be motivated by the intuitive idea that we wish to view a morphism $A \to B$ as something like a 'term' of type B with a 'free variable' of type A, with composition corresponding to substitution. Notice that this condition does *not* amount to imposing the 'η-rule': for example, $\lambda z.f$ and $\lambda z.\lambda x.fx$ may still be distinguished as morphisms $C \to (A \Rightarrow B)$.

We call $\mathscr{C}(\mathbf{A})$ the *intensional category* of $\mathbf{A}$. It comes with a forgetful functor $\Delta_{\mathbf{A}} : \mathscr{C}(\mathbf{A}) \to \mathscr{S}et$ defined by $\Delta_{\mathbf{A}}(A) = A$, $\Delta_{\mathbf{A}}(f) = (\Lambda a.f \cdot a)$, and this is faithful if $\mathbf{A}$ is extensional. Note too that the image of $\Delta_{\mathbf{A}}$ in $\mathscr{S}et$ is exactly the computability model corresponding to $\mathbf{A}$.

In an extensional model, there is of course no difference between a morphism $A \to B$ in $\mathscr{C}(\mathbf{A})$ and a function $A \to B$ computable in $\mathbf{A}$. In non-extensional models, however, one should consider a morphism f in $\mathscr{C}(\mathbf{A})$ as an intensional operation or 'algorithm', and $\Delta_{\mathbf{A}}(f)$ as the function it computes. In the model SP^0, for instance, where every $f \in (A \Rightarrow B)$ satisfies $f = \lambda x.fx$, morphisms $A \to B$ are just nested sequential procedures of the appropriate type. In B, they are in effect forests of decision trees of the kind described in Subsection 3.2.4, mildly standardized by the imposition of the condition $f = \lambda x.fx$. We will see in Section 12.1 that this yields exactly the *irredundant* forests—those in which no query is ever repeated on any path—and that these correspond precisely to *sequential algorithms* of the appropriate type. In the meantime, we simply remark that the algorithms represented by $\lambda xy.x+y$ and $\lambda xy.y+x$ are still distinguishable as morphisms in $\mathscr{C}(\mathsf{B})$.

We now work towards an abstract characterization of those categories that arise as $\mathscr{C}(\mathbf{A})$ for some $\mathbf{A}$, along with the associated functors $\Delta_{\mathbf{A}}$. Such a characterization works most smoothly when suitable *finite products* are present, so we shall begin with this situation. We will then extend to the general case by exploiting the fact that any λ-algebra can be embedded in one with products.

For the case with finite products, we wish to establish a bijective correspondence between a certain class of λ-algebras and a certain class of categorical structures $(\mathscr{C},\Delta)$. For the former, the appropriate definition is the following; note that it strengthens the notion of standard products from Subsection 3.1.1:

Definition 4.1.7. A total relative λ-algebra $(\mathbf{A}^\circ;\mathbf{A}^\sharp)$ *has strong (finite) products* if

- $\mathbf{A}$ has a type $1 = 1^\sharp = \{i\}$ such that $\lambda x^1.x = \lambda x^1.i \in (1 \to 1)^\sharp$,
- for any types A,B there is a type $A \times B$ in $\mathbf{A}$ equipped with elements

  $$pair \in (A \Rightarrow B \Rightarrow A \times B)^\sharp\,, \qquad fst \in (A \times B \Rightarrow A)^\sharp\,, \qquad snd \in (A \times B \Rightarrow B)^\sharp$$

  such that

$$\begin{aligned}
\lambda xy.fst\,(pair\,x\,y) &= \lambda xy.x\,,\\
\lambda xy.snd\,(pair\,x\,y) &= \lambda xy.y\,,\\
\lambda z.pair\,(fst\,z)(snd\,z) &= \lambda z.z\,.
\end{aligned}$$

On the categorical side, we introduce the following notion.

Definition 4.1.8. A *categorical $\lambda\times$-algebra* is a category $\mathscr{C}$ with finite products equipped with the following data for all objects A,B:

- a choice of object $A \Rightarrow B$ along with a morphism $app_{AB} : (A \Rightarrow B) \times A \to B$,
- an operation assigning to each $f : C \times A \to B$ a morphism $\widehat{f} : C \to (A \Rightarrow B)$,

such that the following conditions are satisfied:

1. For any $f : C \times A \to B$ and any morphism $g : D \to C$, we have

$$\widehat{f \circ (g \times id_A)} = \widehat{f} \circ g \;:\; D \to (A \Rightarrow B)\,.$$

2. For any $f : C \times A \to B$, we have $app_{AB} \circ (\widehat{f} \times id_A) = f$.
3. In the case $C = 1$, the mapping $f \mapsto \widehat{f}$ yields a bijection between morphisms $A \to B$ and morphisms $1 \to (A \Rightarrow B)$.

We call $\mathscr{C}$ a *categorical $\lambda\eta\times$-algebra* if additionally we have:

4. For any $g : C \to (A \Rightarrow B)$, we have $\widehat{app_{AB} \circ (g \times id_A)} = g$.

We pause here to note that the latter of these notions coincides with something much more familiar:

Proposition 4.1.9. *$\mathscr{C}$ is a categorical $\lambda\eta\times$-algebra iff $\mathscr{C}$ is cartesian closed.*

Proof. We must show that the morphisms app_{AB} and mappings $f \mapsto \widehat{f}$ satisfy conditions 1–4 of Definition 4.1.8 iff for all $f : C \times A \to B$, $\widehat{f}$ is the *unique* morphism such that $app_{AB} \circ (\widehat{f} \times id_A) = f$. For the forwards implication, condition 2 says that $\widehat{f}$ satisfies this equation, and for the uniqueness we have that if $app \circ (h \times id) = f$

then $h = app \circ \widehat{(h \times id)} = \widehat{f}$ by condition 4. For the reverse implication, suppose $\widehat{f}$ uniquely satisfies the equation for every f; then conditions 2 and 3 are immediate. For condition 1, we have that

$$app_{AB} \circ ((\widehat{f} \circ g) \times id_A) \;=\; app_{AB} \circ (\widehat{f} \circ id_A) \circ (g \times id_A) \;=\; f \circ (g \times id_A)\,,$$

so that $\widehat{f} \circ g$ satisfies the defining property of $f \circ \widehat{(g \times id_A)}$. A similar calculation also establishes condition 4. □

Since we wish to consider categorical $\lambda\times$-algebras $\mathscr{C}$ in conjunction with functors $\mathscr{C} \to \mathscr{S}et$, the following abstract definition will also be needed:

Definition 4.1.10. Suppose $\mathscr{C}$ is any category with a terminal object. A functor $\Delta : \mathscr{C} \to \mathscr{S}et$ is called a *concretion functor* if Δ preserves finite products and acts faithfully on morphisms $1 \to A$ for each A.

We will show that there is a precise correspondence between relative λ-algebras with strong products and categorical $\lambda\times$-algebras. To this end, we first show that any categorical $\lambda\times$-algebra $\mathscr{C}$ admits a compositional interpretation of λ-terms. Suppose T is the type world consisting of objects of $\mathscr{C}$: let us write $[\![\sigma]\!]$ for σ considered as an object of $\mathscr{C}$, and $[\![\Gamma]\!]$ for $[\![\sigma_0]\!] \times \cdots \times [\![\sigma_{r-1}]\!]$ where $\Gamma = x_0^{\sigma_0}, \ldots, x_{r-1}^{\sigma_{r-1}}$. We now define an interpretation $[\![-]\!]^c_-$ assigning to each well-typed λ-term $\Gamma \vdash M : \tau$ over T a morphism $[\![M]\!]^c_\Gamma : [\![\Gamma]\!] \to [\![\tau]\!]$:

- $[\![x_i]\!]^c_\Gamma = \pi_i$, where $\Gamma = x_0, \ldots, x_{r-1}$.
- $[\![MN]\!]^c_\Gamma = app \circ \langle [\![M]\!]^c_\Gamma, [\![N]\!]^c_\Gamma \rangle$.
- $[\![\lambda x^\sigma . M]\!]^c_\Gamma = \widehat{[\![M]\!]^c_{\Gamma, x^\sigma}}$.

The proofs of the following facts are easy exercises in categorical reasoning, by induction on term structure:

Proposition 4.1.11. *Suppose $\mathscr{C}$ is a categorical $\lambda\times$-algebra with $[\![-]\!]^c_-$ the corresponding interpretation.*

(i) If $\Gamma' \vdash M : \tau$, every variable of Γ' appears in Γ, and $\pi_{\Gamma'} : [\![\Gamma]\!] \to [\![\Gamma']\!]$ is the evident projection, then $[\![M]\!]^c_{\Gamma'} = [\![M]\!]^c_\Gamma \circ \pi_{\Gamma'}$.

(ii) If $\Gamma \vdash N : \sigma$ and $\Gamma, x^\sigma \vdash M : \tau$, then $[\![M[x \mapsto N]]\!]^c_\Gamma = [\![M]\!]^c_{\Gamma,x} \circ \langle id_{[\![\Gamma]\!]}, [\![N]\!]^c_\Gamma \rangle$.

(iii) If $\Gamma \vdash M : \tau$, $\Gamma \vdash M' : \tau$ and $M =_\beta M'$, then $[\![M]\!]^c_\Gamma = [\![M']\!]^c_\Gamma$.

(iv) In the case of a $\lambda\eta\times$-algebra, (iii) holds with $=_{\beta\eta}$ in place of $=_\beta$. □

We are now ready to establish our correspondence result. We have already seen how a relative λ-algebra $\mathbf{A}$ gives rise to a category $\mathscr{C}(\mathbf{A})$ with a functor $\Delta_{\mathbf{A}} : \mathscr{C}(\mathbf{A}) \to \mathscr{S}et$. In the other direction, if Δ is a concretion functor on a categorical $\lambda\times$-algebra $\mathscr{C}$, we may define a relative applicative structure $\Delta[\mathscr{C}]$ as follows: $\Delta[\mathscr{C}]^\circ$ consists of the sets $\Delta(A)$ (indexed by objects A of $\mathscr{C}$) with $\Rightarrow$ and application induced from $\mathscr{C}$, and $\Delta[\mathscr{C}]^\sharp$ is the substructure consisting of elements of the form $\Delta(f)(*)$ for some $f : 1 \to A$.

Exercise 4.1.12. In the above situation, we also obtain a computability model $\mathbf{C}$ more directly as the image of Δ in $\mathscr{S}et$. Using condition 3 of Definition 4.1.8, show that $\mathbf{C}$ corresponds to $\Delta[\mathbf{C}]$ as in the proof of Theorem 3.1.18: that is, $|\mathbf{C}| = |\Delta[\mathbf{C}]^\circ|$, and $\mathbf{C}[A,B]$ consists of all functions $\Lambda a. f \cdot a$ where $f \in (A \Rightarrow B)^\sharp$.

The following theorem shows that the above constructions do what we wish:

Theorem 4.1.13. *(i) If* $\mathbf{A}$ *is a relative* λ*-algebra with strong products, then* $\mathscr{C}(\mathbf{A})$ *has the structure of a categorical* $\lambda\times$*-algebra, and* $\Delta_{\mathbf{A}}$ *is a concretion functor.*

(ii) Conversely, if $\mathscr{C}$ *is a categorical* $\lambda\times$*-algebra and* $\Delta : \mathscr{C} \to \mathscr{S}et$ *is any concretion functor, then* $\Delta[\mathscr{C}]$ *has the structure of a relative* λ*-algebra with strong products.*

(iii) These constructions constitute a bijection between relative λ*-algebras with strong products and categorical* $\lambda\times$*-algebras (modulo isomorphism) equipped with a concretion functor. Moreover,* $\mathbf{A}$ *is a relative* $\lambda\eta$*-algebra iff the corresponding* $\mathscr{C}$ *is a categorical* $\lambda\eta\times$*-algebra.*

Proof. (i) It is routine to check that strong products in $\mathbf{A}$ give rise to categorical finite products in $\mathscr{C}(\mathbf{A})$. For any A,B, the morphism $app_{AB} : (A \Rightarrow B) \times A \to B$ is given by $\lambda z.(fst\, z)(snd\, z)$, and for any $f : C \times A \to B$ we may take $\widehat{f} = \lambda y.\lambda x.f(pair\, y\, x)$.

The conditions of Definition 4.1.8 are now easy to verify. For condition 1, it suffices to check that

$$\widehat{f \circ (g \times id_A)} \;=_\beta\; \lambda y.\lambda x.f(pair(gy)x) \;=_\beta\; \widehat{f} \circ g\,,$$

while for condition 2 we note that

$$app \circ (\widehat{f} \times id) \;=_\beta\; \lambda z.f(pair(fst\, z)(snd\, z))\,,$$

which by the axioms for *pair* has the same denotation as $\lambda z.fz$. Condition 3 is readily verified using the axiom $\lambda x^1.x = \lambda x^1.i$. The definition of strong products also readily implies that $\Delta_{\mathbf{A}}$ is a concretion functor on $\mathscr{C}(\mathbf{A})$.

(ii) Given a categorical $\lambda\times$-algebra $\mathscr{C}$ and a concretion functor $\Delta : \mathscr{C} \to \mathscr{S}et$, we may define an internal interpretation of λ-terms in $\Delta[\mathscr{C}]$ by

$$[\![M]\!]^i_{x_0,\ldots,x_{r-1}} \;=\; \Delta[\![\lambda x_0 \ldots x_{r-1}.M]\!]^c(i)\,.$$

Conditions 1–3 of Definition 4.1.5 are immediate from the definition of $[\![-]\!]^c_-$; conditions 4 and 5 follow easily from Proposition 4.1.11.

(iii) It is immediate by construction that $\Delta_{\mathbf{A}}[\mathscr{C}(\mathbf{A})] = \mathbf{A}$. For the converse, note that a morphism $A \to B$ in $\mathscr{C}(\Delta[\mathscr{C}])$ is by definition the image under Δ of some morphism $1 \to (A \Rightarrow B)$ in $\mathscr{C}$. But Δ is faithful on such morphisms, which in turn correspond bijectively to morphisms $A \to B$ in $\mathscr{C}$. This shows that $\mathscr{C}(\Delta[\mathscr{C}]) \cong \mathscr{C}$, and the same considerations also yield that $\Delta_{\Delta[\mathscr{C}]} = \Delta$. We leave the clause on $\lambda\eta$-algebras as an exercise. □

The intuition, then, is that a categorical $\lambda\times$-algebra $\mathscr{C}$ can be seen as a kind of abstract λ-algebra which may be incarnated as various concrete relative λ-algebras

$\mathbf{A}$, all sharing the same substructure $\mathbf{A}^\sharp$. A significant point here (also emphasized by Cockett and Hofstra [56]) is that $\mathscr{C}$ is really the place where β-equality is satisfied, and the validity of $=_\beta$ in various concretions of $\mathscr{C}$ is simply an external manifestation of this. Note that $\mathrm{Hom}(1,-)$ is always an example of a concretion functor; such concretions correspond to *full* λ-algebras $\mathbf{A}$ (those in which $\mathbf{A}^\sharp = \mathbf{A}^\circ$).

It remains to extend our correspondence to λ-algebras without products. If $\mathscr{D}$ is a categorical $\lambda\times$-algebra, by a $\Rightarrow$*-subcategory* of $\mathscr{D}$ we shall mean a full subcategory $\mathscr{C}$ of $\mathscr{D}$ whose objects are closed under $\Rightarrow$. In this situation, any concretion functor Δ' on $\mathscr{D}$ will then readily induce a certain structure on $\mathscr{C}$ essentially making it into a 'λ-algebra'. The structure on $\mathscr{C}$ that we shall present here will be somewhat ad hoc and not of mathematical significance in itself, but it nevertheless serves as a stepping-stone towards other important results below.

Definition 4.1.14. Suppose $\mathscr{C}$ is a category equipped with a functor $\Delta : \mathscr{C} \to \mathscr{S}et$ plus the following additional structure:

- an operation $\Rightarrow$ on the objects of $\mathscr{C}$,
- for each A, B in $\mathscr{C}$, an application operation $\cdot_{AB} : \Delta(A \Rightarrow B) \times \Delta(A) \to \Delta(B)$,
- for each A, B, C in $\mathscr{C}$, a choice of elements $k_{AB} \in \Delta(A \Rightarrow B \Rightarrow A)$ and $s_{ABC} \in \Delta((A \Rightarrow B \Rightarrow C) \Rightarrow (A \Rightarrow B) \Rightarrow (A \Rightarrow C))$.

(i) We say $(\mathscr{C}, \Delta)$ (endowed with the above structure) is a *concrete categorical λ-algebra* if there is some categorical $\lambda\times$-algebra $\mathscr{D}$ with concretion Δ' such that $\mathscr{C}$ is a $\Rightarrow$-subcategory of $\mathscr{D}$ and $\Delta, \Rightarrow, \cdot, k, s$ are induced by $\mathscr{D}$ and Δ' in the obvious way (defining k_{AB}, s_{ABC} via the usual λ-terms).

(ii) In this situation, if $\mathscr{D}$ may moreover be taken to be a categorical $\lambda\eta\times$-algebra, we say $(\mathscr{C}, \Delta)$ (with the above structure) is a *concrete categorical $\lambda\eta$-algebra.*

Note that it does not really make sense to consider 'categorical λ-algebras' in isolation from their concretions: only when all the above structure is present is it possible to specify an interpretation of λ-terms in $\mathscr{C}$. Note also that the $\lambda\times$-algebra $\mathscr{D}$ in Definition 4.1.14 will not be unique.

In any case, from any concrete categorical λ-algebra $(\mathscr{C}, \Delta, \Rightarrow, \cdot, k, s)$, we may obtain a relative applicative structure $\Delta[\mathscr{C}]$ as follows. First, let $\Delta[\mathscr{C}]^\circ$ be the family of sets $\Delta(A)$ indexed by objects $A \in \mathscr{C}$ and endowed with the given $\Rightarrow$ and $\cdot$. To define the substructure $\Delta[\mathscr{C}]^\sharp$ in the absence of a terminal object, we may fix an object $B \in \mathscr{C}$, and let $\Delta[\mathscr{C}]^\sharp$ consist of (say) all elements of the form $\Delta(f)(k_{BB})$ for some $f : (B \Rightarrow B \Rightarrow B) \to A$. Clearly, the same substructure would result from using any other 'computable' element in place of k_{BB} here (cf. the discussion of weak terminals in Subsection 3.1.4). To complete the data required for a relative λ-algebra, we endow $\Delta[\mathscr{C}]$ with the given choice of elements $k_{AB}, s_{ABC} \in \Delta[\mathscr{C}]^\sharp$.

Once again, it is easy to see at this stage that $\Delta[\mathscr{C}]$, viewed as a computability model, is nothing other than the image of Δ in $\mathscr{S}et$.

Theorem 4.1.15. *(i) If $\mathbf{A}$ is any relative λ-algebra, then $(\mathscr{C}(\mathbf{A}), \Delta_{\mathbf{A}})$ carries the structure of a concrete categorical λ-algebra.*

(ii) Conversely, if $(\mathscr{C},\Delta)$ is a concrete categorical λ-algebra, then $\Delta[\mathscr{C}]$ is a relative λ-algebra.

(iii) These constructions constitute a bijection between relative λ-algebras and concrete categorical λ-algebras (modulo isomorphism). Moreover, **A** *is a relative $\lambda\eta$-algebra iff $(\mathscr{C}(\mathbf{A}),\Delta_{\mathbf{A}})$ is a concrete categorical $\lambda\eta$-algebra.*

Proof. (i) Let $\mathbf{A}^\times$ be the relative TPCA obtained from **A** using the product completion construction from the proof of Theorem 3.1.15, via the bijection of Theorem 3.1.18. Rather than showing directly that $\mathbf{A}^\times$ is a relative λ-algebra, we shall use it to build a categorical $\lambda\times$-algebra $\mathscr{C}^\times$ within which $\mathscr{C}(\mathbf{A})$ sits as a $\Rightarrow$-subcategory.

We define our category $\mathscr{C}^\times$ by an analogue of the $\mathscr{C}(-)$ construction: objects are datatypes of $\mathbf{A}^\times$, and morphisms $D \to E$ are tuples $(f_0,\ldots,f_{n-1}) \in (D \to E)^\sharp$ where $f_j = \lambda x_0 \ldots x_{m-1}.f_j x_0 \ldots x_{m-1}$ for each j. The following steps may now be routinely performed by writing suitable vectors of λ-terms and checking the relevant properties via calculations in **A**:

- Define identities and composition in $\mathscr{C}^\times$, and check that $\mathscr{C}^\times$ is a category.
- Check that $\mathscr{C}^\times$ has finite products (the terminal object is the empty product).
- Define suitable morphisms *app* and mappings $f \mapsto \widehat{f}$, and check that they satisfy conditions 1 and 2 of Definition 4.1.8 (condition 3 is immediate by construction).

This shows that $\mathscr{C}^\times$ is a categorical $\lambda\times$-algebra. It is now clear that $\mathscr{C}(\mathbf{A})$ is a $\Rightarrow$-subcategory of $\mathscr{C}^\times$ and that $\Delta_{\mathbf{A}}$ is the restriction of the evident concretion functor on $\mathscr{C}^\times$, so the structure of a concrete categorical λ-algebra is induced on $(\mathscr{C}(\mathbf{A}),\Delta_{\mathbf{A}})$.

(ii) Suppose the structure on $\mathscr{C}$ is induced by some $\lambda\times$-algebra $\mathscr{D}$ with concretion Δ' as in Definition 4.1.14. Clearly $\Delta[\mathscr{C}]$ is a full substructure of $\Delta'[\mathscr{D}]$: that is, $\Delta[\mathscr{C}]^\circ$ and $\Delta[\mathscr{C}]^\sharp$ are the restrictions of $\Delta[\mathscr{C}]^\circ$ and $\Delta[\mathscr{C}]^\sharp$ respectively to the types of $\mathscr{C}$. But $\Delta'[\mathscr{D}]$ is a relative λ-algebra by Theorem 4.1.15(ii), and so admits an internal interpretation of λ-terms as in Definition 4.1.5. Since the objects of $\mathscr{C}$ are closed under $\Rightarrow$, this restricts to an internal interpretation in $\Delta[\mathscr{C}]$, and this shows that $\Delta[\mathscr{C}]$ is a relative λ-algebra.

(iii) As before, it is immediate by construction that $\Delta_{\mathbf{A}}[\mathscr{C}(\mathbf{A})] = \mathbf{A}$. For the other direction, suppose $\mathscr{D}$ and Δ' are as above; then it is easy to check that the isomorphism $\mathscr{C}(\Delta'[\mathscr{D}]) \cong \mathscr{D}$ established by the proof of Theorem 4.1.15(ii) restricts to an isomorphism $\mathscr{C}(\Delta[\mathscr{C}]) \cong \mathscr{C}$. The clause concerning $\lambda\eta$-algebras is also easy. □

4.1.4 The Karoubi Envelope

Even when a categorical λ-algebra $\mathscr{C}$ contains only a limited collection of datatypes explicitly, it is often the case that many other datatypes are already implicitly present in $\mathscr{C}$ in a 'coded' form. In this section we present a simple construction for inflating $\mathscr{C}$ to a λ-algebra $\mathscr{K}(\mathscr{C})$ that renders all these datatypes explicit.

The $\mathscr{K}$ construction will play both a conceptual and a technical role in our theory. In Subsection 4.1.5 we shall use it to establish a connection between λ-algebras and cartesian closed categories. We shall also argue that it is reasonable to think of

$\mathscr{C}$ and $\mathscr{D}$ as 'equivalent' λ-algebras iff $\mathscr{K}(\mathscr{C}) \simeq \mathscr{K}(\mathscr{D})$. Later, in Section 4.2, we shall explore the rich repertoire of datatypes that can be obtained via this deceptively simple construction.

In any category $\mathscr{C}$ whatever, we say an object B is a *retract* of an object A if we have morphisms $t : B \to A$ and $r : A \to B$ such that $r \circ t = id_B$. In this setting we shall write $(t,r) : B \lhd A$, and call t a *section* and r a *retraction*; the pair (t,r) will be called a *retraction pair*. The intuition is that these give a way of 'coding up' B inside A: any morphism $f : C \to B$ may be represented by, and indeed recovered from, the morphism $t \circ f : C \to A$, while any morphism $g : B \to D$ may be recovered from $g \circ r : A \to D$.

Note also that if $(t,r) : B \lhd A$ then the morphism $e = t \circ r$ is an *idempotent* on A: that is, $e \circ e = e$. Conversely, given any idempotent $e : A \to A$, we may notionally imagine an object B (which may or may not be actually present in $\mathscr{C}$) playing the role of the 'image' of e, so that the morphism e splits as $A \xrightarrow{r} B \xrightarrow{t} A$ where $r \circ t = id_B$. The idea behind the following construction is that it completes any given category by formally adding in such idempotent splittings.

Definition 4.1.16. The *Karoubi envelope* of a category $\mathscr{C}$ is the category $\mathscr{K}(\mathscr{C})$ constructed as follows:

- Objects are pairs (A,e), where A is an object of $\mathscr{C}$ and e is an *idempotent* on A.
- A morphism $(A,e) \to (B,d)$ is a morphism $f : A \to B$ of $\mathscr{C}$ such that $d \circ f \circ e = f$.
- The identity on (A,e) is e itself, and the composition of $f : (A,e) \to (B,d)$ with $g : (B,d) \to (C,c)$ is simply their composition $g \circ f$ in $\mathscr{C}$.

Note here that $g \circ f$ is indeed a morphism in $\mathscr{K}(\mathscr{C})$, since

$$c \circ (g \circ f) \circ e \;=\; c \circ g \circ (d \circ f \circ e) \circ e \;=\; (c \circ g \circ d) \circ (d \circ f \circ e) \;=\; g \circ f\,.$$

It is easy to see that composition in $\mathscr{K}(\mathscr{C})$ satisfies the unit laws, and associativity is inherited from $\mathscr{C}$.

Clearly, we have a full embedding $I : \mathscr{C} \to \mathscr{K}(\mathscr{C})$ given by $I(A) = (A, id_A)$ and $I(f) = f$. Moreover, this embedding enjoys an important universal property. Let us say a category $\mathscr{D}$ *has idempotent splittings* if for each idempotent $e : A \to A$ in $\mathscr{D}$ we are given an object A_e along with morphisms $r : A \to A_e$, $t : A_e \to A$ in $\mathscr{D}$ such that $t \circ r = e$ and $r \circ t = id_{A_e}$. The following proposition says that $\mathscr{K}(\mathscr{C})$ represents the 'universal' way of splitting idempotents in $\mathscr{C}$:

Proposition 4.1.17. *(i) $\mathscr{K}(\mathscr{C})$ has idempotent splittings.*

(ii) If $\mathscr{D}$ has idempotent splittings, then any functor $F : \mathscr{C} \to \mathscr{D}$ can be extended, uniquely up to natural isomorphism, to a functor $\widehat{F} : \mathscr{K}(\mathscr{C}) \to \mathscr{D}$ with $I \circ \widehat{F} \cong F$.

(iii) Likewise, given $F, G : \mathscr{C} \to \mathscr{D}$ where $\mathscr{D}$ has idempotent splittings, a natural transformation $\alpha : F \to G$ extends uniquely to a natural transformation $\widehat{\alpha} : \widehat{F} \to \widehat{G}$. Indeed, the construction $\widehat{-}$ yields a functor from the functor category $[\mathscr{C}, \mathscr{K}(\mathscr{D})]$ to $[\mathscr{K}(\mathscr{C}), \mathscr{K}(\mathscr{D})]$.

Proof. (i) We first show that $\mathscr{K}(\mathscr{C})$ has idempotent splittings. Given any morphism $d : (A,e) \to (A,e)$ in $\mathscr{K}(\mathscr{C})$ such that $d \circ d = d$, we have

$$d = d \circ d = (e \circ d \circ e) \circ (e \circ d \circ e) = e \circ d \circ e \circ d \circ e = d \circ d \circ e = d \circ e,$$

and likewise $d = e \circ d$. Thus (A,d) is an object in $\mathcal{K}(\mathcal{C})$, and d is both a morphism $d_0 : (A,e) \to (A,d)$ and a morphism $d_1 : (A,d) \to (A,e)$, where $d_0 \circ d_1 = id_{(A,d)}$ and $d_1 \circ d_0 = d : (A,e) \to (A,e)$.

(ii) Given $F : \mathcal{C} \to \mathcal{D}$ where $\mathcal{D}$ has idempotent splittings, define $\widehat{F}$ as follows. For each object (A,e) in $\mathcal{K}(\mathcal{C})$, let (A_e, r_e, t_e) be the designated splitting for the idempotent $Fe : FA \to FA$ in $\mathcal{D}$, and set $\widehat{F}(A,e) = A_e$. For each morphism $f : (A,e) \to (B,d)$ in $\mathcal{K}(\mathcal{C})$, let (A_e, r_e, t_e) and (B_d, r_d, t_d) be the respective idempotent splittings, and set $\widehat{F}(f) = r_d \circ f \circ t_e$. It is easy to check that $\widehat{F}$ is a functor and that $I \circ \widehat{F} \cong F$, where the natural isomorphisms $F \to I \circ \widehat{F}$ and $I \circ \widehat{F} \to F$ are constituted by the morphisms r_{id_A} and t_{id_A} respectively. Moreover, if $G : \mathcal{K}(\mathcal{C}) \to \mathcal{D}$ is any functor whatever with $G \circ I \cong F$, then applying G to the morphisms $(A, id_A) \xrightarrow{e} (A,e) \xrightarrow{e} (A, id_A)$ we obtain an idempotent splitting $G(A, id_A) \to A'_e \to G(A, id_A)$, which is clearly isomorphic to the given splitting $FA \xrightarrow{r_e} A_e \xrightarrow{t_e} FA$ for Fe in $\mathcal{D}$. Naturality of the isomorphisms is now straightforward.

(iii) Suppose given $\alpha : F \to G$ and take $\widehat{F}, \widehat{G}$ as above. For each (A,e) in $\mathcal{K}(\mathcal{C})$, let (A_e, r_e, t_e) and (A'_e, r'_e, t'_e) be the given splittings of Fe and Ge respectively, and define $\widehat{\alpha}_{(A,e)} = r'_e \circ \alpha_A \circ t_e$. It is easy to see that these constitute the unique natural transformation $\widehat{\alpha}$ with the required properties, and moreover that the passage from α to $\widehat{\alpha}$ is functorial. □

Exercise 4.1.18. Show that $\widehat{F}$ is faithful if F is, and $\widehat{F}$ is full if F is. Thus, any category with idempotent splittings that contains $\mathcal{C}$ as a [full] subcategory will also contain $\mathcal{K}(\mathcal{C})$ as a [full] subcategory.

As a special case of Proposition 4.1.17, since $\mathcal{S}et$ clearly has idempotent splittings, any functor $\Delta : \mathcal{C} \to \mathcal{S}et$ extends uniquely to a functor $\widehat{\Delta} : \mathcal{K}(\mathcal{C}) \to \mathcal{S}et$.

Exercise 4.1.19. Recall that the image $\Delta(\mathcal{C})$ of any functor $\Delta : \mathcal{C} \to \mathcal{S}et$ may be viewed as a total computability model.

(i) Show that $\Delta(\mathcal{C})$ and $\widehat{\Delta}(\mathcal{K}(\mathcal{C}))$ are *equivalent* computability models in the sense of Definition 3.4.4.

(ii) Show that there is a canonical full functor $\mathcal{K}(\mathcal{C}) \to \mathcal{P}er(\Delta(\mathcal{C}))$, which is faithful when Δ is.

We now consider what may be said about the Karoubi envelopes of categorical λ-algebras. Notice that the $\mathcal{K}$ construction 'improves' the structure by bringing in the η-rule:

Theorem 4.1.20. *(i) If $\mathcal{C}$ is a categorical $\lambda\times$-algebra, then $\mathcal{K}(\mathcal{C})$ is a categorical $\lambda\eta\times$-algebra, and if Δ is a concretion functor on $\mathcal{C}$ then $\widehat{\Delta}$ is one on $\mathcal{K}(\mathcal{C})$.*

(ii) If $(\mathcal{C}, \Delta)$ carries a concrete categorical λ-algebra structure, then $(\mathcal{K}(\mathcal{C}), \widehat{\Delta})$ carries a concrete categorical $\lambda\eta$-algebra structure.

Proof. (i) Since $\mathcal{C}$ has finite products, it is easy to see that $\mathcal{K}(\mathcal{C})$ does too: its terminal object is $(1, id_1)$, and the product of (A,e) and (B,d) is $(A \times B, e \times d)$. To endow $\mathcal{K}(\mathcal{C})$ with a categorical $\lambda\eta\times$-algebra structure, take

$$(A,e) \Rightarrow (B,d) \;=\; (A \Rightarrow B, \lambda fx.d(f(ex)))$$

(using the interpretation $[\![-]\!]^c$ of λ-terms in $\mathscr{C}$), then define

$$app_{(A,e)(B,d)} \;=\; \lambda z.d((fst\,z)(e(snd\,z))) \;:\; ((A,e) \Rightarrow (B,d)) \times (A,e) \to (B,d)\,,$$

and for $f : (C,c) \times (A,e) \to (B,d)$, define

$$\widehat{f} \;=\; \lambda x.\lambda y.f(pair\,x\,y) \;:\; (C,c) \to ((A,e) \Rightarrow (B,d))\,.$$

It is easy to check that these are indeed morphisms in $\mathscr{K}(\mathscr{C})$, and that they satisfy conditions 1–3 of Definition 4.1.8 by virtue of the corresponding conditions in $\mathscr{C}$. For condition 4, suppose $g : (C,c) \to ((A,e) \Rightarrow (B,d))$; then $app \circ (g \times id) = \lambda z.d(g(fst\,z))(e(snd\,z))$, whence

$$\widehat{app \circ (g \times id)} \;=\; \lambda xy.d(gx)(ey) \;=\; (\lambda fy.d(f(ey))) \circ g \;=\; g\,.$$

Now suppose $\Delta : \mathscr{C} \to \mathscr{S}et$ is a concretion functor. As noted above, Δ extends to a functor $\widehat{\Delta} : \mathscr{K}(\mathscr{C}) \to \mathscr{S}et$, and clearly $\widehat{\Delta}$ is still a concretion functor.

(ii) Given $(\mathscr{C},\Delta)$ a concrete categorical λ-algebra, let $(\mathscr{D},\Delta')$ be some categorical $\lambda\times$-algebra with concretion in which it embeds. By part (i), $\mathscr{K}(\mathscr{D})$ is then a $\lambda\eta\times$-algebra. But $\mathscr{C}$ is a full subcategory of $\mathscr{D}$, so $\mathscr{K}(\mathscr{C})$ is a full subcategory of $\mathscr{K}(\mathscr{D})$ by Exercise 4.1.18, and the inclusion clearly preserves the $\Rightarrow$ structure. Thus $\mathscr{K}(\mathscr{C})$ is a $\Rightarrow$-subcategory of $\mathscr{K}(\mathscr{D})$. Moreover, $\widehat{\Delta'}$ is a concretion functor on $\mathscr{K}(\mathscr{D})$ extending $\widehat{\Delta}$, so $(\mathscr{K}(\mathscr{C}),\widehat{\Delta})$ carries the structure of a categorical $\lambda\eta$-algebra with completion. □

Exercise 4.1.21. Show that if $(\mathscr{C},\Delta)$ is already a concrete categorical $\lambda\eta$-algebra then the inclusion $I : \mathscr{C} \to \mathscr{K}(\mathscr{C})$ preserves the $\Rightarrow$ structure.

Translating this to the setting of concrete λ-algebras, to any relative λ-algebra **A** we may associate a relative $\lambda\eta$-algebra $\mathbf{K}(\mathbf{A})$ by first passing to the concrete categorical λ-algebra $(\mathbf{C}(\mathbf{A}),\Delta_{\mathbf{A}})$, taking the Karoubi envelope $(\mathscr{K}(\mathbf{C}(\mathbf{A})),\widehat{\Delta_{\mathbf{A}}})$, and transforming this back to a concrete model $\widehat{\Delta_{\mathbf{A}}}[\mathscr{K}(\mathbf{C}(\mathbf{A}))]$ as in Theorem 4.1.15. This is what we shall mean when we refer to the *Karoubi envelope* of a (set-theoretic) relative λ-algebra **A**.

Remark 4.1.22. It is worth observing that if **C** is the computability model corresponding to a concrete λ-algebra **A** (viewed as a category of sets and functions), then the Karoubi envelope of **C** need *not* yield a λ-algebra or even a higher-order model in general, because there may be extensional idempotents $e \in \mathbf{A}[A,A]$ that are not matched by any intensional idempotents in $\mathbf{A}^\sharp(A \Rightarrow A)$. This is one illustration of the importance of the categorical perspective.

Insofar as $\mathbf{K}(\mathbf{A})$ plays the role of a canonical 'completion' of **A**, the following definition naturally suggests itself:

Definition 4.1.23. We say $\mathbf{A}, \mathbf{B}$ are *equivalent as (relative) λ-algebras* if $\mathbf{K}(\mathbf{A})$, $\mathbf{K}(\mathbf{B})$ are equivalent as concrete categorical $\lambda\eta$-algebras, i.e. if $(\mathscr{K}(\mathbf{C}(\mathbf{A})), \widehat{\Delta_{\mathbf{A}}}) \simeq (\mathscr{K}(\mathbf{C}(\mathbf{B})), \widehat{\Delta_{\mathbf{B}}})$.

Thus, $\mathbf{A}$ and $\mathbf{B}$ are equivalent when every type in either of them is implicitly present within the other. Intuitively, we can view $\mathbf{K}(\mathbf{A})$ as abstracting away from merely presentational aspects of $\mathbf{A}$ to yield the essential underlying structure. The fact that $\mathbf{A}$ is equivalent to $\mathbf{K}(\mathbf{A})$ says, in a sense, that little of interest is lost if we restrict attention to $\lambda\eta$-algebras rather than λ-algebras.

The above notion of λ-algebra equivalence is strictly stronger than the TPCA equivalence of Definition 3.4.4. In Subsection 12.1.7 we shall mention some models that are equivalent as TPCAs, but not as λ-algebras.

4.1.5 Retraction Subalgebras

So far, we have characterized λ- and $\lambda\eta$-algebras categorically in terms of subalgebras of $\lambda\times$- and $\lambda\eta\times$-algebras respectively. Whilst the notion of $\lambda\eta\times$-algebra coincides with the mathematically clean notion of cartesian closed category (Proposition 4.1.9), our definition of $\lambda\times$-algebras is not as simple as one might hope for. However, we are now in a position to improve on this, exploiting the fact that every $\lambda\times$-algebra can be embedded in a $\lambda\eta\times$-algebra, namely its Karoubi envelope. This will lead to our last and most abstract characterization of λ-algebras. We will also spell out explicitly what this means in the untyped setting, which is where this characterization will mostly be used in practice.

Let us reconsider the relationship between $\mathscr{C}$ and $\mathscr{K}(\mathscr{C})$ in the setting of Theorem 4.1.20. By inspection of the definition of $\Rightarrow$ within $\mathscr{K}(\mathscr{C})$, it is clear that for any objects A, B in $\mathscr{C}$ we have $I(A) \Rightarrow I(B) \lhd I(A \Rightarrow B)$ in $\mathscr{K}(\mathscr{C})$; moreover, this retraction pair is an isomorphism if $\mathscr{C}$ is already a $\lambda\eta$-algebra. This suggests the following definition. By a *$\Rightarrow$-category* we will mean simply a category with a binary operation $\Rightarrow$ on its objects.

Definition 4.1.24. Suppose $\mathscr{E}$ is a $\Rightarrow$-category. A *retraction subalgebra* of $\mathscr{E}$ consists of a $\Rightarrow$-category $\mathscr{C}$ with a full and faithful functor $J : \mathscr{C} \to \mathscr{E}$, and for each A, B in $\mathscr{C}$ a retraction

$$(t_{AB}, r_{AB}) \;:\; J(A) \Rightarrow_{\mathscr{E}} J(B) \;\lhd\; J(A \Rightarrow_{\mathscr{C}} B)\,.$$

The foregoing remarks show that every $\lambda\times$-algebra $\mathscr{C}$ is a retraction subalgebra of $\mathscr{K}(\mathscr{C})$; hence so is every $\Rightarrow$-subcategory of $(\mathscr{C}, \Delta)$.

Proposition 4.1.25. *Suppose $(\mathscr{E}, \Delta_{\mathscr{E}})$ carries the structure of a concrete categorical λ-algebra. Then any retraction subalgebra $\mathscr{C}$ of $\mathscr{E}$ also inherits such a structure.*

Proof. In the light of Definition 4.1.14, we may assume without loss of generality that $\mathscr{E}$ is a $\lambda\times$-algebra. Suppose J is the embedding of $\mathscr{C}$ in $\mathscr{E}$, with t_{AB}, r_{AB} as in Definition 4.1.24, and define $\Delta_{\mathscr{C}} = \Delta_{\mathscr{E}} \circ J : \mathscr{C} \to \mathscr{S}et$. To make $(\mathscr{C}, \Delta_{\mathscr{C}})$ into a concrete categorical λ-algebra, we must exhibit a categorical $\lambda\times$-algebra in which this structure embeds.

In the case where $\mathscr{C}$ has finite products and J preserves them, it is easy to see that $\mathscr{C}$ itself is already a $\lambda\times$-algebra: we may define

$$app^{\mathscr{C}}_{AB} = J^{-1}(app^{\mathscr{E}}_{J(A)J(B)} \circ (r_{AB} \times id_{J(A)}))\,,$$
$$\widehat{f} = J^{-1}(t_{AB} \circ \widehat{Jf})\,.$$

Conditions 1–3 of Definition 4.1.8 follow easily from the corresponding conditions for $\mathscr{E}$ together with the equation $r \circ t = id$. It is also easy to see that $\Delta_{\mathscr{C}}$ is a concretion functor, so in particular, $(\mathscr{C}, \Delta_{\mathscr{C}})$ is a concrete categorical λ-algebra.

For the general case, we may form a 'finite product closure' $\mathscr{C}^{\times}$ of $\mathscr{C}$ in the following way. Let the objects of $\mathscr{C}^{\times}$ be finite tuples of objects from $\mathscr{C}$, and let the morphisms $(A_0, \ldots, A_{m-1}) \to (B_0, \ldots, B_{n-1})$ be precisely the morphisms

$$J(A_0) \times \cdots \times J(A_{m-1}) \;\to\; J(B_0) \times \cdots \times J(B_{n-1})$$

in $\mathscr{E}$. Clearly $\mathscr{C}^{\times}$ is a category with finite products, and identifying $\mathscr{C}$ with a full subcategory of $\mathscr{C}^{\times}$, the functor J clearly extends to a finite product preserving functor $J^{\times} : \mathscr{C}^{\times} \to \mathscr{E}$.

We may also define an operation $\Rightarrow_{\mathscr{C}^{\times}}$ on the objects of $\mathscr{C}^{\times}$:

$$(A_0, \ldots, A_{m-1}) \Rightarrow_{\mathscr{C}^{\times}} (B_0, \ldots, B_{n-1}) \;=\; (C_0, \ldots, C_{n-1})\,,$$
$$\text{where } C_j = A_0 \Rightarrow_{\mathscr{C}} \cdots \Rightarrow_{\mathscr{C}} A_{m-1} \Rightarrow_{\mathscr{C}} B_j\,.$$

We now claim that $J^{\times}(\vec{A}) \Rightarrow_{\mathscr{E}} J^{\times}(\vec{B}) \lhd J^{\times}(\vec{A} \Rightarrow_{\mathscr{C}^{\times}} \vec{B})$. Indeed, this follows easily from the fact that $J(A) \Rightarrow_{\mathscr{E}} J(B) \lhd J(A \Rightarrow_{\mathscr{C}} B)$ for any A, B along with two readily checked properties of $\Rightarrow_{\mathscr{E}}$:

$$A \Rightarrow (B \times B') \;\cong\; (A \Rightarrow B) \times (A \Rightarrow B')\,, \qquad (A \times A') \Rightarrow B \;\lhd\; A \Rightarrow A' \Rightarrow B\,.$$

The previous argument now applies, and we may conclude that $\mathscr{C}^{\times}$ carries a $\lambda\times$-algebra structure. Since $\mathscr{C}$ is clearly a $\Rightarrow$-subcategory of $\mathscr{C}^{\times}$, we have by Theorem 4.1.15 that $\mathscr{C}$ inherits the structure of a categorical λ-algebra with completion; again, this is easily shown to agree with the structure specified above. □

Exercise 4.1.26. Show explicitly how $(\mathscr{C}, \Delta_{\mathscr{C}})$ may be directly endowed with the data of Definition 4.1.14 using J, r_{AB}, t_{AB} and the relevant structure in $\mathscr{E}$. Check that this structure agrees with that obtained via the above proof.

We may now give our cleanest mathematical characterization of λ-algebras:

Theorem 4.1.27. *A $\Rightarrow$-category admits the structure of a concrete categorical λ-algebra iff it arises as a retraction subalgebra of a cartesian closed category.*

Proof. If $\mathscr{C}$ is a retraction subalgebra of a cartesian closed category $\mathscr{E}$, then any concretion functor on $\mathscr{E}$ (the functor $\mathrm{Hom}(1, -)$ is always one such) makes $\mathscr{E}$ into a concrete categorical λ-algebra, so $\mathscr{C}$ also inherits such a structure by Proposition 4.1.25. Conversely, any concrete categorical λ-algebra structure on $\mathscr{C}$ is inherited from some categorical $\lambda\times$-algebra $\mathscr{D}$ (with concretion $\Delta_{\mathbf{D}}$), which itself is a retraction subalgebra of the cartesian closed category $\mathscr{K}(\mathbf{D})$ (with concretion $\widehat{\Delta_{\mathbf{D}}}$). Thus $\mathscr{C}$ is a retraction subalgebra of $\mathscr{K}(\mathbf{D})$, and the λ-algebra structure on $\mathscr{C}$ is induced from $(\mathscr{K}(\mathbf{D}), \widehat{\Delta_{\mathbf{D}}})$. $\square$

Combining this with Theorem 4.1.15, we obtain the following 'typed' version of the so-called *Scott–Koymans theorem*:

Theorem 4.1.28. *A relative total applicative structure* $\mathbf{A}$ *is a relative* λ*-algebra iff* $\mathbf{A}$ *arises from a retraction subalgebra of a cartesian closed category* $\mathscr{E}$ *via some concretion functor on* $\mathscr{E}$. *The concretion functor may be taken to be* $\mathrm{Hom}(1,-)$ *iff* $\mathbf{A}^\circ = \mathbf{A}^\sharp$. $\square$

The original Scott–Koymans theorem may be read off by specializing Theorem 4.1.28 to untyped λ-algebras $\mathbf{A}$ (i.e. those with just a single datatype). By a *reflexive object* in a cartesian closed category $\mathscr{E}$, we mean an object U equipped with a retraction pair $(t,r) : U^U \lhd U$ in $\mathscr{E}$.

Theorem 4.1.29 (Scott–Koymans). *A relative untyped total applicative structure* $\mathbf{A}$ *is a relative untyped* λ*-algebra iff* $\mathbf{A}$ *arises from a reflexive object in some cartesian closed category* $\mathscr{E}$ *via some concretion on* $\mathscr{E}$. *Again, the concretion may be taken to be* $\mathrm{Hom}(1,-)$ *iff* $\mathbf{A}^\circ = \mathbf{A}^\sharp$. $\square$

If U is a reflexive object in $\mathscr{E}$, we may write $\mathbf{A}[U,\Delta]$ for the relative untyped λ-algebra obtained from the concretion Δ. Explicitly, we have $\mathbf{A}[U,\Delta]^\circ = \Delta(U)$ with application $a \cdot b = \Delta(app)(\Delta(r)(a), b)$, and $\mathbf{A}[U,\Delta]^\sharp = \{\widehat{f} \mid f : 1 \to U\}$. When $\Delta = \mathrm{Hom}(1,-)$, we may write $\mathbf{A}[U,\Delta]$ simply as $\mathbf{A}[U]$.

Reflexive objects will be considered a little further in Subsection 4.2.5.

4.2 Types and Retractions

Having introduced the general concept of retractions and the Karoubi envelope, we now turn to the more specific question of which datatypes arise as retracts of which other datatypes. On the one hand, the Karoubi envelope of any $\lambda\eta$-algebra turns out (under mild conditions) to support a rich collection of datatypes going well beyond the simple types constructed via $\to$ from a given base type. The import of this is that much of the theory of simple types that we develop in this book can be extended 'for free' to these other types. On the other hand, even the class of simple types is more than the bare minimum that we need: one can reconstruct these from just the *pure types* $\overline{k}$ (Section 4.2.1) and the operations between them. This means that for many purposes, nothing of significance is lost by concentrating on the pure types.

The natural setting for this theory is that of $\lambda\eta$-algebras. (We have already noted that any λ-algebra is 'equivalent' to a $\lambda\eta$-algebra.) We begin in Subsections 4.2.1

and 4.2.2 with some results that hold in the very general setting of a $\lambda\eta$-algebra with *booleans* (as in Subsection 3.3.3). This covers typical models with base type $\mathbb{N}$ or $\mathbb{N}_\perp$, as well as those with finite base types such as 2 or $2_\perp$, and even, under mild conditions, those with base type $\mathbb{R}$ (cf. Subsection 13.2.1). In this setting, we are able to show, for instance, that every type constructed from the base type is a definable retract of some pure type. We then move on to some results holding for more specialized classes of models. In Subsection 4.2.3 and the preceding Theorem 4.2.9, we obtain some information specific to 'total' models (e.g. $\lambda\eta$-algebras over $\mathbb{N}$), while in Subsection 4.2.4 we explore some structure typically available in 'partial' models (e.g. $\lambda\eta$-algebras over $\mathbb{N}_\perp$). Even more structure is typically available in models that happen to possess a *universal type*, and we touch on this in Subsection 4.2.5.

Let us marshal some notation and terminology before we begin. We will work in the setting of a relative λ-algebra $\mathbf{A}$ over a type world T; we will typically use $A, B, \ldots$ to range over datatypes of $\mathbf{A}$, and $X, Y, \ldots$ for datatypes of $\mathbf{K}(\mathbf{A})$. We shall move freely between the set-theoretic and categorical perspectives on λ-algebras, often leaving the passage between them implicit. By a *computable element* of $\mathbf{A}$, we will in general mean an element of $\mathbf{A}^\sharp$—or categorically, a morphism in $\mathscr{C}(\mathbf{A})$.

As in the previous section, we will freely use λ-terms (possibly involving constants with designated meanings) as a notation for computable elements, omitting semantic brackets. We allow ourselves the use of composition notation in λ-terms, so that $M \circ N$ denotes $\lambda x.M(N(x))$. We shall also avail ourselves of simple 'pattern matching' syntax in terms involving product types, for instance writing $\lambda(x,f).fx$ for $\lambda z.(snd\,z)(fst\,z)$. If M, M' are λ-terms with free variables among $\vec{x}$, we say the equation $M = M'$ *holds uniformly* in a given λ-algebra if $\lambda\vec{x}.M = \lambda\vec{x}.M'$ holds there.

We will use notations such as $f : A \to B$, $(g,h) : A \cong B$ and $(t,r) : A \lhd B$ with their evident meanings as statements about $\mathscr{C}(\mathbf{A})$, where f,g,h,t,r are computable elements. We also adapt these notations to allow us to talk about *definable* isomorphisms and retractions. If K is some set of constant symbols with designated interpretations as computable elements of $\mathbf{A}$, we will write $(t,r) : A \lhd_K B$ if $(t,r) : A \lhd B$ and both t,r are definable by λ-terms with constants drawn from K; similarly for $(g,h) : A \cong_K B$. In addition, when referring to $\mathbf{K}(\mathbf{A})$, we shall write $(t,r) : (A,e) \lhd_K^+ (B,d)$ if the morphisms e,d,t,r in $\mathscr{K}(\mathscr{C}(\mathbf{A}))$ are all λ-definable relative to K; likewise for $\cong_K^+$. Clearly $A \lhd_K B$ in $\mathbf{A}$ iff $(A,id) \lhd_K^+ (B,id)$ in $\mathbf{K}(\mathbf{A})$.

We will make pervasive use of the following simple observations:

Proposition 4.2.1. *(i) Within any categorical $\lambda\eta$-algebra, the following hold for any objects A,B,C,A',B':*

1. *$A \lhd A$.*
2. *If $A \lhd B$ and $B \lhd C$ then $A \lhd C$.*
3. *If $A \lhd B$ and $A' \lhd B'$, then $A \Rightarrow A' \lhd B \Rightarrow B'$.*

(ii) If K is any set of constants with assigned interpretations in a $\lambda\eta$-algebra $\mathbf{A}$, then properties 1–3 above hold in $\mathbf{A}$ with $\lhd_K$ in place of $\lhd$.

(iii) If K is a set of constants with assigned interpretations in a λ-algebra $\mathbf{A}$, then properties 1–3 above hold in $\mathbf{K}(\mathbf{A})$ with $\lhd_K^+$ in place of $\lhd$.

Proof. (i) Properties 1 and 2 are trivial exercises. For property 3, suppose (t,r) : $A \lhd B$ and $(t',r') : A' \lhd B'$. Then

$$(t'',r'') = (\lambda f.t' \circ f \circ r,\ \lambda g.r' \circ g \circ t) : A \Rightarrow A' \lhd B \Rightarrow B'$$

is also a retraction pair, since

$$\lambda f.r''(t''f) =_\beta \lambda f.r' \circ t' \circ f \circ r \circ t = \lambda fx.fx =_\eta \lambda f.f .$$

Part (ii) is clear from the proof of (i). For (iii), the only further point to note is that if (A,e) and (A',e') are K-definable objects in the sense that e,e' are K-definable, then $(A,e) \Rightarrow (A',e')$ is also K-definable. □

4.2.1 Pure Types

For the rest of this section, we suppose $\mathbf{A}$ is a relative $\lambda\eta$-algebra with *booleans* of some type B in the sense of Subsection 3.3.3: that is, there are elements $tt, ff \in B^\sharp$ and $if_A \in (B \Rightarrow A \Rightarrow A \Rightarrow A)^\sharp$ for each type A such that $if_A \cdot tt \cdot x \cdot y = x$ and $if_A \cdot ff \cdot x \cdot y = y$. The type B will also serve as our base type; in most cases of interest, it will contain more than just tt, ff, and will usually be either $\mathbb{N}$ or $\mathbb{N}_\perp$.

A useful first observation is that any such model actually contains elements $if'_A \in (B \Rightarrow A \Rightarrow A \Rightarrow A)^\sharp$ satisfying the stronger conditions $if' \cdot tt = \lambda xy.x$ and $if' \cdot ff = \lambda xy.y$: simply take $if'_A = \lambda z.\, if_{A \Rightarrow A \Rightarrow A}\, z\,(\lambda xy.x)\,(\lambda xy.y)$.

Another useful observation is that a conditional operator for A automatically yields one for any type $C \Rightarrow A$ via

$$if_{C \Rightarrow A} = \lambda z^B f^{C \Rightarrow A} g^{C \Rightarrow A}.\, \lambda x^C.\, if_A\, z\,(fx)(gx) .$$

In this way, if_B suffices for defining conditionals for all simple types over B.

We will write *Bool* for the set consisting of the constants tt, ff, if_B, and more generally $Bool_A$ for the set $\{tt, ff, if_A\}$. Our first goal is to show that every simple type over B is a *Bool*-definable retract of a pure type over B. Later, we will use this to reconstruct the structure of simple types from just the pure ones.

In various contexts, embeddings of simple types in pure types have been worked out many times over in the literature (see for example [138, 233, 288, 124]). We will here present the results in greater generality than is usually done, and in a slightly sharper form. It is convenient to prove facts about $\mathbf{A}$ by working partly in $\mathbf{K}(\mathbf{A})$.

Proposition 4.2.2. *(i)* $\mathbf{K}(\mathbf{A})$ *has* cartesian squares*: that is, for any object* $X = (A,e)$*, the categorical product* $X \times X$ *exists in* $\mathscr{K}(\mathscr{C}(\mathbf{A}))$.

(ii) For any X, Y *in* $\mathbf{K}(\mathbf{A})$ *where* $X = (A,e)$ *we have*

$$(X \times X) \Rightarrow Y \cong^+_{Bool(A)} X \Rightarrow (X \Rightarrow Y) ,$$

$$Y \Rightarrow (X \times X) \cong^+_{Bool(A)} (Y \Rightarrow X) \times (Y \Rightarrow X) .$$

Proof. (i) Given $X = (A,e)$, let $C = B \Rightarrow A$, and define

$$pair = \lambda xy.\lambda z.\, if'_A\, z\,(ex)\,(ey) \;:\; A \Rightarrow A \Rightarrow C\,,$$
$$fst = \lambda p.e(p\,tt) \;:\; C \Rightarrow A\,,$$
$$snd = \lambda p.e(p\,ff) \;:\; C \Rightarrow A\,.$$

Now let $c = \lambda p.pair\,(fst\,p)(snd\,p)$. It is easy to check that $c \circ c = c$, $e \circ fst \circ c = fst$, and $e \circ snd \circ c = snd$; thus $Z = (C,c)$ is an object of $\mathscr{K}(\mathscr{C}(\mathbf{A}))$, and fst, snd represent morphisms $Z \to X$. It is now a routine exercise to verify that Z serves as the product $X \times X$, with fst and snd as projections.

(ii) For the first isomorphism, the two halves are represented by the elements

$$\lambda f.\lambda xx'.f(pair\,x\,x')\,, \qquad \lambda g.\lambda p.g(fst\;p)(snd\;p)\,.$$

These are $Bool(A)$-definable, and it is routine to check that they represent morphisms of the required types and are mutually inverse. We leave the second isomorphism as an exercise. □

On the syntactic side, we shall work initially with type names in $\mathsf{T}^{\to}(\beta)$, where we interpret β by B. Since $\mathbf{A}$ is a $\lambda\eta$-algebra, the inclusion $\mathbf{A} \hookrightarrow \mathbf{K}(\mathbf{A})$ preserves the meaning of $\Rightarrow$, so that type names $\sigma \in \mathsf{T}^{\to}(\beta)$ may be used to refer to types in either $\mathbf{A}$ or $\mathbf{K}(\mathbf{A})$ without risk of confusion. For convenience, we shall allow ourselves to write simply σ in place of $\mathbf{A}(\sigma)$; we may also write $\sigma \times \sigma$ for the cartesian square in $\mathbf{K}(\mathbf{A})$ given by the above proposition.

Proposition 4.2.3. *(i) If $k \leq l$ then $\overline{k} \lhd_{Bool} \overline{l}$ within* $\mathbf{A}$.
(ii) For any k we have $\overline{k} \times \overline{k} \lhd^{+}_{Bool} \overline{k+1}$ within $\mathbf{K}(\mathbf{A})$.
(iii) For any k we have $\overline{k} \to (\overline{k} \to \overline{0}) \lhd_{Bool} \overline{k+2}$ within $\mathbf{A}$.
(iv) For any type $\sigma \in \mathsf{T}^{\to}(\beta)$, there exists k such that $\sigma \lhd_{Bool} \overline{k}$ within $\mathbf{A}$.

Proof. (i) We first show by induction on k that $\overline{k} \lhd_{Bool} \overline{k+1}$. For $k = 0$, we have $(\lambda x.\lambda y.x, \lambda f.f\;tt) : \overline{0} \lhd_{Bool} \overline{1}$. Also if $\overline{k} \lhd_{Bool} \overline{k+1}$ then by Proposition 4.2.1 we have

$$\overline{k+1} \;=\; \overline{k} \to \overline{0} \;\lhd_{Bool}\; \overline{k+1} \to \overline{0}\,.$$

It now follows by Proposition 4.2.1 that $\overline{k} \lhd_{Bool} \overline{l}$ when $k \leq l$.

(ii) Again we argue by induction on k. For $k = 0$, the proof of Proposition 4.2.2 immediately gives us $\overline{0} \times \overline{0} \lhd^{+}_{Bool} \overline{0} \to \overline{0}$. Now assume $\overline{k} \times \overline{k} \lhd^{+}_{Bool} \overline{k+1}$. Using Proposition 4.2.2 and the fact that $if_{\overline{k}}$ is $Bool$-definable for every k, we have

$$\begin{aligned}
\overline{k+1} \times \overline{k+1} \;&=\; (\overline{k} \to \overline{0}) \times (\overline{k} \to \overline{0}) \\
&\cong^{+}_{Bool}\; \overline{k} \to (\overline{0} \times \overline{0}) \\
&\lhd^{+}_{Bool}\; \overline{k} \to (\overline{0} \to \overline{0}) \\
&\lhd^{+}_{Bool}\; \overline{k} \to (\overline{k} \to \overline{0}) \\
&\cong^{+}_{Bool}\; (\overline{k} \times \overline{k}) \to \overline{0} \\
&\lhd^{+}_{Bool}\; \overline{k+1} \to \overline{0} \;=\; \overline{k+2}\,.
\end{aligned}$$

(iii) Still within $\mathbf{K}(\mathbf{A})$, we now have

$$\begin{aligned}\overline{k} \rightarrow (\overline{k} \rightarrow \overline{0}) &\cong^{+}_{Bool} (\overline{k} \times \overline{k}) \rightarrow \overline{0} \\ &\lhd^{+}_{Bool} \overline{k+1} \rightarrow \overline{0} \;=\; \overline{k+2}\,.\end{aligned}$$

But since $\mathbf{A} \hookrightarrow \mathbf{K}(\mathbf{A})$ is full and respects $\rightarrow$, this means that $\overline{k} \rightarrow \overline{k+1} \lhd_{Bool} \overline{k+2}$ within $\mathbf{A}$.

(iv) By induction on the structure of $\sigma \in \mathsf{T}^{\rightarrow}(\beta)$. For the base type β we may clearly take $k=0$. For a type $\sigma \rightarrow \tau$, suppose inductively that $\sigma \lhd_{Bool} \overline{j}$ and $\tau \lhd_{Bool} \overline{k}$. Take l such that $j \leq l$ and $k \leq l+1$. Then we have

$$\sigma \rightarrow \tau \;\lhd_{Bool}\; \overline{j} \rightarrow \overline{k} \;\lhd_{Bool}\; \overline{l} \rightarrow \overline{l+1} \;\lhd_{Bool}\; \overline{l+2}\,. \qquad \square$$

By inspection of the above proof, we may obtain for each simple type σ a natural number $k(\sigma)$ such that $\sigma \lhd_{Bool} \overline{k(\sigma)}$ as follows:

$$k(\beta) \;=\; 0\,, \qquad k(\sigma \rightarrow \tau) \;=\; 1+\max\,(k(\sigma)+1, k(\tau))\,.$$

Indeed, for each $\sigma \in \mathsf{T}^{\rightarrow}(\beta)$, explicit λ-terms over *Bool* manifesting $\sigma \lhd \overline{k(\sigma)}$ can be extracted from the above proof; these terms are independent of $\mathbf{A}$.

Under stronger hypotheses on $\mathbf{A}$, a lower value for $k(\sigma)$ may be possible. For instance, suppose $\mathbf{A}^{\sharp}(\beta) = \mathbb{N}$ and $\mathbf{A}$ contains computable elements $code : \beta, \beta \rightarrow \beta$ and $first, second : \beta \rightarrow \beta$ such that the following hold uniformly:

$$\begin{aligned} first \cdot (code \cdot n \cdot m) &= n\,, \\ second \cdot (code \cdot n \cdot m) &= m\,, \\ code \cdot (first \cdot p) \cdot (second \cdot p) &= p\,. \end{aligned}$$

(We use these names to avoid confusion with the ordinary pairing and projections associated with product types.) For instance, *code* might represent the pairing operation defined in Example 3.1.5. Letting $Code = Bool \cup \{code, first, second\}$, we then have:

Proposition 4.2.4. *(i) For any k we have* $\overline{k} \times \overline{k} \cong^{+}_{Code} \overline{k}$ *within* $\mathbf{K}(\mathbf{A})$.

(ii) For any $\sigma \in \mathsf{T}^{\rightarrow}(\beta)$, *we have* $\sigma \lhd_{Code} \overline{\mathrm{lv}(\sigma)}$ *within* $\mathbf{A}$, *where* lv *is defined as in Subsection 3.1.7.*

We leave the proof as an exercise.

If furthermore $\mathbf{A}$ is an *extensional* λ-algebra over $\mathbb{N}$, the situation is better still: again under mild assumptions, it is possible to define an *isomorphism* $\sigma \cong \overline{\mathrm{lv}(\sigma)}$. The construction is non-trivial—we will give it in the next subsection once some further machinery for product types is available.

In the general setting of a $\lambda\eta$-algebra $\mathbf{A}$ with booleans, it follows from Proposition 4.2.3 that the pure types 'generate' all the simple types, in the sense that if $\mathscr{C}_{\mathrm{pure}}(\mathbf{A})$ is the full subcategory of $\mathscr{C}(\mathbf{A})$ consisting of the pure types, then all the simple types are present in $\mathscr{K}(\mathscr{C}_{\mathrm{pure}}(\mathbf{A}))$, and indeed $\mathscr{K}(\mathscr{C}_{\mathrm{pure}}(\mathbf{A})) = \mathscr{K}(\mathscr{C}_{\mathrm{simp}}(\mathbf{A}))$

where $\mathscr{C}_{\mathrm{simp}}(\mathbf{A})$ is the full subcategory consisting of the simple types. However, it does not immediately follow that one can reconstruct the whole simple type structure from $\mathscr{C}_{\mathrm{pure}}(\mathbf{A})$, since it is not quite trivial at this stage to define the $\Rightarrow$ operation on $\mathscr{K}(\mathscr{C}_{\mathrm{pure}}(\mathbf{A}))$. We will be able to fill this small gap easily once we have a better grasp of product types.

4.2.2 Product Types

To make further progress, in addition to supposing that $\mathbf{A}$ is a relative $\lambda\eta$-algebra with booleans of type B, we shall assume that $\mathbf{A}$ has the following property: for any types C,D, there is a third type E such that $C \lhd E$ and $D \lhd E$ within $\mathscr{C}(\mathbf{A})$. This is certainly the case when $\mathbf{A}$ is a $\lambda\eta$-algebra over $\mathsf{T}^{\rightarrow}(\beta)$, since Proposition 4.2.3 shows that for any $\sigma,\tau \in \mathsf{T}^{\rightarrow}(\beta)$ there is a pure type $\overline{k}$ with $\sigma,\tau \lhd \overline{k}$.

Under this hypothesis, we may strengthen Proposition 4.2.2 considerably:

Proposition 4.2.5. *(i)* $\mathbf{K}(\mathbf{A})$ *(or more precisely* $\mathscr{K}(\mathscr{C}(\mathbf{A}))$*) has finite cartesian products.*

(ii) $\mathbf{K}(\mathbf{A})$ *is a* $\lambda\eta\times$*-algebra (in other words,* $\mathscr{K}(\mathscr{C}(\mathbf{A}))$ *is cartesian closed).*

Proof. (i) Clearly $(B, \lambda x.tt)$ serves as a terminal object in $\mathbf{K}(\mathbf{A})$. For binary products, suppose (C,c) and (D,d) are in $\mathbf{K}(\mathbf{A})$. Take E such that there are retractions $C \lhd E$ and $D \lhd E$ within $\mathbf{A}$, so that $(C,c) \lhd E$ and $(D,d) \lhd E$ within $\mathbf{K}(\mathbf{A})$, and let $e_c, e_d : E \to E$ be the idempotents associated with the latter retractions. Using Proposition 4.2.2(i), construct the product $E \times E$ in $\mathbf{K}(\mathbf{A})$, and consider the idempotent $\lambda(x,y).(e_c x, e_d y)$ on $E \times E$. It is easy to check that splitting this idempotent yields an object that can serve as $(C,c) \times (D,d)$.

(ii) We first show that for any objects X,Y,W in $\mathscr{K}(\mathscr{C}(\mathbf{A}))$ we have a canonical isomorphism

$$(X \times Y) \Rightarrow W \;\cong\; X \Rightarrow (Y \Rightarrow W)\,.$$

As in part (i), we may take Z such that $X \lhd Z$ and $Y \lhd Z$. Using Proposition 4.2.2(ii), we now have

$$(X \times Y) \Rightarrow W \;\lhd\; (Z \times Z) \Rightarrow W \;\cong\; Z \Rightarrow (Z \Rightarrow W) \;\rhd\; X \Rightarrow (Y \Rightarrow W)\,.$$

Moreover, a routine calculation with the λ-terms given in the proof of Proposition 4.2.2(ii) shows that the composite morphisms are mutually inverse as required.

Specializing to the case $X = Y \Rightarrow W$, the transpose of $id_{Y \Rightarrow W}$ now yields a morphism $app_{Y,W} : (Y \Rightarrow W) \times Y \to W$ for any Y,W, represented by the element $\lambda p.(\mathit{fst}\ p)(\mathit{snd}\ p)$. It is now easy to check that in the general case, the transpose of any $g : X \to (Y \Rightarrow W)$ via the above isomorphism agrees with $app_{Y,W} \circ (g \circ id_Y) : X \times Y \to W$. The requirements for a cartesian closed category are thus satisfied. □

We may now plug the gap we left at the end of Subsection 4.2.1. Let $\mathbf{A}$ be any $\lambda\eta$-algebra over $\mathsf{T}^{\rightarrow}(\beta)$ with booleans of type β, and as in the previous section let $\mathscr{C}_{\mathrm{pure}}$ be the full subcategory of $\mathscr{C}(\mathbf{A})$ consisting of the pure types. We now know that $\mathscr{K}(\mathscr{C}_{\mathrm{pure}}(\mathbf{A})) = \mathscr{K}(\mathscr{C}(\mathbf{A}))$ is cartesian closed. But in any cartesian closed category, both products and exponentials are determined up to isomorphism by their universal properties, so it follows easily that the whole $\lambda\eta$-algebra structure of $\mathbf{A}$ may be recovered (up to isomorphism) from $\mathscr{C}_{\mathrm{pure}}(\mathbf{A})$ along with its concretion $\Delta_{\mathbf{A}}$. In particular, if $(\mathscr{C}_{\mathrm{pure}}(\mathbf{A}), \Delta_{\mathbf{A}}) \simeq (\mathscr{C}_{\mathrm{pure}}(\mathbf{B}), \Delta_{\mathbf{B}})$, then $\mathbf{A} \cong \mathbf{B}$.

In the extensional setting, we may improve on this further:

Theorem 4.2.6. *An extensional λ-algebra $\mathbf{A}$ over $\mathsf{T}^{\to}(\beta)$ with booleans of type β is determined up to isomorphism by its pure type part: that is, by the sets $\mathbf{A}(k)$ and the application operations $\mathbf{A}(k+1) \times \mathbf{A}(k) \to \mathbf{A}(0)$.*

Proof. Suppose we are given the pure type part of $\mathbf{A}$. Re-appropriating the proof of Proposition 4.2.3, for $k \leq l$ we may construct functions

$$\phi_{kl} \;:\; \mathbf{A}(k) \to \mathbf{A}(l)\,, \qquad \psi_{kl} \;:\; \mathbf{A}(l) \to \mathbf{A}(k)\,,$$

external to the structure of $\mathbf{A}$, such that $\psi_{kl} \circ \phi_{kl} = id$, and thence, for any k, construct functions

$$\chi_k \;:\; \mathbf{A}(k) \times \mathbf{A}(k) \to \mathbf{A}(k+1)\,, \qquad \theta_k : \mathbf{A}(k+1) \to \mathbf{A}(k) \times \mathbf{A}(k)$$

such that $\theta_k \circ \chi_k = id$. For example, in the induction step of Proposition 4.2.3(ii), any pair $(f,g) \in \mathbf{A}(k+1) \times \mathbf{A}(k+1)$ uniquely determines a function $\mathbf{A}(k) \to \mathbf{A}(0) \times \mathbf{A}(0)$ and thence a function $\mathbf{A}(k) \times \mathbf{A}(k) \to \mathbf{A}(0)$, which via θ_k yields a function $\mathbf{A}(k+1) \to \mathbf{A}(0)$. Since $\mathbf{A}$ is extensional, this specifies an element of $\mathbf{A}(k+2)$ which we may designate as $\chi_{k+1}(f,g)$.

We may now, for any k,l, construct the set of functions $\mathbf{A}(k) \to \mathbf{A}(l)$ present in $\mathbf{A}$. If $l = 0$, these are just the elements of $\mathbf{A}(k+1)$. Otherwise, they correspond to the functions $\mathbf{A}(k \times (l-1)) \to \mathbf{A}(0)$ in $\mathbf{A}$, and in the light of the above, these can be determined from the elements of $\mathbf{A}(m+1)$ where $m = \max(k, l-1)+1$. Since $\mathbf{A}$ is extensional, this gives us the category $\mathscr{C}_{\mathrm{pure}}(\mathbf{A})$ along with its concretion, and as already shown, this suffices to determine $\mathbf{A}$. □

In a similar vein, we may show that a $\lambda\eta$-algebra over $\mathsf{T}^{\to}(\beta)$ extends *uniquely* to one over $\mathsf{T}^{\to\times 1}(\beta)$, if we require that 1 and $\times$ be interpreted by strong finite products as in Definition 4.1.7. (Recall that these correspond to categorical products.) We will call $\mathbf{B}$ a *$\lambda\eta$-algebra over* $\mathsf{T}^{\to\times 1}(\beta)$ if this is indeed the case.

By Proposition 4.2.5, if $\mathbf{A}$ is a $\lambda\eta$-algebra with booleans over $\mathsf{T}^{\to}(\beta)$, then $\mathbf{K}(\mathbf{A})$ is a $\lambda\eta$-algebra with strong products, and so restricts to a $\lambda\eta$-algebra over $\mathsf{T}^{\to\times 1}(\beta)$, which we call $\mathbf{A}^{\times 1}$. The following facts may now be proved as exercises:

Proposition 4.2.7. *(i) For any $\tau \in \mathsf{T}^{\to\times 1}(\beta)$ we may choose a type $\rho_\tau \in \mathsf{T}^{\to}(\beta)$ and a λ-term $E_\tau : \rho_\tau \to \rho_\tau$ over Bool, independent of $\mathbf{A}$, such that E_τ denotes an idempotent with a splitting $\tau \lhd \rho_\tau$ in $\mathbf{K}(\mathbf{A})$.*

(ii) If $\mathbf{B}$ is any $\lambda\eta$-algebra over $\mathsf{T}^{\to\times 1}(\beta)$ whose restriction to $\mathsf{T}^{\to}(\beta)$ is isomorphic to $\mathbf{A}$, then $\mathbf{B} \cong \mathbf{A}^{\times 1}$.

This shows that it makes little essential difference whether we choose to study models over $\mathsf{T}^{\rightarrow}(\beta)$ or over $\mathsf{T}^{\rightarrow\times 1}(\beta)$: in particular, there is no danger that a single computability notion for types of $\mathsf{T}^{\rightarrow}(\beta)$ will split into several notions when types of $\mathsf{T}^{\rightarrow\times 1}(\beta)$ are considered. Even when we are officially studying models over $\mathsf{T}^{\rightarrow}(\beta)$, we shall sometimes allow ourselves the use of product types for convenience, with the understanding that these are to be interpreted in the Karoubi envelope as above.

We now fulfil a promise from Subsection 4.2.1 by proving a more substantial theorem for *extensional* models over $\mathbb{N}$: under mild assumptions, every type $\sigma \in \mathsf{T}^{\rightarrow\times}(\beta)$ is *isomorphic* to the pure type of the same level. This was noted in unpublished work by Escardó, Lawson and Schröder in connection with the model Ct; however, the proof we give here is new.

Let $\mathbf{A}$ be an extensional (relative) λ-algebra over $\mathsf{T}^{\rightarrow\times 1}(\beta = \mathtt{N})$; recall that all such models are $\lambda\eta$-algebras. We will assume $\mathbf{A}$ contains booleans as in Subsection 3.3.3, bijective pairing and projection operations *code*, *first*, *second* as in Subsection 4.2.1, and an equality tester $eq \in \mathbf{A}^{\sharp}(\beta,\beta \rightarrow \beta)$ such that $eq \cdot n \cdot n = tt$ and $eq \cdot m \cdot n = ff$ for $m \neq n$. We will write $Nat = Bool \cup \{code, first, second, eq\}$. We will also abbreviate $code\,M\,N$ to $\langle M,N \rangle$.

We first define a term $swap : \overline{0},\overline{0} \rightarrow \overline{0} \rightarrow \overline{0}$ as follows:

$$swap \;=\; \lambda x^0 y^0.\lambda z^0.\, if\,(eq\,z\,x)\,y\,(if\,(eq\,z\,y)\,x\,z)\,.$$

Note that for any $x,y,z \in \mathbb{N}$ we have

$$(swap\,x\,y)\,x \;=\; y\,, \qquad (swap\,x\,y)\,y \;=\; x\,, \qquad (swap\,x\,y)((swap\,y\,x)\,z) \;=\; z\,.$$

The non-trivial part of the argument is the following:

Lemma 4.2.8. *For any $k \leq l$, we have $\overline{k} \times \overline{l} \cong_{Nat} \overline{l}$ within* $\mathbf{A}$.

Proof. The case $k = l = 0$ is trivial using the pairing and projection operations on β, so assume $l \geq 1$. We begin with the case $k = 0$: here we wish to encode a pair $(z \in \mathbb{N}, g \in \mathbf{A}(l))$ as an element $G \in \mathbf{A}(l)$. Let U be a λ-term for some fixed element of $\mathbf{A}(l-1)$: e.g. we may take $U = tt$ if $l = 1$, or $U = \lambda y^{l-2}.tt$ if $l > 1$. The idea is to use the value of G at U to smuggle in the value of z along with that of $g(U)$; however, this needs to be done in a rather subtle way to ensure bijectivity.

Let $Q : \overline{k},\overline{l} \rightarrow \overline{l}$, $P_0 : \overline{l} \rightarrow \overline{k}$ and $P_1 : \overline{l} \rightarrow \overline{l}$ be λ-terms over *Nat* as follows:

$$\begin{aligned}
Q &= \lambda z^0 g^l.\, \lambda x^{l-1}.\; swap\,(gU)\,\langle z, gU\rangle\,(gx)\,,\\
P_0 &= \lambda G^l.\, first\,(GU)\,,\\
P_1 &= \lambda G^l.\, \lambda x^{l-1}.\; swap\,(GU)\,(second\,(GU))\,(Gx)\,.
\end{aligned}$$

Clearly these define computable operations $q : \overline{0} \times \overline{l} \rightarrow \overline{l}$ and $p : \overline{l} \rightarrow \overline{0} \times \overline{l}$ in $\mathbf{A}$. It remains to check that these are mutually inverse. Using the equations for *code*, *first*, *second*, *eq* from the previous section and the properties of *swap* listed above, one may verify by straightforward equational reasoning that

$$P_0(Qzg) = z, \quad P_1(Qzg)x = gx, \quad Q(P_0G)(P_1G)x = Gx.$$

Hence by extensionality we have

$$P_0(Qzg) = z, \quad P_1(Qzg) = g, \quad Q(P_0G)(P_1G) = G,$$

so p,q are mutually inverse as required.

For the case $k > 0$, we replace the single element U by a subset of $\mathbf{A}(l-1)$ isomorphic to $\overline{k-1}$ using a retraction pair $(t,r) : \overline{k-1} \lhd \overline{l-1}$. To encode a pair $(f \in \mathbf{A}(k), g \in \mathbf{A}(l))$ as $G \in \mathbf{A}(l)$, we mimic the above argument, smuggling the value of $f(y)$ into $G(ty)$ along with the value of $g(ty)$. Specifically, take terms T,R over *Nat* denoting t,r, and define terms $Q : \overline{k}, \overline{l} \to \overline{l}$, $P_0 : \overline{l} \to \overline{k}$ and $P_1 : \overline{l} \to \overline{l}$ by

$$\begin{aligned}
Q &= \lambda f^k g^l.\, \lambda x^{l-1}.\, swap\ (g(T(Rx)))\ \langle f(Rx), g(T(Rx))\rangle\ (gx),\\
P_0 &= \lambda G^l.\, \lambda y^{k-1}.\, first(G(Ty)),\\
P_1 &= \lambda G^l.\, \lambda x^{l-1}.\, swap\ (G(T(Rx)))\ (second(G(T(Rx))))\ (Gx).
\end{aligned}$$

Clearly these define computable operations $q : \overline{k} \times \overline{l} \to \overline{l}$ and $p : \overline{l} \to \overline{k} \times \overline{l}$ in $\mathbf{K}(\mathbf{A})$. It remains to check that these are mutually inverse. As before, we use simple equational reasoning (now with the help of the property $R(Tx) = x$) to check that

$$P_0(Qfg)x = fx, \quad P_1(Qfg)x = gx, \quad Q(P_0G)(P_1G)x = Gx,$$

so by extensionality we have a definable isomorphism $\overline{k} \times \overline{l} \cong \overline{l}$ as required. □

Theorem 4.2.9. *Let* $\mathbf{A}$ *be an extensional relative* λ*-algebra over* $\mathsf{T}^{\to}(\beta = \mathbb{N})$*, or one over* $\mathsf{T}^{\to\times}(\beta = \mathbb{N})$ *with standard products, equipped with suitable interpretations for the constants in Nat as above. Then for any type* σ *we have* $\sigma \cong_{Nat} \overline{\mathrm{lv}(\sigma)}$ *in* $\mathbf{A}$.

Proof. We first show the result for $\mathsf{T}^{\to}(\beta = \mathbb{N})$, reasoning by argument induction on σ. For β the result is trivial, so suppose inductively we have $\sigma_i \cong_{Nat} \overline{l_i}$ where $l_i = \mathrm{lv}(\sigma_i)$ for each $i < r$. Let $l = \max l_i$. Working in $\mathbf{K}(\mathbf{A})$, we have by Lemma 4.2.8 that $\overline{l_0} \times \cdots \times \overline{l_{r-1}} \cong^+_{Nat} \overline{l}$ and so

$$\sigma_0, \ldots, \sigma_{r-1} \to \beta \cong^+_{Nat} \overline{l_0}, \ldots, \overline{l_{r-1}} \to \beta \cong^+_{Nat} (\overline{l_0} \times \cdots \times \overline{l_{r-1}}) \to \beta \cong^+_{Nat} \overline{l} \to \beta.$$

It follows that $\sigma_0, \ldots, \sigma_{r-1} \to \beta \lhd_{Nat} \overline{l+1}$ within $\mathbf{A}$ itself.

For $\mathsf{T}^{\to\times}(\beta = \mathbb{N})$, we first note that any type τ is definably isomorphic to one of the form $\tau_0 \times \cdots \times \tau_{n-1}$ where $n \geq 1$ and $\tau_0, \ldots, \tau_{n-1} \in \mathsf{T}^{\to}(\beta)$, using the construction of Theorem 3.1.15. By the above argument, each τ_i is definably isomorphic to some $\overline{k_i}$, and the result follows by another appeal to Lemma 4.2.8. □

The above theorem shows that in extensional models over $\mathbb{N}$, the stratification of types according to their structural complexity is quite coarse: intuitively, there is a quantum leap between each type level and the next, but no finer distinctions within type levels. The situation is quite different for extensional models over $\mathbb{N}_\perp$: here, for instance, we will typically have $\mathtt{N}^r \lhd \mathtt{N}^{r+1}$ for any r, but not vice versa.

4.2.3 Further Type Structure in Total Settings

Most of the above theory for pure types and product types works in the very general setting of a (possibly extensional) model over $\mathsf{T}^{\rightarrow}(\beta)$ with booleans of type β. To proceed further with our study of datatypes in $\mathbf{K}(\mathbf{A})$, we shall now specialize to particular kinds of models $\mathbf{A}$: first, to those embodying 'total' computability notions (e.g. λ-algebras over the base type 2 or $\mathbb{N}$), and then, in the next subsection, to those embodying 'partial' notions. In each of these settings, we shall briefly indicate some of the further type structure that is typically available within $\mathbf{K}(\mathbf{A})$, leaving the detailed outworking to the reader.

We start by investigating some consequences of the following simple condition:

Definition 4.2.10. A relative λ-algebra $\mathbf{A}$ has *strong booleans* if it contains computable elements $tt, f\!f, if_A$ of suitable type for which the following hold uniformly.

$$if\ tt \;=\; \lambda xy.x\,, \qquad if\ f\!f \;=\; \lambda xy.y\,, \qquad f \;=\; \lambda z.if\ z(f\ tt)(f\ f\!f)\,.$$

In the cases of interest, $\mathbf{K}(\mathbf{A})$ will have strong booleans when $\mathbf{A}$ contains a type corresponding to 2 or $\mathbb{N}$ (because 2 arises as a retract of $\mathbb{N}$ under mild conditions), but not when the base type is $2_{\perp}$ or $\mathbb{N}_{\perp}$.

We here write $\mathsf{T} = \mathsf{T}^{\rightarrow\times 1+}(\beta)$ for the type world whose types are generated syntactically from β under $\rightarrow, \times, 1, +$. By a *$\lambda\eta\times+$-algebra* over T, we shall mean a $\lambda\eta\times$-algebra over T in which types $\sigma+\tau$ are interpreted as categorical sums.

The following proposition is the analogue for sums of Proposition 4.2.7:

Proposition 4.2.11. *Let* $\mathbf{A}$ *be a* $\lambda\eta$*-algebra over* $\mathsf{T}^{\rightarrow}(\beta)$ *with strong booleans of type* β*. Then*

(i) $\mathbf{K}(\mathbf{A})$ *has binary sums, and thus restricts to a* $\lambda\eta\times+$*-algebra* $\mathbf{A}^{\times 1+}$ *over* T.

(ii) For any $\tau \in \mathsf{T}$ *we may pick a type* $\rho_\tau \in \mathsf{T}^{\rightarrow}(\beta)$ *and a term* $E_\tau : \rho_\tau \rightarrow \rho_\tau$ *over Bool, independent of* $\mathbf{A}$*, such that* E_τ *defines an idempotent with splitting* $\tau \lhd \rho_\tau$.

(iii) If $\mathbf{B}$ *is any* $\lambda\eta\times+$*-algebra over* T *whose restriction to* $\mathsf{T}^{\rightarrow}(\beta)$ *is isomorphic to* $\mathbf{A}$*, then* $\mathbf{B} \cong \mathbf{A}^{\times 1+}$.

We leave the proof as an exercise; the key idea is that a sum type $\sigma+\tau$ may be represented as a retract of $\beta\times\sigma\times\tau$.

Even for extensional models over $\mathbb{N}$, the addition of sum types leads to a more fine-grained hierarchy of type levels than the one indicated by Theorem 4.2.9. For example, if $\mathtt{B}$ denotes the type $1+1$ (a type of strong booleans) and we write $\mathtt{N}$ in place of β, then clearly $\mathtt{N}\rightarrow\mathtt{B}$ is a definable retract of $\mathtt{N}\rightarrow\mathtt{N}$, but in general $\mathtt{N}\rightarrow\mathtt{B}$ is not isomorphic to either $\mathtt{N}$ or $\mathtt{N}\rightarrow\mathtt{N}$.[3]

Although we will officially be working mainly with λ-algebras over $\mathsf{T}^{\rightarrow}(\mathtt{N})$, we shall sometimes make reference to the type $\mathtt{B}$ of booleans and other types constructed from it, understood as living in the Karoubi envelope, in order to pinpoint

[3] This can be readily seen by considering the model Ct to be studied in Chapter 8: within this model, the type $\mathtt{N}\rightarrow\mathtt{B}$ is *compact*, whilst $\mathtt{N}$ and $\mathtt{N}\rightarrow\mathtt{N}$ are not.

more precisely the natural 'type level' at which some phenomenon arises. For example, within the model Ct, the well-known *fan functional* (see Section 8.3) must be assigned the type $((\mathtt{N} \to \mathtt{N}) \to \mathtt{N}) \to \mathtt{N}$ if only the simple types over $\mathtt{N}$ are available, but it is more informative to say that it lives naturally at the type $((\mathtt{N} \to \mathtt{B}) \to \mathtt{N}) \to \mathtt{N}$, since the second '$\mathtt{N}$' serves no other function than to represent the booleans.

In models with base type $\mathbb{N}$, we can typically go further and show that $\mathbf{K}(\mathbf{A})$ admits interpretations of *finitary (positive) inductive types*. We shall suppose here that $\mathbf{A}$ contains a type N of numerals in the sense of Subsection 3.3.4; for sanity, we shall also assume that $N \cong \mathbb{N}$ and that $\mathbf{A}$ is extensional. These conditions imply the existence of strong booleans in $\mathbf{K}(\mathbf{A})$, so we already know that the type operations $\times, +$ and the type symbol 1 may be naturally interpreted there.

Suppose now that $F(\alpha)$ is a formal expression built from types of $\mathsf{T}^{\to\times 1+}(\mathtt{N})$ and the type variable symbol α by means of the binary operators $\times, +$. (For example, $F(\alpha) = (\mathtt{N} \to \mathtt{N}) + (\mathtt{B} \times \alpha \times \alpha)$, where $\mathtt{B} = 1+1$.) Clearly, any such F induces both an endofunctor $\widehat{F}$ on $\mathscr{K}(\mathscr{C}(\mathbf{A}))$ and an endofunctor $\widetilde{F}$ on $\mathscr{S}et$ such that $\Delta(\widehat{F}(X)) = \widetilde{F}(\Delta X)$ for any concretion Δ. What is more, one can show:

Proposition 4.2.12. *Assume F is proper, i.e. $\widetilde{\emptyset}$ is non-empty. Then $\mathscr{K}(\mathscr{C}(\mathbf{A}))$ contains an* initial algebra *for $\widehat{F}$: that is, an object X together with a morphism $\iota : \widehat{F}(X) \to X$ (actually an isomorphism) such that for any object Y and morphism $\iota' : \widehat{F}(Y) \to Y$ there is a unique morphism $h : X \to Y$ with $h \circ \iota = \iota' \circ \widehat{F}(h)$.*

Proof. (Sketch.) One can show that each of the sets $\widetilde{F}^k(\emptyset)$ $(k \geq 1)$ arises as the image of some X_k in $\mathscr{K}(\mathscr{C}(\mathbf{A}))$, and moreover all these X_k may be obtained as retracts of some single type τ (uniformly in k). Specifically, if $F(\alpha)$ is constructed from $\sigma_0, \ldots, \sigma_{r-1} \in \mathsf{T}^{\to\times 1+}(\mathtt{N})$ and α via $\times$ and $+$, we may choose some σ of which all the σ_i are definable retracts, and it then suffices to take $\tau = \mathtt{N} \times (\mathtt{N} \to (\mathtt{N} \times \sigma))$. Intuitively, an element of any X_k will be a finite tree with nodes labelled by elements of the σ_i and other bookkeeping information. Such a structure may be represented by an element of type τ, the first component of type $\mathtt{N}$ being used to record the size of the tree. Using this, we may construct an object X such that $\Delta(X) \cong \bigsqcup_{k \geq 1} \widetilde{F}^k(\emptyset)$ as a retract of τ, and show that this X is an initial algebra as required. We leave the details as a substantial exercise. □

We call this initial algebra the *inductive type* defined by F, and denote it by $\mu\alpha.F(\alpha)$. For example, if $F(\alpha) = (\mathtt{N} \to \mathtt{N}) + (\mathtt{B} \times \alpha \times \alpha)$, then $\mu\alpha.F(\alpha)$ is intuitively the type of well-founded binary trees with leaves labelled by operations $\mathbb{N} \to \mathbb{N}$ and with internal nodes labelled by booleans.

Such inductive types can be shown to have properties akin to those of Propositions 4.2.7 and 4.2.11. In particular, if two models agree on the meaning of simple types, they will agree on the meaning of finitary positive inductive types. This rules out the possibility that a single notion of computability for operations of simple type might bifurcate into several notions once these inductive types are added.

However, this is about as far as we can travel in this direction, at least in the total setting. For typical models over $\mathsf{T}^{\to}(\mathtt{N} = \mathbb{N})$, the Karoubi envelope does *not* admit a good interpretation for infinitary inductive types such as $\mu\alpha.1 + (\mathtt{N} \to \alpha)$, the datatype of countably branching well-founded trees.[4] In fact, finding a home for such a datatype would lead us into the realm of *transfinite types* (see Subsection 13.2.5).

Finally, the following exercise indicates a quite different direction in which the exploration of types in $\mathbf{K}(\mathbf{A})$ might be pursued.

Exercise 4.2.13. In a $\lambda\eta\times$-algebra, by a *uniform family of types* indexed by σ we shall mean a type τ equipped with a 'uniform σ-indexed family of idempotents': that is, an operation $E : \sigma \to (\tau \to \tau)$ such that $Ex = (Ex) \circ (Ex)$ uniformly in $x : \sigma$.

Show that if $\mathbf{A}$ has products and $E : (\sigma_0 \times \cdots \times \sigma_r) \to (\tau \to \tau)$ is a uniform family of types in $\mathbf{K}(\mathbf{A})$, then $\mathbf{K}(\mathbf{A})$ also contains *dependent* product and sum types

$$\lambda(x_0,\ldots,x_{r-1}).\, \Pi y : \sigma_r.E(\vec{x},y) \,:\, (\sigma_0 \times \cdots \times \sigma_{r-1}) \to (\sigma_r \to \tau) \to (\sigma_r \to \tau)\,,$$
$$\lambda(x_0,\ldots,x_{r-1}).\, \Sigma y : \sigma_r.E(\vec{x},y) \,:\, (\sigma_0 \times \cdots \times \sigma_{r-1}) \to (\sigma_r \times \tau) \to (\sigma_r \times \tau)\,.$$

(considered as uniform families indexed by $\sigma_0 \times \cdots \times \sigma_r$), where

$$\Pi y : \sigma_r.E(\vec{x},y) \text{ has image } \{f : \sigma_r \to \tau \mid f = \lambda y.E(\vec{x},y)(fy)\}\,,$$
$$\Sigma y : \sigma_r.E(\vec{x},y) \text{ has image } \{(y,z) : \sigma_r \times \tau \mid z = E(\vec{x},y)z\}\,.$$

To summarize, our discussion so far implies that much of the theory to be developed later for particular models over $\mathsf{T}^{\to}(\mathtt{N} = \mathbb{N})$ (notably in Chapters 5, 8 and 9) can be extended 'for free' to a wide repertoire of other types, including products, sums, finitary inductive types and dependent products and sums.

4.2.4 Partiality, Lifting and Strictness

We now undertake a somewhat similar investigation of models capturing *partial* computability notions—typically, these will be (technically total) λ-algebras over base types such as $2_\perp$ or $\mathbb{N}_\perp$. We shall first identify some basic structure that is frequently present in such models (e.g. in all extensional models of interest, as well as some moderately intensional ones), then show what this implies for the type structure in the Karoubi envelope. Our primary interest here will not be in the kinds of type constructors considered in the previous subsection, but in notions such as *lifting* of types and strict function spaces. The material here will be used in Section 4.3 to show how total models over $\mathbb{N}_\perp$ relate to genuinely partial models over $\mathbb{N}$; it will also inform our discussion of *call-by-value* computation in Section 7.2 and occasionally later in the book.

---

[4] This can be naturally seen for the model HEO to be studied in Chapter 9, for instance. Somewhat loosely, the realization over K_1 of any type in $\mathbf{K}(\mathsf{HEO})$ can be seen to be *arithmetical* in character (i.e. describable by a first-order arithmetical formula), whereas the canonical interpretation of the above type in $\mathscr{P}er(K_1)$ is Π^1_1 complete, and hence non-arithmetical. We leave it to the interested reader to make this argument more precise.

Throughout this subsection, $\mathbf{A}$ is an arbitrary relative λ-algebra. It turns out that a reasonable amount of structure in $\mathbf{K}(\mathbf{A})$ can be derived from the following axiomatic definition:

Definition 4.2.14. A *partiality structure* on $\mathbf{A}$ consists of a type D together with computable elements

$$\top \in D^{\sharp}, \qquad 1_A \in (A \Rightarrow D \Rightarrow A)^{\sharp} \text{ for each } A$$

for which the following axioms hold uniformly. For readability we treat 1 as a left-associative infix symbol.

1. $x \mathbin{1} \top = x$.
2. $(fx) \mathbin{1} z = (f(x \mathbin{1} z)) \mathbin{1} z$.
3. $(x \mathbin{1} z) \mathbin{1} w = x \mathbin{1} (z \mathbin{1} w)$.

Intuitively, D is a type used to represent degrees of 'definedness' or termination for computations.[5] In most cases of interest, all we shall need is a value $\top$ for 'defined' and a value $\bot$ for 'undefined'; however, the above definition in principle allows for more subtle degrees of definedness. An expression $x \mathbin{1_A} z$ should be read as 'x when z is defined' or 'x restricted by z'; in many models, we can understand it operationally as the value of type A obtained by first evaluating z, and then (if this terminates) discarding the result and evaluating x. In terms of this intuition, axiom 2 says roughly that if the result of fx is in any case going to be restricted by z, it makes no difference if a further check on the definedness of z is added to the argument x. Note that this axiom yields $x \mathbin{1} z \mathbin{1} z = x \mathbin{1} z$ by taking $f = id$. Axiom 3 has a somewhat similar justification. Beyond this, however, it is hard to give much intuitive motivation for the particular selection of axioms beyond the fact that they do what we require.

Example 4.2.15. Suppose $\mathbf{A}$ is an extensional TPCA over $\mathsf{T}^{\rightarrow}(\mathtt{U} = \{\top, \bot\}, \mathtt{N} = \mathbb{N}_{\bot})$, such as PC, SF or SR (expanded with a unit type in the obvious way). Take $D = \mathbf{A}(\mathtt{U})$, and define 1_A by

$$x \mathbin{1_A} z = \begin{cases} \bot_A & \text{if } z = \bot, \\ x & \text{if } z = \top, \end{cases}$$

where $\bot_A = \Lambda\vec{x}.\bot$. In all naturally arising cases, 1_A will be present as an element of $(A \Rightarrow D \Rightarrow A)^{\sharp}$, and axioms 1–3 above are easy consequences of extensionality. We thus obtain a partiality structure on $\mathbf{A}$.

Even if a unit type is absent from $\mathbf{A}$, we may (less canonically) take $D = \mathbf{A}(\mathtt{N})$, $\top = 0$, and adapt the definition of 1_A accordingly.

Example 4.2.16. As a non-extensional example, consider the van Oosten algebra B of Subsection 3.2.4. This has a strong product structure, with pairing defined by

[5] The role of D is akin to that of the semidecidable subobject classifier or *dominance* Σ in synthetic domain theory—see for example Rosolini [246] or Hyland [122].

$$\langle f,g\rangle(inl(n)) \;=\; f(n)\,, \qquad \langle f,g\rangle(inr(n)) \;=\; g(n)$$

where $inl(n) = 2n$, $inr(n) = 2n+1$. We may define $\top \in \mathsf{B}$ by $\top(0) = 0$, $\top(n+1)\uparrow$, and define an element $r \in \mathsf{B}$ as follows (here $\langle\cdots\rangle$ denotes coding of finite sequences over $\mathbb{N}$):

$$r(\langle n\rangle) \;=\; ?(inr(0))\,, \qquad r(\langle n,0\rangle) \;=\; ?(inl(n))\,, \qquad r(\langle n,0,m\rangle) \;=\; !m\,.$$

It is easy to see that the computation of $r\cdot\langle x,z\rangle$ at any $n \in \mathbb{N}$ first interrogates z at 0, then if this yields 0 gives the value of x at n. The λ-algebra structure of B now allows us to transpose this to obtain an element $\upharpoonright = \lambda fg.r\langle f,g\rangle$.

The verifications of axioms 1–3 are routine exercises once the λ-algebra structure of B is available. Note that axiom 2 relies on the fact that morphisms correspond to *irredundant* decision forests: for instance, the morphism for $f(x \upharpoonright z) \upharpoonright z$ will perform only a single interrogation of $z(0)$ along any given path. (This point will be made clearer in Section 12.1, where we study B as a reflexive object within the category of sequential algorithms.)

This last observation highlights an important point: a more finely intensional λ-algebra than B that kept track of repeated interrogations would *not* typically carry a partiality structure in the sense of Definition 4.2.14, and so would fall outside the scope of the present theory. The model SP^0 is of this more intensional kind, as are many *game models*: see Section 7.4 and Subsection 12.1.7.

Let $\mathbf{A}^\times$ denote the product completion of $\mathbf{A}$ as in Theorem 3.1.15. It is easy to see that a partiality structure on $\mathbf{A}$ gives rise to one on $\mathbf{A}^\times$ via the following equation, which it will be convenient to regard as a fourth axiom for partiality structures:

4. $(x_0,\ldots,x_{n-1}) \upharpoonright z = (x_0 \upharpoonright z,\ldots,x_{n-1} \upharpoonright z)$.

One reason for our interest in partiality structures on $\mathbf{A}$ is that they give rise to a *lifting operation* on the types of $\mathscr{K}(\mathscr{C}(\mathbf{A}^\times))$. Informally, the *lift* $L(A)$ of a type A plays the role of a type of partial functions $1 \rightharpoonup A$; considered at the set-theoretic level, $L(A)$ will typically extend A by adding a single new element $\perp$ (representing 'undefined') in a suitable way. More specifically, we obtain a structure very typical of lifting operations in domain-theoretic settings, namely that of a *strong monad.* (The definition of this notion is embedded in the proof of the following proposition.) We here let $\mathscr{C}^\times = \mathscr{C}(\mathbf{A}^\times)$, and write $(\times)$ for the product functor $\mathscr{C}^\times \times \mathscr{C}^\times \to \mathscr{C}^\times$.

Proposition 4.2.17. *A partiality structure induces a functor* $L : \mathscr{K}(\mathscr{C}^\times) \to \mathscr{K}(\mathscr{C}^\times)$, *together with natural transformations*

$$\eta : id \to L\,, \qquad \mu : L^2 \to L\,, \qquad t : (\times)\circ(id \times L) \to L\circ(\times)$$

making L into a strong monad.

Proof. Suppose $(D,\top,\upharpoonright)$ is a partiality structure on $\mathbf{A}$. We define the functor L by

$$L(A,e) \;=\; (A \times D,\; \lambda(x,z).(ex \upharpoonright z,\, z))\,,$$
$$L(f : (A,e) \to (B,d)) \;=\; \lambda(x,z).(fx \upharpoonright z,\, z)\,.$$

Using axiom 2, it is easy to check that $\lambda(x,z).(ex \upharpoonright z, z)$ is indeed an idempotent, and that $L(f)$ is indeed a morphism $L(A,e) \to L(B,d)$. The functoriality of L is also easily verified.

For any object $X = (A,e)$, we have a morphism

$$\eta_X \;=\; \lambda x.\,(x,\top) \;:\; X \to L(X)\,.$$

Using axiom 1, it is easily checked that these are morphisms of the stated type and that they constitute a natural transformation $\eta : id \to L$. Likewise, for any $X = (A,e)$ we have a morphism

$$\mu_X \;=\; \lambda xzw.\,(ex \upharpoonright (z \upharpoonright w), z \upharpoonright w) \;:\; L^2(X) \to L(X)\,.$$

To see that this is a morphism of the stated type, use axioms 3 and 4 to check that

$$L^2(A,e) \;=\; (A \times D \times D,\; \lambda(x,z,w).\,(ex \upharpoonright (z \upharpoonright w), z \upharpoonright w, w))\,.$$

It now follows using axiom 2 that the μ_X are indeed morphisms and constitute a natural transformation $L^2 \to L$. Moreover, the monad laws

$$\mu \circ L\eta \;=\; \mu \circ \eta L \;=\; id \;:\; L \to L\,, \qquad \mu \circ L\mu \;=\; \mu \circ \mu L \;:\; L^3 \to L^2$$

can be routinely checked by using all four axioms.

Next, for any $X = (A,e)$ and $Y = (B,d)$, we have a morphism

$$t_{XY} \;=\; \lambda(x,(y,z)).\,((x \upharpoonright z, y \upharpoonright z), z) \;:\; X \times L(Y) \to L(X \times Y)\,,$$

and it is routine to check these yield a natural transformation t as required. The following axioms express that t is a *strength* for the monad (L,η,μ); here we write $\cong$ to mean that two morphisms agree modulo the canonical isomorphisms between their respective domains and codomains.

$$t_{1Y} \cong id_Y\,, \qquad t_{XY} \circ (id_X \times \eta_Y) = \eta_{X \times Y}\,,$$
$$t_{(X \times Y)Z} \cong t_{X(Y \times Z)} \circ (id_X \times t_{YZ})\,, \qquad \mu_{X \times Y} \circ L(t_{XY}) \circ t_{X(LY)} = t_{XY} \circ (id_X \times \mu_Y)\,.$$

Again, the verifications of these axioms are tedious but routine exercises. □

Strong lifting monads may also be identified, on a case-by-case basis, in many other intensional models that lack a partiality structure. In such models, however, even $L(\beta)$ will not in general be a retract of any simple type over β, so the presence of lifting requires a genuinely broader repertoire of types than those of $\mathsf{T}^{\to}(\beta)$ and retracts thereof.

In any case, once a strong lifting monad is available, we may obtain a 'category of partial morphisms' from $\mathscr{C}^{\times}$ by general abstract means. By a partial morphism $f : X \rightharpoonup Y$ we shall mean an ordinary morphism $f : X \rightharpoonup L(Y)$; its composition with a partial morphism $g : Y \rightharpoonup Z$ may be defined to be $\mu_Z \circ L(g) \circ f$. It is easy to check that such partial morphisms form a category purely by virtue of the monad laws; it is known as the *Kleisli category* of the monad (L, η, μ). Likewise, we may regard $(X \Rightarrow L(Y))$ as the type of partial morphisms $X \rightharpoonup Y$.

Exercise 4.2.18. (i) Show how the strength t makes it possible to combine partial morphisms $f : X \rightharpoonup Y$, $f' : X \rightharpoonup Y'$ to obtain a partial morphism $X \rightharpoonup Y \times Y'$. Show however that there are in general two ways to do this, depending (informally) on whether f or f' is evaluated first.

(ii) Using the strength, show how the action of L on morphisms $X \to Y$ may be represented internally by a morphism $(X \Rightarrow Y) \to (L(X) \Rightarrow L(Y))$.

We may also use a partiality structure to articulate the notion of a *strict* morphism $X \to_s Y$ for certain objects X. Informally, a morphism f is strict if $f(x)$ is defined only when x is defined. Strict morphisms arise, for instance, in connection with call-by-value styles of computation, in which the evaluation of a term $f(M)$ always begins with an evaluation of M (see Subsection 7.2.2).

To capture this idea, we introduce a tiny piece of additional structure:

Definition 4.2.19. If $(D, \top, \upharpoonright)$ is a partiality structure on $\mathbf{A}$, a *definedness predicate* on a type A of $\mathbf{A}$ is simply an element $\delta \in (A \Rightarrow D)^\sharp$.

Informally, if $\top \in D$ is considered as the reference point for the concept of definedness, a definedness predicate on a datatype A will say which elements of A we wish to regard as defined. As our leading examples, the type $\mathbb{N}_\perp$ (denoting $\mathtt{N}$) will typically have a definedness predicate sending each n to $\top$ and $\perp$ to $\perp$; and any type $L(A,e)$ will have a definedness predicate sending (x,z) to z. By contrast, a type such as $\mathtt{N} \to \mathtt{N}$ will not normally be considered as carrying a definedness predicate: the question of definedness arises for values of $f(n)$, but not for f itself.

Clearly, a definedness predicate δ on A in $\mathbf{A}$ induces a definedness predicate $\lambda x.\delta(ex)$ on any (A,e) in $\mathbf{K}(\mathbf{A})$. If δ is a definedness predicate on X in $\mathbf{K}(\mathbf{A})$, we may say $f : X \to Y$ is *strict* (and write $f : X \to_s Y$) if $f = \lambda x. fx \upharpoonright \delta x$ within $\mathscr{K}(\mathscr{C})$. Under these conditions, the term $\lambda fx. fx \upharpoonright \delta x$ (interpreted in $\mathscr{K}(\mathscr{C})$) defines an idempotent str_{XY} on $X \Rightarrow Y$, known as *strictification*. We write $X \Rightarrow_s Y$ for the corresponding retract of $X \Rightarrow Y$ within $\mathbf{K}(\mathbf{A})$; by a slight overloading, we also write (inc, str) for the associated retraction pair. Intuitively we may regard $X \Rightarrow_s Y$ as a slightly 'simpler' type than $X \Rightarrow Y$, since we do not in general have $X \Rightarrow Y \lhd X \Rightarrow_s Y$.

As with sum types, we shall sometimes take advantage of the presence of such strict exponentials in the Karoubi envelope to provide more specific identification of the essential type level of certain operators. This will be useful especially when X corresponds to $\mathbb{N}_\perp$, so that morphisms $X \to_s Y$ can typically be identified with operations $\mathbb{N} \to Y$. For instance, the *minimization* operator of Definition 3.3.18, when present, conceptually has type $(\mathtt{N} \to_s \mathtt{B}) \to \mathtt{B}$, since the definition of $min(f)$ makes no reference to the behaviour of f other than on genuine numerals n.

Under mild but fussy additional hypotheses on $\mathbf{A}$, one may also show that $\mathscr{K}(\mathbf{A})$ possesses further type structure along the lines of Subsection 4.2.3. For instance, one can (with a little care) formulate the notion of a type of *partial strong booleans* (intuitively playing the role of $2_\perp$), and show that these give rise to *smash sums*

$X \oplus Y$ for any objects X, Y equipped with definedness predicates. (Intuitively, $X \oplus Y$ is the disjoint union of X and Y with their bottom elements identified.)

Even more type structure is available if we assume $\mathbf{A}$ possesses *fixed point* operators with certain properties. However, at this point we move into the realm of *models of* PCF, which will form the subject of Section 7.1. The type structure arising in the Karoubi envelopes of such models will be briefly indicated in Subsection 7.2.1.

4.2.5 Universal Types

To conclude this section, we mention another kind of structure that can be found within many models of interest:

Definition 4.2.20. (i) A *universal object* in a category $\mathscr{C}$ is an object U such that $X \lhd U$ for all objects X in $\mathscr{C}$. In this case, we write $\mathscr{C}_U$ for the full subcategory of $\mathscr{C}$ on the single object U.

(ii) If $\mathbf{A}$ is a relative λ-algebra over T, a type $\upsilon \in \mathsf{T}$ is a *universal type* if $\mathbf{A}(\upsilon)$ is a universal object in $\mathscr{C}(\mathbf{A})$.

It is easy to see that if U is universal in $\mathscr{C}$, there is a full and faithful functor $I : \mathscr{C} \to \mathscr{K}(\mathscr{C}_U)$: for each object A of $\mathscr{C}$, choose a retraction pair $(t_A, r_A) : A \lhd U$, and set $I(A) = (U, t_A r_A)$. For a similar reason, we have $\mathscr{K}(\mathscr{C}) \simeq \mathscr{K}(\mathscr{C}_U)$.

Now suppose $\mathscr{C}$ itself is cartesian closed. If U is universal in $\mathscr{C}$, then in particular U is a reflexive object (that is, $U \Rightarrow U \lhd U$), so as in Subsection 4.1.5 we obtain a relative untyped λ-algebra $\mathbf{A}[U, \Delta]$ for any concretion Δ on $\mathscr{C}$. (We may write simply $\mathbf{A}[U]$ when $\Delta = \mathrm{Hom}(1, -)$.) Thus, if $(\mathscr{C}, \Delta) = (\mathscr{C}(\mathbf{A}), \Delta_{\mathbf{A}})$ for some $\mathbf{A}$, then $\mathbf{A}$ is equivalent to the untyped λ-algebra $\mathbf{A}[U, \Delta]$ in the sense of Definition 4.1.23, since $\mathscr{C}(\mathbf{A}[U, \Delta]) = \mathscr{C}_U$ and $\mathscr{K}(\mathscr{C}_U) \simeq \mathscr{K}(\mathscr{C})$. Furthermore:

Proposition 4.2.21. *$\mathscr{K}(\mathscr{C}_U)$ is cartesian closed, and $I : \mathscr{C} \to \mathscr{K}(\mathscr{C}_U)$ preserves the cartesian closed structure.*

Proof. By Proposition 4.2.2, the cartesian product $U \times U$ exists within $\mathscr{K}(\mathscr{C}_U)$; hence for any X, Y in $\mathscr{K}(\mathscr{C}_U)$ we obtain an object $X \times Y \lhd U \times U$ within $\mathscr{K}(\mathscr{C}_U)$, so $\mathscr{K}(\mathscr{C}_U)$ has cartesian products. But this translates readily to the statement that $\mathbf{K}(\mathbf{A}[U])$ has strong products; and since $\mathbf{K}(\mathbf{A}[U])$ is certainly a $\lambda\eta$-algebra, this means by Theorem 4.1.13 that $\mathscr{C}(\mathbf{K}(\mathbf{A}[U])) \simeq \mathscr{K}(\mathscr{C}_U)$ is a $\lambda\eta\times$-algebra, and hence a cartesian closed category by Proposition 4.1.9.

To see that I preserves the cartesian structure, we note that the equivalence $\mathscr{K}(\mathscr{C}_U) \simeq \mathscr{K}(\mathscr{C})$ identifies I with the canonical inclusion $\mathscr{C} \hookrightarrow \mathscr{K}(\mathscr{C})$ as defined in Subsection 4.1.4, and it is easy to check that in general this inclusion preserves any products and exponentials that exist in $\mathscr{C}$. $\square$

A cartesian closed category $\mathbf{C}$ with a universal object U also enjoys other pleasant properties. For instance, since $\mathbf{A}[U]$ is a total combinatory algebra, we know from Subsection 3.3.5 that every morphism $f : U \to U$ has a fixed point $Y(f) : 1 \to U$, and it follows easily that for any X in $\mathbf{C}$, every morphism $f : X \to X$ has a fixed point $Y(f) : 1 \to X$.

Of particular interest from the perspective of this book is that a universal object is often present even among the simple types. The following proposition gives a convenient way of establishing that a type is universal:

Proposition 4.2.22. *(i) If* $\mathbf{A}$ *is a relative* $\lambda\eta$*-algebra over* $\mathsf{T}^{\to}(\beta)$*, a type* υ *is universal in* $\mathbf{A}$ *iff* $\beta \lhd \upsilon$ *and* $\upsilon \Rightarrow \upsilon \lhd \upsilon$.

(ii) If $\mathbf{A}$ *is a relative* $\lambda\eta$*-algebra over* $\mathsf{T}^{\to\times}(\beta)$*, then* υ *is universal iff* $\beta \lhd \upsilon$*,* $\upsilon \Rightarrow \upsilon \lhd \upsilon$ *and* $\upsilon \times \upsilon \lhd \upsilon$.

Proof. An easy induction on types using Proposition 4.2.1. □

Conceptually, the significance of a universal type υ is that once we have 'understood' the type $\mathbf{A}(\upsilon)$ and its monoid of endomorphisms, we have in some sense understood the whole of $\mathbf{A}$: in principle, no higher types yield any further structural complexity. In later chapters we will show the following (here $\to_s$ is the strict function type constructor as introduced in Subsection 4.2.4):

- The type $\mathtt{B} \to_s \mathtt{N}$ is universal in (the Karoubi envelope of) PC (Section 10.2).
- $(\mathtt{N} \to_s \mathtt{N}) \to \mathtt{N}$ is universal for SR (Section 11.2).
- $\mathtt{N} \to_s \mathtt{N}$ is universal for the sequential algorithm model SA (Section 12.1).
- The models SP^0 and SF have no universal type (Section 7.7).

Exercise 4.2.23. Show by a diagonal argument that if $\mathbf{A}$ is a total λ-algebra over $\mathsf{T}^{\to}(\mathtt{N} = \mathbb{N})$ in which *suc* is computable, then $\mathbf{A}$ cannot contain a universal type.

In cases where a universal type υ is present, it can often be exploited to advantage to give relatively simple proofs of results of interest. For example, to prove that all computable elements of a simply typed λ-algebra $\mathscr{C}$ are definable in some language $\mathscr{L}$ (say, a typed λ-calculus with certain constants), it is sufficient to show that all computable elements of type υ are $\mathscr{L}$-definable, and that both halves of the above retractions are $\mathscr{L}$-definable. We shall apply this method in Sections 10.3, 11.2 and 12.1; see also Longley [177, 182] and McCusker [192].

4.3 Partial λ-Algebras

Our theory so far has concentrated on *total* λ-algebras. However, many natural attempts to define a class of 'partial computable operations' give rise to genuinely partial structures (for instance, we may consider functionals of types such as $[\mathbb{N} \rightharpoonup \mathbb{N}] \rightharpoonup \mathbb{N}$), so it is natural to ask whether our focus on total models means that we are excluding important notions of partial computability from consideration.

In this section, we offer a partial answer to this question by showing that, at least in the *extensional* setting, there is an exact correspondence between non-total type structures over a set X and total type structures over $X_\perp$ possessing a *ground type restrictor* (Theorem 4.3.5). We are also able to say something about the situation for non-extensional models, although our treatment here is perhaps less definitive and may leave room for future improvements. The results of this section are technically new, though the ideas behind them are familiar from the literature in various guises (see Sieber [270], Riecke [244] and Longley [173, Chapter 6]). As we shall see in Subsection 7.2.2, the present results bear closely on the relationship between *call-by-name* and *call-by-value* styles of computation.

We begin by introducing a mild form of partial λ-calculus which we call the *pλ-calculus*, with significantly weaker conversion rules than the λ_p-calculus of Moggi [195], for instance. The point is not that this system gives rise to an especially good theory in itself, but simply that it provides us with a weak notion of *pλ-algebra* that encompasses a wide class of candidates for 'partial λ-algebras'.

To define the set of well-typed $p\lambda$-terms, we augment the typing rules for λ-terms in Subsection 4.1.1 with the rule

$$\frac{\Gamma \vdash M : \sigma \qquad \Gamma \vdash N : \tau}{\Gamma \vdash M_{\upharpoonright N} : \sigma}$$

In contrast to ordinary λ-terms, the denotation of a $p\lambda$-term (under a given valuation in a model) may sometimes be undefined, owing to the fact that application may be non-total. The construct $-_{\upharpoonright -}$ is a syntactic counterpart to the operator $\upharpoonright$ of Subsection 4.2.4: we think of $M_{\upharpoonright N}$ informally as meaning 'evaluate N; if this succeeds, discard the result and behave like M'. In place of the usual β-equality, we now take $=_{p\beta}$ to be the equivalence relation on $p\lambda$-terms generated by the following clauses, where $C[-]$ ranges over the basic syntactic contexts $-P$, $P-$, $-_{\upharpoonright P}$, $P_{\upharpoonright -}$ and $\lambda x.-$.

$$\begin{array}{ll} M_{\upharpoonright x} =_{p\beta} M\,, & (\lambda x.M)N =_{p\beta} (M[x \mapsto N])_{\upharpoonright N}\,, \\ M_{\upharpoonright \lambda x.N} =_{p\beta} M\,, & M =_{p\beta} N \;\Rightarrow\; C[M] =_{p\beta} C[N]\,. \end{array}$$

The intention here is to establish a set of equations $M =_{p\beta} N$ that can be expected to 'hold uniformly' in any partial λ-algebra. In the case of closed terms M,N, this will mean that the interpretation of M will be defined iff the interpretation of N is, and in this case they will be equal. The equations in the left column above respectively capture the idea that the interpretations of x and $\lambda x.N$ are always defined. One could in principle eliminate $-_{\upharpoonright -}$ from the system since $M_{\upharpoonright N} =_{p\beta} (\lambda x.M)N$, but it is conceptually preferable to treat restriction as a primitive rather than as dependent on function application.

We may now define our concept of partial λ-algebra, using an axiomatization analogous to that of Definition 4.1.5:

Definition 4.3.1. Let $\mathbf{A}$ be a partial applicative structure over T, with $* \in \mathbf{A}(\mathtt{U})$ some arbitrarily chosen element. We say $\mathbf{A}$ is a *pλ-algebra* if it is equipped with a compositional interpretation assigning to each well-typed $p\lambda$-term $x_0^{\sigma_0}, \ldots, x_{r-1}^{\sigma_{r-1}} \vdash M : \tau$ an element $[\![M]\!]_\Gamma^i \in \mathbf{A}(\mathtt{U} \rightarrow \sigma_0, \ldots, \sigma_{r-1} \rightarrow \tau)$, in such a way that the following axioms hold. Here $\simeq$ denotes Kleene equality, and if ν is a valuation for $\Gamma = x_0, \ldots, x_{r-1}$, we set $[\![M]\!]_\nu \simeq [\![M]\!]_\Gamma^i \cdot * \cdot \nu(x_0) \cdot \ldots \cdot \nu(x_{r-1})$.

1. $[\![x]\!]_\nu = \nu(x)$.
2. $[\![MN]\!]_\nu \simeq [\![M]\!]_\nu \cdot [\![N]\!]_\nu$.
3. $[\![\lambda x.M]\!]^i_\Gamma = [\![M]\!]^i_{\Gamma,x}$, and $[\![\lambda x.M]\!]_\nu \downarrow$ always.
4. $[\![M_{\upharpoonright N}]\!]_\nu \simeq [\![M]\!]_\nu$ if $[\![N]\!]_\nu \downarrow$; otherwise $[\![M_{\upharpoonright N}]\!]_\nu$ is undefined.
5. If $\Gamma \vdash M : \sigma$ and $\nu' : \Gamma'$ extends $\nu : \Gamma$, then $[\![M]\!]_\nu \simeq [\![M]\!]'_\nu$.
6. If $M =_{p\beta} M'$ then $[\![M]\!]^i_\Gamma = [\![M']\!]^i_\Gamma$.

The purpose of U and $*$ here is simply to provide a suspension in order to ensure that $[\![M]\!]^i_\Gamma$, unlike $[\![M]\!]_\nu$, is always defined (cf. Subsection 3.3.3).

Whilst this notion covers a wide range of attempts to define a model of partial computable operations, there is an important caveat to be noted. Just as with our notion of partiality structure in Subsection 4.2.4, our definition does not encompass highly intensional models that explicitly track repetitions of evaluations: in particular, the 'β-rule' above will not typically hold in models of call-by-value computation if the number of evaluations of N is deemed significant. Of course, this is not an issue for the extensional models that are our primary concern here.

Clearly, any $p\lambda$-algebra is a TPCA, with $k = [\![\lambda xy.x]\!]$ and $s = [\![\lambda xyz.xz(yz)]\!]$ as usual. Moreover, by mimicking the proof of Proposition 4.1.3, it is straightforward to check that any *extensional* TPCA is a $p\lambda$-algebra, and in this case the partial interpretation is uniquely determined.

We will show that extensional, non-total $p\lambda$-algebras over a base set X correspond exactly to ordinary extensional λ-algebras over $X_\perp$ satisfying some mild conditions; we shall also acquire some information on non-extensional models along the way. Specifically, we will work initially with extensional $p\lambda$-algebras **A** over the type world $\mathsf{T}^{\rightarrow}(\alpha = X, \mathtt{U} = \{*\})$, and with ordinary extensional λ-algebras **B** over $\mathsf{T}^{\rightarrow}(\beta = X_\perp, \mathtt{V} = D)$, where we let $D = \{\top, \perp\}$. The types U and V allow us to exhibit clearly the machinery that we use for representing undefinedness, although in the end we shall dispense with these. Our main interest is in the case $X = \mathbb{N}$.

The technical conditions on λ-algebras that we shall consider are as follows:

Definition 4.3.2. (i) An extensional λ-algebra **B** with $\mathbf{B}(\beta) = X_\perp$ has a *ground type restrictor* if there is an element $-\upharpoonright^0 - \in \mathbf{B}(\beta, \beta \rightarrow \beta)$ such that

$$a \upharpoonright^0 b = \begin{cases} \perp \text{ if } b = \perp\,, \\ a \text{ otherwise}\,. \end{cases}$$

(ii) We say **B** is *standard over* $X_\perp$ if $\mathbf{B}(\beta) = X_\perp$, $\mathbf{B}(\mathtt{V}) = D$, and $\mathtt{V} \lhd \beta$ in **B** via a retraction (t, r), where $r(\perp) = \perp$ and $r(x) = \top$ for all $x \in X$.

In any standard extensional λ-algebra with a ground type restrictor $\upharpoonright^0$, we obtain a partiality structure $(D, \top, \upharpoonright)$, where $x \upharpoonright_\beta z = x \upharpoonright^0 (tz)$, $x \upharpoonright_\mathtt{V} z = r((tx) \upharpoonright^0 (tz))$, and in general $f \upharpoonright_{\vec{\sigma} \rightarrow \gamma} z = \lambda \vec{x}.((f\vec{x}) \upharpoonright_\gamma z)$. The retraction $r : \beta \rightarrow \mathtt{V}$ then serves as a definedness predicate on β.

For the first half of our correspondence, we show how a simply typed $p\lambda$-algebra **A** over $\mathsf{T}^{\rightarrow}(\alpha = X, \mathtt{U} = \{*\})$ gives rise to an ordinary $\lambda\eta$-algebra **B** over $\mathsf{T}^{\rightarrow}(\beta, \mathtt{V})$, with other good properties following when **A** is extensional. To construct **B**, we shall make use of a translation $-^\circ$ from $\mathsf{T}^{\rightarrow}(\beta, \mathtt{V})$ to $\mathsf{T}^{\rightarrow}(\alpha, \mathtt{U})$:

$$\beta^\circ = \mathtt{U} \to \alpha\,, \qquad \mathtt{V}^\circ = \mathtt{U} \to \mathtt{U}\,, \qquad (\sigma \to \tau)^\circ = \sigma^\circ \to \tau^\circ\,.$$

We extend this notation to type environments Γ in the obvious way. We now wish to translate λ-terms $\Gamma \vdash M : \sigma$ to $p\lambda$-terms $\Gamma^\circ \vdash M^\circ : \sigma^\circ$. For this, we require the notion of *hereditary η-expansion*: if x is a variable of type σ° where $\sigma = \sigma_0, \ldots, \sigma_{r-1} \to \gamma$, we (recursively) define

$$x^\eta = \lambda y_0 \ldots y_{r-1} v^{\mathtt{U}}.\, x y_0^\eta \cdots y_{r-1}^\eta v\,.$$

(We will also use the same notation for semantic elements of $\mathbf{A}(\sigma^\circ)$.) We may now define our syntactic translation $M \mapsto M^\circ$ by

$$x^\circ = x^\eta\,, \qquad (MN)^\circ = M^\circ N^\circ\,, \qquad (\lambda x.M)^\circ = \lambda x.(M^\circ)\,.$$

The following properties are easily checked:

1. $x^\eta[x \mapsto x^\eta] =_{p\beta} x^\eta$ (by induction on the type of x); hence $M^\circ[x \mapsto x^\circ] =_{p\beta} M^\circ$ for any M.
2. $M^\circ =_{p\beta} \lambda \vec{y} v. M^\circ \vec{y}^\circ v$ (by induction on the structure of M).

Proposition 4.3.3. *Suppose* $\mathbf{A}$ *is a* $p\lambda$*-algebra over* $\mathsf{T}^{\to}(\alpha = X, \mathtt{U} = \{*\})$. *Define a total applicative structure* $\mathbf{B}$ *over* $\mathsf{T}^{\to}(\beta, \mathtt{V})$ *as follows:*

$$\mathbf{B}(\sigma) = \{f \in \mathbf{A}(\sigma^\circ) \mid f = f^\eta\}\,, \qquad f \cdot^{\mathbf{B}}_{\sigma\tau} x = f \cdot^{\mathbf{A}}_{\sigma^\circ\tau^\circ} x\,.$$

Then

(i) $\mathbf{B}$ *is a* $\lambda\eta$*-algebra over* $\mathsf{T}^{\to}(\beta, \mathtt{V})$.

(ii) If $\mathbf{A}$ *is extensional and non-total, then* $\mathbf{B}$ *is extensional and standard over* $X_\perp$ *and contains a ground type restrictor.*

Proof. (i) We use the translation $-^\circ$ to define an internal interpretation of λ-terms in $\mathbf{B}$: if $\Gamma \vdash M : \sigma$ and ν is a valuation for Γ in $\mathbf{B}$, we may take $[\![M]\!]^{i\mathbf{B}}_\Gamma = [\![M^\circ]\!]^{i\mathbf{A}}_\Gamma \cdot *$. Using property 2 above, the resulting value $[\![M]\!]^{\mathbf{B}}_\nu$ is always defined and is in $\mathbf{B}(\sigma)$.

Since $M^\circ =_{p\beta} \lambda \vec{y} v. M^\circ \vec{y}^\circ v$, we have that $P_{\restriction M^\circ} =_{p\beta} P$ for any $p\lambda$-term P. It is also clear from property 1 above that $M^\circ =_{p\beta} (\lambda x.Mx)^\circ$. Using these facts, a derivation of $M =_{\beta\eta} M'$ routinely translates to a derivation of $M^\circ =_{p\beta} M'^\circ$. By condition 6 of Definition 4.3.1, it follows that if $M =_{\beta\eta} M'$ then $[\![M]\!]^{i\mathbf{B}}_\nu = [\![M']\!]^{i\mathbf{B}}_\nu$. This establishes condition 5 of Definition 4.1.5; the other conditions 1–4 are easily checked.

(ii) Now assume $\mathbf{A}$ is extensional. To see that $\mathbf{B}$ is extensional, suppose $f, f' \in \mathbf{B}(\sigma \to \tau)$ agree on all elements of $\mathbf{B}(\sigma)$. Since $f = f^\eta$ and $f' = f'^\eta$, it is easy to see that they agree on the whole of $\mathbf{A}(\sigma^\circ)$, so by the extensionality of $\mathbf{A}$ we have $f = f'$. If furthermore $\mathbf{A}$ is non-total, it is clear that $\mathbf{B}(\mathtt{V})$ contains just two elements corresponding to 'defined' and 'undefined', and likewise that $\mathbf{B}$ is standard over $X_\perp$.

For the ground type restrictor, we define (in $\mathbf{A}$)

$$\mathsf{1}^0 = \lambda x^{\mathtt{U}\to\alpha} y^{\mathtt{U}\to\alpha}.\lambda v^{\mathtt{U}}.\,(xv)\,\mathsf{1}_{(yv)}\,.$$

It is easily checked by pointwise reasoning that this has the required property. □

Going in the other direction, we will show how a $\lambda\eta$-algebra **B** with a well-behaved partiality structure and definedness predicate on the base type gives rise to a $p\lambda$-algebra **A**. We will make use of the following technical conditions:

- A partiality structure $(D,\top,\upharpoonright)$ on **B** is *strict* if for all $w,z \in D$, $w \upharpoonright z = \top$ implies $z = \top$.
- A definedness predicate $\delta \in (X_\perp \Rightarrow D)$ in **B** is *proper* if $\delta(x \upharpoonright z) = \delta(x) \upharpoonright z$ holds uniformly.

Note that these conditions automatically hold in the extensional setting if $(D,\top,\upharpoonright)$ and δ arise from a ground type restrictor as indicated after Definition 4.3.2. We also use the machinery of lifting and strict function spaces from Subsection 4.2.4, and write ξ_A for the obvious definedness predicate on $L(A)$. Recall that I denotes the inclusion $\mathbf{B} \to \mathbf{K}(\mathbf{B})$.

Proposition 4.3.4. *Suppose* **B** *is a* $\lambda\eta$*-algebra over* $\mathsf{T}^{\rightarrow}(\beta = X_\perp, \mathtt{V} = D)$*, with* $(D,\top,\upharpoonright)$ *a strict partiality structure and* $\delta \in X_\perp \Rightarrow D$ *a proper definedness predicate. For each type* $\sigma \in \mathsf{T}^{\rightarrow}(\alpha,\mathtt{U})$*, define an object* C_σ *in* $\mathbf{K}(\mathbf{B})$ *and a morphism* $\delta_\sigma : C_\sigma \to I(D)$*:*

$$
\begin{aligned}
(C_\alpha, \delta_\alpha) &= (I(X_\perp),\ \delta)\ ,\\
(C_{\mathtt{U}}, \delta_{\mathtt{U}}) &= (I(D),\ id)\ ,\\
(C_{\sigma\to\tau}, \delta_{\sigma\to\tau}) &= (L(C_\sigma \Rightarrow_s C_\tau),\ \xi_{C_\sigma \Rightarrow_s C_\tau})\ .
\end{aligned}
$$

Now take $\mathbf{A}(\sigma) = \delta_\sigma^{-1}(\top)$ *for each* σ*, and define partial application operations* $\cdot_{\sigma\tau} : \mathbf{A}(\sigma \to \tau) \times \mathbf{A}(\sigma) \rightharpoonup \mathbf{A}(\tau)$ *by*

$$
f \cdot_{\sigma\tau} x \;=\; y \;\; \mathit{iff} \;\; f(x) = y \;\in\; \mathbf{A}(\tau)\ .
$$

Then

(i) **A** *is a* $p\lambda$*-algebra over* $\mathsf{T}^{\rightarrow}(\alpha,\mathtt{U})$*.*

(ii) If **B** *is extensional and standard over* $X_\perp$ *and* $(D,\top,\upharpoonright)$ *and* δ *arise from a ground type restrictor* $\upharpoonright$*, then* **A** *is extensional.*

Proof. (i) We sketch the construction of an internal interpretation of $p\lambda$-terms in **A** as in Definition 4.3.1, leaving detailed checking to the reader.

Writing $t' : (\times) \circ (L \times id) \to L \circ (\times)$ for the obvious dual of the natural transformation t from Proposition 4.2.17, and h_τ for the evident map $L(C_\tau) \to C_\tau$ available for each τ, we may define for each σ,τ a strict application operation

$$
app'_{\sigma\tau} \;=\; h_\tau \circ L(\lambda(f,x).inc\, f\, x) \circ t'_{C_\sigma \Rightarrow_s C_\tau, C_\sigma} \;:\; C_{\sigma\to\tau} \times C_\sigma \to C_\tau\ .
$$

We also note that an evident restriction operator $\upharpoonright$ is available on each C_τ, and if $\vec{y} = y_0^{\sigma_0},\ldots,y_{r-1}^{\sigma_{r-1}}$, we use the notation

$$
M \upharpoonright \delta(\vec{y}) \;\equiv\; M \upharpoonright \delta_{\sigma_{r-1}}(y_{r-1}) \upharpoonright \cdots \upharpoonright \delta_{\sigma_0}(y_0)\ .
$$

For each $p\lambda$-term $\Gamma \vdash M : \tau$ (where $\Gamma = x_0^{\sigma_0},\ldots,x_{r-1}^{\sigma_{r-1}}$), we may now define an interpretation $[\![M]\!]'_\Gamma \in C_{\sigma_0} \Rightarrow \cdots \Rightarrow C_{\sigma_{r-1}} \Rightarrow C_\tau$ compositionally by

$$\begin{aligned} [\![x_j]\!]'_\Gamma &= \lambda y_0 \ldots y_{r-1}.(y_j \upharpoonright \delta(\vec{y}))\,, \\ [\![MN]\!]'_\Gamma &= \lambda \vec{y}.\, app'([\![M]\!]'_\Gamma \vec{y},\, [\![N]\!]'_\Gamma \vec{y})\,, \\ [\![\lambda x^\sigma.M]\!]'_\Gamma &= \lambda \vec{y}.\, \eta(str_{C_\sigma C_\tau}([\![M]\!]'_{\Gamma,x}\vec{y})) \upharpoonright \delta(\vec{y})\,, \\ [\![M_{\upharpoonright N}]\!]'_\Gamma &= \lambda \vec{y}.\, ([\![M]\!]'_\Gamma \vec{y}) \upharpoonright \delta_\tau([\![N]\!]'_\Gamma \vec{y})\,. \end{aligned}$$

One can verify by induction on the structure of M that $[\![M]\!]'_\Gamma = \lambda \vec{y}.([\![M]\!]'_\Gamma \vec{y}) \upharpoonright \delta(\vec{y})$, using the fact that δ_α and consequently all the δ_σ are proper. Given this, one can show by induction on derivations that if $M =_{p\beta} M'$ then $[\![M]\!]'_\Gamma = [\![M']\!]'_\Gamma$.

Since $[\![M]\!]'_\Gamma$ is strict in each of its arguments, we can transform it to an element

$$[\![M]\!]^i_\Gamma \in \mathbf{A}(\mathtt{U} \to \sigma_0,\ldots,\sigma_{r-1} \to \tau)$$

from which $[\![M]\!]'_\Gamma$ itself can be recovered; thus $[\![-]\!]^i_-$ is also compositional. The preceding discussion verifies condition 6 of Definition 4.3.1, and the remaining conditions 1–5 are also easily checked. Note that the strictness of $\upharpoonright$ is used for condition 2 (to show that if $[\![MN]\!]_\nu\downarrow$ then $[\![M]\!]_\nu\downarrow$ and $[\![N]\!]_\nu\downarrow$) and for condition 4 (to show that if $[\![M_{\upharpoonright N}]\!]_\nu\downarrow$ then $[\![N]\!]_\nu\downarrow$).

(ii) Suppose the hypotheses apply; then each C_σ contains just a single element $\bot_\sigma$ besides the elements of $\mathbf{A}(\sigma)$, where $\delta_\sigma(\bot_\sigma) = \bot$. To see that $\mathbf{A}$ is extensional, suppose $f,g \in \mathbf{A}(\sigma \to \tau)$ with $f \cdot x \simeq g \cdot x$ for all $x \in \mathbf{A}(\sigma)$. Let f',g' be elements of $C_\sigma \Rightarrow_s C_\tau$ representing f,g respectively; then clearly $f'(x) = g'(x)$ for all $x \in \mathbf{A}(\sigma)$. But since f',g' are strict, we also have $f'(\bot_\sigma) = g'(\bot_\sigma) = \bot_\tau$. Thus $f'(x) = g'(x)$ for all $x \in C_\sigma$, so by extensionality of $\mathbf{B}$ we have $f' = g'$, whence $f = g$. □

It is now routine to check that in the extensional setting, the constructions of Propositions 4.3.3(ii) and 4.3.4(ii) are mutually inverse up to isomorphism: that is, there is an exact correspondence between extensional non-total $p\lambda$-algebras over $\mathsf{T}^{\to}(\alpha,\mathtt{U}=\{*\})$ and extensional λ-algebras over $\mathsf{T}^{\to}(\beta,\mathtt{V})$ with a standard definedness structure. Moreover, it is easy to dispense with the types U and V, since (in effect) they arise by splitting idempotents on α and β respectively. Finally, we recall from Subsection 3.6.4 that every extensional type structure is isomorphic to a unique *canonical* one. Putting all this together, we have

Theorem 4.3.5. *For any non-empty set X, there is a canonical bijection between the following classes of structures:*

- *non-total canonical type structures over X,*
- *total canonical type structures over $X_\bot$ containing a ground type restrictor.* □

In fact, our proof also shows that each extensional non-total $\mathbf{A}$ is *equivalent* to the corresponding total $\mathbf{B}$ in the sense of Definition 3.4.4; thus, every extensional simply typed TPCA is equivalent to a total one.

The above theorem shows that in the simply typed extensional setting, nothing essential is lost by concentrating on total λ-algebras. Even in non-extensional settings, we have by Proposition 4.3.3(i) that every non-total TPCA leaves its mark on the world of total λ-algebras, though there may in principle be distinctions among non-total models that are not reflected there.

Most of the simply typed models we shall study in depth in this book are extensional. The main exceptions are the sequential procedure model SP^0 of Chapters 6 and 7, and the sequential algorithm model SA of Section 12.1. Of these, SA at least satisfies the hypotheses of Propositions 4.3.4(i), so a partial analogue is available. For SP^0 (as distinct from its extensional quotient SF), the question of an appropriate call-by-value counterpart will be commented on in Subsection 6.1.6.

4.4 Prelogical Relations and Their Applications

We conclude this chapter by introducing a powerful proof tool that will form an important part of our arsenal: the technique of *prelogical* (and logical) *relations*. Very broadly, these offer a means by which properties of elements at base type can be coherently extended to properties of elements at all types. This is useful, for instance, when we wish to prove something about λ-terms M by induction on the structure of M, since even a term of base type may contain subterms of arbitrarily high type.

The method of logical relations was first implicitly used by Tait [285] to prove that Gödel's System T is strongly terminating (see Theorem 4.4.6 below). Similar ideas were then used to obtain many other results in λ-calculus and proof theory. In [234, 238], Plotkin observed that many of these applications followed a common pattern, and introduced the general concept of a *logical relation* and the associated basic lemma (cf. Lemma 4.4.3). The theory of logical relations was taken considerably further by Statman in [278]. Later, Honsell and Sannella [119] noted that several known proofs in a similar spirit used relations that were not in fact logical, but could be accounted for within the broader framework of *prelogical relations*, which improves on the theory of logical relations in many mathematical respects. Our exposition here closely follows that of [119].

The theory of prelogical relations applies not just to λ-algebras, but to higher-order structures of many other kinds (they need not even be higher-order models in general). Nonetheless, the theory belongs naturally within the present chapter since it is intrinsically concerned with interpretations of λ-calculus (albeit in a weak sense). In Subsection 4.4.1 we give the basic definitions and fundamental results for prelogical and logical relations; in Subsection 4.4.2 we illustrate the technique by obtaining some key structural results about Gödel's System T. In Subsection 4.4.3 we extract from these some technical results which we shall use in Chapter 5.

4.4.1 Prelogical and Logical Relations

We work in the setting of a type world T with arrow structure, and a set C of constant symbols c each with an assigned type $\tau_c \in \mathsf{T}$. We shall write $\Lambda(C)$ for the set of well-typed λ-terms over T and C defined as in Section 4.1.1.

The following generalizes the notion of pre-interpretation given by Definition 4.1.2.

Definition 4.4.1. Let $\mathbf{A}$ be a total applicative structure. A *weak interpretation* of λ-terms in $\mathbf{A}$ is a function $[\![-]\!]_-$ assigning to each well-typed term $\Gamma \vdash M : \sigma$ and each valuation $\nu : \Gamma$ an element $[\![M]\!]_\nu \in \mathbf{A}(\sigma)$, such that the following conditions are satisfied:

- $[\![x]\!]_\nu = \nu(x)$.
- $[\![MN]\!]_\nu = [\![M]\!]_\nu \cdot [\![N]\!]_\nu$.
- If $\Gamma \vdash M : \sigma$, Γ' extends Γ, and $\nu' : \Gamma'$ extends $\nu : \Gamma$, then $[\![M]\!]_{\nu'} = [\![M]\!]_\nu$.

In contrast to Definition 4.1.2(ii), we do not require $[\![\lambda x.M]\!]_\nu \cdot a = [\![M]\!]_{\nu(x \mapsto a)}$.

We now suppose that $\mathbf{A}_0, \ldots, \mathbf{A}_{n-1}$ are total applicative structures over T, and that each $\mathbf{A}_i$ is equipped with a weak interpretation $[\![-]\!]^i_-$ of well-typed λ-terms over C as in Definition 4.1.2. As usual, we abbreviate $[\![M]\!]^i_\emptyset$ to $[\![M]\!]^i$. Our basic definition is as follows:

Definition 4.4.2. A *prelogical relation* R over $\mathbf{A}_0, \ldots, \mathbf{A}_{n-1}$ consists of a T-indexed family of relations $R_\sigma \subseteq \mathbf{A}_0(\sigma) \times \cdots \times \mathbf{A}_{n-1}(\sigma)$ satisfying the following conditions:

1. For each constant $c \in C$, we have $R_{\tau_c}([\![c]\!]^0, \ldots, [\![c]\!]^{n-1})$.
2. If $R_{\sigma \to \tau}(f_0, \ldots, f_{n-1})$ and $R_\sigma(a_0, \ldots, a_{n-1})$, then $R_\tau(f_0 \cdot a_0, \ldots, f_{n-1} \cdot a_{n-1})$.
3. Suppose $\Gamma, x^\sigma \vdash M : \tau$, and we have valuations ν_i for Γ in $\mathbf{A}^i$ such that

   - $R_\Gamma(\nu_0, \ldots, \nu_{n-1})$ (that is, $R_\rho(\nu_0(y), \ldots, \nu_{n-1}(y))$ for all $y^\rho \in \Gamma$),
   - whenever $R_\sigma(a_0, \ldots, a_{n-1})$, we have $R_\tau([\![M]\!]^0_{\nu_0(x \mapsto a_0)}, \ldots, [\![M]\!]^{n-1}_{\nu_{n-1}(x \mapsto a_{n-1})})$.

   Then $R_{\sigma \to \tau}([\![\lambda x.M]\!]^0_{\nu_0}, \ldots, [\![\lambda x.M]\!]^{n-1}_{\nu_{n-1}})$.

The number n here is called the *arity* of the prelogical relation R. We shall mostly be concerned with the cases $n = 1, 2$, although relations of higher arity will occasionally be useful (see Proposition 7.5.14).

The principal reason for being interested in prelogical relations is spelt out by the following fundamental lemma. We use the notation $R_\Gamma(\nu_0, \ldots, \nu_{n-1})$ as in Definition 4.4.2 above.

Lemma 4.4.3 (Basic lemma). *Suppose R is a prelogical relation over $\mathbf{A}_0, \ldots, \mathbf{A}_{n-1}$ as above, and $\Gamma \vdash M : \tau$ is any well-typed λ-term over C. If $\nu_i : \Gamma$ is a valuation in $\mathbf{A}_i$ for each i, then*

$$R_\Gamma(\nu_0, \ldots, \nu_{n-1}) \;\Rightarrow\; R_\tau([\![M]\!]_{\nu_0}, \ldots, [\![M]\!]_{\nu_{n-1}})\,.$$

The proof is by induction on the structure of M; it is practically a triviality given the way Definition 4.4.2 is set up. In particular, the basic lemma implies that for any *closed* term $M : \tau$ we have $R_\tau([\![M]\!]^0, \ldots, [\![M]\!]^{n-1})$.

The idea behind the use of prelogical relations is as follows. To show that some property P holds for (say) the interpretations of all closed λ-terms of a certain type, it suffices to exhibit some prelogical relation R that guarantees P for such terms. In effect, the relation R plays the role of a generalization of P to terms of arbitrary type, which can serve as a suitable 'induction hypothesis' for a proof by induction on term structure. The power of this method lies in the fact that even relations R that are easy to define may often be of a high logical complexity, so that the use of the basic lemma embodies an induction principle of high proof-theoretic strength. This will be illustrated in Subsection 4.4.2.

The class of prelogical relations enjoys many pleasant properties. For instance, the class of binary prelogical relations contains identity relations and is closed under relational composition; more generally, prelogical relations are closed under intersections, products, projections, permutations and universal quantification. We refer to [119] for further details. Note also that if $\mathbf{A}, \mathbf{B}$ are TPCAs, any prelogical relation R over $\mathbf{A}, \mathbf{B}$ that is total on $\mathbf{A}$ (i.e. such that $\forall a \in \mathbf{A}(\sigma). \exists b \in \mathbf{B}(\sigma). R_\sigma(a, b)$) is a fortiori a type-respecting applicative simulation $\mathbf{A} \longrightarrow\!\!\triangleright\, \mathbf{B}$.

The stronger notion of an (admissible) *logical relation* is in effect obtained by strengthening condition 2 of Definition 4.4.2 to a necessary and sufficient condition for $R_{\sigma\to\tau}(\vec{f})$. Our terminology here is compatible with that of [194] and [119].

Definition 4.4.4. A *logical relation* R over $\mathbf{A}_0, \ldots, \mathbf{A}_{n-1}$ consists of a T-indexed family of relations $R_\sigma \subseteq \mathbf{A}_0(\sigma) \times \cdots \times \mathbf{A}_{n-1}(\sigma)$ satisfying the following conditions:

1. For each constant $c \in C$, we have $R_{\tau_c}([\![c]\!]^0, \ldots, [\![c]\!]^{n-1})$.
2. If $f_i \in \mathbf{A}^i(\sigma \to \tau)$ for each i, then $R_{\sigma\to\tau}(f_0, \ldots, f_{n-1})$ if and only if

$$\forall a_0, \ldots, a_{n-1}.\ R_\sigma(a_0, \ldots, a_{n-1}) \Rightarrow R_\tau(f_0 \cdot a_0, \ldots, f_{n-1} \cdot a_{n-1})\,.$$

A logical relation R is *admissible* if it also satisfies condition 3 of Definition 4.4.2.

Clearly an admissible logical relation is prelogical, and hence subject to the basic lemma. Furthermore, if our weak interpretations $[\![-]\!]^i_-$ satisfy the condition $[\![\lambda x.M]\!]_\nu \cdot a = [\![M]\!]_{\nu(x\mapsto a)}$ (as is the case for interpretations in a λ-algebra, for instance), then condition 3 is automatically satisfied, so that every logical relation is prelogical.

The main interest in logical relations is that they are often easy to specify, since if R is logical then $R_{\sigma\to\tau}$ is completely determined by R_σ and R_τ via condition 2. Suppose for instance that $\mathsf{T} = \mathsf{T}^{\to}(\beta)$; then by induction on types, we see that any logical relation R among structures over T is uniquely determined by R_β. Moreover, given any relation R_β whatever, we can use condition 2 to define a family $R = \{R_\sigma \mid \sigma \in \mathsf{T}\}$, which we refer to as the *logical lifting* of R_β. A typical pattern of argument is therefore that we will simply specify a base type relation R_β, and satisfy ourselves that its logical lifting satisfies conditions 1 and 3 of Definition 4.4.2.

We shall often appeal to the following characterization of logical liftings in terms of the argument form of a type. If $\sigma = \sigma_0, \ldots, \sigma_{r-1} \to \beta$, then $R_\sigma(f_0, \ldots, f_{n-1})$ holds iff whenever $a_{ij} \in \mathbf{A}^i(\sigma_j)$ for all $i < n$ and $j < r$, we have

$$R_{\sigma_0}(a_{00}, \ldots, a_{(n-1)0}) \wedge \cdots \wedge R_{\sigma_{r-1}}(a_{0(r-1)}, \ldots, a_{(n-1)(r-1)}) \\ \Rightarrow R_\beta(f_0 a_{00} \ldots a_{0(r-1)}, \cdots, f_{n-1} a_{(n-1)0} \cdots a_{(n-1)(r-1)}) .$$

Furthermore, for binary relations, the following facts are easily checked:

Proposition 4.4.5. *Suppose that* $\mathbf{A}$ *is a total higher-order structure over* $\mathsf{T}^{\to}(\beta)$, *and that* $R_\beta \subseteq \mathbf{A}(\beta) \times \mathbf{A}(\beta)$ *has logical lifting* $R = \{R_\sigma \mid \sigma \in \mathsf{T}\}$.
(i) If R_β *is symmetric, so is each* R_σ.
(ii) If R_β *is transitive, so is each* R_σ.

Thus, if R_β is a partial equivalence relation, then so are the R_σ, and indeed these are exactly the partial equivalence relations featuring in the extensional collapse construction (see Section 3.6.3). Note, however, that logical lifting does not in general preserve reflexivity: for instance, if R_β is the identity relation, then R is reflexive at all types iff $\mathbf{A}$ is *pre-extensional* in the sense of Exercise 3.6.13, to wit,

$$\forall f, g \in \mathbf{A}(\sigma \to \tau). \forall h \in \mathbf{A}((\sigma \to \tau) \to \mathtt{N}). \; (\forall x. f \cdot x = g \cdot x) \Rightarrow h \cdot f = h \cdot g .$$

As pointed out in [119], the class of logical relations fails to satisfy many of the closure properties enjoyed by prelogical relations: for instance, binary logical relations are not closed under relational composition, and logical relations in general are not closed under intersections or projections. Thus, prelogical relations arguably form a mathematical more natural class than logical ones. Nonetheless, logical relations suffice for most of the applications we shall encounter in this book, and as noted already, their particular virtue is that they can be readily specified as the logical lifting of a base type relation.

4.4.2 Applications to System T

We now give a non-trivial example of the use of logical relations. We use this as an opportunity to establish some of the key structural properties of Gödel's System T, which plays a significant role in our subject although it is weaker than any of the 'general' computability notions to be considered in later chapters (cf. Subsection 5.1.5). For the purpose of this section, we let $\mathsf{T} = \mathsf{T}^{\to}(\mathtt{N})$ and take C to be the set of constants of System T as in Subsection 3.2.3:

$$\begin{aligned} \widehat{0} &: \mathtt{N} , \\ suc &: \mathtt{N} \to \mathtt{N} , \\ rec_\sigma &: \sigma \to (\sigma \to \mathtt{N} \to \sigma) \to \mathtt{N} \to \sigma \qquad \text{for each } \sigma \in \mathsf{T} . \end{aligned}$$

We shall write $\widehat{k}$ for $suc^k \, \widehat{0}$, and shall refer to terms of this form as *numerals*.

We define $\mathbf{T}$ to be the total higher-order structure over T consisting of closed terms of System T modulo α-equivalence, with application operations given by juxtaposition of terms. (We do not quotient by β-equality or any other reduction relation.) A weak interpretation of System T terms within $\mathbf{T}$ itself may now be defined using simultaneous substitution:

$$[\![M]\!]_{(x_0 \mapsto N_0, \ldots, x_{n-1} \mapsto N_{n-1})} = M[x_0 \mapsto N_0, \ldots, x_{n-1} \mapsto N_{n-1}] .$$

It is easy to see that this satisfies the conditions of Definition 4.4.1.

In Subsection 3.2.3, closed System T terms are equipped with a deterministic reduction relation $\rightsquigarrow$. Here, we shall in the first instance consider a more generous non-deterministic reduction relation $\rightsquigarrow_n$ generated as follows:

- $(\lambda x.M)N \rightsquigarrow_n M[x \mapsto N]$.
- $rec_\sigma\, P\, Q\, \widehat{0} \rightsquigarrow_n P$.
- $rec_\sigma\, P\, Q\, (suc\, N) \rightsquigarrow_n Q\, (rec_\sigma\, P\, Q\, N)\, N$.
- If $M \rightsquigarrow_n M'$ then $C[M] \rightsquigarrow_n C[M']$ for any term context $C[-]$.

We write $\rightsquigarrow_n^*$ for the reflexive-transitive closure of $\rightsquigarrow_n$. We say a term $M : \sigma$ is *strongly terminating* if every $\rightsquigarrow_n$-reduction sequence starting from M is finite.

The following classic application of logical relations is due to Tait [285]; the proof we give is essentially the same as his.

Theorem 4.4.6. *Every term in System T is strongly terminating.*

Proof. Clearly M is strongly terminating iff its closure $\lambda \vec{x}.M$ is, so it suffices to show that every closed term is strongly terminating.

Let $R_{\mathtt{N}}(M)$ be the unary predicate 'M is strongly terminating', and let R be its logical lifting. We first show that at any type $\sigma = \sigma_0, \ldots, \sigma_{r-1} \to \mathtt{N}$, $R_\sigma(M)$ implies that M is strongly terminating. For each i, let $N_i : \sigma_i$ be the term $\widehat{0}$ or $\lambda \vec{x}.\widehat{0}$ as appropriate; then clearly $R_{\sigma_i}(N_i)$ for each i. So if $R_\sigma(M)$ then $R_{\mathtt{N}}(MN_0 \ldots N_{r-1})$, so $MN_0 \ldots N_{r-1}$ is strongly terminating. But this means that M is strongly terminating, since any reduction sequence for M yields one for $MN_0 \ldots N_{r-1}$.

To show that all closed terms satisfy R, it now suffices to verify conditions 1 and 3 of Definition 4.4.2.

For condition 1, we note that $R_{\mathtt{N}}(\widehat{0})$ is trivial, and $R_{\mathtt{N} \to \mathtt{N}}(suc)$ is clear since reduction paths for $suc\, M$ correspond precisely to reduction paths for M; hence if M is strongly terminating then so is $suc\, M$. For rec_σ, a little more work is required. First, we check the following three facts relating to 'backwards reductions':

1. If P and Q satisfy R then so does $rec_\sigma(PQ\widehat{0})$.
2. If P, Q and $rec_\sigma PQ\widehat{l}$ satisfy R, then so does $rec_\sigma PQ(\widehat{l+1})$.
3. If P, Q, N satisfy R, $N \rightsquigarrow_n^* \widehat{k}$ and $rec_\sigma PQ\widehat{k}$ satisfies R, then so does $rec_\sigma PQN$.

The reason is similar in each case; we show the third as an example. Suppose $\sigma = \sigma_0, \ldots, \sigma_{r-1} \to \mathtt{N}$; we wish to show that under the given conditions, if $R_{\sigma_i}(T_i)$ for each i then $rec_\sigma PQN\vec{T}$ is strongly terminating. Let π be any reduction sequence starting from $rec_\sigma PQN\vec{T}$, and let π_0 be the maximal initial portion of π that consists entirely of reductions within the subterms P,Q,N,T_i or their residuals. Since P,Q,N,T_i all satisfy R, they are strongly terminating; hence π_0 is finite, and concludes with some term $rec_\sigma P'Q'N'\vec{T}'$, where $P \rightsquigarrow_n^* P'$, $Q \rightsquigarrow_n^* Q'$, $N \rightsquigarrow_n^* N'$ and $T_i \rightsquigarrow_n^* T_i'$ for each i. If π_0 exhausts the whole of π, then π is finite. Otherwise, π must continue with one of the δ-rules for rec_σ, so we must have $N' = \widehat{k}$ since this is the unique normal form for N. But now $rec_\sigma PQ\widehat{k}\vec{T}$ is strongly terminating and reduces by $\rightsquigarrow_n^*$ to $rec_\sigma P'Q'\widehat{k}\vec{T}'$ so the latter is strongly terminating. Hence π is finite. The arguments for the first two facts are similar.

From facts 1 and 2 above, we now see by induction that if P, Q satisfy R then so does $rec_\sigma PQ\widehat{k}$ for any k. But now if P, Q and $N : \mathtt{N}$ satisfy R then $N \rightsquigarrow_n^* \widehat{k}$ where $rec_\sigma PQ\widehat{k}$ satisfies R, hence $rec_\sigma PQN$ by fact 3. Thus rec_σ satisfies R as required.

For condition 3 of Definition 4.4.2, suppose we are given a term $\Gamma, x^\sigma \vdash M : \tau$ and a valuation $\nu : \Gamma$ such that $R_\Gamma(\nu)$ and for all closed $N : \sigma$ we have

$$R_\sigma(N) \;\Rightarrow\; R_\tau(M^*[x \mapsto N]) \,,$$

where we write M^* for the term $[\![M]\!]_\nu$ defined by substitution as above. Assuming $\tau = \tau_0, \ldots, \tau_{r-1} \to \mathtt{N}$, and considering each R_ρ as a subset of $\mathbf{T}(\rho)$, this means that

$$\forall N \in R_\sigma, T_0 \in R_{\tau_0}, \ldots, T_{r-1} \in R_{\tau_{r-1}}.\ M^*[x \mapsto N]T_0 \ldots T_{r-1} \text{ is strongly terminating.}$$

We will show that this implies

$$\forall N \in R_\sigma, T_0 \in R_{\tau_0}, \ldots, T_{r-1} \in R_{\tau_{r-1}}.\ (\lambda x.M^*)NT_0 \ldots T_{r-1} \text{ is strongly terminating.}$$

Since $\lambda x.M^* = (\lambda x.M)^*$, this says precisely that $R_{\sigma\to\tau}((\lambda x.M)^*)$ as required.

The argument is similar to that for fact 3 above. Suppose given $N \in R_\sigma$ and $T_i \in R_{\tau_i}$ for each i. Then we know that the terms $N, T_i, M^*[x \mapsto N]$ and hence M^* itself are all strongly terminating. Hence any reduction sequence π for $(\lambda x.M^*)N\vec{T}$ is either finite or will eventually have to reduce (some residual of) the head β-redex. This is easily seen to result in a term to which $M^*[x \mapsto N]\vec{T}$ reduces; hence π is finite as before. □

The above theorem provides a striking demonstration of the proof-theoretic power of logical relations. It is known that strong termination of System T is equivalent to the consistency of Peano Arithmetic (PA), the equivalence itself being provable within PA. Since by Gödel's incompleteness theorem PA cannot prove its own consistency, Theorem 4.4.6 goes beyond what can be proved in PA. The explanation for this is that although each relation R_σ can be expressed within the language of PA, the family R taken as a whole cannot; the proof thus makes use of an induction principle for a *hyperarithmetical* predicate of level Π^0_ω. We shall informally refer to relations R on λ-terms in the above spirit as *Tait predicates*; they are often known in the literature as *reducibility predicates*, or (somewhat misleadingly from our present point of view) as *computability predicates*.

One can also improve the above theorem to the statement that System T is *strongly normalizing*: that is, every term reduces to a unique normal form. For closed terms $M : \mathtt{N}$ this is easy to see: if $M \rightsquigarrow_n M'$ then clearly M, M' have the same interpretation in the full set-theoretic model $\mathsf{S}(\{\mathbb{N}\})$, so it follows immediately that if $M : \mathtt{N}$ is closed and strongly terminating then all reduction paths lead to the same normal form $\widehat{k}$. For terms of higher type, strong normalization follows (for example) from Theorem 4.4.6 along with the *weak Church–Rosser property*: if $M \rightsquigarrow_n M_0$ and $M \rightsquigarrow_n M_1$, there exists M' such that $M_0 \rightsquigarrow_n^* M'$ and $M_1 \rightsquigarrow_n^* M'$.[6] This latter fact can be established by a tedious though essentially routine case analysis of the possible forms of pairs of redexes within M. We shall not give the details here since they appear in many other texts, and the result itself will not be needed in this book.

We can build a λ-algebra out of closed terms of System T as follows. Let $=_n$ be the equivalence relation on closed terms generated by $\rightsquigarrow_n$. It is easy to see that $M =_n M'$ and $N =_n N'$ imply $MN =_n M'N'$, so we obtain a total applicative structure $T^0/{=_n}$ consisting of closed System T terms modulo $=_n$. This is clearly a TPCA since it contains suitable elements k, s; moreover, by Proposition 4.1.1 we see that the induced interpretation $[\![-]\!]^*_-$ of pure λ-terms interprets each closed term M by its own $=_n$-equivalence class, so $T^0/{=_n}$ is a λ-algebra. Finally, we see from Theorem 4.4.6 that $T^0(\mathtt{N})/{=_n} \cong \mathbb{N}$, so $T^0/{=_n}$ is a simply typed λ-algebra over $\mathbb{N}$.

Exercise 4.4.7. Consider the more restricted (deterministic) reduction relation $\rightsquigarrow$ on System T terms defined in Subsection 3.2.3. Show by induction on term size that if $M : \mathtt{N}$ is closed and irreducible via $\rightsquigarrow$ (i.e. there is no M' such that $M \rightsquigarrow M'$) then M is a numeral $\widehat{k}$. Deduce using Theorem 4.4.6 that $M \rightsquigarrow_n^* \widehat{k}$ iff $M \rightsquigarrow^* \widehat{k}$, and hence that every closed $M : \mathtt{N}$ reduces via $\rightsquigarrow^*$ to a numeral.

The next exercise shows that the structure $\mathbf{T}$ is *pre-extensional*: that is, it has a well-defined extensional quotient that respects the values of ground type terms. It also illustrates the use of binary logical relations. For each type σ we define the *extensional equivalence* relation $\sim_{e\sigma}$ on System T terms of type σ as follows.

$$\begin{aligned} M \sim_{e\mathtt{N}} M' \quad &\text{iff} \quad M, M' \text{ evaluate to the same numeral } \widehat{k}\,, \\ M \sim_{e(\sigma_0,\ldots,\sigma_{r-1}\to\mathtt{N})} M' \quad &\text{iff} \quad \forall N_0,\ldots,N_{r-1}.\, MN_0\ldots N_{r-1} \sim_{e\mathtt{N}} M'N_0\ldots N_{r-1}\,. \end{aligned}$$

Exercise 4.4.8. Define $E_{\mathtt{N}} \subseteq \mathbf{T}(\mathtt{N}) \times \mathbf{T}(\mathtt{N})$ by: $E_{\mathtt{N}}(M, M')$ iff M, M' evaluate to the same numeral $\widehat{k}$ (so that $E_{\mathtt{N}}$ is the restriction of $=_n$ to $\mathbf{T}(\mathtt{N})$), and let E be the logical lifting of $E_{\mathtt{N}}$.

(i) Show that E satisfies conditions 1 and 3 of Definition 4.4.2, and so is a binary logical relation. Deduce that $E_\sigma(M, M)$ for all closed $M : \sigma$.

(ii) Show that $E_\sigma = \sim_{e\sigma}$ for each σ. Deduce that if $M \sim_{e(\sigma\to\tau)} M'$ and $N \sim_{e\sigma} N'$ then $MN \sim_{e\tau} M'N'$.

---

[6] This approach to proving the uniqueness of normal forms relies on König's lemma: from Theorem 4.4.6 we deduce that the tree of all reduction sequences starting from M is finite. There are also other proofs of a more 'elementary' character using Church–Rosser-style properties, though they tend to be combinatorially more complex (see Section 2.2 of Troelstra [288] for discussion).

(iii) Show that we obtain an extensional type structure $T^0/{\sim_e}$ over $\mathbb{N}$ which is a proper quotient of $T^0/{=_n}$.

(iv) Show that $\sim_e$ coincides with the relation $\sim_{obs}$ defined by

$$M \sim_{obs} M' : \sigma \text{ iff } \forall P : \sigma \to \mathtt{N}.\ \forall k \in \mathbb{N}.\ PM =_n \widehat{k} \Leftrightarrow PM' =_n \widehat{k}\,.$$

Finally, we note that Theorem 4.4.6 holds a fortiori for pure simply typed λ-terms, which can be viewed as System T terms not involving any of the constants $\widehat{0}$, *suc*, rec_σ. For such terms, the relation $\rightsquigarrow_n$ may be defined simply by

- $(\lambda x.M)N \rightsquigarrow_n M[x \mapsto N]$,
- if $M \rightsquigarrow_n M'$ then $C[M] \rightsquigarrow_n C[M']$ for any pure λ-term context $C[-]$.

Since reductions of pure terms cannot introduce constants, we immediately have:

Corollary 4.4.9. *Every term of pure λ-calculus is strongly terminating with respect to $\rightsquigarrow_n$. In particular, every term is β-equivalent to a term in β-normal form (that is, a term containing no subterms $(\lambda x.M)N$).* □

However, Corollary 4.4.9 can also be proved within weak systems such as Primitive Recursive Arithmetic, albeit by more delicate means. Naturally, by combining this result with a Church–Rosser property, one can also show that *strong normalization* for pure simply typed λ-terms is provable within weak systems.

4.4.3 Pure Types Revisited

With the help of Corollary 4.4.9, we now obtain some specialized information regarding terms of *pure* type: namely, that under certain conditions on Γ, any well-typed λ-term $\Gamma \vdash\colon \overline{k_0}, \cdots, \overline{k_{r-1}} \to \overline{l}$ is β-equivalent to one of a certain restricted syntactic form. This is of interest because of the special role played by pure types in our theory: as we have seen, not only may a large class of datatypes σ be encoded as retracts of pure types $\overline{k}$, but the λ-terms $M : \overline{k} \to \overline{k}$ that define the corresponding idempotents are of precisely the above kind. Besides being of some independent interest, the information obtained here will play a technical role in Chapter 5.

We shall actually prove our results for typed λ-calculi possibly extended with some additional syntactic constructs whose typing rules are of a certain special form, as explained by the following definition.

Definition 4.4.10. (i) A type $\tau \in \mathsf{T}^{\to}(\beta)$ is *regular* if it has the form $\overline{k_0}, \cdots, \overline{k_{r-1}} \to \overline{l}$, or equivalently has the form $\overline{k_0}, \cdots, \overline{k_{r-1}} \to \overline{0}$ (recall that $\overline{l} = \overline{l-1} \to \overline{0}$ if $l > 0$).

(ii) A syntactic construct $Z(M_0, \ldots, M_{n-1})$ is called *regular* if its typing rule has the form

$$\frac{\Gamma \vdash M_0 : \tau_0 \qquad \Gamma \vdash M_{n-1} : \tau_{n-1}}{\Gamma \vdash Z(M_0, \ldots, M_{n-1}) : \tau}$$

where the τ_i and τ are all regular. We shall write $Z :: \tau_0, \ldots, \tau_{n-1} \to \tau$ to indicate the type signature of Z. In the case $n = 0$, we may refer to Z as a *regular construct.*

(iii) Suppose $x_0^{\sigma_0}, \ldots, x_{r-1}^{\sigma_{r-1}} \vdash M : \tau$ in the λ-calculus over $\mathsf{T}^{\rightarrow}(\beta)$ augmented with some set $\mathscr{Z}$ of regular constructs. We call such a typing judgement *regular* if τ is regular and all the σ_i are pure.

Reviewing the selection of constants introduced in the course of this chapter, we see that *tt*, *ff* and *if* are regular (and we have seen at the start of Section 4.2 that the higher type conditionals if_σ are definable from *if*). Likewise, the constants for numerals $\widehat{n}$ and the numeral operations *suc*, *eq*, *code*, *first*, *second* are regular. (Note that the System T recursors rec_σ are *not* regular.)

The key property of regular terms is the following:

Lemma 4.4.11. *Suppose $\mathscr{Z}$ is a set of regular constructs. Then all regular terms-in-context over $\mathsf{T}^{\rightarrow}(\beta)$ and $\mathscr{Z}$ in β-normal form are generated by the following restricted typing rules (where Γ ranges over pure contexts and τ over regular types)*

$$\Gamma \vdash x : \overline{k} \;\; (x^{\overline{k}} \in \Gamma) \qquad \frac{\Gamma \vdash N : \overline{k}}{\Gamma \vdash xN : \overline{0}} \;\; (x^{\overline{k+1}} \in \Gamma) \qquad \frac{\Gamma, x^{\overline{k}} \vdash M : \tau}{\Gamma \vdash \lambda x^{\overline{k}}.M : \overline{k} \rightarrow \tau}$$

$$\frac{\Gamma \vdash M_0 : \tau_0 \;\; \cdots \;\; \Gamma \vdash M_{n-1} : \tau_{n-1} \quad \Gamma \vdash N_0 : \overline{k_0} \;\; \cdots \;\; \Gamma \vdash N_{r-1} : \overline{k_{r-1}}}{\Gamma \vdash Z(M_0, \ldots, M_{n-1}) N_0 \ldots N_{r-1} : \tau'}$$

$$(r \geq 0, \;\; Z :: \tau_0, \ldots, \tau_{n-1} \rightarrow \tau, \;\; \tau = \overline{k_0}, \ldots, \overline{k_{r-1}} \rightarrow \tau')$$

In particular, every subterm of a regular term is itself regular.

Proof. By induction on term structure. For variables and constants the lemma is trivial (note that we permit $r = 0$ in the last of the above typing rules). For terms-in-context $\Gamma \vdash \lambda x.M : \tau'$ with τ' regular, we have that x is necessarily of some pure type $\overline{k}$ and M is of some regular type τ, so by the induction hypothesis $\Gamma, x^{\overline{k}} \vdash M : \tau$ is derivable via the above rules and we may apply the rule for λ-abstraction.

It remains to consider application terms. Any such term may be written as $MN_0 \ldots N_{r-1}$ where $r \geq 1$ and M is not an application. Since the whole term is assumed to be β-normal, M is not an abstraction either. There are thus two cases:

- M is a variable x of some pure type $\overline{k+1}$. In this case, $r = 1$ and N_0 is of type $\overline{k}$, so we may invoke the induction hypothesis and apply the typing rule for xN_0.
- M is a term $Z(M_0, \ldots, M_{n-1})$ for some regular Z. It follows that each M_i is of regular type and each N_j is of pure type, so again we may invoke the induction hypothesis and then apply the typing rule for $Z(M_0, \ldots, M_{n-1}) N_0 \ldots N_{r-1}$. □

Combining this with Corollary 4.4.9, we have:

Corollary 4.4.12. *Any regular term-in-context over $\mathsf{T}^{\rightarrow}(\beta)$ and C is β-equivalent to one generated by the typing rules of Lemma 4.4.11.* □

This information will be used in Subsection 5.1.3 for establishing the agreement between two different approaches to defining Kleene computability.

Part II
Particular Models

Having developed our general framework in Part I, we now proceed to consider some specific concepts of higher-order computability and the models that embody them. For each of these concepts, we present a variety of mathematical characterizations, establish the (sometimes non-trivial) equivalences between them, analyse the mathematical structure of the models in question, and discuss some important examples (or non-examples) of computable operations.

Roughly speaking, we start with models of computation that treat functions as 'pure extensions'—in effect, as *oracles*—then work our way towards models in which computations may have access to more 'intensional' information such as algorithms, strategies or programs of some kind. This will correlate broadly with a progression from weaker to stronger computability notions—although we cannot follow such a scheme exactly, since the notions we wish to consider do not form a neat linear order.

In Chapters 5–7, we study a group of closely related concepts that in various ways embody an ideal of purely extensional (and sequential) higher-order computation. Chapter 5 is dedicated to Kleene's definition of computability in the context of *total* functionals, using his schemes S1–S9, while Chapter 7 is devoted to Plotkin's PCF—a language for *partial* computable functionals at higher type. Both S1–S9 and PCF are in principle 'programming languages', and there is nothing particularly canonical about their detailed formulations—however, a more abstract and canonical concept of 'algorithm' for sequential higher-order computation is given by *nested sequential procedures* (also known in the literature as PCF *Böhm trees*), which we study in some depth in Chapter 6. The viewpoint we advocate is that nested sequential procedures encapsulate the underlying notion of algorithm that is common to both S1–S9 and PCF.

In Chapters 8 and 9, we consider two important type structures consisting of *total* higher-order functionals over the natural numbers: respectively, the model Ct of *total continuous functionals*, and the model HEO of *hereditarily effective operations*. The former turns out to be the canonical class of total functionals based on the idea of 'continuous operations acting on continuous data', whilst the latter is the canonical class of 'effective operations on effective data'. These models have a certain amount of structural theory in common, although neither can be said to 'contain' the other. In each model, there is a natural and robust notion of effective computability that exceeds the power of Kleene's S1–S9.

Chapters 10 and 11 study two classes of *partial* computable functional that exceed the power of PCF: respectively, the model PC of *partial continuous functionals*, and the model SR of *sequentially realizable functionals*. The latter, in particular, illustrates the additional computational power that can arise from the availability of intensional information. Once again, it turns out that these two models are incomparable with one another, and in fact there can be no 'largest' class of partial computable functionals over $\mathbb{N}$.

In Chapter 12 we consider a small selection of models that are genuinely 'intensional' in that their elements can no longer be considered simply as higher-order functions (over $\mathbb{N}$ or $\mathbb{N}_\perp$). Finally, in Chapter 13, we survey some related areas and possible avenues for further research.

Chapter 5
Kleene Computability in a Total Setting

We begin our study of particular computability notions with the one introduced in Kleene's landmark paper [138] via the computation schemes known as S1–S9. As we shall see, this has a particular conceptual significance within the subject as a whole, as it is in some sense the 'weakest' natural candidate for a general notion of effective computability at higher types. In the present chapter we shall consider this notion in the context of *total* type structures, following Kleene's original approach. In the following chapter, we will investigate ways of adapting this concept of computability to a partial setting, focussing in particular on nested sequential procedures (see Subsection 3.2.5), which serve as a nexus between Kleene computability and Plotkin's PCF.

The general spirit of Kleene's notion is easily grasped. Given some type structure $\mathbf{A}$, we may consider computations in which elements of $\mathbf{A}$ are treated simply as *oracles* or *black boxes*. Thus, if we wish to perform a computation involving an input $f \in \mathbf{A}(\sigma \to \mathtt{N})$, the only way in which we may use f is to feed it with some element of $\mathbf{A}(\sigma)$ that we have already constructed (perhaps with recourse to f itself), and observe the numerical result. In this way, Kleene's notion embodies the ideal of treating functions abstractly as 'pure extensions', without reference to how they might be represented or implemented. Subject to this basic prescription, however, Kleene's definition offers a reasonable candidate for a *general* notion of effective computability, leading to a plausible analogue of Church's thesis for higher-order functions. As regards the formalization of this notion, computable operations are represented in Kleene's definition by numerical *indices* e, and the schemes S1–S9 amount to an inductive definition of an evaluation relation $\{e\}(\Phi_0, \ldots, \Phi_{n-1}) = a$, where the Φ_i are functionals in $\mathbf{A}$ and a is a numerical result.

Perhaps the particular appeal of Kleene computability lies in its purity: the only way to interact with a function is the way that arises from its *being* a function. Kleene computability thus provides a natural baseline for our study of higher-order computability notions: any operation that is computable in Kleene's sense can also be expected to be 'computable' in any other reasonable sense.

Two aspects of Kleene's original work may initially appear puzzling to the modern reader. First, his focus on (hereditarily) *total* functionals might appear odd, given that general notions of computability naturally give rise to partial functionals. Secondly, Kleene developed his theory in the context of the full set-theoretic model S, and one might wonder whether such wildly infinitistic objects as arbitrary elements of S(2) or S(3) are at all amenable to a theory of 'effective' computation—indeed, we shall see that 'computations' in this context are typically infinitely branching trees of transfinite depth.

Both of these features can be explained with reference to Kleene's original motivations. The main goal of Kleene's project was to extend his theory of *logical complexity* for first- and second-order logic (as embodied in the arithmetic and analytic hierarchies) to higher-order systems. This naturally accounts for Kleene's preoccupation with S, and also for why subsequent work on S1–S9 focussed largely on computability relative to 'non-computable' objects such as quantifiers of various type levels. We will touch only lightly on these topics here, our main emphasis being on Kleene computability as a candidate for an 'effective' computability notion to the extent that the objects in question allow.

In our exposition, we shall follow Kleene in restricting attention initially to total objects. As we shall see, working with total functionals entails certain sacrifices, but we nevertheless obtain a class of computable total functionals with a coherent and satisfying structure. We shall also follow Kleene in basing our definition on the schemes S1–S9, which in effect yield a 'programming language' for such functionals. Whilst the definition of this language admittedly has a somewhat arbitrary character, the approach it offers is relatively direct, and mathematically cheaper to set up than the more 'canonical' approach via nested sequential procedures which we shall describe in Chapter 6. The emphasis of the present chapter, then, is on showing how much of the theory can be developed directly on this basis, and without assuming the presence of partial objects in our models.

Unlike Kleene, however, we will not work exclusively in the setting of S, but will interpret S1–S9 in the context of an arbitrary total type structure **A**, although the model S will turn out to play a special role in the theory. (Further results on Kleene computability in the models Ct and HEO will appear in Chapters 8 and 9.)

In Section 5.1, we introduce Kleene computability via Kleene's original S1–S9 definition, along with a closely related λ-calculus presentation which helps to clarify how Kleene computability fits into the framework of Chapters 3 and 4. We develop the basic theory to the point of showing that the Kleene computable functionals over any (Kleene closed) type structure constitute a $\lambda\eta$-algebra, and also briefly consider the more restricted classes of *primitive recursive* and μ*-computable* functionals.

In Section 5.2, we develop a deeper understanding of the key concepts of *termination* and *totality* from the viewpoint of logical complexity. (The necessary prerequisites from the theory of logical complexity are reviewed in Subsection 5.2.1.) For our analysis of termination, we work with a notion of *top-down computation tree*, which provides a model for the computational process underlying Kleene computations which may or may not terminate. As regards totality, we show that if a Kleene index e represents a total functional over S then it does so over any (Kleene closed) type structure **A**; moreover, the set of such indices turns out to be relatively tame in its logical complexity, despite the fact that the definition of totality in S superficially involves sets much larger than the continuum.

In Section 5.3, we present some non-trivial examples of Kleene computable operations, and show how these may be used to chart the frontier between the 'continuous' and 'discontinuous' worlds. Specifically, following ideas of Grilliot [111], we use certain computable 'search functionals' to give an analysis of the 'continuity' or otherwise of an arbitrary $\Phi \in \mathsf{S}(k+2)$ at an arbitrary argument $\xi \in \mathsf{S}(k+1)$. We will see that a sharp dichotomy arises between 'effectively continuous' functionals, for which suitable moduli of convergence are explicitly Kleene computable, and 'effectively discontinuous' ones, from which quantification over $\mathbb{N}$ is systematically computable. For the sake of orientation, we take a brief glance in Section 5.4 at computability relative to discontinuous objects, although a detailed study of this topic would fall outside the scope of this book (cf. Subsections 1.3.1 and 2.2.1).

In Chapter 6 we shall extend our study of Kleene computability into the realm of *partial* objects, with particular emphasis on nested sequential procedures, but also taking in earlier approaches due to Platek and Moschovakis. Most of this will be formally independent of the present chapter, except that Section 5.1 is needed for Section 6.2, where the connection between the two approaches is made. Section 5.1 is also a prerequisite for some parts of Chapters 8 and 9.

Throughout this chapter, we shall often need to refer to familiar concepts from standard first- or second-order computability theory. We shall refer to these as *classical* concepts, in order to distinguish them from concepts associated with the higher-order theory that is our main focus of study.

5.1 Definitions and Basic Theory

In this section we develop the basic theory of Kleene computability in total type structures. We begin in Subsection 5.1.1 with Kleene's original definition for the pure types via S1–S9. This allows us (in Subsection 5.1.2) to establish some basic properties, but in order to give an insightful proof of the crucial *substitution theorem*, we shall resort to an equivalent λ-calculus presentation (Subsection 5.1.3), which also shows how Kleene computability naturally extends to arbitrary simple types. In Subsection 5.1.4, we show how all of this fits within the framework of TPCAs and λ-algebras expounded in Chapters 3 and 4. Finally, in Subsection 5.1.5, we briefly consider two more restricted computability notions that will feature later in the book—the notions of *primitive recursive* and *μ-computable* functional.

5.1.1 Kleene's Definition via S1–S9

We start with Kleene's definition of computability for n-ary functionals over the pure types. We shall work in the setting of an arbitrary *pure type structure* $\mathbf{A}$, consisting of a set $\mathbf{A}(k)$ for each pure type $\bar{k}$, where $\mathbf{A}(0) = \mathbb{N}$ and $\mathbf{A}(k+1)$ is a set of total functions $\mathbf{A}(k) \to \mathbb{N}$. (Further conditions on $\mathbf{A}$ will be adopted as required.) The prototypical example is the full set-theoretic model S, where $\mathsf{S}(k+1)$ consists of all total functions $\mathsf{S}(k) \to \mathbb{N}$.

Specifically, for each choice of $k_0, \ldots, k_{n-1} \in \mathbb{N}$ simultaneously, we shall define a class of partial computable functionals $\mathbf{A}(k_0) \times \cdots \times \mathbf{A}(k_{n-1}) \rightharpoonup \mathbb{N}$. Just as in ordinary first-order computability theory, our notion of general computation naturally gives rise to partial functionals, although our main interest here will be in the total functionals thus generated.

In Kleene's setup, a procedure for computing a functional F is specified by a finitary 'program' that is itself encoded by a natural number e, which we call a *Kleene index* for F. However, this indexing system is not simply some post hoc encoding of possible programs, but itself plays an integral role in the definitions of such procedures via the crucial scheme S9.

Following Kleene, we shall write $\{e\}$ for the partial functional indexed by e. Strictly speaking, we have a separate coding system for each choice of argument types $k_0, \ldots, k_{n-1}$, so we should really write $\{e\}_{k_0 \ldots k_{n-1}}$ for the partial functional on $\mathbf{A}(k_0) \times \cdots \times \mathbf{A}(k_{n-1})$ coded by e; however, we shall usually suppress these type subscripts since they may typically be determined by the context. (Alternatively, one can incorporate the choice of $k_0, \ldots, k_{n-1}$ into the index e, but this too would needlessly burden the notation.)

In fact, Kleene's definition proceeds by defining a ternary *evaluation relation*

$$\{e\}(\vec{\Phi}) = a \,,$$

where $e, a \in \mathbb{N}$ and $\vec{\Phi}$ is a finite list of elements of pure type in $\mathbf{A}$. Once such a relation is in place, we naturally obtain a partial functional $\{e\}_{k_0 \ldots k_{n-1}}$ for each e and each $k_0, \ldots k_{n-1}$. The definition of the evaluation relation is given by a grand inductive definition using the schemes S1–S9.

Whilst various superficial aspects of these schemes may appear somewhat eccentric, we shall see that the notion of computability they give rise to is a very natural and robust one. We here follow Kleene's original definition closely, since his use of the labels S1,...,S9 is widely followed in the literature.[1] As usual, we suppose $\langle \cdots \rangle$ is some fixed primitive recursive coding of finite sequences over $\mathbb{N}$.

Definition 5.1.1. The relation $\{e\}(\Phi_0, \ldots, \Phi_{n-1}) = a$ is defined inductively by the following clauses. Here we write x for arguments that must be of type 0, and f for arguments that must be of type 1. We may also write Φ^k to indicate that an argument Φ must be of type k.

[1] Aside from the fact that we are working over a general structure $\mathbf{A}$ rather than S, our definition differs from Kleene's in two inessential respects. Firstly, Kleene incorporated more explicit type information into his indices than we do, although this makes no difference to the resulting class of functionals. Secondly, Kleene adopted the peculiar convention that the order of listing of arguments was immaterial except for arguments of the same type level, which meant that his version of the permutation scheme S6 was somewhat idiosyncratic (and superficially less general than ours).

S1 (*Successor*): If $e = \langle 1 \rangle$, then $\{e\}(x, \vec{\Phi}) = x+1$.

S2 (*Constants*): If $e = \langle 2, q \rangle$, then $\{e\}(\vec{\Phi}) = q$.

S3 (*Projection*): If $e = \langle 3 \rangle$, then $\{e\}(x, \vec{\Phi}) = x$.

S4 (*First-order composition*): If $e = \langle 4, e_0, e_1 \rangle$, then $\{e\}(\vec{\Phi}) = a$ if for some b we have $\{e_1\}(\vec{\Phi}) = b$ and $\{e_0\}(b, \vec{\Phi}) = a$.

S5 (*Primitive recursion*): If $e = \langle 5, e_0, e_1 \rangle$, then

- $\{e\}(0, \vec{\Phi}) = a$ if $\{e_0\}(\vec{\Phi}) = a$,
- $\{e\}(x+1, \vec{\Phi}) = a$ if for some b we have $\{e\}(x, \vec{\Phi}) = b$ and $\{e_1\}(x, b, \vec{\Phi}) = a$.

S6 (*Permutation*): If $e = \langle 6, d, \pi(0), \ldots, \pi(n-1) \rangle$ where π is a permutation of $\{0, \ldots, n-1\}$, then $\{e\}(\Phi_0, \ldots, \Phi_{n-1}) = a$ if $\{d\}(\Phi_{\pi(0)}, \ldots, \Phi_{\pi(n-1)}) = a$.

S7 (*First-order application*): If $e = \langle 7 \rangle$, then $\{e\}(f, x, \vec{\Phi}) = f(x)$.

S8 (*Higher-order application*): If $e = \langle 8, d \rangle$, then $\{e\}(\Psi^{k+2}, \vec{\Phi}) = \Psi(F)$ if $F \in \mathbf{A}(k+1)$ and for all $\xi \in \mathbf{A}(k)$ we have $\{d\}(\xi, \Psi, \vec{\Phi}) = F(\xi)$.

S9 (*Index invocation*): If $e = \langle 9, m \rangle$ where $m \leq n$, then $\{e\}(x, \Phi_0, \ldots, \Phi_{n-1}) = a$ if $\{x\}(\Phi_0, \ldots, \Phi_{m-1}) = a$.

The scheme S9 says that an argument x can itself be used as an index for a computation. In combination with S4, this means that the numerical result of one subcomputation may be used as the index for a further subcomputation.

We may (inductively) declare e to be a *Kleene index* for the argument types $k_0, \ldots, k_{n-1}$ if e has one of the nine forms listed above and is appropriate to $k_0, \ldots, k_{n-1}$, where e_0, e_1, d are themselves Kleene indices for the appropriate argument types. We assume here that our coding $\langle \cdots \rangle$ has the property

$$\langle i, z_0, \ldots, z_{r-1} \rangle \; > \; z_j \quad \text{for } i > 0 \text{ and } j < r\,;$$

indeed, we shall often make use of this assumption in giving definitions and proofs by induction on indices. Under this assumption, our definition of being a Kleene index is indeed well-founded, and it is primitively recursively decidable whether a given e is a Kleene index for given types $k_0, \ldots, k_{n-1}$.

As usual with such strictly positive inductive definitions, there will be a unique well-founded tree, called the *Kleene computation tree*, verifying that $\{e\}(\vec{\Phi}) = a$ whenever this is the case. Each node of this tree may be labelled with a triple $(e', \vec{\Phi}', a')$ giving the instance of the evaluation relation generated at the corresponding point. Applications of S1, S2, S3 and S7 give us the leaf nodes of the tree, and we call them *initial computations*. A terminating computation $\{e\}(\vec{\Phi}) = a$, where e is an index constructed as in S4, S5, S6 or S9, will have either one or two immediate subcomputations.

In the case of S8, however, there will be an immediate subcomputation $\{d\}(\xi, \vec{\Phi})$ for each $\xi \in \mathbf{A}(k)$, so there will typically be an infinite branching in the computation tree at this point. Thus, in this higher-order setting, 'computation trees' will in general be infinite objects. However, since these trees are well-founded, we may assign to any computation tree T an *ordinal height* $\| T \|$ recursively as follows:

$$\|T\| \;=\; \sup\{\|U\| \mid U \text{ is an immediate subcomputation of } T\} + 1\,.$$

In particular, the height of an initial computation is 1. Proofs by induction on the generation of the evaluation relation are thus tantamount to proofs by induction on the ordinal height of computation trees.

As a first application of these ideas, it is trivial to check by induction on the ordinal rank of computation trees that if $\{e\}(\vec{\Phi}) = a$ and $\{e\}(\vec{\Phi}) = b$ then $a = b$, so we may use $\{e\}(\vec{\Phi})$ as a mathematical expression without ambiguity. Note, however, that the value of $\{e\}(\vec{\Phi})$ will be undefined if $\{e\}$ is not a Kleene index for the appropriate types, or even when it is if no triple $\{e\}(\vec{\Phi}) = a$ is generated by our inductive definition. For example, the value of $\{\langle 9,0\rangle\}(x)$ will be undefined if x is not a Kleene index for the empty argument list.

As in earlier chapters, we shall write $\downarrow$ for definedness and $\simeq$ for Kleene equality, so that (for instance) the schemes S8 and S9 may now be rephrased as

$$\begin{aligned}\{\langle 8,d\rangle\}(\Psi^{k+2},\vec{\Phi}) &\simeq \Psi\,(\Lambda\xi \in \mathbf{A}(k).\,\{d\}(\xi,\Psi,\vec{\Phi}))\,,\\ \{\langle 9,m\rangle\}(x,\Phi_0,\ldots,\Phi_{n-1}) &\simeq \{x\}(\Phi_0,\ldots,\Phi_{m-1})\,,\end{aligned}$$

and this perhaps renders the intuition behind these schemes more clearly. If e is a Kleene index for $k_0,\ldots,k_{n-1}$, we may also write $\{e\}_{k_0\ldots k_{n-1}}$, or usually just $\{e\}$, for the partial function

$$\Lambda(\Phi_0,\ldots,\Phi_{n-1}).\,\{e\}(\Phi_0,\ldots,\Phi_{n-1}) \;:\; \mathbf{A}(k_0)\times\cdots\times\mathbf{A}(k_{n-1}) \rightharpoonup \mathbb{N}\,.$$

We say e is a *total Kleene index* for $k_0,\ldots,k_{n-1}$ if $\{e\}_{k_0\ldots k_{n-1}}$ is a total function. Note that an index $\langle 5,e_0,e_1\rangle$ (for example) may be total even when e_1 is not.

We are now in a position to define the notions of relative and absolute Kleene computability:

Definition 5.1.2. (i) If $\Phi_0,\ldots,\Phi_{n-1}$ are functionals in $\mathbf{A}$ and Ψ is a partial function $\mathbf{A}(k_0)\times\cdots\times\mathbf{A}(k_{m-1}) \rightharpoonup \mathbb{N}$, we say Ψ is *Kleene computable in* $\Phi_0,\ldots,\Phi_{n-1}$ if there is an index e such that

$$\{e\}(\xi_0,\ldots,\xi_{m-1},\Phi_0,\ldots,\Phi_{n-1}) \;\simeq\; \Psi(\xi_0,\ldots,\xi_{m-1})$$

for all $\xi_0 \in \mathbf{A}(k_0),\ldots,\xi_{m-1} \in \mathbf{A}(k_{m-1})$.

(ii) Ψ is *Kleene computable* if it is Kleene computable in the empty list.

(iii) If $\Phi,\Psi \in \mathbf{A}$, we write $\Psi \leq_{\mathrm{Kl}} \Phi$ if Ψ is Kleene computable in Φ.

The Kleene computable (partial) functionals are also known in the older literature as the *Kleene (partial) recursive* functionals.

The relation $\leq_{\mathrm{Kl}}$ induces an equivalence relation $\equiv_{\mathrm{Kl}}$ on elements of $\mathbf{A}$, and the equivalence classes are called *Kleene degrees*. We shall also write $\leq_{\mathrm{Kl}}$ for the induced ordering of the Kleene degrees.

Some informal discussion of the schemes S1–S9 is in order. First, we have noted that the computation tree for a triple $\{\langle d,8\rangle\}(\Psi,\vec{\Phi}) = a$ will be infinitely branching. There are broadly two ways of conceiving this intuitively. The first way (to which

Kleene himself inclined) is to suppose that in order to compute $\{\langle d,8\rangle\}(\Psi,\vec{\Phi})$ we should construct the complete graph of the function $\Lambda\xi.\{d\}(\xi,\Psi,\vec{\Phi})$—that is, compute $\{d\}(\xi,\Psi,\vec{\Phi})$ for every $\xi \in \mathbf{A}(k)$—and then submit this graph to the oracle Ψ. 'Computations' in this setting are thus infinite objects of transfinite depth. Taken by itself, this might give the impression that S1–S9 has nothing to do with computability in any genuinely effective sense. However, an alternative interpretation is possible, in which we assume nothing at all about how Ψ 'works'—for all we know, it may (depending on the constitution of $\mathbf{A}$ itself) be underpinned by perfectly finitary processes. We can then regard the tree witnessing the generation of $\{e\}(\vec{\Phi}) = a$ not as the computation itself, but simply as an infinitary *proof* of this fact purely on the basis of the extensions of the Φ_i. This latter interpretation fits well with the view of Kleene computability as 'common core' for all reasonable computability notions rather than as a good notion in its own right.

Secondly, we note that S9 postulates that there is a *universal algorithm* for any tuple of types $k_0,\ldots,k_{n-1}$, something that in other contexts is seen as a theorem derived from a more 'primitive' definition. Whilst the above inductive definition is mathematically sound, there is intuitively an element of circularity in the fact that S9 states the existence of a universal algorithm for a notion of computability of whose definition S9 is itself a part. This is the main conceptual problem associated with Kleene's definition. We shall later meet some alternative characterizations of Kleene computability (e.g. via nested sequential procedures) that do not suffer from this peculiarity.

Thirdly, in connection with S8, we note that $\{\langle 8,d\rangle\}(\Psi^{k+2},\vec{\Phi})$ will be undefined if $\Lambda\xi^k.\{d\}(\xi,\Psi,\vec{\Phi})$ is not a *total* function (or if it is a total function not present in $\mathbf{A}(k+1)$). This might seem worrying, since it will not even be semidecidable in general whether this particular computation step succeeds—indeed, it might appear quite unrealistic to suppose the oracle Ψ has the magical ability to recognize total inputs in $\mathbf{A}(k+1)$ and 'reject' everything else. On the other hand, we note that one need not assume that our oracles possess any such ability in order to develop a good theory for those computations that *do* terminate, since here the ability to reject unsuitable arguments is not put to the test. (The situation is somewhat akin to the treatment of definedness in Kleene's second model—see Section 3.2.1.) For potentially non-terminating computations, a possible way of reconciling ourselves to this apparent eccentricity arising from S8 will be suggested in Subsection 6.4.3.

For a well-behaved theory, it is also advantageous to work with structures $\mathbf{A}$ satisfying the following condition:

Definition 5.1.3. A pure type structure $\mathbf{A}$ is *Kleene closed* if $\Lambda\xi^k.\{e\}(\xi,\vec{\Phi})$ is present in $\mathbf{A}(k+1)$ whenever it is a total function $\mathbf{A}(k) \to \mathbb{N}$, where the Φ_i are drawn from $\mathbf{A}$.

All our main models of interest will turn out to be Kleene closed; this is of course a triviality in the case of S.

5.1.2 Basic Properties

We now develop some of the basic theory of S1–S9 computability along the lines of [138]. The results in this subsection do not yet require $\mathbf{A}$ to be Kleene closed.

First, we note that an analogue of the classical s-m-n theorem can be obtained trivially in the presence of S4, and an analogue of the *second recursion theorem* follows from this.

Theorem 5.1.4. *For any $m \geq 1$ and any list of types $k_0, \ldots, k_{n-1}$, there is a primitive recursive function $S^m_{k_0,\ldots,k_{n-1}} : \mathbb{N}^{m+1} \to \mathbb{N}$ such that for all $e, x_0, \ldots, x_{m-1} \in \mathbb{N}$ and all $\Phi_0 \in \mathbf{A}(k_0), \ldots, \Phi_{n-1} \in \mathbf{A}(k_{n-1})$ we have*

$$\{e\}(x_0, \ldots, x_{m-1}, \vec{\Phi}) \simeq \{S^m_{k_0,\ldots,k_{n-1}}(e, x_0, \ldots, x_{m-1})\}(\vec{\Phi}) .$$

Indeed, $S^m_{k_0,\ldots,k_{n-1}}(e, x_0, \ldots, x_{m-1})$ may be obtained uniformly as a primitive recursive function of $m, \langle k_0, \ldots, k_{n-1}\rangle, e$ and $\langle x_0, \ldots, x_{m-1}\rangle$.

Proof. For $m = 1$, define $S^1_{k_0,\ldots,k_{n-1}}(e, x) = \langle 4, e, \langle 2, x\rangle\rangle$. Now inductively define

$$S^{m+1}_{k_0,\ldots,k_{n-1}}(e, x_0, \ldots, x_m) = S^1_{k_0,\ldots,k_{n-1}}(S^{m-1}_{k_0,\ldots,k_{n-1}}(e, x_0, \ldots, x_{m-1}), x_m) .$$

The uniformity in m and $k_0, \ldots, k_{n-1}$ are then obvious. □

Following the usage of classical computability theory, we shall typically abbreviate $S^m_{k_0,\ldots,k_{n-1}}$ to S^m_n.

It is essential in Theorem 5.1.4 that the arguments x_i are of ground type. For instance, suppose $e = \langle 7\rangle$ and consider $\{e\}(f, x) = f(x)$. If f is not computable, there is no way we could compute an index for $\Lambda x.f(x)$ from f, which is what the theorem would imply if we allowed a type 1 argument in place of x_0.

Theorem 5.1.5 (Second recursion theorem). *For any $k_0, \ldots, k_{n-1}$ and any index e, there is an index d such that for all $\Phi_0 \in \mathbf{A}(k_0), \ldots, \Phi_{n-1} \in \mathbf{A}(k_{n-1})$ we have*

$$\{d\}(\Phi_0, \ldots, \Phi_{n-1}) \simeq \{e\}(d, \Phi_0, \ldots, \Phi_{n-1}) .$$

Furthermore, such an index d may be computed primitively recursively from e and $\langle k_0, \ldots, k_{n-1}\rangle$.

Proof. Fix $k_0, \ldots, k_{n-1}$ and e. Clearly the schemes S1–S6 generate all primitive recursive first-order functions, so S^1_n is Kleene computable. Using S4, let c be a Kleene index for $\{e\}(S^1_n(x, x), \vec{\Phi})$ as a function of x and $\vec{\Phi}$, and take $d = S^1_n(c, c)$. It is routine to check that d has the required property, and it is now clear that d is primitive recursive in $k_0, \ldots, k_{n-1}$ and e. (Notice that this is essentially the same as the construction of the recursion operator Y in Section 3.3.5.) □

By contrast, the natural analogue of the *first recursion theorem* fails in general because of the eccentric properties of partial functions in the S1–S9 setting: see [139] or Subsection 6.4.4 below.

Interestingly, Theorem 5.1.5 does not itself rely on the scheme S9. However, many natural applications of the theorem will do so, since the argument d typically features in the computation of $\{e\}(d,\vec{\Phi})$ via the induced operation $\{d\}$. We illustrate this with some important examples. As a preliminary, we require the following fact:

Lemma 5.1.6 (Strong definition by cases). *Given Kleene indices d, e_0, e_1 for types $k_0, \ldots, k_{n-1}$, there is an index c such that for any $\vec{\Phi}$ we have*

$$\{c\}(\vec{\Phi}) \;\simeq\; \begin{cases} \{e_0\}(\vec{\Phi}) \text{ if } \{d\}(\vec{\Phi}) = 0\,, \\ \{e_1\}(\vec{\Phi}) \text{ if } \{d\}(\vec{\Phi}) = 1\,. \end{cases}$$

Furthermore, a suitable c may be computed primitively recursively from d, e_0, e_1 and $\langle k_0, \ldots, k_{n-1} \rangle$.

Proof. Given d, e_0, e_1, we may readily construct an index b such that

$$\{b\}(\vec{\Phi}) \;\simeq\; \begin{cases} e_0 \text{ if } \{d\}(\vec{\Phi}) = 0\,, \\ e_1 \text{ if } \{d\}(\vec{\Phi}) = 1\,. \end{cases}$$

Using S4 and S9, we may thence construct c so that $\{c\}(\vec{\Phi}) \simeq \{\{b\}(\vec{\Phi})\}(\vec{\Phi})$. The primitive recursiveness clause is obvious. □

We now give two simple applications of Theorem 5.1.5. We leave the proofs as easy exercises.

Proposition 5.1.7 (Minimization). *If F is Kleene computable over $0, k_0, \ldots, k_{n-1}$, then $min(F)$ is Kleene computable over $k_0, \ldots, k_{n-1}$, where*

$$min(F)(\vec{\Phi}) \;\simeq\; \mu a.\, F(a, \vec{\Phi}) = 0\,.$$

Furthermore, an index for $min(F)$ may be computed primitively recursively from one for F.

Proposition 5.1.8 (Higher-order primitive recursion). *Given Kleene computable functions F over $k, k_0, \ldots, k_{n-1}$ and G over $0, k+1, k, k_0, \ldots, k_{n-1}$, there is a Kleene computable function H over $0, k, k_0, \ldots, k_{n-1}$ such that*

$$\begin{aligned} H(0, \Psi, \vec{\Phi}) &\simeq F(\Psi, \vec{\Phi})\,, \\ H(x+1, \Psi, \vec{\Phi}) &\simeq G(x, (\Lambda \Psi'. H(x, \Psi', \vec{\Phi})), \Psi, \vec{\Phi})\,. \end{aligned}$$

Along with S1–S6, Proposition 5.1.7 tells us that the class of *first-order* Kleene computable functions (those whose arguments are all of type 0) contains the necessary basic functions and is closed under composition, primitive recursion and minimization. It follows that all partial computable functions $\mathbb{N}^n \rightharpoonup \mathbb{N}$ in the classical sense are Kleene computable. Likewise, at second order, all uniformly partial computable functions $(\mathbb{N}^{\mathbb{N}})^m \times \mathbb{N}^n \rightharpoonup \mathbb{N}$ in the classical sense are Kleene computable, by virtue of S7. Proposition 5.1.8 essentially tells us that higher-order recursors in the style of System T are also computable via S1–S9.

More sophisticated applications of the second recursion theorem in effect use it to define functions by recursion on the ordinal rank of computation trees or other well-founded relations—a technique known as *effective transfinite recursion.* We shall illustrate this method in Exercise 5.1.12, and also in Subsection 5.1.3, where it will feature in our proof of the following crucial result:

Theorem 5.1.9 (Substitution theorem). *Suppose F and G are Kleene computable partial functionals over $j, k_0, \ldots, k_{n-1}$ and $j+1, l_0, \ldots, l_{p-1}$ respectively. Then there is a Kleene computable partial functional H over $k_0, \ldots, k_{n-1}, l_0, \ldots, l_{p-1}$ such that for all $\vec{\Phi}, \vec{\Psi}$ we have*

$$H(\vec{\Phi}, \vec{\Psi}) \succeq G(\Lambda \zeta^j . F(\zeta, \vec{\Phi}), \vec{\Psi}) .$$

Furthermore, an index for such an H may be effectively computed from indices for F and G along with j, $\langle k_0, \ldots, k_{n-1} \rangle$ and $\langle l_0, \ldots, l_{p-1} \rangle$.

Our proof will proceed via a λ-calculus presentation of Kleene computability. A direct proof (which is somewhat demanding) can be found in Kleene [138] or in Kechris and Moschovakis [130].

Unfortunately, the Kleene inequality $\succeq$ in Theorem 5.1.9 cannot in general be strengthened to $\simeq$, even under the assumption that F is total. The following notorious counterexample is due to Kleene [139]:

Example 5.1.10. Suppose $\mathbf{A}$ is the full set-theoretic model S, and let $\bullet$ denote classical Kleene application as in Example 3.1.9. Let $F(g^1) = 0$ for all $g^1 \in \mathsf{S}(1)$, and let $G(\Phi^2, e^0) = \Phi(\Lambda y. \theta(e, y))$, where

$$\theta(e,y) = \begin{cases} \text{undefined} & \text{if } y \text{ codes a course of computation for `} e \bullet 0\text{'}, \\ 0 & \text{otherwise.} \end{cases}$$

Clearly both F and G are Kleene computable. However, for any e we have $G(F, e) = 0$ iff $(\Lambda y. \theta(e, y))$ is total, which holds iff $e \bullet 0$ is undefined. Thus, the function $(\Lambda e. G(F, e))$ is not classically computable, and hence not Kleene computable (see Proposition 5.1.13 below).

Thus, the Kleene computable *partial* functionals are not even closed under specialization of total arguments to total computable functionals. At its root, this is a consequence of the eccentricity whereby oracles can 'semidecide' the suitability of their arguments, as discussed in Section 5.1.1.

However, Theorem 5.1.9 does of course yield the following:

Corollary 5.1.11. *If F, G are* total *Kleene computable functions over $j, k_0, \ldots, k_{n-1}$ and $j+1, l_0, \ldots, l_{p-1}$ respectively, and $\Lambda \zeta^j . F(\zeta, \vec{\Phi}) \in \mathbf{A}^{j+1}$ for all $\vec{\Phi}$, then the composition*

$$H = \Lambda \vec{\Phi} \vec{\Psi} . G((\Lambda \zeta . F(\zeta, \vec{\Phi})), \vec{\Psi}$$

is also total and Kleene computable, and an index for H is effectively computable in indices for F, G along with j, $\langle k_0, \ldots, k_{n-1} \rangle$ and $\langle l_0, \ldots, l_{p-1} \rangle$.

In other words, total Kleene computable functions are closed under composition. This is the basic reason why they yield a better theory than the partial ones.

Exercise 5.1.12. (i) Show using S9 that if $m < n$ and F is Kleene computable over $k_0,\ldots,k_{m-1}$ then so is F' over $k_0,\ldots,k_{n-1}$, where

$$F'(\Phi_0,\ldots,\Phi_{n-1}) \;=\; F(\Phi_0,\ldots,\Phi_{m-1}) \,.$$

(ii) Show using Theorem 5.1.5 that if F is Kleene computable over $k_0,k_0,\ldots,k_{n-1}$ then so is F'' over $k_0,\ldots,k_{n-1}$, where

$$F''(\Phi_0,\ldots,\Phi_{n-1}) \;\simeq\; F(\Phi_0,\Phi_0,\Phi_1,\ldots,\Phi_{n-1}) \,.$$

(Hint: show how F'' may be computed assuming we already have a Kleene index z for F'', and apply the second recursion theorem. This is a simple example of effective transfinite recursion.)

To conclude this subsection, we show that Kleene computations at first and second order are relatively easy to analyse. First, note if $\mathscr{U}$ is a computation tree for $\{e\}(\Phi_0,\ldots,\Phi_{n-1}) = a$ where the Φ_i all have type level $\leq k$, then by inspection of S1–S9, all arguments appearing anywhere within $\mathscr{U}$ also have type $\leq k$. In particular, if all the Φ_i are of type 0 or 1, then the same is true throughout $\mathscr{U}$. This means that $\mathscr{U}$ cannot involve S8, which requires an argument of type ≥ 2. Thus $\mathscr{U}$ is finitely branching and well-founded, whence $\mathscr{U}$ is finite by König's lemma.

If all the Φ_i have type 0, then $\mathscr{U}$ is a finite tree whose labels are simply tuples of natural numbers. Such a labelled tree can obviously be coded in an effective way by a single natural number t; moreover, it is primitively recursively decidable whether a given number t codes a correctly formed computation tree of this kind. We may therefore conclude:

Proposition 5.1.13. *For each $n \in \mathbb{N}$ there is a classically computable partial function $\varepsilon^n : \mathbb{N}^{n+1} \rightharpoonup \mathbb{N}$ such that for every $e,x_0,\ldots,x_{n-1} \in \mathbb{N}$ we have*

$$\varepsilon^n(e,x_0,\ldots,x_{n-1}) \;\simeq\; \{e\}(x_0,\ldots,x_{n-1}) \,.$$

Proof. A crude procedure for computing ε^n is as follows: given $e,x_0,\ldots,x_{n-1}$, search for a number t that codes a well-formed computation tree for some triple $\{e\}(x_0,\ldots,x_{n-1}) = a$; if we find one, return a. □

The situation is similar if all the arguments Φ_i are of type 0 or 1. In this case, every type 1 argument appearing anywhere in $\mathscr{U}$ must be one of the Φ_i; thus, $\mathscr{U}$ may be coded by a single natural number t together with the list $\vec{\Phi}$. Moreover, it is decidable via a second-order primitive recursive function whether t and $\vec{\Phi}$ represent a correctly formed computation tree. Thus:

Proposition 5.1.14. *If each of $k_0,\ldots,k_{n-1}$ is 0 or 1, there is a classically computable second-order partial function $\varepsilon^{k_0,\ldots,k_{n-1}}$ such that for every $e \in \mathbb{N}$ and all $\Phi_0,\ldots,\Phi_{n-1}$ of appropriate type we have*

$$\varepsilon^{k_0,\ldots,k_{n-1}}(e,\Phi_0,\ldots,\Phi_{n-1}) \simeq \{e\}(\Phi_0,\ldots,\Phi_{n-1}) .$$

Proof. Given e and $\vec{\Phi}$, search for a number t such that t and $\vec{\Phi}$ code a well-formed computation tree for a triple $\{e\}(\Phi_0,\ldots,\Phi_{n-1}) = a$. If we find one, return a. □

Of course, a much more natural way to 'compute' $\{e\}$ for a given tuple of arguments (whether of low or high type) is to attempt to build a computation tree systematically in a top-down way, starting from the root. This idea will be developed in Section 5.2.2.

5.1.3 A λ-Calculus Presentation

We now offer an alternative presentation of Kleene computability in terms of a suitable λ-calculus. Our purpose in doing so is threefold. Firstly, the perspective of λ-calculus allows us to give a relatively perspicuous proof of the substitution theorem mentioned above (Theorem 5.1.9). Secondly, it offers a convenient way of showing how Kleene computability may be defined for arbitrary simple types, not just the pure ones. Thirdly, the λ-calculus presentation makes clear how Kleene computability fits into the general framework of Chapters 3 and 4.

We shall here work in the context of an arbitrary model $\mathbf{A}$, which for sanity we assume to be a total extensional TPCA over $\mathsf{T} = \mathsf{T}^{\rightarrow}(\mathtt{N} = \mathbb{N})$, and hence a $\lambda\eta$-algebra. (To begin with, we do not assume that $\mathbf{A}$ is Kleene closed as in Definition 5.1.3.) We shall consider the extended λ-calculus over T whose terms are generated by the following typing rules.

$$\Gamma \vdash x^\sigma : \sigma \;\; (x^\sigma \in \Gamma) \qquad \frac{\Gamma, x^\sigma \vdash M : \tau}{\Gamma \vdash \lambda x^\sigma . M : \sigma \rightarrow \tau} \qquad \frac{\Gamma \vdash M : \sigma \rightarrow \tau \quad \Gamma \vdash N : \sigma}{\Gamma \vdash MN : \tau}$$

$$\Gamma \vdash \widehat{0} : \mathtt{N} \qquad \frac{\Gamma \vdash M : \mathtt{N}}{\Gamma \vdash Suc(M) : \mathtt{N}} \qquad \frac{\Gamma \vdash M : \mathtt{N} \quad \Gamma \vdash N : \mathtt{N} \quad \Gamma \vdash P : \mathtt{N},\mathtt{N} \rightarrow \mathtt{N}}{\Gamma \vdash Primrec(M,N,P) : \mathtt{N}}$$

$$\frac{\Gamma \vdash M : \mathtt{N} \quad \Gamma \vdash N_0 : \overline{k_0} \quad \cdots \quad \Gamma \vdash N_{n-1} : \overline{k_{n-1}}}{\Gamma \vdash Eval_\sigma(M,N_0,\ldots,N_{n-1}) : \mathtt{N}} \qquad (\sigma = \overline{k_0},\ldots,\overline{k_{n-1}} \rightarrow \mathtt{N})$$

We shall write Klex for the language thus obtained; we may refer to terms of this language as *Kleene expressions*, or *Kleenexes*. We will see in Section 6.2 that Klex is intimately related to Plotkin's language PCF.

The *Primrec* construct plays the role of the scheme S5, whilst *Eval* plays the role of S9. The latter works in conjunction with an effective Gödel numbering $\lceil - \rceil$ for Kleene expressions: the idea is that if M evaluates to a number $m = \lceil P \rceil$ where $P : \sigma$, then $Eval_\sigma(M,N_0,\ldots,N_{n-1})$ will compute the value of $PN_0 \ldots N_{n-1}$.[2] The

[2] The name *Eval* comes from the `eval` operator in the Lisp programming language, which likewise takes a program as a syntactic object and dynamically renders it as a live operation.

apparent bias here in favour of types of the form $\overline{k_0}, \ldots, \overline{k_{n-1}} \to \mathtt{N}$ is only a temporary measure—we shall see in Exercise 5.1.27 that suitable operators $Eval_\sigma$ for arbitrary σ are definable from the constructs above. Note that the typing rules for $\widehat{0}$, *Suc*, *Primrec*, *Eval* are all *regular* in the sense of Section 4.4.3.

Our primary interest will be in Kleene expressions in *β-normal form.* Expressions not in normal form will play an auxiliary role, though they need not define genuinely computable operations. For instance, if $M : \overline{2}$ and $N : \overline{2} \to \overline{0} \to \overline{0}$ respectively correspond to the functionals F, G of Example 5.1.10, then the function represented by $\lambda e.NMe$ is not computable.

We next define a denotational semantics for Kleene expressions in the model $\mathbf{A}$. We will take a small liberty and treat elements of type $\tau_0, \ldots, \tau_{n-1} \to \mathtt{N}$ as though they were uncurried functions of type $\tau_0 \times \cdots \times \tau_{n-1} \to \mathtt{N}$. Specifically, by analogy with our earlier evaluation relation, we shall inductively define an *interpretation relation*

$$[\![M]\!]_\nu(\Phi_0, \ldots, \Phi_{n-1}) \;=\; a\,,$$

where:

- $\Gamma \vdash M : \tau_0, \ldots, \tau_{n-1} \to \mathtt{N}$ is a well-typed Kleene expression,
- ν is a valuation assigning to each $x^\sigma \in \Gamma$ an element $\nu(x) \in \mathbf{A}(\sigma)$,
- $\Phi_i \in \mathbf{A}(\tau_i)$ for each i,
- $a \in \mathbb{N}$.

This will in effect define a partial function $[\![M]\!]_\nu : \mathbf{A}(\tau_0) \times \cdots \times \mathbf{A}(\tau_{n-1}) \rightharpoonup \mathbb{N}$ whenever $\Gamma \vdash M : \tau_0, \ldots, \tau_{n-1} \to \mathtt{N}$. In tandem with this, we shall define $[\![M]\!]^*_\nu = F \in \mathbf{A}(\tau_0, \ldots, \tau_{n-1} \to \mathtt{N})$ whenever $[\![M]\!]_\nu$ is total and agrees with F. Notice that we cannot simply give a direct compositional definition of $[\![M]\!]^*_\nu$, since a total Kleene expression may contain non-total subexpressions.

Definition 5.1.15. The interpretation relations $[\![M]\!]_\nu(\vec{\Phi}) = a$, $[\![M]\!]^*_\nu = F$ are defined inductively by means of the following clauses (we leave it to the reader to supply the types).

- If $\nu(x)(\vec{\Phi}) = a$, then $[\![x]\!]_\nu(\vec{\Phi}) = a$.
- If $[\![N]\!]^*_\nu = F$ and $[\![M]\!]_\nu(F, \vec{\Phi}) = a$, then $[\![MN]\!]_\nu(\vec{\Phi}) = a$.
- If $[\![M]\!]_{\nu, x \mapsto \Theta}(\vec{\Phi}) = a$, then $[\![\lambda x.M]\!]_\nu(\Theta, \vec{\Phi}) = a$.
- $[\![\widehat{0}]\!]_\nu() = 0$.
- If $[\![M]\!]_\nu() = a$, then $[\![Suc(M)]\!]_\nu() = a + 1$.
- If $[\![M]\!]_\nu() = 0$ and $[\![N]\!]_\nu() = a$, then $[\![Primrec(M,N,P)]\!]_\nu() = a$.
- If $[\![M]\!]_\nu() = b + 1$, $[\![Primrec(b,N,P)]\!]_\nu() = c$ and $[\![P]\!]_\nu(b, c) = a$, then $[\![Primrec(M,N,P)]\!]_\nu() = a$.
- If $[\![M]\!]_\nu() = m = \lceil P \rceil$ where $\Gamma \vdash P : \sigma$, and if $PN_0 \ldots N_{n-1}$ has β-normal form Q where $[\![Q]\!]_\nu() = a$, then $[\![Eval_\sigma(M, N_0, \ldots, N_{n-1})]\!]_\nu() = a$.
- If $F \in \mathbf{A}(\tau_0, \ldots, \tau_{n-1} \to \mathtt{N})$ and $[\![M]\!]_\nu(\vec{\Phi}) = F(\vec{\Phi})$ for all $\vec{\Phi}$, then $[\![M]\!]^*_\nu = F$.

In the clause for $Eval_\sigma$, the step from $PN_0 \ldots N_{n-1}$ to its β-normal form is a technical manoeuvre needed to overcome the problems with substitution mentioned in Section 5.1.2.

Exercise 5.1.16. By adapting ideas from Section 5.1.2, show directly that the minimization operator of Proposition 5.1.7 is representable by a Kleene expression, and hence that the partial functions $\mathbb{N}^r \rightharpoonup \mathbb{N}$ represented by Kleene expressions include all the classically computable partial functions.

We now work towards showing that for pure argument types $k_0, \ldots, k_{n-1}$, the partial functionals $\mathbf{A}(k_0) \times \cdots \times \mathbf{A}(k_{n-1}) \rightharpoonup \mathbb{N}$ representable by closed β-normal Kleene expressions coincide exactly with the Kleene computable functionals as defined by S1–S9. We do this by showing that any S1–S9 description can be translated into an equivalent Kleene expression in *β-normal form*, and conversely, any β-normal Kleene expression can be translated into an equivalent S1–S9 description. To implement this idea, we first appeal to Lemma 4.4.11 to give a more restricted set of typing rules that suffice to generate all closed β-normal Kleene expressions of types $k_0, \ldots, k_{n-1} \rightarrow \mathtt{N}$. Specifically, the rules we need are

$$\Gamma \vdash x : k \;\; (x^k \in \Gamma) \qquad \frac{\Gamma \vdash M : k}{\Gamma \vdash xM : \mathtt{N}} \;\; (x^{k+1} \in \Gamma) \qquad \frac{\Gamma, x^k \vdash M : \tau}{\Gamma \vdash \lambda x^k.M : k \rightarrow \tau}$$

along with the rules for $\widehat{0}$, *Suc*, *Primrec*, $Eval_\sigma$ given earlier. We shall refer to all these as the *regular typing rules* for Kleene expressions; by Lemma 4.4.11, every closed β-normal expression of regular type is generated by these rules.

We may now present the correspondence between Kleene expressions and S1–S9 indices. One obstacle to be negotiated here is that the coding convention for Kleene indices will in general bear no relation to our choice of Gödel numbering for Kleene expressions. This leads us to present our correspondences in terms of *effective translations* between Kleene indices and Kleene expressions. We shall outline a proof of the following technical lemma, leaving detailed checking to the reader. The proof of part (ii) provides a typical illustration of *effective transfinite recursion*.

Lemma 5.1.17. *(i) There is a primitive recursive function $\theta : \mathbb{N} \times \mathbb{N} \rightarrow \mathbb{N}$ such that whenever e is a Kleene index for $k_0, \ldots, k_{n-1}$, we have $\theta(\langle k_0, \ldots, k_{n-1} \rangle, e) = \lceil M \rceil$ where $M[x_0^{k_0}, \ldots, x_{n-1}^{k_{n-1}}] : \mathtt{N}$ is a β-normal Kleene expression such that*

$$\forall \Phi_0, \ldots, \Phi_{n-1}.\;\; [\![M]\!]_{\vec{x} \mapsto \vec{\Phi}} \;\simeq\; \{e\}(\vec{\Phi}) \,.$$

(ii) There is a primitive recursive function $\phi : \mathbb{N} \rightarrow \mathbb{N}$ such that whenever $\tau = l_0, \ldots, l_{m-1} \rightarrow \mathtt{N}$ and $M[x_0^{k_0}, \ldots, x_{n-1}^{k_{n-1}}] : \tau$ is a β-normal Kleene expression, $\phi(\lceil M \rceil)$ is a Kleene index e for $k_0, \ldots, k_{n-1}, l_0, \ldots, l_{p-1}$ such that

$$\forall \Phi_0, \ldots, \Phi_{n-1}, \Psi_0, \ldots, \Psi_{p-1}.\;\; \{e\}(\vec{\Phi}, \vec{\Psi}) \;\simeq\; [\![M]\!]_{\vec{x} \mapsto \vec{\Phi}}(\vec{\Psi}) \,.$$

Proof. (i) The mapping θ may be defined by cases on the syntactic form of e (with a case for each of S1–S9) under the assumption that we already have a (classical)

Kleene index z for θ itself. By the classical second recursion theorem, we thence obtain a computable function θ as required.

Most of the cases are straightforward exercises; the only two that merit attention are those for S4 and S9. For S4, we first define the abbreviation

$$(M \upharpoonright N) \equiv \mathit{Primrec}(N, M, \lambda yz.M) ,$$

and note that $(M \upharpoonright N)$ has value a iff M has value a and the value of N is defined. We now define

$$\theta(\langle k_0, \dots, k_{n-1} \rangle, \langle 4, e_0, e_1 \rangle) = (\theta(e_0)[x_0 \mapsto \theta(e_1), x_1 \mapsto x_0, \dots, x_n \mapsto x_{n-1}] \upharpoonright \theta(e_1))$$

(pretending for convenience that θ returns actual Kleene expressions rather than Gödel numbers for them). It is easy to see that the Kleene expression on the right-hand side is in β-normal form if both $\theta(e_0)$ and $\theta(e_1)$ are. The use of $\upharpoonright$ ensures the correct behaviour even when x_0 does not appear in $\theta(e_0)$.

For S9, we may use Exercise 5.1.16 along with the magical index z to construct a Kleene expression $\mathit{Translate}[w, y] : \mathtt{N}$ such that

$$[\![\mathit{Translate}[\langle k_1, \dots, k_n \rangle, e]]\!]() = \theta(\langle k_1, \dots, k_n \rangle, e)$$

for any $k_1, \dots, k_n$ and e. Using some obvious syntactic sugar for case splits, we may now define

$$\begin{aligned} \theta(\langle k_0, \dots, k_n \rangle, \langle 9, m \rangle) = {} & \mathtt{case}\ m\ \mathtt{of} \\ & 0 \Rightarrow \mathit{Eval}_{\mathtt{N}}(\mathit{Translate}[\langle\rangle, x_0]) \\ & \dots \\ & \mid n \Rightarrow \mathit{Eval}_{k_1,\dots,k_n \to \mathtt{N}}(\mathit{Translate}[\langle k_1, \dots, k_n \rangle, x_0], x_1, \dots, x_n) . \end{aligned}$$

To show that θ has the stated property, one then checks by transfinite inductions on the generation of $\{e\}(\vec{\Phi})$ and $[\![M]\!]_{\vec{x} \mapsto \vec{\Phi}}$ that if either of these is defined then so is the other and they are equal. We omit the tedious details.

(ii) The mapping ϕ may be defined by recursion on the regular typing derivation for M (with a case for each regular typing rule) under the assumption that we already have an (S1–S9) Kleene index z' for ϕ. By Theorem 5.1.5, we thence obtain a computable function ϕ as required.

All the cases are easy exercises except the one for Eval_σ. Note that in this case we have $\tau = \mathtt{N}$ so $p = 0$. To define $\phi(\lceil \mathit{Eval}_\sigma(M, N_0, \dots, N_{n-1}) \rceil)$, we must build a Kleene index e which, given an argument tuple $\vec{\Phi}$, performs the following steps:

- Compute $c = \phi(\lceil M \rceil)$.
- Evaluate $\{c\}$ at $\vec{\Phi}$.
- If this yields a number $m = \lceil P \rceil$, say, compute $\lceil Q \rceil$ where Q is a β-normal form of $PN_0 \dots N_{n-1}$; otherwise diverge.
- Compute $d = \phi(\lceil Q \rceil)$.
- Evaluate $\{d\}$ at $\vec{\Phi}$, and return the result if there is one.

Using S4 and S9 along with the effective computability of β-normal forms, it is easy to see how such an e can be effectively constructed from $\ulcorner Eval_\sigma(M, N_0, \ldots, N_{n-1}) \urcorner$ and the magical index z'.

As in (i), the verification of the required Kleene equality now consists of tedious transfinite inductions on the generation of its left- and right-hand sides. □

Corollary 5.1.18. *For any $k_0, \ldots, k_{n-1}$, the Kleene computable partial functionals $F : \mathbf{A}(k_0) \times \cdots \times \mathbf{A}(k_{n-1}) \rightharpoonup \mathbb{N}$ as defined by S1–S9 are exactly the partial functionals represented by β-normal Kleene expressions $M[x_0^{k_0}, \ldots, x_{n-1}^{k_{n-1}}] : \mathtt{N}$.* □

Note in passing that the above establishes the independence of the class of Kleene computable functionals from the details of the particular coding systems used (whether for Kleene indices or for the Gödel numbering of Kleene expressions).

We may now fill an earlier gap and supply a proof of the substitution theorem 5.1.9. Indeed, the corresponding property for our lambda calculus formulation is fairly easy to establish. The key fact we shall need is the following; here we write $[\![M']\!] \sqsupseteq [\![M]\!]$ to mean that $[\![M']\!]_\nu(\vec{\Psi}) \succeq [\![M]\!]_\nu(\vec{\Psi})$ for all suitable ν and $\vec{\Psi}$.

Lemma 5.1.19. *If M, M' are Kleene expressions with $M \rightsquigarrow_\beta M'$, then $[\![M']\!] \sqsupseteq [\![M]\!]$.*

Proof. First note that if $[\![(\lambda x.N)P]\!]_\nu \cdot \vec{\Phi} = a$ for some $\nu, \vec{\Phi}$, then in particular $F = [\![P]\!]_\nu^*$ is defined, and $[\![(\lambda x.N)P]\!]_\nu(\vec{\Phi}) = [\![N]\!]_{\nu, x \mapsto F}(\vec{\Phi})$ by the definition of $[\![-]\!]$. But also $[\![N]\!]_{\nu, x \mapsto F}(\vec{\Phi}) \simeq [\![N[x \mapsto P]]\!]_\nu(\vec{\Phi})$ by an easy induction on the structure of N. Thus $[\![N[x \mapsto P]]\!] \sqsupseteq [\![(\lambda x.N)P]\!]$.

Next, we show by induction on single-hole expression contexts C that

$$[\![C[N[x \mapsto P]]]\!] \sqsupseteq [\![C[(\lambda x.N)P]]\!] ,$$

the base case being given by the above. For most of our expression constructs, this follows easily from our induction hypothesis using the definition of $[\![-]\!]$. For contexts built using $Eval_\sigma$, this must be supplemented by the observation that the β-normal form of an expression $P'N_0 \ldots N_{n-1}$ is not affected by replacing $(\lambda x.N)P$ by $N[x \mapsto P]$ in some subexpression.

Since all single β-reduction steps have the above form, the lemma is proved. □

Theorem 5.1.20 (Substitution theorem). *Suppose $M[\vec{x}] : \sigma$ and $N[z^\sigma, \vec{y}] : \tau$ are β-normal Kleene expressions. Then there is a β-normal Kleene expression $P[\vec{x}, \vec{y}] : \tau$ such that*

$$[\![P]\!]_{\mu,\nu}(\vec{\Phi}) \succeq [\![N]\!]_{\nu, z \mapsto [\![M]\!]_\mu^*}(\vec{\Phi})$$

whenever μ, ν are valuations for $\vec{x}, \vec{y}$ respectively and the arguments $\vec{\Phi}$ are of suitable type. Furthermore, the Gödel number $\ulcorner P \urcorner$ for a suitable P may be effectively computed from $\ulcorner M \urcorner$ and $\ulcorner N \urcorner$.

Proof. Let P be the β-normal form of $N[z \mapsto M]$; then $[\![P]\!] \sqsupseteq [\![N[z \mapsto M]]\!]$ by Lemma 5.1.19. But if $[\![N]\!]_{\nu, z \mapsto [\![M]\!]_\mu^*}(\vec{\Phi})$ is defined then in particular $[\![M]\!]_\nu^*$ is defined, and so $[\![N[z \mapsto M]]\!]_{\mu,\nu}(\vec{\Phi}) \simeq [\![N]\!]_{\nu, z \mapsto [\![M]\!]_\mu^*}(\vec{\Phi})$ by induction on the structure of N as in the proof of Lemma 5.1.19. The effectivity claim is obvious. □

Exercise 5.1.21. Show that Theorem 5.1.9 follows easily from Theorem 5.1.20 using the effective translations given by Lemma 5.1.17. (Hint: translate Kleene indices for F,G into Kleene expressions M,N respectively, and consider $M_0 = \lambda x_0.M$.)

Remark 5.1.22. If unravelled far enough, our proof of Theorem 5.1.9 is not too different from the direct proofs given in [138, 130]. However, our treatment separates out the use of effective transfinite recursion from those ingredients that are in effect concerned with normalization of λ-terms.

As noted at the end of Subsection 4.4.2, the normalization theorem for simply typed λ-terms is actually provable in Primitive Recursive Arithmetic, so that the operation of normalization is primitive recursive (although our proof in Section 4.4 does not show this). In the light of this, one may strengthen 'effectively' to 'primitively recursively' in Theorem 5.1.20, and hence in Theorem 5.1.9 and Corollary 5.1.11.

We can also use our λ-calculus presentation to give a reasonably abstract and purely finitary 'operational semantics' for Kleene computability via an effective reduction system. This will account for what happens when one Kleene computable functional is applied to another, although of course it has nothing to say about computing with arbitrary set-theoretic functionals. Since the ideas will not be required in the sequel, we give only a brief outline.

We start by giving a purely 'first-order' reduction system for closed expressions of ground type in β-normal form. We here view the reduction of a Kleene expression M to its normal form $\beta nf(M)$ as a single 'computation step'. Let $\rightsquigarrow_0$ be the small-step reduction given inductively by:

- $\mathit{Primrec}\,(\widehat{0},N,\lambda yz.P) \;\rightsquigarrow_0\; N$,
- $\mathit{Primrec}\,(\widehat{b+1},N,\lambda yz.P) \;\rightsquigarrow_0\; P[y \mapsto \widehat{b},\, z \mapsto \mathit{Primrec}\,(\widehat{b},N,\lambda yz.P)]$,
- $\mathit{Eval}_\sigma(\lceil P \rceil, N_0,\ldots,N_{r-1}) \;\rightsquigarrow_0\; \beta nf(PN_0\ldots N_{r-1})$,
- if $M \rightsquigarrow_0 M'$ then $C[M] \rightsquigarrow_0 C[M']$, where $C[-]$ has any of the forms

$$\mathit{Suc}(-)\,, \qquad \mathit{Primrec}\,(-,N,P)\,, \qquad \mathit{Eval}_\sigma(-,N_0,\ldots,N_{r-1})\,.$$

Theorem 5.1.23. *If $M : \mathtt{N}$ is a closed β-normal Kleene expression, then*

$$M \rightsquigarrow_0^* \widehat{a} \;\; \mathit{iff} \;\; [\![M]\!]() = a\,.$$

Proof. Given a reduction sequence $R : M \rightsquigarrow_0^* \widehat{a}$, it is easy to build a derivation of $[\![M]\!]() = a$ via the defining clauses for $[\![-]\!]$, by induction on the length of R.

For the converse, observe that if $\Gamma \vdash M : \mathtt{N}$ where all variables in Γ are of type $\mathtt{N}$, then whenever $[\![M]\!]_\nu() = a$, the derivation of this fact takes the form of a finitely branching tree. (This is easily seen by induction on the definition of $[\![M]\!]_\nu() = a$.) By König's Lemma, such derivations are therefore finite. It is now easy to see that if M is closed and $[\![M]\!]() = a$, then either $M = \widehat{a}$ or $M \rightsquigarrow_0 M'$ where $[\![M']\!]() = a$ via a smaller derivation. It follows that if $[\![M]\!]() = a$ then $M \rightsquigarrow_0^* \widehat{a}$. □

Thus, for example, if $M : \sigma \to \mathtt{N}$ and $N : \sigma$ are closed Kleene expressions and $[\![MN]\!]() = a$, then $[\![\beta nf(MN)]\!]() = a$ by Lemma 5.1.19, and so the value a may be computed by β-reducing MN to normal form and then applying $\rightsquigarrow_0$. However, the converse is not true, for the reason illustrated by Example 5.1.10: even if $[\![N]\!]^*$ is defined, we may have $\beta nf(MN) \rightsquigarrow_0^* \widehat{a}$ but not $[\![MN]\!]() = a$.

We may now combine $\rightsquigarrow_0$ with β-reduction to obtain a somewhat more natural system. Specifically, for arbitrary closed terms $M : \mathtt{N}$, we define $\rightsquigarrow_1$ inductively by:

- $(\lambda x.M)N \rightsquigarrow_1 M[x \mapsto N]$,
- $Primrec\,(\widehat{0},N,P) \rightsquigarrow_1 N$,
- $Primrec\,(\widehat{b+1},N,P) \rightsquigarrow_1 P\widehat{b}(Primrec\,(\widehat{b},N,P))$,
- $Eval_\sigma(\lceil P \rceil, N_0, \ldots, N_{r-1}) \rightsquigarrow_1 PN_0 \ldots N_{r-1}$,
- if $M \rightsquigarrow_1 M'$ then $C[M] \rightsquigarrow_1 C[M']$, where $C[-]$ has any of the forms

$$(-)N\,, \qquad Suc(-)\,, \qquad Primrec\,(-,N,P)\,, \qquad Eval_\sigma(-,N_0,\ldots,N_{r-1})\,.$$

Theorem 5.1.24. *If $M : \mathtt{N}$ is closed and $[\![M]\!] = a$, then $M \rightsquigarrow_1^* \widehat{a}$.*

The proof proceeds by a detailed comparison between $\rightsquigarrow_1$-reductions for M and $\rightsquigarrow_0$-reductions for $\beta nf(M)$. We omit the unilluminating details.

5.1.4 Models Arising from Kleene Computability

Next, we raise our level of abstraction somewhat and consider how Kleene computability fits into the framework of Chapters 3 and 4.

First, we note the obvious fact that closed Kleene expressions modulo β-equality form a λ-algebra, which we shall call Klex^0/β; it is clearly endowed with a simulation in K_1 via Gödel numbering. This model will play a role in Section 6.2, where we will also see that it may be consistently quotiented by the equivalence generated by $\rightsquigarrow_0$ or $\rightsquigarrow_1$.

Our main interest, however, is in the class of Kleene computable functionals over a given type structure. Again, we start with a total extensional TPCA $\mathbf{A}$ over $\mathsf{T}^{\rightarrow}(\mathtt{N} = \mathbb{N})$, which we do not initially assume to be Kleene closed.

We may say a partial function $F : \mathbf{A}(\sigma_0) \times \cdots \times \mathbf{A}(\sigma_{r-1}) \rightharpoonup \mathbf{A}(\tau)$ is *Kleene computable* if there is a β-normal Kleene expression $x_0^{\sigma_0}, \ldots, x_{r-1}^{\sigma_{r-1}} \vdash M : \tau$ such that for all $\vec{\Phi} \in \mathbf{A}(\sigma_0) \times \mathbf{A}(\sigma_{r-1})$ we have $F(\vec{\Phi}) \simeq [\![M]\!]^*_{\vec{x} \mapsto \vec{\Phi}}$. It is easy to see that the identity on $\mathbf{A}(\sigma)$ is Kleene computable; moreover, by Theorem 5.1.20, if $F : \mathbf{A}(\sigma) \rightharpoonup \mathbf{A}(\tau)$ and $G : \mathbf{A}(\tau) \rightharpoonup \mathbf{A}(\upsilon)$ are Kleene computable, there is some Kleene computable $H : \mathbf{A}(\sigma) \rightharpoonup \mathbf{A}(\upsilon)$ such that $H(\Phi) \succeq G(F(\Phi))$ for all Φ. Thus, the datatypes $\mathbf{A}(\sigma)$ and Kleene computable partial functions between them constitute a lax computability model $\mathsf{Kl}_p(\mathbf{A})$ (see Section 3.1.6). We also write $\mathsf{Kl}_t(\mathbf{A})$ for the submodel of $\mathsf{Kl}_p(\mathbf{A})$ consisting of total Kleene computable functions $\mathbf{A}(\sigma) \rightarrow \mathbf{A}(\tau)$; this is of course a computability model in the strict sense of Section 3.1.1.

Even in well-behaved cases, we cannot expect $\mathsf{Kl}_p(\mathbf{A})$ to be a higher-order model, since the datatypes contain only total functionals. On the other hand, $\mathsf{Kl}_t(\mathbf{A})$ will be a higher-order model (with the evident operation $\Rightarrow$ on types) whenever **A** is *Kleene closed*. In the TPCA setting, this notion may be suitably formulated as follows:

Definition 5.1.25. A total extensional TPCA **A** is *Kleene closed* if for every partial Kleene computable $F : \mathbf{A}(\sigma_0) \times \cdots \times \mathbf{A}(\sigma_{r-1}) \rightharpoonup \mathbf{A}(\tau)$ and every $\Phi_1 \in \mathbf{A}(\sigma_1), \ldots, \Phi_{r-1} \in \mathbf{A}(\sigma_{r-1})$ where $r > 0$, the function $\Lambda\xi \in \mathbf{A}(\sigma_0).F(\xi, \vec{\Phi})$, if total, is present in $\mathbf{A}(\sigma_0 \to \tau)$.

It is easy to check that if **A** is Kleene closed then $\mathsf{Kl}_t(\mathbf{A})$ is a higher-order computability model in the sense of Definition 3.1.13. (In fact, this only requires the above condition to hold for *total* computable F.) We shall write $\mathbf{A}^{\mathrm{Kl}}$ for the corresponding substructure of **A**, so that $\mathbf{A}^{\mathrm{Kl}}(\mathtt{N}) = \mathbb{N}$ and $\mathbf{A}^{\mathrm{Kl}}(\sigma \to \tau)$ is the set of Kleene computable total functions $\mathbf{A}(\sigma) \to \mathbf{A}(\tau)$. The higher-order model $\mathsf{Kl}_t(\mathbf{A})$ thus corresponds to the relative TPCA $(\mathbf{A};\mathbf{A}^{\mathrm{Kl}})$ via Theorem 3.1.18. Moreover, since **A** is automatically a $\lambda\eta$-algebra, it is now a triviality that $(\mathbf{A};\mathbf{A}^{\mathrm{Kl}})$ is a *relative* $\lambda\eta$-*algebra* in the sense of Definition 4.1.2(ii).

It is also easy to see that $(\mathbf{A};\mathbf{A}^{\mathrm{Kl}})$ has booleans in the sense of Section 3.3.3; thus, the whole theory of retractions from Section 4.2 is available for this structure. In particular, each type $\mathbf{A}(\sigma)$ is a retract of some pure type $\mathbf{A}(k)$ via total Kleene computable functions $\mathbf{A}(\sigma) \to \mathbf{A}(k)$ and $\mathbf{A}(k) \to \mathbf{A}(\sigma)$.

This means that, with hindsight, one can obtain the whole structure $(\mathbf{A};\mathbf{A}^{\mathrm{Kl}})$ from just the S1–S9 definable functions on pure types—for instance, by constructing the category of computable functions $\mathbf{A}(k) \to \mathbf{A}(l)$ (represented by S1–S9 definable functions over $\overline{k}, \overline{l-1}$ when $l > 0$) and then taking its Karoubi envelope. However, in order to define composition in this category, one already needs Corollary 5.1.11, which is awkward to prove directly in terms of S1–S9.

Exercise 5.1.26. Suppose **A** is a total extensional TPCA over $\mathsf{T}^{\to}(\mathtt{N} = \mathbb{N})$. Show that the pure type part of **A** is Kleene closed in the sense of Definition 5.1.3 iff **A** is Kleene closed in the sense of Definition 5.1.25(i). (Hint: in such a structure **A**, the standard definition of the retractions $\mathbf{A}(\sigma) \lhd \mathbf{A}(k)$ requires only *tt*, *ff* and $\mathit{if}_{\mathtt{N}}$.)

Exercise 5.1.27. (i) Show that for any type $\sigma = \sigma_0, \ldots, \sigma_{n-1} \to \mathtt{N}$, regular or not, and any Kleene expressions $M : \sigma, N_0 : \sigma_0, \ldots, N_{n-1} : \sigma_{n-1}$, a term $\mathit{Eval}_\sigma(M, \vec{N})$ can be constructed with the property given in Definition 5.1.15.

(ii) Hence (or otherwise) show that $(\mathbf{A};\mathbf{A}^{\mathrm{Kl}})$ contains System T recursors as in Proposition 3.3.11, so that $0, 1, 2, \ldots \in \mathbf{A}^{\mathrm{Kl}}(\mathtt{N})$ is a system of numerals.

We may also note that $\mathbf{A}^{\mathrm{Kl}}$, considered as a model in its own right, is an *effective* TPCA in the sense of Section 3.5.1. Specifically, we may define a simulation $\gamma : \mathbf{A}^{\mathrm{Kl}} \longrightarrow\!\!\!\!\triangleright K_1$ by

$$n \Vdash^{\gamma}_{\sigma} x \ \text{ iff } \ n = \ulcorner M \urcorner \text{ for some closed } M : \sigma \text{ with } [\![M]\!] = x\,.$$

The effectivity clause of Theorem 5.1.20 tells us that application in $\mathbf{A}^{\mathrm{Kl}}$ is effective in K_1, that is, γ is a simulation in the sense of Definition 3.4.7. Furthermore, γ is numeral-respecting: combining Lemma 5.1.17(ii) for $\tau = \mathtt{N}$ with Proposition 5.1.13 for $n = 0$, we obtain a K_1 index d such that $n \Vdash^{\gamma}_{\mathtt{N}} m$ implies $d \bullet n = m$.

Exercise 5.1.28. Define $\delta : K_1 \longrightarrow\!\!\!\triangleright \mathbf{A}^{\mathrm{Kl}}$ by $x \Vdash^{\delta} n$ iff $x = n$. Show that δ is an applicative simulation and $\gamma\delta \sim id_{K_1}$. Show also that $\gamma\delta \preceq id_{\mathbf{A}^{\mathrm{Kl}}}$ but not vice versa.

5.1.5 Restricted Computability Notions

We conclude this section with a brief discussion of two more restricted classes of functionals that give rise to substructures of $\mathbf{A}^{\mathrm{Kl}}$. These will play significant roles both in this chapter and later in the book.

First, if we drop S9 from our definition of Kleene computability or equivalently the construct *Eval* from our language of Kleene expressions, we obtain the class of *primitive recursive functionals*. In contrast to full Kleene computability, it is easy to see that this class contains only *total* functionals (thus, the definition of $[\![-]\!]_{-}$ in Section 5.1.3 can be considerably simplified in this case). Clearly, the effective translations of Lemma 5.1.17 still work under these restrictions, and the discussion of Section 5.1.4 goes through, so that we obtain a relative $\lambda\eta$-algebra $(\mathbf{A};\mathbf{A}^{\mathrm{prim}})$, where $\mathbf{A}^{\mathrm{prim}} \subseteq \mathbf{A}^{\mathrm{Kl}}$.

We shall write $\mathsf{Klex}^{\mathrm{prim}}$ for the language of Kleene expressions without *Eval*. By considering the possible shapes of β-normal expressions of type level 1 in this system, it is easy to show that at first and second order, the functions present in $\mathbf{A}^{\mathrm{prim}}$ are precisely the primitive recursive ones in the classical sense. Furthermore, for arbitrary types $\sigma = \sigma_0, \ldots, \sigma_{r-1} \to \mathtt{N}$, we can define weak versions of the recursors rec_σ from Gödel's System T (see Section 3.2.3). Specifically, we define

$$\widehat{rec}_\sigma \ : \ \sigma \to (\mathtt{N} \to \mathtt{N} \to \sigma) \to \mathtt{N} \to \sigma$$

by

$$\widehat{rec}_\sigma \ = \ \lambda xfn.\lambda z_0 \ldots z_{r-1}.\, Primrec(n,\ x\vec{z},\ \lambda yv.fyv\vec{z})\ .$$

This satisfies the equations

$$\widehat{rec}_\sigma x f 0 \vec{z} = x\vec{z}\,,$$
$$\widehat{rec}_\sigma x f (n+1)\vec{z} = f n (\widehat{rec}_\sigma x f n \vec{z})\vec{z}\,.$$

However, the language $\mathsf{Klex}^{\mathrm{prim}}$ is strictly weaker than Gödel's System T, even at type 1. For the inclusion, it is easy to see that in total type structures over $\mathbb{N}$, the operator *Primrec* is equivalent in power to the System T recursor rec_0.[3] For the

[3] The reader is warned here of a small notational mismatch: $Primrec(M,N,P)$ corresponds broadly to $rec_0\ N\ (\lambda xy.Pyx)\ M$. For rec_0 we have retained Gödel's notation; the difference in the order of the two arguments of P stems from a difference between this and Kleene's presentation. As regards the three arguments of *Primrec*, our convention is designed to emphasize that M is envisaged as being evaluated first.

Whilst these mismatches might appear annoying, the notational distinction is in principle significant. Indeed, in the presence of *partial* functions, there is a crucial computational difference between the two forms of primitive recursion, as will be explained at the end of Subsection 6.3.3.

strictness, there is a well-known example of a function $h : \mathbb{N} \times \mathbb{N} \to \mathbb{N}$ (the *Ackermann function*) which is not primitive recursive but is definable using the operator rec_1 of System T (see e.g. [275, Chapter 29]). Alternatively, it is an interesting exercise to use the operator rec_1 to construct a *universal* program for ordinary primitive recursive functions $\mathbb{N} \to \mathbb{N}$; a simple diagonalization argument shows that no such program can itself be primitive recursive.

Note in passing that the absence of System T recursors in $\mathbf{A}^{\mathrm{prim}}$ means that in this model $0, 1, \ldots,$ are not actually a system of *numerals* in the sense of Section 3.3.4.

To summarize, both S1–S8 and System T offer reasonable candidates for a higher-order generalization of primitive recursion; however, the former is conservative over the first-order notion, whilst the latter is not.

For our other restricted notion of computability, we have already noted that the classical *minimization* operator is S1–S9 computable, and is likewise definable via a Kleene expression (Proposition 5.1.7; Exercise 5.1.16). We may therefore obtain a notion of computability intermediate between $\mathbf{A}^{\mathrm{prim}}$ and $\mathbf{A}^{\mathrm{Kl}}$ by considering S1–S8 together with the following scheme.

S10 (*Minimization*): If $e = \langle 10, d \rangle$ then $\{e\}(\vec{\Phi}) = a$ if $a = \mu x.\{d\}(x, \vec{\Phi}) = 0$.

(Note that if $a = \mu x.\{d\}(x, \vec{\Phi}) = 0$ then we must have $\{d\}(x, \vec{\Phi})$ defined for all $x \leq a$.) Equivalently, we may extend $\mathsf{Klex}^{\mathrm{prim}}$ to a new language $\mathsf{Klex}^{\mathrm{min}}$ by adding the construct

$$\frac{\Gamma \vdash M : \mathtt{N} \to \mathtt{N}}{\Gamma \vdash Min(M) : \mathtt{N}}$$

which we endow with a denotational semantics as follows:

- If $[\![M]\!]_\nu(a) = 0$ and $[\![M]\!]_\nu(b) > 0$ for all $b < a$, then $[\![Min(M)]\!]_\nu() = a$.

Again, the translations of Lemma 5.1.17 and the discussion of Section 5.1.4 go through; we hence obtain a lax computability model $Min_p(\mathbf{A})$ of partial computable functionals, and a relative $\lambda\eta$-algebra $(\mathbf{A}; \mathbf{A}^{\mathrm{min}})$ where $\mathbf{A}^{\mathrm{prim}} \subseteq \mathbf{A}^{\mathrm{min}} \subseteq \mathbf{A}^{\mathrm{Kl}}$.

We shall refer to the computable operations present in $Min_p(\mathbf{A})$ as the *μ-computable* partial functionals (the term *μ-recursive* is also used in the literature). Clearly, this class includes all classically computable partial functionals $\mathbb{N}^n \rightharpoonup \mathbb{N}$, and indeed all classically computable *type 2 partial functionals* $\mathbb{N}^n \times (\mathbb{N}^{\mathbb{N}})^m \rightharpoonup \mathbb{N}$. It follows that the inclusion $\mathbf{A}^{\mathrm{prim}}(1) \subseteq \mathbf{A}^{\mathrm{min}}(1)$ is strict, since $\mathbf{A}^{\mathrm{min}}(1)$ contains all classically computable total functions, and that $\mathbf{A}^{\mathrm{min}}(\sigma) = \mathbf{A}^{\mathrm{Kl}}(\sigma)$ if $\mathrm{lv}(\sigma) \leq 2$. We shall see in Section 6.3.5, however, that there are total functionals $\Phi \in \mathbf{A}^{\mathrm{Kl}}(3)$ that are not μ-computable.

We shall see in Chapter 6 that both Kleene computability and μ-computability have good credentials as mathematically natural notions: the Kleene computable functionals are those represented by *nested sequential procedures*, while the μ-computable functionals are exactly those represented by *left-bounded* procedures.

5.2 Computations, Totality and Logical Complexity

Having established the basic properties of the Kleene computable functionals, we now proceed to a somewhat deeper analysis of the concepts of computation, termination and totality, examining these concepts in particular from the standpoint of *logical complexity*. We start in Subsection 5.2.1 with a brief summary of the key ideas of logical complexity for the benefit of readers unfamiliar with them—this provides some relevant conceptual background for Subsection 5.2.2, as well as some necessary technical prerequisites for Subsection 5.2.3 and later parts of the book.

In Subsection 5.2.2, we introduce the notion of *top-down* computation trees, picking up on an idea from the end of Section 5.1.2. We show how such trees may be used to analyse the logical complexity of termination in Kleene's setting, and to obtain other structural information.

In Subsection 5.2.3, we consider a closely related question: which Kleene indices e define *total* computable functionals of a given type σ? In general, this will depend on the type structure we are working in; but an important observation is that since the set-theoretic model S can be viewed as the *maximal* type structure, any Kleene index that defines a total functional in S will do so in any other Kleene closed model. This class of 'absolutely total' indices turns out to be relatively benign in terms of its logical complexity: we show that totality in S is a Π^1_1 property.

5.2.1 Logical Complexity Classes

We begin with a brief review of the theory of logical complexity. The core of the theory is due to Kleene [134, 137]; the basic idea is to classify predicates according to the structural complexity of the logical formulae needed to define them. We shall here introduce the definitions of the complexity classes known as Π^0_n, Σ^0_n, Δ^0_n and Π^1_n, Σ^1_n, Δ^1_n over products of copies of $\mathbb{N}$ and $\mathbb{N}^{\mathbb{N}}$; readers already familiar with these classes may skip this subsection. For a fuller treatment, see Rogers [245].

Throughout this subsection, X will denote a set of the form $\mathbb{N}^n \times (\mathbb{N}^{\mathbb{N}})^m$; we shall refer to such sets as *pointsets*. By a *predicate* on X we shall mean simply a subset $A \subseteq X$. As the bottom level in our classification scheme, we let $\Delta^0_0 = \Pi^0_0 = \Delta^0_0$ be the class of all (classically) primitive recursive predicates on pointsets. Clearly this class is closed under the boolean operations of union, intersection and complement, under inverse images of primitive recursive functions, and under definitions via bounded number quantifiers $\exists x < y$ and $\forall x < y$.

We shall also consider a logical language for defining predicates. For convenience, we choose a language with two sorts, one for $\mathbb{N}$ and one for $\mathbb{N}^{\mathbb{N}}$, a predicate symbol for each primitive recursive predicate, and a constant or function symbol for every primitive recursive function or functional. Our language is also equipped with the usual boolean operators, and with quantifiers $\forall, \exists$ for both $\mathbb{N}$ and $\mathbb{N}^{\mathbb{N}}$; we call these *number* and *function* quantifiers respectively.

Definition 5.2.1. (i) A predicate is *arithmetical* if it is definable from the predicate symbols using boolean operators and number quantifiers.

(ii) We define the complexity classes Π^0_n and Σ^0_n for $n > 0$ by mutual recursion:

- A predicate $A \subseteq X$ is Π^0_{n+1} if there is a Σ^0_n predicate $B \subseteq \mathbb{N} \times X$ such that for all $\vec{x} \in X$ we have
$$\vec{x} \in A \;\Leftrightarrow\; \forall y \in \mathbb{N}.\,(y,\vec{x}) \in B\,.$$
- A predicate $A \subseteq X$ is Σ^0_{n+1} if its complement $X - A$ is Π^0_{n+1}.

(iii) A predicate is Δ^0_n if it is both Π^0_n and Σ^0_n.

Proposition 5.2.2. *(i) Any Σ^0_n predicate A may be defined by a formula of the form*

$$\vec{x} \in A \;\Leftrightarrow\; \exists y_{n-1}.\forall y_{n-2}.\cdots R(y_{n-1}, y_{n-2}, \ldots, y_0, \vec{x})$$

with n alternating number quantifiers, where R is primitive recursive. Likewise, any Π^0_n predicate B may be defined by a formula

$$\vec{x} \in B \;\Leftrightarrow\; \forall y_{n-1}.\exists x_{n-2}.\cdots S(y_{n-1}, y_{n-2}, \ldots, y_0, \vec{x})$$

with n alternating number quantifiers, where S is primitive recursive.

(ii) Let $n > 0$. Then the class of Σ^0_n predicates is closed under finite unions and intersections, under bounded number quantification, and under existential number quantifications $\exists x \in \mathbb{N}$. Likewise, the Π^0_n predicates are closed under finite unions and intersections, bounded number quantification, and universal number quantifications $\forall x \in \mathbb{N}$. Hence every arithmetical predicate is Σ^0_n or Π^0_n for some n.

(iii) Each of the classes Π^0_n, Σ^0_n and Δ^0_n is closed under preimages via primitive recursive (functions and) functionals.

Proof. Part (i) is easy and is left as an exercise for the reader.

For (ii), recall that we have a primitive recursive pairing function $pair : \mathbb{N}^2 \to \mathbb{N}$ with projections $(-)_0, (-)_1 : \mathbb{N} \to \mathbb{N}$ such that $pair((n)_0, (n)_1) = n$ for all n. Using this, we can collapse a pair of number quantifiers of the same type into one:

$$\exists x.\exists y.\Phi(x,y) \mapsto \exists z.\Phi((z)_0, (z)_1)\,,$$
$$\forall x.\forall y.\Phi(x,y) \mapsto \forall z.\Phi((z)_0, (z)_1)\,.$$

The remaining details can now be left to the reader.

Part (iii) is proved by an easy induction on the length of the quantifier prefix. □

Definition 5.2.3. If Γ consists of a class Γ_X of predicates on X for each pointset X, we call $U \in \Gamma_{\mathbb{N}\times X}$ a *universal* for Γ at X if for every predicate $A \in \Gamma_X$ there exists $e \in \mathbb{N}$ such that $\vec{x} \in A$ iff $(e,\vec{x}) \in U$.

Proposition 5.2.4. *If $n > 0$ and Γ is one of the classes Σ^0_n or Π^0_n, then for each pointset X there is a universal predicate for Γ at X.*

Proof. We prove this for Σ^0_1. The rest follows easily using induction on n, and is left to the reader.

A predicate A is Σ^0_1 if and only if A is the domain of a partial computable functional. Using the existence of universal algorithms and the appropriate variant of Kleene's T-predicate, we obtain a universal predicate for Σ^0_1 at any given X. □

In the notations Π^0_n, Σ^0_n, Δ^0_n, the upper index indicates that we are only using quantifiers of type 0. We now turn our attention to the hierarchy obtained when we allow quantifiers of type 1 as well; this is known as the *analytic hierarchy*.

Definition 5.2.5. (i) A predicate is Π^1_0 (equivalently Σ^1_0) iff it is arithmetical.

(ii) A predicate A is Σ^1_{n+1} if it is of the form

$$\vec{x} \in A \;\Leftrightarrow\; \exists f \in \mathbb{N}^{\mathbb{N}}.\,(f,\vec{x}) \in B$$

where B is Π^1_n. A predicate A is Π^1_{n+1} if it is the complement of a Σ^1_{n+1} predicate.

(iii) A predicate is Δ^1_n if it is both Σ^1_n and Π^1_n.

In order to simplify or permute quantifier prefixes, we shall make use of the fact that there are primitive recursive bijections:

$$\mathbb{N}^{\mathbb{N}} \cong \mathbb{N} \times \mathbb{N}^{\mathbb{N}}\,, \quad \mathbb{N}^{\mathbb{N}} \cong (\mathbb{N}^{\mathbb{N}})^2\,, \quad \mathbb{N}^{\mathbb{N}} \cong (\mathbb{N}^{\mathbb{N}})^{\mathbb{N}}\,.$$

These enable us to perform the following translations and their duals:

$$\begin{aligned}
\exists f \in \mathbb{N}^{\mathbb{N}}.\exists g \in \mathbb{N}^{\mathbb{N}}.\Phi(f,g) &\mapsto \exists h.\Phi((h)_0,(h)_1)\,,\\
\forall x \in \mathbb{N}.\exists f \in \mathbb{N}^{\mathbb{N}}.\Phi(x,f) &\mapsto \exists f.\forall x.\Phi(x,(f)_x)\,,\\
\exists x \in \mathbb{N}.\forall y \in \mathbb{N}.\Phi(x,y) &\mapsto \forall f \in \mathbb{N}^{\mathbb{N}}.\exists x \in \mathbb{N}.\Phi(x,f(x))\,.
\end{aligned}$$

The details are left to the reader (note that the countable axiom of choice is required). We can also collapse $\exists x \in \mathbb{N}.\exists f \in \mathbb{N}^{\mathbb{N}}$ into one existential quantifier over $\mathbb{N}^{\mathbb{N}}$.

These quantifier operations may be used to show the following:

Proposition 5.2.6. *(i) For $n > 0$, the class of Σ^1_n predicates is closed under finite unions and intersections, under existential and universal number quantification, and under existential function quantification.*

(ii) For $n > 0$, the class of Π^1_n predicates is closed under finite unions and intersections, existential and universal number quantification, and universal function quantification.

(iii) Every Π^1_1 predicate A may be defined by an expression

$$\vec{x} \in A \;\Leftrightarrow\; \forall f \in \mathbb{N}^{\mathbb{N}}.\exists z \in \mathbb{N}.\,R(z,f,\vec{x})$$

where R is primitive recursive.

(iv) If $n > 0$, the classes Σ^1_n and Π^1_n have universal predicates at each pointset X.

(v) The classes Π^1_n, Σ^1_n and Δ^1_n are closed under preimages via computable functionals.

One important consequence of the existence of universal sets is that Π^0_n and Σ^0_n differ for each $n > 0$, and so do Π^1_n and Σ^1_n for each n. A further consequence is that the classes form proper hierarchies: the inclusions $\Sigma^0_n \subseteq \Sigma^0_{n+1}$ and $\Sigma^1_n \subseteq \Sigma^1_{n+1}$ are strict, and likewise for Π.

Definition 5.2.7. Let Γ be one of the classes Π^0_n, Σ^0_n, Π^1_n or Σ^1_n for $n > 0$. Suppose $A \subseteq \mathbb{N}$ and $B \subseteq \mathbb{N}^{\mathbb{N}}$ are in Γ.

(i) We say A is *Γ-complete* if for all X of the form $\mathbb{N}^n$ and all $C \in \Gamma_X$, there is a total computable function $f : X \to \mathbb{N}$ such that

$$\forall \vec{x} \in X.\, \vec{x} \in C \;\Leftrightarrow\; f(\vec{x}) \in A\,.$$

(ii) We say B is Γ-complete if for all X of the form $\mathbb{N}^n \times (\mathbb{N}^{\mathbb{N}})^m$ and all $D \in \Gamma_X$, there is a total computable functional $F : X \to \mathbb{N}^{\mathbb{N}}$ such that

$$\forall \vec{x} \in X.\, \vec{x} \in D \;\Leftrightarrow\; F(\vec{x}) \in B\,.$$

We end this subsection with an important classical result. We leave it to the reader to supply the coding details necessary to make formal sense of the statement.

Theorem 5.2.8. *The set of well-founded trees of finite sequences over $\mathbb{N}$ is Π^1_1-complete.*

Proof. First, let T be a tree of finite sequences. Then T is well founded iff

$$\forall f \in \mathbb{N}^{\mathbb{N}}.\exists n.\, \langle f(0),\ldots,f(n-1)\rangle \notin T\,.$$

Thus the set of well-founded trees is Π^1_1.

Conversely, suppose $A \subseteq X$ is any Π^1_1 predicate. By Proposition 5.2.6, we may write A in the form

$$\vec{x} \in A \;\Leftrightarrow\; \forall f.\exists n.\, R(f,n,\vec{x})$$

where R is primitive recursive. For any $\vec{y} = y_0,\ldots,y_{m-1}$, let us write

$$U_{\vec{y}} \;=\; \{f \in \mathbb{N}^{\mathbb{N}} \mid f(0) = y_0 \wedge \cdots \wedge f(m-1) = y_{m-1}\}\,.$$

By suitably designing R, we may assume that for all f, $\vec{x}$ and n, if $R(f,n,\vec{x})$ then the necessary information about f will be contained in $f(0),\ldots,f(n-1)$. Moreover, we may assume that $R(f,n,\vec{x}) \wedge n < n' \Rightarrow R(f,n',\vec{x})$. Thus there is a computable predicate H such that

$$H(\langle y_0,\ldots,y_{m-1}\rangle,n,\vec{x}) \;\Leftrightarrow\; \forall f.\,(f \in U_{\vec{y}} \Rightarrow R(f,n,\vec{x}))\,.$$

We may now define a set of finite sequences $T_{\vec{x}}$ for any $\vec{x}$ by

$$\vec{y} \in T_{\vec{x}} \;\Leftrightarrow\; \neg\exists n < m.\, H(\langle\vec{y}\rangle,n,\vec{x})\,.$$

Clearly $T_{\vec{x}}$ is always a tree and is computable uniformly in $\vec{x}$. We leave it to the reader to check that $T_{\vec{x}}$ is well-founded iff $\vec{x} \in A$. $\square$

Exercise 5.2.9. Show that the least fixed point of a Π^1_1 operator on $\mathbb{N}$ is Π^1_1. Specifically, show using quantifier manipulations that if an operator $\Theta : \mathscr{P}(\mathbb{N}) \to \mathscr{P}(\mathbb{N})$ is monotone and Π^1_1 (that is, the relation $y \in \Theta(X)$ is Π^1_1), then so is the set

$$\{x \in \mathbb{N} \mid \forall I \subseteq \mathbb{N}.\, (\forall y \in \mathbb{N}.\, y \in \Theta(I) \Rightarrow y \in I) \Rightarrow x \in I\}\,.$$

5.2.2 Top-Down Computation Trees

In Section 5.1.1, 'computations' were defined as Kleene computation trees. This is somewhat unsatisfactory in terms of the intuitive sense of 'computation': to say that the computation of $\{e\}(\vec{\Phi})$ 'terminates' is to say that there happens to exist a well-founded computation tree whose root is some triple $\{e\}(\vec{\Phi}) = a$, whilst to say the computation 'diverges' is simply to say there is no such tree. What this fails to provide is a good model of a *computational process* which will exist for any e and $\vec{\Phi}$, and which may or may not terminate.

We shall here refer to the Kleene computation trees of Section 5.1.1 as *bottom-up* computation trees, since they are inductively generated from initial computations by plugging together subtrees. Recall that the nodes of a bottom-up tree are labelled with tuples $(e', \vec{\Phi}', a)$. By contrast, we shall here define for any given $e, \vec{\Phi}$ a *top-down* computation tree $\mathscr{T}(e, \vec{\Phi})$, whose node labels have the simpler form $(e', \vec{\Phi}')$. Such trees were introduced implicitly by Kleene in [138, Section 5], and more explicitly in [139, Section 9].

The idea is to build such a tree starting from the root $(e, \vec{\Phi})$; the presence of a node labelled $(e', \vec{\Phi}')$ signals that the computation involves a subtask of computing $\{e'\}(\vec{\Phi}')$, which may or may not be completed. We shall refer to the nodes of the top-down tree (somewhat loosely) as *subcomputations*.

Definition 5.2.10. (i) For any tuple $(e, \vec{\Phi})$ where $\vec{\Phi} = \Phi_0, \ldots, \Phi_{n-1}$, we define a family of *immediate subcomputations* of $(e, \vec{\Phi})$ as follows:

- If e is an index for an initial computation (via S1, S2, S3 or S7), and $\vec{\Phi}$ are of the appropriate types for e, then $(e, \vec{\Phi})$ has no immediate subcomputations.
- If $e = \langle 4, e_0, e_1 \rangle$, then $(e_1, \vec{\Phi})$ is designated as the *first* immediate subcomputation of $(e, \vec{\Phi})$. If $\{e_1\}(\vec{\Phi})$ is undefined, there are no other immediate subcomputations, whilst if $\{e_1\}(\vec{\Phi}) = b$ then $\langle e_0, b, \vec{\Phi} \rangle$ is designated as the *second* immediate subcomputation of $(e, \vec{\Phi})$.
- If $e = \langle 5, e_0, e_1 \rangle$ and $\Phi_0 = 0$, then the sole immediate subcomputation of $(e, \vec{\Phi})$ is $(e_0, \vec{\Phi})$.
- If $e = \langle 5, e_0, e_1 \rangle$ and $\Phi_0 = x + 1$, then $(e, x, \Phi_1, \ldots, \Phi_{n-1})$ is designated as the first immediate subcomputation of $\langle e, \vec{\Phi} \rangle$. If $\{e\}(x, \Phi_1, \ldots, \Phi_{n-1})$ is undefined, there are no other immediate subcomputations, whilst if $\{e\}(x, \Phi_1, \ldots, \Phi_{n-1}) = b$ then $(e_1, x, b, \Phi_1, \ldots, \Phi_{n-1})$ is designated as the second immediate subcomputation.
- If $e = \langle 6, d, \pi(0), \ldots, \pi(n-1) \rangle$, then the sole immediate subcomputation of $(e, \vec{\Phi})$ is $(d, \Phi_{\pi(0)}, \ldots, \Phi_{\pi(n-1)})$.

- If $e = \langle 8,d\rangle$ and Φ_0 is of type $k+2$, then for each $\xi \in \mathbf{A}(k)$ we have an *immediate subcomputation of* $(e,\vec{\Phi})$ *at* ξ, namely $(d,\xi,\vec{\Phi})$.
- If $e = \langle 9,m\rangle$ where $m \leq n$ and $\Phi_0 = d$, then the sole immediate subcomputation of $(e,\vec{\Phi})$ is $(d,\Phi_1,\ldots,\Phi_{m-1})$.
- If none of the above clauses apply, then by convention we take $(e,\vec{\Phi})$ to be the sole immediate subcomputation of itself.

(ii) The *top-down computation tree* for $(e,\vec{\Phi})$, denoted $\mathscr{T}(e,\vec{\Phi})$, is the unique labelled tree rooted at $(e,\vec{\Phi})$ in which the successors of any given node correspond precisely to its immediate subcomputations. A *leaf node* in this tree is one with no successors (corresponding to an instance of S1, S2, S3 or S7).

In principle, an analogous notion of top-down tree could be given for our λ-calculus presentation of Kleene computability, thereby showing that the concept is not restricted to the pure types. However, we shall prefer here to work with Kleene's presentation via indices, both for continuity with the literature and because the more general case adds only a notational burden.

Note that $\mathscr{T}(e,\vec{\Phi})$ exists whether or not $\{e\}(\vec{\Phi})$ is defined. For the sake of comparison, in cases where $\{e\}(\vec{\Phi})\downarrow$, we may also define the *bottom-up computation tree* $\mathscr{U}(e,\vec{\Phi})$ to be the Kleene computation tree as defined in Section 5.1.1, i.e. the well-founded tree embodying the inductive generation of the relevant triple $(e,\vec{\Phi},a)$.

Proposition 5.2.11. *$\mathscr{T}(e,\vec{\Phi})$ is well-founded iff $\{e\}(\vec{\Phi})\downarrow$. In this case, $\mathscr{T}(e,\vec{\Phi})$ may be obtained from $\mathscr{U}(e,\vec{\Phi})$ simply by omitting the result value a' from each node label.*

Proof. First suppose $\{e\}(\vec{\Phi})\downarrow$. Then the well-founded tree $\mathscr{U}(e,\vec{\Phi})$ exists, and it is easy to see that by removing the result values we get $\mathscr{T}(e,\vec{\Phi})$. Conversely, if $\mathscr{T}(e,\vec{\Phi})$ is well founded, we can fill in the correct value at each node by recursion on the rank of the node, yielding a tree $\mathscr{U}(e,\vec{\Phi})$ witnessing that $\{e\}(\vec{\Phi})\downarrow$. □

An immediate corollary of the above proposition is that $\{e\}(\vec{\Phi})$ is undefined iff there is an infinite path through $\mathscr{T}(e,\vec{\Phi})$. Such a path is called a *Moschovakis witness* to the non-termination of the computation in question.

Our first main result is that the tree $\mathscr{T}(e,\vec{\Phi})$ is *semicomputable* uniformly in e and $\vec{\Phi}$. To make this more precise, we should indicate how $\mathscr{T}(e,\vec{\Phi})$ can itself be represented by an object of finite type. We give the idea here without spelling out all the details, and leaning heavily on the use of computable retractions in the style of Section 4.2.

First, we require a way of labelling the *branches* emanating from any given node in a top-down tree. Suppose for the time being that e is a Kleene index at some type level $k \geq 3$, and $\vec{\Phi}$ a list of arguments with maximum level $k-1$. Then the branches of any given node in $\mathscr{T}(e,\vec{\Phi})$ may be labelled either with $0,1$ (in the case of a node with at most two successors) or with an element of $\mathbf{A}(l)$ for some $l \leq k-3$ (in the case of an S8 node). In either case, with the help of suitable retractions as in Section 4.2, such a branch may be labelled by an element of $\mathbf{A}(k-3)$. Such retractions also allow us to construe any element of $\mathbf{A}(k-3)$ as a branch from any given node—it will be easier to live with the existence of many elements of $\mathbf{A}(k-3)$ representing the same branch than with the stricture that only certain elements are valid branch representations in a given context.

It follows easily that a *position* in the tree $\mathscr{T}(e,\vec{\Phi})$, being specified by a finite sequence of branches starting from the root, may also be represented by an element of $\mathbf{A}(k-3)$. The tree itself can now be represented as a partial function from positions to *labels*. Clearly, within $\mathscr{T}(e,\vec{\Phi})$, every node has a label of the form $(e',\vec{\Phi}')$, where each Φ'_i is either one of the Φ_j or is given by the label on some S8-branch along the finite path leading to the node in question. Thus, in the presence of $\vec{\Phi}$ and the node position in the tree, the identity and order of the arguments $\vec{\Phi}'$ may be specified by finitary information encodable as a natural number. In this way, the tree $\mathscr{T}(e,\vec{\Phi})$ may be represented by a partial function $\gamma_{e,\vec{\Phi}} : \mathbf{A}(k-3) \rightharpoonup \mathbb{N}$, where $\gamma_{e,\vec{\Phi}}(\zeta)$ is defined iff $\zeta \in \mathbf{A}(k-3)$ represents the position of a genuine node in $\mathscr{T}(e,\vec{\Phi})$.

In the same way, the bottom-up tree $\mathscr{U}(e,\vec{\Phi})$, when it exists, may be represented by a partial function $\delta_{e,\vec{\Phi}} : \mathbf{A}(k-3) \rightharpoonup \mathbb{N}$; in this case, the label at each node includes the local result value a'. The following is now readily established:

Theorem 5.2.12. *Suppose $k \geq 3$, and $k_0,\ldots,k_{n-1}$ is a list of argument types with maximum $k-1$. Then there are Kleene computable partial functions*

$$C,D \;:\; \mathbf{A}(0) \times \mathbf{A}(k_0) \times \cdots \times \mathbf{A}(k_{n-1}) \times \mathbf{A}(k-3) \rightharpoonup \mathbb{N}$$

such that for any suitably typed $e,\vec{\Phi},\zeta$ we have

1. $C(e,\vec{\Phi},\zeta) \;\simeq\; \gamma_{e,\vec{\Phi}}(\zeta)$,
2. *if* $\{e\}(\vec{\Phi})\downarrow$, *then* $D(e,\vec{\Phi},\zeta) \;\simeq\; \delta_{e,\vec{\Phi}}(\zeta)$.

Proof. We may compute $C(e,\vec{\Phi},\zeta)$ by recursion on the length of the path represented by ζ. If ζ is the root position, the value of $\gamma_{e,\vec{\Phi}}(\zeta)$ is trivial to determine. Otherwise, if ζ is a successor of some position ζ' via the branch designated by some $\xi \in \mathbf{A}(k-3)$, it suffices to note that $\gamma_{e,\vec{\Phi}}(\zeta)$ is Kleene computable from $\vec{\Phi}$, $\gamma_{e,\vec{\Phi}}(\zeta')$ and ζ itself. In the case that the node at ζ is the second immediate subcomputation of the one at ζ' via S4 or S5, we must of course execute the relevant first subcomputation $\{e'\}(\vec{\Phi}')$ in order to obtain the necessary value of b. This can be done using S9, since the relevant arguments $\vec{\Phi}'$ may be retrieved from $\vec{\Phi}$ and ζ.

Likewise, having computed $\gamma_{e,\vec{\Phi}}(\zeta)$, we may attempt to compute the local value a' at each node ζ using S9. In the case that $\{e\}(\vec{\Phi})\downarrow$, this will indeed yield $\delta_{e,\vec{\Phi}}(\zeta)$ as required. □

In the simpler case $k<3$, the tree $\mathscr{T}(e,\vec{\Phi})$ is finitely branching (see the end of Section 5.1.2), so Theorem 5.2.12 will hold if we use the type 0 in place of $k-3$.

Much of the interest in these trees lies in the fact that a top-down tree equipped with a Moschovakis witness can serve as a *certificate* for the non-termination of $\{e\}(\vec{\Phi})$, much as a bottom-up tree can serve as a certificate for its termination.

There is a loose analogy here with the T-predicate from classical computability theory: a course of computation y can certify that $\{e\}(n)$ terminates, and the predicate $T(e,n,y)$ asserting that it does so may be verified by restricted means (it is primitive recursive). We shall show how something similar works at higher types: both top-down and bottom-up trees, with suitable auxiliary information, can function as certificates whose correctness predicate is of low logical complexity.

For simplicity, we shall restrict attention at this point to the full set-theoretic model S. We have seen that a top-down tree $\mathscr{T}(e,\vec{\Phi})$ is naturally represented by a *partial* functional $\gamma_{e,\vec{\Phi}}$. On the other hand, for the purpose of verifying their correctness, we would prefer our certificates to be *total* objects. To allow for some flexibility in this regard, we adopt the following definitions.

Definition 5.2.13. Suppose $k > 3$.

(i) We say that $G : \mathsf{S}(k-3) \rightharpoonup \mathbb{N}$ *certifies* $\{e\}(\vec{\Phi})\uparrow$ if G encodes a triple (γ,ε,μ), where

- $\gamma : \mathsf{S}(k-3) \rightharpoonup \mathbb{N}$ extends $\gamma_{e,\vec{\Phi}}$,
- $\varepsilon : \mathsf{S}(k-3) \rightharpoonup \mathbb{N}$ is such that $\varepsilon(\zeta) = a'$ whenever $\mathscr{T}(e,\vec{\Phi})$ has some label $(e',\vec{\Phi}')$ at position ζ and $\{e'\}(\vec{\Phi}') = a'$,
- $\mu : \mathbb{N} \to \mathsf{S}(k-3)$ represents a Moschovakis witness within $\mathscr{T}(e,\vec{\Phi})$.

(ii) Likewise, we shall say that $F : \mathsf{S}(k-3) \rightharpoonup \mathbb{N}$ *certifies* $\{e\}(\vec{\Phi})\downarrow$ if indeed $\{e\}(\vec{\Phi})\downarrow$ and F encodes a pair (δ,θ), where

- $\delta : \mathsf{S}(k-3) \rightharpoonup \mathbb{N}$ extends $\delta_{e,\vec{\Phi}}$,
- $\theta : \mathsf{S}(0 \to (k-3)) \rightharpoonup \mathbb{N}$ witnesses the well-foundedness of $\mathscr{U}(e,\vec{\Phi})$ in the following sense: whenever $\pi \in \mathsf{S}(0 \to (k-3))$ represents an infinite sequence of branches, $\theta(\pi)$ is defined and the sequence $\pi(0),\dots,\pi(\theta(\pi)-1)$ represents a leaf node position in $\mathscr{U}(e,\vec{\Phi})$. (We shall say in this case that $\theta(\pi)$ is a *leaf node index* for π.)

We have already seen that the minimal candidates for γ and (in the case of termination) δ are partial Kleene computable uniformly in e and $\vec{\Phi}$. This is clearly also true for ε and (in the case of termination) for θ, though not for μ.

Exercise 5.2.14. Construct a type 2 Kleene index e and a computable $f \in \mathsf{S}(1)$ such that the computation of $\{e\}(f)$ diverges, but the (unique) Moschovakis witness in $\mathscr{T}(e,f)$ is non-computable. (Hint: use the *Kleene tree* as described in Subsection 9.1.2.)

Clearly, whether $\{e\}(\Phi)\downarrow$ or $\{e\}(\Phi)\uparrow$, a *total* certificate for the relevant fact exists in S, though it may not be computable. The low logical complexity of the correctness of total certificates is established by the following theorem (essentially a repackaging of ideas from Kleene [138]):

Theorem 5.2.15. *Suppose $k > 3$ and the argument types k_i have maximum $k-1$. Then there exist primitive recursive functionals*

$$I,J \;:\; \mathsf{S}(0)\times\mathsf{S}(k_0)\times\cdots\times\mathsf{S}(k_{n-1})\times\mathsf{S}(k-2)\times\mathsf{S}(k-3)\to\mathbb{N}$$

such that for any e, $\vec{\Phi}$ and $F,G\in\mathsf{S}(k-2)$ we have

1. *F certifies $\{e\}(\vec{\Phi})\downarrow$ iff $\forall z\in\mathsf{S}(k-3).\, I(e,\vec{\Phi},F,z)=0$,*
2. *G certifies $\{e\}(\vec{\Phi})\uparrow$ iff $\forall z\in\mathsf{S}(k-3).\, J(e,\vec{\Phi},G,z)=0$.*

Proof. For termination certificates, the idea is that an element $z\in\mathsf{S}(k-3)$ can equally well be construed as coding a node position ζ in $\mathscr{U}(e,\vec{\Phi})$ or an infinite path π. We would therefore like a functional I such that if F codes (δ,θ) and z represents the position ζ and the path π, then $I(e,\vec{\Phi},F,z)=0$ asserts that δ is 'locally correct' at ζ *and* $\theta(\pi)$ is a leaf node index for π in δ. Here 'locally correct' means that either some ancestor of ζ is already a leaf node in δ (in which case the label at ζ does not matter), or, if $\delta(\zeta)$ represents the label $(e',\vec{\Phi}',a')$, then $e',\vec{\Phi}'$ are as expected on the basis of the predecessor node of ζ (and on its sister node in the case of a second subcomputation), while a' is as expected on the basis of any immediate successor nodes. We leave it as an exercise to firm up the definition of a primitive recursive I verifying these properties.

For non-termination certificates, the argument is similar, except that information required for checking local correctness is now distributed between γ and ε. We construe an element z as coding both a node position ζ and an index $i\in\mathbb{N}$, and construct J so that $J(e,\vec{\Phi},F,z)=0$ iff γ and ε are locally correct at ζ *and* the $(i+1)$th node along μ is a genuine successor to the ith node according to γ. □

To obtain the corresponding results for $k\leq 3$, the type for termination certificates must be increased to $\mathsf{S}(1)\rightharpoonup\mathbb{N}$ in order to be able to accommodate the component θ, and the type of I adjusted accordingly. For non-termination certificates, the above theorem still applies when $k=3$, but if $k\leq 2$ the type for non-termination certificates should be taken to be $\mathbb{N}\rightharpoonup\mathbb{N}$.

Results of this kind were successfully used by Kleene as the basis of his extension of the theory of logical complexity to higher types (see [138, Sections 6–8]). Related ideas will appear in Section 8.5, when we analyse the complexity of termination in the model Ct.

Remark 5.2.16. In Theorem 5.2.15, both correctness conditions for certificates involve a quantification over $\mathsf{S}(k-3)$. We may conclude that the correctness of certificates is actually semidecidable, if we are willing to exploit the morally questionable ability of oracles to semidecide the totality of their arguments. For instance, to test the correctness of non-termination certificates, we may define

$$J'(e,\vec{\Phi},F)\;\simeq\;\Phi_i(\lambda z.\text{ if }J(e,\vec{\Phi},F,z)=0\text{ then }0\text{ else undefined})$$

with J as above, where Φ_i is an argument of type $k-1$. Then G certifies $\{e\}(\vec{\Phi})\uparrow$ iff $J'(e,\vec{\Phi},G)\downarrow$.

The following shows that if we are not concerned to limit ourselves to a 'primitive recursive' style of verification as in Theorem 5.2.15, then even a Moschovakis witness μ by itself can play the role of a non-termination certificate:

Proposition 5.2.17. *In the situation of Theorem 5.2.15, there is a Kleene partial computable functional*

$$K : \mathsf{S}(0) \times \mathsf{S}(k_0) \times \cdots \times \mathsf{S}(k_{n-1}) \times \mathsf{S}(k-3) \rightharpoonup \mathbb{N}$$

such that for any $e, \vec{\Phi}$ and $\nu \in \mathsf{S}(k-3)$, we have $K(e,\vec{\Phi},\nu)\downarrow$ iff ν codes a Moschovakis witness within $\mathscr{T}(e,\vec{\Phi})$.

Proof. We consider ν as coding a function $\mu : \mathbb{N} \to \mathsf{S}(k-3)$. To confirm that μ represents a Moschovakis witness in $\mathscr{T}(e,\vec{\Phi})$, it suffices to check that for each $i \in \mathbb{N}$, the node specified by the path $\mu(0),\ldots,\mu(i-1)$ does indeed exist in $\mathscr{T}(e,\vec{\Phi})$. But for each i, this can be readily verified by the termination of a suitable Kleene computation (cf. the construction of $\gamma_{e,\vec{\Phi}}$ in the proof of Theorem 5.2.12). The fact that this holds for all i thus comes down to the totality of some Kleene computable function $\mathbb{N} \rightharpoonup \mathbb{N}$, and as in Remark 5.2.16, this can be tested via a suitable call to some argument of type $k-1$. □

It might seem counterintuitive that the non-termination of a computation can be verified by the *termination* of a certain test, and indeed that our predicates for termination and non-termination have the same logical form. The explanation is that in order to verify non-termination, we would need to 'guess' (or be given) a suitable Moschovakis witness μ, which is not in general computable from the data. Indeed, using precisely this observation, we can turn the whole situation around to obtain the following negative result. Notice that the value of k is here offset by 3 relative to the discussion so far.

Theorem 5.2.18. *Suppose $k > 0$ and $\Psi \in \mathsf{S}(k+2)$. Then the Ψ-semidecidable predicates, even on sets $\mathsf{S}(l_0) \times \cdots \mathsf{S}(l_{r-1})$ where $l_0,\ldots,l_{r-1} \leq k$, are not closed under existential quantification over $\mathsf{S}(k)$.*

Proof. Let K be as in Proposition 5.2.17, taking $n=0$ and $k_0 = k+2$. Then the predicate $P(e) \equiv \exists z.K(e,\Psi,z)\downarrow$ asserts that there exists a Moschovakis witness within $\mathscr{T}(e,\Psi)$, i.e. that $\{e\}(\Psi)\uparrow$. We can now use a diagonal argument to show that P is not Ψ-semidecidable. Suppose $P(e) \Leftrightarrow \{p\}(e,\Phi)\downarrow$, and use the second recursion theorem to construct q such that $\{q\}(\Psi) \simeq \{p\}(q,\Psi)$. Then $\{q\}(\Psi)\downarrow$ iff $P(q)$ iff $\{q\}(\Psi)\uparrow$, a contradiction. □

This slightly strengthens a theorem of Moschovakis [198, 130], in which Ψ was additionally required to be *normal* (see Definition 5.4.3). Such results are of interest because it is a crucial classical property of semidecidable predicates on sets $\mathbb{N}^r$ that they are closed under $\exists_{\mathbb{N}}$ (see also Corollary 5.4.6). The above theorem shows that this does not extend to existential quantification over a type $k > 0$, even in the presence of a potentially powerful functional of type $k+2$. On the other hand, a result beyond the scope of this book states that the Ψ-semidecidable predicates on types of level $\leq k+1$ *are* closed under $\exists_{\mathsf{S}(k)}$ if Ψ is normal of type $k+3$ (see Grilliot [110], Harrington and MacQueen [114]).

5.2.3 *Totality and the Maximal Model*

So far, we have been considering the possibility of interpreting Kleene indices in a range of models $\mathbf{A}$, but have not said much about how the interpretations in different models might be related. We now turn our attention to this question, with a particular interest in the special role played by S as the *maximal* type structure.

Mathematically, it matters little whether we work just with the pure types as in Section 5.1.1, or with arbitrary simple types as in Section 5.1.3. We find the former approach to be marginally more convenient.

To begin, we need a suitable notion of when one type structure is 'larger' than another. It is clear that the evident notion of inclusion is not appropriate here: if two type structures differ at type $\overline{k}$, they are strictly speaking not even comparable at type $\overline{k+1}$. A much better way of comparing type structures is via the concept of a prelogical relation as in Section 4.4. This notion may be adapted to the setting of pure type structures over $\mathbb{N}$ as follows:

Definition 5.2.19. Let $\mathbf{A} = \{\mathbf{A}(k) \mid k \in \mathbb{N}\}$ and $\mathbf{B} = \{\mathbf{B}(k) \mid k \in \mathbb{N}\}$ be two (total) pure type structures over $\mathbb{N}$. A *total prelogical relation* from $\mathbf{A}$ to $\mathbf{B}$ is a family $R = \{R(k) \subseteq \mathbf{A}(k) \times \mathbf{B}(k) \mid k \in \mathbb{N}\}$ such that

1. $R(0) = \{(n,n) \mid n \in \mathbb{N}\}$,
2. if $(f,g) \in R(k+1)$ and $(a,b) \in R(k)$, then $f(a) = g(b)$,
3. for each $a \in \mathbf{A}(k)$ there is at least one $b \in \mathbf{B}(k)$ such that $(a,b) \in R(k)$.

Such a relation is *logical* if additionally

4. if $f \in \mathbf{A}(k+1)$, $g \in \mathbf{B}(k+1)$, and $f(a) = g(b)$ for all $(a,b) \in R(k)$, then $(f,g) \in R(k+1)$.

The following recapitulates some relevant facts from Section 4.4:

Exercise 5.2.20. (i) Suppose R is a total prelogical relation from $\mathbf{A}$ to $\mathbf{B}$. Show that if $(a,b) \in R(k)$ and $(a',b) \in R(k)$ then $a = a'$.

(ii) Show that the only total prelogical relation from $\mathbf{A}$ to $\mathbf{A}$ is the identity.

(iii) Suppose R is a total prelogical relation from $\mathbf{A}$ to $\mathbf{B}$, and S a total prelogical relation from $\mathbf{B}$ to $\mathbf{C}$. Show that the typewise relational composition $S \circ R$ is a total prelogical relation from $\mathbf{A}$ to $\mathbf{C}$.

(iv) Show that if there is a total logical relation from $\mathbf{A}$ to $\mathbf{B}$, then it is unique.

(v) Let $\mathbf{A}$ be any pure type structure. Show that there is a total logical relation from $\mathbf{A}$ to the full set-theoretic type structure S.

Let us write $\mathbf{A} \preceq \mathbf{B}$ if there exists a total prelogical relation from $\mathbf{A}$ to $\mathbf{B}$. The above facts show that $\preceq$ is a partial ordering on pure type structures, and that (as might be expected) S is the unique maximal pure type structure under this ordering.

As explained in Section 5.1.1, the schemes S1–S9 may be interpreted in any pure type structure; we write $\{e\}_{\mathbf{A}}$ to denote the interpretation of the index e in $\mathbf{A}$. We next show how prelogical relations allow us to relate Kleene computations in different type structures.

Lemma 5.2.21. *Let* **A** *and* **B** *be Kleene closed pure type structures as in Definition 5.1.3. Let R be a total logical relation from A to B, and suppose* $(\Phi_i, \Psi_i) \in R(k_i)$ *for each* $i < r$. *If* $\{e\}_{\mathbf{B}}(\Psi_0, \ldots, \Psi_{r-1}) = a$ *then* $\{e\}_{\mathbf{A}}(\Phi_0, \ldots, \Phi_{r-1}) = a$.

Proof. By induction on the ordinal rank of the computation tree for $\{e\}_B(\vec{\Psi}) = a$. We consider only the case for S8, all other cases being trivial. Suppose $e = \langle 8, d \rangle$ and $k = k_0 - 2$, so that

$$\{e\}_{\mathbf{B}}(\Psi_0, \ldots, \Psi_{r-1}) \;=\; \Psi_0(G), \;\text{ where } G \;=\; \Lambda\xi \in \mathbf{B}(k).\, \{d\}_{\mathbf{B}}(\xi, \Psi_0, \ldots, \Psi_{r-1})\,.$$

For each $\zeta \in \mathbf{A}(k)$ there exists $\xi \in \mathbf{B}(k)$ with $(\zeta, \xi) \in R(k)$, so by the induction hypothesis, the functional

$$F \;=\; \Lambda\zeta \in \mathbf{A}(k).\, \{d\}(\xi, \Phi_0, \ldots, \Phi_{r-1})$$

is total; hence in $F \in \mathbf{A}(k+1)$. This establishes that $\{e\}_{\mathbf{A}}(\Phi_0, \ldots, \Phi_{r-1})\downarrow$. Furthermore, from the induction hypothesis and condition 4 of Definition 5.2.19, we see that $(F, G) \in R(k+1)$, and it follows that $\Phi_0(F) = \Psi_0(G)$, whence $\{e\}_{\mathbf{A}}(\vec{\Phi}) = a$. □

The converse is not true in general. For example, let Ct be the type structure of total continuous functionals as in Section 3.7. We shall see in Chapter 8 that Ct is Kleene closed, and by Exercise 5.2.20 we have Ct $\preceq$ S. Now define $f_n : \mathbb{N} \to \mathbb{N}$ by $f_n(m) = 1$ if $m = n$, $f_n(m) = 0$ otherwise, and define $f_\infty(m) = 0$ for all m. In Ct, though not in S, it is the case that every type $\overline{2}$ functional is *continuous*, so that for every $G \in \mathsf{Ct}(2)$ there exists n with $G(f_n) = G(f_\infty)$. Take d a Kleene index such that in any Kleene closed **A** we have

$$\forall G \in \mathbf{A}(2).\; \{d\}(G) \;\simeq\; \mu n.\, G(f_n) = G(f_\infty)\,.$$

Then d defines a total Kleene computable function in Ct, but not in S. Using S8, we may now obtain a Kleene index e with argument type $\overline{4}$ such that $\{e\}(\Lambda\Phi.0)$ has value 0 in Ct, but is undefined in S.

This shows that non-termination does not in general transfer downwards with respect to $\preceq$. However, the following exercise (not required in the sequel) gives a tool for transferring it under certain conditions:

Exercise 5.2.22. Let **A** be a Kleene closed pure type structure, and let R be the unique total logical relation from **A** to S. Suppose $\Psi_j \in \mathsf{S}(k_j)$ for each j and $\{e\}(\Psi_0, \ldots, \Psi_{r-1})\uparrow$, and let

$$(e_0, \vec{\Psi}_0),\; (e_1, \vec{\Psi}_1),\; \cdots$$

be a Moschovakis witness in $\mathscr{T}(e, \vec{\Psi})$, where $\Psi_{ij} \in \mathsf{S}(k_{ij})$ for each i, j. Suppose moreover that $\vec{\Phi}_0, \vec{\Phi}_1, \ldots$ are finite sequences of elements from **A** such that $(\Phi_{ij}, \Psi_{ij}) \in R(k_{ij})$ for each i, j. Show that

$$(e_0, \vec{\Phi}_0),\; (e_1, \vec{\Phi}_1),\; \cdots$$

is a Moschovakis witness in $\mathscr{T}(e,\vec{\Phi})$, and hence that $\{e\}(\vec{\Phi}))\uparrow$ within $\mathbf{A}$.

Recall that a Kleene index e is *total* with respect to $\mathbf{A}$ if $\{e\}_{\mathbf{A}}(\Phi_0,\ldots,\Phi_{r-1})\downarrow$ for all $\Phi_0,\ldots,\Phi_{r-1}$ in $\mathbf{A}$ of suitable types. The following is now immediate from Lemma 5.2.21 and Exercise 5.2.20:

Theorem 5.2.23. *(i) A Kleene index e (for specified argument types k_i) is total with respect to all Kleene closed pure type structures* $\mathbf{A}$ *iff it is total with respect to* S.

(ii) Two Kleene indices e,e' represent the same total computable functional in all Kleene closed $\mathbf{A}$ *iff they do so in* S. □

Thus, whilst the set of total indices will in general vary from one type structure to another, the indices that are total with respect to S have a special status in that they can be regarded as defining 'absolutely total' operations. We now proceed to analyse the logical complexity of this class of indices. Superficially, one might expect the notion of totality to have a high complexity which moreover increases with the type level, since the obvious formula asserting the totality of e over $\mathsf{S}(k)$ involves a quantification over $\mathsf{S}(k)$. In fact, the situation is much more benign than this: at all type levels, totality turns out to be a Π^1_1 property. We prove this with the help of the following lemma, which shows how the 'uncountable' structure S may for present purposes be replaced by a countable one. The ideas here were essentially introduced in [197].

Lemma 5.2.24. *There is a countable Kleene closed pure type structure* $\mathbf{A}$ *such that*

(i) for any index e with any choice of argument types, e is total with respect to S *iff e is total with respect to* $\mathbf{A}$*;*

(ii) two indices e,e' represent the same total computable functional in S *iff they do so in* $\mathbf{A}$.

Proof. The argument is inspired by the proof of the Löwenheim–Skolem theorem in classical model theory. In a sense, we shall construct an elementary substructure of S, then take its extensional collapse.

Let $\mathbf{B}=\{\mathbf{B}(k)\mid k\in\mathbb{N}\}$ be such that each $\mathbf{B}(k)$ is a countable subset of $\mathsf{S}(k)$ and the following conditions are satisfied:

1. If $\Phi\in\mathsf{S}(k)$ is Kleene computable relative to $\Phi_0\in\mathbf{B}(k_0),\ldots,\Phi_{r-1}\in\mathbf{B}(k_{r-1})$, then $\Phi\in\mathbf{B}(k)$.
2. If $\Phi,\Phi'\in\mathbf{B}(k+1)$ and $\Phi\neq\Phi'$, there exists $\xi\in\mathbf{B}(k)$ such that $\Phi(\xi)\neq\Phi'(\xi)$.
3. If $\Psi_0,\ldots,\Psi_{s-1}\in\mathsf{S}$, $\Phi_s,\ldots,\Phi_{r-1}\in\mathbf{B}$ and $\{e\}(\Psi_0,\ldots,\Psi_{s-1},\Phi_s,\ldots,\Phi_{r-1})\uparrow$, then there exist $\Phi_0,\ldots,\Phi_{s-1}\in\mathbf{B}$ such that $\{e\}(\Phi_0,\ldots,\Phi_{r-1})\uparrow$.

Such a family of countable sets can easily be constructed by an iteration up to ω, treating the above properties 1–3 as closure conditions.

We next define sets $\mathbf{A}(k)$ together with bijections $P_k:\mathbf{B}(k)\to\mathbf{A}(k)$ inductively as follows:

- $\mathbf{A}(0)=\mathbf{B}(0)=\mathbb{N}$, $P_0(n)=n$,
- $\mathbf{A}(k+1)=\{P_{k+1}(\Phi)\mid\Phi\in\mathbf{B}(k+1)\}$,
  where $P_{k+1}(\Phi)=(\Lambda\zeta\in\mathbf{A}(n).\Phi(P_k^{-1}(\zeta)))$.

Assuming that P_k is bijective, it is immediate from the definition of $\mathbf{A}(k+1)$ that P_{k+1} is surjective, and that it is injective follows from condition 3 above. We thus obtain a total pure type structure $\mathbf{A}$ in which we can interpret S1–S9 as in Section 5.1.1.

We now state two claims and then prove them. In both these claims, we assume $\Phi_i \in \mathbf{B}(k_i)$ and $\Psi_i = P_{k_i}(\Phi_i) \in \mathbf{A}(k_i)$ for all $i < r$. We also set $k = k_0 - 2$.

Claim 1: If $\{e\}_{\mathbf{A}}(\Psi_0, \ldots, \Psi_{r-1}) = a$ then $\{e\}_{\mathsf{S}}(\Phi_0, \ldots, \Phi_{r-1}) = a$.

Claim 2: $\mathbf{A}$ is Kleene closed.

It is easy to see that the lemma follows from these claims. For part (i) of the lemma, if e is total with respect to S then it is total with respect to $\mathbf{A}$ by claim 2 and Theorem 5.2.23(i), whilst if e is total for $\mathbf{A}$ then it is total for S by claim 1 and condition 3 for $\mathbf{B}$; For part (ii), if e, e' define the same computable functional in S then they do so in $\mathbf{A}$ by claim 2 and Theorem 5.2.23(ii), whilst if e, e' define the same functional in $\mathbf{A}$ then they do so in S by claim 1 and condition 2 for $\mathbf{B}$.

Proof of claim 1: By induction on the computation tree for $\{e\}_{\mathbf{A}}(\Psi_0, \ldots, \Psi_{r-1}) = a$. Only the case for S8 merits consideration, the others being easy. Suppose $e = \langle 8, d \rangle$ and

$$\{e\}_{\mathbf{A}}(\vec{\Psi}) \;=\; F, \text{ where } F \;=\; (\Lambda \zeta \in \mathbf{A}(k).\, \{d\}_{\mathbf{A}}(\zeta, \vec{\Psi}))\,.$$

If for some $\xi \in \mathsf{S}(k)$ we have that $\{d\}_{\mathsf{S}}(\xi, \vec{\Phi})\uparrow$, then by condition 3 for $\mathbf{B}$ there will be some such ξ in $\mathbf{B}(k)$. Let $\zeta = P_k(\xi)$. Since we are assuming $\{e\}_{\mathbf{A}}(\vec{\Phi})\downarrow$, we have that F is total, so in particular $\{d\}_{\mathbf{A}}(\zeta, \vec{\Psi})\downarrow$. Now by the induction hypothesis we obtain $\{d\}_{\mathsf{S}}(\xi, \vec{\Phi})\downarrow$, contradicting the choice of ξ. It follows that the functional

$$G = (\Lambda \xi \in \mathsf{S}(k).\, \{d\}_{\mathsf{S}}(\xi, \vec{\Phi}))$$

is total, and hence that $G \in \mathbf{B}(k+1)$ by condition 1 for $\mathbf{B}$. By the induction hypothesis and the definition of P_{k+1}, it follows that $P_{k+1}(G) = F$, so by definition of P_{k_0} we have $\{e\}_{\mathbf{A}}(\vec{\Psi}) = \{e\}_{\mathsf{S}}(\vec{\Phi})$.

Proof of claim 2: Suppose the functional

$$F \;=\; (\Lambda \zeta \in \mathbf{A}(k).\, \{d\}_{\mathbf{A}}(\zeta, \vec{\Psi}))$$

is total. As in the proof of claim 1, we have that

$$G = (\Lambda \xi \in \mathsf{S}(k).\, \{d\}_{\mathsf{S}}(\xi, \vec{\Phi}))$$

is total, and hence that $G \in \mathbf{B}(k+1)$ and that $F = P_{k+1}(G)$. It follows that $F \in \mathbf{A}(k+1)$ by definition of $\mathbf{A}(k+1)$; thus $\mathbf{A}$ is Kleene closed. □

We may now use this lemma to show that the property of being an index for a total functional is of a relatively low complexity.

Theorem 5.2.25. *The sets*

$$Z_0 = \{(e,k) \mid \forall \Phi \in \mathsf{S}(k).\, \{e\}(\Phi)\downarrow\}\,,$$

$$Z_1 = \{(e,e',k) \mid \forall \Phi \in \mathsf{S}(k).\ \{e\}(\Phi) = \{e'\}(\Phi)\}$$

are Π^1_1.

Proof. By Lemmas 5.2.21 and 5.2.24, we know the following are equivalent for any e,k:

1. $(e,k) \in Z_0$.
2. For all countable Kleene closed pure type structures $\mathbf{A}$ and all $\Psi \in \mathbf{A}(k)$, $\{e\}_{\mathbf{A}}(\Phi)$ is defined.

As it stands, the complexity of statement 2 is Π^1_2, but this statement may be refined with the help of the following notion.

Let $\mathbf{A}$ be any countable pure type structure, and let C be a set of sequences $(e,\Psi_0,\ldots,\Psi_{r-1},a)$ where the Ψ_i are in $\mathbf{A}$. Call C a *computation relation* on $\mathbf{A}$ if

(i) C is a fixed point of the inductive definition given by S1–S9 interpreted in $\mathbf{A}$,
(ii) for each e and $\Psi_0,\ldots,\Psi_{r-1}$, there is at most one $a \in \mathbb{N}$ such that $(e,\vec{\Psi},a) \in C$,
(iii) $\mathbf{A}$ is *C-closed* in the following sense: if for some $\Psi_0,\ldots,\Psi_{r-1} \in \mathbf{A}$ we have

$$\forall \zeta \in \mathbf{A}(k).\ \exists a \in \mathbb{N}.\ (e,\zeta,\Psi_0,\ldots,\Psi_{r-1},a) \in C\,,$$

then there exists $F \in \mathbf{A}(k+1)$ with

$$\forall \zeta \in \mathbf{A}(k).\ (e,\zeta,\Psi_0,\ldots,\Psi_{r-1},F(\zeta)) \in C\,.$$

Clearly, $\mathbf{A}$ is Kleene closed if and only if there is a computation relation C on $\mathbf{A}$. Moreover, if $\mathbf{A}$ and C are coded up as subsets of $\mathbb{N}$, conditions (ii) and (iii) above are clearly expressible by arithmetical formulae. Furthermore, in the presence of (iii), condition (i) is arithmetically expressible, since all quantifications may be taken to be over elements of $\mathbf{A}$ or finite sequences thereof. Thus, the property 'C is a computation relation on $\mathbf{A}$' is itself arithmetical.

We now see that statement 2 above is equivalent to

3. For all countable pure type structures $\mathbf{A}$ and all computation relations C on $\mathbf{A}$, we have
$$\forall \Psi \in \mathbf{A}(k).\ \exists a \in \mathbb{N}.\ (e,\Psi,a) \in C\,,$$

and it is clear that the complexity of this is Π^1_1.

The verification that Z_1 is a Π^1_1 set is an easy adaptation of this argument, bearing in mind Lemma 5.2.24(ii). □

In essence, the above theorem says that the simulation $\gamma : \mathsf{S}^{\mathrm{Kl}} \longrightarrow\!\!\!\triangleright\, K_1$ from Section 5.1.4 is Π^1_1 in the sense of Section 3.5.1. The logical complexity of the corresponding simulations $\mathsf{Ct}^{\mathrm{Kl}} \longrightarrow\!\!\!\triangleright\, K_1$ and $\mathsf{HEO}^{\mathrm{Kl}} \longrightarrow\!\!\!\triangleright\, K_1$ will be considered in Chapters 8 and 9 respectively.

5.3 Existential Quantifiers and Effective Continuity

So far, our discussion has been rather thin on interesting examples of Kleene computable functionals. Part of our purpose in this final section is to redress this by pointing out some of the surprising phenomena that arise even from such a relatively weak notion as Kleene computability. We shall focus here on computable functionals within S; examples specific to Ct or HEO will be mentioned in Chapters 8 and 9 respectively.

A second purpose of this section is to explore a little more deeply the notion of Kleene computability relative to some $\Phi \in \mathsf{S}$ as introduced in Definition 5.1.2. A key role here is played by a certain object ${}^2\exists \in \mathsf{S}(2)$ embodying existential quantification over $\mathbb{N}$, and which we treat as the prototype of a *discontinuous* functional. We shall see that there a sharp dichotomy emerges: any functional Φ is either powerful enough that ${}^2\exists \leq_{\mathrm{Kl}} \Phi$, or Φ and all functionals computable from it manifest some remarkable *effective continuity* properties.

As explained in Chapter 1, computability relative to ${}^2\exists$ and stronger functionals really lies beyond the scope of this book, although this was an important part of Kleene's original motivation for introducing S1–S9, and will be briefly touched on in Section 5.4. Our present interest in ${}^2\exists$ is as a litmus test of effective continuity.

We begin with the following general definition:

Definition 5.3.1. Given any set $S \subseteq \mathsf{S}(\sigma)$, we say a functional $E \in \mathsf{S}((\sigma \to \mathtt{N}) \to \mathtt{N})$ is an *existential quantifier for* S if for all $f \in \mathsf{S}(\sigma \to \mathtt{N})$ we have

$$E(f) \;=\; \begin{cases} 0 \text{ if } \forall x \in S.\, f(x) = 0\,, \\ 1 \text{ if } \exists x \in S.\, f(x) > 0\,. \end{cases}$$

If $S = \mathsf{S}(\mathtt{N}) = \mathbb{N}$, there is a unique such functional, which we denote by ${}^2\exists$.

Alternatively, if one prefers to work in the Karoubi envelope of S as in Chapter 4, then E may be considered as having the type $(\sigma \to \mathtt{B}) \to \mathtt{B}$. A mild variant of the above definition also features in Section 7.3.

If S is any *finite* set whose elements are all Kleene computable, then trivially some existential quantifier for S is Kleene computable. By contrast, ${}^2\exists$ is not Kleene computable: it is easy to see that if it were, the characteristic function of the *halting set* from classical computability theory would be Kleene computable, contradicting Proposition 5.1.13. One might naturally suppose that no infinite set admitted a computable existential quantifier; however, this is refuted by the following remarkable observations due to Escardó [83]. For this purpose, we define $g_n : \mathbb{N} \to \mathbb{N}$ by $g_n(m) = 0$ if $m < n$, $g_n(m) = 1$ otherwise, and take $g_\infty(m) = 0$ for all m. We write $N_\infty \subseteq \mathsf{S}(1)$ for the set $\{g_0, g_1, \ldots, g_\infty\}$, and N for the set $\{g_0, g_1, \ldots\}$ excluding g_∞.

Theorem 5.3.2. *(i) There is an existential quantifier E for $N_\infty \subseteq \mathsf{S}(1)$ that is Kleene primitive recursive in the sense of Section 5.1.5.*

(ii) There is even a primitive recursive existential quantifier E' for $N \subseteq \mathsf{S}(1)$.

Proof. (i) For any $F \in \mathsf{S}(1 \to 0)$ and any $n \in \mathbb{N}$, define

$$p(F,n) = \begin{cases} 0 \text{ if } \forall m \le n. F(g_m) = 0 , \\ 1 \text{ if } \exists m \le n. F(g_m) > 0 . \end{cases}$$

Now define

$$E(F) = \begin{cases} 0 \text{ if } F(\Lambda n. p(F,n)) = 0 , \\ 1 \text{ otherwise.} \end{cases}$$

Clearly E is primitive recursive. To see that E is an existential quantifier for N_∞, note that if $n \in \mathbb{N}$ is minimal such that $F(g_n) > 0$ then $\Lambda n.p(F,n) = g_n$ so $E(F) = 1$; on the other hand, if $F(g_n) = 0$ for all $n \in \mathbb{N}$ then $\Lambda n.p(F,n) = g_\infty$ so $E(F) = \min(1, F(g_\infty))$; but in this case $\exists g \in N_\infty.F(g) > 0$ iff $F(g_\infty) > 0$.

(ii) Define $S \in \mathsf{S}(1 \to 1)$ by $S(g)(0) = 0$ and $S(g)(n+1) = g(n)$; clearly S is primitive recursive and $S(g_n) = g_{n+1}$, $S(g_\infty) = g_\infty$. Suppose $F \in \mathsf{S}(2)$; by composing with the map $0 \mapsto 0$, $n+1 \mapsto 1$, we may assume $F(g) \le 1$ for all g. We may now use the existential quantifier E from part (i) to test whether $\exists g \in N_\infty. F(g) \neq F(S(g))$. Since $F(g_\infty) = F(S(g_\infty))$ always, this amounts to testing whether $\exists n \in \mathbb{N}. F(g_n) \neq F(S(g_n))$, which is clearly equivalent to $\exists n \in \mathbb{N}. F(g_n) \neq F(g_0)$. If this is true, then $\exists n. F(g_n) = 0$; if not, we may decide $\exists n. F(g_n) = 0$ simply by testing $F(g_0)$. □

In the interests of brevity and simplicity, we have availed ourselves here of classical case splits on whether some given property holds for all $n \in \mathbb{N}$. With a little extra care, however, a fully constructive presentation of the above theorem is possible: see Escardó [83] for details.

Whilst the functional E' above might seem to come precariously close to computing ${}^2\exists$, it does not really do so. Essentially this is because there is no *computable* way of extending an arbitrary function $N \to \mathbb{N}$ to a function $N_\infty \to \mathbb{N}$.

The basic idea of exploiting nested calls to F in this way is sometimes known as the 'Grilliot trick', after Grilliot [111] (see Remark 5.3.9 below). As an aside, we mention a somewhat similar example appearing on page 581 of Barendregt [9], there attributed to Kreisel. This will feature in Subsection 8.1.4, where we compare the interpretations of System T in the models S and Ct.

Exercise 5.3.3. Let g_n, g_∞ be as in Theorem 5.3.2, and for $F \in \mathsf{S}(2)$ and $n \in \mathbb{N}$ define

$$q(F,n) = \begin{cases} 0 \ \text{ if } \forall m \le n+1.\ F(g_m) \neq F(g_\infty) , \\ 1 \text{ otherwise,} \end{cases}$$

$$D(F) = \begin{cases} 0 \ \text{ if } F(g_0) = F(g_\infty) \vee F(\Lambda n.q(F,n)) \neq F(g_\infty) , \\ 1 \text{ otherwise.} \end{cases}$$

Show that $D(F) = 0$ for all *continuous* F, though not for all $F \in \mathsf{S}(2)$.

We now work towards obtaining the dichotomy mentioned above. Our main result (Theorem 5.3.5) is based on one obtained by the second author, extending ideas from [206], and subsequently improved in the light of insights from Escardó [83]. Much of the essence of this is contained in the following lemma, which marshals information from [83] in a form suited to our purpose. Here, if $\{a_n\}_n$ is a sequence of integers with limit a, we write

$$\mathsf{mod}_n\, a_n \;=\; \mu n.\forall m \geq n.\, a_m = a\,.$$

Lemma 5.3.4. *Suppose $F \in \mathsf{S}(2)$. Then it is primitively recursively decidable uniformly in F whether or not $F(g_\infty) = \lim_{m\to\infty} F(g_m)$ or not. Moreover, there exist a μ-computable $M \in \mathsf{S}(2 \to 0)$ and a primitive recursive $D \in \mathsf{S}(2 \to 2)$, independent of F, such that:*

1. *if $F(g_\infty) = \lim_{m\to\infty} F(g_m)$, then $M(F) = \mathsf{mod}_m F(g_m)$,*
2. *if $F(g_\infty) \neq \lim_{m\to\infty} F(g_m)$, then $D(F) = {}^2\exists$.*

Proof. Using the functional E' from Theorem 5.3.2, the following predicate $\Xi(F,n)$ may be primitively recursively decided uniformly in $F \in \mathsf{S}(2)$ and $n \in \mathbb{N}$:

$$\Xi(F,n) \;\Leftrightarrow\; \forall m \in \mathbb{N}.\, m \geq n \Rightarrow F(g_m) = F(g_\infty)\,.$$

By applying E' again to $\Lambda x.\Xi(F,x)$, we may decide primitively recursively in F whether

$$\exists n \in \mathbb{N}.\forall m \in \mathbb{N}.\, m > n \Rightarrow F(g_m) = F(g_\infty)$$

which is the logical formula for $F(g_\infty) = \lim_{m\to\infty} F(g_m)$.

By defining $M(F) \simeq \mu n.\Xi(F,n)$, we obtain a μ-computable functional M with the required property. As regards D, let us first define $C \in \mathsf{S}(2,1 \to 0 \to 0)$ by

$$C(F,h)(n) \;=\; \begin{cases} 1 & \text{if } \exists m \leq m' \leq n.\, h(m) > 0 \wedge F(g_{m'}) \neq F(g_\infty)\,, \\ 0 & \text{otherwise.} \end{cases}$$

Clearly, if $h(m) = 0$ for all m then $C(F,h) = \Lambda n.0 = g_\infty$, whereas if m is minimal with $h(m) > 0$ then we may take $m' \geq m$ minimal with $F(g_{m'}) \neq F(g_\infty)$, and then $C(F,h) = g_{m'}$. We may now define

$$D(F,h) \;=\; \begin{cases} 0 & \text{if } F(C(F,h)) = F(g_\infty)\,, \\ 1 & \text{otherwise.} \end{cases}$$

Then clearly $D(F,h) = {}^2\exists(h)$ provided $F(g_m) \not\to F(g_\infty)$. □

We will use this to show that we can test whether an arbitrary functional Φ is 'continuous' at an arbitrary argument ξ, at least with respect to a standard sequence of approximations. To formulate this precisely, we shall use a naive way of 'approximating' any given functional by means of very simple ones. Specifically, for each $n \in \mathbb{N}$, we define the *nth approximation* $(z)^0_n$ to an element $z \in \mathsf{S}(0) = \mathbb{N}$ to be simply $\min\{z,n\}$; we then inductively define the nth approximation $(f)^{\sigma\to\tau}_n$ to an element $f \in \mathsf{S}(\sigma \to \tau)$ to be the function $\Lambda x.(f((x)^\sigma_n))^\tau_n$. It is easy to see that the mappings $(-)^\sigma$ are idempotents whose images correspond to the finite type structure over the set $\{0,\ldots,n\}$. These approximations will also play an important role in Chapter 8.

The idea now is to attempt to characterize an arbitrary functional $\Phi \in \mathsf{S}$ as the 'limit' of the sequence $(\Phi)_0, (\Phi)_1, \ldots$. Clearly, the set of functionals that can be characterized in this way can have at most the cardinality of the continuum, because the set of approximating functionals is countable altogether. Nevertheless, the following theorem indicates when it is justified to call Φ a 'limit' of the $(\Phi)_n$. For notational simplicity, we concentrate here on functionals of pure type.

Theorem 5.3.5. *(i) If $x \in \mathbb{N}$, then $x = \lim_{n\to\infty}(x)_n$, and $\mathsf{mod}_n(x)_n = x$.*

(ii) If $f \in \mathsf{S}(1)$ and $x \in \mathbb{N}$, then $f(x) = \lim_{n\to\infty}(f)_n(x)$, and $\mathsf{mod}_n(f)_n(x)$ is μ-computable from f and x.

(iii) If $\Phi \in \mathsf{S}(k+2)$ and $\xi \in \mathsf{S}(k+1)$, then it is primitively recursively computable uniformly in Φ and ξ whether $\Phi(\xi) = \lim_{n\to\infty}(\Phi)_n(\xi)$ or not. Moreover, there exist a μ-computable functional M_{k+2} and a primitive recursive functional D_{k+2}, independent of Φ and ξ, such that

1. if $\Phi(\xi) = \lim_{n\to\infty}(\Phi)_n(\xi)$, then $M_{k+2}(\Phi,\xi) = \mathsf{mod}_n(\Phi)_n(\xi)$,
2. otherwise, we have $D_{k+2}(\Phi,\xi,h) = {}^2\exists(h)$ for all $h \in \mathsf{S}(1)$.

Proof. Parts (i) and (ii) are easy exercises. For (iii), first define $B_0 \in \mathsf{S}(1,0 \to 0)$ by

$$B_0(g)(x) = \begin{cases} x & \text{if } g(x) = 0\,, \\ \mu m \leq x.\, g(m) \neq 0 & \text{otherwise,} \end{cases}$$

and note that $B_0(g_m) = (-)^0_m$ for all $m \in \mathbb{N}$ and $B_0(g_\infty) = id$. From this, we may obtain operations $B_k \in \mathsf{S}(1,k \to k)$ for all $k \in \mathbb{N}$ via

$$B_{k+1}(g)(H) = B_0(g) \circ H \circ B_k(g)\,,$$

and note that $B_k(g_m) = (-)^k_m$ for all m and $B_0(g_\infty) = id$. Clearly the B_k are Kleene primitive recursive.

The question of whether $\Phi(\xi) = \lim_{n\to\infty}(\Phi)_n(\xi)$ is now precisely that of whether $F_{\Phi,\xi}(g_\infty) = \lim_{n\to\infty} F_{\Phi,\xi}(g_n)$, where $F_{\Phi,\xi} = \Lambda g.B_{k+1}(g,\Phi)(\xi)$. From this it is clear that all parts of the theorem follow immediately from Lemma 5.3.4. □

We may also adapt these ideas to provide a notion of approximation for *computation trees*, and obtain a similar dichotomy: either the 'approximation' is successful and suitable moduli are computable, or we can compute ${}^2\exists$ uniformly in the data.

Definition 5.3.6. We define the ith approximation $\{e\}_i(\Phi_0,\ldots,\Phi_{n-1})$ to the (possibly non-terminating) computation of $\{e\}(\Phi_0,\ldots,\Phi_{n-1})$ by induction on i, and by cases on the scheme in question:

- If e is an index for S1, S2, S3 or S7, we let $\{e\}_i(\vec{\Phi}) = \{e\}(\vec{\Phi})$.
- In all other cases, we let $\{e\}_0(\vec{\Phi}) = 0$.
- If $e = \langle 4, e_1, e_2\rangle$ we let

  $$\{e\}_{i+1}(\vec{\Phi}) = \{e_1\}_i(\{e_2\}_i(\vec{\Phi}),\vec{\Phi})\,.$$

  If e is an index for S5, S6, S8 or S9, we define $\{e\}_{i+1}(\vec{\Phi})$ in a similar way, adding the subscript i to the right-hand side of the original scheme.
- If e is not an index for any scheme, or the typing of the arguments does not fit in with the requirements of the index, we let $\{e\}_i(\vec{\Phi}) = 0$ for all i.

Theorem 5.3.7. *(i) For each sequence $\vec{\sigma}$ of pure types, there is a partial Kleene computable $F_{\vec{\sigma}}$ of type $1,0,\vec{\sigma} \to 0$ such that for all $i,e,\vec{\Phi}$ we have*

$$F_{\vec{\sigma}}(g_i,e,\vec{\Phi}) = \{e\}_i(\vec{\Phi}), \qquad F_{\vec{\sigma}}(g_\infty,e,\Phi) \simeq \{e\}(\vec{\Phi}).$$

(ii) There exist primitive recursive functions α, β, γ such that if $\Phi_j \in \mathsf{S}(\sigma_j)$ for each j, $\{e\}(\vec{\Phi})\downarrow$, and $t \in \mathbb{N}$ codes the list $\vec{\sigma}$, then

1. *$\{\alpha(e,t)\}(\vec{\Phi})\downarrow$, and $\{\alpha(e,t)\}(\vec{\Phi}) = 0$ iff $\{e\}(\vec{\Phi}) = \lim_{i\to\infty}\{e\}_i(\vec{\Phi})$,*
2. *if $\{\alpha(e,t)\}(\vec{\Phi}) = 0$, then $\{\beta(e,t)\}(\vec{\Phi})\downarrow$ and has value $\mathsf{mod}_i(\{e\}_i(\vec{\Phi}))$,*
3. *if $\{\alpha(e,t)\}(\vec{\Phi}) > 0$, then $\{\gamma(e,t)\}(\vec{\Phi})\downarrow$, and its value will be an index for computing ${}^2\exists$ from $\vec{\Phi}$.*

Proof. (Sketch.) For (i), first construct $P \in \mathsf{S}(1 \to 1)$ primitive recursive such that $P(g_{n+1}) = g_n$ and $P(g_\infty) = g_\infty$. The idea is then to simulate Definition 5.3.6, replacing i with g_i and the predecessor operation by P. Note that the base case can be detected since for $g \in N_\infty$ we have $g = g_0$ iff $g(0) = 1$. Broadly speaking, this enables us to construct $F_{\vec{\sigma}}$ using the recursion theorem for Kleene computability.

A seeming obstacle here is that in view of S8, we actually need to define $F_{\vec{\tau}}$ simultaneously for infinitely many sequences $\vec{\tau}$ extending $\vec{\sigma}$. We may overcome this by working with an argument type k sufficiently high that all relevant argument tuples $\vec{\Psi} : \vec{\tau}$ may be embedded in $\mathsf{S}(k)$ by means of standard retractions, computable uniformly in a code for $\vec{\tau}$. We leave the further details to the reader.

(ii) For fixed t, the existence of suitable primitive recursive functions $\alpha(-,t)$, $\beta(-,t)$, $\gamma(-,t)$ is now an immediate application of Lemma 5.3.4. An inspection of the construction confirms that the primitive recursiveness is also uniform in t. □

An important consequence of this theorem is the following result, which gives an insight into the nature of *1-sections*. By the 1-section of Φ we mean the set $1\mathrm{-sc}(\Phi)$ consisting of all $f \in \mathsf{S}(1)$ such that $f \leq_{\mathsf{Kl}} \Phi$. We shall also require the idea of the *Turing jump*: given $f \in \mathsf{S}(1)$, we write $f' \in \mathsf{S}(1)$ for the characteristic function of the halting set for (classical) Turing machines with oracle f.

Corollary 5.3.8. *Let $\Phi \in \mathsf{S}(k)$ for some k. Then exactly one of the following holds:*

1. ${}^2\exists \leq_{\mathsf{Kl}} \Phi$.
2. *There exists $h \in 1\mathrm{-sc}(\Phi)$ such that every $f \in 1\mathrm{-sc}(\Phi)$ is computable in h'.*

Proof. If ${}^2\exists \leq_{\mathsf{Kl}} \Phi$, it is easy to see that $1\mathrm{-sc}(\Phi)$ is closed under the operation $-'$, so condition 2 cannot hold.

If not, then we know from Theorem 5.3.7 that whenever $\{e\}(x,\Phi)$ terminates, we have

$$\{e\}(x,\Phi) = \lim_{i\to\infty}\{e\}_i(x,\Phi).$$

Define $h(\langle e,x,i\rangle) = \{e\}_i(x,\Phi)$, and consider the set

$$G = \{\langle e,x,i\rangle \mid \exists j > i.\, h(\langle e,x,i\rangle) \neq h(\langle e,x,j\rangle)\}.$$

Clearly $h \in 1\!-\!\mathrm{sc}(\Phi)$, and G is decidable in h'. Finally, if $f \in 1\!-\!\mathrm{sc}(\Phi)$, take e such that $f = (\Lambda x.\{e\}(x,\Phi))$; then $f(x) = \{e\}_i(x,\Phi)$ for the smallest i such that $\langle e,x,i\rangle \notin G$. Thus f is computable in G, and hence in h'. □

The force of this corollary is that if $1\!-\!\mathrm{sc}(\Phi)$ is closed under Turing jump, then we are in case 1 and it follows that $1\!-\!\mathrm{sc}(\Phi)$ is closed under very much more. This will be illustrated by Corollary 5.4.2 below.

Remark 5.3.9. The history of the above results is somewhat complex. Grilliot [111] proved the dichotomy at type 2: for any F of type 2, either ${}^2\exists$ is computable in F or the 1-section of F is not closed under jump. This was generalized to type 3 functionals by Bergstra [31] and to all finite types by Normann [206].

For any functional Φ of type ≥ 2, let the *trace* $h_\Phi : \mathbb{N} \to \mathbb{N}$ of Φ be defined by applying Φ to a canonical enumeration of the finitary functionals of type $k-1$. Wainer [301] showed that if a type 2 functional F is not effectively discontinuous, then the 1-section of F may be computably generated from sets semidecidable in h_F, much as in the proof of Corollary 5.3.8. Furthermore, F induces a 'pseudo-jump' operation on the class of sets semidecidable in h_F, which generates the whole of $1\!-\!\mathrm{sc}(F)$. The extensions of Grilliot's trick to higher types also show, with the same caveat, that for any functional Φ of any finite type, either ${}^2\exists$ is computable in Φ or the 1-section of Φ is computably generated from sets semidecidable relative to h_Φ.

5.4 Computability in Normal Functionals

One of Kleene's original motivations for introducing S1–S9 computability was to investigate the computational power of higher-order quantifiers, and thereby find higher-order generalizations of the hyperarithmetical hierarchy. This led in the 1960s and 1970s to a deep study of computability relative to so-called *normal* functionals. Although, as we have stressed, this theory falls properly outside the scope of this book, we conclude this chapter with a brief glance at this substantial field of research.

We begin with the following remarkable result of Kleene [138].

Theorem 5.4.1. *The 1-section of* ${}^2\exists$ *contains exactly the* Δ^1_1 *or* hyperarithmetical *functions.*

Proof. (Sketch.) Recall that a set is Δ^1_1 if it is definable both by a Π^1_1 formula and by a Σ^1_1 formula. A function is Δ^1_1 if the graph is Δ^1_1. The fact that any $f \leq_{\mathrm{Kl}} {}^2\exists$ is Δ^1_1 can be read off fairly easily from Theorem 5.2.15. For the converse, we need the fact that the Δ^1_1 functions can be stratified into a *hyperarithmetical hierarchy*: every Δ^1_1 function f is computable in some $\mathbf{0}^{(d)}$, where $\mathbf{0}$ is the Turing degree of $(\Lambda x.0)$, d is a Kleene-style notation for some computable ordinal, and $-^{(d)}$ denotes the d-fold iterate of the Turing jump. However, an easy transfinite induction on notations shows that $\mathbf{0}^{(d)}$ is ${}^2\exists$-computable for every computable ordinal notation d. □

Corollary 5.4.2. *The set of* Π^0_ω *or* arithmetical *functions (i.e. those definable by a first-order arithmetical formula) is not the 1-section of any* $\Phi \in \mathsf{S}$.

Proof. The set in question is closed under Turing jump, but does not contain all hyperarithmetical functions. Thus neither alternative of Corollary 5.3.8 applies. □

The idea now is to generalize the theory of ${}^2\exists$ to quantifiers of higher type, and indeed to other functionals from which such quantifiers may be computed:

Definition 5.4.3. (i) If $k \geq 2$, we write ${}^k\exists \in \mathsf{S}(k)$ for the unique existential quantifier over $\mathsf{S}(k-2)$ in the sense of Definition 5.3.1.

(ii) A functional $F \in \mathsf{S}(k)$ is *normal* if ${}^k\exists$ is Kleene computable in F.

In general, Kleene computability relative to a normal $F \in \mathsf{S}(k)$ enjoys much better properties than pure Kleene computability: for instance, the F-computable partial functionals over types $\leq k$ are properly closed under substitution (cf. Theorem 5.1.9 and Example 5.1.10). Most of the work on S1–S9 in the 1960s and 1970s therefore focussed on computability relative to normal functionals.

Two of the key results are *stage comparison* and *Gandy selection*. Informally, stage comparison plays a role which in ordinary recursion theory is fulfilled by interleaving arguments: given two computations of which at least one terminates, there is a way of telling which of them terminates 'sooner' (in terms of the ordinal height of computations). The idea was introduced by Gandy [100] in the case $k = 2$, and generalized to higher types by Platek and Moschovakis.

Theorem 5.4.4 (Stage comparison). *Let* $F \in \mathsf{S}(k)$ *be normal. Then there is a partial Kleene computable functional* χ *such that whenever* e *and* d *are indices for* F*-computable partial functionals of level* $\leq k$ *and* $\vec{\xi}$ *and* $\vec{\eta}$ *are suitably typed sequences of arguments of level* $\leq k+1$, *we have:*

- $\chi(F, \langle e, \vec{\xi} \rangle, \langle d, \vec{\eta} \rangle)\downarrow$ *whenever* $\{e\}(F, \vec{\xi})\downarrow$ *or* $\{d\}(F, \vec{\eta})\downarrow$,
- *if* $\chi(F, \langle e, \vec{\xi} \rangle, \langle d, \vec{\eta} \rangle) = 0$ *then* $\{e\}(F, \vec{\xi})\downarrow$ *and* $\{d\}(F, \vec{\eta})$ *does not have a computation tree of lower rank than that of* $\{e\}(F, \vec{\xi})$,
- *if* $\chi(F, \langle e, \vec{\xi} \rangle, \langle d, \vec{\eta} \rangle) > 0$ *then* $\{d\}(F, \vec{\eta})\downarrow$ *and* $\{e\}(F, \vec{\xi})$ *does not have a computation tree of rank as low as that of* $\{d\}(F, \vec{\eta})$.

Moreover, a Kleene index for such a χ *may be computed primitively recursively from an index for computing* ${}^k\exists$ *from* F.

Here we assume that $\langle e, \vec{\xi} \rangle \in \mathsf{S}(k+1)$ codes up the values $e, \vec{\xi}$, together with information on the number and type of the ξ_i. Stage comparison may be used to show, for example, that if $F \in \mathsf{S}(k+2)$ is normal, the F-semidecidable subsets of $\mathsf{S}(k+2)$ are closed under binary unions, and a subset of $\mathsf{S}(k+2)$ is F-decidable if both it and its complement are F-semidecidable.

Gandy selection says that in the presence of a normal functional, we can not only tell whether there exists a natural number passing a certain test, but can actually find one computably in the data:

Theorem 5.4.5 (Gandy selection). *Let $F \in \mathsf{S}(k)$ be normal. Then there is a partial Kleene computable* selection function ν *such that for all finite sequences $\vec{\xi}$ of objects of types $\leq k$ and all $e \in \mathbb{N}$, we have:*

- $\nu(F,\vec{\xi},e)\downarrow$ *iff there exists $n \in \mathbb{N}$ such that $\{e\}(F,\vec{\xi},n)\downarrow$,*
- *if $\nu(F,\vec{\xi},e) = n$ then $\{e\}(F,\vec{\xi},n)\downarrow$.*

Moreover, an index for such a ν may be found primitively recursively in an index for computing ${}^{k}\exists$ from F.

Corollary 5.4.6. *If $F \in \mathsf{S}(k)$ is normal, the F-semidecidable predicates on products of types $\leq k$ are closed under existential quantification over $\mathbb{N}$.*

For example, if $F \in \mathsf{S}(k)$ is normal, c is an index for an F-semidecidable subset $C \subseteq \mathsf{S}(k-2)$, and $d \in \mathbb{N}$, then it is F-semidecidable in c whether d is the index of a non-empty F-decidable subset $D \subseteq C$. So by Gandy selection, it is F-semidecidable in c whether C has any non-empty F-decidable subset.

We may apply this fact as follows. If $k > 2$ and $A \subseteq \mathbb{N}$ is any set whose complement is F-semidecidable, then by the proof of Theorem 5.2.18 there will be an F-semidecidable set $B \subseteq \mathbb{N} \times \mathsf{S}(k-2)$ whose projection to $\mathbb{N}$ equals A. Let B_n be the section of B over n; then by the above observation, the set E of all n such that B_n has a non-empty F-decidable subset is itself F-semidecidable. But if A itself is chosen not to be F-semidecidable, then E cannot be all of A, so we may draw the following conclusion:

Lemma 5.4.7. *If $F \in \mathsf{S}(k)$ is normal where $k > 2$, there is a non-empty F-semidecidable set $C \subseteq \mathsf{S}(k-2)$ that has no non-empty F-decidable subset.*

By contrast, Grilliot [110] proposed a selection theorem for subsets of $\mathsf{S}(k-3)$: uniformly in an index for a nonempty F-semidecidable set $A \subseteq \mathsf{S}(k-3)$ (where $F \in \mathsf{S}(k)$ is normal), we can find an index for a nonempty F-decidable set $B \subseteq A$.[4] For $k = 3$ this follows trivially from Gandy selection, but for $k > 3$ it yields something new, since Gandy-style selection functions need no longer exist here. This is an example of a phenomenon that manifests itself only at types ≥ 4.

A rich and sophisticated mathematical theory of Kleene computations relative to normal functionals, later generalized to *set recursion* (see Normann [205]), was developed on the basis of Gandy selection, Grilliot selection and the lack of full selection for F-semidecidable subsets of $\mathsf{S}(k-2)$ (with $k \geq 3$ and $F \in \mathsf{S}(k)$ normal). A full treatment would require another volume, so we leave the topic here. For further information on this line of research, and for detailed references to the literature, see Moldestad [196], Sacks [248, 249], or the recent survey paper by Normann [218].

---

4 Grilliot's proof of this theorem in [110] was incorrect, but a correct proof was provided by Harrington and MacQueen [114].

Chapter 6
Nested Sequential Procedures

In this chapter, we study in depth the *nested sequential procedure* (NSP) model which we introduced briefly in Subsection 3.2.5. The point of view we shall adopt is that NSPs offer an abstract formulation of an underlying concept of 'algorithm' that is common to both Kleene computability (as studied in the previous chapter) and Plotkin's PCF (the subject of the next chapter). We shall consider NSPs both as forming a self-contained model of computation in their own right, and as a calculus of algorithms that can be interpreted in many other (total and partial) models.

As already indicated in Chapter 2, the ideas behind nested sequential procedures have a complex history. The general idea of sequential computation via nested oracle calls was the driving force behind Kleene's later papers (e.g. [141, 144]), although the concept did not receive a particularly transparent or definitive formulation there. Many of the essential ideas of NSPs can be found in early work of Sazonov [252], in which a notion of *Turing machine with oracles* was used to characterize the 'sequentially computable' elements of the model PC. NSPs as studied here were first explicitly introduced in work on game semantics for PCF—both by Abramsky, Jagadeesan and Malacaria [5] (under the name of *evaluation trees*) and by Hyland and Ong [124] (under the name of *canonical forms*). In these papers, NSPs played only an ancillary role; however, it was shown by Amadio and Curien [6] how (under the name of PCF *Böhm trees*) they could be made into a model of PCF in their own right. Similar ideas were employed again by Sazonov [253] to give a standalone characterization of the class of sequentially computable *functionals*. More recently, Normann and Sazonov [221] gave an explicit construction of the NSP model in a somewhat more semantic spirit than [6], using the name *sequential procedures*. We shall adopt this name here, often adding the epithet 'nested' to emphasize the contrast with other flavours of sequential computation.

Whilst the model we shall study is identical with that of [6] and [221], our method of construction will be distinctive in several ways. First, our approach focusses on the *dynamic process* whereby one NSP operates on another, thereby directly manifesting its 'sequential' character—this is somewhat less clear in the treatments in [6] and [221], where application of NSPs is arrived at indirectly via finite approximations. Secondly, in contrast to previous approaches, our construction of the relevant applicative structure does not depend on any hard theorems—although the proof that this structure is a λ-algebra still requires some real work. Thirdly, our approach adapts immediately to an *untyped* setting, indicating that types are not the real reason why the λ-algebra axioms hold. Finally, our treatment of this basic theory is purely 'finitistic' in character, in contrast to those of [6] and [221] which require Π_2^0 induction.

In Section 6.1, we present our construction of the NSP model SP^0 as a standalone computation model, fleshing out the description in Subsection 3.2.5, and establishing its fundamental properties. We also define the model SF of (PCF-)sequential functionals as the extensional quotient of SP^0. In Section 6.2, we show how Kleene computability may be conveniently recast in terms of NSPs: any Kleene expression gives rise to an NSP representing the abstract algorithm it embodies, and this NSP can itself be directly interpreted in suitable total models **A**.

In Section 6.3, we show what the NSP perspective is good for. In particular, combinatorial properties of NSPs can be used to capture some interesting expressivity distinctions: we shall identify several substructures of SP^0, and use these to analyse the expressive power of systems such as $\mathsf{T}+min$ or the more restricted $\mathsf{Klex}^{\mathrm{min}}$. In this way, we obtain some new non-definability results for which no other methods of proof are currently known.

In Section 6.4, we ask how our treatment of Kleene computability might be adapted to *partial* type structures. We shall in fact present three rival interpretations of NSPs in a partial setting, paying close attention to the conceptual differences between them and to their respective computational strengths. The last, and most fruitful, of these interpretations turns out to be well-behaved in precisely those models possessing *least fixed points*. This can be seen as a motivation for this important concept, and paves the way for our study of PCF in which least fixed points are taken as basic. In the course of this section, we also briefly review earlier treatments of Kleene computability due to Platek [233] and Moschovakis [200, 130].

6.1 The NSP Model: Basic Theory

We here present the construction of the NSP model SP^0 and the verification that it is a $\lambda\eta$-algebra. We also look briefly at its observational quotient SF and at analogous call-by-value and untyped models, and exhibit the 'sequential' character of computation in SP^0 via a simulation in van Oosten's B. The reader is advised to study the motivating examples given in Section 3.2.5 before tackling this material.

6.1.1 Construction of the Model

We begin by building the model SP^0 as an applicative structure. Let T be the type world $\mathsf{T}^{\rightarrow}(\mathtt{N})$, and assume we are given an infinite supply of typed variables x^{σ} for each $\sigma \in \mathsf{T}$. As in Section 3.2.5, we shall consider NSPs to be trees of possibly infinite breadth and depth as defined coinductively by the following grammar:

$$\begin{array}{rl} \textit{Procedures:} & p,q ::= \lambda x_0 \ldots x_{r-1}.e \\ \textit{Expressions:} & d,e ::= \bot \mid n \mid \texttt{case}\, a\, \texttt{of}\, (i \Rightarrow e_i \mid i \in \mathbb{N}) \\ \textit{Applications:} & a ::= x q_0 \ldots q_{r-1} \end{array}$$

In an abstraction $\lambda x_0 \ldots x_{r-1}$, the variables x_i are required to be distinct. We shall frequently use vector notation $\vec{x}, \vec{q}$ for sequences $x_0 \ldots x_{r-1}$ and $q_0 \ldots q_{r-1}$. We also use t as a metavariable ranging over *terms* of any of the above three kinds. The notions of free variable and (infinitary) α-equivalence are defined in the expected way, and we shall work with terms only up to α-equivalence.

We wish to restrict our attention to *well-typed* terms. Informally, a term will be well-typed unless a typing violation occurs at some specific point within its syntax tree. More precisely, t is well-typed if for every application $xq_0 \ldots q_{r-1}$ appearing within t, the type of x has the form $\sigma_0, \ldots, \sigma_{r-1} \to \mathtt{N}$, and for each $i < r$, the procedure q_i has the form $\lambda x_{i0} \cdots x_{i(s_i-1)}.\, e_i$, where each x_{ij} has some type τ_{ij} and $\sigma_i = \tau_{i0}, \ldots, \tau_{i(s_i-1)} \to \mathtt{N}$. If Γ is any set of variables, we write $\Gamma \vdash e$ and $\Gamma \vdash a$ to mean that e,a respectively are well-typed with free variables in Γ; we also write $\Gamma \vdash p : \sigma$ when p is well-typed in Γ and of the form $\lambda x_0^{\sigma_0} \cdots x_{r-1}^{\sigma_{r-1}}.e$, where $\sigma = \sigma_0, \ldots, \sigma_{r-1} \to \mathtt{N}$.

For each type σ, we now define

$$\mathsf{SP}(\sigma) = \{p \mid \Gamma \vdash p : \sigma \text{ for some } \Gamma\}, \qquad \mathsf{SP}^0(\sigma) = \{p \mid \emptyset \vdash p : \sigma\}.$$

To give a computational calculus for NSPs, we expand our language to a system of *meta-terms* defined coinductively as follows:

$$\begin{array}{rl} \textit{Meta-procedures:} & P,Q ::= \lambda x_0 \ldots x_{r-1}.E \\ \textit{Meta-expressions:} & D,E ::= \bot \mid n \mid \texttt{case}\, G\, \texttt{of}\, (i \Rightarrow E_i \mid i \in \mathbb{N}) \\ \textit{Ground meta-terms:} & G ::= E \mid xQ_0 \ldots Q_{r-1} \mid PQ_0 \ldots Q_{r-1} \end{array}$$

We use T to range over arbitrary meta-terms. We also have an evident notion of simultaneous capture-avoiding substitution $T[\vec{x} \mapsto \vec{Q}]$ for meta-terms.

The typing rules for meta-terms extend those for ordinary terms in an obvious way. Specifically, we say that T is well-typed if for every ground meta-term $XQ_0 \ldots Q_{r-1}$ appearing within T, X is either a variable of type $\sigma_0, \ldots, \sigma_{r-1} \to \mathtt{N}$ or a meta-procedure of the form $\lambda y_0^{\sigma_0} \ldots y_{r-1}^{\sigma_{r-1}}.D$, and in either case, for each $i < r$, Q_i has the form $\lambda x_{i0} \cdots x_{i(s_i-1)}.E_i$, where each x_{ij} has some type τ_{ij} and $\sigma_i = \tau_{i0}, \ldots, \tau_{i(s_i-1)} \to \mathtt{N}$. Again, we write $\Gamma \vdash E$, $\Gamma \vdash G$ to mean that E,G are well-typed in Γ, and $\Gamma \vdash P : \sigma$ to mean that P is well-typed in Γ and of the form $\lambda x_0^{\sigma_0} \ldots x_{r-1}^{\sigma_{r-1}}.E$ where $\sigma = \sigma_0, \ldots, \sigma_{r-1} \to \mathtt{N}$.

It is easy to see that if $\Gamma, \vec{x} \vdash T(:\sigma)$, and for each $j < r$ we have $\Gamma \vdash Q_j : \tau_j$ where τ_j is the type of x_j, then also $\Gamma \vdash T[\vec{x} \mapsto \vec{Q}](:\sigma)$.

The next stage is to introduce a *basic reduction* relation $\rightsquigarrow_b$ for ground meta-terms, which we do by the following rules:

(b1) $(\lambda x_0 \ldots x_{r-1}.E)Q_0 \ldots Q_{r-1} \rightsquigarrow_b E[\vec{x} \mapsto \vec{Q}]$ (*β-rule*).
(b2) $\mathtt{case}\ \bot\ \mathtt{of}\ (i \Rightarrow E_i) \rightsquigarrow_b \bot$.
(b3) $\mathtt{case}\ n\ \mathtt{of}\ (i \Rightarrow E_i) \rightsquigarrow_b E_n$.
(b4) $\mathtt{case}\ (\mathtt{case}\ G\ \mathtt{of}\ (i \Rightarrow E_i))\ \mathtt{of}\ (j \Rightarrow F_j) \rightsquigarrow_b$
$\mathtt{case}\ G\ \mathtt{of}\ (i \Rightarrow \mathtt{case}\ E_i\ \mathtt{of}\ (j \Rightarrow F_j))$.

Note that the β-rule applies even when $r = 0$: thus $\lambda.2 \rightsquigarrow_b 2$.

From this, a *head reduction* relation $\rightsquigarrow_h$ on meta-terms is defined inductively:

(h1) If $G \rightsquigarrow_b G'$ then $G \rightsquigarrow_h G'$.
(h2) If $G \rightsquigarrow_h G'$ and G is not a case meta-term, then

$$\mathtt{case}\ G\ \mathtt{of}\ (i \Rightarrow E_i) \ \rightsquigarrow_h\ \mathtt{case}\ G'\ \mathtt{of}\ (i \Rightarrow E_i)\ .$$

(h3) If $E \rightsquigarrow_h E'$ then $\lambda\vec{x}.E \ \rightsquigarrow_h\ \lambda\vec{x}.E'$.

The following facts are easily verified:

Proposition 6.1.1. *(i) For any meta-term T, there is at most one T' with $T \rightsquigarrow_h T'$.*
(ii) If $\Gamma \vdash T(:\sigma)$ and $T \rightsquigarrow_h T'$, then $\Gamma \vdash T'(:\sigma)$.
(iii) If $T \rightsquigarrow_h T'$ then $T[\vec{x} \mapsto \vec{Q}] \rightsquigarrow_h T'[\vec{x} \mapsto \vec{Q}]$. □

Note in particular that a substitution never blocks an application of (h2), since if G is not a case meta-term then neither is $G[\vec{x} \mapsto \vec{Q}]$.

We call a meta-term a *head normal form* if it cannot be further reduced using $\rightsquigarrow_h$. The possible shapes of head normal forms are $\bot$, n, $\mathtt{case}\ y\vec{Q}\ \mathtt{of}\ (i \Rightarrow E_i)$ and $y\vec{Q}$, the first three optionally prefixed by $\lambda\vec{x}$ (where $\vec{x}$ may or may not contain y).

We now define the *general reduction* relation $\rightsquigarrow$ inductively as follows:

(g1) If $T \rightsquigarrow_h T'$ then $T \rightsquigarrow T'$.
(g2) If $E \rightsquigarrow E'$ then $\lambda\vec{x}.E \ \rightsquigarrow\ \lambda\vec{x}.E'$.
(g3) If $Q_j = Q'_j$ except at $j = k$ where $Q_k \rightsquigarrow Q'_k$, then

$$xQ_0 \ldots Q_{r-1} \rightsquigarrow xQ'_0 \ldots Q'_{r-1}\ ,$$
$$\mathtt{case}\ xQ_0 \ldots Q_{r-1}\ \mathtt{of}\ (i \Rightarrow E_i) \rightsquigarrow \mathtt{case}\ xQ'_0 \ldots Q'_{r-1}\ \mathtt{of}\ (i \Rightarrow E_i)\ .$$

(g4) If $E_i = E'_i$ except at $i = k$ where $E_k \rightsquigarrow E'_k$, then

$$\mathtt{case}\ xQ_0 \ldots Q_{r-1}\ \mathtt{of}\ (i \Rightarrow E_i) \ \rightsquigarrow\ \mathtt{case}\ xQ_0 \ldots Q_{r-1}\ \mathtt{of}\ (i \Rightarrow E'_i)\ .$$

An important point to note is that the terms t are precisely the meta-terms in normal form, i.e. those that cannot be reduced using $\rightsquigarrow$.

We write $\rightsquigarrow^*$ for the reflexive-transitive closure of $\rightsquigarrow$. We also write $\sqsubseteq$ for the evident syntactic orderings on meta-procedures and on ground meta-terms: thus, $T \sqsubseteq U$ iff T may be obtained from U by replacing zero or more subterms (possibly infinitely many) by $\bot$. It is easy to see that each set $\mathsf{SP}(\sigma)$ or $\mathsf{SP}^0(\sigma)$ forms a directed-complete partial order under $\sqsubseteq$, with least element $\lambda\vec{x}.\bot$.

By a *finite* term t we shall mean one generated by the following grammar, this time construed inductively:

$$\begin{array}{lrl} \textit{Procedures:} & p,q ::= & \lambda x_0 \ldots x_{r-1}.e \\ \textit{Expressions:} & d,e ::= & \bot \mid n \mid \texttt{case}\ a\ \texttt{of}\ (0 \Rightarrow e_0 \mid \cdots \mid r-1 \Rightarrow e_{r-1}) \\ \textit{Applications:} & a ::= & x q_0 \ldots q_{r-1} \end{array}$$

We regard finite terms as a subset of general terms by identifying the conditional branching $(0 \Rightarrow e_0 \mid \cdots \mid r-1 \Rightarrow e_{r-1})$ with

$$(0 \Rightarrow e_0 \mid \cdots \mid r-1 \Rightarrow e_{r-1} \mid r \Rightarrow \bot \mid r+1 \Rightarrow \bot \mid \ldots).$$

We now explain how a general meta-term T *evaluates* to a term t. Since t may be infinite, this will in general be an infinite process, but we can capture the value of t as the limit of the finite portions that become visible at finite stages in the reduction. To this end, for any meta-term T we define

$$\Downarrow_{\mathrm{fin}} T \;=\; \{t \text{ finite} \mid \exists T'.\, T \rightsquigarrow^* T' \wedge t \sqsubseteq T'\}.$$

Intuitively, to evaluate T, we start by reducing T to head normal form, then recursively evaluate any remaining syntactic constituents Q_i, E_j so that an NSP term gradually crystallizes out from the top downwards. At any stage, the portion of the tree consisting of nodes in head normal form will be embodied by some finite t, which will serve as a finite approximation to the eventual value of T. The key property of $\Downarrow_{\mathrm{fin}} T$ is the following:

Proposition 6.1.2. *For any meta-term T, the set $\Downarrow_{\mathrm{fin}} T$ is directed with respect to $\sqsubseteq$. That is, $\Downarrow_{\mathrm{fin}} T$ is non-empty, and any $t,t' \in \Downarrow_{\mathrm{fin}} T$ have an upper bound in $\Downarrow_{\mathrm{fin}} T$.*

Proof. Clearly $\Downarrow_{\mathrm{fin}} T$ contains $\bot$ or $\lambda\vec{x}.\bot$. Now suppose $t_0, t_1 \in \Downarrow_{\mathrm{fin}} T$, so that there exist T_0', T_1' with $T \rightsquigarrow^* T_i'$ and $t_i \sqsubseteq T_i'$. We will show by induction on the combined depth of t_0 and t_1 that $t_0 \sqcup t_1$ exists and is in $\Downarrow_{\mathrm{fin}} T$. We consider here a sample of the cases, the others being similar or easier.

If t_0 or t_1 is $\bot$, then clearly either $t_0 \sqsubseteq t_1$ or $t_1 \sqsubseteq t_0$ and so $t_0 \sqcup t_1 \in \Downarrow_{\mathrm{fin}} T$. If $t_0 = n_0$ and $t_1 = n_1$, then $T_0' = n_0$ and $T_1' = n_1$, and clearly these are head reducts of T since no step generated via (g3) or (g4) can yield just a numeral. But head reduction is deterministic, so $n_0 = n_1$ and again $t_0 \sqcup t_1 \in \Downarrow_{\mathrm{fin}} T$.

Now suppose for example $t_i = \texttt{case}\ x q_{i0} \ldots q_{i(r-1)}\ \texttt{of}\ (j \Rightarrow e_{ij})$, where the head normal form of T is $\texttt{case}\ x Q_0 \ldots Q_{r-1}\ \texttt{of}\ (j \Rightarrow E_j)$. Then for each $j < r$ we have $q_{0j}, q_{1j} \in \Downarrow_{\mathrm{fin}} Q_j$, whence $q_{0j} \sqcup q_{1j} \in \Downarrow_{\mathrm{fin}} Q_j$ by the induction hypothesis; likewise for each j we have $e_{0j}, e_{1j} \in \Downarrow_{\mathrm{fin}} E_j$, whence $e_{0j} \sqcup e_{1j} \in \Downarrow_{\mathrm{fin}} E_j$. It now readily follows that $t_0 \sqcup t_1 \in \Downarrow_{\mathrm{fin}} T$ as required. □

It is also easy to see that each $\mathsf{SP}(\sigma)$ is *directed-complete*: that is, any directed subset of $\mathsf{SP}(\sigma)$ has a least upper bound in $\mathsf{SP}(\sigma)$. The following definition is therefore justified.

Definition 6.1.3. For any meta-term T, we define $\ll T \gg$, the *value* of T, to be the ordinary term

$$\ll T \gg \;=\; \bigsqcup(\Downarrow_{\mathrm{fin}} T)\,.$$

We may relate this as follows to the mildly different approach taken in Section 3.2.5, which bypasses the notion of finiteness. It is easy to see that any $p \in \mathsf{SP}(\sigma)$ is the least upper bound of the finite procedures below it; hence if $p \sqsubseteq \bigsqcup(\Downarrow T)$ then $\Downarrow_{\mathrm{fin}} p \subseteq \Downarrow_{\mathrm{fin}} T$ and so $p \sqsubseteq \bigsqcup(\Downarrow_{\mathrm{fin}} T)$; but also $\Downarrow_{\mathrm{fin}} T \subseteq \Downarrow T$ so $\bigsqcup(\Downarrow_{\mathrm{fin}} T)$ is also the least upper bound of $\Downarrow T$.

Note in passing that the value $\ll G \gg$ of a ground meta-term G may be either an expression or an application. In either case, it is certainly a ground meta-term.

The following easy observations will frequently be useful:

Proposition 6.1.4. *(i)* $\ll \lambda \vec{x}.E \gg = \lambda \vec{x}. \ll E \gg$.
(ii) If $T \rightsquigarrow^* T'$ *then* $\ll T \gg = \ll T' \gg$.

Proof. (i) Trivial.
(ii) That $\Downarrow_{\mathrm{fin}} T' \subseteq \Downarrow_{\mathrm{fin}} T$ is immediate; that $\Downarrow_{\mathrm{fin}} T \subseteq \Downarrow_{\mathrm{fin}} T'$ is shown by an easy adaptation of the proof of Proposition 6.1.2. □

The sets $\mathsf{SP}(\sigma)$ may now be made into a total applicative structure SP by defining

$$(\lambda x_0 \cdots x_r.e)\cdot q \;=\; \lambda x_1 \cdots x_r. \ll e[x_0 \mapsto q] \gg \;.$$

Clearly the sets $\mathsf{SP}^0(\sigma)$ are closed under this application operation, so we also obtain an applicative substructure SP^0.

Exercise 6.1.5. Verify that application in SP is *monotone* and *continuous*, that is:
(i) if $p \sqsubseteq p'$ and $q \sqsubseteq q'$, then $p\cdot q \sqsubseteq p'\cdot q'$;
(ii) if $p\cdot q = r$ and $r_0 \sqsubseteq r$ is finite, there exist finite procedures $p_0 \sqsubseteq p$ and $q_0 \sqsubseteq q$ such that $r_0 \sqsubseteq p_0 \cdot q_0$.

Remark 6.1.6. Our construction of these applicative structures has required only relatively easy facts such as Propositions 6.1.1 and 6.1.2. In particular, in contrast to [6, 221], our approach bypasses the need to prove at the outset that finite NSPs are closed under application, a 'deeper' result which we shall establish in Subsection 6.3.1.

6.1.2 The Evaluation Theorem

We now embark on the proof that the structures SP, SP^0 are $\lambda\eta$-algebras. Most of the serious work comes in the proof of the following syntactic result, to which this subsection is dedicated:

Theorem 6.1.7 (Evaluation theorem). *Let $C[-]$ be any meta-term context with a hole for ground meta-terms. Then for any well-typed ground meta-term G such that $C[G]$ is well-formed, we have*

$$\ll C[G] \gg \;=\; \ll C[\ll G \gg] \gg .$$

Loosely speaking, this says that order of evaluation is not too important: it makes no difference whether we first evaluate the subterm G or simply evaluate $C[G]$ as it stands. The reason, intuitively, is that the reduction sequence needed to compute a given node in $\ll C[G] \gg$ has essentially the same structure as that required for the corresponding node in $\ll C[\ll G \gg] \gg$: the difference is just that in the former case, the required parts of $\ll G \gg$ are computed on the fly, whilst in the latter case, the whole of $\ll G \gg$ is precomputed. Our proof will work by exhibiting the skeletal structure that is common to both of these reductions, and hence showing that they yield the same result. Owing to the coinductive nature of the definitions, reductions have to be treated here as 'top-down' processes; the reader may wish to work out the possible first steps of the relevant reductions in order to appreciate the difficulties involved.

The outworking of this idea is bureaucratically complex, and the proof may be safely skipped by the reader if desired—only the statement of Theorem 6.1.7, and those of the ensuing Corollaries 6.1.11, 6.1.12 and Lemma 6.1.13, will be used in what follows. The reader may also find alternative proofs in the literature of closely related results, using induction on types (see e.g. [6, 221]). However, there are at least two possible reasons for preferring our present approach. The first is that, unlike these other proofs, our method carries over immediately to the *untyped* setting (see Subsection 6.1.7); this in itself indicates that types are not the real reason why the evaluation theorem is true. The second is proof-theoretic: whereas the other proofs require at least Π^0_2 induction, our present proof involves only 'finitistic' styles of reasoning and so manifests the essentially elementary character of the theorem (although we shall not attempt to formalize this claim here).

To explain the idea, we begin with a concrete example. We shall consider the following finite term G (with free variable $f^{\mathrm{N}\to\mathrm{N}}$) and its reduction to $\ll G \gg$. We shall omit some parentheses to reduce clutter, and abbreviate $\lambda.n$ to $\widetilde{n}$. We number the reduction steps for future reference.

$$\begin{aligned}
G \;&=\; (\lambda x.\,\texttt{case}\; f(\lambda.\texttt{case}\; x \;\texttt{of}\; 0 \Rightarrow 1) \;\texttt{of}\; 2 \Rightarrow \texttt{case}\; 3 \;\texttt{of}\; 3 \Rightarrow 4)\; \widetilde{0} \\
&\rightsquigarrow^0 \texttt{case}\; f(\lambda.\texttt{case}\; \widetilde{0} \;\texttt{of}\; 0 \Rightarrow 1) \;\texttt{of}\; 2 \Rightarrow \texttt{case}\; 3 \;\texttt{of}\; 3 \Rightarrow 4 \\
&\rightsquigarrow^1 \texttt{case}\; f(\lambda.\texttt{case}\; 0 \;\texttt{of}\; 0 \Rightarrow 1) \;\texttt{of}\; 2 \Rightarrow \texttt{case}\; 3 \;\texttt{of}\; 3 \Rightarrow 4 \\
&\rightsquigarrow^2 \texttt{case}\; f\, \widetilde{1} \;\texttt{of}\; 2 \Rightarrow \texttt{case}\; 3 \;\texttt{of}\; 3 \Rightarrow 4 \\
&\rightsquigarrow^3 \texttt{case}\; f\, \widetilde{1} \;\texttt{of}\; 2 \Rightarrow 4 \;=\; \ll G \gg .
\end{aligned}$$

We shall consider the evaluations of $C[G]$ and $C[\ll G \gg]$, where

$$C[-] \;=\; (\lambda f.\texttt{case}\; - \;\texttt{of}\; 4 \Rightarrow 5)\; p\,, \qquad p \;=\; \lambda y.\texttt{case}\; y \;\texttt{of}\; 1 \Rightarrow 2\,.$$

The evaluation of $C[\ll G \gg]$ may now be performed directly in eight steps, but we shall here present this evaluation in a way that makes explicit which portions of $\ll G \gg$ are required when. To this end, we introduce labels for the expression nodes within $\ll G \gg$: the whole of $\ll G \gg$ is labelled by the empty sequence ε, and the occurrences of 1 and 4 are labelled by L_0 and R_2 respectively. (The general convention for labelling will be explained in the proof below.) These labels may be expanded using the rules:

$$\varepsilon \rightsquigarrow \texttt{case}\ f\,(\lambda.L_0)\ \texttt{of}\ 2 \Rightarrow R_2\,, \qquad L_0 \rightsquigarrow 1\,, \qquad R_2 \rightsquigarrow 4\,,$$

so that the whole of $\ll G \gg$ may be obtained from ε by repeated expansion. We can now present the evaluation of $C[\ll G \gg]$ as an evaluation of $C[\varepsilon]$, expanding labels only when required:

$$\begin{aligned}
C[\varepsilon] &= (\lambda f.\texttt{case}\ \varepsilon\ \texttt{of}\ 4 \Rightarrow 5)\ p \\
&\rightsquigarrow \texttt{case}\ \varepsilon\langle f \mapsto p\rangle\ \texttt{of}\ 4 \Rightarrow 5 \\
&\rightsquigarrow^{\dagger} \texttt{case}\ (\texttt{case}\ p(\lambda.L_0\langle f \mapsto p\rangle)\ \texttt{of}\ 2 \Rightarrow R_2\langle f \mapsto p\rangle)\ \texttt{of}\ 4 \Rightarrow 5 \\
&\rightsquigarrow \texttt{case}\ p(\lambda.L_0\langle f \mapsto p\rangle)\ \texttt{of}\ 2 \Rightarrow\ \texttt{case}\ R_2\langle f \mapsto p\rangle\ \texttt{of}\ 4 \Rightarrow 5 \\
&\rightsquigarrow \texttt{case}\ (\texttt{case}\ \lambda.L_0\langle f \mapsto p\rangle\ \texttt{of}\ 1 \Rightarrow 2)\ \texttt{of}\ 2 \Rightarrow\ \texttt{case}\ R_2\langle f \mapsto p\rangle\ \texttt{of}\ 4 \Rightarrow 5 \\
&\rightsquigarrow \texttt{case}\ (\lambda.L_0\langle f \mapsto p\rangle)\ \texttt{of}\ 1 \Rightarrow \texttt{case}\ 2\ \texttt{of}\ 2 \Rightarrow\ \texttt{case}\ R_2\langle f \mapsto p\rangle\ \texttt{of}\ 4 \Rightarrow 5 \\
&\rightsquigarrow \texttt{case}\ L_0\langle f \mapsto p\rangle\ \texttt{of}\ 1 \Rightarrow \texttt{case}\ 2\ \texttt{of}\ 2 \Rightarrow\ \texttt{case}\ R_2\langle f \mapsto p\rangle\ \texttt{of}\ 4 \Rightarrow 5 \\
&\rightsquigarrow^{\dagger} \texttt{case}\ 1\ \texttt{of}\ 1 \Rightarrow \texttt{case}\ 2\ \texttt{of}\ 2 \Rightarrow\ \texttt{case}\ R_2\langle f \mapsto p\rangle\ \texttt{of}\ 4 \Rightarrow 5 \\
&\rightsquigarrow^{*} \texttt{case}\ R_2\langle f \mapsto p\rangle\ \texttt{of}\ 4 \Rightarrow 5 \\
&\rightsquigarrow^{\dagger} \texttt{case}\ 4\ \texttt{of}\ 4 \Rightarrow 5 \quad \rightsquigarrow 5\,.
\end{aligned}$$

The first step here is a β-reduction that does not depend on the values of the labels. In the second line, $\langle f \mapsto p\rangle$ denotes an *explicit substitution*, indicating that once ε has been expanded, the substitution $f \mapsto p$ needs to be applied. In fact, this happens in the very next step, where the expansion of $\varepsilon\langle f \mapsto p\rangle$ yields

$$\texttt{case}\ p\,(\lambda.L_0\langle f \mapsto p\rangle)\ \texttt{of}\ 2 \Rightarrow R_2\langle f \mapsto p\rangle\,.$$

Note that the subexpression labels L_0 and R_2 inherit the explicit substitution. Later on, when the values of L_0 and R_2 are required, the corresponding expansions are invoked (in this instance, the substitutions have no effect since f does not appear in the relevant expressions). Such 'expansion-substitution' steps are marked above with '$\dagger$', the other steps being ordinary NSP reductions.

It is clear that this *skeletal reduction* in some sense embodies the evaluation of $C[\ll G \gg]$. Formally, the latter may be retrieved from the above by instantiating all labels as follows and performing any prescribed substitutions:

$$\varepsilon \mapsto \ \ll G \gg\,, \qquad L_0 \mapsto 1\,, \qquad R_2 \mapsto 4\,.$$

When this is done, the '$\dagger$' steps leave the meta-term unchanged, whilst the others remain as ordinary reductions.

The key idea is that the same skeleton can be interpreted in another way as a template for the reduction of $C[G]$, thus manifesting that $C[G]$ also evaluates to 5. To a first approximation, this works as follows. Suppose that we instantiate the labels in our skeletal reduction via the following alternative substitutions:

$$\varepsilon \mapsto G\,, \qquad L_0 \mapsto \texttt{case}\,\widetilde{0}\,\texttt{of}\;0 \Rightarrow 1\,, \qquad R_2 \mapsto \texttt{case}\;3\;\texttt{of}\;3 \Rightarrow 4\,.$$

Roughly speaking, viewing the evaluation of G as a successive evaluation of subterms to head normal form, these give, for each node of $\ll G \gg$, the subterm that needs to be thus evaluated. Thus, G itself is reduced to head normal form by step 0 of the evaluation, leaving the subterm $\texttt{case}\,\widetilde{0}\,\texttt{of}\;0 \Rightarrow 1$ to be evaluated in steps 1 and 2, and $\texttt{case}\;3\;\texttt{of}\;3 \Rightarrow 4$ to be evaluated in step 3.

We shall denote this latter substitution operation by $^\bullet$. Formally, the relevant fact is that if $\alpha \rightsquigarrow v_\alpha$ is one of our expansion rules, then $\alpha^\bullet \rightsquigarrow_h^* v_\alpha^\bullet$. Thus, if our skeletal reduction is instantiated via $^\bullet$, we obtain (roughly) an evaluation of $C[G]$, where each expansion step now stands for the entire head evaluation of $\alpha^\bullet$ to $v_\alpha^\bullet$.

There is, however, a further subtlety. In the true evaluation of $C[G]$, the head evaluation of $\alpha^\bullet$ need not happen 'in place' as the above discussion might suggest, but may be interfered with by the case commuting rule (b4). For instance, if we apply $\bullet$ to the seventh line of the skeletal reduction, the subsequent evaluation steps will not have the form

$$\texttt{case}\,(\texttt{case}\,\widetilde{0}\,\texttt{of}\;0 \Rightarrow 1)\,\texttt{of}\cdots \rightsquigarrow \texttt{case}\,(\texttt{case}\,0\,\texttt{of}\;0 \Rightarrow 1)\,\texttt{of}\cdots \rightsquigarrow \texttt{case}\;1\;\texttt{of}\cdots\,,$$

but rather

$$\begin{aligned} &\texttt{case}\,(\texttt{case}\,\widetilde{0}\,\texttt{of}\;0 \Rightarrow 1)\,\texttt{of}\cdots \rightsquigarrow \texttt{case}\,\widetilde{0}\,\texttt{of}\;0 \Rightarrow \texttt{case}\;1\;\texttt{of}\cdots \\ &\rightsquigarrow \texttt{case}\;0\;\texttt{of}\;0 \Rightarrow \texttt{case}\;1\;\texttt{of}\cdots \rightsquigarrow \texttt{case}\;1\;\texttt{of}\cdots\,. \end{aligned}$$

However, it turns out that once some simple reshuffling of case expressions is allowed for, the evaluation of $C[G]$ indeed follows the structure of the skeleton.

Whilst this example does not exhibit all the complexities of the general situation, it serves to illustrate the general direction of our argument. We now proceed to the proof of Theorem 6.1.7 itself.

Proof. We begin by assigning an *address* α to each expression node within the syntax tree of $\ll G \gg$; we shall write g_α for the expression at position α. Addresses will be finite sequences of *tags*, each of which is either a left tag L_i or a right tag R_j $(i, j \in \mathbb{N})$.

As a special case, we start by assigning the address ε (the empty sequence) to the root node of $\ll G \gg$, even if $\ll G \gg$ is an application rather than an expression, so that $g_\varepsilon = \ll G \gg$. Assuming g_α has already been defined, we assign addresses to the immediate subexpressions of α as follows:

- If g_α is $\bot$ or a numeral n, there is nothing to do.
- If $g_\alpha = \texttt{case}\; y(\lambda \vec{z}_0.d_0)\cdots(\lambda \vec{z}_{r-1}.d_{r-1})\;\texttt{of}\;(j \Rightarrow e_j)$, we assign addresses αL_i, αR_j to the respective occurrences of d_i and e_j, so that $g_{\alpha L_i} = d_i$ and $g_{\alpha R_j} = e_j$.
- If $g_\alpha = y(\lambda \vec{z}_0.d_0)\cdots(\lambda \vec{z}_{r-1}.d_{r-1})$, we again take $g_{\alpha L_i} = d_i$ for each i (this can only happen when $\alpha = \varepsilon$).

We work with an extended calculus of meta-terms in which addresses α may appear, and may also be subject to explicit substitutions $\langle \vec{x} \mapsto \vec{P} \rangle$. Specifically, we extend the grammar for meta-terms from Subsection 6.1.1 with the clause

$$E \ ::= \ \alpha \langle \vec{x}_0 \mapsto \vec{P}_0 \rangle \cdots \langle \vec{x}_{s-1} \mapsto \vec{P}_{s-1} \rangle \ .$$

For each value of i, we think of the substitutions $x_{i0} \mapsto P_{i0}$, $x_{i1} \mapsto P_{i1}$, ... as being performed simultaneously, whilst the clusters of substitutions for $i = 0, 1, \ldots$ are performed in sequence. We also add the technical stipulation that a meta-term must come with a finite global bound on the nesting depth of substitutions $\langle x \mapsto P \rangle$, although meta-terms may in other respects be infinitely deep as usual.

We now define a reduction system for our new meta-terms. Here we write * to stand for a (possibly empty) sequence of substitutions $\langle \vec{x}_0 \mapsto \vec{P}_0 \rangle \cdots \langle \vec{x}_{s-1} \mapsto \vec{P}_{s-1} \rangle$ applied to an address α. For any variable y, we also define

$$y^* \ = \ \begin{cases} P_{kl}[\vec{x}_{k+1} \mapsto \vec{P}_{k+1}] \cdots [\vec{x}_{s-1} \mapsto \vec{P}_{s-1}] & \text{if } y = x_{kl} \ , \\ y & \text{if } y \text{ is not among the } x_{kl} \ . \end{cases}$$

We adopt the clauses for $\rightsquigarrow_b$ from Subsection 6.1.1, noting that in the β-rule the usual definition of substitution should now be extended with the clause

$$(\alpha \langle \vec{x}_0 \mapsto \vec{P}_0 \rangle \cdots \langle \vec{x}_{s-1} \mapsto \vec{P}_{s-1} \rangle)[\vec{y} \mapsto \vec{Q}] \ = \ \alpha \langle \vec{x}_0 \mapsto \vec{P}_0 \rangle \cdots \langle \vec{x}_{s-1} \mapsto \vec{P}_{s-1} \rangle \langle \vec{y} \mapsto \vec{Q} \rangle \ .$$

We also augment these reduction rules with the following new clauses:

$$\begin{aligned} \alpha^* &\rightsquigarrow_b g_\alpha \ \text{ if } g_\alpha = \bot \text{ or } g_\alpha = n \ , \\ \alpha^* &\rightsquigarrow_b \mathtt{case}\ y^* \, (\lambda \vec{z}_0 . \alpha L_0^*) \cdots (\lambda \vec{z}_{r-1} . \alpha L_{r-1}^*) \ \mathtt{of}\ i \Rightarrow \alpha R_i^* \\ &\qquad\qquad \text{if } g_\alpha = \mathtt{case}\ y \, (\lambda \vec{z}_0 . d_0) \cdots (\lambda \vec{z}_{r-1} . d_{r-1}) \ \mathtt{of}\ (i \Rightarrow e_i) \ , \\ \varepsilon^* &\rightsquigarrow_b y^* \, (\vec{z}_0 . \alpha L_0^*) \cdots (\vec{z}_{r-1} . \alpha L_{r-1}^*) \ \text{ if } g = y \, (\lambda \vec{z}_0 . e_0) \cdots (\lambda \vec{z}_{r-1} . e_{r-1}) \ . \end{aligned}$$

We refer to these new rules as *expansions*. Subject to this change to the definition of $\rightsquigarrow_b$, we define $\rightsquigarrow_h$, $\rightsquigarrow$ and $\rightsquigarrow^*$ as before, and again set

$$\Downarrow_{\mathrm{fin}} T \ = \ \{ t \text{ finite} \mid \exists T' .\, T \rightsquigarrow^* T' \wedge t \sqsubseteq T' \} \ .$$

Now consider the set $\Downarrow_{\mathrm{fin}} C[\varepsilon]$. We shall show in two lemmas below that this set coincides with both $\Downarrow_{\mathrm{fin}} C[G]$ and $\Downarrow_{\mathrm{fin}} C[\ll G \gg]$. Since $\ll T \gg = \bigsqcup(\Downarrow_{\mathrm{fin}} T)$ for any T, it follows that $\ll C[G] \gg = \ll C[\ll G \gg] \gg$. $\square$

In the context of the above reduction system, notice that if $T \rightsquigarrow_h T'$ by expanding some α^* and then $T' \rightsquigarrow_h T''$ by expanding some $\beta^\star$ (where * is some other sequence of substitutions), then $\beta^\star$ must appear within the substitutions represented by *. It follows that a consecutive sequence of head expansions must be bounded in length. We write $\exp(T)$ for the result of applying head expansions to T until no longer possible.

Note too that $T \rightsquigarrow_h T'$ still implies $T[\vec{x} \mapsto \vec{Q}] \rightsquigarrow_h T'[\vec{x} \mapsto \vec{Q}]$.

The coincidence between our sets of finite approximations is established by analysing the evaluation of $C[\varepsilon]$ in two different ways. The details are somewhat tedious, and the reader may prefer to skip them. Let us consider $C[\ll G \gg]$ first.

Lemma 6.1.8. $\Downarrow_{\mathrm{fin}} C[\varepsilon] = \Downarrow_{\mathrm{fin}} C[\ll G \gg]$.

Proof. We shall define an operation $T \mapsto T^\circ$ on our extended meta-terms which replaces each address α by the corresponding g_α, and carries out any pending substitutions on this subterm. The crucial clause in the definition of T° is

$$(\alpha\langle\vec{x}_0 \mapsto \vec{P}_0\rangle \cdots \langle\vec{x}_{s-1} \mapsto \vec{P}_{s-1}\rangle)^\circ \;=\; g_\alpha[\vec{x}_0 \mapsto \vec{P}_0^\circ] \cdots [\vec{x}_{s-1} \mapsto \vec{P}_{s-1}^\circ] \,,$$

the remaining clauses being as expected. The definition is well-founded because of our stipulation on nesting of substitutions. Note that $(T[\vec{x} \mapsto \vec{P}])^\circ = T^\circ[\vec{x} \mapsto \vec{P}^\circ]$.

The following facts may be routinely checked by inspection of our reduction rules and definitions. Here $\rightsquigarrow_h^*$ denotes the reflexive-transitive closure of $\rightsquigarrow_h$, and T, T' range over extended meta-terms.

- If $T \rightsquigarrow_b T'$ via an expansion rule, then $T^\circ = T'^\circ$. Thus $(\exp(T))^\circ = T^\circ$.
- If $T \rightsquigarrow_b T'$ via a non-expansion rule, then $T^\circ \rightsquigarrow_b T'^\circ$.
- Hence if $T \rightsquigarrow_h T'$ then $T^\circ \rightsquigarrow_h^* T'^\circ$ in at most one step. (The only subtlety here is that if $G \rightsquigarrow_h G'$ where G is not a `case` meta-term but G° is, then G has the form α^* so that G is an expansion rule.)
- Conversely, if $T^\circ \rightsquigarrow_h T^\dagger$, then for some T' we have that $\exp(T) \rightsquigarrow T'$ via a non-expansion rule and $T'^\circ = T^\dagger$. (This requires some easy case analysis of the ways in which a redex can arise within T°.) Thus $T \rightsquigarrow_h^* T'$ in at least one step.

As before, we call T a *head normal form* if it cannot be reduced by $\rightsquigarrow_h$. Let $\mathrm{hnf}(T)$ denote the unique head normal form T' such that $T \rightsquigarrow_h^* T'$, if there is one. From the above facts we conclude the following:

- If T is a head normal form then so is T°.
- If T° is a head normal form then so is $\exp(T)$.
- Thus, $\mathrm{hnf}(T)$ exists iff $\mathrm{hnf}(T^\circ)$ exists, and in this case $\mathrm{hnf}(T^\circ) = \mathrm{hnf}(T)^\circ$.
- In particular, if $\mathrm{hnf}(T) = \mathtt{case}\, x\,(\lambda\vec{z}_0.D_0)\cdots(\lambda\vec{z}_{r-1}.D_{r-1})\ \mathtt{of}\ (i \Rightarrow E_i)$, then

$$\mathrm{hnf}(T^\circ) = \mathtt{case}\, x\,(\lambda\vec{z}_0.D_0^\circ)\cdots(\lambda\vec{z}_{r-1}.D_{r-1}^\circ)\ \mathtt{of}\ (i \Rightarrow E_i^\circ)\,,$$

  and likewise for $\mathrm{hnf}(T) = x\,(\lambda\vec{z}_0.D_0)\cdots(\lambda\vec{z}_{r-1}.D_{r-1})$.

Applying this last fact recursively, it follows easily that $\Downarrow_{\mathrm{fin}} T = \Downarrow_{\mathrm{fin}} T^\circ$ for any T. Taking $T = C[\varepsilon]$, we thus have that

$$\Downarrow_{\mathrm{fin}} C[\varepsilon] \;=\; \Downarrow_{\mathrm{fin}} C[\varepsilon]^\circ \;=\; \Downarrow_{\mathrm{fin}} C[\ll G \gg]\,. \qquad \square$$

We now wish to show by a similar analysis that a different instantiation of the skeletal reduction yields a reduction sequence for $C[G]$. However, this is only true modulo some reshuffling of `case` expressions, and some definitions and a further lemma are needed to make this precise.

Let us write $\rightsquigarrow_c$ for the reduction on ground meta-terms given by just (b4) and (b2), and $\rightsquigarrow_c^*$ for its reflexive-transitive closure. A reduction step $G \rightsquigarrow_h G'$ will be called *proper* if $G \not\rightsquigarrow_c G'$. We define a relation $\succeq$ inductively as follows:

1. If $T \rightsquigarrow_c^* U$ then $T \succeq U$.
2. If $T \rightsquigarrow_c^* \mathtt{case}\ G\ \mathtt{of}\ (i \Rightarrow T_i)$ and $T_i \succeq U_i$ for each i, then $T \succeq \mathtt{case}\ G\ \mathtt{of}\ (i \Rightarrow U_i)$.

Notice that if $T \succeq U$ and $U \rightsquigarrow_c^* U'$ then $T \succeq U'$.

We shall call $C[-]$ a *head context* if $C[-]$ is obtained by nesting zero or more contexts of the form $\mathtt{case}\ -\ \mathtt{of}\ (k \Rightarrow F_k)$. The following lemma says that 'in-place' reductions of a meta-term G in such a context are tracked by honest reductions of $C[G]$, modulo the slack admitted by $\succeq$.

Lemma 6.1.9. *Suppose $C[-]$ is a head context and $C[G] \succeq U$.*

(i) If $G \rightsquigarrow_h G'$, there exists U' with $U \rightsquigarrow_h^* U'$ and $C[G'] \succeq U'$.

(ii) Moreover, if $G \rightsquigarrow_h G'$ is proper then $U \rightsquigarrow_h^* U'$ contains exactly one proper step; otherwise it contains none.

Proof. (i) First suppose $C[-]$ is the identity context. We reason by cases on the possible forms of G:

- If $G = (\lambda \vec{x}.E)\vec{Q}$, then $U = G$ so we may take $U' = G'$.
- If $G = \mathtt{case}\ n\ \mathtt{of}\ (i \Rightarrow E_i)$ then $U = \mathtt{case}\ n\ \mathtt{of}\ (i \Rightarrow U_i)$ where $E_i \succeq U_i$ for all i, so $G' = E_i$ and we may take $U' = U_i$.
- If $G = \mathtt{case}\ (\lambda \vec{x}.E)\vec{Q}\ \mathtt{of}\ (\cdots)$, a similar argument applies.
- If $G \rightsquigarrow_c G'$, clearly either $C[G'] \succeq U$ already or $U \rightsquigarrow_c U'$ with $C[G'] \succeq U'$.

Next, suppose $C[-]$ is a single-level context $\mathtt{case}\ -\ \mathtt{of}\ (k \Rightarrow F_k)$. Again, we argue by cases on the form of G:

- If $G = (\lambda \vec{x}.E)\vec{Q}$, then $U = \mathtt{case}\ G\ \mathtt{of}\ (k \Rightarrow U_k)$ where $F_k \succeq U_k$; hence we may take $U' = \mathtt{case}\ G'\ \mathtt{of}\ (k \Rightarrow U_k)$.
- If $G = \mathtt{case}\ n\ \mathtt{of}\ (i \Rightarrow E_i)$, then $U \rightsquigarrow_c^* \mathtt{case}\ n\ \mathtt{of}\ (i \Rightarrow U_i)$ in at most one step, where $C[E_i] \succeq U_i$; hence we may take $U' = U_n$.
- If $G = \mathtt{case}\ (\lambda \vec{x}.E)\vec{Q}\ \mathtt{of}\ (\cdots)$, the argument is similar.
- If $G = \mathtt{case}\ \bot\ \mathtt{of}\ (\cdots)$, then $U \rightsquigarrow_c^* \bot$ in at most two steps; also $G' = \bot$ so $C[G] \rightsquigarrow_c \bot$.
- Suppose $G = \mathtt{case}\ (\mathtt{case}\ C\ \mathtt{of}\ (i \Rightarrow D_i))\ \mathtt{of}\ (j \Rightarrow E_j)$. Then we have

$$U \rightsquigarrow_c^* \mathtt{case}\ C\ \mathtt{of}\ (i \Rightarrow \mathtt{case}\ D_i\ \mathtt{of}\ (j \Rightarrow \mathtt{case}\ E_j\ \mathtt{of}\ (k \Rightarrow U_k)))$$

  in at most two steps, and it is easy to verify that the latter term may serve as U'.

Finally, suppose $C[-]$ is a head context of depth $r > 1$. Then clearly $C[-] \rightsquigarrow_c^* C'[-]$ in $r-1$ steps, where $C'[-]$ is a head context of depth 1. So if $C[G] \succeq U$, then $C[G'] \succeq U''$ where $U \rightsquigarrow_c^* U''$ in at most $r-1$ steps. But now by the above we can find $U'' \rightsquigarrow_h^* U'$ with $C'[G'] \succeq U'$, whence also $C[G'] \succeq U'$.

(ii) is now immediate by inspection of the above. □

We may now proceed to the analogue of Lemma 6.1.8 for $C[G]$:

Lemma 6.1.10. $\Downarrow_{\mathsf{fin}} C[\varepsilon] = \Downarrow_{\mathsf{fin}} C[G]$.

Proof. The proof broadly parallels that of Lemma 6.1.8, but uses a different way of expanding the addresses, derived from the process by which G evaluates to g. To each node α in g we assign a meta-expression $\alpha^\bullet$ by induction on the length of α:

- Set $\varepsilon^\bullet = G$.
- If $\alpha^\bullet$ is defined and reduces via $\rightsquigarrow_h^*$ to the head normal form

$$\mathtt{case}\ x(\lambda\vec{z}_0.D_0)\cdots(\lambda\vec{z}_{r-1}.D_{r-1})\ \mathtt{of}\ (i \Rightarrow E_i)$$

  then set $\alpha L_i^\bullet = D_i$ for each i, and $\alpha R_j^\bullet = E_j$ for each j.
- If $\mathrm{hnf}(\alpha^\bullet) = x(\lambda\vec{z}_0.D_0)\cdots(\lambda\vec{z}_{r-1}.D_{r-1})$, set $\alpha L_i^\bullet = D_i$ for each i (this can only happen when $\alpha = \varepsilon$).

Notice that this defines $\alpha^\bullet$ for exactly those nodes α that appear in the syntax tree of g, and that for all such α we have $\ll \alpha^\bullet \gg = g_\alpha$. We hence obtain an operation $T \mapsto T^\bullet$ by analogy with the definition of T°; note that $(T[\vec{x} \mapsto \vec{P}])^\bullet = T^\bullet[\vec{x} \mapsto \vec{P}^\bullet]$.

Let us write v_α for the head normal form to which α expands by the rules given earlier. From the way these expansion rules and the instantiation $^\bullet$ are constructed, we see that $\alpha^\bullet \rightsquigarrow_h^* v_\alpha^\bullet$.

We now relate reductions of T to those of any U such that $T^\bullet \succeq U$.

- Suppose $T \rightsquigarrow_h T'$ via an expansion rule: say $T = C[\alpha^*]$ where $C[-]$ is a head context. Then clearly $\alpha^* \rightsquigarrow v_\alpha^*$ (defining v_α^* in the obvious way), so $T' = C[v_\alpha^*]$. Thus $T^\bullet = C^\bullet[G]$ where $G = (\alpha^*)^\bullet$, and $T'^\bullet = C^\bullet[G']$ where $G' = (v_\alpha^*)^\bullet$. Moreover, since $\alpha^\bullet \rightsquigarrow_h^* v_\alpha^\bullet$, we have

$$G \;=\; (\alpha^\bullet)^{(*\bullet)} \;\rightsquigarrow_h^*\; (v_\alpha^\bullet)^{(*\bullet)} \;=\; G'$$

  for the obvious meaning of $^{(*\bullet)}$. So if $T^\bullet \succeq U$, then by iterated application of Lemma 6.1.9 we have $T'^\bullet \succeq U'$.
- If $T \rightsquigarrow_h T'$ via a non-expansion rule, then $T^\bullet \rightsquigarrow_h T'^\bullet$ via the same rule. So if $T^\bullet \succeq U$, then by Lemma 6.1.9 we again have $T'^\bullet \succeq U'$ where $U \rightsquigarrow_h^* U'$.
- If T is a head normal form for the expanded definition of $\rightsquigarrow_h$, then T cannot contain an address in head position, so $T^\bullet$ is a head normal form. Hence so is any U such that $T^\bullet \succeq U$. Combining this with the preceding points, if T has a head normal form T' and $T^\bullet \succeq U$, then U has a head normal form U' and $T'^\bullet \succeq U'$.
- Conversely, if $T^\bullet \succeq U$ and U has a head normal form U', the following may be noted about the head reduction sequence of T:
  - The number of proper non-expansion steps cannot exceed the number of proper steps in $U \rightsquigarrow_h^* U'$.
  - The number of consecutive expansions possible is bounded.
  - If $T \rightsquigarrow_h^* T^\dagger$, the number of consecutive $\rightsquigarrow_c$ steps starting from $T^\dagger$ is bounded. This is because we may find $U \rightsquigarrow_h^* U^\dagger$ and a reduction $T^\dagger \rightsquigarrow_c^* V^\dagger$ such that either $V^\dagger = U^\dagger$ or $V^\dagger = \mathtt{case}\ H\ \mathtt{of}\ (i \Rightarrow V_i)$ and $U^\dagger = \mathtt{case}\ H\ \mathtt{of}\ (i \Rightarrow U_i)$ where $V_i \succeq U_i$. The $\rightsquigarrow_c$ reduction path of $V^\dagger$ now mirrors that of $U^\dagger$, and hence is finite since $U^\dagger$ has a head normal form.

It follows that the head reduction of T terminates in a head normal form.

We now see that if $T^\bullet \succeq U$, then T has a head normal form iff U does; moreover, if hnf(T) is n or $\perp$ then $\mathrm{hnf}(U) = \mathrm{hnf}(T)$; and if $\mathrm{hnf}(T) = \mathtt{case}\ x\vec{Q}\ \mathtt{of}\ (j \Rightarrow T_j)$ where $Q_i = \lambda\vec{z}_i.E_i$ for each i, then $\mathrm{hnf}(U) = \mathtt{case}\ x\vec{Q}^\bullet\ \mathtt{of}\ (j \Rightarrow U_j)$ where $Q_i^\bullet = \lambda\vec{z}_i.E_i^\bullet$, $E_i \succeq E_i^\bullet$ and $T_j \succeq U_j$. Applying this recursively, we see that $\Downarrow_{\mathrm{fin}} T = \Downarrow_{\mathrm{fin}} U$. In particular, since $C[\varepsilon]^\bullet = C[G]$, the lemma follows. □

This completes the proof of Theorem 6.1.7. We next note some useful corollaries and extensions.

Corollary 6.1.11. *(i) If $C[-]$ is any meta-term context and T any meta-term such that $C[T]$ is well-formed, then $\ll C[T] \gg = \ll C[\ll T \gg] \gg$.*

(ii) If $C[-^0, -^1, \ldots]$ is any meta-term context with countably many holes and $C[T^0, T^1, \ldots]$ is well-formed, then

$$\ll C[T^0, T^1, \ldots] \gg \;=\; \ll C[\ll T^0 \gg, \ll T^1 \gg, \ldots] \gg .$$

Proof. (i) If T is a meta-expression or ground meta-term then this is already covered by Theorem 6.1.7. If T is a meta-procedure $\lambda\vec{x}.E$, the result follows from the theorem since $\ll T \gg = \lambda\vec{x}. \ll E \gg$.

(ii) This is strictly a corollary to the proof of Theorem 6.1.7 rather than its statement. We simply adapt the proof by using disjoint sets of addresses α^i for each of the holes $-^i$. □

Corollary 6.1.12. *(i) $\ll E[\vec{y} \mapsto \vec{q}] \gg = \ll \ll E \gg [\vec{y} \mapsto \vec{q}] \gg$.*

(ii) $(\lambda x_0 \ldots x_{r-1}.e) \cdot q_0 \cdot \ldots \cdot q_{i-1} = \ll \lambda x_i \ldots x_{r-1}.e[x_0 \mapsto q_0] \cdots [x_{i-1} \mapsto q_{i-1}] \gg$.

Proof. (i) By Theorem 6.1.7 and Proposition 6.1.4(ii), we have

$$\ll E[\vec{y} \mapsto \vec{q}] \gg = \ll (\lambda\vec{y}.E)\vec{q} \gg = \ll (\lambda\vec{y}. \ll E \gg)\vec{q} \gg = \ll \ll E \gg [\vec{y} \mapsto \vec{q}] \gg .$$

(ii) An easy induction on i using (i) and the definition of application. □

We also have the following lemma, which makes explicit some quantitative information that will be useful later on.

Lemma 6.1.13. *Suppose $c = \mathtt{case}\ p\vec{q}\ \mathtt{of}\ (i \Rightarrow e_i)$ and $\ll c \gg = n$. Then $\ll p\vec{q} \gg$ is some numeral m where $\ll e_m \gg = n$. Moreover, each of the reduction sequences $p\vec{q} \rightsquigarrow_h^* m$, $e_m \rightsquigarrow_h^* n$ contains fewer proper reduction steps than $c \rightsquigarrow_h^* n$.*

Proof. By Theorem 6.1.7 we have $\ll c \gg = \ll \mathtt{case} \ll p\vec{q} \gg \mathtt{of}\ (i \Rightarrow e_i) \gg = n$. It follows that $\ll p\vec{q} \gg$ must have some value $m \in \mathbb{N}$, and that $\ll e_m \gg = n$. Moreover, Lemma 6.1.9 tells us that we have $c \rightsquigarrow_h^* \mathtt{case}\ m\ \mathtt{of}\ (i \Rightarrow e_i')$ with the same number of proper steps as $p\vec{q} \rightsquigarrow_h^* m$, where $e_i' \succeq e_i$; also $e_m' \rightsquigarrow_h^* n$ with the same number of proper steps as $e_m \rightsquigarrow_h^* n$. Finally, the reduction $c \rightsquigarrow_h^* n$ is made up from these two reductions along with the additional proper step $\mathtt{case}\ m\ \mathtt{of}\ (i \Rightarrow e_i) \rightsquigarrow e_m$. □

6.1.3 Interpretation of Typed λ-Calculus

Next, we work towards showing that SP admits a good interpretation of the ordinary λ-calculus as in Chapter 4. Clearly, not every closed term of the λ-calculus is a sequential procedure as it stands: for instance, $\lambda xy.xy$ is not a syntactically valid procedure term, since y itself is not a procedure. In order to treat variables as procedures, we require the notion of *η-expansion*. For x^σ a typed variable, we define a procedure $x^{\sigma\eta}$ by argument induction on σ:

$$x^{(\sigma_0,\dots,\sigma_{r-1}\to\mathtt{N})\eta} \;=\; \lambda z_0\dots z_{r-1}.\,\mathtt{case}\,x z_0^{\sigma_0\eta}\dots z_{r-1}^{\sigma_{r-1}\eta}\;\mathtt{of}\;(i\Rightarrow i)\,.$$

In particular, $x^{\mathtt{N}\eta} = \lambda.\mathtt{case}\,x\;\mathtt{of}\;(i\Rightarrow i)$. We shall sometimes write $x^{\sigma\eta}$ simply as x^η, but the reader should bear in mind that the meaning of η-expansion depends crucially on the type of x. If $\vec{z}$ is the sequence $z_0\cdots z_{r-1}$, we shall write $\vec{z}^\eta$ for $z_0^\eta\dots z_{r-1}^\eta$. For convenience, we will often write $\mathtt{case}\,G\,\mathtt{of}\;(i\Rightarrow i)$ simply as G.

The key properties of η-expansions require some significant work:

Lemma 6.1.14. *Let* $\sigma = \sigma_0,\dots,\sigma_{r-1}\to\mathtt{N}$.
(i) For any $x:\sigma$ *and* $q_i\in\mathsf{SP}(\sigma_i)$, *we have* $\ll x^{\sigma\eta}\,\vec{q}\gg = \mathtt{case}\,x\vec{q}\;\mathtt{of}\;(i\Rightarrow i)$.
(ii) For any $p\in\mathsf{SP}(\sigma)$, *we have* $\ll \lambda y_0^{\sigma_0}\cdots y_{r-1}^{\sigma_{r-1}}.p\vec{y}^\eta \gg = p$.

Proof. We reason in a way that minimizes reliance on the type structure, instead showing that both the above equations hold to arbitrary *depth* k. To this end, we introduce equivalence relations $=_k$ on expressions as follows:

- $e =_0 e'$ iff $e=e'=\bot$ or $e=e'=n$ or both e,e' are $\mathtt{case}$ expressions with the same head variable x.
- $e =_{k+1} e'$ iff $e =_0 e'$ and in the situation

$$\begin{aligned} e &= \mathtt{case}\,x(\lambda\vec{y}_0.d_0)\cdots(\lambda\vec{y}_{r-1}.d_{r-1})\;\mathtt{of}\;(i\Rightarrow e_i)\,,\\ e' &= \mathtt{case}\,x(\lambda\vec{y}_0.d'_0)\cdots(\lambda\vec{y}_{r-1}.d'_{r-1})\;\mathtt{of}\;(i\Rightarrow e'_i)\,,\end{aligned}$$

  we have $d_j =_k d'_j$ for each $j<r$ and $e_i =_k e'_i$ for each i.

We also set $\lambda\vec{x}.e =_k \lambda\vec{x}.e'$ iff $e =_k e'$. An equation $t=t'$ *holds to depth* k if $t =_k t'$.

For the present lemma, it will suffice to establish the following claims:

1. Both (i) and (ii) hold to depth 0.
2. If (ii) holds to depth k, then (i) holds to depth $k+1$.
3. If (ii) holds to depth k, then (ii) holds to depth $k+1$.

For claim 1, equation (i) at depth 0 is trivial, and (ii) at depth 0 follows easily. For claim 2, assuming (ii) at depth k, we have:

$$\begin{aligned} \ll x^\eta\,\vec{q}\gg &= \ll(\lambda\vec{y}.\mathtt{case}\,x(\lambda\vec{z}.y_0\vec{z}^\eta)\cdots(\lambda\vec{z}.y_{r-1}\vec{z}^\eta)\;\mathtt{of}\;(i\Rightarrow i))\vec{q}\gg\\ &= \ll\mathtt{case}\,x\lambda\vec{z}.q_0\vec{z}^\eta\cdots\lambda\vec{z}.q_{r-1}\vec{z}^\eta\;\mathtt{of}\;(i\Rightarrow i)\gg\\ &= \mathtt{case}\,x\ll\lambda\vec{z}.q_0\vec{z}^\eta\gg\cdots\ll\lambda\vec{z}.q_{r-1}\vec{z}^\eta\gg\;\mathtt{of}\;(i\Rightarrow i)\\ &=_{k+1} \mathtt{case}\,xq_0\dots q_{r-1}\;\mathtt{of}\;(i\Rightarrow i)\,.\end{aligned}$$

For claim 3, suppose (ii) holds to depth k, so that (i) holds to depth $k+1$ by claim 2. Let $p = \lambda\vec{y}.e$; then

$$\ll \lambda\vec{y}.\, p\vec{y}^{\eta} \gg \;=\; \lambda\vec{y}.\, \ll e[\vec{y} \mapsto \vec{y}^{\eta}] \gg \;.$$

Let * denote the substitution $[\vec{y} \mapsto \vec{y}^{\eta}]$. It will suffice to prove that $\ll e^* \gg =_{k+1} e$.

We may assume $e = \mathtt{case}\; x\vec{q}\; \mathtt{of}\; (j \Rightarrow d_j)$. If x is not among the y_i, then $\ll e^* \gg = \mathtt{case}\; x \ll \vec{q}^* \gg \mathtt{of}\; (j \Rightarrow \ll d_j^* \gg)$, and the claim follows easily from (ii) at depth k. If $x = y_i$, then by Corollary 6.1.11(ii) we have

$$\begin{aligned} \ll e^* \gg &= \ll \mathtt{case}\; y_i^{\eta}\vec{q}^*\; \mathtt{of}\; (j \Rightarrow d_j^*) \gg \\ &= \ll \mathtt{case}\; \ll y_i^{\eta}\vec{q}^* \gg \mathtt{of}\; (j \Rightarrow \ll d_j^* \gg) \gg \\ &= \ll \mathtt{case}\; \ll y_i^{\eta} \ll \vec{q}^* \gg\gg \mathtt{of}\; (j \Rightarrow \ll d_j^* \gg) \gg \;. \end{aligned}$$

But $\ll y_i^{\eta} \ll \vec{q}^* \gg\gg =_{k+1} \mathtt{case}\; y \ll \vec{q}^* \gg \mathtt{of}\; (i \Rightarrow i)$ by (i) at depth $k+1$; it follows that $\ll y_i^{\eta} \ll \vec{q}^* \gg\gg = \mathtt{case}\; y\vec{r}\; \mathtt{of}\; (i \Rightarrow i)$ where $\vec{r} =_k \ll \vec{q}^* \gg$. Thus

$$\begin{aligned} \ll e^* \gg &= \ll \mathtt{case}\; (\mathtt{case}\; y\vec{r}\; \mathtt{of}\; (i \Rightarrow i))\; \mathtt{of}\; (j \Rightarrow \ll d_j^* \gg) \gg \\ &= \mathtt{case}\; y\vec{r}\; \mathtt{of}\; (j \Rightarrow \ll d_j^* \gg) \;. \end{aligned}$$

Now using (ii) at depth k we see that $r_i =_k \ll q_i^* \gg =_k q_i$ and $\ll d_j^* \gg =_k d_j$ for each i, j, whence $\ll e^* \gg =_{k+1} \mathtt{case}\; x\vec{q}\; \mathtt{of}\; (j \Rightarrow d_j) = e$ as required. □

A simple consequence of Lemma 6.1.14 is the following alternative formula for application:

Lemma 6.1.15. $p \cdot q = \lambda\vec{v}.\, \ll pq\vec{v}^{\eta} \gg$.

Proof. Assume by α-conversion that p has the form $\lambda y\vec{v}.e$. Then by the definition of application, $p \cdot q = \lambda\vec{v}.\, \ll e[y \mapsto q] \gg$. But now

$$\begin{aligned} p \cdot q &= \ll \lambda\vec{v}.(p \cdot q)\vec{v}^{\eta} \gg \quad \text{by Lemma 6.1.14(ii)} \\ &= \lambda\vec{v}.\, \ll \ll e[y \mapsto q] \gg [\vec{v} \mapsto \vec{v}^{\eta}] \gg \quad \text{by } \beta\text{-equality} \\ &= \lambda\vec{v}.\, \ll e[y \mapsto q][\vec{v} \mapsto \vec{v}^{\eta}] \gg \quad \text{by Corollary 6.1.12(i)} \\ &= \lambda\vec{v}.\, \ll pq\vec{v}^{\eta} \gg \quad \text{by } \beta\text{-equality.} \end{aligned}$$ □

We are now ready to exhibit an interpretation of typed λ-calculus (over the type world $\mathsf{T}^{\rightarrow}(\mathtt{N})$) in SP. We do this along the lines of an 'internal interpretation' as described in Section 4.1.2. To reduce notational confusion, we shall use the symbols $\underline{\lambda}$ and $\bullet$ for abstraction and application in this meta-level calculus, reserving λ and juxtaposition for the syntax of NSPs themselves. At the level of NSPs, we shall allow ourselves to write $\lambda x_0 \ldots x_{r-1}.p$ for $\lambda x_0 \ldots x_{r-1}\vec{y}.e$ when $p = \lambda\vec{y}.e$.

We first give a simple compositional translation $[-]$ from typed λ-terms to sequential procedures:

$$[x] = x^{\eta}\,, \qquad [M \bullet N] = [M] \cdot [N]\,, \qquad [\underline{\lambda}x.M] = \lambda x.[M]\,.$$

For Γ any finite list of typed variables $x_0 \dots x_{r-1}$ and M any typed λ-term in context Γ, we now simply define

$$[\![M]\!]^i_\Gamma = \lambda x_0 \dots x_{r-1}.[M] .$$

Clearly $[\![M]\!]^i_\Gamma$ is a closed procedure, so $[\![M]\!]^i_\Gamma \in \mathsf{SP}^0$. For any fixed Γ, we may readily obtain $[\![M]\!]^i_\Gamma$ from $[M]$ and vice versa just by adding or removing λ-abstractions; hence $[\![-]\!]^i_-$ itself is compositional. Moreover:

Proposition 6.1.16. *For any typed λ-term M in context $\Gamma = x_0^{\sigma_0}, \dots, x_{r-1}^{\sigma_{r-1}}$ and any elements $a_i \in \mathsf{SP}(\sigma_i)$, we have $[\![M]\!]^i_\Gamma \cdot a_0 \cdot \dots \cdot a_{r-1} = \ll [M][\vec{x} \mapsto \vec{a}] \gg$.*

Proof. Let $[M] = \lambda \vec{y}.e$, where we assume the variables $\vec{y}$ have been chosen to avoid clashes. Then using Corollary 6.1.12(ii) we have

$$[\![M]\!]^i_\Gamma \cdot \vec{a} = (\lambda \vec{x}.\lambda \vec{y}.e) \cdot \vec{a} = \ll \lambda \vec{y}.e[\vec{x} \mapsto \vec{a}] \gg = \ll [M][\vec{x} \mapsto \vec{a}] \gg . \quad \square$$

Proposition 6.1.17. *$[\![-]\!]^i_-$ is an internal interpretation of typed λ-terms in SP satisfying the conditions of Definition 4.1.5.*

Proof. We have seen that $[\![-]\!]^i_-$ is compositional, so it suffices to verify conditions 1–5 of Definition 4.1.5. We assume $\Gamma = x_0, \dots, x_{r-1}$.

For condition 1, Proposition 6.1.16 and Lemma 6.1.14(ii) give

$$[\![x_i]\!]^i_\Gamma \cdot a_0 \cdot \dots \cdot a_{r-1} = \ll x_i^\eta [\vec{x} \mapsto \vec{a}] \gg = \ll \lambda \vec{y}.a_i \vec{y}^\eta \gg = a_i .$$

For condition 2, suppose M, N are λ-terms in context Γ. Let $p = \lambda y \vec{z}.e = [M]$, $q = [N]$, and write * for the substitution $[\vec{x} \mapsto \vec{a}]$. Then

$$\begin{aligned}
[\![M \bullet N]\!]^i_\Gamma \cdot a_0 \cdot \dots \cdot a_{r-1} &= \ll [M \bullet N]^* \gg \;=\; \ll (p \cdot q)^* \gg \quad \text{by Proposition 6.1.16} \\
&= \ll \ll \lambda \vec{z}.e[y \mapsto q] \gg^* \gg \quad \text{by Proposition 6.1.4(i)} \\
&= \ll (\lambda \vec{z}.e[y \mapsto q])^* \gg \quad \text{by Corollary 6.1.12(i)} \\
&= \ll \lambda \vec{z}.e^*[y \mapsto q^*] \gg \\
&= \ll \lambda \vec{z}. \ll e^* \gg [y \mapsto \ll q^* \gg] \gg \quad \text{using Theorem 6.1.7} \\
&= \ll p^* \gg \cdot \ll q^* \gg \\
&= ([\![M]\!]^i_\Gamma \cdot \vec{a}) \cdot ([\![N]\!]^i_\Gamma \cdot \vec{a}) \quad \text{by Proposition 6.1.16.}
\end{aligned}$$

Condition 3 says that $[\![\underline{\lambda} x.M]\!]^i_\Gamma = [\![M]\!]^i_{\Gamma,x}$, which is immediate from the definitions of $[-]$ and $[\![-]\!]^i_-$.

Condition 4 (the weakening rule) follows immediately from Proposition 6.1.16: in the equation

$$[\![M]\!]^i_{\Delta'} \cdot a \cdot \dots \cdot a_{t-1} = [\![M]\!]^i_\Delta \cdot a_{\theta(0)} \cdot \dots \cdot a_{\theta(r-1)} ,$$

both sides are easily seen to equal $\ll [M][x_{\theta(0)} \mapsto a_{\theta(0)}, \dots, x_{\theta(r-1)} \mapsto a_{\theta(r-1)}] \gg$.

Condition 5 is the β-rule: $[\![(\underline{\lambda} y.M) \bullet N]\!]^i_\Gamma = [\![M[y \mapsto N]]\!]^i_\Gamma$. By the definitions of $[\![-]\!]^i_-$ and $[-]$, it suffices to show that

$$(\lambda y.[M]) \cdot [N] \;=\; [M[y \mapsto N]] \,.$$

For this, we argue by induction on the structure of M:

- If $M \equiv y$, then using Lemma 6.1.14(ii) we have

$$(\lambda y.[M]) \cdot [N] \;=\; (\lambda y.y^\eta) \cdot [N] \;=\; \lambda \vec{v}. \ll [N]\vec{v}^\eta \gg \;=\; [N] \;=\; [y[y \mapsto N]] \,.$$

- If $M \equiv z \neq y$, then

$$(\lambda y.[M]) \cdot [N] \;=\; (\lambda y.z^\eta) \cdot [N] \;=\; \lambda \vec{v}. \ll z\vec{v}^\eta \gg \;=\; z^\eta \;=\; [z[y \mapsto N]] \,.$$

- If $M = P \centerdot Q$, let $p = [P]$, $q = [Q]$ and write * for the substitution $[y \mapsto [N]]$. Then

$$\begin{aligned}
(\lambda y.[M]) \cdot [N] &= \ll (p \cdot q)^* \gg \\
&= \ll p^* \gg \cdot \ll q^* \gg \quad \text{as in the proof of condition 2} \\
&= ((\lambda y.[P]) \cdot [N]) \cdot ((\lambda y.[Q]) \cdot [N]) \\
&= [P[y \mapsto N]] \cdot [Q[y \mapsto N]] \quad \text{by induction hypothesis} \\
&= [(P[y \mapsto N]) \centerdot (Q[y \mapsto N])] \\
&= [M[y \mapsto N]] \,.
\end{aligned}$$

- If $M = \underline{\lambda} z.M'$, let $p = [M']$ and $q = [N]$. Then

$$\begin{aligned}
(\lambda y.[M]) \cdot [N] &= (\lambda y.\lambda z.p) \cdot q \\
&= \ll \lambda z.p[y \mapsto q] \gg \\
&= \lambda z.((\lambda y.p) \cdot q) \\
&= \lambda z.[M'[y \mapsto N]] \quad \text{by induction hypothesis} \\
&= [M[y \mapsto N]] \,. \qquad \square
\end{aligned}$$

Theorem 6.1.18. SP *and* SP^0 *are* $\lambda\eta$*-algebras.*

Proof. Combining the above with Proposition 4.1.6 gives us that both SP and SP^0 are λ-algebras, since the elements a_i may range over SP but the elements $[\![M]\!]^i_\Gamma$ themselves are in SP^0. To see that SP, SP^0 both satisfy the η-rule, note that

$$\begin{aligned}
[\![\underline{\lambda} y.M \centerdot y]\!]^i_\Gamma &= \lambda \vec{x}.\lambda y.[M] \cdot y^\eta \\
&= \lambda \vec{x}.\lambda y.\lambda \vec{v}. \ll [M]y^\eta \vec{v}^\eta \gg \quad \text{by Lemma 6.1.15} \\
&= \lambda \vec{x}.[M] \quad \text{by Lemma 6.1.14(ii)} \\
&= [\![M]\!]^i_\Gamma \,. \qquad \square
\end{aligned}$$

It follows a fortiori that SP and SP^0 are higher-order models. Indeed, suitable combinators k, s may be given explicitly by

$$k \;=\; \lambda xy\vec{z}.\, x\vec{z}^\eta \,, \qquad s \;=\; \lambda xyz\vec{w}.\, xz^\eta\, (\lambda \vec{v}.\, yz^\eta \vec{v}^\eta)\, \vec{w}^\eta \,.$$

6.1.4 The Extensional Quotient

Nested sequential procedures in principle provide an *intensional* model of computation, in that they represent algorithms for computing sequential operations as distinct from the functions they compute: witness the existence of multiple NSPs for computing addition, as mentioned in Subsection 3.2.5. We may, however, obtain a model of sequential *functions* by taking a suitable extensional quotient of SP^0.

We start by defining a relation $\sim_{\rm app}$ called *applicative equivalence* on each set $\mathsf{SP}^0(\sigma)$. If $p, p' \in \mathsf{SP}^0(\sigma_0, \ldots, \sigma_{t-1} \to \mathtt{N})$, we let

$$p \sim_{\rm app} p' \;\text{ iff }\; \forall q_0, \ldots, q_{t-1} \in \mathsf{SP}^0.\; p \cdot q_0 \cdot \ldots \cdot q_{t-1} \;=\; p' \cdot q_0 \cdot \ldots \cdot q_{t-1}\,.$$

Note that $\sim_{\rm app}$ is the identity relation on $\mathsf{SP}^0(\mathtt{N})$.

The key properties of $\sim_{\rm app}$ are as follows:

Lemma 6.1.19. *(i) If $p \sim_{\rm app} p'$ then $p \cdot r \sim_{\rm app} p' \cdot r$.*
(ii) If $p \sim_{\rm app} p'$ then $r \cdot p \sim_{\rm app} r \cdot p'$.

Proof. (i) is immediate.

For (ii), suppose $p \sim_{\rm app} p'$ and $r = \lambda x\vec{y}.d$. By Corollary 6.1.12, we have

$$r \cdot p \cdot q_0 \cdot \ldots q_{t-1} \;=\; \ll d[x \mapsto p, \vec{y} \mapsto \vec{q}] \gg$$

for any $\vec{q}$ of suitable type, so it is enough to show

$$\ll d[x \mapsto p, \vec{y} \mapsto \vec{q}] \gg \;=\; \ll d[x \mapsto p', \vec{y} \mapsto \vec{q}] \gg\,.$$

In fact, we shall prove more generally that if e is any expression with $\mathrm{FV}(e) \subseteq \{x\}$, then $\ll e[x \mapsto p] \gg = \ll e[x \mapsto p'] \gg$. For this, it will be sufficient to prove that if $e[x \mapsto p] \rightsquigarrow_h^* n$ then $e[x \mapsto p'] \rightsquigarrow_h^* n$; we reason by induction on the number of proper reduction steps in $\ll e[x \mapsto p] \gg \rightsquigarrow_h^* n$.

Consider the head reduction sequence for e itself. There are two possibilities:

1. $e \rightsquigarrow_h^* n$. Then substituting p' for x, we obtain $e[x \mapsto p'] \rightsquigarrow_h^* n$.
2. $e \rightsquigarrow_h^* c = \mathtt{case}\, x\vec{t}\; \mathtt{of}\; (i \Rightarrow e_i)$. Since $c[x \mapsto p] \rightsquigarrow_h^* n$, say with w proper reduction steps, we have by Lemma 6.1.13 that $(x\vec{t})[x \mapsto p]$ reduces to some numeral m with fewer than w proper steps, and also that $e_m[x \mapsto p] \rightsquigarrow_h^* n$ with fewer than w proper steps. Hence by the induction hypothesis we have

$$(x\vec{t})[x \mapsto p'] \rightsquigarrow_h^* m\,, \qquad e_m[x \mapsto p'] \rightsquigarrow_h^* n\,.$$

Thus $c[x \mapsto p'] \rightsquigarrow_h^* n$ by Theorem 6.1.7, whence $e[x \mapsto p'] \rightsquigarrow_h^* n$ as required. □

Part (ii) of the above is an example of a *context lemma*: a statement that observational equivalence with respect to some restricted class of contexts (in this case, applicative contexts) implies observational equivalence with respect to arbitrary contexts. We will see in Section 7.1 that the above result readily implies the well-known context lemma for PCF.

Lemma 6.1.19 says that application is well-defined on $\sim_{\text{app}}$-classes; we therefore obtain an applicative structure $\mathsf{SP}^0/\sim_{\text{app}}$. Clearly this is a $\lambda\eta$-algebra, since composing the quotient map with the interpretation of typed λ-terms given in Subsection 6.1.3 yields an interpretation in $\mathsf{SP}^0/\sim_{\text{app}}$ satisfying the conditions of Definition 4.1.5 along with the η-rule.

The following facts about $\sim_{\text{app}}$ are now straightforward:

Proposition 6.1.20. *(i) If $p, p' \in \mathsf{SP}^0(\sigma \to \tau)$, then $p \sim_{\text{app}} p'$ iff $p \cdot r \sim_{\text{app}} p' \cdot r$ for all $r \in \mathsf{SP}^0(\sigma)$.*

(ii) If $p, p' \in \mathsf{SP}^0(\sigma)$, then $p \sim_{\text{app}} p'$ iff $r \cdot p = r \cdot p'$ for all $r \in \mathsf{SP}^0(\sigma \to \mathtt{N})$.

Proof. (i) is immediate. For (ii), one implication is given by Lemma 6.1.19(ii). For the converse, note that if $\sigma = \sigma_0, \ldots, \sigma_{t-1} \to \mathtt{N}$ and $q_i \in \mathsf{SP}(\sigma_i)$ for each i, then $r \cdot p = r \cdot p'$ with $r = \lambda x^\sigma . x\vec{q}$ gives us $p \cdot q_0 \cdot \ldots \cdot q_{t-1} = p' \cdot q_0 \cdot \ldots \cdot q_{t-1}$. □

Thus, the applicative structure SP^0 is *pre-extensional* in the sense of Exercise 3.6.13, and $\mathsf{SP}^0/\sim_{\text{app}}$ coincides with its extensional collapse with respect to the identity relation on $\mathbb{N}_\perp$. We henceforth refer to $\mathsf{SP}^0/\sim_{\text{app}}$ as the type structure of *(PCF-)sequential functionals*, and denote it by SF. We shall study SF as a model in its own right in Sections 7.5–7.7.

Exercise 6.1.21. Define a preorder $\preceq_{\text{app}}$ on each $\mathsf{SP}^0(\sigma)$ by

$$p \preceq_{\text{app}} p' \quad \text{iff} \quad \forall q_0, \ldots, q_{t-1}.\ p \cdot q_0 \cdot \ldots \cdot q_{t-1} \sqsubseteq p' \cdot q_0 \cdot \ldots \cdot q_{t-1} \,,$$

where $\sqsubseteq$ is the usual partial ordering on $\mathsf{SP}^0(\mathtt{N}) \cong \mathbb{N}_\perp$. By mimicking the proof of Lemma 6.1.19, show that if $p \preceq_{\text{app}} p'$ then $r \cdot p \preceq_{app} r \cdot p'$ for all $r \in \mathsf{SP}^0(\sigma \to \tau)$, and conversely if $r \cdot p \preceq r \cdot p'$ for all $r \in \mathsf{SP}^0(\sigma \to \mathtt{N})$ then $p \preceq_{\text{app}} p'$. Conclude that $\preceq_{\text{app}}$ induces a partial order structure $\preceq$ on SF which is pointwise at types $\sigma \to \tau$, and that all functions in SF are *monotone* with respect to $\preceq$.

6.1.5 Simulating NSPs in B

It is integral to our conception of the models $\mathsf{SP}, \mathsf{SP}^0$ that the application operation is not merely mathematically well-defined, but is itself 'sequentially computable' in some natural sense. One way of expressing this is to say that SP admits a simulation in the van Oosten model B (see Subsection 3.2.4), and one advantage of our presentation of NSPs is that it makes this fact relatively transparent. We now outline how such a simulation works, omitting tedious coding details. This also provides a convenient way of identifying the *effective submodels* of SP and SP^0.

We first indicate how NSP meta-procedures (and hence procedures) may be represented by partial functions $\mathbb{N} \rightharpoonup \mathbb{N}$ (that is, elements of B). Clearly, the syntax tree of a procedure P may be represented by a partial function f_P from *addresses* (i.e. positions in the tree) to *contents* (i.e. what appears in the tree at a given position).

Addresses may be taken to be finite sequences of *tags* much as in Subsection 6.1.2, although we shall here take our addresses to refer to arbitrary meta-term nodes in the syntax tree. (We should also ensure that we have a tag to indicate the subterm P within a meta-term $P\vec{Q}$.) The content at a given address should contain the following information:

- Whether the node in question is a meta-procedure, a meta-expression or a ground meta-term.
- If it is a meta-procedure node, the finite list of variables bound at this node.
- If it is a meta-expression node, whether it is a `case` meta-expression or a numeral n, and the value of n in the latter case.
- If it is a ground meta-term node, whether it consists of an expression or an application, whether it has a head variable x, and the identity of x in this case.

We do not include a content code for $\bot$; rather, an occurrence of $\bot$ in the tree is represented by the value of f_P being undefined at the relevant address. (In addition, f_P will be undefined on addresses that do not exist in the tree.) Modulo tedious coding, such a function f_P may be considered as a partial function $\mathbb{N} \rightharpoonup \mathbb{N}$.

Recall that the *head normal forms* are the meta-terms of shape $\bot$, n, $yQ_0 \ldots Q_{r-1}$ or `case` $yQ_0 \ldots Q_{r-1}$ `of` $(i \Rightarrow E_i)$, possibly prefixed by $\lambda\vec{x}$. By inspection of the reduction rules for meta-terms, it is informally clear that the following are computable within B (indeed within $\mathsf{B}^{\mathrm{eff}}$) given a code f_P along with an address α:

1. Whether the meta-term T at address α in P is already in head normal form. (This test may sometimes fail to terminate, but only in cases where T has no head normal form or has head normal form $\bot$.)
2. If T is not a head normal form, the code $f_{P'}$ of the meta-term P' obtained from P by performing a head reduction $T \rightsquigarrow_h T'$ at address α.
3. By iterating 1 and 2, the code $f_{P''}$ for the meta-term P'' obtained by reducing the meta-term T at α to head normal form. (The result of this step, being itself an element of B, will always be 'defined', though its value at α will be undefined if T has no head normal form.)

Given codes f_p, f_q for procedures $p \in \mathsf{SP}(\sigma \to \tau)$ and $q \in \mathsf{SP}(\sigma)$, we may now (within $\mathsf{B}^{\mathrm{eff}}$) compute the code f_r for $r = p \cdot q$ as follows. It suffices to say how to compute $f_r(\alpha)$ for some specified address α.

1. Build the code for the meta-procedure $\lambda x_1 \ldots x_r.e[x_0 \mapsto q]$, where $p = \lambda x_0 \ldots x_r.e$.
2. For each prefix β of α in turn (starting from $\beta = \varepsilon$), reduce the subterm at address β to head normal form.
3. Return the content at address α in the resulting code.

We leave the finer details to the conscientious reader. In summary, we have constructed a single-valued simulation $\gamma : \mathsf{SP} \longrightarrow\!\!\triangleright\, \mathsf{B}$ (where $f_p \Vdash^{\gamma} p$), with application at all types σ, τ realized by a single element of $\mathsf{B}^{\mathrm{eff}}$. Composing with the inclusion $\mathsf{SP}^0 \hookrightarrow \mathsf{SP}$, we obtain a simulation $\gamma^0 : \mathsf{SP}^0 \longrightarrow\!\!\triangleright\, \mathsf{B}$.

We say a sequential procedure p is *effective* if f_p is a computable partial function (that is, an element of $\mathsf{B}^{\mathrm{eff}}$). Since $\mathsf{B}^{\mathrm{eff}}$ is closed under application, so are the effective sequential procedures. We write $\mathsf{SP}^{\mathrm{eff}}$, $\mathsf{SP}^{0,\mathrm{eff}}$ for the effective substructures of

$\mathsf{SP}, \mathsf{SP}^0$ respectively. It is also easy to see that the interpretation of any λ-term (as in Subsection 6.1.3) is an effective NSP; hence $\mathsf{SP}^{\mathrm{eff}}$, $\mathsf{SP}^{0,\mathrm{eff}}$ are themselves $\lambda\eta$-algebras that come with simulations in $\mathsf{B}^{\mathrm{eff}}$ (and hence also in K_1).

We also obtain an effective submodel of the extensional quotient SF. Since every NSP is a supremum of effective ones, the definition of $\sim_{\mathrm{app}}$ in Subsection 6.1.4 retains the same meaning when the arguments q_i are restricted to $\mathsf{SP}^{0,\mathrm{eff}}$. All the results of Subsection 6.1.4 thus go through in the effective setting. We write $\mathsf{SF}^{\mathrm{eff}}$ for the extensional collapse of $\mathsf{SP}^{0,\mathrm{eff}}$; clearly this is an extensional λ-algebra in its own right, but we shall more typically view it as a substructure of SF.

6.1.6 Call-by-Value Types

We now comment briefly on the 'partial' or 'call-by-value' analogues of SP^0 and SF in the sense indicated in Section 4.3. The model SF, being extensional, falls within the scope of Theorem 4.3.5, and so has a canonical call-by-value counterpart for general reasons. The model SP^0, however, is too intensionally fine-grained for this general theory to be applicable. Indeed, SP^0 does not even carry a suitable partiality structure in the sense of Definition 4.2.14: the obvious candidate fails to satisfy the axiom $(fx) \upharpoonright z = (f(x \upharpoonright z)) \upharpoonright z$, because the difference between one and two evaluations of z is visible in the corresponding NSPs.

Certain types do nevertheless admit a natural 'call-by-value' interpretation within the Karoubi envelope of SP^0. For example, the following strictification operator, which will play an important role in Chapter 7, yields an idempotent on $\mathsf{SP}^0(1)$:

$$byval \;=\; \lambda f^{\mathtt{N}\to\sigma}.\lambda x^{\mathtt{N}}.\ \mathtt{case}\ x\ \mathtt{of}\ (i \Rightarrow \mathtt{case}\ f(\lambda.i)\ \mathtt{of}\ (j \Rightarrow j))\,.$$

The corresponding splitting gives us a call-by-value interpretation $\mathsf{SP}^0_v(1)$ for the type $\overline{1}$; we may thence obtain call-by-value interpretations for all *pure* types via $\mathsf{SP}^0_v(k+1) = \mathsf{SP}^0_v(k) \Rightarrow \mathsf{SP}^0(\mathtt{N})$.

To give a call-by-value interpretation for $\mathtt{N}^{(2)} \to \mathtt{N}$, however, a non-trival extension of the NSP model is needed: procedures must be allowed to continue with a further computation after returning some ground type information. This can indeed be achieved by suitably modifying our definition of NSPs, although space does not permit us to develop the idea here.

6.1.7 An Untyped Approach

To conclude this section, we briefly outline an alternative route to the construction of our models: we may build a single *untyped* λ-algebra of NSPs which is of some interest in its own right, then recover the typed model from its Karoubi envelope.

Untyped NSPs are defined coinductively by the following grammar, where x is understood to range over untyped variables. Note that every λ now binds an *infinite* sequence of variables x_i (where $x_0x_1\cdots$ stands for an arbitrary infinite sequence), and every application involves an infinite sequence of arguments q_i.

$$\begin{array}{lrl} \textit{Procedures:} & p,q ::= & \lambda x_0 x_1 \cdots .e \\ \textit{Expressions:} & d,e ::= & \bot \mid n \mid \texttt{case}\ a\ \texttt{of}\ (i \Rightarrow e_i \mid i \in \mathbb{N}) \\ \textit{Applications:} & a ::= & x q_0 q_1 \cdots \end{array}$$

The definition of meta-terms is adapted similarly. Since all variables are untyped, no typing constraints are needed to ensure that reduction works smoothly.

The definition of reduction is given as for typed NSPs, except that the β-rule is replaced by the following infinitary version whereby a λ-abstraction consumes an infinite sequence of arguments in a single step:

$$(\lambda x_0 x_1 \cdots .E) Q_0 Q_1 \cdots \rightsquigarrow_b E[\vec{x} \mapsto \vec{Q}] \,.$$

With these changes, the construction proceeds as before, and yields an untyped applicative structure $(\mathsf{USP},\cdot)$ with a substructure $(\mathsf{USP}^0,\cdot)$ of closed NSPs.

The proof of Theorem 6.1.7 now goes through with trivial adjustments, as does the verification that USP and USP^0 are (untyped) $\lambda\eta$-algebras. The only significant point is that in the absence of types, we have just a single notion of (infinitary) η-expansion for variables, given corecursively by

$$x^\eta = \lambda z_0 z_1 \cdots .\ \texttt{case}\ x z_0^\eta z_1^\eta \cdots\ \texttt{of}\ (i \Rightarrow i) \,.$$

(Our proof of Lemma 6.1.14 adapts immediately to the infinitary setting.)

We may now turn to the Karoubi envelope of USP^0 as defined in Chapter 4. Consider the element

$$r = \lambda x y_0 y_1 \cdots .\ x \bot \bot \cdots \,.$$

It is routine to check that $\lambda x. r \cdot (r \cdot x^\eta) = r$ in USP^0; also that $r \cdot (\lambda \vec{y}.n) = \lambda \vec{y}.n$, $r \cdot (\lambda \vec{y}.\bot) = \lambda \vec{y}.\bot$, and $r \cdot (\lambda \vec{y}.\texttt{case}\ y_i\ \texttt{of}\ (\cdots)) = \lambda \vec{y}.\bot$. Thus r is an idempotent whose fixed points correspond to elements of $\mathsf{SP}^0(\mathtt{N}) \cong \mathbb{N}_\bot$. Starting from this, we may obtain a typed $\lambda\eta$-algebra by repeated exponentiation in $\mathscr{K}(\mathsf{USP}^0)$. With a little effort, it can be shown that this coincides exactly with SP^0 as defined in Subsection 3.2.5. This shows that the typing constraints of Subsection 3.2.5 are not an ad hoc imposition, but fall out inevitably from a general construction.

A surprising amount of the theory of NSPs can be developed in terms of USP^0 with no reference to a type structure. For example, USP^0 admits a natural definition of applicative equivalence $\sim_{\mathrm{app}}$, and the context lemma goes through as before; we thus obtain a quotient $\mathsf{USP}^0/\sim_{\mathrm{app}}$, which turns out to be an extensional lambda algebra and hence a $\lambda\eta$-model (see [9, chapter 5]). The Karoubi envelope of this algebra, in turn, yields an extensional typed model which can be shown to coincide with SF. Thus, the equivalence relations $\sim_{\mathrm{app}}$ on the sets $\mathsf{SP}^0(\sigma)$ can all be seen as restrictions of a single equivalence relation.

6.2 NSPs and Kleene Computability

We now make the connection with Kleene computability as studied in Chapter 5. We shall work here with the λ-calculus presentation of Kleene computability, both because λ-terms mesh relatively well with the calculus of meta-terms from Subsection 6.1.1, and so that our results are applicable to arbitrary simple types. In particular, we shall show that the interpretation of Kleene expressions in total type structures **A** as described in Chapter 5 can be factored through an interpretation in NSPs (given independently of **A**). One may therefore think of the NSP arising from a Kleene expression as an abstract embodiment of the computational procedure or 'algorithm' that the expression represents.

As in Subsection 5.1.3, we shall write Klex for the language of ordinary Kleene expressions. We shall also augment this to a language Klex^{Ω} of *oracle Kleene expressions* by adding a constant $C_f : \mathtt{N} \rightarrow \mathtt{N}$ for every mathematical partial function $f : \mathbb{N} \rightharpoonup \mathbb{N}$ with the obvious interpretation clause: $[\![C_f]\!](n) = m$ iff $f(n) = m$. Note, however, that the Gödel numbering $\lceil - \rceil$ used in connection with $Eval_{\sigma}$ is applicable only to ordinary Kleene expressions.

6.2.1 NSPs and Kleene Expressions

We now show how terms of Klex or Klex^{Ω} may be interpreted as sequential procedures. In Subsection 7.1.1 a corresponding interpretation for PCF will be given, and we will show how it relates to the present interpretation.

The idea is to give a recipe for 'growing' an NSP from a given Kleene expression from the root downwards, much in the spirit of Böhm trees for the untyped λ-calculus. We make use here of the notion of η-expansion of NSP variables from Subsection 6.1.3. Note that the first three clauses in the following definition simply recapitulate the interpretation of λ-terms from Subsection 6.1.3.

Definition 6.2.1. Given an expression $\Gamma \vdash M : \sigma$ of Klex^{Ω}, we coinductively generate a well-typed sequential procedure $\Gamma \vdash [M]^S : \sigma$ as follows (for readability we omit empty λ-abstractions):

- $[x]^S = x^{\eta}$.
- $[\lambda x.M]^S = \lambda x.[M]^S$.
- $[MN]^S = [M]^S \cdot [N]^S$.
- $[\widehat{0}]^S = 0$.
- $[Suc(M)]^S = \ll \mathtt{case}\ [M]^S\ \mathtt{of}\ (i \Rightarrow i+1) \gg$.
- $[Primrec(M,N,P)]^S = \ll \mathtt{case}\ [M]^S\ \mathtt{of}\ (j \Rightarrow R'_j\, [N]^S\, [P]^S) \gg$, where

$$R'_j \;=\; \lambda xf.\, \mathtt{case}\, x\, \mathtt{of}\; i_0 \Rightarrow \mathtt{case}\; f0i_0\; \mathtt{of}\; i_1 \Rightarrow \cdots\, \mathtt{case}\; f(j-1)i_{j-1}\; \mathtt{of}\; i_j \Rightarrow i_j\,.$$

- $[Eval_\sigma(M,N_0,\ldots,N_{n-1})]^S = \ll \mathtt{case}\ [M]^S\ \mathtt{of}\ (\lceil P \rceil \Rightarrow [PN_0 \ldots N_{n-1}]^S) \gg$.
- $[C_f]^S = \lambda x.\mathtt{case}\ x\ \mathtt{of}\ (i \Rightarrow f(i))$.

Clearly, if M is a term of Klex, then $[M]^S$ will be an effective NSP.

As noted in Subsection 5.1.4, ordinary closed Kleene expressions modulo β-equality constitute a λ-algebra Klex^0/β. We also write $\mathsf{Klex}^{0,\Omega}/\beta$ for the λ-algebra of closed oracle Kleene expressions modulo β-equality. Since SP^0 is also a λ-algebra (Theorem 6.1.18), we see immediately from the first three clauses above that $[-]^S$ yields a λ-algebra homomorphism (and hence a type-respecting simulation) $h : \mathsf{Klex}^{0,\Omega}/\beta \to \mathsf{SP}^0$; this restricts to a homomorphism $\mathsf{Klex}^0/\beta \to \mathsf{SP}^{0,\mathrm{eff}}$.

The following exercises relate this interpretation to the reduction system for Kleene expressions which we introduced briefly in Subsection 5.1.3. They will not play a role in what follows.

Exercise 6.2.2. Using Theorem 6.1.7, check that $[-]^S$ respects the reduction relations $\rightsquigarrow_0$, $\rightsquigarrow_1$ of Subsection 5.1.3. Deduce that the quotient $\mathsf{Klex}^0/\sim$ is non-degenerate, where $\sim$ is the congruence generated by $\rightsquigarrow_0$.

Exercise 6.2.3. Verify that the induced homomorphism $h^\sim : \mathsf{Klex}^0/\sim \to \mathsf{SP}^{0,\mathrm{eff}}$ is itself realizable in K_1: that is, if $\gamma : \mathsf{Klex}^0/\sim \longrightarrow\!\triangleright K_1$ is the simulation given by Gödel numbering and $\delta : \mathsf{SP}^{0,\mathrm{eff}} \to K_1$ a simulation as in Subsection 6.1.5, then $\gamma \preceq \delta \circ h^\sim$. Deduce that γ respects numerals in the sense of Definition 3.4.9.

Show also that $h : \mathsf{Klex}^{0,\Omega}/\sim \to \mathsf{SP}^0$ is realizable in $\mathsf{B}^{\mathrm{eff}}$, with respect to the simulation of Subsection 6.1.5 along with an evident simulation $\mathsf{Klex}^{0,\Omega}/\sim \longrightarrow\!\triangleright \mathsf{B}$.

In Subsection 7.1.6 we will obtain the following important theorem:

Theorem 6.2.4 (Definability). *(i) Every $p \in \mathsf{SP}^0(\sigma)$ arises as $[M]^S$ for some closed $M : \sigma$ in Klex^Ω.*

(ii) Every $p \in \mathsf{SP}^{0,\mathrm{eff}}(\sigma)$ arises as $[M]^S$ for some closed $M : \sigma$ in Klex. *Moreover, a suitable term M may be computed from a program generating p.*

This can be proved directly from the definition of $[-]^S$, but it will be somewhat more convenient to obtain it via the corresponding result for PCF. Specifically, we shall prove that every NSP is definable in a certain language $(\mathrm{PCF} + \mathit{byval})^\Omega$ (and likewise in the effective setting), then establish a translation from $(\mathrm{PCF} + \mathit{byval})^\Omega$ to Klex^Ω that preserves the NSP semantics. Theorem 6.2.4 will not be used in the rest of this chapter, except tangentially in Exercise 6.4.11.

6.2.2 Interpreting NSPs in Total Models

Next, we show how the interpretation of Kleene expressions in a total type structure $\mathbf{A}$ (as in Subsection 5.1.3) may be factored through the above interpretation $[-]^S$. We do this by giving a direct (partial) interpretation of NSP meta-terms in $\mathbf{A}$.

Definition 6.2.5. Let $\mathbf{A}$ be any total type structure over $\mathbb{N}$. We define the following relations by simultaneous induction:

$$[\![E]\!]_\nu = n\,, \qquad [\![G]\!]_\nu = n\,, \qquad [\![P]\!]_\nu(\vec{\Phi}) = n\,, \qquad [\![P]\!]^*_\nu = F\,,$$

where E, G, P range over well-typed meta-expressions, ground meta-terms and meta-procedures respectively; ν ranges over finite valuations of variables in $\mathbf{A}$; $n \in \mathbb{N}$; $\Phi_i \in \mathbf{A}(\sigma_i)$ and $F \in \mathbf{A}(\sigma)$ where P has type $\sigma = \sigma_0, \ldots, \sigma_{r-1} \to \mathtt{N}$.

- $[\![n]\!]_\nu = n$.
- If $\nu(x) = \Phi$, $[\![Q_j]\!]^*_\nu = F_j$ for $j < r$ and $\Phi(F_0, \ldots, F_{r-1}) = n$, then

$$[\![xQ_0 \ldots Q_{r-1}]\!]_\nu = n\,.$$

- If $[\![Q_j]\!]^*_\nu = F_j$ for $j < r$ and $[\![P]\!]_\nu(F_0, \ldots, F_{r-1}) = n$, then

$$[\![PQ_0 \ldots Q_{r-1}]\!]_\nu = n\,.$$

- If $[\![G]\!]_\nu = m$ and $[\![E_m]\!]_\nu = n$, then

$$[\![\mathtt{case}\ G\ \mathtt{of}\ (i \Rightarrow E_i)]\!]_\nu = n\,.$$

- If $[\![E]\!]_\nu = n$ regarding E as a meta-expression, then also $[\![E]\!]_\nu = n$ regarding E as a ground meta-term.
- If $P = \lambda x_0, \ldots, x_{r-1}.E$ and $[\![E]\!]_{\nu, x_0 \mapsto \Phi_0, \ldots, x_{r-1} \mapsto \Phi_{r-1}} = n$, then $[\![P]\!]_\nu(\vec{\Phi}) = n$.
- If $F \in \mathbf{A}(\sigma)$ and $[\![P]\!]_\nu(\vec{\Phi}) = F(\vec{\Phi})$ for all $\vec{\Phi}$, then $[\![P]\!]^*_\nu = F$.

A fortiori, we thus have a partial interpretation of NSPs in $\mathbf{A}$. Combining this with the translation of Definition 6.2.1, we may now recover our original interpretation of Kleene expressions from Subsection 5.1.3. The full proof of the following is tedious and lengthy; we give here only the essential ingredients.

Theorem 6.2.6. *Suppose $\Gamma \vdash M : \sigma$ is a β-normal Kleene expression, and ν a valuation for Γ within $\mathbf{A}$.*
(i) For any $\vec{\Phi}$, we have $[\![M]\!]_\nu(\vec{\Phi}) = a$ iff $[\![[M]^S]\!]_\nu(\vec{\Phi}) = a$.
(ii) $[\![M]\!]^*_\nu = F$ iff $[\![[M]^S]\!]^*_\nu = F$.

Proof. (Sketch.) The left-to-right implications are proved simultaneously by induction on the generation of statements $[\![M]\!]_\nu(\vec{\Phi}) = a$, $[\![M]\!]^*_\nu = F$ as in Definition 5.1.15. Likewise, the right-to-left implications are established simultaneously by induction on the generation of statements $[\![[M]^S]\!]_\nu(\vec{\Phi}) = a$, $[\![[M]^S]\!]^*_\nu = F$ as in Definition 6.2.5, allied with a case split on the syntactic form of M. In both cases, we use that if M is β-normal, any applications within M are of the form $xN_0 \ldots N_{r-1}$. We omit the details, contenting ourselves with marshalling some key facts that are needed:

1. $[\![x^\eta]\!]^*_\nu = \nu(x)$. This is established by an easy induction on the type of x.
2. $x^\eta \cdot q_0 \cdot \ldots \cdot q_{r-1} = \lambda \vec{z}.\,\mathtt{case}\ xq_0 \ldots q_{r-1}\vec{z}^\eta\ \mathtt{of}\ (j \Rightarrow j)$. This is an easy exercise in the use of Lemmas 6.1.14 and 6.1.15.

3. If $[\![e]\!]_\nu = n$ and $[\![d_n]\!]_\nu = a$, then $[\![\ll \mathtt{case}\ e\ \mathtt{of}\ (j \Rightarrow d_j) \gg]\!]_\nu = a$. This is readily seen from the definition of $[\![-]\!]_\nu$ plus the fact that $\ll \mathtt{case}\ e\ \mathtt{of}\ (j \Rightarrow d_j) \gg$ is obtained from e by replacing each leaf j by d_j.
4. If $\Gamma \vdash p : \overline{0},\overline{0} \to \overline{0}$ and $m,n \in \mathbb{N}$, then $[\![p \cdot (\lambda.m) \cdot (\lambda.n)]\!]_\nu() \simeq [\![p]\!]_\nu(m,n)$. This is easily shown by comparing the generation of the values of the two sides. □

Exercise 6.2.7. Check that Theorem 6.2.6 still holds if Kleene expressions are allowed to contain the construct *Min* of Subsection 5.1.5, where we define

$$[Min(M)]^S = \ll \mathtt{case}\ [M\widehat{0}]^S\ \mathtt{of}\ (0 \Rightarrow 0 \mid - \Rightarrow \mathtt{case}\ [M\widehat{1}]^S\ \mathtt{of}\ (0 \Rightarrow 1 \mid - \Rightarrow \cdots)) \gg .$$

Corollary 6.2.8 (Adequacy). *If $M : \mathtt{N}$ is a closed Kleene expression, then $M \rightsquigarrow_0^* \widehat{n}$ iff $[M]^S = n$.* □

Proof. By Theorem 5.1.23, we have $M \rightsquigarrow_0^* \widehat{n}$ iff $[\![M]\!]() = n$. By Theorem 6.2.6, this holds iff $[\![[M]^S]\!]() = n$, and by Definition 6.2.5, this is true iff $[M]^S = n$. □

To a large extent, the above results allow us to dispense with Kleene expressions and work instead at the more abstract level of NSPs. However, we shall still wish to refer on occasion to Kleene expressions, e.g. in order to identify the class of such expressions that corresponds to some *substructure* of SP^0.

To complete the picture, we show that the interpretation in $\mathbf{A}$ interacts well with meta-term evaluation, and in particular with NSP application. We shall say that a meta-term T has *order* $\delta \in \mathbb{N}$ if for every subterm $P\vec{Q}$ within T, the $\vec{Q}$ have type level $< \delta$ (so that P has type level $\leq \delta$).

Theorem 6.2.9. *(i) Let G be a ground meta-term of finite order, and ν a valuation in $\mathbf{A}$ for the free variables of G. Then $[\![G]\!]_\nu \preceq [\![\ll G \gg]\!]_\nu$.*

(ii) If $p \in \mathsf{SP}^0(\sigma_0,\ldots,\sigma_{r-1} \to \mathtt{N})$, $q \in \mathsf{SP}^0(\sigma_0)$ and $[\![q]\!]^*$ is defined, then for any $\vec{\Phi}$ we have

$$[\![p \cdot q]\!](\Phi_1,\ldots,\Phi_{r-1}) \succeq [\![p]\!]([\![q]\!]^*, \Phi_1,\ldots,\Phi_{r-1}) .$$

In particular, if also $[\![p]\!]^*$ is defined then $[\![p \cdot q]\!]^* = [\![p]\!]^*([\![q]\!]^*)$.

Proof. (i) The key idea is that any true assertion $[\![G]\!]_\nu = n$ will have a well-founded derivation via the inductive clauses of Definition 6.2.5, whether or not G itself is well-founded. We shall write $[\![G]\!]_\nu =_\xi n$ (respectively $[\![G]\!]_\nu =_{<\xi} n$) to mean that $[\![G]\!]_\nu = n$ via a derivation of ordinal height ξ (respectively $< \xi$); we also use similar annotations for assertions $[\![P]\!]_\nu^* = F$.

As a preliminary lemma, we claim that if $C = \mathtt{case}\ g\ \mathtt{of}\ (i \Rightarrow e_i)$ where g, e_i are terms, then $[\![C]\!]_\nu \preceq [\![\ll C \gg]\!]_\nu$. To show this, we note that if $[\![C]\!]_\nu = n$ then $[\![g]\!]_\nu =_\xi m$ for some m and ξ, and we argue by induction on ξ. If g is a numeral (necessarily m) then $[\![\ll C \gg]\!]_\nu = [\![e_m]\!]_\nu = n$, and if $g = x\vec{Q}$ then C is already a normal form. So suppose $g = \mathtt{case}\ y\vec{q}\ \mathtt{of}\ (j \Rightarrow d_j)$, where $[\![y\vec{q}]\!]_\nu = l$. Then $[\![d_l]\!]_\nu =_{<\xi} m$, so by the induction hypothesis and the definition of $[\![-]\!]_-$, we have

$$\begin{aligned}[\![C]\!]_\nu &\simeq [\![\mathtt{case}\ d_l\ \mathtt{of}\ (i \Rightarrow e_i)]\!]_\nu \simeq [\![\ll \mathtt{case}\ d_l\ \mathtt{of}\ (i \Rightarrow e_i) \gg]\!]_\nu \\ &\preceq [\![\mathtt{case}\ y\vec{q}\ \mathtt{of}\ (j \Rightarrow \ll \mathtt{case}\ d_j\ \mathtt{of}\ (i \Rightarrow e_i) \gg)]\!]_\nu \simeq [\![\ll C \gg]\!]_\nu .\end{aligned}$$

We now establish the following claims by simultaneous induction on $\delta \in \mathbb{N}$:

1. If G is a ground meta-term of order δ, then $[\![G]\!]_\nu = n$ implies $[\![\ll G \gg]\!]_\nu = n$.
2. If P is a meta-procedure of order δ, then $[\![P]\!]^*_\nu = F$ implies $[\![\ll P \gg]\!]^*_\nu = F$.

We shall prove these claims for δ assuming that they hold for all $\varepsilon < \delta$, so that there is no need for a separate base case. We argue by an inner induction on the height ξ of the derivations of assertions $[\![G]\!]_\nu = n$ and $[\![P]\!]^*_\nu = F$.

Suppose both claims hold for all meta-terms of order $< \delta$, and for all order δ meta-terms yielding assertions of height $< \xi$. To establish claim 1 for δ and ξ, we suppose $[\![G]\!]_\nu =_\xi n$ where G is of order δ, and show by cases that $[\![\ll G \gg]\!]_\nu = n$.

If G is a numeral (necessarily n), this is trivial, and the case $G = \bot$ is impossible.

If $G = \texttt{case}\ G'\ \texttt{of}\ (i \Rightarrow E_i)$, then $[\![G']\!]_\nu =_{<\xi} m$ for some m and $[\![E_m]\!]_\nu =_{<\xi} n$. Let $g = \ll G' \gg$ and $e_m = \ll E_m \gg$; then by the inner induction hypothesis we have $[\![g]\!]_\nu = m$ and $[\![e_m]\!]_\nu = n$. Thus $[\![\texttt{case}\ g\ \texttt{of}\ (i \Rightarrow e_i)]\!]_\nu = n$, so by our lemma we have $[\![\ll \texttt{case}\ g\ \texttt{of}\ (i \Rightarrow e_i) \gg]\!]_\nu = n$. But $\ll \texttt{case}\ g\ \texttt{of}\ (i \Rightarrow e_i) \gg = \ll G \gg$ by Corollary 6.1.11, so the conclusion is established.

If $G = x\vec{Q}$ where $\nu(x) = \Psi$, then for each j we have an assertion $[\![Q_j]\!]^*_\nu =_{<\xi} F_j$, where $\Psi(\vec{F}) = n$. So by the inner induction hypothesis we have $[\![\ll Q_j \gg]\!]^*_\nu = F_j$ for each j, whence $[\![\ll G \gg]\!]_\nu \simeq [\![x \ll Q_0 \gg \cdots \ll Q_{r-1} \gg]\!]_\nu = \Psi(\vec{F}) = n$.

If $G = P\vec{Q}$ where $P = \lambda\vec{x}.E$, then we have $[\![Q_j]\!]^*_\nu =_{<\xi} F_j$ for each j, and $[\![E]\!]_{\nu,\vec{x}\mapsto\vec{F}} =_{<\xi} n$. Let $q_j = \ll Q_j \gg$ for each j and $e = \ll E \gg$; then by the inner induction hypothesis we have $[\![q_j]\!]^*_\nu = F_j$ and $[\![e]\!]_{\nu,\vec{x}\mapsto\vec{F}} = n$. But also by Corollary 6.1.11 we have

$$[\![\ll G \gg]\!]_\nu \ \simeq\ [\![\ll E[\vec{x} \mapsto \vec{Q}] \gg]\!]_\nu \ \simeq\ [\![\ll e[\vec{x} \mapsto \vec{q}] \gg]\!]_\nu \,.$$

Here $e[\vec{x} \mapsto \vec{q}]$ is of order $< \delta$, since $e, \vec{q}$ are normal forms and the q_j have type level $< \delta$. So by the outer induction hypothesis and compositionality of $[\![-]\!]_-$, we have

$$[\![\ll e[\vec{x} \mapsto \vec{q}] \gg]\!]_\nu \ \succeq\ [\![e[\vec{x} \mapsto \vec{q}]]\!]_\nu \ \simeq\ [\![e]\!]_{\nu,\vec{x}\mapsto\vec{F}} = n \,.$$

This completes the argument for claim 1.

For claim 2, suppose $[\![P]\!]^*_\nu =_\xi F$ where $P = \lambda\vec{x}.E$ is of order δ. Then for any argument list $\vec{\Phi}$ we have $[\![E]\!]_{\nu,\vec{x}\mapsto\vec{\Phi}} =_{<\xi} F(\vec{\Phi})$, so by the inner induction hypothesis we have $[\![\ll P \gg]\!]_\nu(\vec{\Phi}) = [\![\ll E \gg]\!]_{\nu,\vec{x}\mapsto\vec{\Phi}} = F(\vec{\Phi})$. But since $\vec{\Phi}$ is arbitrary, this shows that $[\![\ll P \gg]\!]^*_\nu = F$. The induction is now complete.

(ii) Suppose $p = \lambda\vec{x}.e$. Then by the definition of $p \cdot q$ we have

$$[\![p \cdot q]\!](\Phi_1, \ldots, \Phi_{r-1}) \ \simeq\ [\![\ll e[x_0 \mapsto q] \gg]\!]_\nu$$

where $\nu : x_i \mapsto \Phi_i$ for $1 \le i < r$. But also

$$[\![p]\!]([\![q]\!]^*, \Phi_1, \ldots, \Phi_{r-1}) \ \simeq\ [\![e]\!]_{\nu,x_0\mapsto[\![q]\!]^*} \ \simeq\ [\![e[x_0 \mapsto q]]\!]_\nu \,,$$

and the claimed results follow using part (i). □

6.3 Substructures of SP^0

The NSP model offers a considerable richness of structure which can be exploited to analyse aspects of Kleene computability (and indeed PCF) that are not readily accessible by other known means. In this section, we illustrate this by considering several natural *substructures* that can be discovered within SP^0. By a substructure, we here mean simply a class of procedures closed under application; in many cases, this class will form a higher-order model in its own right. Such substructures can represent clearly demarcated levels of computational or expressive power; a typical use for them is to show that some operation f is not definable in some sublanguage $\mathscr{L}$, by exhibiting a substructure that forms a model for $\mathscr{L}$ but does not contain f.

The principal substructures we shall consider are those consisting of:

- *finite* procedures, defined as in Section 6.1.1,
- *well-founded* sequential procedures,
- *left-well-founded* procedures: i.e. those in which infinite sequencing of `case` branchings is permitted, but infinitely deep nesting of applications is not,
- *left-bounded* procedures: those for which there is a global finite bound on the depth of nestings.

We shall establish the existence of these substructures in Subsections 6.3.1–6.3.3, then use them to obtain some non-trivial expressivity results in Subsections 6.3.4 and 6.3.5. For another result in this vein, see Theorem 7.7.2; we also anticipate that many other interesting substructures of SP^0 will be forthcoming.

6.3.1 Finite and Well-Founded Procedures

The *finite* NSP terms were defined in Section 6.1 as those generated by a certain finitary grammar, construed inductively. As mentioned there, we may construe such terms as ordinary NSP terms by identifying the conditional branching $(0 \Rightarrow e_0 \mid \cdots \mid r-1 \Rightarrow e_{r-1})$ with

$$(0 \Rightarrow e_0 \mid \cdots \mid r-1 \Rightarrow e_{r-1} \mid r \Rightarrow \bot \mid r+1 \Rightarrow \bot \mid \ldots)\,.$$

We shall write $\mathsf{SP}^{\mathrm{fin}}$, $\mathsf{SP}^{0,\mathrm{fin}}$ for the classes of finite procedures in SP, SP^0 respectively. The notion of a finite *meta-term* may also be defined in the analogous way.

Likewise, the *well-founded* NSP terms [resp. meta-terms] can be defined as those generated by the usual infinitary grammar for NSP terms [resp. meta-terms], but construed inductively rather than coinductively. Equivalently, the well-founded meta-terms may be characterized as those whose syntax tree contains no infinite descending paths. We shall write $\mathsf{SP}^{\mathrm{wf}}$, $\mathsf{SP}^{0,\mathrm{wf}}$ for the classes of well-founded procedures in SP, SP^0.

It is natural to consider finiteness and well-foundedness together, since the difference is merely that in the latter case, infinitary `case` branchings are admitted. We shall concentrate here on proving that the well-founded procedures are closed under application—it will be easily seen that the same proof (with mild simplifications) applies to finite procedures.

Recall that in the setting of typed NSPs, variables are considered as explicitly typed. Given any well-typed NSP meta-term T, each syntactic subterm of T may therefore be assigned a unique type from $\mathtt{T}^{\rightarrow}(\mathtt{N})$. By a *redex* within T we shall mean a subterm $P\vec{Q}$. As in Subsection 6.2.2, we will say T is of *order* $\delta \in \mathbb{N}$ if for every redex $P\vec{Q}$ within T, P has type level at most δ.

We shall need the following preliminary fact:

Lemma 6.3.1. *Suppose* $C = \mathtt{case}\ g\ \mathtt{of}\ (i \Rightarrow e_i)$, *where* g *is a well-founded expression or application and each* e_i *is well-founded. Then* $\ll C \gg$ *is well-founded. Furthermore, if* C *is finite then so is* $\ll C \gg$.

Proof. By induction on the structure of g. If $g = y\vec{q}$, then already $\ll C \gg = C$; and if g is $\perp$ or n, then C reduces in a single step to some well-founded e_i. So suppose $g = \mathtt{case}\ y\vec{q}\ \mathtt{of}\ (j \Rightarrow d_j)$ where $y, \vec{q}, d_j$ are well-founded; then C reduces in one step to

$$\mathtt{case}\ y\vec{q}\ \mathtt{of}\ (j \Rightarrow \mathtt{case}\ d_j\ \mathtt{of}\ (i \Rightarrow e_i))\ .$$

Now for each j we have that $\ll \mathtt{case}\ d_j\ \mathtt{of}\ (i \Rightarrow e_i) \gg$ is some well-founded expression f_j by the induction hypothesis. Hence $\ll C \gg = \mathtt{case}\ y\vec{q}\ \mathtt{of}\ (j \Rightarrow f_j)$ is well-founded. The finiteness claim is now clear by inspection. □

The crux of our analysis is the proof of the following lemma. Much of the content of this already featured in the proof of Theorem 6.2.9.

Lemma 6.3.2. *If* T *is any well-founded and well-typed meta-term of finite order, then* $\ll T \gg$ *is well-founded.*

Proof. By induction on the order of T. For $\delta \in \mathbb{N}$, let $\Phi(\delta)$ be the statement:

- Whenever $\Gamma \vdash T (: \tau)$, if T is well-founded of order δ, then $\ll T \gg$ is well-founded.

We shall show that if $\Phi(\varepsilon)$ for all $\varepsilon < \delta$, then $\Phi(\delta)$ (so that there is no need for a separate base case).

Assuming $\Phi(\varepsilon)$ for all $\varepsilon < \delta$, suppose $\Gamma \vdash T (: \tau)$ where T is well-founded of order δ. We will establish that $\ll T \gg$ is well-founded by an inner induction on the structure of T.

If T is of form $\perp$ or n, the result is trivial.

If $T = \mathtt{case}\ G\ \mathtt{of}\ (i \Rightarrow E_i)$, then by the inner induction hypothesis both $\ll G \gg$ and every $\ll E_i \gg$ are well-founded. Let $C = \mathtt{case}\ \ll G \gg\ \mathtt{of}\ (i \Rightarrow \ll E_i \gg)$. Then by Lemma 6.3.1, $\ll C \gg$ is well-founded; but $\ll C \gg = \ll T \gg$ using Corollary 6.1.11.

If $T = x\vec{Q}$, then by the inner induction hypothesis each $q_i = \ll Q_i \gg$ is well-founded, and clearly $\ll T \gg = x\vec{q}$.

If $T = P\vec{Q}$ (the critical case), then again $p = \ll P \gg$ and each $q_i = \ll Q_i \gg$ are well-founded by the inner induction hypothesis. Suppose $p = \lambda \vec{x}.e$. Since $P\vec{Q}$ is well-typed, so is $p\vec{q}$ and hence so is $e[\vec{x} \mapsto \vec{q}]$. Note also that the q_i have level $< \delta$. We now see that $e[\vec{x} \mapsto \vec{q}]$ has order $< \delta$: indeed, both e and the q_i are normal forms, so the only redexes within $e[\vec{x} \mapsto \vec{q}]$ are of the form $q_i\vec{r}$, where $\mathrm{lv}(q_i) < \delta$. Furthermore, $e[\vec{x} \mapsto \vec{q}]$ is evidently well-founded. So by the outer induction hypothesis, its value is well-founded. But clearly $\ll T \gg = \ll e[\vec{x} \mapsto \vec{q}] \gg$ using Theorem 6.1.7. Thus $\ll T \gg$ is well-founded as required.

Finally, if $T = (\lambda\vec{x}.E)$ is well-founded of order δ then so is E, so $\ll E \gg$ is well-founded by the inner induction hypothesis. Hence $\ll T \gg = \lambda\vec{x}. \ll E \gg$ is well-founded. □

Theorem 6.3.3. $\mathsf{SP}^{0,\mathrm{wf}}$ *is closed under application.*

Proof. If $p,q \in \mathsf{SP}^0$ are well-founded and of suitable type then so is $\lambda\vec{v}.pq\vec{v}$. Moreover, this meta-term is of finite order, since the only redex appearing within it is $pq\vec{v}$ itself. Hence by Lemma 6.3.2 and Proposition 6.1.15, $p \cdot q = \lambda\vec{v}. \ll pq\vec{v} \gg$ is well-founded. □

Analogous but slightly simpler arguments apply to *finite* sequential procedures. Indeed, it is easy to see that Lemma 6.3.2 goes through with 'well-founded' replaced everywhere by 'finite'. We thus obtain:

Theorem 6.3.4. $\mathsf{SP}^{0,\mathrm{fin}}$ *is closed under application.* □

The substructure $\mathsf{SP}^{0,\mathrm{fin}}$ is perhaps of limited interest by itself, since the elements k,s of SP^0 are clearly not finite. In the well-founded case, however, we have:

Theorem 6.3.5. *(i)* $\mathsf{SP}^{0,\mathrm{wf}}$ *is a higher-order submodel of* SP^0.

(ii) $\mathsf{SP}^{0,\mathrm{wf}}$ *admits an interpretation of closed System* T *terms such that if* $M \rightsquigarrow M'$ *then* M, M' *denote the same procedure.*

Proof. (i) Immediate from Theorem 6.3.3, since the procedures $k_{\sigma\tau}$ and $s_{\rho\sigma\tau}$ in SP^0 given at the end of Subsection 6.1.3 are well-founded. (Note that the η-expansion $x^{\sigma\eta}$ of a typed variable x^σ is clearly well-founded.)

(ii) To define the interpretation, it suffices to specify well-founded procedures for the System T constants $\hat{0}$, suc, rec_σ. For $\hat{0}$ and suc these are obvious. For rec_σ, we may inductively define a well-founded procedure $R_{\sigma,j}$ for each $j \in \mathbb{N}$ as follows:

$$R_{\sigma,0}[x,f,\vec{z}] = x\vec{z}^{\eta} ,$$
$$R_{\sigma,j+1}[x,f,\vec{z}] = f(\lambda\vec{w}.R_{\sigma,j}[x,f,\vec{w}])(\lambda.j)\vec{z}^{\eta} .$$

We may then define

$$R_\sigma = \lambda xfn\vec{z}.\ \mathtt{case}\ n\ \mathtt{of}\ (j \Rightarrow R_{\sigma,j}[x,f,\vec{z}]) .$$

Clearly R_σ is well-founded (although not of finite depth), and suffices for the interpretation of rec_σ. The soundness with respect to the reduction rules from Example 3.2.2 is now easily checked. □

If desired, one can obviously extend the above theorem to the language T+$\bot$, where $\bot : \mathtt{N}$ is a constant denoting a non-terminating expression, for which we may add the reduction rule $\bot \rightsquigarrow \bot$.

Remark 6.3.6. (i) Our proof of the closure of finite procedures under application uses the principle of induction for Π^0_2 statements, since the finite analogue of $\Phi(\varepsilon, \delta)$ (where ε and δ may be taken to be finite) is an example of such a statement. Whilst this appears to be the 'natural' method of proof, we believe that a more finitistic treatment is possible, although the challenge of giving a palatable such treatment is left for future work. As with pure simply typed λ-terms, it appears that the natural bound on the size of $p \cdot q$ in terms of that of p and q falls just outside the class of Kalmár elementary functions.

(ii) Whilst it might seem that our proofs of Theorems 6.3.3 and 6.3.4 rely heavily on the simple type structure, this is not in fact essential to the results themselves. Indeed, one can prove that even in the untyped model USP^0 (see Subsection 6.1.7), the finite and well-founded procedures each form a class closed under application.[1] This is done by showing that all such procedures can be annotated with countable ordinals that play a role analogous to their type level.

6.3.2 Left-Well-Founded Procedures

Suppose $p \in \mathsf{SP}$ and π is some path through its syntax tree. If $\mathtt{case}\ a\ \mathtt{of}\ (i \Rightarrow e_i)$ is a subterm appearing along π, we shall say π takes a *left branch* at this subterm if it descends into a, and a *right branch* if it descends into some e_i. Left and right branches within meta-terms may be defined analogously.

Clearly, if p is not well-founded, then p contains an infinite path π which must involve either infinitely many left branches or infinitely many right branches (perhaps both). Intuitively, there is a considerable difference between these two situations: an infinite sequence of right branches simply means that an infinite sequence of subcomputations may be performed sequentially, whilst an infinite sequence of left branches corresponds to the somewhat more striking phenomenon of infinitely deep nesting. The following condition on NSPs is thus a natural one to consider:

Definition 6.3.7. We say $p \in \mathsf{SP}^0$ is *left-well-founded*, or LWF, if no path π through the syntax tree of p involves infinitely many left branches. Equivalently, p is LWF if the tree of application subterms within p is well-founded.

LWF procedures thus represent a level of computational power intermediate between that of general procedures and well-founded ones. We shall write $\mathsf{SP}^{0,\mathrm{lwf}}$ for the class of left-well-founded elements of SP^0. Our main result in this subsection will be that $\mathsf{SP}^{0,\mathrm{lwf}}$ is closed under application, and hence is a higher-order submodel

[1] This contrasts sharply with the situation for Böhm trees in untyped lambda calculus (see [9]): the term $M = \lambda x.f(xx)$ has a finite Böhm tree, but the Böhm tree of MM is infinite.

of SP^0. It will follow easily that $\mathsf{SP}^{0,\mathrm{lwf}}$ is a model for System T extended with the minimization operator.

For the purpose of this subsection, we shall extend our usual (coinductive) grammar of meta-terms to allow *metavariables* X to appear as meta-expressions:

$$\textit{Meta-expressions:}\quad D,E ::= X \mid \bot \mid n \mid \mathtt{case}\ G\ \mathtt{of}\ (i \Rightarrow E_i \mid i \in \mathbb{N})$$

However, we also impose a restriction on where these metavariables may occur. By a *rightward position* in the syntax tree of a meta-term, we shall mean a node reached by a path involving no ground meta-term nodes. A *well-formed* meta-term will be one in which every occurrence of any metavariable X is at a rightward position.

For non-well-formed meta-terms such as $(\lambda x.X)(\lambda.0)$, there is a difficulty with β-reduction, as we have given no definition of substitution for metavariables. However, by inspection of the reduction rules, it is easy to check that the usual definition of $\rightsquigarrow^*$ is sound for well-formed meta-terms. Moreover:

Proposition 6.3.8. *If T is well-formed and $T \rightsquigarrow^* T'$, then T' is well-formed. Hence $\ll T \gg$ is well-defined and well-formed for any well-formed T.* □

We shall now define a *left-well-founded meta-term* to be a well-formed meta-term in which no path has infinitely many left branches. A meta-term is *ordinary* if it contains no metavariables.

As mentioned earlier, we may generate all *well-founded* meta-terms by construing our grammar inductively rather than coinductively, regarding each production as a 'formation rule' for meta-terms. To generate all *left-well-founded* meta-terms, however, we need to add one further formation rule, which we now explain.

Given a family $\mathscr{F}$ of well-formed meta-expressions, along with a chosen *root* $F \in \mathscr{F}$ and a function ξ mapping certain metavariables X to meta-expressions $\xi(X) \in \mathscr{F}$, we may define the possibly infinite *plugging* $\Pi(\mathscr{F},F,\xi)$ as follows: begin with F, then repeatedly expand each metavariable X to $\xi(X)$ whenever $X \in \mathrm{dom}\,\xi$. Somewhat more formally, we may take $F_0 = F$, then inductively obtain F_{i+1} from F_i by expanding X to $\xi(X)$ whenever $X \in \mathrm{dom}\,\xi$. Finally, we define $\Pi(\mathscr{F},F,\xi) = \bigsqcup F_i^-$, where each F_i^- is obtained from F_i by replacing all metavariable occurrences by $\bot$. It is easy to see that $\Pi(\mathscr{F},F,\xi)$ is well-formed.

In conjunction with the usual grammar rules, this plugging operation suffices to generate all left-well-founded terms:

Proposition 6.3.9. *The class of LWF meta-terms (with metavariables) is inductively generated by means of the following formation rules.*

1. *X, $\bot$ and n are always LWF as meta-expressions.*
2. *If E is LWF, then so is $\lambda\vec{x}.E$.*
3. *If G and each E_i are LWF, then $\mathtt{case}\ G\ \mathtt{of}\ (i \Rightarrow E_i)$ is LWF as a meta-expression.*
4. *If E is LWF as a meta-expression and is ordinary, then E is LWF as a ground meta-term.*
5. *If P and each Q_i are LWF and ordinary, then $x\vec{Q}$ and $P\vec{Q}$ are LWF.*
6. *(Plugging rule.) If $\mathscr{F}$ consists of LWF meta-expressions, $F \in \mathscr{F}$ and $\mathrm{ran}\,\xi \subseteq \mathscr{F}$, then $\Pi(\mathscr{F},F,\xi)$ is LWF.*

Proof. Clearly the class of LWF meta-terms is closed under the formation rules 1–5. To see that it is closed under rule 6, note that if $\mathscr{F}$ consists of LWF meta-terms and π is an infinite path through $\Pi(\mathscr{F},F,\xi)$, then either π is eventually a path entirely within some $E \in \mathscr{F}$, in which case π involves only finitely many left branches since E is LWF, or π passes through infinitely many meta-terms in $\mathscr{F}$, each appearing at a rightward position in $\Pi(\mathscr{F},F,\xi)$, whence π involves no left branches. Thus, every meta-term generated by the above formation rules is LWF.

For the converse, suppose T is some well-formed meta-term not generated by rules 1–6. We will show how to construct a path through T with infinitely many left branches. Clearly, it suffices to consider the case where T is itself a meta-expression. First, assign a separate metavariable to every rightward meta-expression node within T except the root, and use these to disconnect T into a set $\mathscr{F}$ of well-formed fragments of the form $\mathtt{case}\ G\ \mathtt{of}\ (i \Rightarrow X_i)$, with a valuation ξ for metavariables and a distinguished root fragment F, so that $T_0 = \Pi(\mathscr{F},F,\xi)$. Since this is an application of rule 6, it follows that there must be some fragment $\mathtt{case}\ G\ \mathtt{of}\ (i \Rightarrow X_i)$ in $\mathscr{F}$ that is not generated by rules 1–6. But this must mean that G is not generated by these rules, which in turn means that G must contain a meta-expression T' that is not so generated; note that the position of T' in T involves a left branch. By recursively applying the same construction starting from T', we hence obtain a path with infinitely many left branches in T. □

In principle, clause 3 in the above definition is now subsumed by clause 5, but it is pleasant to retain it anyway. The main point is that our inductive generation of left-well-founded terms will allow us to carry out a syntactic analysis similar to the one for well-founded terms. Obviously, in the presence of the plugging rule, the same left-well-founded term may be generated in many different ways, but this will not matter for our purposes.

As in Subsection 6.3.1, we say a meta-term T has *order* δ if for every redex $P\vec{Q}$ within T, the type level of P is at most δ. Next, we note that Lemma 6.3.1 may be strengthened as follows:

Lemma 6.3.10. *Suppose* $C = \mathtt{case}\ g\ \mathtt{of}\ (i \Rightarrow e_i)$, *where* g, e_i *are LWF normal forms. Then* $\ll C \gg$ *is LWF.*

Proof. Let g' be obtained from g by replacing each numeral occurrence n in rightmost position with e_n. Clearly g' is LWF. To see that $g' = \ll C \gg$, note first that if g is well-founded then this can be proved as in Lemma 6.3.1. For the general case, write g as a limit $\bigsqcup_n g_n$ of well-founded terms, and let $C_n = \mathtt{case}\ g_n\ \mathtt{of}\ (i \Rightarrow e_i)$ so that $C = \bigsqcup_n C_n$; then $g'_n = \ll C_n \gg$ for each n, whence $g' = \ll C \gg$ by the evident continuity of $\ll - \gg$. □

We also require a new lemma specific to our plugging apparatus:

Lemma 6.3.11. *If* $\mathscr{F},F,\xi$ *are as above and* $\ll E \gg$ *is left-well-founded for each* $E \in \mathscr{F}$, *then* $\ll \Pi(\mathscr{F},F,\xi) \gg$ *is left-well-founded.*

Proof. Let $D = \Pi(\mathscr{F},F,\xi)$. Suppose π is some infinite path through $\ll D \gg$, and let ρ be the evident reduction sequence starting from D that evaluates the nodes of

$\ll D \gg$ along π in turn, in the sense of successively reducing the relevant subterms to head normal form. Clearly, for any node α appearing in π, some finite initial portion of ρ suffices to evaluate the nodes up to and including α, and the remainder of ρ then concerns the evaluation of nodes below α.

Moreover, it is easy to see that ρ is shadowed in an obvious way by some reduction ρ' starting from F, in which we allow ourselves the additional *expansion* rules $X \rightsquigarrow \xi(X)$. Using Proposition 6.3.8 and arguing by induction on initial segments of ρ', it is clear that every meta-term appearing in ρ' is a meta-expression fragment; hence every expansion occurs at a rightward position.

We wish to show that π does not involve infinitely many left branches. There are two cases to consider.

Case 1: ρ' involves infinitely many expansion steps. Then given any node α of $\ll D \gg$ appearing along π, we may find an initial portion of ρ' that suffices to evaluate D up to α, and an occurrence of an expansion after this portion. But this expansion occurs at a rightward position and below α, so α itself is at a rightward position. Since α was arbitrary, this shows that π consists entirely of right branches.

Case 2: ρ' involves only finitely many expansion steps. Set $E = F$ if ρ' contains no expansions at all, or $E = \xi(X)$ if $X \rightsquigarrow \xi(X)$ is the last expansion featuring in ρ'. Then clearly, all subsequent reductions within ρ' simply amount to reductions of E not involving any expansions. So beyond a certain point, π is simply a path through $\ll E \gg$. But $\ll E \gg$ is LWF by hypothesis, and so π contains only finitely many left branches. □

A related but much easier fact is the following:

Lemma 6.3.12. *If $e, q_0, q_1, \ldots$ are ordinary LWF normal forms, then the meta-term $e[\vec{x} \mapsto \vec{q}]$ is LWF.*

Proof. Similar to the proof that LWF meta-terms are closed under rule 6 (see Proposition 6.3.9). □

With all these facts in hand, the analogue of Lemma 6.3.2 now goes through:

Lemma 6.3.13. *If T is any left-well-founded and well-typed meta-term of finite order, then $\ll T \gg$ is left-well-founded.*

Proof. Closely analogous to Lemma 6.3.2. Let $\Psi(\delta)$ be the statement:

> Whenever $\Gamma \vdash T(:\tau)$, if T is left-well-founded of order δ, then $\ll T \gg$ is left-well-founded.

We show that if $\Psi(\varepsilon)$ for all $\varepsilon < \delta$ then $\Psi(\delta)$, and the result follows by induction.

To establish this, we argue by an inner induction on the generation of LWF terms according to Proposition 6.3.9. The argument is parallel to that in Lemma 6.3.2, augmented with an induction step for plugging as supplied by Lemma 6.3.11. Note that the induction case for `case` meta-expressions uses Lemma 6.3.10, and the one for meta-terms $P\vec{Q}$ involves an appeal to Lemma 6.3.12. □

By analogy with Theorem 6.3.3, we now have:

Theorem 6.3.14. $\mathsf{SP}^{0,\mathrm{lwf}}$ *is closed under application.* □

Let us write $\mathsf{SP}^{0,\mathrm{lwf}}$ for the substructure of SP^0 consisting of left-well-founded procedures. Since $\mathsf{SP}^{0,\mathrm{wf}}$ contains k and s combinators and interpretations of the System T constants (Theorem 6.3.5), the same is true for $\mathsf{SP}^{0,\mathrm{lwf}}$. Furthermore, there is an evident LWF (but non-well-founded) procedure for the minimization operator $min : \overline{2}$ (cf. Exercise 6.2.7):

$$\lambda f.\,\mathtt{case}\ f0\ \mathtt{of}\ (0 \Rightarrow 0 \mid - \Rightarrow \mathtt{case}\ f1\ \mathtt{of}\ (0 \Rightarrow 1 \mid - \Rightarrow \cdots))\,.$$

We thus have:

Theorem 6.3.15. $\mathsf{SP}^{0,\mathrm{lwf}}$ *is a higher-order submodel of* SP^0 *and admits a natural interpretation of System* $\mathrm{T}+min$. □

Exercise 6.3.16. Extend the operational semantics of System T given in Example 3.2.2 to one for $\mathrm{T}+min$, and show that our interpretation in $\mathsf{SP}^{0,\mathrm{lwf}}$ is sound with respect to this semantics.

6.3.3 Left-Bounded Procedures

Continuing in the same vein, we now draw attention to another submodel of SP^0, related to $\mathsf{SP}^{0,\mathrm{lwf}}$ but smaller. Let $\mathsf{SP}^{0,\mathrm{lbd}}$ consist of all *left-bounded* closed sequential procedures, i.e. those procedures p for which there is some global bound $d \in \mathbb{N}$ on the nesting depth of applications within p. We refer to the least such bound as the *left-depth* of p, or $ld(p)$. We shall show that left-bounded procedures form a well-behaved submodel that supports *μ-computable* operations in the sense of Subsection 5.1.5.

Theorem 6.3.17. $\mathsf{SP}^{0,\mathrm{lbd}}$ *is a higher-order submodel of* SP^0.

Proof. It is easy to show by induction on σ that the infinitary η-expansion $x^{\sigma\eta}$ of x^σ is left-bounded, and hence that k,s are left-bounded. For closure under application, we shall proceed by assigning a 'depth measure' to certain NSP meta-terms, and then showing that NSP reduction is non-increasing with respect to this measure.

First, define a function $D : \mathbb{N}^3 \to \mathbb{N}$ by the following double recursion:

$$\begin{aligned} D(0,a,b) &= a+b\,, \\ D(k+1,0,b) &= b\,, \\ D(k+1,a+1,b) &= D(k,b,D(k+1,a,b)) + D(k+1,a,b) + 1\,. \end{aligned}$$

It is routine to check that D is monotone in each of its arguments.

We now define a left-depth measure $ld(T)$ for certain NSP *meta-terms* T (possibly with plugging metavariables X) by means of the following clauses. Strictly speaking, the measure is associated with a particular way of generating T via the inductive clauses of Proposition 6.3.9 (we shall not need the fact that different ways of generating the same T yield the same measure).

$$\begin{aligned}
ld(\bot) \;=\; ld(n) \;=\; ld(X) &= 0\,, \\
ld(\lambda\vec{x}.E) &\simeq ld(E)\,, \\
ld(\texttt{case}\; G\; \texttt{of}\; (i \Rightarrow E_i)) &\simeq \max(ld(G), \max_i ld(E_i)) \;\text{ if this is finite,} \\
ld(xQ_0 \ldots Q_{r-1}) &\simeq 1 + \max_i ld(Q_i)\,, \\
ld(pQ_0 \ldots Q_{r-1}) &\simeq D(k, ld(p), \max_i ld(Q_i))\,, \\
&\qquad \text{where } k \text{ is the maximum type level of the } Q_i\,, \\
ld(\Pi(\mathscr{F}, F, \xi)) &\simeq \max_{E \in \mathscr{F}} ld(E) \;\text{ if this is finite.}
\end{aligned}$$

It is easy to check that if p is a procedure, this definition of $ld(p)$ agrees with the one given at the start of this subsection.

Clearly, if $p, \vec{q}$ are left-bounded procedures and $p\vec{q}$ is well-typed, then $ld(p\vec{q})$ is defined. We shall show that in this case

$$ld(\ll p\vec{q} \gg) \;\leq\; ld(p\vec{q})\,.$$

The closure of $\mathsf{SP}^{0,\mathrm{lbd}}$ under application then follows easily. We argue by induction on the maximum type level of the q_i, which we call k.

For $k = 0$, suppose $p = \lambda\vec{x}.e$. Then the x_i all have type $\overline{0}$, so any application subterm within e involving some x_i is simply x_i itself. It is thus clear that

$$ld(e[\vec{x} \mapsto \vec{q}]) \;\leq\; ld(e) + \max_i ld(q_i) \;=\; D(0, ld(e), \max_i ld(q_i)) \;=\; ld(p\vec{q})\,.$$

Moreover, $e[\vec{x} \mapsto \vec{q}]$ contains no proper (i.e. non-nullary) β-redexes; hence neither does any $\rightsquigarrow^*$-reduct of it, since the remaining rules cannot create such redexes. But if $T \rightsquigarrow T'$ via a reduction that is not a proper β-redex, it is easy to see that $ld(T') \leq ld(T)$. It follows that $ld(\ll e[\vec{x} \mapsto \vec{q}] \gg) \leq ld(e[\vec{x} \mapsto \vec{q}])$, since if $\ll e[\vec{x} \mapsto \vec{q}] \gg$ were to contain a nested sequence of (say) d applications, then so would some finite reduct of $e[\vec{x} \mapsto \vec{q}]$. But $\ll e[\vec{x} \mapsto \vec{q}] \gg = \ll p\vec{q} \gg$, so $ld(\ll p\vec{q} \gg) \;\leq\; ld(p\vec{q})$.

Now assume the induction claim holds for k, and suppose $p\vec{q}$ is a β-redex where the q_i have maximum type level $k+1$. Again let $p = \lambda\vec{x}.e$; we shall also write * for the substitution $[\vec{x} \mapsto \vec{q}]$. Let $b = \max_i ld(q_i)$. We first show by induction on the bottom-up generation of left-bounded terms that if t is left-bounded then $ld(t^*) \leq D(k+1, ld(t), b)$. The non-trivial cases are the following:

- If $t = y\vec{p'}$ where y is not among the x_i, we have

$$\begin{aligned}
ld(t^*) = ld(y(\vec{p'}^*)) \;&=\; 1 + \max_j ld(p'^*_j) \\
&\leq 1 + \max_j D(k+1, ld(p'_j), b) \;\text{ by induction hypothesis} \\
&= 1 + D(k+1, \max_j ld(p'_j), b) \;\text{ since } D \text{ is monotone} \\
&\leq D(k+1, 1 + \max_j ld(p'_j), b) \;\text{ by the definition of } D \\
&= D(k+1, ld(t), b)\,.
\end{aligned}$$

- If $t = x_i\vec{p'}$, we have

$$\begin{aligned} ld(t^*) &= ld(q_i(\vec{p'}^*)) \\ &\leq D(k, ld(q_i), \max_j ld(p'^*_j)) \;\text{ since each } p'^*_j \text{ has type level } \leq k \\ &\leq D(k, b, \max_j D(k+1, ld(p'_j), b)) \;\text{ by induction hypothesis} \\ &= D(k, b, D(k+1, \max_j ld(p'_j), b)) \;\text{ since } D \text{ is monotone} \\ &\leq D(k+1, 1+\max_j ld(p'_j), b) \;\text{ by definition of } D \\ &= D(k+1, ld(t), b)\,. \end{aligned}$$

Since $ld(p\vec{q}) = D(k+1, ld(e), b)$, we may therefore conclude that $ld(e^*) \leq ld(p\vec{q})$.

Since $\ll p\vec{q} \gg = \ll e^* \gg$, it now suffices to show that

$$ld(\ll e^* \gg) \;\leq\; ld(e^*)\,.$$

Here we argue by induction on $ld(e)$. If $ld(e) = 0$ then $e[\vec{x} \mapsto \vec{q}] = e = \ll e \gg$ so the claim is trivial. Otherwise, consider an arbitrary outermost subterm of form $x_i\vec{p'}$ within e. By Theorem 6.1.7, the outer and inner induction hypotheses, and the monotonicity of D, we have

$$ld(\ll q_i\vec{p'}^* \gg) \;=\; ld(\ll q_i \ll \vec{p'}^* \gg\gg) \;\leq\; ld(q_i \ll \vec{p'}^* \gg) \;\leq\; ld(q_i\vec{p'}^*)\,.$$

Let $e^\dagger$ be obtained from e^* by replacing all such outermost subterms $q_i\vec{p'}^*$ with $\ll q_i\vec{p'}^* \gg$. Then the above yields $ld(e^\dagger) \leq ld(e^*)$ via an inductive construction of e from the relevant subterms $x_i\vec{p'}$. But $\ll e^* \gg = \ll e^\dagger \gg$ using Corollary 6.1.11, so it now suffices to show that $ld(\ll e^\dagger \gg) \leq ld(e^\dagger)$. But this is clear since $e^\dagger$ contains no β-redexes, hence neither do any of its reducts, so the argument we used in the case $k = 0$ again applies here. □

Clearly the standard interpretations of $\widehat{0}$, *suc* are left-bounded, as are $k_{\sigma\tau}$, $s_{\rho\sigma\tau}$ and the element $min \in \mathsf{SP}^0(\overline{2})$ as given by Exercise 6.2.7. By contrast, the procedures R_σ for rec_σ given in the proof of Theorem 6.3.5 are not left-bounded. However, for the purpose of modelling Kleene's *Primrec*, the following form of primitive recursion (already introduced in Definition 6.2.1) provides a satisfactory left-bounded alternative:

$$\begin{aligned} R'_{0,j} &= \lambda xf.\, \mathtt{case}\, x\, \mathtt{of}\; i_0 \Rightarrow\; \mathtt{case}\; f0i_0\; \mathtt{of}\; i_1 \Rightarrow \cdots \mathtt{case}\; f(j-1)i_{j-1}\; \mathtt{of}\; i_j \Rightarrow i_j\,, \\ R'_0 &= \lambda xfn.\, \mathtt{case}\, n\, \mathtt{of}\; (j \Rightarrow R'_{0,j}xf)\,. \end{aligned}$$

We thus have everything we need to interpret $\mathsf{Klex}^{\mathrm{min}}$, the language of μ-computable functionals (see Subsection 5.1.5):

Theorem 6.3.18. *Let* $[\![-]\!]$ *be the interpretation of* $\mathsf{Klex}^{\mathrm{min}}$ *in* SP^0 *given as in Definition 6.2.1 and Exercise 6.2.7. Then* $[\![M]\!] \in \mathsf{SP}^{0,\mathrm{lbd}}$ *for every closed* M. □

By Theorem 6.2.6 and Exercise 6.2.7, the term M and the procedure $[\![M]\!]$ have the same denotation in any total type structure $\mathbf{A}$.

In fact, the relationship between left-boundedness and μ-computability is much tighter than this: we will show here that in a somewhat innocuous extension of $\mathsf{Klex}^{\mathrm{min}}$, *every* effective left-bounded procedure is definable. This result will be used in Subsection 6.4.3 to show that in the type structures of primary interest, the μ-computable elements are precisely those denotable by effective left-bounded procedures.

For this purpose, we introduce an element $plug \in \mathsf{SP}^0(2)$, which may be regarded as an internal counterpart of the plugging rule in Proposition 6.3.9. Informally, *plug* assembles a right-branching tree from a function f that codes what each individual node should look like:

$$\begin{aligned} plug = \lambda f.\ & \mathtt{case}\ f\langle\rangle\ \mathtt{of}\ (2i_0 \Rightarrow i_0 \mid 2i_0+1 \Rightarrow \\ & \mathtt{case}\ f\langle i_0\rangle\ \mathtt{of}\ (2i_1 \Rightarrow i_1 \mid 2i_1+1 \Rightarrow \\ & \mathtt{case}\ f\langle i_0,i_1\rangle\ \mathtt{of}\ (2i_2 \Rightarrow i_2 \mid 2i_2+1 \Rightarrow \cdots)))\ . \end{aligned}$$

Clearly, *plug* is effective and left-bounded with depth 1, but not well-founded. An intuition for the use of *plug* may be gained from the following exercise.

Exercise 6.3.19. Show that the procedure $min \in \mathsf{SP}^0(\overline{2})$ is definable from *plug* in $\mathsf{Klex}^{\mathrm{prim}}$, the language of Kleene primitive recursion (see Subsection 5.1.5).

We write $\mathsf{Klex}^{\mathrm{plug}}$ for the extension of $\mathsf{Klex}^{\mathrm{prim}}$ with a constant *plug*, and $[\![-]\!]$ for its interpretation in SP. Although *plug* itself does not appear to be definable in $\mathsf{Klex}^{\mathrm{min}}$, we will see in Subsection 6.4.3 that the denotation of *plug* in any *extensional* partial type structure over $\mathbb{N}_\bot$ is $\mathsf{Klex}^{\mathrm{min}}$ definable. This is the sense in which $\mathsf{Klex}^{\mathrm{plug}}$ is an innocuous extension of $\mathsf{Klex}^{\mathrm{min}}$.

We now give the main theorem. By the *code* $\lceil p \rceil$ for an NSP $p \in \mathsf{SP}(\sigma)$, we shall here mean the element of $\mathsf{SP}^0(1)$ representing the partial function $f_p : \mathbb{N} \rightharpoonup \mathbb{N}$ that codes the syntax tree of p, as defined in Subsection 6.1.5.

Theorem 6.3.20. *Every element $p \in \mathsf{SP}^{0,\mathrm{lbd}}(\sigma)$ is $\mathsf{Klex}^{\mathrm{plug}}$ definable relative to $\lceil p \rceil$.*

Proof. We will prove the following claim by induction on d: for each type σ, there is a $\mathsf{Klex}^{\mathrm{plug}}$ program $E^d_\sigma : \overline{1} \to \sigma$ such that for any procedure $p \in \mathsf{SP}^{0,\mathrm{lbd}}(\sigma)$ of left-depth $\leq d$, we have $[\![E^d_\sigma]\!] \cdot \lceil p \rceil = p$.

For $d = 0$, the construction of E^d_σ is trivial, since the only procedures of left-depth 0 are of the form $\lambda\vec{x}.\bot$ or $\lambda\vec{x}.n$. So suppose $d > 0$ where we have already constructed suitable terms E^{d-1}_τ for all τ, and consider an arbitrary $p = \lambda\vec{x}.e \in \mathsf{SP}^{0,\mathrm{lbd}}(\sigma)$ of left-depth $\leq d$. By considering the rightward branching structure of e and using the plugging machinery of Subsection 6.3.2, we may express e as a plugging of meta-expression fragments $F_{\vec{i}}$, one for each rightward path $\vec{i} = i_0, \ldots, i_{t-1}$ in e, where each $F_{\vec{i}}$ is either a constant n or of the form

$$\mathtt{case}\ x_m q_0 \ldots q_{s-1}\ \mathtt{of}\ (j \Rightarrow X_{\vec{i},j})$$

where $q_0, \ldots, q_{s-1}$ are of left-depth $\leq d-1$ and may contain free variables from $\vec{x}$. (The intention, of course, is that we expand each metavariable $X_{\vec{i}}$ to $F_{\vec{i}}$, starting from the root X_ε.) Moreover, the following information is clearly $\mathsf{Klex}^{\mathrm{prim}}$ computable from $\lceil p \rceil$ and $\vec{i}$:

- Whether $F_{\vec{i}}$ is a constant or a case expression, and the identity of the corresponding number n or variable x_m as appropriate.
- For a case expression, the list of codes $\lceil \lambda\vec{x}.q_0 \rceil, \ldots, \lceil \lambda\vec{x}.q_{s-1} \rceil$.

Using this along with suitable programs $E^{d-1}_{\vec{\tau}}$ and a case split on the finitely many possible variables x_m, we now see that the following procedure $H(\lceil p \rceil, \vec{i}, \vec{x})$ is $\mathsf{Klex}^{\mathrm{plug}}$ computable uniformly in $\lceil p \rceil$, $\vec{i}$ and the variables $\vec{x}$:

$$H(\lceil p \rceil, \vec{i}, \vec{x}) \;=\; \begin{cases} 2n & \text{if } F_{\vec{i}} \equiv n\,, \\ \mathtt{case}\; x_m\vec{q}\; \mathtt{of}\; (j \Rightarrow 2j+1) & \text{if } F_{\vec{i}} \equiv \mathtt{case}\; x_m\vec{q}\; \mathtt{of}\; (j \Rightarrow X_{\vec{i},j})\,. \end{cases}$$

It follows that p itself is $\mathsf{Klex}^{\mathrm{plug}}$ computable uniformly in $\lceil p \rceil$ via the program

$$\lambda\vec{x}.\, plug\, (\lambda\langle\vec{i}\rangle.\, H(\lceil p \rceil, \vec{i}, \vec{x}))\,. \qquad \square$$

In particular, every *effective* left-bounded procedure is $\mathsf{Klex}^{\mathrm{plug}}$ definable.

Finally, we draw attention to a subtlety pertaining to the semantics of $\mathsf{Klex}^{\mathrm{min}}$. The terms of $\mathsf{Klex}^{\mathrm{min}}$ correspond superficially to those of $\mathrm{T}_0 + min$, where T_0 is the sublanguage of System T with rec_0 as the sole primitive recursor. However, it would be misleading to say we have an interpretation of $\mathrm{T}_0 + min$ in $\mathsf{SP}^{0,\mathrm{lbd}}$: whereas the procedure R_0 of Theorem 6.3.5 accords with the operational semantics of rec_0 as given in Example 3.2.2, the above procedure R_0' does not. For example, in $\mathrm{T}_0 +$ min we have $rec_0 \perp (\lambda xy.\widehat{0})\,\widehat{1} \rightsquigarrow^* 0$ where $\perp = min(\lambda z.\widehat{1})$, but the corresponding $\mathsf{Klex}^{\mathrm{min}}$ computation diverges. The relationship between these two kinds of primitive recursion will be further clarified by Theorem 6.3.23.

Exercise 6.3.21. (i) Let $\mathsf{SP}^{0,\mathrm{prim}}$ denote the class of *Kleene primitive recursive* closed NSPs, i.e. those arising from terms of $\mathsf{Klex}^{\mathrm{prim}}$. Show that $\mathsf{SP}^{0,\mathrm{prim}}$ is closed under application, and that all procedures in $\mathsf{SP}^{0,\mathrm{prim}}$ are well-founded, left-bounded and $\perp$-free. Note that $\mathsf{SP}^{0,\mathrm{prim}}$ is a *total* computability model over $\mathbb{N}$.

(ii) Show that these properties also hold for the class $\mathsf{SP}^{0,\mathrm{prim+}}$ of closed NSPs that are Kleene primitive recursive relative to the following 'strong definition by cases' operator $D \in \mathsf{SP}^0(0,1,1 \to 0)$:

$$D \;=\; \lambda xfg.\, \mathtt{case}\; x\; \mathtt{of}\; (0 \Rightarrow \mathtt{case}\; f0\; \mathtt{of}\; (i \Rightarrow i) \mid 1 \Rightarrow \mathtt{case}\; g0\; \mathtt{of}\; (i \Rightarrow i))\,.$$

6.3.4 *Expressive Power of* $\mathrm{T} + min$ *and* $\mathsf{Klex}^{\mathrm{min}}$

We are now ready to present some applications of these substructures of SP^0, showing that they shed some interesting light on the computational strength of certain programming languages. We have seen that SP^0 supports an interpretation of the language Klex (which will turn out to be equivalent to PCF), whilst the more restricted models $\mathsf{SP}^{0,\mathrm{lwf}}$ and $\mathsf{SP}^{0,\mathrm{lbd}}$ admit interpretations of the weaker languages $\mathrm{T}+min$ and $\mathsf{Klex}^{\min}$ respectively. Here we shall compare the expressive power of these three languages, and the following picture will emerge:

- For first-order types $\mathtt{N}^{(r)} \rightarrow \mathtt{N}$, all three languages have the same power: even $\mathsf{Klex}^{\min}$ suffices for defining all effective sequential functions.
- At second order, there are sequential functionals definable in T but not in $\mathsf{Klex}^{\min}$, and functionals definable in Klex but not in $\mathrm{T}+min$. However, if we are only interested in the behaviour of second-order functionals F on (strict) *total* arguments g, then again all three languages have the same power.
- At third order, there are even 'total' functionals definable in T but not in $\mathsf{Klex}^{\min}$, and 'total' functionals definable in Klex but not $\mathrm{T}+min$.

We will concentrate here on questions of definability within SF itself, though our results will be formulated in such a way that they transfer easily to other models of interest, both total and partial. We shall reap some of the consequences for total models in the next subsection.

For types σ of level 1 (i.e. those of the form $\mathtt{N}^{(r)} \rightarrow \mathtt{N}$), the following establishes that our three languages are equally powerful. The theorem as stated is close to a result of Trakhtenbrot [287]. The proof anticipates a technique of 'simulating NSPs' that will play a major role in Subsection 7.1.5. The argument we give avoids reliance on the machinery of Theorem 6.3.20.

Theorem 6.3.22. *Every* $f \in \mathsf{SF}^{\mathrm{eff}}(\mathtt{N}^{(r)} \rightarrow \mathtt{N})$ *is definable in* $\mathsf{Klex}^{\min}$.

Proof. (Sketch.) Suppose $p \in \mathsf{SP}^0(\mathtt{N}^{(r)} \rightarrow \mathtt{N})$ is effective and represents f. Then p has the form $\lambda x_0 \ldots x_{r-1}.e$, where every expression node within e has the form n, $\bot$, or $\mathtt{case}\, x_i\, \mathtt{of}\, (\cdots)$ for some $i < r$. Moreover, a suitable code $\lceil p \rceil \in \mathsf{SP}(1)$ for p is programmable in $\mathsf{Klex}^{\min}$, say by a program $C : \overline{1}$. We shall use C to construct a $\mathsf{Klex}^{\min}$ program $P : \mathtt{N}^{(r)} \rightarrow \mathtt{N}$ that simulates p on any given tuple of arguments $\vec{x}$.

The task of P is to track the path through p that will be followed when we apply p to $\vec{x}$. First note that given any $j \in \mathbb{N}$, we may use primitive recursion to track this path for j steps, evaluating arguments x_i as and when necessary to determine which branch to choose, and returning a code for the node in p at which we end up (if any). We may now use *min* to search for the least j for which this node is a leaf n, and return n as the final result of P. It is then easy to check that P denotes f. □

A more delicate construction, which we do not give here, shows that in fact every effective *procedure* $p \in \mathsf{SP}^0(\mathtt{N}^{(r)} \rightarrow \mathtt{N})$ is the denotation of some $\mathsf{Klex}^{\min}$ program P according to the semantics given by Definition 6.2.1 and Exercise 6.2.7 (see also Theorem 6.3.18). This means that our three languages have the same expressive power even at the more fine-grained level of SP^0.

We now turn to the situation at second order. Here we shall show that all three of our languages differ in terms of the sequential functionals they define. We begin with a simple application of $\mathsf{SP}^{0,\mathrm{lbd}}$ which clarifies the relationship between the two versions of primitive recursion discussed at the end of Subsection 6.3.3.

Theorem 6.3.23. *The System* T *recursor* rec_0 *(regarded as an element of* SF*) is not definable in* $\mathsf{Klex}^{\min}$.

Proof. We first show that for any n, if the functional

$$F_n \;=\; \Lambda f_0 \ldots f_{n-1}.\, f_{n-1}(f_{n-2}(\cdots(f_0 0)\cdots)) \;\in\; \mathsf{SF}(1^{(n)} \to 0)$$

is *semi-represented* by an NSP p in the sense that $p \cdot \vec{q} \sqsupseteq F(\vec{f}) \in \mathbb{N}_\perp$ whenever $\vec{q}$ represents $\vec{f}$, then $ld(p) \geq n$. For $n = 0$ this is trivial. Assuming the claim holds at n, suppose $p = \lambda \vec{f}.e$ semi-represents F_{n+1}. Since $F_{n+1}(\lambda x.\bot)\cdots(\lambda x.\bot)(\lambda x.m) = m$ for any m, it is easy to see that e has the form $\texttt{case}\; f_n e' \;\texttt{of}\; (\cdots)$. We shall show that $p' = \lambda f_0 \ldots f_{n-1}.e'$ semi-represents F_n. Indeed, if for some $f_0, \ldots, f_{n-1} \in \mathsf{SF}(1)$ represented by $q_0, \ldots, q_{n-1}$ we had $p' \cdot \vec{q} \not\sqsupseteq F_n(\vec{f})$, then we could choose f_n such that $f_n(p' \cdot \vec{q}) = \bot$ but $f_n(F_n(\vec{f})) = F_n(\vec{f}) \neq \bot$. Hence if q_n represents f_n then $p \cdot \vec{q} = \bot$ whereas $F_{n+1}(\vec{f}) \neq \bot$, a contradiction. Now by the induction hypothesis, p' has depth $\geq n$, whence p has depth $\geq n+1$.

Since any tuple $f_0, \ldots, f_{n-1}$ may be assembled into a single function f where $f_i = f(i)$ for each i, the above shows that the function $G_n = \Lambda f.rec_0 \widehat{0} f \widehat{n}$ cannot be represented by a procedure of left-depth $< n$. But if rec_0 itself were represented by a procedure of left-depth d, this would readily yield a procedure for G_{d+1} of left-depth d, a contradiction. $\Box$

To show that $\mathrm{T} + min$ is less powerful than Klex at second order, we resort to a more complex argument using the model $\mathsf{SP}^{0,\mathrm{lwf}}$. We begin by adapting an idea from Berger [22] to obtain a useful property of LWF-representable functionals:

Theorem 6.3.24. *Let* $p \in \mathsf{SP}^{0,\mathrm{lwf}}((\mathtt{N}^{(r)} \to \mathtt{N}) \to \mathtt{N})$, *with* $[p]$ *the corresponding sequential functional. Then either* $[p]$ *is a constant function, or there exists* k *such that for all* $g \in \mathsf{SF}(\mathtt{N}^{(r)} \to \mathtt{N})$, *if* $[p](g)\downarrow$ *then for some* $m_0, \ldots, m_{r-1} < k$ *we have* $g(m_0, \ldots, m_{r-1})\downarrow$.

Proof. Let Δ be the tree consisting of all leftward paths through p (i.e. paths consisting entirely of left branches). Since p is LWF, all such paths terminate; moreover, any non-leaf expression node of Δ has the form $\texttt{case}\; g(q_0, \ldots, q_{r-1}) \;\texttt{of}\; (\cdots)$, so Δ is finitely branching of arity r. So by König's lemma, Δ is finite. We argue by induction on the height of Δ, associating a suitable bound $k(p)$ to every such $p \in \mathsf{SP}^{0,\mathrm{lwf}}((\mathtt{N}^{(r)} \to \mathtt{N}) \to \mathtt{N})$.

Suppose $p = \lambda g.e$. If e is a numeral j, set $k(p) = j+1$, and if $e = \bot$, set $k(p) = 1$. In either of these cases, $[p]$ is a constant function so the theorem holds.

If $e = \texttt{case}\; g(p_0, \ldots, p_{r-1}) \;\texttt{of}\; (\cdots)$, set $k(p) = \max_{i<r} k(\lambda g.e_i)$ where $p_i = \lambda.e_i$ for each i. We claim that if $[p]$ is not a constant function, then $k(p)$ has the property required by the theorem. Take any $g \in \mathsf{SF}(\mathtt{N}^{(r)} \to \mathtt{N})$ such that $[p](g)\downarrow$, and take $q \in \mathsf{SP}^{0,\mathrm{lwf}}(\mathtt{N}^{(r)} \to \mathtt{N})$ with $[q] = g$, so that pq evaluates to some numeral; then clearly $(g(p_0, \ldots, p_{r-1}))[g \mapsto q]$ evaluates to some numeral n. For each $i < r$, take $p_i' = p_i$ if p_i is of the form $\lambda.j$ or $\lambda.\bot$; otherwise take $p_i' = \lambda.x_i$ where x_i is a free variable of type $\mathtt{N}$. There are now two cases to consider:

1. $(g(p'_0,\ldots,p'_{r-1}))[g \mapsto q]$ evaluates to n. In this case, pick $m_0,\ldots,m_{r-1}$ by taking $m_i = j_i$ whenever p'_i has the form $\lambda.j_i$, and $m_i = 0$ otherwise. Then clearly $g(m_0,\ldots,m_{r-1}) = n$ with each $m_i < k$.
2. $(g(p'_0,\ldots,p'_{r-1}))[g \mapsto q]$ reduces to some head normal form $\mathtt{case}\ x_i\ \mathtt{of}\ (\cdots)$. Then clearly $(g(p_0,\ldots,p_{r-1}))[g \mapsto q]$ head reduces to $\mathtt{case}\ e_i[g \mapsto q]\ \mathtt{of}\ (\cdots)$, so $(\lambda g.e_i)q$ evaluates to a numeral. So by the induction hypothesis, there exist $m_0,\ldots,m_{r-1} < k(\lambda g.e_i) \le k$ with $g(m_0,\ldots,m_{r-1})\downarrow$. □

Corollary 6.3.25 (Berger). *Let* $H \in \mathsf{SF}((\mathtt{N},\mathtt{N} \to \mathtt{N}) \to \mathtt{N})$ *be the sequential function arising from the sequential procedure*

$$\lambda g.\ g(0,g(1,g(2,\cdots)))$$

(abbreviating $\mathtt{case}\ a\ \mathtt{of}\ (i \Rightarrow i)$ *by a). Then H is not definable in* $(\mathrm{T}+min)^\Omega$.

Proof. Suppose H were definable in $(\mathrm{T}+min)^\Omega$; then there would be an LWF procedure p with $H = [\![p]\!]$, so by Theorem 6.3.24 there would exist k such that for all $g \in \mathsf{SF}(\mathtt{N},\mathtt{N} \to \mathtt{N})$ we have

$$H(g)\downarrow\ \Rightarrow\ \exists m_0,m_1 < k.\ g(m_0,m_1)\downarrow\ .$$

Now define $g \in \mathsf{SF}(\mathtt{N},\mathtt{N} \to \mathtt{N})$ by

$$g(m_0,m_1)\ =\ \mathit{if}\ m_0 \ge k\ \mathit{then}\ k\ \mathit{else}\ (\mathit{if}\ m_1 \ge k\ \mathit{then}\ k\ \mathit{else}\ \bot)$$

where $\ge$ is understood to be strict. Then $H(g) = g(0,g(1,\ldots g(k,\bot)\ldots)) = k$, a contradiction. □

On the other hand, an easy exercise in the use of the second recursion theorem shows that the functional H above is definable in Klex. Alternatively, it will be trivial to define such functionals in PCF once the interpretation of the latter is in place (see Subsection 7.1.1).

In the example of Corollary 6.3.25, the fact that H can be distinguished from any LWF-representable functional depends crucially on the possibility of applying H to *partial* arguments g. However, if we are only interested in the behaviour of second-order functionals on (strict) *total* arguments, it turns out that $\mathsf{Klex}^{\mathrm{min}}$ is as good as Klex or even the much richer model $\mathsf{PC}^{\mathrm{eff}}$. The following fact is shown by Berger [22]; the proof can be seen as a miniature version of the proof of Theorem 8.6.1.

Proposition 6.3.26. *For any* $F \in \mathsf{PC}^{\mathrm{eff}}((\mathtt{N}^{(r)} \to \mathtt{N}) \to \mathtt{N})$*, there exists a functional* F' *definable in* $\mathsf{Klex}^{\mathrm{min}}$ *such that* $F'(g) = F(g)$ *whenever g is strict and total.*

Proof. Let $\lceil - \rceil$ be some natural coding of finite strict functions $h \in \mathsf{PC}(\mathtt{N}^{(r)} \to \mathtt{N})$ as natural numbers. We say $g \in \mathsf{PC}(\mathtt{N}^{(r)} \to \mathtt{N})$ *matches* $\lceil h \rceil \in \mathbb{N}$ iff $h \sqsubseteq g$; note that if g is total then the condition 'g matches $\lceil h \rceil$' can be tested within $\mathsf{Klex}^{\mathrm{prim}}$. Given F as above, let $\mathscr{F}$ denote the set of pairs (g,n) where $g \in \mathsf{PC}(\mathtt{N}^{(r)} \to \mathtt{N})$ is a finite strict function and $F(g) = n$. Let $c : \mathbb{N} \to \mathbb{N}$ be such that $c(0), c(1), \ldots$ is a computable enumeration of codes $\langle \lceil g \rceil, n \rangle$ for $(g,n) \in \mathscr{F}$. Now define F' by

$$F'(g) \;=\; snd(c(\mu i.\, g \text{ matches } fst(c(i))))\,.$$

It is easy to see that F' is definable in $(\mathsf{Klex}^{\min})^{\Omega}$, or $\mathsf{Klex}^{\min}$ in the effective case, and agrees with F on all strict and total g. □

Note that if the functional F in the above proposition is hereditarily extensional with respect to $\mathbb{N}$ (that is, $F \in \mathsf{Ext}(\mathsf{PC};\mathbb{N})$ in the notation of Subsection 3.6.3), then the associated F' will satisfy $F'(g) = F(g)$ for all total g, whether strict or not. Thus, every second-order functional in $\mathsf{Ct} = \mathsf{EC}(\mathsf{PC};\mathbb{N})$ with a representative in $\mathsf{PC}^{\mathrm{eff}}$ has a representative definable in $\mathsf{Klex}^{\min}$.

Let us now move on to the situation at third order, where our languages diverge even for 'total' objects. We shall discuss the following results (writing T_k for the sublanguage of T with just the recursors $rec_0, \ldots, rec_k$):

1. There are third-order 'total' functionals definable in T_1 but not in $\mathsf{Klex}^{\min}$.
2. There are third-order 'total' functionals definable in Klex but not in $\mathrm{T}+min$.

We shall prove statement 1 in detail by showing that rec_1 is such a functional, using the substructure $\mathsf{SP}^{0,\mathrm{lbd}}$ to establish the non-definability. The proof is complex, but illustrates a method which appears to be quite general and powerful, and which we expect to have further applications in the future. (The reader may wish to prepare for this proof by studying other non-definability arguments of a somewhat similar character, e.g. the proofs of Theorems 8.5.3 and 8.5.8.) We shall then briefly indicate how statement 2 can be established by a similar method, using the substructure $\mathsf{SP}^{0,\mathrm{lwf}}$ and taking a well-known *bar recursion* operator as our example.

We have so far not specified the precise concept of 'totality' we have in mind. The notions of *hereditarily total* and *hereditarily extensional* elements from Section 3.6 both offer reasonable candidates for such a concept; moreover, these in principle have a different meaning for each model over $\mathbb{N}_\perp$ in which our languages may be interpreted. In order to present our theorem in a strong form that will apply to all reasonable concepts of totality, we shall state it with reference to a certain core class of *strongly total* procedures that admit a well-behaved interpretation in all models of interest, and will be 'total' in all reasonable senses. For convenience, we take this to be the class $\mathsf{SP}^{0,\mathrm{prim}+}$ as defined in Exercise 6.3.21. We will often leave it to the reader to check that specific procedures asserted to belong to this class really do so.

Going beyond $\mathsf{SP}^{0,\mathrm{prim}+}$, one may also say that the System T_1 recursor rec_1 of type $\overline{1},(\overline{1},\overline{0} \to \overline{1}),\overline{0} \to \overline{1}$ is present in SP^0 and is total in all reasonable senses, e.g. $rec_1 \in \mathsf{Tot}(\mathsf{SP}^0;\mathbb{N}) \cap \mathsf{Ext}(\mathsf{SP}^0;\mathbb{N})$. However:

Theorem 6.3.27. *The functional rec_1 is not definable in $\mathsf{Klex}^{\min}$. More precisely, there is no $\Psi \in \mathsf{SP}^0$ definable in $\mathsf{Klex}^{\min}$ such that $\Psi \cdot f \cdot F \cdot m \cdot n = rec_1 \cdot f \cdot F \cdot m \cdot n$ even for all strongly total f, F, m, n.*

Proof. It will suffice to show that a simpler total functional $\Phi \in \mathsf{SP}^0(0,2 \to 0)$, readily definable from rec_1 in $\mathsf{Klex}^{\mathrm{prim}}$, does not agree with any $\mathsf{Klex}^{\min}$ definable Ψ on all arguments from $\mathsf{SP}^{0,\mathrm{prim}+}$. We define elements $\Phi_n \in \mathsf{SP}^0(3)$ and Φ by:

$$\begin{aligned}\Phi_n(F) &= F(\Lambda x_0.F(\Lambda x_1.\cdots F(\Lambda x_n.\langle x_0,\ldots,x_n\rangle)\cdots))\,,\\ \Phi(n)(F) &= \Phi_n(F)\,.\end{aligned}$$

It is an easy exercise to formalize this definition in T_1 using the recursor rec_1, and thence to interpret it in SP^0 as per Theorem 6.3.5. Note that each procedure Φ_n has left-depth $n+1$, so that Φ itself is not left-bounded.

We shall show that there is no left-bounded procedure Ψ that agrees with Φ on all $n \in \mathbb{N}$ and $F \in \mathsf{SP}^{0,\mathrm{prim}+}(2)$. This will suffice, since by Theorem 6.3.18 the interpretation of any $\mathsf{Klex}^{\mathrm{min}}$ term is left-bounded.

For this, it will be enough to show that Φ_d is not extensionally equivalent to any procedure of left-depth d. Suppose for contradiction Ψ_d were such a procedure; let F_0 be the procedure $\lambda g.g(0)$, and suppose $\Psi_d \cdot F_0 = c$. Our strategy will be to find a neigbourhood $\mathscr{F}$ of F_0 such that $\Psi_d \cdot F = c$ for all $F \in \mathscr{F}$, and an $F_1 \in \mathscr{F}$ such that $\Phi_d \cdot F_1 \neq c$.

Consider the computation of $\Psi_d \cdot F_0$. At top level, this follows a path through Ψ_d of the form

$$\lambda nF.\ \mathtt{case}\ F(g_0^0)\ \mathtt{of}\ u_0^0 \Rightarrow \cdots \Rightarrow \mathtt{case}\ F(g_{n^0-1}^0)\ \mathtt{of}\ u_{n^0-1}^0 \Rightarrow c\,,$$

where the g_i^0 are themselves type 1 procedures that may contain F free. (The superscript 0 indicates that we are analysing the computation at 'level 0'.) In order to define a suitable 'neighbourhood' of F_0 within which to work, choose some modulus $m^0 > n^0+1$, and consider the computations of $(\lambda F.g_i^0) \cdot F_0 \cdot z$ for all $i < n^0$ and $z < m^0$. (The condition on m^0 is a little hard to motivate at this point; its purpose will emerge in the construction of F_1 below.) Each of these computations must yield some natural number r_{iz}^0; note that $r_{i0}^0 = u_i^0$. For suppose $(\lambda F.g_i^0) \cdot F_0 \cdot z = \bot$, and let F' be the procedure $\lambda g.\ \mathtt{case}\ g(z)\ \mathtt{of}\ (j \Rightarrow F_0(g))$. Then it is easy to see that $\Psi_d \cdot F'$ is undefined, contradicting the fact that $\Phi_d \cdot F'$ is defined for all $F' \in \mathsf{SP}^{0,\mathrm{prim}+}(2)$.

At this point, it is convenient to record the information gleaned so far in the form of certain *neighbourhoods* of F_0. For each $i < n^0$, define

$$\begin{aligned}V_i^0 &= \{g \in \mathsf{SP}^0(1) \mid \forall z < m^0.\ g \cdot z = r_{iz}^0\}\,,\\ \mathscr{F}_i^0 &= \{F \in \mathsf{SP}^0(2) \mid \forall g \in V_i^0.\ F \cdot g = u_i^0\}\,.\end{aligned}$$

Clearly $F_0 \in \mathscr{F}_i^0$ for each i. The idea is that the $\mathscr{F}_i^0$ will form part of a system of neighbourhoods which suffice to ensure $\Psi_d \cdot F = c$ for all F in their intersection.

Proceeding to the next level, we repeat the above analysis for each of the computations of $(\lambda F.g_i^0) \cdot F_0 \cdot z$ where $i < n^0$ and $z < m^0$ (so that we are now considering subterms of left-depth $\leq d-1$). At top level, each of these traces a path through g_i^0 consisting of a finite sequence of applications of F, leading to the result r_{iz}^0. Taking all these computations together, let $g_0^1, \ldots, g_{n^1-1}^1$ denote the (occurrences of) type 1 procedures to which F is applied (which may contain F free), with $u_0^1, \ldots, u_{n^1-1}^1$ the corresponding outcomes. Although we will not explicitly track the fact in our notation, we should consider each of the g_i^1 as a 'child' of the procedure g_j^0 from which it arose; note that each occurrence g_i^1 corresponds to a path through the syntax tree of Ψ_d with two left branches.

As a suitable modulus for our analysis of the g_i^1, take $m^1 > n_0 + n_1 + 2$ with $m^1 \geq m^0$ (again, the reason for this choice will emerge later). As before, it is easy to see that $(\lambda F.g_i^1) \cdot F_0 \cdot z$ yields some number r_{iz}^1 for each $i < n^1$ and $z < m^1$. We now augment our collection of neighbourhoods by defining

$$V_i^1 = \{g \in \mathsf{SP}^0(1) \mid \forall z < m^1 .\, g \cdot z = r_{iz}^1\} ,$$
$$\mathscr{F}_i^1 = \{F \in \mathsf{SP}^0(2) \mid \forall g \in V_i^1 .\, F \cdot g = u_i^1\}$$

for each $i < n^1$; note once again that $F_0 \in \mathscr{F}_i^1$ because $r_{i0}^1 = u_i^1$.

We now repeat this analysis for each of the computations of $(\lambda f.g_i^1) \cdot F_0 \cdot z$ where $i < n^1$ and $z < m^1$ (so that we are now considering subterms of left-depth $\leq d-2$). Having identified the relevant type 1 procedures $g_0^2, \ldots, g_{n^2-1}^2$, we pick a modulus $m^2 > n^0 + n^1 + n^2 + 3$ with $m^2 \geq m^1$, and proceed as before. By this point, it is clear how to continue the construction to arbitrary depth, ending with a set of neighbourhoods $\mathscr{F}_i^{d-1}$ for $i < n^{d-1}$.

We may now define the *critical neighbourhood* $\mathscr{F} \subseteq \mathsf{SP}^0(2)$ by

$$\mathscr{F} = \bigcap_{w<d,\ i<n^w} \mathscr{F}_i^w .$$

Note that $F_0 \in \mathscr{F}$. We claim that the following hold for all $F \in \mathscr{F}$ and all $w < d$:

1. $(\lambda F.g_i^w) \cdot F \cdot z = r_{iz}^w$ for all $i < n^w$ and $z < m^w$.
2. $(\lambda F.F(g_i^w)) \cdot F = u_i^w$ for all $i < n^w$.
3. $\Psi_d \cdot F = c$.

Claims 1 and 2 here are shown simultaneously by reverse induction on w, and one more application of the same argument also establishes claim 3.

We now embark on the construction of our counterexample $F_1 \in \mathscr{F}$. First some notation: for any $F \in \mathsf{SP}^0(2)$ and any $x_0, \ldots, x_{w-1} \in \mathbb{N}$ where $w \leq d$, define $g_{(x_0,\ldots,x_{w-1})}^F \in \mathsf{SP}^0(1)$ as follows:

$$g_{(x_0,\ldots,x_{d-1})}^F = \Lambda x_d . \langle x_0, \ldots, x_d \rangle ,$$
$$g_{(x_0,\ldots,x_{w-1})}^F = \Lambda x_w . F(g_{x_0,\ldots,x_w}^F) \quad \text{if } w < d .$$

Note that $\Phi_d \cdot F = F(g_{()}^F)$. Note also that $g_{(x_0,\ldots,x_{w-1})}^{F_0}(0) = \langle x_0, \ldots, x_{w-1}, 0_w, \ldots, 0_d \rangle$ (adding subscripts to the zeros to show how many there are), so for each w the mapping $(x_0, \ldots, x_{w-1}) \mapsto g_{(x_0,\ldots,x_{w-1})}^{F_0}(0)$ is injective.

We choose a path through the tree for Ψ_d in the following way. First, in view of our choice of m^0, we may pick $0 < z_0 < m^0$ so that $y_1 = g_{(z_0)}^{F_0}(0)$ differs from r_{i0}^0 for all $i < n^0$. Next, in view of our choice of m^1, we may pick $0 < z_1 < m^1$ so that $y_2 = g_{(z_0,z_1)}^{F_0}(0)$ differs from r_{i0}^0 and r_{j0}^1 for all $i < n^0$ and $j < n^1$, and also from y_0. In general, for $0 < w \leq d$, we inductively pick $0 < z_{w-1} < m^{w-1}$ so that $y_w =$

$g^{F_0}_{(z_0,\ldots,z_{w-1})}(0)$ differs from r^u_{i0} for all $u < w$ and $i < n^u$, and also from $y_1, \ldots, y_{w-1}$. Notice that for each w we have $y_w = \langle z_0, \ldots, z_{w-1}, 0_w, \ldots, 0_d \rangle$.

Let K be a number larger than any that has featured in the construction so far, and larger than all numbers $\langle x_0, \ldots, x_d \rangle$ where $x_0, \ldots, x_d \in \{0, z_0, \ldots, z_{d-1}\}$. Define $F_1 \in \mathsf{SP}^{0,\mathrm{prim}+}(2)$ by

$$
\begin{array}{rcl}
F_1 & = & \lambda g.\ \mathtt{case}\ g(0)\ \mathtt{of}\ (\\
 & & \quad y_d \Rightarrow K \\
 & & \quad \mid y_{d-1} \Rightarrow \mathtt{case}\ g(z_{d-1})\ \mathtt{of}\ (K \Rightarrow K \mid j \Rightarrow y_{d-1}) \\
 & & \quad \mid \cdots \\
 & & \quad \mid y_1 \Rightarrow \mathtt{case}\ g(z_1)\ \mathtt{of}\ (K \Rightarrow K \mid j \Rightarrow y_1) \\
 & & \quad \mid i \Rightarrow \mathtt{case}\ g(z_0)\ \mathtt{of}\ (K \Rightarrow K \mid j \Rightarrow i) \\
 & & \quad).
\end{array}
$$

Here i, j function as 'pattern variables' that catch all cases not handled by the preceding clauses.

Let us check that $F_1 \in \mathscr{F}$. Suppose $w < d$, $i < n^w$ and $g \in V^w_i$, so that $g \cdot z = r^w_{iz}$ for all $z < m^w$; we wish to show that $F_1 \cdot g = u^w_i$. First note that $y_d, y_{d-1}, \ldots, y_{w+1}$ were chosen to be different from $g(0) = r^w_{i0}$, and also that for each $v \leq w$, K was chosen to be larger than $g \cdot z_v = r^w_{iz_v}$. So by the definition of F_1 we see that $F_1 \cdot g = g(0) = r^w_{i0} = u^w_i$ as required. Thus $F_1 \in \mathscr{F}$, whence $\Psi_d \cdot F_1 = c$ by claim 3 above.

On the other hand, we shall show that $\Phi_d \cdot F_1 = K$. First, we claim that for any $w < u \leq d$ and any $x_{w+1}, \ldots, x_u$ we have

$$
g^{F_1}_{(z_0,\ldots,z_{w-1},0,x_{w+1},\ldots,x_{u-1})}(x_u) \;=\; \langle \vec{z}, 0, x_{w+1}, \ldots, x_u, 0_{u+1}, \ldots, 0_d \rangle \,.
$$

To show this, we fix $w < d$, and argue by reverse induction on u. For $u = d$, this is immediate from the general definition of $g^{F_1}_{(x_0,\ldots,x_{d-1})}$ above. Assuming the claim holds at u, for any $x_{w+1}, \ldots, x_{u-1}$ we have

$$
g^{F_1}_{(\vec{z},0,x_{w+1},\ldots,x_{u-2})}(x_{u-1}) \;=\; F_1(g^{F_1}_{(\vec{z},0,x_{w+1},\ldots,x_{u-1})}) \,.
$$

To compute this, note that the induction hypothesis gives $g^{F_1}_{(\vec{z},0,\vec{x})}(0) = \langle \vec{z}, 0, \vec{x}, \vec{0} \rangle$, and observe that this differs from y_d. Next note that for $0 \leq j < d$ we have $g^{F_1}_{(\vec{z},0,\vec{x})}(z_j) = \langle \vec{z}, 0, \vec{x}, z_j, \vec{0} \rangle$, again by the induction hypothesis, and observe that this differs from K. It thus follows by the definition of F_1 that

$$
F_1(g^{F_1}_{(\vec{z},0,x_{w+1},\ldots,x_{u-1})}) \;=\; \langle \vec{z}, 0, x_{w+1}, \ldots, x_{u-1}, \vec{0} \rangle \,,
$$

yielding the induction claim for $u-1$ as required. In particular, when $u = w+1$ so that $\vec{x}$ is empty, we have $g^{F_1}_{(\vec{z},0)}(0) = \langle \vec{z}, \vec{0} \rangle = y_w$; and by applying a similar argument once again we deduce $g^{F_1}_{(\vec{z})}(0) = y_w$.

We can now show by reverse induction that for $0 < w \leq d$ we have

$$F_1 \cdot g^{F_1}_{(z_0,\ldots,z_{w-1})} = K\,.$$

For $w = d$, this holds because $g^{F_1}_{(z_0,\ldots,z_{d-1})} = \langle z_0,\ldots,z_{d-1},0\rangle = y_d$. For $w < d$, we have shown $g^{F_1}_{(z_0,\ldots,z_{w-1})}(0) = y_w$, and by the induction hypothesis we have $g^{F_1}_{(z_0,\ldots,z_{w-1})}(z_w) = F_1(g_{(z_0,\ldots,z_w)}) = K$, so the claim follows by the definition of F_1. In particular we have $F_1 \cdot g^{F_1}_{(z_0)} = K$, and applying a similar argument one last time yields $\Phi_d \cdot F_1 = F_1(g^{F_1}_{()}) = K$. But $K \neq c$, contradicting the assumption that $\Psi_d \cdot F = \Phi_d \cdot F$ for all $F \in \mathsf{SP}^{0,\mathrm{prim}+}(2)$. $\square$

We shall see in the next subsection how this also implies that rec_1 cannot be μ-computable in any total extensional model.

We conclude the present subsection with a formal statement of the other expressivity result mentioned earlier, to the effect that a well-known *bar recursion* operator provides an example of a third-order total function that is Kleene computable but not definable in $\mathrm{T}+min$. Bar recursion will be discussed in more depth in Subsections 7.3.3 and 8.3.1, but the statement we give here will be technically self-contained. We will in fact claim that even a certain specialized instance of bar recursion is not LWF-representable.

A little notation and terminology will be needed. If $x = \langle x_0,\ldots,x_{r-1}\rangle$ where $\langle\cdots\rangle$ is a coding for finite sequences of natural numbers, we shall write $x.z$ for $\langle x_0,\ldots,x_{r-1},z\rangle$; we also write $x_0\ldots x_{r-1}0^\omega$ or $\vec{x}0^\omega$ for the function $\mathbb{N} \to \mathbb{N}$ sending i to x_i if $i < r$, and 0 if $i \geq r$.

Now suppose $\mathbf{A}$ is any higher-order model over $\mathbb{N}$ or $\mathbb{N}_\bot$ equipped with an interpretation of the language $\mathsf{Klex}^{\mathrm{prim}+} = \mathsf{Klex}^{\mathrm{prim}} + D$ (see Exercise 6.3.21). We call an element $B \in \mathbf{A}(2,2 \to 1)$ a *restricted bar recursor* if whenever $F,G \in \mathbf{A}(2)$ are $\mathsf{Klex}^{\mathrm{prim}+}$ definable, the following conditions hold for all $x = \langle x_0,\ldots,x_{r-1}\rangle$:

1. $B(F,G)(x) = 2x+1$ if $F(\vec{x}0^\omega) < r$.
2. $B(F,G)(x) = G(\lambda z.B(F,G)(x.z))$ otherwise.

(In the terminology of Section 7.3.3, this will correspond to bar recursion with the 'leaf function' L specialized to the function $x \mapsto 2x+1$.)

On the one hand, the above two conditions can be read as a recursive definition of B which can be readily expressed in Klex; hence restricted bar recursors do exist within SP^0 and many other models. Moreover, we will see in Subsection 7.3.3 that such operators are 'total' in the senses of interest. On the other hand, we have:

Theorem 6.3.28. *Within* SP^0 *itself, a restricted bar recursor cannot be LWF, and hence cannot be* $\mathrm{T}+min$ *definable.*

The proof is a more elaborate application of the method used in Theorem 6.3.27 above; see Longley [184] for details. Again, we will see in the next subsection what this theorem implies for the status of bar recursion in total models.

We anticipate that many further applications of NSPs to definability questions will be possible. For instance, we conjecture that for any k, a substructure of SP^0

may be found that distinguishes $\mathrm{T}_k + min$ from $\mathrm{T}_{k+1} + min$ (cf. Berger [22]). The relative power of $\mathsf{Klex}^{\min}$ and $\mathrm{T}_0 + min$ for representing total functionals also awaits clarification. Another recent application of these methods will be mentioned as Theorem 7.7.2.

6.3.5 Total Models Revisited

We may now revisit the interpretation of NSPs in total type structures from Section 6.2 in the light of our study of substructures of SP^0. We start with some observations concerning well-founded procedures.

Proposition 6.3.29. *Suppose $p \in \mathsf{SP}^0(\sigma)$ is well-founded and $\perp$-free. Then $[\![p]\!]^* \in \mathbf{A}(\sigma)$ is defined for any Kleene closed $\mathbf{A}$.*

Proof. An easy transfinite induction on the height of p. □

From Theorem 5.2.8, we know that the class of well-founded procedures (coded as functions $\mathbb{N} \to \mathbb{N}$) is Π^1_1-complete, and by Theorem 5.2.25, this is also the complexity of the class of total Kleene indices themselves. One might therefore be tempted to conjecture that every Kleene computable element of S was denotable by some well-founded procedure. That this is not the case is shown by the following counterexample:

Example 6.3.30. Consider the function $\Theta_0 \in \mathsf{S}(3)$ defined by

$$\begin{aligned} \Theta_0(F^2) \;=\; & \textit{if } F(\Lambda n.F(\Lambda x.n+1)) = F(\Lambda n.0) \textit{ then } 0 \\ & \textit{else } \mu n.F(\Lambda x.n+1) \neq 0\,. \end{aligned}$$

This is clearly total and Kleene computable, but there is no *well-founded* sequential procedure p such that $[\![p]\!]^* = \Theta_0$. For suppose p were such a procedure, and consider the binary subtree of p whose nodes are top-level calls to F, and whose branches are given by case branchings $0 \Rightarrow \cdots$ and $1 \Rightarrow \cdots$. By König's lemma, this tree is finite, and so has a finite set L of leaf nodes. Now define $F_m \in \mathsf{S}(2)$ for each $m \in \mathbb{N}$ by

$$F_m(g) \;=\; \begin{cases} 1 \text{ if } g = (\Lambda x.0) \text{ or } g = (\Lambda x.m+1)\,, \\ 0 \text{ otherwise}\,. \end{cases}$$

Clearly, for each m, the value of $\Theta_0(F_m)$ will be the value of one of the nodes in L, since the range of F_m is $\{0,1\}$. But this is impossible, since L is finite, whereas it is easy to see that $\Theta_0(F_m) = m$ for each m.

There are also sequential procedures that are not even left-well-founded, but which nevertheless give rise to total Kleene computable functionals. For instance, consider the functional $\Theta_1 \in \mathsf{S}(3)$ defined recursively by:

$$\begin{aligned} \Theta_1(F^2) &= \textit{if } F(\Lambda n.F(\Lambda x.n+1)) = F(\Lambda n.0) \textit{ then } 0 \textit{ else } \Theta_1'(0,F)\,, \\ \Theta_1'(n,F) &= F(\Lambda y.\textit{ if } F(\Lambda x.n+1) \neq 0 \text{ then } 0 \textit{ else } \Theta_1'(n+1,F))\,. \end{aligned}$$

This definition may be readily translated into an S1–S9 index (or Kleene expression) using the second recursion theorem. It is clear that the corresponding sequential procedure is not LWF, as it involves an infinite nesting of calls to F. Nonetheless, Θ_1 is a total functional: for any $F \in \mathsf{S}(2)$, either $F(\Lambda n.F(\Lambda x.n+1)) = F(\Lambda n.0)$, in which case $\Theta_1(F) = 0$, or there is some least m such that $F(\Lambda x.m+1) \neq 0$, in which case the recursion in the above definition will bottom out at $\Theta_1'(m,F)$.

Note, however, that Θ_1 is not such a 'strong' counterexample as Θ_0, since we may readily find other LWF procedures (and Kleene expressions for them) that give rise to the functional Θ_1. Indeed, the following natural question is currently open:

Question 6.3.31. Are there Kleene computable elements of S that are not denotable by any LWF procedure?

Both Θ_0 and Θ_1 highlight an important conceptual point about the nature of Kleene computability. For any particular choice of arguments $\vec{\Phi}$, a Kleene expression M will give rise to a well-founded computation tree, and this computation will of course involve only a well-founded portion of the procedure $[\![M]\!]^S$. However, it is impossible in general to isolate a single well-founded (or even left-well-founded) portion of $[\![M]\!]^S$ that suffices for *all* possible arguments $\vec{\Phi}$: the tree that is potentially available for exploration must be allowed to be non-well-founded. This reinforces an idea from Subsection 5.2.2: Kleene computations have to be thought of as proceeding in a top-down or 'demand driven' manner, the oracle calls being generated on the fly as the computation progresses.

Finally, we show how the non-definability results of Subsection 6.3.4 yield information on computability in total type structures. First, a consequence of the proof of Theorem 6.3.27:

Corollary 6.3.32. *Let* **A** *be a total type structure over* $\mathbb{N}$ *supporting Kleene primitive recursion. Then the operation* Φ *given by*

$$\Phi_n(F) = F(\Lambda x_0.F(\Lambda x_1.\cdots F(\Lambda x_n.\langle x_0,\ldots,x_n\rangle)\cdots)) \,,$$
$$\Phi(n)(F) = \Phi_n(F) \,,$$

if present at all in **A**, *is not* μ*-computable in the sense of Subsection 5.1.5. A fortiori, the same applies to the System* T *recursor* rec_1.

Proof. Suppose Φ is μ-computable; then Φ is the denotation of some expression M of $\mathsf{Klex}^{\min}$. By Theorem 6.3.18, the corresponding NSP $p = [\![M]\!]^S$ is left-bounded. We will show that p itself satisfies the defining clauses for Φ for all $n \in \mathbb{N}$ and $F \in \mathsf{SP}^{0,\mathrm{prim}+}(2)$, which the proof of Theorem 6.3.27 has shown to be impossible.

Let Z be the logical relation between SP^0 and **A** which, at ground type, relates $q \in \mathsf{SP}^0(\mathtt{N})$ and $n \in \mathbf{A}(\mathtt{N})$ iff $q = \lambda.n$. By a straightforward induction on terms of $\mathsf{Klex}^{\min}$, one can show that $Z(p,\Phi)$. Moreover, for any $F = [\![N]\!]^S$ where $N : \overline{2}$ in $\mathsf{Klex}^{\mathrm{prim}+}$, it is easy to check that there exists $\widehat{F} \in \mathbf{A}(2)$ with $Z(F,\widehat{F})$ (noting that the operator D of Exercise 6.3.21 has a $\mathsf{Klex}^{\mathrm{prim}}$ definable counterpart in **A**). We thus have that $Z(p \cdot n \cdot F, \Phi(n)(\widehat{F}))$ for any n, and the relevant equations for p now follow readily from those for Φ. □

This result is clearly applicable to the models S, Ct and HEO, for instance. A third-order functional in S that is Kleene computable but not μ-computable was first given by Kleene [138, Section 8] using ideas from degree theory; however, Kleene's example relies on the presence of discontinuous arguments F. A more subtle argument due to Bergstra [31], which we present in Subsection 8.5.2, establishes the existence of such functionals within Ct, but this still relies on the presence of non-computable arguments. The proof we have given is more complex than these, but more 'robust' in that it works also for HEO and other effective type structures.

Similarly, we may obtain the following as a consequence of Theorem 6.3.28 (for proof details see Longley [184]):

Corollary 6.3.33. *In a total type structure* **A** *over* $\mathbb{N}$ *supporting Kleene primitive recursion, no restricted bar recursor can be* μ*-computable, or even* μ*-computable relative to the System T recursors if these exist in* **A**. □

For instance, this shows that the bar recursion operator in Ct is not μ-computable. (We will see in Subsection 8.3.1 that such an operator exists and is Kleene computable over Ct.) This answers affirmatively the analogue for Ct of Question 6.3.31.

6.4 Kleene Computability in Partial Models

So far, we have considered Kleene's concept of higher-order computability (and its reformulation via NSPs) only for total type structures over $\mathbb{N}$. We now move on to consider how this concept might be extended to partial models (technically, total type structures over $\mathbb{N}_\perp$). We have in mind three leading examples of such models:

1. The *full monotone* type structure M defined as follows: $\mathsf{M}(0) = \mathbb{N}_\perp$ (with the usual partial ordering), and $\mathsf{M}(\sigma \to \tau)$ is the set of all monotone functions $\mathsf{M}(\sigma) \to \mathsf{M}(\tau)$ (endowed with the pointwise ordering).
2. The type structure PC of *partial continuous* functionals (see Subsection 3.2.2).
3. The effective submodel $\mathsf{PC}^{\mathrm{eff}}$ of PC.

There are at least two motivations for extending Kleene computability to such settings. Firstly, we have seen how, in total models, the need for a 'totality' side-condition in the scheme S8 gives rise to some apparent oddities, such as the fact that partial computable functionals do not compose in general. Since definitions of computability naturally give rise to partial functionals as well as total ones, one might reasonably expect that working in a partial setting should result in a smoother mathematical theory. Secondly, many important total models **A** naturally come with a type-respecting simulation in some partial type structure **P**: indeed, the full models S, Ct and HEO respectively arise as the extensional collapses of M, PC and $\mathsf{PC}^{\mathrm{eff}}$ relative to $\mathbb{N} \subset \mathbb{N}_\perp$ (see Section 3.7). Assuming we have a good theory of Kleene computability in the partial model **P**, we can then single out the elements of **A** that have computable representatives in **P**, and ask whether this gives us a useful handle on Kleene computability within **A** itself.

In fact, there are several competing ways in which the definition of Kleene computability can be adapted to partial models, and the differences between them highlight some important conceptual points. In Subsection 6.4.1, we will offer three possible 'partial' versions of Kleene computability, which we call the *strict*, *intermediate* and *liberal* interpretations. Of these, it is the liberal interpretation that is mathematically the most satisfactory, and this is the one we shall mainly focus on (both here and in Chapter 7). On the other hand, there is a sense in which it is problematic to regard the resulting partial functionals as 'computable' in general, and a comparison with the other two interpretations is useful for clarifying the subtle conceptual issues at stake here.

In Subsection 6.4.2, we show that the models in which the liberal interpretation works well are exactly those with *least fixed point* operators. This can be seen as a motivation for this latter concept, and in the following chapter we shall pursue an approach to higher type computability that takes this important notion as primitive. We shall see, indeed, that the interpretations of PCF which we shall study there naturally generalize the liberal interpretation that we consider here.

We then turn in Subsection 6.4.3 to consider the situation of a total model **A** simulated by a partial model **P**. We shall see that in certain cases, though not all, the meanings of Kleene computability in **A** and **P** agree; once again, a comparison between our three interpretations in **P** is useful for pinpointing the exact source of the various phenomena that arise. Finally, having introduced the idea of least fixed points, we are well placed to review another classical approach to Kleene computability, namely that of Moschovakis via *inductive definitions*, which we discuss briefly in Subsection 6.4.4.

6.4.1 Three Interpretations of Kleene Computability

We here present three plausible interpretations of Kleene computability in partial models, in order of increasing computational power. We work in the setting of a total type structure **P** over $\mathbb{N}_\perp$. (Recall that the term 'total' here refers to the fact that the application operations are mathematically total functions, although their value may sometimes be $\perp$.) For notational convenience, we shall assume that **P** is canonical, and write applications in **P** as ordinary function applications.

All three of our interpretations factor naturally through NSPs, so to make this fact manifest, we shall present them as adaptations of Definition 6.2.5 above. From this, it is routine to obtain more direct formulations of our interpretations in the style of S1–S9, using the translation from Kleene indices to NSPs given by Lemma 5.1.17 and Definition 6.2.1. (See however Exercise 6.4.11 for some associated subtleties.)

Our three interpretations differ in their treatment of the new element $\perp \in \mathbb{N}_\perp$. Our account of the first two interpretations will be rather cursory, since our main interest is simply in the contrast with the third one that they offer. We start with a very literal-minded or *strict* adaptation of Definition 6.2.5, in which $\perp$ is in essence treated no differently from any other element of $\mathbb{N}_\perp$.

Definition 6.4.1. (i) We define the following relations by simultaneous induction:

$$[\![E]\!]^s_\nu = u\,, \qquad [\![G]\!]^s_\nu = u\,, \qquad [\![P]\!]^s_\nu(\vec{\Phi}) = u\,, \qquad [\![P]\!]^{s*}_\nu = F\,,$$

where E, G, P range over well-typed meta-expressions, ground meta-terms and meta-procedures respectively; ν ranges over finite valuations of variables in $\mathbf{P}$; $u \in \mathbb{N}_\perp$; and $\Phi_i \in \mathbf{P}(\sigma_i)$, $F \in \mathbf{P}(\sigma)$, where P has type $\sigma = \sigma_0, \ldots, \sigma_{r-1} \to \mathtt{N}$.

- $[\![u]\!]^s_\nu = u$.
- If $\nu(x) = \Phi$, $[\![Q_j]\!]^{s*}_\nu = F_j$ for $j < r$ and $\Phi(\vec{F}) = u$, then $[\![xQ_0 \ldots Q_{r-1}]\!]^s_\nu = u$.
- If $[\![Q_j]\!]^{s*}_\nu = F_j$ for $j < r$ and $[\![P]\!]^s_\nu(\vec{F}) = u$, then $[\![PQ_0 \ldots Q_{r-1}]\!]^s_\nu = u$.
- If $[\![G]\!]^s_\nu = m \in \mathbb{N}$ and $[\![E_m]\!]^s_\nu = u$, then $[\![\mathtt{case}\ G\ \mathtt{of}\ (i \Rightarrow E_i)]\!]^s_\nu = u$.
- If $[\![G]\!]^s_\nu = \perp$, then $[\![\mathtt{case}\ G\ \mathtt{of}\ (i \Rightarrow E_i)]\!]^s_\nu = \perp$.
- If $[\![E]\!]^s_\nu = u$ regarding E as a meta-expression, then also $[\![E]\!]^s_\nu = u$ regarding E as a ground meta-term.
- If $P = \lambda x_0, \ldots, x_{r-1}.E$ and $[\![E]\!]^s_{\nu, x_0 \mapsto \Phi_0, \ldots, x_{r-1} \mapsto \Phi_{r-1}} = u$, then $[\![P]\!]^s_\nu(\vec{\Phi}) = u$.
- If $F \in \mathbf{P}(\sigma)$ and $[\![P]\!]^s_\nu(\vec{\Phi}) = F(\vec{\Phi})$ for all $\vec{\Phi}$, then $[\![P]\!]^{s*}_\nu = F$.

An element of $\mathbf{P}$ is called *strictly Kleene computable* if it is of the form $[\![p]\!]^{s*}$ for some $p \in \mathsf{SP}^{0,\mathrm{eff}}$.

(ii) We say $\mathbf{P}$ is *strictly Kleene closed* if whenever $\Gamma \vdash p : \sigma$ is effective and $[\![p]\!]^{s*}_\nu = F$, we have $F \in \mathbf{P}(\sigma)$.

It is straightforward to show that our leading examples M, PC, $\mathsf{PC}^{\mathrm{eff}}$ are strictly Kleene closed. One can also adapt the proof of Theorem 6.2.9 to show that, in general, the strictly Kleene computable elements of $\mathbf{P}$ are closed under application: if $p \in \mathsf{SP}^0(\sigma \to \tau)$, $q \in \mathsf{SP}^0(\sigma)$ and $[\![p]\!]^{s*}$, $[\![q]\!]^{s*}$ are defined, then $[\![p]\!]^{s*}([\![q]\!]^{s*}) = [\![p \cdot q]\!]^{s*}$.

Clearly, the above is in some sense a direct adaptation of Definition 6.2.5 to the partial setting. Indeed, the interpretation $[\![-]\!]^s$ shares many familiar properties of classical Kleene computability: for instance, any valid triple $[\![P]\!]^s_\nu(\vec{\Phi}) = u$ is witnessed by a unique well-founded *computation tree* corresponding to its inductive generation, and such trees may also be systematically constructed in a 'top-down' manner reflecting a dynamic computation process as described in Subsection 5.2.2. Furthermore, the definition of $[\![-]\!]^s$ makes sense for any $\mathbf{P}$ whatever, and always yields a *deterministic* evaluation relation (although of course there will be pairs $(P, \vec{\Phi})$ for which no value of u is generated). This accords well with the idea of computing with 'pure extensions' without any reference to their specific presentation or properties.

To see what is distinctive about the strict interpretation, one must note carefully the difference between saying that $[\![e]\!]^s_\nu = \perp$ and saying that $[\![e]\!]^s_\nu$ does not have a specified value at all: we have $[\![e]\!]^s_\nu = \perp$ only if this fact is explicitly generated by our inductive definition. In order for an application $xq_0 \ldots q_{r-1}$ to have a defined value at some ν, it is necessary that the q_i denote everywhere-specified functions relative to ν—that is, that the value of $[\![q_i]\!]^s_\nu(\vec{\Psi})$ is specified (even if $\perp$) for every

$\vec{\Psi}$ of suitable type. Again, this fits with the idea that all we know about a functional $\Phi : \sigma_0, \ldots, \sigma_{r-1} \to \mathtt{N}$ is how it will behave when presented with fully specified arguments $F_i : \sigma_i$; nothing is guaranteed for arguments that lack a complete, well-founded specification. Indeed, the definition of $[\![-]\!]^s$ naturally provides what we need in order to interpret all *well-founded* procedures:

Exercise 6.4.2. Show that if **P** is strictly Kleene closed and $p \in \mathsf{SP}^0$ is well-founded, then $[\![p]\!]^{s*}_{\mathbf{P}}$ is well-defined. Show that this property fails in general for left-well-founded procedures.

On the other hand, some major deficiencies of $[\![-]\!]^s$ should be noted. Firstly, we have done nothing to eliminate the need for an 'S8 side-condition': in the above presentation, this takes the form of the requirement in the final clause that $[\![P]\!]^s_{\mathrm{V}}(\vec{\Phi})$ is specified for all $\vec{\Phi}$, and even in our leading models this will frequently fail to be satisfied. Secondly, this strict interpretation offers almost no scope for interesting computations with non-total functionals: this is because a result $\bot$ can only ever arise from the result of an oracle call or from an explicit occurrence of $\bot$ in a procedure, and never from a genuinely infinite computation process.

The latter of these deficiencies is largely remedied by our second interpretation, which allows certain infinite interactions to give rise to $\bot$ whilst maintaining the desirable properties of $[\![-]\!]^s$. The idea here is that if the attempt to compute $[\![P]\!]^s_{\mathrm{V}}(\vec{\Phi})$ gives rise to an infinite sequence of subcomputations, this in itself can be treated as a 'proof' that $[\![e]\!]_{\mathrm{V}} = \bot$.

Definition 6.4.3. (i) We define the following relations by simultaneous induction:

$$[\![E]\!]^i_{\mathrm{V}} = u\,, \qquad [\![G]\!]^i_{\mathrm{V}} = u\,, \qquad [\![P]\!]^i_{\mathrm{V}}(\vec{\Phi}) = u\,, \qquad [\![P]\!]^{i*}_{\mathrm{V}} = F\,.$$

The inductive clauses are those of Definition 6.4.1 augmented by the following:

- If $E^0, E^1, \ldots$ is an infinite sequence of NSP meta-expressions and $n^0, n^1, \ldots$ a sequence of natural numbers such that each E^j has the form $\mathtt{case}\ G^j\ \mathtt{of}\ (i \Rightarrow E^j_i)$ where $E^j_{n^j} = E^{j+1}$, and if $[\![G^j]\!]^i_{\mathrm{V}} = n_j$ for each j, then $[\![E^0]\!]^i_{\mathrm{V}} = \bot$.

An element of **P** is called *intermediately Kleene computable* if it is of the form $[\![p]\!]^{i*}$ for some $p \in \mathsf{SP}^{0,\mathrm{eff}}$.

(ii) We say **P** is *intermediately Kleene closed* if whenever $\Gamma \vdash p : \sigma$ is effective and $[\![p]\!]^{i*}_{\mathrm{V}} = F$, we have $F \in \mathbf{P}(\sigma)$.

Once again, our leading models are easily shown to be intermediately Kleene closed, and in general the intermediately Kleene computable elements of **P** are closed under application, again by an adaptation of the proof of Theorem 6.2.9.

All the good properties mentioned above for $[\![-]\!]^s$ are still maintained by $[\![-]\!]^i$. Furthermore, $[\![-]\!]^i$ allows for non-trivial computations with partial functionals: for instance, it is easily seen that the minimization operator *min* is $[\![-]\!]^i$-computable (in models where it exists). Indeed, the definition of $[\![-]\!]^i$ naturally gives what we need for interpreting *left-well-founded* procedures:

Exercise 6.4.4. Show that if $\mathbf{P}$ is intermediately Kleene closed and $p \in \mathsf{SP}^0$ is left-well-founded, then $[\![p]\!]^{i*}_{\mathbf{P}}$ is well-defined.

However, the definition of $[\![-]\!]^i$ still does not dispense with an S8-style side-condition, and even in our leading models, there will be procedures that receive no interpretation (see for example the non-well-founded procedure F_σ given in the proof of Proposition 6.4.14).

Our third candidate, the *liberal* interpretation, addresses these issues by means of a very different approach to $\bot$. It has the pleasing feature that *every* sequential procedure is assigned an interpretation in the models of interest, resulting in a much smoother mathematical theory unhampered by tiresome side-conditions. This is the notion that is usually intended when Kleene computability over partial type structures is discussed in the literature (although the subtleties we are considering here are not always made explicit). As we shall see in Section 7.1, it also coincides in typical models with the notion of PCF-computability. Nonetheless, we shall see that the liberal interpretation is significantly different in spirit from $[\![-]\!]^s$ and $[\![-]\!]^i$: it lacks some of the familiar characteristics of Kleene computability, with the consequence that it is conceptually problematic to view the resulting elements of $\mathbf{P}$ as 'computable' in the traditional sense of Kleene computation.

The key idea behind the liberal interpretation is that in order to establish (say) that $[\![xq]\!]_\nu = n$ where $\nu(x) = \Phi$, it is not necessary that the function $[\![q]\!]^*_\nu$ be specified in its entirety, but only enough of it to ensure that $\Phi([\![q]\!]^*_\nu) = n$ 'whatever $[\![q]\!]^*_\nu$ might be'. In fact, it is not even necessary to justify that the mathematical expression '$[\![q]\!]^*_\nu$' is meaningful at all before we can assign a value to $[\![xq]\!]_\nu$—although in hindsight it will turn out that *all* procedures are assigned a meaning by the interpretation.

For the liberal interpretation of Kleene computability to make sense, our model $\mathbf{P}$ must satisfy certain conditions:

Definition 6.4.5. Let $\mathbf{P}$ be a total type structure over $\mathbb{N}_\bot$.

(i) We say $\mathbf{P}$ is *pointwise monotone* if every $f \in \mathbf{P}(\sigma \to \tau)$ is monotone with respect to the pointwise order on $\mathbf{P}(\sigma)$ and $\mathbf{P}(\tau)$.

(ii) We say $\mathbf{P}$ *has consistent joins* if whenever $f, f' \in \mathbf{P}(\sigma_0, \ldots, \sigma_{r-1} \to \mathtt{N})$ are consistent (that is, there are no $x_i \in \mathbf{P}(\sigma_i)$ such that $f\vec{x}$ and $f'\vec{x}$ are distinct natural numbers), they have a pointwise upper bound within $\mathbf{P}$.

It is easy to see that our three leading models all satisfy these conditions. However, the fact that such conditions are needed at all is already an indicator that we are no longer computing with pure, unconstrained oracles in quite the same way as before. We may now give the definition of our liberal interpretation as follows:

Definition 6.4.6. Assume $\mathbf{P}$ is a total type structure over $\mathbb{N}_\bot$ that is pointwise monotone and has consistent joins.

(i) We define the following relations by simultaneous induction:

$$[\![E]\!]^l_\nu = n\,, \qquad [\![G]\!]^l_\nu = n\,, \qquad [\![P]\!]^l_\nu(\vec{\Phi}) = n\,, \qquad [\![P]\!]^{l*}_\nu \sqsupseteq F\,,$$

where E, G, P range over well-typed meta-expressions, ground meta-terms and meta-procedures respectively; ν ranges over finite valuations of variables in $\mathbf{P}$; $n \in \mathbb{N}$; and $\Phi_i \in \mathbf{P}(\sigma_i)$, $F \in \mathbf{P}(\sigma)$, where P has type $\sigma = \sigma_0, \ldots, \sigma_{r-1} \to \mathtt{N}$.

- $[\![n]\!]^l_\nu = n$.
- If $\nu(x) = \Phi$, $[\![Q_j]\!]^{l*}_\nu \sqsupseteq F_j$ for $j < r$ and $\Phi(\vec{F}) = n$, then $[\![xQ_0 \ldots Q_{r-1}]\!]^l_\nu \;=\; n$.
- If $[\![Q_j]\!]^{l*}_\nu \sqsupseteq F_j$ for $j < r$ and $[\![P]\!]^l_\nu(\vec{F}) = n$, then $[\![PQ_0 \ldots Q_{r-1}]\!]^l_\nu \;=\; n$.
- If $[\![G]\!]^l_\nu = m \in \mathbb{N}$ and $[\![E_m]\!]^l_\nu = n$, then $[\![\mathtt{case}\; G \;\mathtt{of}\; (i \Rightarrow E_i)]\!]^l_\nu \;=\; n$.
- If $[\![E]\!]^l_\nu = n$ regarding E as a meta-expression, then $[\![E]\!]^l_\nu = n$ regarding E as a ground meta-term.
- If $P = \lambda x_0, \ldots, x_{r-1}.E$ and $[\![E]\!]^l_{\nu, x_0 \mapsto \Phi_0, \ldots, x_{r-1} \mapsto \Phi_{r-1}} = n$, then $[\![P]\!]^l_\nu(\vec{\Phi}) = n$.
- If $F \in \mathbf{P}(\sigma)$ and $[\![P]\!]^s_\nu(\vec{\Phi}) = F(\vec{\Phi})$ whenever $F(\vec{\Phi}) \in \mathbb{N}$, then $[\![P]\!]^{l*}_\nu \sqsupseteq F$.

(ii) For any P and ν as above, we may now define a function $[\![P]\!]^{l*}_\nu$ by setting $[\![P]\!]^{l*}_\nu(\vec{\Phi}) = n$ if $[\![P]\!]^l_\nu(\vec{\Phi}) = n$, and $[\![P]\!]^{l*}_\nu(\vec{\Phi}) = \bot$ if no value for $[\![P]\!]^l_\nu(\vec{\Phi})$ is generated by the inductive definition above. An element of $\mathbf{P}$ is called *liberally Kleene computable* if it is of the form $[\![p]\!]^{l*}$ for some $p \in \mathsf{SP}^{0,\mathrm{eff}}$.

(iii) We say $\mathbf{P}$ is *liberally Kleene closed* if whenever $\Gamma \vdash p : \sigma$ is effective and ν is a valuation for Γ, we have $[\![p]\!]^{l*}_\nu \in \mathbf{P}(\sigma)$.

Notice that the element $\bot$ does not feature at all in part (i) of this definition—as in Definition 5.1.1, our inductive process generates only definite numerical results for computations. However, the relations $[\![P]\!]^{l*}_\nu \sqsupseteq F$ introduce a new ingredient not present in Kleene's original definition—an ingredient that has profound consequences for the computational strength of this notion. The key point to note is that in the second clause of part (i), we will have

$$\Phi([\![Q_0]\!]^{l*}_\nu, \ldots, [\![Q_{r-1}]\!]^{l*}_\nu) \;=\; n$$

whatever the $[\![Q_j]\!]^{l*}_\nu \sqsupseteq F_j$ turn out to be, assuming that $\mathbf{P}$ is pointwise monotone. Indeed, the hypotheses of monotonicity and consistent joins are needed to ensure that evaluation is deterministic:

Exercise 6.4.7. Using the fact that $\mathbf{P}$ is pointwise monotone and has consistent joins, show by simultaneous induction on the generation of $[\![-]\!]^l$ that:

- if $[\![E]\!]^l_\nu = n$ and $[\![E]\!]^l_\nu = n'$, then $n = n'$,
- if $[\![P]\!]^{l*}_\nu \sqsupseteq F$ and $[\![P]\!]^{l*}_\nu \sqsupseteq F'$, then F, F' are consistent.

Exercise 6.4.8. (i) Show by induction on derivations that if $[\![E]\!]^i_\nu = n$ then $[\![E]\!]^l_\nu = n$.

(ii) Give an example to show that we may have $[\![E]\!]^i_\nu$ unspecified, but $[\![E]\!]^l_\nu = 0$. (Hint: refer to Proposition 6.4.14 below.) Thus, $[\![-]\!]^l$ is not simply obtained from $[\![-]\!]^i$ by taking $[\![E]\!]^l_\nu = \bot$ whenever $[\![E]\!]^i_\nu$ is unspecified.

Exercise 6.4.9. Check that M, PC, $\mathsf{PC}^{\mathrm{eff}}$ are all liberally Kleene closed.

Exercise 6.4.10. Check that the proof of Theorem 6.2.9 adapts straightforwardly to show that in a liberally Kleene closed model we have $[\![p]\!]^{l*}([\![q]\!]^{l*}) = [\![p \cdot q]\!]^{l*}$ for any $p \in \mathsf{SP}^0(\sigma \to \tau)$, $q \in \mathsf{SP}^0(\sigma)$.

Exercise 6.4.11. (i) Establish how Kleene's schemes S1–S9 (as given in Definition 5.1.1) may be interpreted 'directly in a partial model **P** in the spirit of the liberal interpretation, understanding all variables appearing in S1–S9 to range over the appropriate $\mathbf{P}(\sigma)$.

(ii) Show that this direct reading does not agree on the nose with the one obtained by combining Definition 6.4.6 with our translation from Kleene indices (via Kleene expressions) to NSPs (see Lemma 5.1.17 and Definition 6.2.1). In particular, if $e = \langle 4, \langle 2, n \rangle, e' \rangle$ where $\{e'\}$ is everywhere undefined, show that $\{e\}() = n$ under the direct reading but $\{e\}() \uparrow$ under the indirect one.

(iii) Show, however, that the direct and indirect interpretations give rise to the same class of computable partial functionals, and indeed that the corresponding systems of indices are effectively intertranslatable. (Hint: show that the direct interpretation may also be factored through NSPs, and use Theorem 6.2.4.)

The main mathematical advantage of the liberal interpretation is that it is defined for all meta-terms, so that the complications arising from S8 are entirely avoided. However, this comes at a price: there is no longer in general a *unique* well-founded tree witnessing the generation of a given statement $[\![E]\!]^l_\nu = n$. For example, if $[\![xQ]\!]^l_\nu = n$, there may be many possible choices of F such that $[\![Q]\!]^{l*}_\nu \sqsupseteq F$. What is more, there need not even be a *smallest* such tree in general: e.g. if we have $Q = \lambda z^1.0$ and $\nu(x^2)(F) = 0$ iff $F(0) = 0$ or $F(1) = 0$, then a tree for $[\![xQ]\!]^l_\nu = 0$ may be constructed using either F_0 or F_1 (where $F_i(n)$ is 0 if $n = i$ and $\bot$ otherwise), but not using any smaller function.

For a similar reason, there is a difficulty in formulating a *top-down* procedure for generating computation trees in the spirit of Subsection 5.2.2. For instance, if $\nu(x)$ is as above, a procedure for evaluating $[\![xQ']\!]^l_\nu$ might speculatively launch a subcomputation for $[\![Q']\!]^l_\nu(0)$ or for $[\![Q']\!]^l_\nu(1)$, but in either case it risks the possibility that there is no numerical result so that no well-founded subtree will be generated.[2]

These observations suggest that it is somewhat problematic to regard the $[\![-]\!]^l$-denotable elements of **P** as 'computable' for the kinds of general reasons that suffice for $[\![-]\!]^s$ and $[\![-]\!]^i$. Typically, a justification for deeming them 'computable' will rest instead on the specific constitution of **P** and how it is presented.

6.4.2 Least Fixed Points

In this subsection, we shed some further light on the liberal interpretation by characterizing it in terms of *least fixed point* operators. We shall show, firstly, that any liberally Kleene closed model possesses least fixed points, and secondly, that these

[2] In fact, the problem goes very much deeper than this. Even if we allow top-down trees to contain non-terminating subcomputations and hence to be non-well-founded, and set ourselves the task of generating a top-down tree that *contains* a well-founded computation tree (when there is one) via some transfinite procedure, it appears that this cannot be accomplished by any 'depth-first' computation strategy. We leave the detailed clarification of the situation to future work.

in turn may be used to reconstruct the interpretation $[\![-]\!]^l$. The concept of least fixed points will be one of the chief cornerstones of later chapters, and the present subsection offers one kind of explanation for how this concept enters into our story.

In the previous subsection, the interpretation $[\![-]\!]^l$ was defined for arbitrary meta-terms. This was useful for the purpose of Exercise 6.4.10, but in the present subsection we shall for simplicity restrict attention to $[\![-]\!]^l$ as applied to terms, so that the third clause of Definition 6.4.6(i) may be ignored.

The concept of least fixed points makes sense in a very general setting:

Definition 6.4.12. Let $\mathbf{P}$ be any total applicative structure as in Definition 3.1.16.

(i) A *preorder structure* on $\mathbf{P}$ consists of a preorder $\sqsubseteq_\sigma$ on each $\mathbf{P}(\sigma)$ such that the application operations in $\mathbf{P}$ are monotone with respect to $\sqsubseteq$. (Note that $\sqsubseteq_{\sigma\to\tau}$ need not be the pointwise preorder on $\mathbf{P}(\sigma\to\tau)$.)

(ii) If $\mathbf{P}$ has a preorder structure, we say $\mathbf{P}$ *has (uniform) least fixed points* if for every type σ there exists $Y_\sigma \in \mathbf{P}((\sigma\to\sigma)\to\sigma)$ such that for all $f \in \mathbf{P}(\sigma\to\sigma)$ we have

1. $f\cdot(Y_\sigma\cdot f)) = Y_\sigma\cdot f$,
2. if $z\in\mathbf{P}(\sigma)$ and $f\cdot z = z$, then $Y_\sigma\cdot f \sqsubseteq_\sigma z$.

Example 6.4.13. The following well-known argument, due to Tarski, shows that fixed point operators exist in M. Note first that each $\mathsf{M}(\sigma)$ is closed under (pointwise) least upper bounds of increasing chains. For any $f \in \mathsf{M}(\sigma\to\sigma)$, we may therefore define an ordinal-indexed sequence of elements $x_\alpha \in \mathsf{M}(\sigma)$ as follows:

$$x_0 \;=\; \bot_\sigma\,, \qquad x_{\alpha+1} \;=\; f(x_\alpha)\,, \qquad x_\alpha \;=\; \bigsqcup_{\beta<\alpha} x_\beta \;\; (\alpha \text{ a limit})\,.$$

Clearly the x_α form an increasing sequence, and for cardinality reasons, there must be some α with $x_{\alpha+1} = x_\alpha$. It is now easy to show that x_α must be the least fixed point $\mathit{Fix}(f)$ of f, and moreover that the mapping $f \mapsto \mathit{Fix}(f)$ is itself monotone.

The existence of least fixed point operators in PC and $\mathsf{PC}^{\mathrm{eff}}$ was already noted in Example 3.5.6. Indeed, we shall see in Section 7.1 that all *models of* PCF possess such operators—these include many non-pointwise monotone or non-extensional models of interest, such as the models SR and SA of Chapters 11 and 12 respectively.

Our first task is to show how least fixed point operators arise as an instance of the liberal interpretation:

Proposition 6.4.14. *If* $\mathbf{P}$ *satisfies the conditions of Definition 6.4.5 and is liberally Kleene closed, then* $\mathbf{P}$ *has least fixed points with respect to the pointwise order.*

Proof. Given any type $\sigma = \sigma_0,\ldots,\sigma_{r-1} \to \mathtt{N}$, consider the effective NSP F_σ with free variable $g^{\sigma\to\sigma}$ defined corecursively by

$$F_\sigma \;=\; \lambda x_0^{\sigma_0}\ldots x_{r-1}^{\sigma_{r-1}}.\,\mathtt{case}\; g\,F_\sigma\, x_0^\eta \ldots x_{r-1}^\eta \;\mathtt{of}\; (i \Rightarrow i)\,.$$

Assuming $\mathbf{P}$ is liberally closed, for any $f \in \mathbf{P}(\sigma \to \sigma)$ we may define $Fix(f) = [\![F_\sigma]\!]^{l*}_{g \mapsto f} \in \mathbf{P}(\sigma)$; we shall show that this is the least fixed point of f. This will be sufficient, since we may then define $Y_\sigma = [\![\lambda g.F_\sigma]\!]^{l*}$.

To see that $Fix(f)$ is a fixed point, we apply the definition of $[\![-]\!]^l$ to the above definition of F_σ. This yields that for any $\vec{\Phi}$ we have

$$[\![F_\sigma]\!]^l_{g \mapsto f}(\vec{\Phi}) \;=\; n \;\text{ iff }\; \exists F' \in \mathbf{P}(\sigma \to \sigma).\; [\![F_\sigma]\!]^l_{g \mapsto f} \sqsupseteq F' \;\wedge\; f(F')(\vec{\Phi}) \;=\; n\,.$$

Since the largest candidate for F' here is $[\![F_\sigma]\!]^{l*}_{g \mapsto f}$ itself, this implies

$$Fix(f)(\vec{\Phi}) \;=\; f(Fix(f))(\vec{\Phi})\,,$$

and since the $\vec{\Phi}$ are arbitrary and $\mathbf{P}$ is extensional, this means $Fix(f) = f(Fix(f))$.

To show that this fixed point is the least one, suppose $f(z) = z$, and suppose $Fix(f)(\vec{\Phi}) = n$ via some derivation tree $\mathscr{U}$. By an *F-node* in $\mathscr{U}$, we shall mean a node corresponding to an assertion $[\![F_\sigma]\!]^l_{g \mapsto f}(\vec{\Psi}) = m$ for some $\vec{\Psi}$ and some m. We shall show that $z(\vec{\Phi}) = n$, reasoning by induction on the ordinal height of the tree of F-nodes within $\mathscr{U}$.

If the only F-node is the root node, this means that the trivial approximation $[\![F_\sigma]\!]^{*l}_{g \mapsto f} \sqsupseteq F' = \lambda x.\bot$ is sufficient for generating the assertion $[\![gF_\sigma\vec{x}]\!]^l_{g \mapsto f, \vec{x} \mapsto \vec{\Phi}}$, since otherwise every non-$\bot$ value of F' would demand a subtree headed by another F-node. Thus $f(\lambda x.\bot)(\vec{\Phi}) = n$, whence also $z(\vec{\Phi}) = f(z)(\vec{\Phi}) = n$ by monotonicity.

Otherwise, the value $Fix(f)(\vec{\Phi}) = n$ arises from some assertion $[\![F_\sigma]\!]^{*l}_{g \mapsto f} \sqsupseteq F'$, where for some non-empty set of argument tuples $\vec{\Psi}$ constituting the domain of F', $\mathscr{U}$ contains a subcomputation for some value $[\![F_\sigma]\!]^l_{g \mapsto f}(\vec{\Psi}) = m$, and by the induction hypothesis, we have $z(\vec{\Psi}) = m$. The relevant portion of $\mathscr{U}$ may therefore be adapted to yield a proof that $f(z)(\vec{\Phi}) = n$, whence $z(\vec{\Phi}) = n$. □

In the other direction, we shall show how the whole of the liberal interpretation can itself be constructed as a least fixed point. What is more, this style of definition carries over to a much wider range of models than those covered by Definition 6.4.6—essentially because it has the character of an 'internal interpretation' of λ-calculus as in Subsection 4.1.2, in contrast to the 'pointwise' spirit of Definition 6.4.6. Although the outline of the construction that we give here will be self-contained, the reader may find it helpful to compare it with Subsection 7.1.5, where a very similar construction will be presented in more formal detail (the connection with the present definition will be spelt out in Subsection 7.1.7).

For the remainder of this subsection, we shall adopt the following assumptions on our model $\mathbf{P}$:

1. $\mathbf{P}$ is a typed λ-algebra over $\mathsf{T}^{\to}(\mathtt{N} = \mathbb{N}_\bot)$.
2. $\mathbf{P}$ carries a preorder structure $\sqsubseteq$, where $\sqsubseteq_{\mathtt{N}}$ is the usual partial order on $\mathbb{N}_\bot$.
3. $\mathbf{P}$ has least fixed points as in Definition 6.4.12.
4. The elements $0, 1, \ldots \in \mathbf{P}(\mathtt{N})$ form a system of numerals for $\mathbf{P}$ in the sense of Subsection 3.3.4.

We do not assume that **P** is extensional or pointwise monotone. We shall see in Section 7.1 that every *model of* PCF with $\mathbf{P}(\mathtt{N}) = \mathbb{N}_\perp$ satisfies all the above conditions, although such models also satisfy a continuity condition which is not relevant here. The requirement that **P** has numerals ensures that a modicum of basic computational machinery is available: for instance, that all type 2 primitive recursive operations are representable in **P**. In fact, it will be convenient to work in the Karoubi envelope $\mathbf{K}(\mathbf{P})$; clearly this also satisfies conditions 1–4 above and additionally has product types.

The idea is that for any finite environment Γ and type σ, we shall define our interpretation operation $(p, \nu) \mapsto [\![p]\!]^*_\nu$ (where $\Gamma \vdash p : \sigma$ and ν is a valuation for Γ) as an element

$$\mathbb{I}^p_{\Gamma,\sigma} \;\in\; \mathbf{K}(\mathbf{P})(1, \tau_\Gamma \to \sigma)\,.$$

Here the type $\overline{1}$ is used to code NSP terms t via partial functions $\lceil t \rceil : \mathbb{N} \rightharpoonup \mathbb{N}$ as in Subsection 6.1.5, and τ_Γ is chosen to be capable of representing valuations ν for Γ within **P**. Likewise, our interpretation operations $(e, \nu) \mapsto [\![e]\!]_\nu$ and $(a, \nu) \mapsto [\![a]\!]_\nu$ for expressions and applications will be defined as elements

$$\mathbb{I}^e_\Gamma,\, \mathbb{I}^a_\Gamma \;\in\; \mathbf{K}(\mathbf{P})(1, \tau_\Gamma \to \mathtt{N})\,.$$

To make our construction work, we will actually need to choose τ_Γ so that it is also capable of representing valuations of all larger environments Γ' that arise in the course of interpreting NSP terms in Γ. Specifically, let $\sigma_0, \ldots, \sigma_{t-1}$ be the types of the variables in Γ, and let $\rho_0, \ldots, \rho_{c-1}$ be a non-repetitive enumeration of the types in $\sigma_0, \ldots, \sigma_{t-1}, \sigma$ along with all syntactic constituents of these types. Now set

$$\tau_\Gamma \;=\; \mathtt{N} \times (\mathtt{N} \to \rho_0) \times \cdots \times (\mathtt{N} \to \rho_{c-1})\,.$$

If Γ' is any environment whose variables have types among the ρ_i, we can represent a valuation for Γ' within **P** by an element of $\mathbf{K}(\mathbf{P})(\tau_\Gamma)$: the component of type $\mathtt{N} \to \rho_i$ allows us to represent the values of all the variables of type ρ_i, whilst the component of type $\mathtt{N}$ is used to record how many such variables there are for each i, and which variables are associated with which indices. The desired interpretation map $\mathbb{I}^p_{\Gamma,\sigma}$, for example, will now associate to each (p, ν') its interpretation $[\![p]\!]^*_{\nu'}$, where $\Gamma' \vdash p : \sigma$ for some Γ' as above, and ν' is a valuation for Γ.

We will define by simultaneous recursion suitable interpretation maps $\mathbb{I}^e_\Gamma$, $\mathbb{I}^a_\Gamma$ and $\mathbb{I}^p_{\Gamma,\rho} \in \mathbf{K}(\mathbf{P})(1, \tau_\Gamma \to \rho)$ for each $\rho \in \{\rho_0, \ldots, \rho_{c-1}\}$. Somewhat informally, a 'circular' definition of these interpretation maps may be given as follows. We set $\Gamma = x_0, \ldots, x_{t-1}$, and abuse notation by writing e.g. $[\![q]\!]^p_\nu$ in place of $\mathbb{I}^p_{\Gamma,\rho}(\lceil q \rceil)(\nu)$, where $\lceil - \rceil$ is a coding of NSP terms in $\mathbf{P}(1)$, and ρ is the type of q.

$$\begin{aligned}
[\![n]\!]^e_\nu &= n \;\text{ for } n \in \mathbb{N}\,, \\
[\![\mathtt{case}\; a \;\mathtt{of}\; (i \Rightarrow e_i)]\!]^e_\nu &= (\lambda i.\, [\![e_i]\!]^e_\nu)([\![a]\!]^a_\nu)\,, \\
[\![x_i q_0 \cdots q_{r-1}]\!]^a_\nu &= \nu(x_i)([\![q_0]\!]^p_\nu) \cdots ([\![q_{r-1}]\!]^p_\nu)\,, \\
[\![\lambda \vec{y}.e]\!]^p_\nu &= \lambda \vec{z}.\, [\![e]\!]^e_{\nu, \vec{y} \mapsto \vec{z}}\,.
\end{aligned}$$

The crucial point here is that this 'definition' of the class of interpretation maps of interest does not depend on any interpretation maps outside this class.

Note that the coding $\ulcorner - \urcorner$ is applicable only to those NSP terms that can be represented within $\mathbf{P}(1)$. (For instance, in the case of $\mathsf{PC}^{\mathrm{eff}}$ we can represent just the effective NSP terms, so our interpretation will be limited to these.) We shall assume the coding $\ulcorner - \urcorner$ is sensibly chosen so that if t is representable within $\mathbf{P}(1)$ then so is every subterm of t, and also so that basic operations on syntax trees such as subterm extraction are representable within $\mathbf{P}$ (this is needed to make formal sense of the right-hand sides of the above clauses).

Subject to all this, the above circular definition can be formally understood as a simultaneous least fixed point definition of the finitely many interpretation maps in question. We can now verify that this 'least fixed point interpretation' of NSP terms generalizes the liberal interpretation of Definition 6.4.6:

Theorem 6.4.15. *Suppose* $\mathbf{P}$ *satisfies the conditions of Definition 6.4.5 and is liberally Kleene closed (whence* $\mathbf{P}$ *satisfies assumptions 1–4 above). Then for any procedure* $\Gamma \vdash p : \sigma$ *with a code* $\ulcorner p \urcorner$ *in* $\mathbf{P}(1)$, *any valuation* ν *for* Γ, *any* $\vec{\Phi}$ *in* $\mathbf{P}$ *of appropriate types and any* $n \in \mathbb{N}$, *we have*

$$\mathbb{I}^{p}_{\Gamma,\sigma}(\ulcorner p \urcorner)(\nu)(\vec{\Phi}) \;=\; n \;\; \textit{iff} \;\; [\![p]\!]^{l}_{\nu}(\vec{\Phi}) \;=\; n \,.$$

Likewise, for $\Gamma \vdash e$ *with a code in* $\mathbf{P}(1)$, *we have*

$$\mathbb{I}^{e}_{\Gamma}(\ulcorner e \urcorner)(\nu) \;=\; n \;\; \textit{iff} \;\; [\![e]\!]^{l}_{\nu} \;=\; n \,,$$

and similarly for $\mathbb{I}^{a}$.

Proof. (Sketch.) For the right-to-left implications, we must check that the relations

$$\mathbb{I}^{p}_{\Gamma,\rho}(\ulcorner p \urcorner)(\nu)(\vec{\Phi}) \;=\; n \,, \qquad \mathbb{I}^{e}_{\Gamma}(\ulcorner e \urcorner)(\nu) \;=\; n \,, \qquad \mathbb{I}^{a}_{\Gamma}(\ulcorner a \urcorner)(\nu) \;=\; n$$

satisfy the closure conditions of Definition 6.4.6(i); this is easily done by comparing these conditions with the clauses of the circular definition above. For the converse, we may define alternative interpretation operations $\mathbb{I}'^{p}_{\Gamma,\rho}$ (for all relevant ρ) by

$$\mathbb{I}'^{p}_{\Gamma,\rho}(\ulcorner p \urcorner)(\nu) \;=\; [\![p]\!]^{l*}_{\nu}$$

and likewise for $\mathbb{I}'^{e}$ and $\mathbb{I}'^{a}$. To see that $\mathbb{I}'^{p}_{\Gamma,\rho} \in \mathbf{K}(\mathbf{P})(1, \tau_{\Gamma} \to \rho)$, one may give a direct construction of $\mathbb{I}'^{p}_{\Gamma,\rho}$ (simultaneously for all relevant ρ) as an $[\![-]\!]^{l}$-computable function—this amounts to defining a 'universal' function for liberal computability. We may now check by pointwise calculations that the $\mathbb{I}'$ interpretations satisfy the clauses of the above circular definition, whence $\mathbb{I} \sqsubseteq \mathbb{I}'$. $\square$

We have thus shown that the operators Y_{σ} are in some sense the *prototypical* liberally computable operations: once these are available, the entire interpretation $[\![-]\!]^{l}$, and hence all other liberally computable operations, can be reconstructed from them via relatively basic machinery. One can see this as evidence for the importance of least fixed points as a key concept in higher-order computability theory.

6.4.3 Relating Partial and Total Models

We now return to the question of how all this relates to the original notion of Kleene computability in a *total* model **A** that can be represented in **P**. In the course of this, we will shed some further light on the differences in computational power between the strict, intermediate and liberal interpretations. Our treatment here will be fairly brief, although the remaining details should be within reach as exercises for the interested reader.

The generic situation we have in mind is that of two total type structures, **A** over $\mathbb{N}$ and **P** over $\mathbb{N}_\bot$, together with a type-respecting simulation $\varepsilon : \mathbf{A} \longrightarrow\!\!\!\!\triangleright\, \mathbf{P}$ such that $\varepsilon_{\mathbb{N}}(n,u)$ iff $u = n$. To obtain useful results, it is best to assume that ε is actually a logical relation—that is, $\varepsilon(f,g)$ iff $\varepsilon(f(x),g(y))$ whenever $\varepsilon(x,y)$—so that **A** can be viewed as the extensional collapse of **P** with respect to $\mathbb{N} \subset \mathbb{N}_\bot$. We have three particular scenarios in mind:

$$\mathsf{S} \longrightarrow\!\!\!\!\triangleright\, \mathsf{M}\,, \qquad \mathsf{Ct} \longrightarrow\!\!\!\!\triangleright\, \mathsf{PC}\,, \qquad \mathsf{HEO} \longrightarrow\!\!\!\!\triangleright\, \mathsf{PC}^{\mathrm{eff}}\,.$$

In each of these cases, the model **A** is Kleene closed in the classical sense, and **P** is Kleene closed in the liberal sense (and hence also in the intermediate and strict senses).

We shall consider the strict and intermediate interpretations together, as they present very similar pictures. Our first result is that in each of the above three situations, everything in **A** that is realized in **P** by a strictly Kleene computable function, or even by an intermediately computable one, is Kleene computable in the original sense. This indicates that the possibility of performing detours via partial functionals does not, of itself, result in an ability to compute more total functionals.

Proposition 6.4.16. *Let* $\varepsilon : \mathbf{A} \longrightarrow\!\!\!\!\triangleright\, \mathbf{P}$ *be one of the three situations above. If* $G = [\![p]\!]^{i*} \in \mathbf{P}(\sigma)$ *for* $p \in \mathsf{SP}^0(\sigma)$ *and* $\varepsilon_\sigma(F,G)$, *then* $F = [\![p]\!]^{*} \in \mathbf{A}(\sigma)$. *A fortiori, the same holds with* $[\![-]\!]^{s*}$ *in place of* $[\![-]\!]^{i*}$.

Proof. (Sketch.) By induction on the derivation of values for $[\![p]\!]^{i}(\vec{\Phi})$ (suitably generalizing the induction claim to open NSP terms). The possibility that must be excluded is that $[\![xq_0 \ldots q_{r-1}]\!]^{i}_{\nu} \in \mathbb{N}$ for all relevant ν but that some $[\![q_i]\!]^{i*}_{\nu}$ falls outside the image of ε. For this, we use the fact that each of our three situations enjoys the following 'separation property': if $\varepsilon_{\sigma\to\tau}(\Phi,\Psi)$ and $G \in \mathbf{P}(\sigma)$ is not in the image of ε, there exists $\Psi' \sqsubseteq \Psi$ such that $\varepsilon(\Phi,\Psi')$ but $\Psi'(G) = (\Lambda\vec{y}.\bot) \in \mathbf{P}(\tau)$. (This may be checked with the help of Corollary 8.1.2 and its analogues for M and HEO.) □

In the case of $\mathsf{S} \longrightarrow\!\!\!\!\triangleright\, \mathsf{M}$, the converse to Proposition 6.4.16 also holds:

Proposition 6.4.17. *If* $F = [\![p]\!]^{*} \in \mathsf{S}(\sigma)$ *for* $p \in \mathsf{SP}^0(\sigma)$ *then* $G = [\![p]\!]^{*s} \in \mathsf{M}(\sigma)$ *is well-defined and* $\varepsilon_\sigma(F,G)$. *A fortiori, we also have* $G = [\![p]\!]^{*i}$.

Proof. (Sketch.) By induction on the derivation of values for $[\![-]\!]$. The main task is to show that if $[\![p]\!]^{*}$ is defined and total then the extension $[\![p]\!]^{*s}$ to the whole of M is well-specified. For this, we show that if $\vec{\Phi}' \sqsubseteq \vec{\Phi}$ in M and $[\![p]\!]^{s}(\vec{\Phi})\downarrow$, a computation tree for $[\![p]\!]^s(\vec{\Phi})$ may be inductively lowered to one for $[\![p]\!]^s(\vec{\Phi}')$. We then appeal to a special property of $\mathsf{S} \longrightarrow\!\!\!\triangleright \mathsf{M}$ which is easily checked: any $\Phi' \in \mathsf{M}$ can be extended to some Φ in the image of ε. □

Thus, the Kleene computable elements of S are exactly those with strictly (resp. intermediately) computable realizers in M.

In the case of $\mathsf{Ct} \longrightarrow\!\!\!\triangleright \mathsf{PC}$ and $\mathsf{HEO} \longrightarrow\!\!\!\triangleright \mathsf{PC}^{\mathrm{eff}}$, however, the converse to Proposition 6.4.16 fails for both $[\![-]\!]^s$ and $[\![-]\!]^i$, albeit for an annoying reason. Consider, for instance, the NSP $p \in \mathsf{SP}^0(3)$ naturally associated with the program

$$\lambda F.\,\mu n.\,F(d_n) = F(d_{n+1})\,,$$

where $d_n(n) = 1$ and $d_n(m) = 0$ for $m \neq n$. Then $[\![p]\!]^* \in \mathsf{Ct}(3)$ is defined, since for any $F \in \mathsf{Ct}(2)$ the $F(d_n)$ converge to $F(\Lambda m.0)$, whereas we obtain no well-defined value (not even $\bot$) for $[\![p]\!]^s$ applied to the function $G = (\Lambda g.\mu n.\,g(n) = 1) \in \mathsf{PC}(2)$ (note that $G(d_n) = n$ for each n). It is also easy to adapt this counterexample to the case of $[\![-]\!]^i$. In this sense, the ability to compute with partial functionals is actually a disadvantage where $[\![-]\!]^s$ and $[\![-]\!]^i$ are concerned. These observations suggest that the classes of functionals in Ct or HEO with strictly or intermediately computable representatives may not be very natural ones to consider.

We now turn our attention to the liberal interpretation. Since, for any $p \in \mathsf{SP}^0$, the interpretation $[\![p]\!]^l$ in each of M, PC, $\mathsf{PC}^{\mathrm{eff}}$ is always well-specified, there is this time no difficulty in showing the following by a routine induction on derivations:

Proposition 6.4.18. *Suppose $\varepsilon : \mathbf{A} \longrightarrow\!\!\!\triangleright \mathbf{P}$ is one of our three leading examples. If $[\![p]\!]^* = F$ for $p \in \mathsf{SP}^0$, then $[\![p]\!]^{*l} = G$ where $\varepsilon(F,G)$.* □

In the case of $\mathsf{S} \longrightarrow\!\!\!\triangleright \mathsf{M}$, we once again have perfect accord between the partial and the total settings. This characterization of Kleene computability in S is in essence due to Platek [233]; see also Subsection 7.1.7.

Theorem 6.4.19. *For any $p \in \mathsf{SP}^0(\sigma)$, $F \in \mathsf{S}(\sigma)$, we have $[\![p]\!]^*_{\mathsf{S}} = F$ iff $\varepsilon(F, [\![p]\!]^{*l}_{\mathsf{M}})$.*

Proof. The forwards implication is given by Proposition 6.4.18. For the converse, the key observation is that every element $\Phi \in \mathsf{S}(\sigma)$ has a *minimal* realizer $\widehat{\Phi} \in \mathsf{M}(\sigma)$, characterized by the property that $\Phi(\vec{x}) = \bot$ if some x_i is outside the image of ε. We reason by induction on derivations for $[\![-]\!]^l$, much as in the proof of Proposition 6.4.16. □

In the case of $\mathsf{Ct} \longrightarrow\!\!\!\triangleright \mathsf{PC}$ and $\mathsf{HEO} \longrightarrow\!\!\!\triangleright \mathsf{PC}^{\mathrm{eff}}$, however, the situation is more subtle and interesting. On the one hand, if we restrict our attention to *left-well-founded* procedures, the accord between partial and total settings is maintained:

Theorem 6.4.20. *(i) A procedure $p \in \mathsf{SP}^{0,\mathrm{lwf}}(\sigma)$ denotes an element $F \in \mathsf{Ct}(\sigma)$ iff $[\![p]\!]^{l*} \in \mathsf{PC}(\sigma)$ is an ε-realizer for F.*

(ii) An element $F \in \mathsf{Ct}(\sigma)$ is μ-computable iff some ε-realizer for F is μ-computable (i.e. is of the form $[\![p]\!]^{l*}$ for some $\mathsf{Klex}^{\min}$ definable p).[3]

[3] This can be seen as a corrected version of a result claimed in [220]. The proof in [220] contained a significant error in the treatment of the schemes S4 and S5. See also the discussion at the start of Section 8.6.

Likewise for HEO *and* $\mathsf{PC}^{\mathrm{eff}}$.

Proof. (i) Suppose $F = [\![p]\!]^*$ for some $p \in \mathsf{SP}^{0,\mathrm{lwf}}(\sigma)$. Since the liberal interpretation is always well-specified, we may define $G = [\![p]\!]^{l*}$, and an easy induction on derivations shows that $\varepsilon(F,G)$. Conversely, suppose $\varepsilon(F,G)$ where $G = [\![p]\!]^{l*}$. If p is LWF then $[\![p]\!]^{i*}$ is well-defined by Exercise 6.4.4, and then $[\![p]\!]^{i*} = G$ by Exercise 6.4.8. Hence by Proposition 6.4.16 we have $F = [\![p]\!]^*$, so F is μ-computable.

Part (ii) follows easily, since $\mathsf{Klex}^{\mathrm{min}}$ definable procedures are left-well-founded by Theorem 6.3.18. □

On the other hand, once we pass beyond the realm of LWF procedures and admit the full power of Kleene computation, it turns out that we get genuinely more computable *total* functionals by working in the partial model than in the total one. For example, we will see in Subsection 8.3.2 that the famous *fan functional* is not Kleene computable in Ct, but is representable by a liberally Kleene computable element of PC. What is more, this phenomenon is not restricted to isolated examples, but turns out to be systematic: Theorem 8.6.1 shows that *every* effective element of Ct (that is, every element with a representative in $\mathsf{PC}^{\mathrm{eff}}$) has an $[\![-]\!]^l$-computable representative in PC, even though not every element of $\mathsf{PC}^{\mathrm{eff}}$ is $[\![-]\!]^l$-computable.

Likewise, in Section 9.3, we shall give an example due to Gandy of a functional in HEO that is not Kleene computable but is realized by a $[\![-]\!]^l$-computable element of $\mathsf{PC}^{\mathrm{eff}}$. Once again, the phenomenon is systematic: every element of HEO has a $[\![-]\!]^l$-computable representative in $\mathsf{PC}^{\mathrm{eff}}$ (Theorem 9.2.13).

These phenomena show that a crucial threshold in computational power is crossed once we move to the liberal interpretation of Kleene computability. Indeed, the import of some of our later results will be that once this threshold has been crossed, no further increases or variations in the computational power of **P** (subject to some mild constraints) will affect the class of total computable functionals that arises (see Theorems 8.6.7 and 9.2.15). It is for this reason that we have attached such importance here to pinpointing the exact location of this threshold, and analysing its nature via a comparison of our three variants of Kleene computability.

To summarize the results of our investigation: all three of these variants interpreted in M give rise to the same substructure $\mathsf{S}^{\mathrm{Kl}} \subseteq \mathsf{S}$, and this can be seen as further evidence that Kleene computability in S is a mathematically robust notion. In the case of Ct —▷ PC or HEO —▷ $\mathsf{PC}^{\mathrm{eff}}$, the picture is more complex, but our results at least conspire to pinpoint the root cause of the mismatch between $[\![-]\!]^l_{\mathsf{PC}}$ and $[\![-]\!]_{\mathsf{Ct}}$ as evidenced by the fan functional. As Proposition 6.4.16 indicates, it is not that any expressivity is gained simply by allowing computations to take detours via partial functionals; rather, the phenomenon stems from the particular treatment of undefinedness embodied in the liberal interpretation (cf. Proposition 6.4.16). Furthermore, the difference only becomes visible in models with some kind of continuity constraint (cf. Theorem 6.4.19), and even then only in connection with *infinite nesting* of function calls (see Exercises 6.4.8 and 6.4.4).

As an aside, we are now in a position to fulfil a promise from Subsection 6.3.3 and establish the following fact, which furnishes evidence that μ-computability is a mathematically natural notion in the models of interest:

Theorem 6.4.21. *In any of* S, M, Ct, PC, HEO, $\mathsf{PC}^{\mathrm{eff}}$, *the μ-computable elements are precisely those denoted by effective left-bounded procedures (using $[\![-]\!]^l$ or equivalently $[\![-]\!]^i$ in the case of* M, PC, $\mathsf{PC}^{\mathrm{eff}}$*).*

Proof. First, an easy adaptation of the proof of Theorem 6.3.22 shows that there is a term of $\mathsf{Klex}^{\mathrm{min}}$ whose denotation in any extensional type structure **P** over $\mathbb{N}_\perp$ coincides with that of *plug*. Since by Theorem 6.3.20 every $\mathsf{SP}^{0,\mathrm{lbd,eff}}$-denotable element of **P** is $\mathsf{Klex}^{\mathrm{plug}}$ denotable, the desired property holds for the partial models M, PC, $\mathsf{PC}^{\mathrm{eff}}$. For the total models, the result now follows by Theorem 6.4.20. □

This is also a convenient point at which to revisit Kleene's original definition of computability over total type structures (Definition 5.1.1), and to consider what light our excursion into partial functionals may shed on some of the problematic features of this definition, notably the totality side-condition in S8 and the resulting non-composability of partial computable functionals.

On a naive reading, Kleene's S8 would appear to say that an oracle $\Phi \in \mathbf{A}(k+1)$ will yield a numerical result whenever it is presented with a *total* type $\overline{k}$ argument, but will diverge otherwise, suggesting that oracles by nature have a magical ability to semidecide the totality of their input. However, our experience of presenting total functionals in terms of partial ones now suggests that this is a somewhat spurious idea: it would be more realistic to say that we *do not know* what will happen if we try to apply an oracle to a non-total argument. Thus, for instance, if our oracle Φ happens to be represented by some $\Psi \in \mathbf{P}(k+1)$, there may well be non-total arguments $F \in \mathbf{P}(k)$ for which $\Psi(F)$ yields a value in $\mathbb{N}$—indeed, in continuous settings the existence of such F is inevitable. From this perspective, a computation triple $\{e\}(\vec{\Phi}) = a$ should be viewed as asserting not simply that the operation described by e yields the value a when applied to Φ, but that it is *guaranteed* to do so purely by virtue of the extensions of the Φ_i as elements of **A**. This fits with the idea that Kleene computability constitutes a 'common core' for all reasonable computability notions, as well as making the Π^1_1 nature of termination (when $\mathbf{A} = \mathsf{S}$) somewhat more intelligible.

This, in turn, suggests that Kleene's original definition of partial computable functionals over S might not be the most fruitful one to consider. One might argue that for an index e to compute a partial functional F, it ought to be *known* that the computation of $\{e\}(\vec{\Phi})$ diverges whenever $F(\vec{\Phi})$ is undefined. For instance, one might declare $F : \mathsf{S}(\sigma_0) \times \cdots \times \mathsf{S}(\sigma_{r-1}) \to \mathbb{N}_\perp$ to be computable iff there is some $[\![-]\!]^l$-computable $G \in \mathsf{M}(\vec{\sigma} \to \mathtt{N})$ such that whenever $\vec{\Phi}, \vec{\Psi}$ are suitably typed and $\varepsilon(\Phi_i, \Psi_i)$ for each i, we have $G(\vec{\Psi}) = F(\vec{\Phi}) \in \mathbb{N}_\perp$. It is easy to show that the computable partial functionals thus defined are closed under composition, and that this notion enjoys other pleasing properties not shared by the original definition: for instance, the semidecidable subsets of any $\mathsf{S}(\sigma)$ are closed under binary unions. We leave the further development of a theory of partial computable functionals along these lines as a possible (if not exactly pressing) topic for future research.

6.4.4 Kleene Computability via Inductive Definitions

Having established the key role of fixed points, we now offer another perspective on Kleene computability due to Moschovakis [200, 130]. We shall begin with a version of Moschovakis's ideas somewhat freely adapted to our present setup, and proceed from there to a brief sketch of his original approach. Since these results do not feature later in the book, we will not give detailed proofs.

First, we show that Proposition 6.4.18 can be strengthened considerably: if fixed points in $\mathbf{P}$ are available, then every element of $\mathbf{A}^{\mathrm{Kl}}$ can be computed using just a single invocation of Fix_σ along with a bare minimum of other machinery.

Suppose, in general, that $\mathbf{P}$ is a total type structure over $\mathbb{N}_\perp$ which is liberally Kleene closed, and hence by Proposition 6.4.14 possesses least fixed point operators with respect to the pointwise ordering. Suppose too that $\mathscr{L}$ is some language consisting of terms $M:\tau$ in environments $\Gamma = x_0^{\sigma_0},\ldots,x_{r-1}^{\sigma_{r-1}}$, and endowed with an interpretation $[\![-]\!]$ in $\mathbf{P}$ such that if $f:\sigma \vdash M:\sigma$ then $[\![M]\!]_f$ is a monotone function $\mathbf{P}(\sigma) \to \mathbf{P}(\sigma)$. We may then call an element $\Phi \in \mathbf{P}(\sigma)$ *$\mathscr{L}$-inductive* if there exist a closed $\mathscr{L}$ term $f:\mathtt{N}^{(t)} \to \sigma \vdash M:\mathtt{N}^{(t)} \to \sigma$ and numbers $n_0,\ldots,n_{t-1} \in \mathbb{N}$ such that

$$\Phi \;=\; (Fix_{\mathtt{N}^{(t)}\to\sigma}[\![M]\!]_f)(\vec{n})\,.$$

In particular, let $\mathscr{L}_M$ be the simply typed λ-calculus with constants $0:\mathtt{N}$, $suc:\mathtt{N}\to\mathtt{N}$, $eq:\mathtt{N}^{(2)}\to\mathtt{N}$ and $ifzero:\mathtt{N}^{(3)}\to\mathtt{N}$, and let $[\![-]\!]$ be its evident interpretation in $\mathbf{P}$ (interpreting *ifzero* as 'strong definition by cases'). In the light of Proposition 6.4.14 and Exercise 6.4.10, every $\mathscr{L}_M$-inductive element of $\mathbf{P}$ is liberally Kleene computable. We now show that the converse holds.

The following lemma gives some important closure properties of the inductive elements. We leave the proofs as non-trivial exercises.

Lemma 6.4.22. *(i) The $\mathscr{L}_M$-inductive functionals are closed under composition: if $\Psi \in \mathbf{P}(\tau_0,\ldots,\tau_{r-1}\to\upsilon)$ and $\Phi_i \in \mathbf{P}(\vec{\sigma}\to\tau_i)$ for each $i<r$ are all $\mathscr{L}_M$-inductive, then so is*

$$(\Lambda\vec{z}.\,\Psi(\Phi_0(\vec{z}),\ldots,\Phi_{r-1}(\vec{z}))) \;\in\; \mathbf{P}(\vec{\sigma}\to\upsilon)\,.$$

(ii) If $\Phi \in \mathbf{P}(\vec{\sigma},\tau\to\tau)$ is $\mathscr{L}_M$-inductive, then so is

$$(\Lambda\vec{z}.\,Fix_\tau(\Lambda x.\Phi(\vec{z},x))) \;\in\; \mathbf{P}(\vec{\sigma}\to\tau)\,.$$

These properties allow us to perform complex definitions involving repeated substitutions and fixed point constructions, knowing that we will still remain within the class of $\mathscr{L}_M$-inductive functions. For instance:

Exercise 6.4.23. (i) Show that $min \in \mathbf{P}(2)$ is inductive, where $min(f)=n$ iff $f(n)>0$ and $f(m)=0$ for all $m<n$.

(ii) Show that the predecessor function is inductive. (Hint: if $n\neq 0$ then $pre(n) = \mu m.\,suc(m)=n$.)

(iii) Show that the functionals Fix_σ are inductive.

(iv) Show that any $\mathscr{L}_M$-definable functional is itself $\mathscr{L}_M$-inductive.

It follows readily from Lemma 6.4.22 and Exercise 6.4.23 that every PCF-*definable* element of **P** is $\mathscr{L}_M$-inductive. Since the machinery of Subsection 6.4.2 in effect yields a PCF definition of the liberal interpretation map, and every effective NSP has a PCF-definable code, we may conclude:

Theorem 6.4.24. *Every liberally Kleene computable element of* **P** *is* $\mathscr{L}_M$*-inductive.*

In particular, using Theorem 6.4.19, the elements of S realized by $\mathscr{L}_M$-inductive elements of M are exactly those of S^{Kl}.

Not every inductive element of M represents an element of S^{Kl}, or even a Kleene computable partial functional over S, because in general $Fix_\sigma[\![M]\!]$ may behave non-extensionally on representatives of total functionals, even though $[\![M]\!]$ itself will always behave extensionally on such representatives. For example, if

$$M : \overline{2} \to \overline{2} \;=\; \lambda F^2.\lambda f^1.f(Ff)$$

and $f_0, f_1 \in \mathsf{M}(1)$ both represent $(\Lambda x.0) \in \mathsf{S}(1)$ but with $f_0(\bot) = \bot$, $f_1(\bot) = 0$, then $(Fix[\![M]\!])(f_0) = \bot$ but $(Fix[\![M]\!])(f_1) = 0$. It is thus natural to ask whether we can find some other language $\mathscr{L}$ (interpreted in M) such that the $\mathscr{L}$-inductive functionals exactly represent the Kleene computable partial functionals over S.

This is in effect what the approach of Moschovakis achieves (we here recast the presentation of [130] in the spirit of typed λ-calculi). In place of $\mathscr{L}_M$, we adopt a more tightly controlled calculus $\mathscr{L}_S$ that distinguishes between 'total' objects and 'partial' ones, and also accords special treatment to the ground type $\mathtt{N}$.

In $\mathscr{L}_S$, the *total types* are the usual types $\sigma \in \mathsf{T}^{\to}(\mathtt{N})$, whilst the *partial types* have the form $\sigma_0, \ldots, \sigma_{r-1} \rightharpoonup \mathtt{N}$ where the σ_i are total types. We let σ, τ range over total types, α, β over partial ones, and write $\mathtt{N}_\bot$ for the partial type $\rightharpoonup \mathtt{N}$. Environments Γ are lists of variables which may be of total or partial type; we use the metavariable x for total variables and f for partial ones. Well-typed terms $\Gamma \vdash M : \alpha$ are generated by the typing rules below, which also involve auxiliary judgements of the form $\Gamma \vdash x : \sigma$ for $\sigma \neq \mathtt{N}$. We take C to range over the following typed constants:

$$0 : \mathtt{N}_\bot, \qquad suc : \mathtt{N} \rightharpoonup \mathtt{N}, \qquad eq : \mathtt{N}^{(2)} \rightharpoonup \mathtt{N}, \qquad ifzero : \mathtt{N}^{(3)} \rightharpoonup \mathtt{N}.$$

For σ a total type, we define $\sigma^* = \mathtt{N}_\bot$ if $\sigma = \mathtt{N}$, and $\sigma^* = \sigma$ otherwise. If $\alpha = \sigma_0, \ldots, \sigma_{r-1} \rightharpoonup \mathtt{N}$, we write $\alpha^\dagger$ for the total type $\sigma_0, \ldots, \sigma_{r-1} \to \mathtt{N}$ (note that $\mathtt{N}_\bot^\dagger = \mathtt{N}$).

$$\Gamma \vdash x : \sigma^* \;\; (x^\sigma \in \Gamma) \qquad\qquad \Gamma \vdash f : \alpha \;\; (f^\alpha \in \Gamma) \qquad\qquad \Gamma \vdash C : \alpha \;\; (C : \alpha)$$

$$\frac{\Gamma, x_0^{\sigma_0}, \ldots, x_{r-1}^{\sigma_{r-1}} \vdash M : \mathtt{N}_\bot}{\Gamma \vdash \lambda \vec{x}.M : \sigma_0, \ldots, \sigma_{r-1} \rightharpoonup \mathtt{N}}$$

$$\frac{\Gamma \vdash P : \sigma_0, \ldots, \sigma_{r-1} \rightharpoonup \mathtt{N} \quad \Gamma \vdash M_0 : \sigma_0^* \;\; \cdots \;\; \Gamma \vdash M_{r-1} : \sigma_{r-1}^*}{\Gamma \vdash PM_0 \cdots M_{r-1} : \mathtt{N}_\bot}$$

$$\frac{\Gamma \vdash x : \alpha_0^\dagger, \ldots, \alpha_{s-1}^\dagger \to \mathtt{N} \quad \Gamma \vdash P_0 : \alpha_0 \;\; \cdots \;\; \Gamma \vdash P_{s-1} : \alpha_{s-1}}{\Gamma \vdash x \cdot (P_0, \ldots, P_{s-1}) : \mathtt{N}_\bot}$$

Notice that we have two kinds of application: one for applying a potentially partial term P to terms which must be either total variables x or else of ground type, and another for applying a total variable x to potentially partial objects. It is an easy exercise to represent the application of one total variable to another within this system via hereditary η-expansion, but we may not apply one *partial* term to another unless the latter is of ground type.

The language $\mathscr{L}_S$ may be interpreted in M as follows. Total types are understood to consist of representatives of elements of S, whilst partial types consist of those elements that extensionally represent partial functionals over S. The meaning of an application $PM_0 \dots M_{r-1}$ is taken to be $\perp$ if some M_i of ground type denotes $\perp$. More radically, the meaning of $x \cdot (P_0, \dots, P_{s-1})$ is taken to be $\perp$ if some P_i represents a non-total functional; this means that $\mathscr{L}_S$ can define *discontinuous* functionals. On the other hand, $\mathscr{L}_S$ enforces pleasing extensionality properties (we here write $h \sim h'$ if $h, h' \in$ M represent the same element of S or the same partial functional over S):

Proposition 6.4.25. *(i) If $\Gamma \vdash M : \sigma$ in $\mathscr{L}_S$ and $\nu \sim \nu'$ elementwise, then $[\![M]\!]_\nu \sim [\![M]\!]_{\nu'}$. In particular, $[\![M]\!]_\nu$ represents a partial functional over* S.

(ii) Every $\mathscr{L}_S$-inductive element of M *represents a partial functional over* S.

It is shown in [130], in effect, that the closure properties of Lemma 6.4.22 hold also for $\mathscr{L}_S$, and hence that the Kleene application operations $(e, \vec{\Phi}) \mapsto \{e\}(\vec{\Phi})$ are $\mathscr{L}_S$-inductive. Thus every Kleene computable partial functional over S is realized by an $\mathscr{L}_S$-inductive element. Conversely, it is also shown in [130] that the partial functional represented by any $\mathscr{L}_S$-inductive element is Kleene computable over S. We thus have a characterization of Kleene computability over S via partial elements, but with no junk.

What is more, Lemma 6.4.22(ii) (for $\mathscr{L}_S$) offers perhaps the most satisfactory analogue of the classical *first recursion theorem* that is available for Kleene computability over S.[4] The price to pay is the somewhat artificial character of the language $\mathscr{L}_S$, along with the discontinuity of the application $\cdot$, which in effect reintroduces the problematic side-condition from S8.

Finally, we may apply Occam's razor to the above interpretation to recover the one originally used by Moschovakis. Since everything is now extensional by fiat, we may as well work directly with elements of S rather than their representatives in M. Thus, a total type σ can be interpreted simply by the sets $\mathsf{S}(\sigma)$, and a partial type $\sigma_0, \dots, \sigma_{r-1} \rightharpoonup \mathtt{N}$ by the set of all partial functionals $\mathsf{S}(\sigma_0) \times \cdots \times \mathsf{S}(\sigma_{r-1}) \rightharpoonup \mathbb{N}$. The above interpretation can now be recast in the obvious way, so that terms with free variables are interpreted as monotone operators on these partial function spaces. As before, the $\mathscr{L}_S$-inductive partial functionals over S are defined to be those of the form $(Fix_{\mathtt{N}^{(t)} \to \sigma}[\![M]\!])(\vec{n})$ for some closed $\mathscr{L}_S$ term M, and once again, we obtain a junk-free characterization of the Kleene computable partial functionals over S.

[4] The failure of the most obvious formulation of the first recursion theorem for Kleene computability in S is discussed in [139], where Kleene also presents his own version of the theorem carefully hedged in by suitable restrictions.

Chapter 7
PCF and Its Models

In the previous chapter, we considered NSPs as abstract algorithms for sequential higher-order computation, and showed that the partial type structures admitting a well-behaved (liberal) interpretation of NSPs are exactly those with *least fixed point* operators. In the present chapter, we will develop an alternative perspective on essentially the same computability notion that takes least fixed points as primitive—a perspective conveniently embodied by Plotkin's language PCF as introduced in Subsection 3.2.3. Like Kleene's S1–S9, PCF offers a syntactic calculus for defining NSPs (and elements of other type structures), but in contrast to S1–S9 it is coding-free and hence cleaner to work with as a 'programming language'. Indeed, PCF has in practice proved valuable as a basis for the design of functional programming languages such as Standard ML and Haskell.

As explained in Chapter 2, a version of PCF (in a combinatory form) was first introduced in Platek's Ph.D. thesis [233] as a term language for defining elements in the type structure M of partial monotone functionals (see Section 6.4). Essentially the same language, under the name LCF (Logic of Computable Functions), was used by Scott for denoting elements in the model PC of partial *continuous* functionals, this time with a more explicit emphasis on computability. Soon afterwards, Plotkin [236] endowed LCF with an operational semantics based on purely symbolic rules for evaluating programs; this meant that LCF could be considered as a standalone programming language, independently of any mathematical model or class of functionals in which it might be interpreted. In this capacity, the language was re-christened PCF (Programming language for Computable Functions). Our exposition in the present chapter will be much in the spirit of Plotkin's work and of the theoretical computer science tradition to which it gave rise, with an emphasis on *continuous* models of PCF, and paying close attention to the relationship between operational and denotational perspectives.

In view of its status as a standalone executable programming language, one can distinguish two possible reasons for being interested in mathematical models for PCF, both of which will figure in this chapter. Firstly, one's primary interest might be in the language, and one might hope to learn something about it from the models—for instance, to establish general properties of all PCF programs, or behavioural equivalences between particular programs. Alternatively, one's main interest might be in the models and notions of computability over them—in this case, we might see PCF as conveniently embodying a relatively weak computability notion that makes sense for a large class of models. (This, of course, assumes that we are willing to regard the least fixed point operators as 'computable' for the models in question—see the discussion at the end of Subsection 6.4.1.)

In Section 7.1, we develop the basic theory of PCF and its models, showing how it relates to the theory of NSPs as presented in Chapter 6. After recalling the definition of the language and its operational semantics, we introduce the general notion of a *model of* PCF, and show that the interpretation of PCF in any such model is *adequate* with respect to the operational semantics. We also discuss the concepts of *observational equivalence* of programs and *full abstraction* for models of PCF. We place particular emphasis on the sequential procedure model SP^0, and prove an important definability result: in an innocuous extension of PCF, every element of SP^0 is definable relative to a type 1 oracle (and every effective element is definable outright). Finally, we establish the connection with the language Klex of Kleene expressions, and with the fixed point interpretation of NSPs described in Subsection 6.4.2.

In Section 7.2, we consider a *call-by-value* version of PCF, as well as some extensions with a richer repertoire of datatypes. We show that under some simple conditions, a model of PCF already provides, in essence, the structure needed to give interpretations of these closely related languages.

In Section 7.3, we present some specific examples of PCF-programmable operations, illustrating the sometimes surprising scope of this computability notion. We also mention some important non-examples in order to point out the limitations of the language, and use these non-examples to show that various prominent models of PCF are not fully abstract.

In Section 7.4, we briefly review another perspective on the same computability notion provided by *game semantics*. This yields a model isomorphic to SP^0, although the isomorphism is somewhat non-trivial. A particular attraction of game models is that they allow us to situate PCF-computability within a wider framework, also embracing models for languages with control flow operators, imperative variables and other 'non-functional' features.

In Sections 7.5–7.7, we shift our emphasis from 'intensional' representations of PCF algorithms (via NSPs or games) to the class of functionals that these define. Specifically, we resume our study of the model SF of *(*PCF*)-sequential functions* (the extensional quotient of SP^0) which we began in Subsection 6.1.4. Whilst good mathematical characterizations of the sequential functionals are available at low types, an important undecidability result due to Loader [172] precludes the possibility of a genuinely 'effective' characterization of such functionals in general. We also present some results on the order-theoretic structure of SF, showing for instance that there are increasing chains of sequential functionals whose least upper bound is not present in SF. Finally, we show that the model SF contains no *universal* type, in contrast to the models PC, SR and SA to be considered in later chapters.

7.1 Interpretations of PCF

Recall from Subsection 3.2.3 that PCF is the simply typed λ-calculus (over the type world $\mathsf{T} = \mathsf{T}^{\rightarrow}(\mathtt{N})$) with constants

$$\begin{aligned} \widehat{n} &: \mathtt{N} \quad \text{for each } n \in \mathbb{N} \,, \\ suc, pre &: \mathtt{N} \rightarrow \mathtt{N} \,, \\ ifzero &: \mathtt{N},\mathtt{N},\mathtt{N} \rightarrow \mathtt{N} \,, \\ Y_\sigma &: (\sigma \rightarrow \sigma) \rightarrow \sigma \qquad \text{for each } \sigma \in \mathsf{T} \,, \end{aligned}$$

and reduction rules

$$\begin{aligned} (\lambda x.M)N &\rightsquigarrow M[x \mapsto N] \,, & ifzero\, \widehat{0} &\rightsquigarrow \lambda xy.x \,, \\ suc\, \widehat{n} &\rightsquigarrow \widehat{n+1} \,, & ifzero\, \widehat{n+1} &\rightsquigarrow \lambda xy.y \,, \\ pre\, \widehat{n+1} &\rightsquigarrow \widehat{n} \,, & Y_\sigma f &\rightsquigarrow f(Y_\sigma f) \,, \\ pre\, \widehat{0} &\rightsquigarrow \widehat{0} \,, \end{aligned}$$

which may be applied (for any closed instantiation of the free variable f) in the evaluation contexts generated by

$$[-]N \,, \qquad suc\,[-] \,, \qquad pre\,[-] \,, \qquad ifzero\,[-] \,.$$

Many inessential variations on this definition are possible. For example, the version of PCF introduced by Plotkin [236] included a primitive type of booleans—this and other mild extensions will be considered in Subsection 7.2.1. Other variations in the exact choice of constants may also be found in the literature. We shall not delay here to discuss such differences further, since all these variants are easily seen to be intertranslatable, and all our main results will be equally applicable to any of them.

Exercise 7.1.1. (i) Construct a PCF term $eq : \mathtt{N}^{(2)} \rightarrow \mathtt{N}$ such that $eq\, \widehat{n}\, \widehat{n} \rightsquigarrow^* \widehat{0}$ for all n and $eq\, \widehat{n}\, \widehat{m} \rightsquigarrow^* \widehat{1}$ whenever $n \neq m$.

(ii) Show that every partial computable function $\mathbb{N}^r \rightharpoonup \mathbb{N}$ is (strongly) representable by a PCF term of type $\mathtt{N}^{(r)} \rightarrow \mathtt{N}$.

By analogy with the definition of Klex^Ω (see Section 6.2), we shall also write PCF^Ω ('PCF with oracles') for the language obtained by extending the definition with a constant $C_f : \mathtt{N} \rightarrow \mathtt{N}$ for every mathematical partial function $f : \mathbb{N} \rightharpoonup \mathbb{N}$, along with reduction rules

$$C_f\, \widehat{n} \rightsquigarrow \widehat{m} \quad \text{whenever } f(n) = m \,,$$

and adding $C_f\,[-]$ to our collection of generating evaluation contexts. Note that in the literature, PCF^Ω sometimes denotes the language with oracles only for *total* functions $\mathbb{N} \rightarrow \mathbb{N}$; this minor difference will be discussed in Remark 7.1.35 below.

We shall also be interested in a mild extension of PCF and PCF$^{\Omega}$ with a constant *byval* : $\overline{1} \to \overline{1}$. Intuitively, this will transform any function $f : \overline{1}$ into a version of f that accepts its argument 'by value': whereas the computation of fx might in general evaluate x zero, one or more times, the computation of $(byval\ f)x$ will evaluate x exactly once; if this yields a numeral $\widehat{n}$, then $f\widehat{n}$ is evaluated. To give an operational semantics reflecting this intention, we may add reduction rules

$$byval\ f\ \widehat{n} \rightsquigarrow f\ \widehat{n} \quad \text{for each } n \in \mathbb{N}\ ,$$

along with new evaluation contexts $byval\ M\ [-]$. As we shall see, the addition of *byval* does not lead to any new computable functionals, but it will extend the class of intensional procedures or strategies that we can define.

The section will be structured as follows (for simplicity we here use the name 'PCF' as representative of the above four languages). In Subsection 7.1.1, we shall give general conditions for a λ-algebra to be a *model of* PCF, somewhat strengthening the conditions given by Example 3.5.10. We draw particular attention to the model SP^0 and its quotient SF. In Subsection 7.1.2, we show that in any model of PCF some useful mathematical structure is available—notably an *observational ordering* and a well-behaved notion of *observational limit*. We then turn to various properties of interpretations of PCF in such models. In Subsection 7.1.3 we show that such interpretations are faithful to the operational semantics of PCF, a property we refer to as *adequacy*. In Subsection 7.1.4 we note that if two PCF terms have the same interpretation in some model then they are *observationally equivalent*, and show among other things that the model SF is *fully abstract* (that is, it perfectly captures observational equivalence for PCF). In Subsection 7.1.5 we show how sequential procedures may be 'implemented' within PCF itself, and use this to prove *definability* results for the models SP^0 and SF: e.g. every element of SF is denotable by a term of PCF$^{\Omega}$. We also use these results to show how any PCF-computable operation of type $\overline{k}$ can (roughly speaking) be defined without recourse to subterms of level higher than k. Finally, in Subsections 7.1.6 and 7.1.7 we make the connection with ideas from Chapter 6, making explicit the relationship of PCF to Kleene expressions and to the fixed point interpretation of NSPs.

7.1.1 Models of PCF

Suppose $\mathbf{P}$ is any $\lambda\eta$-algebra over $\mathsf{T}^{\to}(\mathtt{N})$. As explained in Section 4.1, we can interpret simply typed λ-terms in $\mathbf{P}$, so in order to interpret PCF terms, it suffices to specify suitable interpretations for the constants of PCF. These are given by the following definition. We shall overload notation by using the same symbols for the PCF constants and the corresponding elements of $\mathbf{P}$.

Definition 7.1.2. (i) A *pre-model of* PCF is a $\lambda\eta$-algebra $\mathbf{P}$ over $\mathsf{T}^{\to}(\mathtt{N})$ equipped with distinct elements

$$\begin{aligned} \widehat{n} &\in \mathbf{P}(\mathtt{N}) \quad \text{for } n \in \mathbb{N}\,, \\ suc, pre &\in \mathbf{P}(\mathtt{N} \rightarrow \mathtt{N})\,, \\ ifzero &\in \mathbf{P}(\mathtt{N},\mathtt{N},\mathtt{N} \rightarrow \mathtt{N})\,, \\ Y_\sigma &\in \mathbf{P}((\sigma \rightarrow \sigma) \rightarrow \sigma) \end{aligned}$$

satisfying the following equations for all $n \in \mathbb{N}$:

$$\begin{aligned} suc\,\widehat{n} &= \widehat{n+1}\,, & \lambda xy.\, ifzero\,\widehat{0}\,x\,y &= \lambda xy.x\,, \\ pre\,\widehat{n+1} &= \widehat{n}\,, & \lambda xy.\, ifzero\,\widehat{n+1}\,x\,y &= \lambda xy.y\,, \\ pre\,\widehat{0} &= \widehat{0}\,, & \lambda f.\,Y_\sigma f &= \lambda f.\,f(Y_\sigma f)\,. \end{aligned}$$

(ii) A *pre-model of* PCF^Ω is a pre-model of PCF additionally equipped with an element $C_f \in \mathbf{P}(\mathtt{N} \rightarrow \mathtt{N})$ for each $f : \mathbb{N} \rightharpoonup \mathbb{N}$ such that $C_f \cdot \widehat{n} = \widehat{m}$ whenever $f(n) = m$.

(iii) A *pre-model of* PCF + *byval* is a pre-model of PCF equipped with an element $byval \in \mathbf{P}(1 \rightarrow 1)$ such that $\lambda f.\, byval\, f\, \widehat{n} = \lambda f.\, f\, \widehat{n}$ for each n.

For many purposes, we do not strictly require the η-rule in the above definition. However, its presence will smooth our path in a few places, and it does not entail much significant loss of generality (cf. the discussion in Subsection 4.1.4).

Whereas any pre-model will correctly capture the values of all terminating computations (see Proposition 7.1.12 below), it need not faithfully capture *non-termination* in general, nor need it reflect the *finitary* character of terminating computations in any useful way. We therefore introduce a stronger notion of model by adding further conditions of two kinds. First, we require $\mathbf{P}$ to contain a special element $\perp \in \mathbf{P}(\mathtt{N})$ (with certain properties) representing non-termination. Secondly, we focus on the elements

$$Y_\sigma^k \;=\; \lambda f^{\sigma\rightarrow\sigma}.f^k(\lambda\vec{x}.\perp) \;\in\; \mathbf{P}((\sigma \rightarrow \sigma) \rightarrow \sigma)$$

for $k \in \mathbb{N}$, which can be viewed as finite approximations to the fixed point operator Y_σ. We impose a continuity condition on $\mathbf{P}$ saying that Y_σ behaves as a 'limit' of these approximations. Formally:

Definition 7.1.3. (i) A *model of* PCF is a pre-model $\mathbf{P}$ of PCF endowed with an element $\perp \in \mathbf{P}(\mathtt{N})$ such that

$$suc \perp \;=\; \perp\,, \qquad pre \perp \;=\; \perp\,, \qquad \lambda xy.\, ifzero \perp x\,y \;=\; \lambda xy.\perp\,,$$

and satisfying the following (we write $\perp_\sigma$ for $\lambda\vec{x}.\perp$ and Y_σ^k for $\lambda f.f^k\perp_\sigma$).

- *Monotonicity axiom*: for any $g \in \mathbf{P}(\mathtt{N} \rightarrow \mathtt{N})$, if $g \cdot \perp = \widehat{n}$ then $g \cdot \widehat{m} = \widehat{n}$ for any m.
- *Continuity axiom*: for any $h \in \mathbf{P}(((\sigma \rightarrow \sigma) \rightarrow \sigma) \rightarrow \mathtt{N})$, if $h \cdot Y_\sigma = \widehat{n}$ then there exists k such that $h \cdot Y_\sigma^k = \widehat{n}$.

(ii) A *model of* PCF^Ω is a model of PCF which is also a pre-model of PCF^Ω such that $C_f \widehat{n} = \perp$ whenever $f(n)$ is undefined.

(iii) A *model of* PCF + *byval* is a model of PCF which is also a pre-model of PCF + *byval* such that $\lambda f.\, byval\, f \perp = \lambda f.\perp$.

It is an interesting exercise to show that if $\mathbf{P}(\mathtt{N}) = \mathbb{N}_\perp$ then the monotonicity axiom is automatic. Informally, the continuity axiom says that no semantic observation h is able to 'tear away' Y_σ from all of its finite approximants.

Example 7.1.4. The partial continuous model PC of Subsection 3.2.2 is clearly a model of PCF^Ω: as noted under Example 3.5.6, the elements Y_σ are given by the least fixed point operators $\bigsqcup_n(\Lambda f.f^n(\perp))$, and the interpretations of the other constants are easy. The continuity axiom holds because $Y_\sigma = \bigsqcup Y_\sigma^k$ and all functionals in PC preserve least upper bounds. We may also interpret *byval* in a trivial way by taking $\mathit{byval}\ f\ n = f(n)$, $\mathit{byval}\ f\ \perp = \perp$.

Likewise, the effective submodel $\mathsf{PC}^{\mathrm{eff}}$ is a model of $\mathrm{PCF} + \mathit{byval}$.

Example 7.1.5. Another important model of $\mathrm{PCF}^\Omega + \mathit{byval}$ is the sequential procedure model SP^0. The interpretations of the constants are as follows; note that the definition of Y_σ uses a corecursively defined procedure $F_\sigma[g]$ (with free variable $g^{\sigma\to\sigma}$) which also featured in the proof of Proposition 6.4.14.

$$
\begin{aligned}
\widehat{n} &= \lambda.n\,, \\
\mathit{suc} &= \lambda x^{\mathtt{N}}.\ \mathtt{case}\ x\ \mathtt{of}\ (i \Rightarrow i+1)\,, \\
\mathit{pre} &= \lambda x^{\mathtt{N}}.\ \mathtt{case}\ x\ \mathtt{of}\ (0 \Rightarrow 0 \mid i+1 \Rightarrow i)\,, \\
\mathit{ifzero} &= \lambda x^{\mathtt{N}} y^{\mathtt{N}} z^{\mathtt{N}}.\ \mathtt{case}\ x\ \mathtt{of} \\
&\qquad\qquad (0 \Rightarrow \mathtt{case}\ y\ \mathtt{of}\ (j \Rightarrow j) \mid i+1 \Rightarrow \mathtt{case}\ z\ \mathtt{of}\ (j \Rightarrow j))\,, \\
F_\sigma[g] &= \lambda x_0^{\sigma_0} \ldots x_{r-1}^{\sigma_{r-1}}.\ \mathtt{case}\ g\ F_\sigma[g]\ x_0^\eta \cdots x_{r-1}^\eta\ \mathtt{of}\ (i \Rightarrow i)\,, \\
Y_\sigma &= \lambda g^{\sigma\to\sigma}.F_\sigma[g]\,, \\
C_f &= \lambda x^{\mathtt{N}}.\ \mathtt{case}\ x\ \mathtt{of}\ (i \Rightarrow f(i))\,, \\
\mathit{byval} &= \lambda f^{\mathtt{N}\to\mathtt{N}} x^{\mathtt{N}}.\ \mathtt{case}\ x\ \mathtt{of}\ (i \Rightarrow \mathtt{case}\ f(\lambda.i)\ \mathtt{of}\ (j \Rightarrow j))\,.
\end{aligned}
$$

It is a matter of routine calculation to check that these satisfy the required equations: for instance, that the terms $\lambda f.Y_\sigma f$ and $\lambda f.f(Y_\sigma f)$ denote the same sequential procedure. The continuity axiom is an easy consequence of the continuity of application in SP^0 (Exercise 6.1.5). The procedure *byval* embodies the intended idea of evaluating x just once and passing the resulting value to f.

Since the above procedures are obviously effective (discounting the C_f), the effective submodel $\mathsf{SP}^{0,\mathrm{eff}}$ is also a model of $\mathrm{PCF} + \mathit{byval}$.

Example 7.1.6. In addition, we obtain that the model SF of sequential functionals is a model of $\mathrm{PCF}^\Omega + \mathit{byval}$, simply by applying the quotient mapping $\mathsf{SP}^0 \to \mathsf{SF}$ to the above choice of elements. As in PC, the interpretation of *byval* is just the strictification map in this case.

Other models of PCF that will feature later in the book include the *stable* and *strongly stable* models St, SSt (Chapter 11), the *sequential algorithms* model SA (Chapter 12), and a range of models arising from game semantics (Section 7.4). A model in which $\mathbf{P}(\mathtt{N}) \neq \mathbb{N}_\perp$ will be briefly mentioned in Subsection 10.1.2.

7.1.2 Mathematical Structure in Models of PCF

Before proceeding to study the interpretation of PCF within such models, we first draw attention to some aspects of their mathematical structure.

Although our definitions so far have not assumed any order-theoretic structure on our models, we may in fact obtain such a structure for free by borrowing some key ideas from synthetic domain theory.

Definition 7.1.7. Let **P** be any λ-algebra over $\mathsf{T}^{\rightarrow}(\mathtt{N})$ with elements $\widehat{n} \in \mathbf{P}(\mathtt{N})$ for each $n \in \mathbb{N}$.

(i) Given $x, y \in \mathbf{P}(\sigma)$, let us write $x \preceq_{\mathrm{obs}} y$ if for all $g \in \mathbf{P}(\sigma \rightarrow \mathtt{N})$ and $n \in \mathbb{N}$, we have that $g(x) = \widehat{n}$ implies $g(y) = \widehat{n}$.

(ii) Given $x \in \mathbf{P}(\sigma)$ and $S \subseteq \mathbf{P}(\sigma)$, we say x is an *observational limit* of S if for all $g \in \mathbf{P}(\sigma \rightarrow \mathtt{N})$ and $n \in \mathbb{N}$, $g(x) = \widehat{n}$ iff there exists $z \in S$ with $g(z) = \widehat{n}$.

The idea is that the numerals $\widehat{n}$ function as 'observable values', and $x \preceq_{\mathrm{obs}} y$ says that any 'observable property' of x is also satisfied by y. Clearly $\preceq_{\mathrm{obs}}$ is reflexive and transitive, and so constitutes a preorder on each $\mathbf{P}(\sigma)$. The following facts are easy to check:

Proposition 7.1.8. *(i) If x is an observational limit of S, then x is a least upper bound of S with respect to $\preceq_{\mathrm{obs}}$: that is, $z \preceq_{\mathrm{obs}} x$ for all $z \in S$, and if $z \preceq_{\mathrm{obs}} y$ for all $z \in S$ then $x \preceq_{\mathrm{obs}} y$.*

(ii) Every $f \in \mathbf{P}(\sigma \rightarrow \tau)$ is monotone with respect to $\preceq_{\mathrm{obs}}$ and preserves observational limits. □

In many models of interest, $\preceq_{\mathrm{obs}}$ coincides with the pointwise ordering. For SF this is given by Exercise 6.1.21, and for PC it is easily checked. For the stable and strongly stable models, however, $\preceq_{\mathrm{obs}}$ turns out to be a strictly smaller relation than the pointwise order, as we will see in Chapter 11.

If **P** is a model of PCF, more may be said about the observational preorder.

Proposition 7.1.9. *The following hold in any model* **P** *of* PCF*:*

(i) For any $n \in \mathbb{N}$, we have $\bot \preceq_{\mathrm{obs}} \widehat{n}$ but $\widehat{n} \not\preceq_{\mathrm{obs}} \bot$.

(ii) $\bot_\sigma \preceq_{\mathrm{obs}} z$ for any $z \in \mathbf{P}(\sigma)$.

(iii) $Y_\sigma^k \preceq_{\mathrm{obs}} Y_\sigma^{k+1}$ and $Y_\sigma^k \preceq_{\mathrm{obs}} Y_\sigma$ for $k \in \mathbb{N}$.

(iv) Y_σ is an observational limit (and hence a least upper bound) of the Y_σ^k.

Proof. (i) That $\bot \preceq_{\mathrm{obs}} \widehat{n}$ is given by the monotonicity axiom in Definition 7.1.3, and that $\widehat{n} \not\preceq_{\mathrm{obs}} \bot$ is obvious by considering the identity observation.

(ii) Suppose $\sigma = \sigma_0, \ldots, \sigma_{r-1} \rightarrow \mathtt{N}$. Given $z \in \mathbf{P}(\sigma)$, consider the operation

$$F \;=\; \lambda y^{\mathtt{N}}.\lambda x_0 \ldots x_{r-1}.\, \mathit{ifzero}\; y\; (z\vec{x})\; \bot\,.$$

With the help of the η-rule, it is easy to see that $F(\bot) = \bot_\sigma$ and $F(\widehat{0}) = z$ in **P**. But $\bot \preceq_{\mathrm{obs}} \widehat{0}$ by part (i), whence $\bot_\sigma \preceq_{\mathrm{obs}} z$ by Proposition 7.1.8(ii).

(iii) Consider the operation

$$G = \lambda g^{(\sigma\to\sigma)\to\sigma}.\lambda f^{\sigma\to\sigma}.\ f^k(gf)\ .$$

Then clearly $G(\bot) = Y^k_\sigma$, $G(Y^1_\sigma) = Y^{k+1}_\sigma$ and $G(Y_\sigma) = Y_\sigma$ in **P**. But $\bot \preceq_{\mathrm{obs}} Y^1_\sigma$ and $\bot \preceq_{\mathrm{obs}} Y_\sigma$ by part (ii), and the result follows.

(iv) Immediate from (iii) and the continuity axiom in Definition 7.1.3. □

As an aside, it is now easy to see that any model of PCF satisfies conditions 1–4 of Subsection 6.4.2, and hence admits a fixed point interpretation of sequential procedures as described there. We will see in Subsection 7.1.7 that this accords with the interpretation of PCF itself in the model.

We say that a model **P** is *simple* if $\preceq_{\mathrm{obs}}$ is in fact a partial order on each $\mathbf{P}(\sigma)$. Every extensional model is clearly simple, as are some non-extensional models of interest, such as SA. If **P** is not simple, we may readily obtain a simple model by quotienting by the equivalence relation associated with $\preceq_{\mathrm{obs}}$ (cf. Subsection 3.6.1).

In a simple model, the observational limit of a set S is unique if it exists. It follows that given the element $\bot$, the choice of elements Y_σ is uniquely determined, since the Y^k_σ are determined by $\bot$ and the λ-algebra structure. Many other pleasing results from classical domain theory are also available for simple models. We mention here just one example, a classical property of simultaneous least fixed points that has relevance to the interpretation of mutually recursive definitions:

Exercise 7.1.10 (Bekič lemma). Formulate a suitable version of PCF with product types, and a corresponding notion of model. (This will be done more officially in Subsection 7.1.5.) Now suppose **P** is a simple model of PCF with product types, and suppose $f \in \mathbf{P}((\sigma\times\tau)\to(\sigma\times\tau))$. Show that

$$\begin{aligned} \mathit{fst}(Y_{\sigma\times\tau}f) &= Y_\sigma(\lambda x^\sigma.\mathit{fst}(f(x, Y_\tau(\lambda y^\tau.\mathit{snd}(f(x,y)))))) \,, \\ \mathit{snd}(Y_{\sigma\times\tau}f) &= Y_\tau(\lambda y^\tau.\mathit{snd}(f(\mathit{fst}(Y_{\sigma\times\tau}f),y)))\,. \end{aligned}$$

Another property of interest is the following. We say a model **P** is *normalizable at type* $\overline{1}$ if there is an element $\mathit{norm} \in \mathbf{P}(1\to 1)$ such that:

1. $\mathit{norm}(f)(n) = f(n)$ for all $f \in \mathbf{P}(1)$ and $n \in \mathbb{N}$,
2. if $f(n) = g(n)$ for all $n \in \mathbb{N}$, then $\mathit{norm}(f) = \mathit{norm}(g)$.

In this case, we say $f \in \mathbf{P}(1)$ is *normal* if $f = \mathit{norm}(f)$.

Again, every extensional model clearly has this property, and in many intensional models of interest, the element *byval* is just such a normalizer. If **P** is normalizable at type $\overline{1}$, we have some particularly useful examples of observational limits:

Exercise 7.1.11. Suppose **P** is a model of PCF and is normalizable at type $\overline{1}$.

(i) Let $h \in \mathbf{P}(1)$ be normal such that $h(n) = 0$ for all n, and for each k, let h_k be normal such that $h_k(n) = 0$ for $n < k$ and $h_k(n) = \bot$ for $n \geq k$. By constructing an operation mapping each Y_k to h_k and Y to h, show that h is an observational limit of the h_k.

(ii) Show that any normal $f \in \mathbf{P}(1)$ is an observational limit of the f_k, where $f_k(n) = f(n)$ for $n < k$, and $f_k(n) = \bot$ for $n \geq k$.

(iii) Deduce that any $F \in \mathbf{P}(2)$ is continuous on total elements: if g_k and g are normal representatives of *total* functions $\mathbb{N} \to \mathbb{N}$ and the g_k converge to g in the usual sense, then $F(g_k) = F(g)$ for all sufficiently large k.

7.1.3 Soundness and Adequacy

For the purpose of this subsection, let PCF* be any of the four languages PCF, PCF$^{\Omega}$, PCF + *byval*, PCF$^{\Omega}$ + *byval*, and suppose $\mathbf{P}$ is a pre-model of PCF*. Since we have given meanings in $\mathbf{P}$ to the constants of PCF*, the internal interpretation of λ-terms as described in Subsection 4.1.2 immediately yields an interpretation $[\![-]\!]^{\mathbf{P}}_{-}$ of PCF* in $\mathbf{P}$. (Recall that if $\Gamma \vdash M : \tau$ where $\Gamma = x_0^{\sigma_0}, \ldots, x_{r-1}^{\sigma_{r-1}}$, then $[\![M]\!]^{\mathbf{P}}_{\Gamma} \in \mathbf{P}(\sigma_0, \ldots, \sigma_{r-1} \to \tau)$.) Furthermore, the equations in Definition 7.1.2 readily yield the following:

Proposition 7.1.12. *If M is closed and $M \rightsquigarrow M'$ in* PCF*, *then* $[\![M]\!]^{\mathbf{P}} = [\![M']\!]^{\mathbf{P}}$.

Proof. By induction on the generation of $\rightsquigarrow$. If $M \rightsquigarrow M'$ via the β-rule then $[\![M]\!]^{\mathbf{P}} = [\![M']\!]^{\mathbf{P}}$ because $\mathbf{P}$ is a λ-algebra. If $M \rightsquigarrow M'$ via any other reduction rule of PCF* then $[\![M]\!]^{\mathbf{P}} = [\![M']\!]^{\mathbf{P}}$ by the corresponding equation in Definition 7.1.2. If $M \rightsquigarrow M'$ is derived from a reduction of either of these kinds by means of the evaluation contexts, then $[\![M]\!]^{\mathbf{P}} = [\![M']\!]^{\mathbf{P}}$ by compositionality of $[\![-]\!]^{\mathbf{P}}$. □

Corollary 7.1.13 (Soundness). *If $M \rightsquigarrow^* \widehat{n}$, then $[\![M]\!]^{\mathbf{P}} = \widehat{n}$.* □

We say an interpretation of PCF* is *adequate* if it satisfies both Corollary 7.1.13 and its converse. We now show that if $\mathbf{P}$ is a model of PCF* then the interpretation $[\![-]\!]^{\mathbf{P}}$ is adequate. The method of proof here is due to Plotkin [236].

Let PCF*$^{-}$ be obtained from PCF* by deleting the constants Y_σ and their associated reduction rules, and replacing them with constants

$$\bot : \mathtt{N}, \qquad Y_\sigma^0, Y_\sigma^1, \ldots : (\sigma \to \sigma) \to \sigma$$

for each σ. The idea is that the constants Y_σ^k will serve as syntactic approximants to the true fixed point operator Y_σ, just as the *elements* Y_σ^k of Definition 7.1.3 serve as semantic approximants to Y_σ. The reduction rules for these new constants are:

$$\begin{array}{rclrcl} suc\ \bot & \rightsquigarrow & \bot\,, & Y_\sigma^0 f & \rightsquigarrow & \lambda\vec{x}.\bot\,, \\ pre\ \bot & \rightsquigarrow & \bot\,, & Y_\sigma^{k+1} f & \rightsquigarrow & f(Y_\sigma^k f)\,, \\ ifzero\ \bot & \rightsquigarrow & \lambda xy.\bot\,, & C_f\ \bot & \rightsquigarrow & \bot\,, \\ byval\ f\ \bot & \rightsquigarrow & \bot\,, & C_f\ \widehat{n} & \rightsquigarrow & \bot \text{ if } f(n) \text{ is undefined.} \end{array}$$

The main interest of PCF*$^{-}$ lies in the following termination property. We write $\mathbb{V}$ for the set of syntactic values $\{\bot, \widehat{0}, \widehat{1}, \ldots\}$.

Lemma 7.1.14. *For every closed $M : \mathtt{N}$ in* PCF*$^{-}$, *we have $M \rightsquigarrow^* V$ for some $V \in \mathbb{V}$.*

Proof. A standard use of logical relations, similar to the proof of Theorem 4.4.6. Let $R_{\mathtt{N}}$ be the unary predicate on closed PCF^{*-} terms of type N given by

$$R_{\mathtt{N}}(M) \quad \text{iff} \quad M \rightsquigarrow^* V \text{ for some } V \in \mathbb{V}\,,$$

and let R be the logical lifting of $R_{\mathtt{N}}$ to closed PCF^{*-} terms of arbitrary type. By the basic lemma for prelogical relations (Lemma 4.4.3), to show that $R_{\mathtt{N}}(M)$ for all closed $M : \mathtt{N}$ it will suffice to check that R is a prelogical relation as in Definition 4.4.2. Condition 2 of this definition is automatic by the definition of logical lifting, and the verification of condition 3 works exactly as in the proof of Theorem 4.4.6.

For condition 1, we simply need to check that each of the constants of PCF^{*-} satisfies R. For $\bot$, $\widehat{n}$, C_f, *suc*, *pre*, *ifzero*, *byval* this is easy. For the constants Y^k_σ, we argue by induction on k. Let $\sigma = \sigma_0, \ldots, \sigma_{r-1} \to \mathtt{N}$. For $k = 0$, it suffices to note that $Y^0_\sigma M N_0 \ldots N_{r-1} \rightsquigarrow^* \bot$ for any closed $M, N_0, \ldots, N_{r-1}$. For the induction step, suppose $M : \sigma \to \sigma$ satisfies R. Then $Y^{k+1}_\sigma M \rightsquigarrow M(Y^k_\sigma M)$, where Y^k_σ satisfies R by the induction hypothesis. Hence $M(Y^k_\sigma M)$ satisfies R since R is closed under application, and so $Y^{k+1}_\sigma M$ satisfies R since R is easily seen to be closed under backwards reduction. Thus Y^{k+1}_σ satisfies R as required. □

Remark 7.1.15. The use of logical relations is not strictly necessary here: one can prove the above lemma more finitistically via a syntactic translation of PCF^{*-} into the simply typed λ-calculus, which itself can be shown to be normalizing by finitistic means (see Loader [171]). However, this latter proof is syntactically much more intricate than the one given above.

We may now use the language PCF^{*-} to obtain the converse to Corollary 7.1.13.

Theorem 7.1.16 (Adequacy). *Suppose* **P** *is a model of* PCF*. *For every closed* $M : \mathtt{N}$ *we have* $M \rightsquigarrow^* \widehat{n}$ *iff* $[\![M]\!]^{\mathbf{P}} = \widehat{n}$.

Proof. The forwards implication is given by Corollary 7.1.13. For the converse, we first extend our interpretation $[\![-]\!]^{\mathbf{P}}$ to PCF^{*-} simply by interpreting Y^k_σ by Y^k_σ and $\bot$ by $\bot$. The soundness of the new reduction rules of PCF^{*-} is ensured by β-equality and the equations of Definition 7.1.3, so as before we have that if $V \in \mathbb{V}$ and $M \rightsquigarrow^* V$ in PCF^{*-} then $[M] = [V]$.

Next, it follows readily from Proposition 7.1.9(iv) that if M is a closed PCF* term and $[\![M]\!]^{\mathbf{P}} = \widehat{n}$, then for some sufficiently large k we have $[\![M^k]\!] = \widehat{n}$, where M^k is the PCF^{*-} term obtained by replacing all occurrences of constants Y_σ in M by Y^k_σ. But now by Lemma 7.1.14 we have $M^k \rightsquigarrow^* V$ for some $V \in \mathbb{V}$, and by the soundness observation above we have $[\![V]\!]^{\mathbf{P}} = [\![M^k]\!]^{\mathbf{P}}$, whence $V = \widehat{n}$. Finally, by replacing all Y^j_σ by Y_σ throughout the reduction $M^k \rightsquigarrow^* \widehat{n}$, we see that $M \rightsquigarrow^* \widehat{n}$ in PCF*. □

The above proof is a good illustration of our primary reason for focussing on models satisfying the continuity axiom. Such models faithfully reflect the *finitary* character of computations in PCF, in the sense that if a term M denotes a numeral n, then this fact relies only on finite 'portions' of the Y combinators within M.

The following exercise shows that a non-continuous model can also be adequate, although the proof in this case is parasitic on its relationship to a continuous model:

Exercise 7.1.17. Recall from Section 6.4 the *full monotone* type structure M over $\mathbb{N}_\perp$, in which $\mathsf{M}(\sigma \to \tau)$ consists of all monotone functions $\mathsf{M}(\sigma) \to \mathsf{M}(\tau)$ ordered pointwise. Recall also from Subsection 6.4.2 that M possesses fixed point operators: for any $f \in \mathsf{M}(\sigma \to \sigma)$, $Y_\sigma(f)$ is the supremum over all ordinals α of $f^\alpha(\perp_\sigma)$, where $f^0(x) = x$, $f^{\alpha+1}(x) = f(f^\alpha(x))$, and $f^\alpha(x) = \bigsqcup_{\beta<\alpha} f^\beta(x)$ when α is a limit ordinal. It follows easily that M is a pre-model of PCF.

(i) Let R denote the logical relation between M and PC arising from the identity relation on $\mathbb{N}_\perp$. Show that if $R_{\sigma\to\sigma}(f,g)$ then $R(f^k(\perp_\sigma), g^k(\perp_\sigma))$ for all $k \in \mathbb{N}$, and $R(f^\alpha(\perp_\sigma), Y_\sigma(g))$ for all $\alpha \geq \omega$. Deduce that $R_{(\sigma\to\sigma)\to\sigma}(Y_\sigma^{\mathsf{M}}, Y_\sigma^{\mathsf{PC}})$.

(ii) Using this, show that $R_\sigma([\![M]\!]^{\mathsf{M}}, [\![M]\!]^{\mathsf{PC}})$ for any closed PCF term $M : \sigma$. Deduce that the interpretation $[\![-]\!]^{\mathsf{M}}$ is adequate.

Finally, we remark that for certain intensional models such as SP^0, a stronger kind of adequacy statement is possible, saying in effect that the denotational and operational semantics agree not only on the final *results* of computations, but also to some extent on the *process* by which these results are computed. (This is close to the concept of *normal form bisimulation* introduced by Lassen and Levy [168].) The following exercise gives a flavour of this, though there is scope for further investigation in this area.

Exercise 7.1.18. (i) Adapt the proof of Theorem 7.1.16 to show the following. Suppose $x^\sigma \vdash M : \mathtt{N}$ is a PCF + *byval* term and $x^\sigma \vdash E$ is an NSP meta-expression with $[\![M]\!]^{\mathsf{SP}^0} = \ll E \gg$. Then the following are equivalent:

1. $M \rightsquigarrow^* C[x\vec{N}]$ for some terms $\vec{N}$ and evaluation context $C[-]$.
2. $\ll E \gg$ has the form $\mathtt{case}\ x\vec{q}\ \mathtt{of}\ (j \Rightarrow e_j)$ for some $\vec{q}$ and e_j.
3. $E \rightsquigarrow_h^* \mathtt{case}\ x\vec{Q}\ \mathtt{of}\ (j \Rightarrow E_j)$ for some $\vec{Q}$ and E_j, using the head reduction of Subsection 6.1.1.

Moreover, in this case, $[\![N_i]\!] = q_i = \ll Q_i \gg$ and $[\![C[j]]\!] = e_j = \ll E_j \gg$ for each i, j.

(ii) Show that this intensional property would break if the denotation of *byval* were changed to agree with that of $\lambda fx.\, ifzero\ x\,(fx)\,(fx)$, whereas Theorem 7.1.16 itself would still hold.

7.1.4 Observational Equivalence and Full Abstraction

Next, we consider what models of PCF allow us to say about *observational equivalence*. This is a key concept which may in general be defined in either operational or denotational terms:

Definition 7.1.19. Suppose $\mathscr{L}$ is a simply typed λ-calculus with certain constants, with o a designated type of *observations*, and with certain closed terms of type o designated as *values*.

(i) An *observing context* for an environment Γ and type τ is a closed syntactic context $C[-^\tau] : o$ of $\mathscr{L}$ binding all the variables of Γ, and with just a single occurrence of the hole $-^\tau$.

(ii) If $\mathscr{L}$ is equipped with a deterministic *evaluation* relation $M \rightsquigarrow^* V$ on closed terms of type o (where V ranges over values), we say the terms $\Gamma \vdash M, M' : \tau$ of $\mathscr{L}$ are *observationally equivalent in $\mathscr{L}$ (in the operational sense)* if for all observing contexts $C[-]$ for Γ, τ we have

$$C[M] \rightsquigarrow^* V \ \text{ iff } \ C[M'] \rightsquigarrow^* V \, .$$

(iii) If the constants of $\mathscr{L}$ have a type-respecting interpretation in a λ-algebra $\mathbf{A}$, giving rise to a compositional interpretation of $[\![-]\!]$ of $\mathscr{L}$ in $\mathbf{A}$ as in Subsection 4.1.2, we say $\Gamma \vdash M, M' : \tau$ are *observationally equivalent in $\mathscr{L}$ (in the denotational sense)* if for all observing contexts $C[-]$ for Γ, τ we have $[\![C[M]]\!] = [\![C[M']]\!]$.

If we have both an operational and a denotational semantics for $\mathscr{L}$ that are related via the *adequacy* property (that is, $M \rightsquigarrow^* V$ iff $[\![M]\!] = [\![V]\!]$), and if moreover there is just one element in $\mathbf{A}(o)$ that is not denoted by a value, clearly these two notions of observational equivalence coincide.

Specializing to any of the four languages PCF* introduced above, we may set $o = \mathtt{N}$ and take the values to be just the numerals $\widehat{n}$; we may then write $\sim^*_{\mathrm{obs}}$ for the resulting operationally defined notion of observational equivalence in PCF*. By Theorem 7.1.16 and the above remarks, this coincides with denotational observational equivalence in any model $\mathbf{P}$ of PCF* with $\mathbf{P}(\mathtt{N}) = \mathbb{N}_\bot$. We shall see shortly that for all four of our languages, the relations $\sim^*_{\mathrm{obs}}$ are in agreement.

Some simple examples of observational equivalences will be given in Exercise 7.1.28 below; some less obvious ones will be discussed in Subsection 7.3.1.

Exercise 7.1.20. (i) Show that if $M, M' : \tau$ are *closed* terms, then $M \sim^*_{\mathrm{obs}} M'$ iff for all closed $P : \tau \to \mathtt{N}$ in PCF* we have $PM \rightsquigarrow^* \widehat{n}$ iff $PM' \rightsquigarrow^* \widehat{n}$.

(ii) If $\Gamma \vdash M, M' : \tau$ where $\Gamma = x_0^{\sigma_0}, \ldots, x_{r-1}^{\sigma_{r-1}}$, show that $M \sim^*_{\mathrm{obs}} M'$ iff $\lambda\vec{x}.M \sim^*_{\mathrm{obs}} \lambda\vec{x}.M'$. (Hint: given a context $C[-^\tau]$, devise a context $C'[-^{\vec{\sigma}\to\tau}]$ such that $[\![C[M]]\!] = [\![C'[\lambda\vec{x}.M]]\!]$ in any adequate model.)

(iii) Show that it makes no difference to the relation $\sim^*_{\mathrm{obs}}$ if in Definition 7.1.19 the contexts $C[-^\tau]$ are allowed to contain several occurrences of the hole $-^\tau$.

(iv) Recall from Subsection 3.6.1 the two-stage definition of observational equivalence on closed terms via the operational equivalence $\sim_{\mathrm{op}}$. Using the existence of an adequate model, show that this agrees with Definition 7.1.19 for closed terms.

The following is a simple but important consequence of Theorem 7.1.16:

Proposition 7.1.21. *Let* $\mathbf{P}$ *be a model of* PCF*. *Suppose* $\Gamma \vdash M, M' : \tau$ *are* PCF* *terms. If* $[\![M]\!]^{\mathbf{P}}_\Gamma = [\![M']\!]^{\mathbf{P}}_\Gamma$ *then* $M \sim^*_{\mathrm{obs}} M'$.

Proof. Suppose $[\![M]\!] = [\![M']\!]$; then for any observing context $C[-^\tau] : \mathtt{N}$, we have $[\![C[M]]\!] = [\![C[M']]\!]$ by compositionality, whence $C[M] \rightsquigarrow^* V$ iff $C[M'] \rightsquigarrow^* V$ by adequacy. $\square$

The general principle that adequate, compositional interpretations establish observational equivalences in this way is one of the key ideas of denotational semantics. In practice, it is useful to have a selection of models at our disposal, since different models can be used to establish different instances of $\sim_{\mathrm{obs}}$. Some examples illustrating this will be given in Subsection 7.3.1.

It is now natural to ask whether there is some single model **P** of PCF* for which the converse of Proposition 7.1.21 holds, so that **P** by itself suffices for establishing all valid observational equivalences. This motivates the following general concept:

Definition 7.1.22. In the situation of Definition 7.1.19, we say the model **A** of $\mathscr{L}$ is *operationally* [resp. *denotationally*] *fully abstract* if whenever $\Gamma \vdash M, M' : \tau$ are observationally equivalent $\mathscr{L}$-terms in the operational [resp. denotational] sense, we have $[\![M]\!]_\Gamma = [\![M']\!]_\Gamma$ in **A**.

Clearly, the operational and denotational notions of full abstraction coincide in the case of adequate models with at most one non-value at type o. The possibility of framing full abstraction as a purely denotational property (with no reference to an operational semantics) was noted in Milner [193].

Exercise 7.1.23. Show that in both the operational and denotational senses, **A** is fully abstract for $\mathscr{L}$ iff whenever $M, M' : \tau$ are *closed* observationally equivalent $\mathscr{L}$-terms, we have $[\![M]\!] = [\![M']\!]$ in **A**.

We shall now work towards showing that the model SF of sequential functions is indeed fully abstract for all four of our languages. In practice, however, this model will not be particularly useful for establishing specific observational equivalences, since in general it is not easy to tell whether $[\![M]\!]^{\mathsf{SF}} = [\![M']\!]^{\mathsf{SF}}$. (The problem will be discussed more deeply in Section 7.5.) Most of what can be easily retrieved from the model SF is embodied in the *context lemma* given as Theorem 7.1.26 below.

For the remainder of this subsection, we shall write $[-]$ for the interpretation $[\![-]\!]^{\mathsf{SP}^0}$, and $[\![-]\!]$ for $[\![-]\!]^{\mathsf{SF}}$. We start by observing that all *finite* NSPs, at least, are definable in PCF + *byval*:

Lemma 7.1.24. *If* $p \in \mathsf{SP}^{0,\mathrm{fin}}(\tau)$*, there is a* PCF + *byval term* $P : \tau$ *with* $[P] = p$*.*

Proof. We may translate any finite NSP term t into a PCF + *byval* term $\widetilde{t}$ in an obvious way, rendering the construct $\mathtt{case}\ a\ \mathtt{of}\ (0 \Rightarrow e_0 \mid \cdots \mid r-1 \Rightarrow e_{r-1})$ by

$$byval\ (\lambda x.ifzero\ x\ \widetilde{e_0}\ (ifzero\ (pre\,x)\ \widetilde{e_1}\ (\cdots (ifzero\ (pre^{r-1}\,x)\ \widetilde{e_{r-1}}\ \bot)\cdots)))\ \widetilde{a}\,,$$

where $\bot$ abbreviates $Y_{\mathtt{N}}(\lambda x^{\mathtt{N}}.x)$. A routine structural induction now shows that for any finite procedure p in context $\Gamma = x_0, \ldots, x_{r-1}$ we have $[\widetilde{p}]_\Gamma = \lambda \vec{x}.p$. In particular, if p is closed then $[\widetilde{p}] = p$. $\square$

Theorem 7.1.25 (Full abstraction for SF). *Let* PCF* *be any of the four languages introduced above, with* $\sim^*_{\mathrm{obs}}$ *the associated observational equivalence. Then for any closed terms* $M, M' : \sigma$ *in* PCF**, we have*

$$[\![M]\!]^{\mathsf{SF}} = [\![M']\!]^{\mathsf{SF}} \ \textit{iff} \ M \sim^{*}_{\mathrm{obs}} M' .$$

In particular, all four of our relations $\sim^{*}_{\mathrm{obs}}$ *agree wherever their domains overlap.*

Proof. The left-to-right implication is given by Proposition 7.1.21. We first show the converse for the two languages with *byval*. Suppose $M, M' : \sigma$ are closed PCF$^{\Omega}$ terms and $[\![M]\!] \neq [\![M']\!]$ in SF. In view of the definition of SF (see Subsection 6.1.4), this implies that there is some $p \in \mathsf{SP}^0(\sigma \to \mathtt{N})$ such that $p \cdot [M] \neq p \cdot [M']$. Since any NSP is the supremum of the finite procedures below it, by the continuity of NSP application we may conclude that there is some *finite* p_0 with this property. But now by Lemma 7.1.24, we obtain a closed PCF + *byval* term $P : \sigma \to \mathtt{N}$ such that

$$[PM] = [P] \cdot [M] = p_0 \cdot [M] \neq p_0 \cdot [M'] = [P] \cdot [M'] = [PM'] .$$

Using the adequacy of $[-]$ (Theorem 7.1.16), we conclude that $M \not\sim_{\mathrm{obs}} M'$.

For the two languages without *byval*, we note that

$$[\![byval]\!]^{\mathsf{SF}} = [\![\lambda f x.\ ifzero\, x\, (fx)(fx)]\!]^{\mathsf{SF}} ,$$

whence $byval \sim_{\mathrm{obs}} \lambda f x.\ ifzero\, x\, (fx)(fx)$ in PCF$^{\Omega}$ + *byval* by Proposition 7.1.21. Thus, if $C[-]$ were any context in PCF + *byval* distinguishing M and M', we would obtain a distinguishing context $C'[-]$ in PCF by replacing *byval* throughout $C[-]$ by $\lambda f x.\ ifzero\, x\, (fx)(fx)$. Hence observational equivalence in PCF agrees with that in PCF + *byval*, and likewise for the oracle versions. □

In the same vein, we have the following result, due originally to Milner [193].

Theorem 7.1.26 (Context lemma for PCF). *If* $M, M' : \sigma = \sigma_0, \ldots, \sigma_{r-1} \to \mathtt{N}$ *are closed terms of* PCF$^{\Omega}$ + *byval, the following are equivalent:*

1. $M \sim_{\mathrm{obs}} M'$.
2. *For all closed* PCF *terms* $N_0 : \sigma_0, \ldots, N_{r-1} : \sigma_{r-1}$ *and all* $n \in \mathbb{N}$, *we have* $MN_0 \ldots N_{r-1} \rightsquigarrow^{*} \widehat{n}$ *iff* $M'N_0 \ldots N_{r-1} \rightsquigarrow^{*} \widehat{n}$.

Proof. The implication 1 ⇒ 2 is immediate by applying the definition of $\sim_{\mathrm{obs}}$ to the observing context $-N_0 \ldots N_{r-1}$. Conversely, if $M \not\sim_{\mathrm{obs}} M'$, then by full abstraction and extensionality of SF we have $[\![M]\!] \cdot z_0 \cdot \ldots \cdot z_{r-1} \neq [\![M']\!] \cdot z_0 \cdot \ldots \cdot z_{r-1}$ for some $z_i \in \mathsf{SF}(\sigma_i)$. So for some $q_i \in \mathsf{SP}^0(\sigma_i)$ we have $[M] \cdot q_0 \cdot \ldots \cdot q_{r-1} \neq [M'] \cdot q_0 \cdot \ldots \cdot q_{r-1}$, and by continuity of application in SP^0, the q_i here may be taken to be finite. By Lemma 7.1.24 we may find PCF + *byval* terms Q_i denoting these q_i, and as in the proof of Theorem 7.1.25 these may be replaced by PCF terms N_i with $[\![N_i]\!]^{\mathsf{SF}} = [\![Q_i]\!]^{\mathsf{SF}}$. We then have $[\![MN_0 \ldots N_{r-1}]\!]^{\mathsf{SF}} \neq [\![M'N_0 \ldots N_{r-1}]\!]^{\mathsf{SF}}$, and by adequacy of $[\![-]\!]^{\mathsf{SF}}$ this contradicts statement 2 above. □

Remark 7.1.27. The context lemma for PCF can also be proved purely operationally by an analysis of reduction sequences; this was the method employed in [193]. Here we have in effect derived it from the context lemma for NSPs (Theorem 6.1.19), a slightly more abstract version of the same result which was used in the construction of SF.

Exercise 7.1.28. Use the context lemma to establish the following simple observational equivalences in PCF:

$$\begin{aligned}\lambda xy.\, \mathit{ifzero}\; x\; (\mathit{ifzero}\; y\; \widehat{0}\; \widehat{1})\; (\mathit{ifzero}\; y\; \widehat{2}\; \widehat{3}) &\sim \lambda xy.\, \mathit{ifzero}\; y\; (\mathit{ifzero}\; x\; \widehat{0}\; \widehat{2})\; (\mathit{ifzero}\; x\; \widehat{1}\; \widehat{3})\,, \\ \lambda x.\, \mathit{ifzero}\; x\; (\mathit{ifzero}\; x\; \widehat{0}\; \widehat{1})\; (\mathit{ifzero}\; x\; \widehat{2}\; \widehat{3}) &\sim \lambda x.\, \mathit{ifzero}\; x\; \widehat{0}\; \widehat{3}\,, \\ \lambda f.\, f\, (\mathit{ifzero}\; (f\widehat{0})\; \widehat{1}\; \widehat{2}) &\sim \lambda f.\, \mathit{ifzero}\; (f\widehat{0})\; (f\widehat{1})\; (f\widehat{2})\,.\end{aligned}$$

Exercise 7.1.29. One may also formulate inequational versions of the above results as follows. For closed $\mathrm{PCF}^{\Omega} + \mathit{byval}$ terms $M, M' : \sigma$, define $M \preceq^{*}_{\mathrm{obs}} M'$ iff for all closed contexts $C[-]$ of PCF^{*} and all $n \in \mathbb{N}$ we have

$$C[M] \rightsquigarrow^{*} \widehat{n} \;\text{ implies }\; C[M'] \rightsquigarrow^{*} \widehat{n}\,.$$

We also have the pointwise ordering $\preceq_{\mathrm{app}}$ defined on each $\mathsf{SF}(\sigma_0, \ldots, \sigma_{r-1} \to \mathtt{N})$ by

$$f \preceq_{\mathrm{app}} g \;\text{ iff }\; \forall \vec{z}.\, f z_0 \ldots z_{r-1} \sqsubseteq g z_0 \ldots z_{r-1}\,.$$

It follows immediately from Exercise 6.1.21 that $\preceq_{\mathrm{app}}$ coincides with the semantic observational ordering $\preceq_{\mathrm{obs}}$ on SF.

(i) Show that for closed $M, M' : \sigma$ in PCF^{*} we have

$$[\![M]\!]^{\mathsf{SF}} \preceq_{\mathrm{app}} [\![M']\!]^{\mathsf{SF}} \;\text{ iff }\; M \preceq^{*}_{\mathrm{obs}} M'\,,$$

and hence that the relations $\preceq^{*}_{\mathrm{obs}}$ agree for all four languages PCF^{*}. (We say a model with a partial ordering is *inequationally fully abstract* if the above property holds.)

(ii) Show also that for closed M, M' we have $M \preceq_{\mathrm{obs}} M'$ iff for all closed PCF terms $N_0 : \sigma_0, \ldots, N_{r-1} : \sigma_{r-1}$ and all $n \in \mathbb{N}$, $M N_0 \ldots N_{r-1} \rightsquigarrow^{*} \widehat{n}$ implies $M' N_0 \ldots N_{r-1} \rightsquigarrow^{*} \widehat{n}$.

7.1.5 *Completeness of* $\mathrm{PCF}^{\Omega} + \mathit{byval}$ *for Sequential Procedures*

We have already seen in Subsection 6.1.5 how any procedure $p \in \mathsf{SP}^{0}(\sigma)$ may be coded as a partial function $\mathbb{N} \rightharpoonup \mathbb{N}$; this may in turn be canonically represented by an element of $\mathsf{SP}^{0}(1)$, which we shall denote by $\lceil p \rceil$. Our goal in this subsection is to obtain an important definability result: for each type σ, there is a procedure $I_\sigma \in \mathsf{SP}^{0}(1 \to \sigma)$, definable in $\mathrm{PCF} + \mathit{byval}$, which maps any such code $\lceil p \rceil \in \mathsf{SP}^{0}(1)$ to p itself. Since every *effective* code $\lceil p \rceil$ is itself definable in PCF, it follows immediately that every $p \in \mathsf{SP}^{0,\mathrm{eff}}$ is definable in $\mathrm{PCF} + \mathit{byval}$, and also that every element of $\mathsf{SF}^{\mathrm{eff}}$ is PCF-definable.

It is illuminating to present this result in terms of simulations and transformations. As shown in Subsection 6.1.5, the coding of closed procedures as functions $\mathbb{N} \rightharpoonup \mathbb{N}$ gives rise to a simulation $\gamma : \mathsf{SP}^{0} \longrightarrow\!\!\!\!\triangleright\, \mathsf{B}$ of sequential procedures in the van Oosten model. We will also see that the obvious coding of functions $\mathbb{N} \rightharpoonup \mathbb{N}$ in $\mathsf{SP}^{0}(1)$ yields a simulation $\delta : \mathsf{B} \longrightarrow\!\!\!\!\triangleright\, \mathsf{SP}^{0}$ such that $\gamma \circ \delta \sim \mathit{id}_{\mathsf{B}}$ in the sense of Definition 3.4.2. The import of our main result will then be that there are PCF + *byval* definable elements $I_\sigma \in \mathsf{SP}^0(1 \to \sigma)$ that constitute a transformation $\delta \circ \gamma \preceq id_{\mathsf{SP}^0}$.

Let us start with the simulation $\delta : \mathsf{B} \longrightarrow\!\!\triangleright\, \mathsf{SP}^0(1)$. For any $f : \mathbb{N} \to \mathbb{N}_\bot$, let $\delta(f)$ be the type $\overline{1}$ procedure $\lambda n.\mathtt{case}\ n\ \mathtt{of}\ (i \Rightarrow f(i))$. Although not strictly necessary for our definability result, it is worth pausing to establish the following:

Proposition 7.1.30. *The mapping δ is a simulation $\mathsf{B} \longrightarrow\!\!\triangleright\, \mathsf{SP}^0$ in the sense of Section 3.4; indeed, it is tracked by a* PCF*-definable element of* $\mathsf{SP}^0(1,1 \to 1)$.

Proof. We simply need to provide a PCF program that implements the application operation of B as defined in Subsection 3.2.4. We give here the pseudocode for a program $app : \overline{1}, \overline{1} \to \overline{1}$ that does this by mediating the relevant dialogue between two elements $f, g \in \mathsf{B}$ to produce the element $f \cdot g$.

$$\begin{aligned} play\ f^1\ g^1\ \langle m_0, \ldots, m_r \rangle &= case\ f\ \langle m_0, \ldots, m_r \rangle\ of \\ &\qquad (2n \Rightarrow (let\ m = g(n)\ in\ play\ f\ g\ \langle m_0, \ldots, m_r, m \rangle) \\ &\qquad \mid 2n+1 \Rightarrow n) \\ app\ f\ g &= \lambda m.play\ f\ g\ \langle m \rangle \end{aligned}$$

It is routine to check that $[\![app]\!]^{\mathsf{SP}^0}$ is indeed a realizer for δ. $\square$

Since the procedure $[\![app]\!]$ is PCF-definable, it is certainly effective, so δ also restricts to a simulation $\mathsf{B}^{\mathrm{eff}} \longrightarrow\!\!\triangleright\, \mathsf{SP}^{0,\mathrm{eff}}$.

The simulation $\gamma : \mathsf{SP}^0 \longrightarrow\!\!\triangleright\, \mathsf{B}$ has already been introduced in Subsection 6.1.5, and it is easy to verify that $\gamma \circ \delta \sim id_{\mathsf{B}}$. As regards the composite $\delta \circ \gamma$, for any $p \in \mathsf{SP}^0(\sigma)$ we will write $\lceil p \rceil \in \mathsf{SP}^0(1)$ for the procedure $\delta(\gamma(p))$, and call $\lceil p \rceil$ the *code* for p. Clearly we cannot hope for a transformation $id_{\mathsf{SP}^0} \preceq \delta \circ \gamma$: if p, p' are distinct but observationally equivalent procedures then $\lceil p \rceil$ and $\lceil p' \rceil$ are observationally distinct, so a realizer for such a transformation would need to split an observational equivalence class, which cannot be done within SP^0 itself. We shall show, however, that $\delta \circ \gamma \preceq id_{\mathsf{SP}^0}$ via procedures definable in PCF + *byval*.

The idea behind our construction was first employed by Platek [233]; similar methods have also been applied in [5, 124, 186, 253]. One can also see our construction as an intensionally more fine-grained counterpart of the one outlined in Subsection 6.4.2; the connection will be spelt out in Subsection 7.1.7 below.

Informally, for each type $\sigma = \sigma_0, \ldots, \sigma_{r-1} \to \mathtt{N}$, our task is to write an 'interpreter' that renders an arbitrary code $\lceil p \rceil$ as a live operation of type σ, capable of being applied to arguments $q_i \in \mathsf{SP}(\sigma_i)$ without access to codes for the latter. In this scenario, if $p = \lambda \vec{x}.e$, the q_i will become 'bound' to the variables x_i, and intuitively must be 'stored' so as to be available whenever we encounter a subexpression in p with head variable x_i. Suppose $x_i \vec{r}$ is such a subexpression. Since we only have access to q_i as an operation of type σ_i—intuitively, it comes to us as an 'oracle from the outside world'—we have no choice but to interpret the r_j themselves as live operations of the appropriate types σ_{ij}, then feed these to q_i. This will in turn involve simulating the behaviour of r_j on an arbitrary sequence of arguments $\vec{s}$, which again will involve storing these arguments for use when required. Indeed, as we work our way down the original syntax tree $\lceil p \rceil$, more and more variables may come into scope—we will therefore need a facility for storing an unbounded number of 'oracles' received from the outside world. Note, however, that these oracles will be of only finitely many distinct types, all appearing as syntactic constituents of σ itself.

These considerations motivate the following technical machinery. We define the notions of *odd* and *even constituents* of a given type σ inductively as follows. If $\sigma = \sigma_0, \ldots, \sigma_{r-1} \to \mathtt{N}$, then

- σ is an even constituent of itself,
- each even constituent of each σ_i is an odd constituent of σ,
- each odd constituent of each σ_i is an even constituent of σ.

Note that the same type may occur as both an odd and an even constituent of σ.

At this point, it will ease notation to extend PCF with product types. That is, we work over the type world $\mathsf{T}^{\to\times}(\mathtt{N})$, and take as our constants those of ordinary PCF (including constants Y_τ for every $\tau \in \mathsf{T}^{\to\times}(\mathtt{N})$) along with pairing and projection operators. Using the ideas of Section 4.2, we may readily interpret this extended language PCF$^\times$ in the Karoubi envelope $\mathbf{K}(\mathsf{SP}^0)$, and we have by Proposition 4.2.7(ii) that any type $\tau \in \mathsf{T}^{\to\times}(\mathtt{N})$ arises as the splitting of a PCF-definable idempotent on some $\sigma \in \mathsf{T}^{\to}(\mathtt{N})$. From this, it is easily shown that if $\rho \in \mathsf{T}^{\to}(\mathtt{N})$ and $p \in \mathsf{SP}^0(\rho)$ is definable in PCF$^\times + \mathit{byval}$, then p is already definable in PCF $+ \mathit{byval}$. This is because if P is any PCF$^\times + \mathit{byval}$ term denoting p, we may systematically replace all subterms of P by corresponding terms of some PCF type; we leave the details to the reader.

Now fix σ, let $\rho_0, \ldots, \rho_{c-1}$ and $\tau_0, \ldots, \tau_{d-1}$ be non-repetitive enumerations of (respectively) the odd and even constituents of σ, and consider the type

$$\psi \;=\; \mathtt{N} \times (\mathtt{N} \to \rho_0) \times \cdots \times (\mathtt{N} \to \rho_{c-1}) \,.$$

We shall use ψ as the type of the 'memory bank' required for our computation. Here, each component $h_j : \mathtt{N} \to \rho_j$ will be used to record a list $h_j(0), h_j(1), \ldots$ of 'oracles of type ρ_j', of finite but arbitrary length. The first component, of type $\mathtt{N}$, is used to record the following bookkeeping information:

- For each $j < c$, the number of oracles of type ρ_j currently stored—in other words, the next available address in the mapping $h_j : \mathtt{N} \to \rho_j$.
- For each typed variable x^ρ currently in scope, the index j and address k such that $\rho = \rho_j$ and the oracle bound to x^ρ is stored as $h_j(k)$. (Note that typed variables will themselves be represented here by numerical codes as in Subsection 6.1.5.)

It is clear that for each $j < c$ we may define PCF$^\times$ programs

$$\mathit{lookup}_j : \mathtt{N}, \psi \to \rho_j \qquad \mathit{store}_j : \mathtt{N}, \rho_j, \psi \to \psi$$

satisfying the following properties (we here write $[\![-]\!]$ for the interpretation of PCF$^\times$ in $\mathbf{K}(\mathsf{SP}^0)$, and write $\lceil x \rceil$ for the numerical code for a variable x).

$$[\![\mathit{lookup}_j]\!] \cdot \lceil x \rceil \cdot ([\![\mathit{store}_j]\!] \cdot \lceil x \rceil \cdot p \cdot S) = p \,,$$

$$[\![lookup_j]\!] \cdot \lceil x \rceil \cdot ([\![store_j]\!] \cdot \lceil y \rceil \cdot p \cdot S) = [\![lookup_j]\!] \cdot \lceil x \rceil \cdot S \ \text{ if } y \not\equiv x \,.$$

We also require some simple machinery corresponding to syntactic operations on type $\overline{1}$ codes. We extend our notation of codes to arbitrary NSP terms, writing $\lceil t \rceil \in \mathsf{SP}^0(1)$ for the code for t. It is routine to construct PCF programs as follows:

- For NSP expressions, we require programs

  $$isNum, numVal \;:\; \overline{1} \rightarrow \overline{0}\,, \qquad left \;:\; \overline{1} \rightarrow \overline{1}\,, \qquad right \;:\; \overline{0}, \overline{1} \rightarrow \overline{1}$$

  such that

  $$\begin{array}{rlrl} [\![isNum]\!] \cdot \lceil n \rceil &= \widehat{0}\,, & [\![left]\!] \cdot \lceil \mathtt{case}\ a\ \mathtt{of}\ (i \Rightarrow e_i) \rceil &= \lceil a \rceil\,, \\ [\![isNum]\!] \cdot \lceil \mathtt{case} \cdots \rceil &= \widehat{1}\,, & [\![right]\!] \cdot \widehat{j} \cdot \lceil \mathtt{case}\ a\ \mathtt{of}\ (i \Rightarrow e_i) \rceil &= \lceil e_j \rceil\,, \\ [\![numVal]\!] \cdot \lceil n \rceil &= \widehat{n}\,. & & \end{array}$$

- For applications, we require

  $$headVar \;:\; \overline{1} \rightarrow \overline{0}\,, \qquad typeOf \;:\; \overline{0} \rightarrow \overline{0}\,, \qquad arg \;:\; \overline{0}, \overline{1} \rightarrow \overline{1}$$

  such that

  $$\begin{array}{rlrl} [\![headVar]\!] \cdot \lceil x^\rho q_0 \ldots q_{t-1} \rceil &= \lceil x^\rho \rceil\,, & [\![arg]\!] \cdot \widehat{j} \cdot \lceil x^\rho q_0 \ldots q_{t-1} \rceil &= \lceil q_j \rceil\,, \\ [\![typeOf]\!] \cdot \lceil x^\rho \rceil &= \lceil \rho \rceil\,. & & \end{array}$$

- For procedures, we require

  $$var \;:\; \overline{0}, \overline{1} \rightarrow \overline{0}\,, \qquad body \;:\; \overline{1} \rightarrow \overline{1}$$

  such that

  $$[\![var]\!] \cdot \widehat{j} \cdot \lceil \lambda x_0 \ldots x_{t-1}.e \rceil \;=\; \lceil x_j \rceil\,, \qquad [\![body]\!] \cdot \lceil \lambda x_0 \ldots x_{t-1}.e \rceil \;=\; \lceil e \rceil\,.$$

We now come to the core of the construction. We wish to define by simultaneous recursion certain operations

$$\begin{array}{l} eval : \overline{1} \rightarrow \psi \rightarrow \overline{0}\,, \\ app : \overline{1} \rightarrow \psi \rightarrow \overline{0}\,, \\ sim_{\tau_j} : \overline{1} \rightarrow \psi \rightarrow \tau_j \ \text{ for each even constituent } \tau_j \text{ of } \sigma \end{array}$$

such that: *eval* evaluates a given expression code $\lceil e \rceil$ relative to a given memory state S which provides meanings for the free variables appearing in e; *app* does the same for a given application code $\lceil a \rceil$; and sim_τ likewise simulates a code $\lceil p \rceil$ for a given $p \in \mathsf{SP}(\tau_j)$ relative to a given S. All this may be achieved in $\mathrm{PCF}^\times + byval$ via the following pseudocode:

$$
\begin{array}{l}
eval \ \lceil e \rceil \ S \ = \\
\quad \textit{if } isNum(e) \textit{ then } numVal(e) \\
\quad \textit{else } byval \ (\lambda i.\, eval(right\ i\ \lceil e \rceil)\, S)\ (app\,(left\ \lceil e \rceil)\, S)\ , \\
app\ \lceil a \rceil\ S\ = \\
\quad \textit{case } typeOf\,(headVar\ \lceil a \rceil) \textit{ of} \\
\qquad \rho_j \Rightarrow (lookup_j\ (headVar\ \lceil a \rceil)\, S) \\
\qquad\qquad (sim_{\rho_{j0}}\ (arg\ \widehat{0}\ \lceil a \rceil)\, S)\ \cdots\ (sim_{\rho_{j(t-1)}}\ (arg\ \widehat{t-1}\ \lceil a \rceil)\, S) \\
\qquad \text{where } \rho_j \equiv \rho_{j0}, \ldots, \rho_{j(t-1)} \to \mathtt{N}\ , \\
sim_{\tau_j}\ \lceil p \rceil\ S\ = \\
\qquad \lambda x_0 \ldots x_{t-1}.\ eval\ (body\ \lceil p \rceil) \\
\qquad\quad (store\ (var\ \widehat{0}\ \lceil p \rceil)\ x_0\ (\cdots (store\ (var\ \widehat{t-1}\ \lceil p \rceil)\ x_{t-1}\ S) \cdots)) \\
\qquad \text{where } \tau_j \equiv \tau_{j0}, \ldots, \tau_{j(t-1)} \to \mathtt{N}\ .
\end{array}
$$

Note that in the definition of *app*, the code pertaining to ρ_j cannot work 'uniformly in j'; rather, a separate portion of code must be provided for each $j < c$. The use of *byval* should also be noted: to simulate a code $\lceil \mathtt{case}\ a\ \mathtt{of}\ (i \Rightarrow e_i) \rceil$ correctly, the value of a is computed just once, and the resulting value i is 'memoized' so that it need not be recomputed in the course of simulating $\lceil e_i \rceil$.

To express the above simultaneous recursion formally in $\mathsf{PCF}^{\times} + byval$, consider the type

$$
\kappa \ = \ (\overline{1}, \psi \to \mathtt{N}) \ \times \ (\overline{1}, \psi \to \mathtt{N}) \ \times \ (\overline{1}, \psi \to \tau_0) \ \times \ \cdots \ \times \ (\overline{1}, \psi \to \tau_{d-1})\ .
$$

We may now construct the desired tuple $(eval, app, sim_{\tau_0}, \ldots, sim_{\tau_{d-1}})$ by a single ordinary recursion using Y_κ. Alternatively, the above pseudocode may be syntactically expanded into a single recursive definition of *eval*, from which the sim_{τ_j} are then easily defined.

Next, we formulate the key properties of the above programs. It is convenient here to work in SP rather than SP^0, allowing procedures with free variables. If Γ is any finite list of typed variables, let us say $S \in \mathsf{SP}(\psi)$ is *correct for* Γ if $[\![lookup]\!] \cdot \lceil x^\rho \rceil \cdot S = x^{\rho\eta}$ for all $x^\rho \in \Gamma$, where $x^{\rho\eta}$ is defined as in Subsection 6.1.3.

Lemma 7.1.31. *(i) If $\Gamma \vdash e$ and S is correct for Γ, then $[\![eval]\!] \cdot \lceil e \rceil \cdot S = e$.*

(ii) If $\Gamma \vdash p : \tau$ for some even constituent τ of σ, and S is correct for Γ, then $[\![sim_\tau]\!] \cdot \lceil p \rceil \cdot S = p$.

Proof. Identify *eval* and sim_τ with their denotations to ease notation. The following facts follow routinely from the definitions of *eval*, *app*, sim_τ, the definition of $[\![-]\!]$ given by Exercise 7.1.5, and the stated properties of the various syntactic operations on codes, with the help of results on NSP evaluation from Section 6.1.

- If $e \equiv n$ or $e \equiv \bot$ then $[\![eval]\!] \cdot \lceil e \rceil \cdot S = e$ for any S.
- If $\Gamma \vdash e \equiv \mathtt{case}\ x q_0 \ldots q_{r-1}\ \mathtt{of}\ (i \Rightarrow e_i)$, where x has type $\rho_{j0}, \ldots, \rho_{j(r-1)} \to \mathtt{N}$ and S is correct for Γ, then

$$
\begin{array}{l}
[\![eval]\!] \cdot \lceil e \rceil \cdot S \ = \ \mathtt{case}\ x\,([\![sim_{\rho_{j0}}]\!] \cdot \lceil q_0 \rceil \cdot S) \cdots ([\![sim_{\rho_{j(r-1)}}]\!] \cdot \lceil q_{r-1} \rceil \cdot S)\ \mathtt{of} \\
\qquad\qquad\qquad\qquad (i \Rightarrow [\![eval]\!] \cdot \lceil e_i \rceil \cdot S)
\end{array}
$$

- If $\Gamma \vdash p \equiv \lambda x_0 \ldots x_{t-1}.e \ : \tau_j$ where S is correct for Γ, then

$$[\![sim_{\tau_j}]\!] \cdot \lceil p \rceil \cdot S \ = \ \lambda x_0 \ldots x_{t-1}. \ [\![eval]\!] \cdot \lceil e \rceil \cdot S'$$

where S' is correct for $\Gamma, x_0, \ldots, x_{t-1}$.

It now follows by induction that the equations in (i) and (ii) hold to any desired finite depth d, and hence that they hold exactly. □

To conclude our construction, we define the $\mathrm{PCF}^{\times} + byval$ term

$$I'_{\sigma} \ = \ \lambda z.\ sim_{\sigma}\ z\ E \ : \ \overline{1} \rightarrow \sigma \,,$$

where $E : \psi$ is a term defining the initial (empty) memory state. By earlier remarks, we may now find a PCF + *byval* term I_{σ} denoting the same NSP as I'_{σ}. The following is now immediate from the preceding lemma:

Theorem 7.1.32. *For all* $p \in \mathsf{SP}^0(\sigma)$ *we have* $[\![I_{\sigma}]\!] \cdot \lceil p \rceil = p$. *Thus* $\delta \circ \gamma \preceq id_{\mathsf{SP}^0}$. □

Note that $I_{\sigma} \cdot \lceil p \rceil$ yields p on the nose, not just up to observational equivalence. We thus obtain:

Corollary 7.1.33. *(i) Every* $p \in \mathsf{SP}^0(\sigma)$ *is definable in* $\mathrm{PCF}^{\Omega} + byval$.
(ii) Every $p \in \mathsf{SP}^{0,\mathrm{eff}}(\sigma)$ *is definable in* PCF + *byval.*

Proof. (i) For any $p \in \mathsf{SP}^0(\sigma)$, we have $p = [\![I_{\sigma} C_g]\!]$, where C_g is the constant representing $\lceil p \rceil$.
(ii) If p is effective then $\lceil p \rceil$ is partial computable, so it is easy to see that the canonical type 1 procedure for $\lceil p \rceil$ is definable in PCF + *byval*, say by $P : \overline{1}$. We then have that $p = [\![I_{\sigma} P]\!]$. □

Similar results hold for SF, except that here we no longer need *byval*, since it may be replaced by a denotationally equivalent PCF term as in the proof of Theorem 7.1.25:

Corollary 7.1.34. *(i) Every* $f \in \mathsf{SF}(\sigma)$ *is definable in* PCF^{Ω}.
(ii) Every $f \in \mathsf{SF}^{\mathrm{eff}}(\sigma)$ *is definable in* PCF. □

Remark 7.1.35. One may ask whether the same results hold when the oracle constants C_f are restricted to *total* functions $f : \mathbb{N} \rightarrow \mathbb{N}$, as is sometimes done in the literature. In classical logic, this poses no problem for Corollary 7.1.33, since in place of a partial function f we may equally well use the function mapping n to $f(n)+1$ if $f(n)$ is defined, 0 if not. However, it does not hold *constructively* that every element of SP^0 is PCF-definable relative to a total oracle, unless we have in mind some presentation of SP^0 that allows such a 'total code' to be extracted from a procedure p.

In the effective setting, the following variation on Theorem 7.1.32 is available, using $\overline{0}$ rather than $\overline{1}$ as the type of 'codes':

Corollary 7.1.36. *Let* $\ulcorner\!\ulcorner - \urcorner\!\urcorner$ *be an effective coding of closed* PCF *terms as natural numbers. For each type* σ *there is a* PCF + *byval program* $J_\sigma : \overline{0} \to \sigma$ *such that for any closed* $M : \sigma$ *we have* $[\![J_\sigma \ulcorner\!\ulcorner M \urcorner\!\urcorner]\!] = [\![M]\!]$ *in* SP^0.

Proof. Using ideas from Subsection 6.1.5, it is easy to construct a PCF program $K : \overline{0} \to \overline{1}$ such that $[\![K]\!] \cdot \ulcorner\!\ulcorner M \urcorner\!\urcorner = \ulcorner [\![M]\!] \urcorner$ for any closed $M : \sigma$. Composing this with I_σ yields a program J_σ with the required property. □

At an abstract level, one can see this as combining the above simulations δ, γ with simulations $\delta' : K_1 \longrightarrow\!\!\!\!\rhd \mathsf{B}^{\mathrm{eff}}$ and $\gamma' : \mathsf{B}^{\mathrm{eff}} \longrightarrow\!\!\!\!\rhd K_1$ such that $\delta' \circ \gamma' \preceq id_{\mathsf{B}^{\mathrm{eff}}}$.

One can also cast the above results in a purely operational way with no reference to NSPs: for example, every closed PCF$^\Omega$ term $M : \sigma$ is observationally equivalent to one of the form $I_\sigma C_g$, and every closed PCF term $M : \sigma$ is observationally equivalent to one of the form $J_\sigma \widehat{n}$. A direct proof of this latter statement using purely operational concepts (though otherwise similar to the one above) is given in Longley and Plotkin [186].

As further corollaries of the proof of Theorem 7.1.32, we are able to obtain various standardization results to the effect that any $p \in \mathsf{SP}^0(\sigma)$ is definable by a term $M : \sigma$ without recourse to subterms or recursors of types 'higher' than σ; in particular, every closed PCF term $M' : \sigma$ is observationally equivalent to such a term. We here sketch some results of this kind for the *pure* types; similar ideas can also be applied to other types of interest.

Suppose first that $p \in \mathsf{SP}^0(k)$ where $k \geq 3$, so that p has the form $\lambda x^{k-1}.e$. It is easy to see that x is the only variable of level $k-1$ that may appear in p; in fact, all other variables within p have level at most $k-3$. This means that in the proof of Theorem 7.1.32, we can get away without recording oracles of type $\overline{k-1}$ in our memory state S: the only such oracle is the one bound to x, and this can instead be treated as a parameter that remains fixed throughout the entire computation. Thus, in the definition of ψ, we may assume the ρ_j are all pure types of level at most $k-3$. If $k \geq 4$, it follows that the type ψ of memory states has type level $\leq k-3$. An inspection of the pseudocode in Subsection 7.1.5 now reveals the following result:

Theorem 7.1.37. *If* $k \geq 4$, *then every* $p \in \mathsf{SP}(k)$ *is definable by a* PCF$^\Omega$ + *byval program* $M : \overline{k}$ *such that*

- *all recursors* Y_τ *appearing in* M *are for types* τ *of level* $\leq k-2$ *(so that* Y_τ *is itself of level* $\leq k$*),*
- *all other proper subterms of* M *have type level* $\leq k-1$. □

In the light of Theorem 7.7.2, it seems very likely that this result is optimal in terms of type levels. There is also scope for obtaining more general or more refined results, e.g. with tighter constraints on the exact type of the recursors needed.

For $k = 3$, we have only that $\mathrm{lv}(\psi) \leq 1$, so our analysis shows only that Y_2 suffices; we do not know whether Y_1 is already enough. For $k = 2$, we do not need a memory bank at all, and some mild reworking of our machinery for type $\overline{1}$ codes shows that here Y_1 is indeed sufficient.

7.1.6 Relationship to Kleene Expressions

We may now fulfil a promise made in Subsection 6.2.1, showing that the language Klex^{Ω} of oracle Kleene expressions, like PCF + *byval*, suffices for defining all elements of SP^0 (Theorem 6.2.4). We do this by means of a syntactic translation $-^{\dagger}$ from PCF + *byval* to Klex^{Ω} that respects the NSP semantics.

We define our translation by providing a suitable oracle Kleene expression $c^{\dagger}$ for each PCF^{Ω} + *byval* constant c, and translating variables, application and abstraction by themselves. For the constants $\widehat{n}$, *suc*, *pre*, *ifzero*, *byval* and C_f, the translations are straightforward (the notation $\lceil - \rceil$ here refers to Gödel numbering of Kleene expressions as in Subsection 5.1.3):

$$\begin{aligned}
\widehat{n}^{\dagger} &= Suc^{n}(\widehat{0})\,, \\
suc^{\dagger} &= \lambda x.Suc(x)\,, \\
pre^{\dagger} &= \lambda x.Primrec(x,\widehat{0},\lambda yz.y)\,, \\
ifzero^{\dagger} &= \lambda xyz.\,Eval_{\mathrm{N,N\to N}}(ifzero^{+}(x,\lceil \lambda yz.y \rceil, \lceil \lambda yz.z \rceil),y,z)\,, \\
&\qquad \text{where } ifzero^{+}(x,y,z) \equiv Primrec(x,y,\lambda uv.z)\,, \\
byval^{\dagger} &= \lambda fx.Primrec(\widehat{1},x,\lambda yz.fz)\,, \\
C_f^{\dagger} &= C_f\,.
\end{aligned}$$

Notice that we cannot translate *ifzero* simply by $\lambda xyz.\,ifzero^{+}(x,y,z)$, since the latter would be strict in the second argument.

For Y_{σ}, we mimic the standard proof of the second recursion theorem (Theorem 5.1.5). Explicitly, let us define

$$\begin{aligned}
M[y^{\mathrm{N}},F^{\sigma\to\sigma}] &= F\,(\lambda \vec{x}.Eval_{(\sigma\to\sigma)\to\sigma}(y,F,\vec{x}))\,, \\
N &= \lambda z^{\mathrm{N}}.\lambda F^{\sigma\to\sigma}.\,M[\lceil \lambda F\vec{x}.Eval_{\mathrm{N},(\sigma\to\sigma)\to\sigma}(\lfloor z \rfloor, \lfloor z \rfloor, F,\vec{x})\rceil,\ F]\,, \\
Y_{\sigma} &= \lambda F\vec{x}.\,Eval_{\mathrm{N},(\sigma\to\sigma)\to\sigma}(\lceil N \rceil, \lceil N \rceil, F, \vec{x})\,.
\end{aligned}$$

Here $\lfloor z \rfloor$ denotes the ordinary Kleene expression with Gödel number z; notice that $\lceil \lambda F\vec{x}.Eval(\lfloor z \rfloor, \lfloor z \rfloor, F, \vec{x}) \rceil$ is a classically computable function of z.

The key property of our translation is as follows:

Proposition 7.1.38. *For any* PCF^{Ω} + *byval term* M*, we have* $[M^{\dagger}]^{S} = [\![M]\!]^{\mathsf{SP}^0}$*, where* $[-]^{S}$ *is the NSP semantics of* Klex^{Ω} *as defined as in Subsection 6.1.3, and* $[\![-]\!]^{\mathsf{SP}^0}$ *is the interpretation of* PCF^{Ω} + *byval in* SP^0 *as given by Example 7.1.5.*

Proof. Since the definitions of $[-]^{S}$ and $[\![-]\!]^{\mathsf{SP}^0}$ agree in their treatment of variables, applications and abstractions, it suffices to check using Definition 6.2.1 and Example 7.1.5 that $[c^{\dagger}]^{S} = [\![c]\!]^{\mathsf{SP}^0}$ for each constant c of PCF^{Ω}. This is routine for all constants other than the Y_{σ} (and is recommended as an exercise for the reader). For Y_{σ}, the definition of $[-]^{S}$ yields

$$[Y_{\sigma}F\vec{x}]^{S} = [Eval(\lceil N \rceil, \lceil N \rceil, F, \vec{x})]^{S}$$

$$
\begin{aligned}
&= [N\lceil N\rceil F\vec{x}]^S \\
&= [M[\lceil \lambda F\vec{x}.Eval(\lceil N\rceil, \lceil N\rceil, F, \vec{x})\rceil, F]\,\vec{x}]^S \\
&= [M[\lceil Y_\sigma\rceil, F]\,\vec{x}]^S \\
&= [F(\lambda\vec{x}.Eval(\lceil Y_\sigma\rceil, F, \vec{x}))\,\vec{x}]^S \\
&= [F(\lambda\vec{x}.Y_\sigma F\vec{x})\,\vec{x}]^S \\
&= \texttt{case}\ F(\lambda\vec{x}.[Y_\sigma F\vec{x}]^S)\vec{x}^\eta\ \texttt{of}\ (j \Rightarrow j)\,.
\end{aligned}
$$

This serves as a corecursive definition of $[Y_\sigma F\vec{x}]^S$, from which it is easy to see that $[Y_\sigma]^S$ coincides with $[\![Y_\sigma]\!]^{\mathsf{SP}^0}$ as defined in Example 7.1.5. □

Note too that the above restricts to an *effective* translation from PCF + *byval* to ordinary Kleene expressions. One may also explicitly define an effective translation from Kleene expressions to PCF terms that respects the NSP semantics, but we shall have no need for this here.

Combining the above proposition with Corollary 7.1.33, we obtain that every $p \in \mathsf{SP}^0(\sigma)$ [resp. $\mathsf{SP}^{0,\mathrm{eff}}(\sigma)$] arises as $[M]^S$ for some closed $M : \sigma$ in Klex^Ω [resp. Klex], as claimed by Theorem 6.2.4. The effectivity clause of Theorem 6.2.4 also falls out from the above discussion.

As noted earlier, PCF is preferable to Klex as a programming language for partial functionals in that no coding is involved. However, the language of Kleene expressions is not wasted, since there is no good compositional way to interpret PCF in a *total* type structure over $\mathbb{N}$ in the absence of a supporting structure of partial objects.

7.1.7 Interpretation of NSPs Revisited

To conclude this section, we clarify how PCF relates to the interpretation of sequential procedures via least fixed points as outlined in Subsection 6.4.2, and in particular how our PCF programs I_σ relate to the Kleene computable operations constructed there. This leads to two new results. Firstly, we establish that PCF-definability coincides with *liberal Kleene computability* (see Definition 6.4.6) in models where the latter makes sense. Secondly, we sketch a proof that for any model $\mathbf{P}$ of PCF satisfying some mild conditions, the interpretation of PCF in $\mathbf{P}$ can actually be factored through the NSP model—this suggests that NSPs capture most of the information about PCF programs that is likely to be of interest.

In Subsection 6.4.2, the setting was that of a model $\mathbf{P}$ with fixed point operators satisfying certain conditions 1–4. For such a model, we sketched a 'fixed point' definition of a family of operators

$$
\mathbb{I}^p_{\Gamma,\sigma} \in \mathbf{P}(1, \tau_\Gamma \to \sigma)\,, \qquad \mathbb{I}^e_\Gamma,\ \mathbb{I}^a_\Gamma \in \mathbf{P}(1, \tau_\Gamma \to \mathtt{N})
$$

such that $\mathbb{I}^p_{\Gamma,\sigma}(\lceil q\rceil)(\nu)$ yields an interpretation of a procedure $\Gamma \vdash q : \sigma$ with respect to a valuation ν for Γ, and similarly for $\mathbb{I}^e_\Gamma$ and $\mathbb{I}^a_\Gamma$. We may now see that, modulo some minor adjustments to their definition, such operators may be readily retrieved from the constructions of Subsection 7.1.5.

The type τ_Γ here plays the role of the type ψ in Subsection 6.4.2 as a 'type of valuations', both being defined as $\mathtt{N} \times (\mathtt{N} \to \rho_0) \times \cdots \times (\mathtt{N} \to \rho_{c-1})$. An inconsequential difference is that in Subsection 6.4.2 we did not bother to restrict $\rho_0, \ldots, \rho_{c-1}$ to the *odd* constituent types of $\Gamma \to \sigma$, but one may clearly do so, and the results of Subsection 6.4.2 are unaffected. As a further refinement, one may choose an element $byval \in \mathbf{P}(1 \to 1)$ such that $\lambda f. byval\ f\ \widehat{n} = \lambda f. f\ \widehat{n}$ for each n (note that $\lambda fx. ifzero\, x(fx)(fx)$ is always one such element), and in the circular definition of $\mathbb{I}^a$ replace the clause

$$[\![\mathtt{case}\ a\ \mathtt{of}\ (i \Rightarrow e_i)]\!]^e_\nu \;=\; (\lambda i.\, [\![e_i]\!]^e_\nu)([\![a]\!]^a_\nu)$$

by

$$[\![\mathtt{case}\ a\ \mathtt{of}\ (i \Rightarrow e_i)]\!]^e_\nu \;=\; byval(\lambda i.\, [\![e_i]\!]^e_\nu)([\![a]\!]^a_\nu)\,.$$

Note that if $\mathbf{P}$ is an *extensional* model over $\mathbb{N}_\perp$, this makes no difference to the resulting interpretations, since here only one choice of *byval* is possible, and $[\![e_i]\!]^e_\nu$ is strict in i in any case.

In this situation, we see that $\mathbf{P}$ is at least a pre-model of PCF + *byval*, and it can be checked that the operators $\mathbb{I}^e_\Gamma$ and $\mathbb{I}^a_\Gamma$ are nothing other than the denotations in $\mathbf{P}$ of the programs *eval* and *app* of Subsection 7.1.5. From these, we may obtain operators $\mathbb{I}^p_{\Gamma,\rho}$ via

$$\mathbb{I}^p_{\Gamma,\rho}\,\lceil \lambda\vec{y}.e\rceil(\nu) \;=\; \lambda\vec{z}.\,\mathbb{I}^e_{\Gamma,\vec{y}}\,\lceil e\rceil(\nu[\vec{y} \mapsto \vec{z}])\,,$$

and for suitable types ρ this agrees with the denotation of sim_ρ.

We may now draw the following consequence from Theorem 6.4.15:

Theorem 7.1.39. *Suppose* $\mathbf{P}$ *satisfies the conditions of Definition 6.4.5 and is liberally Kleene closed. Then* $\mathbf{P}$ *is a pre-model of* PCF*, and the liberally Kleene computable elements of* $\mathbf{P}$ *are precisely the* PCF*-definable ones.*

Proof. Suppose $\mathbf{P}$ satisfies the conditions, and $x \in \mathbf{P}(\sigma)$ is liberally Kleene computable. Then by Theorem 6.4.15, x is of the form $\mathbb{I}^p_{\emptyset,\sigma}(\lceil p\rceil)(\emptyset)$ for some $p = \lambda\vec{z}.e \in \mathsf{SP}^{0,\mathrm{eff}}(\sigma)$, and since $\mathbf{P}$ is extensional, this coincides with $\lambda\vec{z}.\, eval\,\lceil e\rceil\,(S(\vec{z}^\eta))$, where $S(\vec{z}^\eta)$ represents the valuation mapping each z_i to z_i^η. But the latter is clearly PCF-definable.

For the converse, recall from Proposition 4.1.1 that every closed PCF term M is β-equivalent to, and hence has the same denotation in $\mathbf{P}$ as, one constructed by application from the PCF constants plus the λ-terms for k, s. But the least fixed point operators are liberally Kleene computable by Proposition 6.4.14; the remaining constants and k, s are easily seen to be liberally Kleene computable; and the liberally Kleene computable elements are closed under application by Exercise 6.4.10. □

As already noted in Subsection 6.4.2, one advantage of the above 'fixed point' approach over the liberal interpretation itself is that it is directly applicable to a much wider range of models, including many non-extensional ones. From the perspective of PCF, we may express this as follows. Suppose that $\mathbf{P}$ is a model of PCF + *byval* in the sense of Subsection 7.1.1, and is normalizable at type $\overline{1}$. Then for any environment $\Gamma = x_0,\ldots,x_{n-1}$ and procedure $\Gamma \vdash p \in \mathsf{SP}(\sigma)$, we may define the *fixed point interpretation* $[\![p]\!]^f_\Gamma \in \mathbf{P}(\Gamma \to \sigma)$ simply by

$$[\![p]\!]^f_\Gamma \;=\; [\![I_{\Gamma\to\sigma}]\!]^{\mathbf{P}} \cdot \lceil \lambda\vec{x}.p \rceil \;,$$

where $I_{\Gamma\to\sigma}$ is the PCF + *byval* term as in Theorem 7.1.32, and we write $\lceil \lambda\vec{x}.p\rceil$ for the unique *normal* code for $\lambda\vec{x}.p$.

We now present a result to the effect that, under mild hypotheses, the direct interpretation of PCF + *byval* in $\mathbf{P}$ factors through the above interpretation $[\![-]\!]^f$ via the interpretation $[\![-]\!]^{\mathsf{SP}^0}$ of PCF + *byval* as defined in Example 7.1.5.

Theorem 7.1.40. *Suppose* $\mathbf{P}$ *is a model of* PCF + *byval satisfying the following conditions:*

1. $\mathbf{P}$ *is simple (in the sense of Subsection 7.1.2).*
2. $\mathbf{P}$ *is normalizable at type* $\overline{1}$.
3. *For each of the constants* $c = suc, pre, ifzero, byval$*, we have* $[\![c]\!]^{\mathbf{P}} = [\![\,[\![c]\!]^{\mathsf{SP}^0}]\!]^f$.

Then $[\![M]\!]^{\mathbf{P}} = [\![\,[\![M]\!]^{\mathsf{SP}^0}]\!]^f$ *for any closed term* M *of* PCF + *byval.*

Proof. (Sketch.) As in the proof of Theorem 7.1.39, it will be enough to prove the theorem for terms M constructed by application from the PCF + *byval* constants along with the λ-terms for k,s. For the constants $\widehat{n}$ this is trivial, and for *suc*, *pre*, *ifzero*, *byval* it is true by fiat. For the terms for k,s, we first check by induction on the types of variables that if $\Gamma = x_0,\ldots,x_{n-1}$ then $[\![x_i^\eta]\!]^f_\Gamma = \lambda\vec{x}.x_i$ in $\mathbf{P}$; the assertions for k,s then follow routinely. A similar argument also shows that $[\![\,[\![Y^k_\sigma]\!]^{\mathsf{SP}^0}]\!]^f = Y^k_\sigma$ in $\mathbf{P}$ for each k. By taking observational limits on both sides, and recalling that in a simple model such limits are uniquely determined, we conclude that $[\![\,[\![Y_\sigma]\!]^{\mathsf{SP}^0}]\!] = Y_\sigma$ in $\mathbf{P}$.

It remains to handle the induction step for application. Since both $[\![-]\!]^{\mathbf{P}}$ and $[\![-]\!]^{\mathsf{SP}^0}$ respect application, it is enough to show that for any $p \in \mathsf{SP}^0(\sigma\to\tau)$ and $q \in \mathsf{SP}^0(\sigma)$, we have $[\![p\cdot q]\!]^f = [\![p]\!]^f([\![q]\!]^f)$ in $\mathbf{P}$. By a continuity argument allied with the fact that $\mathbf{P}$ is simple, it will suffice to show this for *finite* p,q. (The normalizability at type $\overline{1}$ is used at this point—cf. Exercise 7.1.11(i).) To establish this, we need to extend our interpretation $[\![-]\!]^f$ to arbitrary NSP meta-terms, which we do by extending the definition of the program $I_{\Gamma\to\sigma}$ so as to simulate codes for meta-procedures in an obvious way. The new ingredient here is an extra clause in the definition of *app* to achieve the following effect:

$$app\;\lceil PQ_0\ldots Q_{r-1}\rceil\; S \;=\; (sim\;\lceil P\rceil\; S)(sim\;\lceil Q_0\rceil\; S)\cdots(sim\;\lceil Q_{r-1}\rceil\; S)\;.$$

It will then be the case that $[\![\lambda\vec{x}.pq\vec{x}^\eta]\!]^f = [\![p]\!]^f([\![q]\!]^f)$. But now $\lambda\vec{x}.pq\vec{x}^\eta \rightsquigarrow^* p\cdot q$ in finitely many steps by Theorem 6.3.4. Moreover, it can be checked (with some effort) that if $M \rightsquigarrow M'$ then $[\![M]\!]^f = [\![M']\!]^f$, whence the result. □

Thus, under moderate hypotheses, interpretations of PCF factor through NSPs in a standard way. This means that although such interpretations may well draw denotational distinctions between observationally equivalent terms, they can never distinguish between terms yielding the same NSP.

If **P** is extensional, the hypotheses 1–3 above are automatically satisfied. They are also in practice satisfied in many intensional models of interest, including the simple quotients of typical *game models* (see Section 7.4). The restriction to simple models is a mild one insofar as every model has a canonical simple quotient; it is admittedly something of an oddity here that SP^0 itself is not simple.

7.2 Variants and Type Extensions of PCF

So far we have been considering PCF in its original *call-by-name* version, and moreover restricting attention to the simple types over $\mathtt{N}$. Before proceeding further with our study of this language, we digress briefly to show that a model of this basic version of PCF will often have all the structure we need to model significantly more besides. In Subsection 7.2.1, we give just a brief glimpse of some of the other datatypes that are typically available 'for free' in the Karoubi envelope of a model of PCF, continuing a line of discussion initiated in Section 4.2. In Subsection 7.2.2, we show that under mild assumptions, a model of ordinary PCF also gives rise to a model of *call-by-value* version PCF_v via the machinery of partial λ-algebras developed in Section 4.3. This call-by-value interpretation will not feature in the rest of this chapter, though it will sometimes play a role in later chapters.

7.2.1 *Further Type Structure in Models of* PCF

Let **P** be any model of PCF. From Section 4.2 we know that the Karoubi envelope $\mathbf{K}(\mathbf{P})$ contains many other useful datatypes besides those of $\mathsf{T}^{\rightarrow}(\mathtt{N})$: for instance, we have already referred in Subsection 7.1.5 to a version of PCF with product types which may be interpreted there. Here we shall pause to fix notation for some other trivial extensions of PCF that will be useful in the remainder of this chapter. We also indicate (very briefly) some of the further type structure available in $\mathbf{K}(\mathbf{P})$ beyond that obtained in Section 4.2; this structure will not be required in the sequel.

First, we may extend PCF with a unit type $\mathtt{U}$ and a boolean type $\mathtt{B}$, along with constants

$$\top : \mathtt{U}\,, \qquad then_\gamma : \mathtt{U}, \gamma \rightarrow \gamma\,,$$
$$tt, ff : \mathtt{B}\,, \qquad if_\gamma : \mathtt{B}, \gamma, \gamma \rightarrow \gamma\,,$$

where γ ranges over the *ground types* (currently $\mathtt{U}$, $\mathtt{B}$ and $\mathtt{N}$). The reduction rules for these new constructs are

$$then\,\top \rightsquigarrow \lambda x.x\,, \qquad if\;tt \rightsquigarrow \lambda xy.x\,, \qquad if\;ff \rightsquigarrow \lambda xy.y\,,$$

and the new primitive evaluation contexts are $then[-]$ and $if_\gamma[-]$. We will usually write *then* as an infix; the meaning of M *then* N essentially corresponds to that of $N\upharpoonright_M$ in Section 4.3. For readability, we will sometimes write $if\, M N P$ as *if* M *then* N *else* P. Clearly this extended language admits an evident interpretation in $\mathbf{K}(\mathbf{P})$, and the adequacy theorem 7.1.16 goes through as before.

Somewhat less formally, it will be useful to add a ground type $\mathtt{N}^*$ of finite lists of natural numbers; we shall display elements of this type using the notation $[n_0,\ldots,n_{l-1}]$, and will adopt notation for standard list operations as required.

Next, it is easily checked that the operation *byval* defined in Example 7.1.5 is idempotent in SP^0, and it follows by the results of Subsection 7.1.7 that *byval* is idempotent in all other models of interest. Using this, we obtain an idempotent *str* on $\mathtt{N}\to\tau$ for every simple type τ: for instance, if $\tau=\tau_0,\ldots,\tau_{r-1}\to\mathtt{N}$ we define

$$str \;=\; \lambda f.\,\lambda x^{\mathtt{N}} y_0\ldots y_{r-1}.\,byval\,(\lambda z.fz\vec{y})\,x\,,$$

and this gives rise to a semantic type of *strict* operations $\mathtt{N}\to\tau$ within $\mathbf{K}(\mathbf{P})$, which we shall denote by $\mathtt{N}\to_s\tau$ following a convention introduced in Subsection 4.2.4. Likewise, we have a type $\gamma\to_s\tau$ for any other ground type γ, though we do not in general have $\sigma\to_s\tau$ for arbitrary σ.

As regards further type structure, it turns out that $\mathbf{K}(\mathbf{P})$ supports a rich repertoire of infinitary inductive types, and sometimes even mixed-variance recursive types, going significantly beyond what we obtained in Section 4.2. Since these extensions are not required elsewhere in the book, we will content ourselves with just a brief glimpse of what is possible, and we do this in the form of a project for the ambitious reader.

Exercise 7.2.1. (i) Show that under mild hypotheses on a model $\mathbf{P}$ of PCF, we naturally obtain a *lifted sum* operation as a bifunctor $\oplus:\mathscr{K}(\mathbf{P})\times\mathscr{K}(\mathbf{P})\to\mathscr{K}(\mathbf{P})$, where $X\oplus Y$ is given as a suitable retract of $\mathbf{P}(\mathtt{B})\times X\times Y$.

(ii) Under the same hypotheses, let $F:\mathscr{K}(\mathbf{P})\to\mathscr{K}(\mathbf{P})$ be any functor constructed via composition and pairing from constant functors together with $\times$, $\oplus$, $A\Rightarrow-$ and $B\Rightarrow_s-$, where A,B are constant objects with $B\lhd\mathbf{P}(\mathtt{N})$. By adapting ideas from Subsection 4.2.3, show that $\mathbf{K}(\mathbf{P})$ contains an *initial algebra* for F in the sense of Proposition 4.2.12. In this way we obtain interpretations for infinitary inductive types such as the type $\mu\alpha.\mathtt{N}\times(\mathtt{N}\to\alpha)$ of infinitely branching lazy trees.

(iii) Now suppose additionally that $\mathbf{P}$ has a *universal* type υ. Let G be any unary operation on objects of $\mathscr{K}(\mathbf{P})$ constructed via composition and pairing from constants together with $\times$, $\oplus$, $\Rightarrow$ and $\Rightarrow_s$. (Note that G will not in general be a functor, since $\Rightarrow$ and $\Rightarrow_s$ are of mixed variance.) By representing types of $\mathbf{K}(\mathbf{P})$ by elements of $\mathbf{P}(\upsilon\to\upsilon)$, show how a sensible meaning may be assigned to the 'least fixed point' of G as an object of $\mathbf{K}(\mathbf{P})$. (The idea here is due to Scott [259].) We thus obtain interpretation for mixed-variance recursive types such as $\mu\alpha.\mathtt{N}\oplus(\alpha\to\alpha)$.

(iv) Investigate conditions under which the initial algebra for a functor F as in (ii) coincides with its least fixed point as in (iii).

7.2.2 Call-by-Value PCF

A natural alternative to PCF is provided by its *call-by-value* counterpart, which we denote by PCF_v. The essential difference is that under call-by-value, an application term MN is evaluated by first evaluating N to a value V, then passing this as the argument to M. (Semantically, it is natural to think here of models $\mathbf{P}$ in which $\mathbf{P}(\mathtt{N})$ is $\mathbb{N}$ rather than $\mathbb{N}_\perp$, and $\mathbf{P}(\sigma \to \tau)$ consists of partial functions from $\mathbf{P}(\sigma)$ to $\mathbf{P}(\tau)$.) On the face of it, this offers an equally plausible candidate for a concept of higher type sequential computation, so it is reasonable to ask whether PCF and PCF_v represent genuinely different computability notions. The main thrust of this subsection will be to show that there is relatively little essential difference between the two variants: any model for PCF will (under mild assumptions) yield an 'equivalent' model for PCF_v, in such a way that properties such as completeness may be transferred from one to the other.

We define PCF_v as the simply typed λ-calculus over the type world $\mathsf{T}^{\to}(\mathtt{N})$ with the same constants as PCF (see Section 7.1), except that a constant Y_σ is included only for arrow types σ (i.e. not for $\sigma = \mathtt{N}$). By a *value* in PCF_v, we shall here mean a closed term that is either a constant or a λ-abstraction $\lambda x.M$; we let V range over values. The reduction rules of PCF_v are

$$\begin{aligned} (\lambda x.M)V &\rightsquigarrow M[x \mapsto V]\,, & \mathit{ifzero}\,\widehat{0} &\rightsquigarrow \lambda xy.x\,,\\ \mathit{suc}\,\widehat{n} &\rightsquigarrow \widehat{n+1}\,, & \mathit{ifzero}\,\widehat{n+1} &\rightsquigarrow \lambda xy.y\,,\\ \mathit{pre}\,\widehat{n+1} &\rightsquigarrow \widehat{n}\,, & Y_{\tau\to\tau'}f &\rightsquigarrow \lambda x^{\tau}.f(Y_{\tau\to\tau'}f)x\,,\\ \mathit{pre}\,\widehat{0} &\rightsquigarrow \widehat{0}\,, \end{aligned}$$

which may be applied (for any closed instantiation of f) in the evaluation contexts generated by

$$[-]N\,, \qquad V[-]\,.$$

To illustrate the difference between PCF and PCF_v, we note that $(\lambda x.\widehat{1})(Y_1(\lambda f.f)\widehat{0})$ evaluates to $\widehat{1}$ in the former, but yields a non-terminating reduction in the latter.

Suppose now that we begin with a model $\mathbf{P}$ of PCF. We saw in Section 4.3 how, under certain conditions, $\mathbf{P}$ gives rise to a partial λ-algebra, which we shall here denote by $\mathbf{P}_v$. Our purpose here is to show that in this situation $\mathbf{P}_v$ admits an adequate interpretation of PCF_v. We shall in principle assume familiarity with the setup of Section 4.3, but will for convenience summarize some of the key ingredients here.

First, we recall from Proposition 4.3.4 that the construction of the partial λ-algebra $\mathbf{P}_v$ requires $\mathbf{P}$ to have certain structure, namely a *strict partiality structure* $(D, \top, \downharpoonleft)$ and a *proper definedness predicate* $\delta \in (\mathbf{P}(\mathtt{N}) \Rightarrow D)$. Here we shall take $D = \mathbf{P}(\mathtt{N})$, $\top = \widehat{0}$, and for each type $\sigma = \sigma_0, \ldots, \sigma_{r-1} \to \mathtt{N}$ define $f \downharpoonleft_\sigma z = \lambda \vec{x}.\,\mathit{ifzero}\; z\; (f\vec{x})\; \perp$. We also define $\delta = \lambda x^{\mathtt{N}}.\,\mathit{ifzero}\; x\; \widehat{0}\,\widehat{0}$. As a hypothesis on $\mathbf{P}$, we adopt the assumption that these do indeed constitute a strict partiality structure and proper definedness predicate. As noted in Section 4.3, these hypotheses hold whenever $\mathbf{P}$ is an extensional model over $\mathbb{N}_\perp$, and also for the sequential algorithms model SA, though not for more fine-grained intensional models such as SP^0.

Next, we recall how $\mathbf{P}_v$ is constructed under these conditions. Whereas $\mathbf{P}$ is a λ-algebra over $\mathsf{T}^{\rightarrow}(\mathtt{N})$, we define $\mathbf{P}_v$ as a partial applicative structure over $\mathsf{T}^{\rightarrow}(\mathtt{N})$. (We can dispense here with the types $\mathtt{V}$ and $\mathtt{U}$ that featured in Section 4.3.) For each type σ, we define an object $C_\sigma \in \mathbf{K}(\mathbf{P})$ and a morphism $\delta_\sigma : C_\sigma \to \mathbf{P}(\mathtt{N})$ by

$$(C_{\mathtt{N}}, \delta_{\mathtt{N}}) = (\mathbf{P}(\mathtt{N}),\ \delta)\ ,$$
$$(C_{\sigma\to\tau}, \delta_{\sigma\to\tau}) = (L(C_\sigma \Rightarrow_s C_\tau),\ \xi_{C_\sigma \Rightarrow_s C_\tau})\ ,$$

where L is the lift functor arising from $(D, \top, 1)$ as in Subsection 4.2.4, and ξ_A is the obvious definedness predicate on $L(A)$. We then take $\mathbf{P}_v(\sigma) = \delta_\sigma^{-1}(\top) \subseteq C_\sigma$ for each σ, and endow $\mathbf{P}_v$ with the evident partial application operations induced by application in $\mathbf{P}$. The content of Proposition 4.3.4(i) is then that $\mathbf{P}_v$ is a $p\lambda$-algebra.

Some further notation will be helpful. If $\Gamma = x_0^{\sigma_0}, \ldots, x_{r-1}^{\sigma_{r-1}}$, we will write C_Γ for $C_{\sigma_0} \times \cdots \times C_{\sigma_{r-1}}$ (note that $\mathbf{K}(\mathbf{P})$ has products since $\mathbf{P}$ has booleans). We use the notation λx with reference to the $\lambda\eta$-algebra structure of $\mathbf{K}(\mathbf{P})$; a superscript σ or Γ on the bound variable will indicate the type C_σ or C_Γ. In addition, we write $\lambda_s x$ for *strict λ-abstraction*: if x has type C_σ and M has type C_τ, then $\lambda_s x.M = \eta(str(\lambda x.M))$ has type $C_{\sigma\to\tau}$. We also recall the morphism $app' : C_{\sigma\to\tau} \times C_\sigma \to C_\tau$ constructed in the proof of Proposition 4.3.4. Finally, if A is any type in $\mathbf{K}(\mathbf{P})$ with $(t,r) : A \lhd \mathbf{P}(\sigma)$ its designated retraction pair, we write $Y_A : (A \Rightarrow A) \Rightarrow A$ for the operator $\lambda g.r(Y_\sigma(\lambda x.r(g(tx))))$; note that $Y_A = \lambda f.f(Y_A f)$ in $\mathbf{K}(\mathbf{P})$.

We are now ready to define an interpretation of PCF$_v$ in $\mathbf{P}_v$. Much as in the proof of Proposition 4.3.4, we will first compositionally define for each term $\Gamma \vdash M : \tau$ of PCF$_v$ an interpretation $[M]_\Gamma : C_\Gamma \to C_\tau$ in $\mathbf{K}(\mathbf{P})$, then use this to obtain a (partial) interpretation $[\![M]\!]_\nu \in \mathbf{P}_v(\tau)$ where ν is any valuation for Γ in $\mathbf{P}_v$.

$$\begin{aligned}
[\widehat{n}]_\Gamma &= \lambda p^\Gamma.\widehat{n}\ ,\\
[suc]_\Gamma &= \lambda p^\Gamma.\lambda_s x.\, suc\, x\ ,\\
[pre]_\Gamma &= \lambda p^\Gamma.\lambda_s x.\, pre\, x\ ,\\
[ifzero]_\Gamma &= \lambda p^\Gamma.\lambda_s z^{\mathtt{N}}.\lambda_s x^{\mathtt{N}}.\lambda_s y^{\mathtt{N}}.\, ifzero\, z x y\ ,\\
[Y_{\tau\to\tau'}]_\Gamma &= \lambda p^\Gamma.\lambda_s f^{(\tau\to\tau')\to(\tau\to\tau')}.\, Y_{C_{\tau\to\tau'}}(\lambda g^{\tau\to\tau'}.\lambda_s x^\tau.app'\langle app'\langle f,g\rangle,x\rangle)\ ,\\
[x_i]_\Gamma &= \lambda(x_0,\ldots,x_{r-1}).x_i\ ,\\
[MN]_\Gamma &= \lambda p^\Gamma.app'\langle [M]_\Gamma p, [N]_\Gamma p\rangle\ ,\\
[\lambda x^\sigma.M]_\Gamma &= \lambda p^\Gamma.\lambda_s x^\sigma.[M]_{\Gamma,x}\langle p,x\rangle\ .
\end{aligned}$$

If $\Gamma \vdash M : \tau$ in PCF$_v$ where $\Gamma = x_0^{\sigma_0}, \ldots, x_{r-1}^{\sigma_{r-1}}$, and ν maps each variable x_i to an element of $\mathbf{P}_v(\sigma_i)$, we may now define

$$[\![M]\!]_\nu \ \simeq\ [M]_\Gamma \cdot \nu(x_0) \cdot \ldots \cdot \nu(x_{r-1})\ .$$

When Γ and ν are empty, we may write simply $[\![M]\!]$ in place of $[\![M]\!]_\nu$. The following theorem now expresses the *adequacy* of this interpretation:

Theorem 7.2.2. *Under the above hypotheses on* $\mathbf{P}$, *for any closed* PCF_v *term* $M : \tau$ *we have* $[\![M]\!]_\Gamma \downarrow$ *in* $\mathbf{P}_v$ *iff* $M \rightsquigarrow^* V$ *for some value* V, *and in this case* $[\![M]\!]_\Gamma = [\![V]\!]_\Gamma$.

Proof. (Sketch.) By tedious calculation, one can verify that if $M \rightsquigarrow M'$ in PCF_v then $[\![M]\!]_\Gamma = [\![M']\!]_\Gamma$, and it is easy to check that $[\![V]\!]_\Gamma \downarrow$ for every value V. It remains to show that if $[\![M]\!]_\Gamma \downarrow$ then the evaluation of M terminates—this is done just as in the proof of Theorem 7.1.16, using the easily verified fact that $[Y_{\tau \to \tau'}]$ is an observational limit of the elements $[\lambda f.\lambda x.f^k(\bot)(x)]$ for $k \in \mathbb{N}$. $\square$

Going in the other direction, one may also give an independent definition of a 'model of PCF_v' and show that any such model yields a model of PCF via the construction of Proposition 4.3.3. However, we shall not pursue this avenue here.

Next, we outline the sense in which $\mathbf{P}$ and $\mathbf{P}_v$ are 'equivalent': call-by-name types may be embedded in call-by-value types and vice versa, with the effect that completeness results for extensions of PCF may readily be transferred between the two versions (this will be useful in Chapters 11 and 12). Working in $\mathbf{K}(\mathbf{P})$, let us here write $[\![\sigma]\!]_n$ for $\mathbf{P}(\sigma)$, and $[\![\tau]\!]_v$ for C_τ as defined above. (In other contexts, it is often more appropriate to regard the subset $\mathbf{P}_v(\tau) \subseteq C_\tau$ as the genuine call-by-value interpretation of τ.) We also define translations $\overline{-}$ and $\widetilde{-}$ on types of $\mathtt{T}^{\to}(\mathtt{N})$ by

$$\overline{\mathtt{N}} = \mathtt{N}\,, \qquad \overline{\sigma \to \tau} = (\mathtt{N} \to \overline{\sigma}) \to \overline{\tau}\,,$$
$$\widetilde{\mathtt{N}} = \mathtt{N}\,, \qquad \widetilde{\sigma \to \tau} = \widetilde{\sigma} \to (\mathtt{N} \to \widetilde{\tau})\,.$$

Informally, the type $\mathtt{N} \to \overline{\sigma}$ will serve to represent the lifting of $\overline{\sigma}$ (we could equally well use the unit type here in place of $\mathtt{N}$ if available). By contrast, the type $\mathtt{N} \to \widetilde{\tau}$ will represent the lifting of $\widetilde{\tau}$ in a different way: the value at $\widehat{0}$ will record the definedness or otherwise of the element of the lifting, while the value at $\widehat{1}$ will record the actual value of type $\widetilde{\tau}$ when defined.

Proposition 7.2.3. *For all types* $\sigma, \tau \in \mathtt{T}^{\to}(\mathtt{N})$ *where* $\tau \neq \mathtt{N}$, *we have* $[\![\sigma]\!]_n \lhd [\![\overline{\sigma}]\!]_v$ *and* $[\![\tau]\!]_v \lhd [\![\widetilde{\tau}]\!]_n$ *within* $\mathbf{K}(\mathbf{P})$. *Moreover, these retractions may be chosen so that both halves of the composite retraction* $[\![\sigma]\!]_n \lhd [\![(\overline{\sigma})^{\sim}]\!]_n$ *are* PCF*-definable, and both halves of the composite retraction* $[\![\tau]\!]_v \lhd [\![(\widetilde{\tau})^{-}]\!]_v$ *are essentially* PCF_v*-definable. (Strictly speaking, it is the induced retraction* $[\![\mathtt{N} \to \tau]\!]_v \lhd [\![\mathtt{N} \to (\widetilde{\tau})^{-}]\!]_v$ *that is* PCF_v*-definable.)*

Proof. Again we work in the $\lambda\eta$-algebra structure of $\mathsf{K}(\mathbf{P})$. Define $(t_\sigma, r_\sigma) : [\![\sigma]\!]_n \lhd [\![\overline{\sigma}]\!]_v$ inductively by

$$t_{\mathtt{N}} = id\,, \qquad t_{\sigma \to \tau} = \lambda f.(\lambda_s y.t_\tau(f(r_\sigma(app'\langle y, \widehat{0}\rangle)))))\,,$$
$$r_{\mathtt{N}} = id\,, \qquad r_{\sigma \to \tau} = \lambda g.(\lambda x.r_\tau(app' g, \lambda_s z.t_\sigma x))\,.$$

Also define $(t'_\tau, r'_\tau) : [\![\tau]\!]_v \lhd [\![\widetilde{\tau}]\!]_n$ for $\tau \neq \mathtt{N}$ by

$$t'_{\mathtt{N}} = id\,, \qquad t'_{\sigma \to \tau} = \lambda g.(\lambda x.\lambda n.\ \mathit{ifzero}\ n\ (\lambda \vec{w}.(\widehat{0} \mid \delta(g)))\ (t'_\tau(app'\langle g, r'_\sigma x\rangle))))\,,$$
$$r'_{\mathtt{N}} = id\,, \qquad r'_{\sigma \to \tau} = \lambda f.\ (\lambda_s y.r'_\tau(f(t'_\sigma y)\widehat{1}))\ \mid\ \delta(f\,d\,\widehat{0}\,\vec{e})\,.$$

Here $\upharpoonright$ and δ have their meanings from Section 4.3, and $d, \vec{e}$ are arbitrary elements of appropriate type so that $f\,d\,\widehat{0}\,\vec{e}$ has type N. It is easily verified by induction that the (t_σ, r_σ) are retraction pairs. For (t'_τ, r'_τ), the induction step boils down to the equation $(\lambda_s y.\, app'\langle g, y\rangle) \upharpoonright \delta(g) = g$ for $g : L(X \Rightarrow_s Y)$; this can itself be verified using the concrete representation of lifting in the proof of Proposition 4.2.17.

The definability clauses are clear once the λ-terms for the composite retractions have been routinely simplified; we leave the details of this to the reader. □

Remark 7.2.4. Along similar lines, it is also easy to show by induction that each call-by-name *pure* type $[\![k]\!]_n$ is a retract of the call-by-value pure type $[\![k+1]\!]_v$. From this together with Proposition 4.2.3 and the above proposition, it follows readily that every call-by-value type $[\![\tau]\!]_v$ is a retract of a pure call-by-value type. Moreover, an explicit computation of the λ-terms in question readily shows both halves of these retractions to be essentially PCF$_v$-definable.

From Proposition 7.2.3 it is easy to read off that the models **P** and $\mathbf{P}_v$ are *equivalent* in the sense of Definition 3.4.4. Furthermore, the above retractions can be used to transfer completeness results from a call-by-value to a call-by-name setting. Suppose, for instance, that in the above situation we have elements $C_i \in [\![\sigma_i]\!]_n$ for $i < r$, and that $C'_i = t_{\sigma_i}(C_i) \in [\![\overline{\sigma_i}]\!]_v$ for each i. We then have the following, which will be useful in Chapters 11 and 12:

Theorem 7.2.5. *If every element of each* $[\![\tau]\!]_v$ *is definable in* PCF$_v + C'_0, \ldots, C'_{r-1}$, *then every element of each* $[\![\sigma]\!]_n$ *is definable in* PCF $+ C_0, \ldots, C_{r-1}$.

Proof. (Sketch.) Suppose $x \in [\![\sigma]\!]_n$. Take $M : \overline{\sigma}$ a closed PCF$_v$ term denoting $t_\sigma(x)$. By induction on the structure of M, it is straightforward to construct a closed PCF term $M' : (\overline{\sigma})^\sim$ denoting $t'_{\overline{\sigma}}(t_\sigma(x))$. But by Proposition 7.2.3, the retraction $r_\sigma \circ r'_{\overline{\sigma}}$ is PCF-definable, so we obtain a closed PCF term denoting x. □

Further proof details may be found in Chapter 6 of Longley [173], which also obtains a similar result in the other direction, and moreover considers a third variant of the language, namely the *lazy* PCF of Bloom and Riecke [38]. For other related results and associated syntactic translations, see Sieber [270] and Riecke [244].

Many other results that we shall obtain for call-by-name languages and models can be easily transferred to the call-by-value world using the above ideas. For example, in Section 7.7 we will see that there is no universal type for call-by-name PCF, and it is then clear from Proposition 7.2.3 that the same will be true of PCF$_v$.

7.3 Examples and Non-examples of PCF-Definable Operations

We now look at some interesting examples and non-examples of operations computable in PCF (or equivalently representable by effective NSPs). We pay attention both to the PCF programs themselves and to the operations they define in important models such as PC.

Although the notion of computability represented by PCF is a relatively weak one in comparison with others considered later in this book, it supports some remarkable and surprising computational phenomena. The universal operators I_σ described in Subsection 7.1.5 already provide one non-trivial demonstration of the power of PCF, and here we shall flesh out the picture with some further examples. Even the weaker language $\mathrm{T}+min$, which suffices for most 'bread-and-butter' PCF-style computation, has a few surprises to offer, although our more far-reaching examples will make essential use of the infinite nesting associated with non-left-well-founded procedures (see Subsection 6.3.2).

Several of our examples will have the character of interesting *total* functionals within Ct for which representatives in PC are PCF-computable. In Theorem 8.6.1 we will see that in fact *all* effective elements of Ct are PCF-computable in this way, but for now we will focus on some specific examples that are illuminating in their own right, and which will pave the way for the proof of the general result.

In order to be more specific about the types at which various phenomena naturally arise, we shall here enrich our language PCF with the ground types $\mathtt{U}$, $\mathtt{B}$ and $\mathtt{N}^*$ as detailed in Subsection 7.2.1. We will also sometimes refer to strict function spaces $\mathtt{N} \rightarrow_s \tau$, although we do not strictly speaking add such types to our language—rather, we view the subscript s simply as an annotation that gives more precise information about the operation in question. For example, in all models of interest, the minimization operator lives naturally at type $(\mathtt{N} \rightarrow_s \mathtt{B}) \rightarrow \mathtt{N}$, a semantically somewhat simpler type than $(\mathtt{N} \rightarrow \mathtt{B}) \rightarrow \mathtt{N}$.

To ease notation, we shall generally mix syntax and semantics rather freely, using the same notation for PCF programs and for elements of the models in question. We will also avail ourselves of various kinds of syntactic 'sugar' for which the translation into PCF proper is evident.

Whilst our main focus is on positive examples, our first subsection will be dedicated to some non-examples highlighting important limitations in the power of PCF.

7.3.1 Counterexamples to Full Abstraction

We begin with the most famous 'deficiency' of PCF-style sequential computation: the absence of operations that involve parallel evaluation. Consider for example the following *parallel convergence* operator $pconv : \mathtt{U},\mathtt{U} \rightarrow \mathtt{U}$, which is easily seen to exist in the model PC (cf. the operator P of Subsection 1.1.6):

$$pconv\, x\, y = \top \text{ if } x = \top \text{ or } y = \top\,,$$
$$pconv\, x\, y = \bot \text{ if } x = y = \bot\,.$$

It is easy to show that there can be no procedure $p \in \mathsf{SP}^0(\mathtt{U},\mathtt{U} \rightarrow \mathtt{U})$ that denotes $pconv$. For suppose $p = \lambda xy.e$ were such a procedure. We argue by cases on the form of e. Clearly we cannot have $e = \top$ or $e = \bot$, since $pconv$ is not a constant function. However, if e has the form $\mathtt{case}\; x \;\mathtt{of}\; (\top \Rightarrow e')$ then $e \cdot \top \cdot \bot$ diverges whereas $pconv\,\top\,\bot$ converges; while if e has the form `case` y `of` $(\top \Rightarrow e')$ then $e\cdot\bot\cdot\top$ diverges whereas $pconv\,\bot\,\top$ converges. It follows that $pconv$ is not definable in PCF or even $\mathrm{PCF}^{\Omega} + byval$.

Next, consider the famous *parallel-or* function $por \in \mathsf{PC}(\mathtt{B},\mathtt{B} \to \mathtt{B})$ given by

$$\begin{aligned} por\,x\,y &= tt \text{ if } x = tt \text{ or } y = tt\,, \\ por\,x\,y &= ff \text{ if } x = ff \text{ and } y = ff\,, \\ por\,x\,y &= \bot \text{ otherwise.} \end{aligned}$$

It is easy enough to write a closed PCF program M such that $[\![M]\!]^{\mathsf{PC}}(por) = pconv$ (we leave this as an exercise). It follows that por is not denotable by an NSP, and hence not definable in PCF. These examples show that even very simple compact elements of PC can fail to be PCF-definable.

The example $pconv$ can also be used to exhibit a counterexample to full abstraction for PC. Consider the two 'detecting programs' $D_0, D_1 : (\mathtt{U},\mathtt{U} \to \mathtt{U}) \to \mathtt{U}$:

$$\begin{aligned} D_0 &= \lambda f.(f\,\top\,\bot)\ then\ (f\,\bot\,\top)\,, \\ D_1 &= \lambda f.(f\,\bot\,\bot)\,. \end{aligned}$$

These have distinct denotations in PC, since $[\![D_0]\!](pconv) = \top$ but $[\![D_1]\!](pconv) = \bot$. However, D_0 and D_1 are observationally equivalent in PCF. Intuitively, this is because the only way to distinguish them would be to apply them to $pconv$. More formally, one can show by case analysis that $[\![D_0]\!]^{\mathsf{SP}^0} \cdot q = [\![D_1]\!]^{\mathsf{SP}^0} \cdot q$ for all $q \in \mathsf{SP}^0(\mathtt{U},\mathtt{U} \to \mathtt{U})$, and appeal to the context lemma. Alternatively, and more cleanly, one can compute the interpretations of D_0 and D_1 in either the *stable* model St or the *strongly stable* model SSt (see Chapter 11); it turns out that in either of these models we have $[\![D_0]\!] = [\![D_1]\!]$, whence $D_0 \sim_{\mathrm{obs}} D_1$.

For the sake of completeness, we briefly mention some similar counterexamples showing that St and SSt are not fully abstract (specific knowledge of these models will not be required here). We first note that in each of PC, St and SSt, the type $\mathtt{U} \to \mathtt{U}$ consists of just three functions, defined by the terms $\lambda x.\top$, $\lambda x.x$ and $\lambda x.\bot$. Moreover, in St and SSt, there is a functional $F : (\mathtt{U} \to \mathtt{U}) \to \mathtt{B}$ such that

$$F(\lambda x.\top) \;=\; tt\,, \qquad F(\lambda x.x) \;=\; ff\,, \qquad F(\lambda x.\bot) \;=\; \bot\,.$$

(This example will play an important role in Chapter 11.) However, it is easy to see that F is not present in $\mathsf{PC}((\mathtt{U} \to \mathtt{U}) \to \mathtt{B})$ since it is not monotone with respect to the pointwise ordering: we have $\lambda x.x \sqsubseteq \lambda x.\top$ in $\mathsf{PC}(\mathtt{U} \to \mathtt{U})$ but $ff \not\sqsubseteq tt$. This also provides a way of seeing that no program meeting the specification for F can be written in PCF.

Mimicking the detecting functions D_0, D_1 above, we may define PCF programs $D'_0, D'_1 : ((\mathtt{U} \to \mathtt{U}) \to \mathtt{B}) \to \mathtt{U}$ as follows:

$$\begin{aligned} D'_0 &= \lambda f.\ if\ f(\lambda x.\top) = tt\ and\ f(\lambda x.x) = ff\ then\ \top\ else\ \bot\,, \\ D'_1 &= \lambda f.\bot\,. \end{aligned}$$

These receive distinct denotations in St or SSt since they behave differently on F. However, they do receive the same denotation in PC, so in this case PC may be used to establish the observational equivalence.

The models St and SSt can likewise be distinguished from each other. For example, there is a function G of type $\mathtt{B},\mathtt{B},\mathtt{B} \to \mathtt{B}$ known as the *Gustave function* (due to Berry) which lives in St though not in SSt or PC:

$$G \perp tt\, ff \;=\; tt\,, \qquad G\, ff \perp tt \;=\; tt\,, \qquad G\, tt\, ff \perp \;=\; tt\,, \qquad G\, ff\, ff\, ff \;=\; ff\,.$$

We leave it as an exercise to write two detecting programs which, informally, could only be distinguished by applying them to the Gustave function. These programs receive distinct denotations in St, but the same denotation in SSt and PC, and hence are observationally equivalent.

Finally, we mention an artificial but ingenious example due to Curien [60], showing that there are even observational equivalences that elude all of the models PC, St, SSt. Consider the following specification for a functional $C : (\mathtt{B},\mathtt{B} \to \mathtt{U}) \to \mathtt{U}$:

$$\begin{aligned} C\,f &= \top \text{ if } f\,tt \perp = \top \text{ or } f \perp tt = \top \text{ or } f \perp ff = f\,ff\,tt = \top\,, \\ C\,f &= \perp \text{ if } f\,x\,y = \perp \text{ for all } x, y\,. \end{aligned}$$

It can be checked that each of our three models contains a functional C with these properties, although no such functional is definable in PCF. (This can be shown by a laborious case analysis of sequential procedures at the type in question; see also Exercise 7.5.15.) Now consider the following detecting programs:

$$\begin{aligned} D_0'' &= \lambda c.\, c(\lambda xy.\, \mathit{if}\, x\, \mathit{then} \top \mathit{else} \perp)\ \mathit{then}\ c(\lambda xy.\, \mathit{if}\, y\, \mathit{then} \top \mathit{else} \perp)\ \mathit{then} \\ &\qquad\qquad c(\lambda xy.\, \mathit{if}\, y\, \mathit{then}\, (\mathit{if}\, x\, \mathit{then} \perp \mathit{else} \top)\, \mathit{else} \top)\,, \\ D_1'' &= \lambda c.\, c(\lambda xy.\perp)\,. \end{aligned}$$

These receive distinct denotations in all three models, since they differ on a function with the specified properties of C. Nonetheless, they are observationally equivalent by virtue of the context lemma, since it is not hard to see that any PCF term on which they differed would have to satisfy the specification of C.

The idea of detecting programs turns out to be very general. An important argument due to Milner (see Theorem 7.5.18) shows that in any model of PCF, if x is any 'finite' element of minimal type level such that x is not PCF-definable, then there are PCF programs D_0, D_1 such that $D_0 \sim_{\mathrm{obs}} D_1$ but $[\![D_0]\!](x) \neq [\![D_1]\!](x)$.

7.3.2 The Power of Nesting

We now turn to some of the surprising and counterintuitive things that *can* be done in PCF. Our first example is taken from Berry, Curien and Lévy [34].

Let $B \in \mathsf{PC}((\mathtt{U},\mathtt{U} \to \mathtt{U}) \to \mathtt{U})$ be defined as follows:

$$B(f) = \top \text{ if } f\top\bot = \top \text{ or } f\bot\top = \top\,,$$
$$B(f) = \bot \text{ otherwise.}$$

One might initially suppose that an implementation of this would require a parallel convergence test. However, it is readily checked that B is defined by the following PCF program (where $\bot$ abbreviates $Y_{\mathtt{U}}(\lambda x.x)$):

$$B = \lambda f.\, f\,(f\top\bot)\,(f\bot\top)\,.$$

This illustrates the remarkable power of *nesting* of function calls, something that has already featured in Section 5.3; we will meet further examples shortly. The functional B will also serve as a useful counterexample later on: see Exercise 7.5.10. Note that B, like all the examples in the preceding subsection, can be implemented even in the weaker language $\mathsf{Klex}^{\mathrm{prim}} + \bot$.

7.3.3 Bar Recursion

All the examples given so far have been concerned with the *finitary* fragment of PCF (i.e. the types constructed from $\mathtt{U}$ and $\mathtt{B}$). We now consider some examples of an infinitary character. We begin with an operator embodying the principle of *bar recursion*, a close relative of the concept of *bar induction* introduced by Brouwer [41].

It is a familiar idea that many important functions on finite trees (for example, their height, or a list of the nodes they contain) can be computed by recursion on their structure. The 'base case' is given by an operation defining the value of the function at leaf nodes (i.e. for single-node trees), while the 'step case' is given by an operation to be applied at branch nodes, which builds the value at such a node using the values at its daughter nodes. Bar recursion extends this principle of definition to infinitely branching but well-founded trees.

Suppose, for example, that we wish to define functions with values of type τ over countably branching trees, i.e. those in which the branches from any node may be indexed by $\mathtt{N}$. (Actually, any other type σ could equally well be used as the branching type here, but countably branching trees are pleasant to visualize.) Any node in such a tree may be identified by a path from the root consisting of a finite list of natural numbers, i.e. by a value of type $\mathtt{N}^*$. A *leaf function*, specifying the value of the desired function at leaf nodes, can then be of type $\mathtt{N}^* \rightarrow_s \tau$, while a *branch function*, specifying how the value at a given node may be computed from the assemblage of values at its daughter nodes, can have type $\mathtt{N}^* \rightarrow_s (\mathtt{N} \rightarrow_s \tau) \rightarrow \tau$.

It remains to explain how the well-founded tree itself is presented. Suppose that F is any operation of type $(\mathtt{N} \rightarrow_s \mathtt{N}) \rightarrow \mathtt{N}$. Suppose too that we can define a predicate p_F on finite sequences $\vec{x} \in \mathbb{N}^*$, decidable uniformly in F, such that any *infinite* sequence $x_0, x_1, \ldots$ has some finite prefix satisfying p_F. The minimal sequences satisfying p_F will then constitute the leaves of a well-founded tree. The key observation is that we can indeed provide such a 'stopping condition' p_F if F is hereditarily total and *continuous*. Spector's original stopping condition [276] was given by

$$F(x_0 \ldots x_{r-1} 0^\omega) \ < \ r \,,$$

where we write $x_0 \ldots x_{r-1} t^\omega$ for the function $\mathbb{N} \to \mathbb{N}$ sending i to x_i if $i < r$, and t if $i \geq r$. Another possible stopping condition (used by Kohlenbach [147]) is

$$F(\vec{x} 0^\omega) \ = \ F(\vec{x} 1^\omega) \,.$$

In either case, it is easy to see that if F is hereditarily total and continuous, then any infinite sequence $x_0, x_1, \ldots$ has a finite initial segment satisfying the stopping condition; the first such initial segment is taken by definition to be a leaf of the tree presented by F. Since, in our models of interest, *all* type 2 operations are continuous, we can regard an arbitrary hereditarily total element of type $(\mathtt{N} \to_s \mathtt{N}) \to \mathtt{N}$ as presenting a well-founded tree.

We shall therefore consider bar recursion as a third-order operator

$$\mathsf{BR}_\tau \ : \ ((\mathtt{N} \to_s \mathtt{N}) \to \mathtt{N}) \ \to \ (\mathtt{N}^* \to_s \tau) \ \to \ (\mathtt{N}^* \to_s (\mathtt{N} \to_s \tau) \to \tau) \ \to \ (\mathtt{N}^* \to_s \tau) \,,$$

where the first argument specifies a well-founded tree, the second and third arguments are the leaf and branch functions respectively, and the result of type $\mathtt{N}^* \to_s \tau$ gives the value of the resulting function at each node of the tree. In the light of the above discussion, the defining clauses for such an operator (using Spector's stopping condition) are as follows. Here we use $\vec{x}$ to range over finite sequences of natural numbers, write $|\vec{x}|$ for the length of $\vec{x}$, and write $\vec{x};z$ for the result of appending the number z to $\vec{x}$.

$$\begin{array}{lll} \mathsf{BR}(F,L,G)(\vec{x}) & = L(\vec{x}) & \text{if } F(\vec{x}0^\omega) < |\vec{x}| \,, \\ \mathsf{BR}(F,L,G)(\vec{x}) & = G(\vec{x})(\Lambda z \in \mathbb{N}.\, \mathsf{BR}(F,L,G)(\vec{x};z)) & \text{otherwise.} \end{array}$$

We understand these as semantic conditions to be satisfied in any model of PCF. However, they also translate fairly directly into a syntactic definition in PCF involving the recursor $Y_{\mathtt{N}^* \to \tau}$:

$$\mathsf{BR}_\tau \ = \ \lambda FLG.\, Y(\lambda B.\, (\lambda \vec{x}.\, \mathit{if}\ F(\vec{x}0^\omega) < |\vec{x}|\ \mathit{then}\ L(\vec{x})\ \mathit{else}\ G(\vec{x})(\lambda z. B(\vec{x};z)))) \,.$$

It is easy to see that in any model of PCF, the interpretation $[\![\mathsf{BR}_\tau]\!]$ satisfies the above semantic conditions.

Note that by taking $\tau = \mathtt{N}$, coding up $\mathtt{N}^*$ as $\mathtt{N}$ and specializing the leaf function to $\lambda x.2x+1$, we obtain a *restricted bar recursor* as defined in Subsection 6.3.4. Thus, restricted bar recursors indeed exist and are PCF-definable (or equivalently Klex-definable) in SP^0 as in all other models of interest.

Bar recursion provides an interesting example of a PCF-computable operation that goes beyond the power of $\mathrm{T} + \mathit{min}$ (this is the import of Theorem 6.3.28). Indeed, it is an illuminating exercise to sketch part of the NSP arising from $\mathsf{BR}_\mathtt{N}$ (or its restricted version) in order to observe the infinitely deep nesting of calls to G, a feature that it shares with the NSP for Y_1 itself.

Unlike Y_1, however, BR_τ has the distinction of being a *total* operation in all models of interest. This may be understood in two ways:

Theorem 7.3.1. *Let* $\mathbf{P}$ *be any extensional model of* PCF^Ω, *and let* BR_τ *denote the bar recursion operator for* τ *in* $\mathbf{P}$ *defined as above.*

(i) BR_τ *is* hereditarily total, *i.e. it is an element of* $\mathsf{Tot}(\mathbf{P};\mathbb{N})$ *as defined in Subsection 3.6.2.*

(ii) BR_τ *is* hereditarily extensional, *i.e. it is an element of* $\mathsf{Ext}(\mathbf{P};\mathbb{N})$ *as defined in Subsection 3.6.3.*

Proof. (i) Suppose F,L,G as above are elements of $\mathsf{Tot}(\mathbf{P};\mathbb{N})$. We will use bar induction in the metatheory to prove that $\mathsf{BR}_\tau FLG\vec{x} \in \mathbb{N}$ for all lists $\vec{x} \in \mathbb{N}^*$; we shall abbreviate $\mathsf{BR}_\tau FLG$ by B. First note that F acts continuously on total arguments by Exercise 7.1.11(iii); hence for any infinite sequence $x_0,x_1,\ldots$ we may take $m = F(\Lambda i.x_i)$ and then find $r > m$ such that $F(x_0\ldots x_{r'-1}0^\omega) = m$ for all $r' \geq r$. It follows that the tree T is well-founded, where T consists of all prefixes of all sequences $x_0,\ldots,x_{r-1}$ such that $F(x_0\ldots x_{r-1}0^\omega) \geq r$. For any $\vec{x} \notin T$, by taking $r = |\vec{x}|$ we see that $F(\vec{x}0^\omega) < |\vec{x}|$, so we immediately have $B(\vec{x}) = L\vec{x} \in \mathbb{N}$ since L is total. We shall show also that $B(\vec{x}) \in \mathbb{N}$ for all $\vec{x} \in T$.

If $\vec{x}$ is a leaf of T, then for every $z \in \mathbb{N}$ we have $B(\vec{x};z) \in \mathbb{N}$ as above; thus $(\lambda z.B(\vec{x};z)) \in \mathsf{Tot}(\mathbf{P},\mathbb{N})$, and since $G \in \mathsf{Tot}(\mathbf{P},\mathbb{N})$ we have $G(\lambda z.B(\vec{x};z)) \in \mathbb{N}$, whence $B(\vec{x}) \in \mathbb{N}$. Now suppose $\vec{x}$ is a node in T such that for all its successors $\vec{x};z$ within T we have $B(\vec{x};z) \in \mathbb{N}$; then once again $(\lambda z.B(\vec{x};z)) \in \mathsf{Tot}(\mathbf{P},\mathbb{N})$, whence $B(\vec{x}) \in \mathbb{N}$. Applying bar induction, we therefore have $B(\vec{x}) \in \mathbb{N}$ for all $\vec{x}$.

(ii) Let $\sim_\sigma$ denote the partial equivalence relation on $\mathbf{P}(\sigma)$ generated by the identity on $\mathbb{N}$ as in Subsection 3.6.3; for simplicity we will omit type subscripts. Assume $F \sim F'$, $L \sim L'$ and $G \sim G'$; then an argument precisely analogous to the above shows that $\mathsf{BR}_\tau FLG\vec{x} = \mathsf{BR}_\tau F'L'G'\vec{x} \in \mathbb{N}$ for all $\vec{x} \in \mathbb{N}^*$. □

The leading example is provided by the model PC. Since $\mathsf{Ct} \cong \mathsf{EC}(\mathsf{PC},\mathbb{N})$, part (ii) of the above theorem tells us that bar recursors are present as elements of Ct. We will revisit this situation in Subsection 8.3.1, where we will see that these bar recursors are actually Kleene computable over Ct directly. Another example is the model SF; indeed, from the totality of BR_τ in SF it follows immediately that BR_τ is total in SP^0, as claimed in Subsection 6.3.4. However, the restriction to models of PCF^Ω is essential: e.g. for $\mathbf{P} = \mathsf{PC}^{\mathrm{eff}}$ the above proof fails, since for arbitrary $x_0,x_1,\ldots$ the function $\Lambda i.x_i$ need not be present in $\mathsf{PC}^{\mathrm{eff}}(1)$.

Some variations on bar recursion are worth noting. In [26], Berger and Oliva introduced a *modified bar recursion* operator

$$\mathsf{MBR}_\tau \;:\; ((\mathtt{N} \rightarrow_s \mathtt{N}) \rightarrow \tau) \;\rightarrow\; (\mathtt{N}^* \rightarrow_s (\mathtt{N} \rightarrow_s \tau) \rightarrow (\mathtt{N} \rightarrow_s \mathtt{N})) \;\rightarrow\; (\mathtt{N}^* \rightarrow_s \tau)\,,$$

definable in PCF by

$$\mathsf{MBR}_\tau \;=\; \lambda FG.\,Y(\lambda B.\,(\lambda \vec{x}.\,F(\vec{x}\,\#\,G(\vec{x})(\lambda z.B(\vec{x};z)))))\,,$$

where $x_0,\dots,x_{r-1}\#f$ denotes the function $\lambda i.\, \mathit{if}\ i<r\ \mathit{then}\ x_i\ \mathit{else}\ f(i-r)$. This differs from ordinary bar recursion in having no explicit stopping condition or 'leaf clause': however, totality is guaranteed by the fact that as one descends the tree of calls to G by any path, the sequences $\vec{x}$ will become progressively longer until they are sufficient to 'satisfy' F. Indeed, the proof of Theorem 7.3.1 can be readily adapted to work for MBR_τ. (In essence, $\mathsf{MBR}_\mathtt{N}$ is equivalent to Gandy's functional Γ which we shall consider in Subsection 8.3.3; see Berger and Oliva [26].)

The definition of MBR_τ was itself inspired by a somewhat wilder functional introduced by Berardi, Bezem and Coquand [17], and consequently known as the *BBC functional*. Here the data that augment as one descends through the call tree are not sequences $\vec{x}\in\mathbb{N}^*$, but sequences $(y_0,x_0)\dots(y_{r-1},x_{r-1})\in(\mathbb{N}\times\mathbb{N})^*$ that serve as graphs for finite portions of functions $\mathbb{N}\to\mathbb{N}$.

$$\mathsf{BBC}_\tau \;:\; ((\mathtt{N}\to_s\mathtt{N})\to\tau) \;\to\; (\mathtt{N}\to_s(\mathtt{N}\to_s\tau)\to\mathtt{N}) \;\to\; ((\mathtt{N}\times\mathtt{N})^*\to_s\tau)\,,$$
$$\mathsf{BBC}_\tau = \lambda FG.\, Y(\lambda B.\,(\lambda\vec{t}.\, F(\vec{t}\star\lambda y.G(y)(\lambda x.B(\vec{t};(y,x))))))\,,$$

where $(y_0,x_0)\dots(y_{r-1},x_{r-1})\star f$ is defined as the function mapping i to x_j if $i=y_j$, and to $f(i)$ if i is not among the y_j.

The reason this is wilder than MBR_τ is that the behaviour can no longer be visualized so readily as an exploration of a countably branching tree laid out in advance. Indeed, as we descend the call tree, the tuples (y,x) may accumulate in any order, e.g. $(0,5)(2,4)(7,1)(3,4)\cdots$. Because of this, the proof of totality is more subtle for BBC_τ than for BR_τ or MBR_τ: see Berger [23].

Although these operators are offered here chiefly as interesting examples of the power of PCF programming, we should also mention the remarkable metamathematical work that gave rise to them. The definition of BR_τ was originally introduced by Spector [276] as the crucial ingredient of a computational interpretation for full classical analysis (i.e. second-order arithmetic) extending Gödel's Dialectica interpretation for first-order arithmetic. Much later, Berardi, Bezem and Coquand [17] introduced the BBC functional in the context of a realizability-style interpretation of classical analysis, using it in particular to give a computational interpretation for dependent and countable choice principles. Since then, Berger, Oliva and Escardó have undertaken detailed studies of these and other variants of bar recursion in connection with a range of computational interpretations for analysis—see [25, 87]. These interpretations have in turn made possible the extraction of non-trivial algorithms from certain classical proofs. This body of work represents one of the most exciting application areas for computable operations of higher type (see also Subsections 1.2.1 and 1.2.2).

7.3.4 The Fan Functional

Our next example, the *fan functional*, is again a total third-order functional arising from early ideas of Brouwer. Interpreted in Ct, the fan functional is the prototypical example of an effective continuous functional that is not Kleene computable. (This property is also shared by MBR_τ and BBC_τ, though not by BR_τ.) We will prove this non-computability result as Theorem 8.5.3 in our study of Ct; our emphasis here is on the definition of the fan functional and its computability in PCF.

Working first at the level of total functions, let us introduce some notation for elements of the *Cantor space* $\mathbb{B}^{\mathbb{N}}$. For convenience, we shall here take $\mathbb{B} = \{0,1\}$, and regard functions $f : \mathbb{N} \to \mathbb{B}$ as infinite sequences $f(0), f(1), f(2) \ldots$ over $\mathbb{B}$. If s is a finite sequence of booleans, we shall write $s0^\omega$ for the sequence consisting of s followed by infinitely many zeros; similarly for $s1^\omega$. We also write U_s for the set of all infinite sequences commencing with s.

Now suppose $G : \mathbb{B}^{\mathbb{N}} \to \mathbb{N}$ is continuous. This means that any infinite sequence f has a finite prefix s such that G is constant on U_s. Writing T for the tree of sequences s such that G is not constant on U_s, this means that T contains no infinite paths. But T is a binary tree, so by König's lemma, it is finite altogether, and hence has some finite depth n, which we may call a *modulus of uniform continuity* for G. The fan functional, by definition, returns this modulus of uniform continuity:

$$\mathsf{FAN}(G) \;=\; \mu n. \forall f, g \in \mathbb{B}^{\mathbb{N}}.\, (\forall i < n. f(i) = g(i)) \;\Rightarrow\; G(f) = G(g)\,.$$

As it stands, this formula involves a quantification over $\mathbb{B}^{\mathbb{N}}$, and so does not directly give us an effective way of computing FAN. What is surprising is that a representative of FAN is computable in PCF. This result is due to Berger [18]. According to a private communication from Martin Hyland, it was also known to Gandy, although he never published a proof.

We shall see later that this is an instance of a very general theorem: *every* effective element of Ct has a PCF-computable representative (Theorem 8.6.1). However, Berger's proof for the case of FAN is worthy of attention in its own right, both because it was an important inspiration for the proof of Theorem 8.6.1 and offers a useful warm-up to it, and because it yields a more efficient way of computing FAN than the general construction does.[1]

Theorem 7.3.2. FAN *is computable in* PCF. *More precisely, there is a* PCF *program* $\mathsf{FAN} : ((\mathtt{N} \to_s \mathtt{B}) \to \mathtt{N}) \to \mathtt{N})$ *such that if* $\mathbf{P}$ *is any extensional model of* PCF^Ω, *then* $[\![\mathsf{FAN}]\!]^{\mathbf{P}}$ *represents an element* $\mathsf{FAN} \in \mathsf{EC}(\mathbf{P}; \mathbb{N})$ *meeting the above specification.*

Proof. We will first define a PCF program $\delta : ((\mathtt{N} \to_s \mathtt{B}) \to \mathtt{N}) \to \mathtt{B}^* \to (\mathtt{N} \to_s \mathtt{B})$. The idea is that $\delta\ G\ s$ will return an example of some $f \in U_s$ such that $G(f) \neq G(s0^\omega)$ if one exists; otherwise it will simply return $s0^\omega$. We can then use $G(\delta\ G\ s)$ as way of testing whether $G(f) = G(s0^\omega)$ for all $f \in U_s$.

We define δ via the following clauses; it is easy to see how these may be translated into a formal PCF definition using the fixed point operator for $\mathtt{B}^* \to (\mathtt{N} \to_s \mathtt{B})$.

1. $\delta\ G\ s\ i = s_i$ if $i < |s|$. Otherwise:

[1] In fact, Berger originally proved his theorem for a certain *extended fan functional*, and his construction was somewhat more elaborate than the one above. We have taken the liberty of reformulating his algorithm in the style of this book.

2. $\delta\, G\, s\, i = (s0^\omega)i = 0$ if $G(\delta\, G\, (s0)) = G(s00^\omega) = G(s10^\omega) = G(\delta\, G\, (s1))$.
3. $\delta\, G\, s\, i = \delta\, G\, s0\, i$ if $G(\delta\, G\, (s0)) \neq G(s00^\omega)$.
4. $\delta\, G\, s\, i = (s10^\omega)i$ if $G(\delta\, G\, (s0)) = G(s00^\omega) \neq G(s10^\omega)$.
5. $\delta\, G\, s\, i = \delta\, G\, s1\, i$ if $G(\delta\, G\, (s0)) = G(s00^\omega) = G(s10^\omega) \neq G(\delta\, G\, (s1))$.

Now interpret this in a model **P** as above, and suppose $G \in \mathbf{P}((\mathtt{N} \to \mathtt{B}) \to \mathtt{N})$ represents a functional $\widehat{G} \in \mathsf{EC}(\mathbf{P};\mathbb{N})$. For $s \in \mathbf{B}^*$, let us write $s\bot^\omega \in \mathbf{P}(1)$ for the strict function such that $(s\bot^\omega)(i) = s_i$ if $i < |s|$ and $(s\bot^\omega)(i) = \bot$ otherwise. It is easy to see that if $G(s\bot^\omega) \in \mathbb{N}$ in **P** then $G(f) = G(s\bot^\omega)$ for all f representing an element of U_s, and $\delta\ G\ s = s0^\omega$ in **P**. Moreover, by appealing to the continuity of G as in the proof of Theorem 7.3.1, we have a well-founded tree T consisting of all sequences s such that $G(s\bot^\omega) = \bot$, and by bar induction on T we may readily show that $\delta\ G\ s$ is total for all s, and that $G(\delta\ G\ s) = G(s0^\omega)$ if and only if $\widehat{G}$ is constant on U_s.

To define the fan functional in PCF, we may now recursively define

1. $\mathsf{FAN}'\ G\ s = 0$ if $G(\delta\ G\ s) = G(s0^\omega)$,
2. $\mathsf{FAN}'\ G\ s = 1 + \max\{\mathsf{FAN}'\ G\ (s0), \mathsf{FAN}'\ G\ (s1)\}$ otherwise,

and take $\mathsf{FAN}\ G = \mathsf{FAN}'\ G\ \varepsilon$. If G represents $\widehat{G} \in \mathsf{EC}(\mathbf{P},\mathbb{N})$, clearly $\mathsf{FAN}(G) \in \mathbb{N}$ and its value coincides with $\mathsf{FAN}(\widehat{G})$ as defined above. □

Again, our leading example is the model PC; the above shows that here there is a PCF-definable representative for $\mathsf{FAN} \in \mathsf{Ct}(((\mathtt{N} \to \mathtt{B}) \to \mathtt{N}) \to \mathtt{N})$. Once again, the requirement that **P** be a model of PCF$^\Omega$ is essential: for example, in the model PC^{eff}, the theorem fails because the *Kleene tree* functional given by Definition 9.1.10 lacks any finite modulus of uniform continuity.

It is well known that König's lemma for binary trees is equivalent to many important theorems in analysis (e.g. in the sense of *Reverse Mathematics*: see [273]). This led Alex Simpson [272] to explore possible applications of Berger's method to computable analysis. We give just a brief outline of the ideas here.

Let us write $\mathbb{D}$ for the set $\{-1,0,1\}$ of *signed binary digits*. A function $r : \mathbb{N} \to \mathbb{D}$ will determine a real number

$$x_r \;=\; \sum_{i=0}^{\infty} 2^{-(i+1)} r(i) \;\in\; [-1,1]\,,$$

and the mapping $r \mapsto x_r : \mathbb{D}^\mathbb{N} \to [-1,1]$ is clearly surjective; we call r a *signed binary representation* of x_r. What is more, continuous functions on $[-1,1]$ can be 'lifted' to continuous functions on the set of representations:

Exercise 7.3.3. Show that if $F : [-1,1] \to [-1,1]$ is continuous, there is a continuous $\overline{F} : \mathbb{D}^\mathbb{N} \to \mathbb{D}^\mathbb{N}$ such that $F(x_r) = x_{\overline{F}(r)}$ for all $r \in \mathbb{D}^\mathbb{N}$.

This lifting $\overline{F}$ can be regarded as an element of $\mathsf{Ct}(\mathtt{rep} \to \mathtt{rep})$, where $\mathtt{rep} = \mathtt{N} \to \mathtt{D}$ is the type of signed binary representations. Since $\mathsf{Ct} = \mathsf{EC}(\mathsf{PC};\mathbb{N})$, any such $\overline{F}$ can in turn be represented by an element $\check{F} \in \mathsf{PC}(\mathtt{rep} \to \mathtt{rep})$.

Ascending to the next type level, we may now ask which operators on the set of continuous functions $[-1,1] \to [-1,1]$ can be lifted to continuous operators on these representations. Simpson [272] showed that there are PCF-definable functionals

$$\mathsf{Sup}, \mathsf{Int} \in \mathsf{PC}((\mathtt{rep} \to \mathtt{rep}) \to \mathtt{rep})$$

such that if $\check{F} \in \mathsf{PC}(\mathtt{rep} \to \mathtt{rep})$ represents $F : [-1,1] \to [-1,1]$ then

$$\mathsf{Sup}(\check{F}) \text{ represents } \sup_{x \in [-1,1]} F(x) , \qquad \mathsf{Int}(\check{F}) \text{ represents } \int_{-1}^{1} F(x)\,\mathrm{d}x .$$

The definition of these operators mimics that of FAN. In order to compute $\mathsf{Sup}(\check{F})$ or $\mathsf{Int}(\check{F})$ to any requested accuracy, we in effect undertake a repeated (and recursive) subdivision of the interval $[-1,1]$ until we arrive at subintervals I sufficiently small that F can be regarded as 'constant' on I to within the relevant error margins. The constancy of F on such subintervals is detected using a version of the operator δ above, and the suprema or integrals for these subintervals are then recursively combined in a straightforward way to yield the supremum or integral over $[-1,1]$.

Note in passing that although any continuous F on $[-1,1]$ attains its supremum, the proof of this fact is famously non-constructive: thus, we cannot in general compute, from $\check{F}$, a representation r of a point x_r at which the supremum is attained.

7.3.5 Selection Functions and Quantifiers

The proof of Theorem 7.3.2 demonstrates the remarkable fact that exhaustive searches over the set $\mathbb{B}^{\mathbb{N}}$ are computationally possible. In particular, the operator δ yields an example of a computable *selection function*: from any given G, it allows us to compute an example of an element f of $\mathbb{B}^{\mathbb{N}}$ satisfying $G(f) \neq G(0^\omega)$ whenever one exists, or the standard element 0^ω if none does. Using this, it is easy to test whether such an f indeed exists.

This surprising observation led Escardó [80] to consider the general question of which infinite sets admit computable exhaustive searches in this way. We shall present some of Escardó's results here, focussing on those aspects that shed interesting light on the power of PCF itself. (The more topological aspects of Escardó's theory will be considered further in Subsection 8.4.1.) For concreteness, we follow Escardó [80] in focussing on the model PC and its hereditarily total elements.

The key notions are as follows.

Definition 7.3.4. Suppose $K \subseteq \mathsf{PC}(\sigma)$ for some type σ.

(i) K is *(PCF-)searchable* if there is a PCF-definable $\varepsilon_K \in \mathsf{PC}((\sigma \to \mathtt{B}) \to \sigma)$ such that whenever $P \in \mathsf{PC}(\sigma \to \mathtt{B})$ is defined on K (i.e. $P(x) \in \{tt, f\!f\}$ for all $x \in K$), we have

1. $\varepsilon_K(P) \in K$,

2. if $P(x) = tt$ for some $x \in K$, then $P(\varepsilon_K(P)) = tt$.

Such an element ε_K is called a *selection function* for K.

(ii) K is *(PCF-)exhaustible* if there is a PCF-definable $\exists_K \in \mathsf{PC}((\sigma \to \mathtt{B}) \to \mathtt{B})$ such that whenever $P \in \mathsf{PC}(\sigma \to \mathtt{B})$ is defined on K we have

1. $\exists_K(P) = tt$ if there exists $x \in K$ with $P(x) = tt$,
2. $\exists_K(P) = ff$ if $P(x) = ff$ for all $x \in K$.

Such an element $\exists_K$ is called an *existential quantifier* for K.

Proposition 7.3.5. *(i) Any searchable set is exhaustible.*

(ii) If $F \in \mathsf{PC}(\sigma \to \tau)$ is PCF*-definable and $K \subseteq \mathsf{PC}(\sigma)$ is searchable [resp. exhaustible], then the image $F(K) \subseteq \mathsf{PC}(\tau)$ is searchable [resp. exhaustible].*

Proof. (i) Take $\exists_K = \lambda P.P(\varepsilon_K(P))$.

(ii) Define respectively $\varepsilon_{F(K)} = \lambda P.F(\varepsilon_K(P \circ F))$ and $\exists_{F(K)} = \lambda P.\exists_K(P \circ F)$. $\square$

Example 7.3.6. (i) Any finite inhabited set is trivially searchable and exhaustible.

(ii) The proof of Theorem 7.3.2 shows that the set of total elements of $\mathbf{P}(\mathtt{N} \to \mathtt{B})$ (essentially Cantor space) is searchable and hence exhaustible. Equivalently, the set of total elements $f \in \mathbf{P}(\mathtt{N} \to \mathtt{N})$ such that $f(n) < 2$ for all n is searchable. Hence all images of this set under PCF-definable functions are searchable.

(iii) In Theorem 5.3.2 we considered the set $N_\infty = \{g_0, g_1, \ldots, g_\infty\}$, where $g_n : \mathbb{N} \to \mathbb{N}$ by $g_n(m) = 0$ if $m < n$, $g_n(m) = 1$ otherwise, and $g_\infty(m) = 0$ for all m. We shall here regard N_∞ as a set of strict elements in $\mathsf{PC}(1)$. As the proof of Theorem 5.3.2 shows, N_∞ admits a selection function and hence an existential quantifier definable even in the weak language $\mathsf{Klex}^{\mathrm{prim}}$; we may thus say that N_∞ is *primitive recursively searchable* (hence exhaustible).

As we have seen, searchable sets are trivially exhaustible. What is more surprising is that for the sets in which we are chiefly interested, the converse holds—indeed, for such sets K, a suitable selection function ε_K is computable uniformly in a quantifier $\exists_K$. We shall show this here for the critical case $\sigma = \mathtt{N} \to \mathtt{N}$; the construction provides an interesting example of a fourth-order PCF-definable operation.

As several of the previous examples show, we are often interested in elements of PC as a way of representing, and computing with, elements of its extensional collapse Ct. It is therefore natural to restrict attention to sets K that represent subsets of Ct:

Definition 7.3.7. A set $K \subseteq \mathsf{PC}(\sigma)$ is *entire* if there is a set $\widehat{K} \subseteq \mathsf{EC}(\mathsf{PC};\mathbb{N})(\sigma)$ such that K is the set of all representatives of elements of $\widehat{K}$.

Lemma 7.3.8. *If $K \subseteq \mathsf{PC}(\mathtt{N} \to \mathtt{N})$ is entire and exhaustible, the corresponding set $\widehat{K}$ is closed in the usual topology on $\mathbb{N}^{\mathbb{N}}$.*

Proof. Suppose for contradiction that $f \notin \widehat{K}$ but that each neighbourhood

$$U_k = \{g \in \mathbb{N}^{\mathbb{N}} \mid \forall i < k.\, g(i) = f(i)\}$$

contains an element of $\widehat{K}$. For each k, it is easy to construct $P_k \in \mathsf{PC}((\mathtt{N} \rightarrow \mathtt{N}) \rightarrow \mathtt{B})$ such that for all $g \in \mathsf{PC}(\mathtt{N} \rightarrow \mathtt{N})$ representing $\widehat{g} \in \mathsf{Ct}(\mathtt{N} \rightarrow \mathtt{N})$, we have $P_k(g) = \mathit{ff}$ if $\widehat{g} \notin U_k$, and $P_k(g) = \bot$ if $\widehat{g} \in U_k$. Since $\bigcap_k U_k = \{f\}$, we have that $(\bigsqcup_k P_k)(g) = \mathit{ff}$ for all $g \in K$. So if $\exists_K$ is an existential quantifier for K then $\exists_K(\bigsqcup_k P_k) = \mathit{ff}$. Hence by continuity $\exists_K P_k = \mathit{ff}$ for some k. Now extend P_k to a predicate Q on K where $Q(g) = \mathit{tt}$ if $\widehat{g} \in U_k$; then $P_k \sqsubseteq Q$, so $\exists_K Q = \mathit{ff}$. But this contradicts the definition of $\exists_K$, since there exists $g \in K$ with $\widehat{g} \in U_k$. □

Theorem 7.3.9. *If $K \subseteq \mathsf{PC}(\mathtt{N} \rightarrow \mathtt{N})$ is inhabited, entire and exhaustible, then it is searchable; moreover, a selection function for K is uniformly* $\mathsf{Klex}^{\min}$ *definable relative to an existential quantifier for K.*

Proof. Suppose $P \in \mathsf{PC}((\mathtt{N} \rightarrow \mathtt{N}) \rightarrow \mathtt{B})$ is a predicate defined on K. To define $h = \varepsilon_K(P) \in \mathsf{PC}(\mathtt{N} \rightarrow \mathtt{N})$, we use course-of-values recursion on the argument n:

$$h(n) \;=\; \mu m.\ \exists_K(\lambda f.\ f(0) = h(0) \wedge \cdots \wedge f(n-1) = h(n-1) \wedge f(n) = m)\,.$$

Since K is inhabited, it is easy to see that h is total, and that for every n there exists $f \in K$ such that $f(i) = h(i)$ for all $i < n$. But by the previous lemma, K represents a closed set in $\mathbb{N}^{\mathbb{N}}$; hence $h \in K$. □

In [80], Theorem 7.3.9 is extended to arbitrary types σ, essentially by showing that entire exhaustible subsets of $\mathsf{PC}(\sigma)$ are 'computable retracts' of entire exhaustible subsets of $\mathsf{PC}(\mathtt{N} \rightarrow \mathtt{N})$ in a suitably loose sense. The technology for this will be discussed in Section 8.4.

As our final source of examples, we turn briefly to the theory of *generalized* selection functions and generalized quantifiers as developed by Escardó and Oliva [85]. The idea here is to replace the type $\mathtt{B}$ in the definitions above by an arbitrary type of 'generalized truth values'; indeed, the following definitions make sense in any λ-algebra $\mathbf{A}$ equipped with a chosen type $R \in \mathbf{A}$.

Definition 7.3.10. (i) A *(generalized) quantifier* for a type $X \in \mathbf{A}$ is simply an element $\phi \in (X \Rightarrow R) \Rightarrow R$.

(ii) A *(generalized) selection function* for X is an element $\varepsilon \in (X \Rightarrow R) \Rightarrow X$. Any such ε gives rise to a quantifier $\overline{\varepsilon} = \lambda p.p(\varepsilon p)$.

If R is the familiar type of booleans, then the usual existential and universal quantifiers over X (if they exist within $\mathbf{A}$) provide examples of generalized quantifiers. In both cases, selection functions for these (if they exist in $\mathbf{A}$) yield 'critical test cases' for predicates: thus, if $\varepsilon, \varepsilon'$ are selection functions yielding $\exists, \forall$ respectively, then εp will satisfy p iff some element of X does, and $\varepsilon' p$ will satisfy p iff all do. Similar remarks hold for quantifiers over some subset $K \subseteq X$ as in Definition 7.3.4.

As an example of another kind, let R be the real interval $[0,1]$ and X some compact space (in a suitable category). The supremum function $\sup : (X \Rightarrow R) \Rightarrow R$ is an example of an R-valued quantifier; a corresponding selection function (if it exists) yields a point at which a function $X \rightarrow R$ attains its supremum. Related examples arise in game theory: if X is a set of possible actions or moves for some agent, and R is a set of possible *outcomes* for the game as a whole, a selection function $\varepsilon \in (X \Rightarrow R) \Rightarrow X$ can be thought of as a *policy function* for the agent, giving a choice of action assuming that the action's effect on the outcome is known. For instance, a typical policy might involve maximizing an agent's *payoff* as determined by some function $R \to \mathbb{R}$.

Certain general constructions on selection functions and quantifiers are of interest, notably the following (asymmetrical) *product* construction. We assume here we are in a λ-algebra with products.

Definition 7.3.11. (i) If ϕ, ψ are R-valued quantifiers over X, Y respectively, their *product* $\phi \otimes \psi$ is the quantifier on $X \times Y$ given by

$$(\phi \otimes \psi)(p) \;=\; \phi(\lambda x.\psi(\lambda y.p(x,y))) \,.$$

(ii) If ε, δ are selection functions for R-valued predicates on X, Y respectively, their product $\varepsilon \otimes \delta$ is the selection function on $X \times Y$ defined by

$$\begin{aligned}(\varepsilon \otimes \delta)(p) &= (a, b(a))\,, \quad \text{where} \\ b(x) &= \delta(\lambda y.p(x,y))\,, \\ a &= \varepsilon(\lambda x.p(x,b(x)))\,.\end{aligned}$$

(Notice the formal resemblance to the *Bekič lemma* from Exercise 7.1.10.)

Both these product operations are conceptually third-order operators definable in pure λ-calculus. An easy unfolding of the definitions shows that $\overline{\varepsilon \otimes \delta} = \overline{\varepsilon} \otimes \overline{\delta}$.

To give some intuition for these, consider the classical boolean-valued quantifiers $\forall_X$ and $\exists_Y$ for some sets X, Y. In this case, the product $\forall_X \otimes \exists_Y$ sends a predicate p on $X \times Y$ to the truth-value $\forall x.\exists y.p(x,y)$. If selection functions ε, δ exist for $\forall_X, \exists_Y$ respectively, the above definition yields a selection function (ε, δ), returning for each predicate p a 'critical pair' (x_0, y_0) such that $p(x_0, y_0)$ iff $\forall x.\exists y.p(x,y)$. In terms of game theory, we may imagine that a value of x is chosen by Player I who is trying to make p false, then y is chosen by Player II who is trying to make p true. Assuming both players play optimally, the result of the play will be just such a critical pair. By generalizing to n-ary products and allowing for more complex game outcomes, we are able to compute optimal strategies in this way for finite games with non-trivial payoff functions.

The next step is to generalize this to *infinite* products; this will enable us to compute optimal strategies even for games of unlimited length. Suppose $\varepsilon_0, \varepsilon_1, \ldots$ is an infinite sequence of selection functions, which for convenience we assume all to be over the same set X. (The situation for infinite products of quantifiers is entirely analogous.) In suitable models, we can define the infinitary product $\bigotimes$ as a least fixed point:

$$\bigotimes_i \varepsilon_i \;=\; \varepsilon_0 \otimes \big(\bigotimes_i \varepsilon_{i+1}\big) \,.$$

To make this more concrete, suppose we are in a model of PCF and are considering selection functions for boolean-valued predicates on a type σ. An infinite sequence of such functions may be represented as an element of type $\mathtt{N} \rightarrow_s ((\sigma \rightarrow \mathtt{B}) \rightarrow \sigma)$; hence the operation $\bigotimes$ itself will be a PCF-definable element of type

$$\Xi(\sigma) \;=\; (\mathtt{N} \rightarrow_s ((\sigma \rightarrow \mathtt{B}) \rightarrow \sigma)) \;\rightarrow\; (((\mathtt{N} \rightarrow_s \sigma) \rightarrow \mathtt{B}) \rightarrow \sigma)\,.$$

This is conceptually a third-order operation (relative to σ). However, the case $\sigma = \mathtt{N} \rightarrow \mathtt{N}$ is of particular interest, since, as we have seen, there are many non-trivial sets $K \subseteq \mathsf{PC}(\mathtt{N} \rightarrow \mathtt{N})$ for which selection functions and quantifiers exist. In this case, $\bigotimes$ will of course be technically of type level 4. Whilst the above construction superficially requires the fixed point operator $Y_{\Xi(\sigma)}$, it is shown in [85] how it may be reformulated in a 'bar recursive' style which requires only $Y_{\sigma^* \rightarrow \sigma}$ (where $\sigma^* \rightarrow \sigma$ is conceptually first-order relative to σ).

The proof that $\bigotimes$ is hereditarily total and extensional (say in PC) requires a bar induction similar to that in the proof of Theorem 7.3.1. Specifically, given any predicate p on infinite sequences over σ, for any such infinite sequence $x_0 x_1 \ldots$ there will be some r such that $p(x_0 \ldots x_{r-1} \bot\bot \ldots)$ is defined, and it is easy to check that the infinite product will then be defined when specialized to sequences beginning with $x_0 \ldots x_{r-1}$. This gives us a bar consisting of sequences where the product is defined, from which we may then inductively work backwards until the empty sequence is reached. For further details, see [85].

The links between various kinds of products and various forms of bar recursion are investigated in detail in [87].

Remark 7.3.12. Some interesting experiments into the practical efficiency of these product operators and related functionals have been undertaken by Escardó. A variation on the above naive definition of infinite products, which appears to yield a significant efficiency gain in practice, is reported at the end of [80].

The examples given in this section probably do not exhaust the possibilities for remarkable and surprising higher type computations in PCF. We end with some informal open questions.

Question 7.3.13. Is there an interesting application for a conceptually fourth-order PCF-definable operation that is not definable in $\mathrm{T} + \mathit{min}$?

Question 7.3.14. Is there an interesting application for a PCF operation that makes essential use of some Y_τ where τ is of type level ≥ 2 relative to the natural 'base types' for the operation? (For instance, in the above example Ξ, we may regard σ as a 'base type', since the program in question is in effect polymorphic in σ.)

7.4 Game Models

We now look briefly at another class of intensional models for PCF closely related to SP^0: those arising from *game semantics*. These models made their debut around 1993, when similar ideas were applied (essentially independently) by Abramsky,

Jagadeesan and Malacaria [5], by Hyland and Ong [124] and by Nickau [203] to construct models of PCF, all giving rise (in effect) to the fully abstract model SF. Since then, the ideas of game semantics have been applied to a wealth of other languages—indeed, one of the major strengths of the approach lies in its capacity for modelling and comparing a wide spectrum of programming languages within a common semantic paradigm. Here we shall describe a version of the Hyland–Ong model as representative of game models of PCF, following the presentation of Abramsky and McCusker [3] to illustrate how this model fits into a broader framework embracing other languages. We shall then discuss its close relationship to the sequential procedure model SP^0.

Like the NSP model, the game model of PCF that we shall present characterizes a purely functional style of nested sequential computation. However, whereas the spirit of the NSP model is to present algorithms as self-contained packages which can be submitted to *oracles* in a single step, game models offer a more fine-grained view of the interaction between a function and its arguments as a kind of conversation or dialogue. The idea may be conveyed by a concrete example (taken from [178]). Consider the following terms of PCF + *byval* (in which '+' denotes an obvious implementation of addition):

$$\begin{aligned} F &\equiv \lambda gh.\, g\widehat{3} + g(Y_1 h) &&: \overline{1},\overline{1} \to \overline{0}\,, \\ g &\equiv \lambda x.\, x + \widehat{2} &&: \overline{1}\,, \\ h &\equiv \lambda x.\, \widehat{4} &&: \overline{1}\,. \end{aligned}$$

Let us see how to model the evaluation of $F(g)(h)$. This can be represented as a dialogue between a Player P, who follows a strategy determined by F, and an Opponent O, whose strategy is determined by g and h. Moves in the game are either questions (written '?') or answers (natural numbers) which are matched to previous questions. Furthermore, each move is associated with some occurrence of the ground type $\overline{0}$ within $\overline{1},\overline{1} \to \overline{0}$, according to its 'meaning' in the context of the game. We display here the dialogue for $F(g)(h)$ along with an informal gloss for each move:

		$(\overline{0}$	$\to \overline{0})$	$\to (\overline{0}$	$\to \overline{0})$	$\to \overline{0}$	
1	O:					?	What is $F(g)(h)$?
2	P:		?				What is $g(x)$?
3	O:	?					What is x here?
4	P:	3					$x = 3$.
5	O:		5				In that case, $g(x) = 5$.
6	P:		?				All right. Now what is $g(x')$?
7	O:	?					What is x' here?
8	P:				?		What is $h(y)$?
9	O:				4		$h(y) = 4$ (whatever y is).
10	P:	4					Then $x' = 4$.
11	O:		6				In that case, $g(x') = 6$.
12	P:					11	Well then, $F(g)(h) = 11$.

Note that there is no notion of winning or losing in this game: the participants simply collaborate in enacting the computation in question.

We now proceed to the formal definition of the game model, closely following [3]. We will only be concerned here with the interpretation of types in $\mathsf{T}^{\rightarrow}(\mathtt{N})$; this permits some slight simplifications in our treatment.

Definition 7.4.1. (i) An *arena* consists of the following data:

1. A set M of *moves*, each labelled as either an Opponent or a Player move, and each labelled as either a question or an answer. (Formally, each $m \in M$ is of one of the kinds OQ, OA, PQ, PA.)
2. A choice of *initial move* i, which is always an Opponent question.
3. An *enabling relation* $\vdash \subseteq M \times M$ such that
   - if $m \vdash n$ and n is an answer, then m is a question,
   - if $m \vdash n$, then m, n have opposite O/P labels and $n \neq i$.

(ii) A (single-threaded) *justified sequence* in an arena $(M, i, \vdash)$ is a sequence s of moves of M equipped with an associated sequence of *pointers*: the first move of s (if there is one) is i, and every other move occurrence n within s is endowed with a pointer to an earlier move occurrence m within s such that $m \vdash n$. In this case, we say that *m justifies n* in s.

For instance, in the example above, the question on line 3 is justified by the one on line 2 (and likewise for lines 7 and 6); this reflects the fact that we are not in a position to evaluate the argument of g before g itself has been called. The questions on lines 2, 6, 8 are all justified by the initial question, and each answer is justified by the question it answers.

The next definition captures the intuition that if either participant asks a question and later receives an answer to it, the intervening portion of dialogue can in some way be regarded as having 'dropped out of view'. This will form an important ingredient in the characterization of purely functional (PCF-style) computation. For all the game models we consider here, we will impose the condition that the justifier of any move is always 'in view' when the move occurs.

Definition 7.4.2. (i) If s is a finite justified sequence, its *P-view* $\ulcorner s \urcorner$ and *O-view* $\llcorner s \lrcorner$ are sequences of moves defined inductively as follows (where m, n range over move occurrences and s, t over finite justified sequences).

$$
\begin{aligned}
\ulcorner \varepsilon \urcorner &= \llcorner \varepsilon \lrcorner = \varepsilon\,, \\
\ulcorner i \urcorner &= i\,, \\
\ulcorner sn \urcorner &= \ulcorner s \urcorner n && \text{if } n \text{ is a } P\text{-move}\,, \\
\ulcorner smtn \urcorner &= \ulcorner s \urcorner mn && \text{if } n \text{ is an } O\text{-move justified by } m\,, \\
\llcorner sn \lrcorner &= \llcorner s \lrcorner n && \text{if } n \text{ is an } O\text{-move}\,, \\
\llcorner smtn \lrcorner &= \llcorner s \lrcorner mn && \text{if } n \text{ is a } P\text{-move justified by } m\,.
\end{aligned}
$$

(ii) A justified sequence s is *legal* if it satisfies the following conditions:

1. *Alternation: O*-moves and *P*-moves alternate throughout s.

2. *Visibility:* If tm is a prefix of s where m is a P-move, the justifier of m occurs in $\lceil t \rceil$; likewise, if tm is a prefix of s where m is a non-initial O-move, the justifier of m occurs in $\lfloor t \rfloor$.

As an example, the P-view of the entire dialogue displayed above will be the sub-dialogue consisting of lines 1, 2, 5, 6, 11, 12. Note that if s is legal, then $\lceil s \rceil, \lfloor s \rfloor$ may themselves be made into justified sequences in an obvious way.

Definition 7.4.3. (i) A *(well-opened) game A* consists of an arena $(M_A, i_A, \vdash_A)$ together with a non-empty, prefix-closed set P_A of legal finite justified sequences for this arena (known as *positions* of the game).

(ii) A *(P-)strategy* for a game A is a non-empty set S of even-length positions from P_A such that

1. if $smn \in S$ then $s \in S$,
2. for any odd-length position t, S contains at most one position of the form tn.

Thus, a Player strategy is in effect a partial function from odd-length positions to Player moves, along with justifications for the resulting moves.

We may now construct the games we need to interpret the PCF types. For the base type, we use the following game N:

- The moves of N are an initial Opponent question written as '?', and a Player answer n for each $n \in \mathbb{N}$. For the enabling relation, we have $? \vdash n$ for each n and nothing else.
- The positions of N are ε, ?, and $?n$ for each n.

Given games A and B, we may define a game $A \Rightarrow B$ as follows:

- The set $M_{A \Rightarrow B}$ is the disjoint union of M_A and M_B, except that moves from M_A have their O/P labelling reversed in $M_{A \Rightarrow B}$.
- The initial move of $A \Rightarrow B$ is just the initial move of B.
- The enabling relation of $A \Rightarrow B$ is given by

$$m \vdash_{A \Rightarrow B} n \quad \text{iff} \quad m \vdash_A n \text{ or } m \vdash_B n \text{ or } (m = i_B \wedge n = i_A) \,.$$

- Positions of $A \Rightarrow B$ are legal finite sequences s such that
  - $s \upharpoonright B \in P_B$, where $s \upharpoonright B$ is the subsequence of s consisting of all moves in B,
  - for any occurrence m of i_A in s, we have $s \upharpoonright m \in P_A$, where $s \upharpoonright m$ is the subsequence of s consisting of move occurrences hereditarily justified by m.

We may now define games $\mathscr{G}(\sigma)$ by $\mathscr{G}(\mathtt{N}) = N$, $\mathscr{G}(\sigma \to \tau) = \mathscr{G}(\sigma) \Rightarrow \mathscr{G}(\tau)$. We write $\mathsf{G}(\sigma)$ for the set of strategies of $\mathscr{G}(\sigma)$; note that $\mathsf{G}(\mathtt{N}) \cong \mathbb{N}_\perp$.

It remains to define a suitable application operation for strategies. Given strategies S for $A \Rightarrow B$ and T for A, we may 'play off' S against T to obtain a strategy $S \cdot T$ for B in the following way. We first define the set $Int(S,T)$ of *interaction sequences* compatible with both S and T to consist of all $s \in P_{A \Rightarrow B}$ such that

1. $s \in S$,

2. for all occurrences m of i_A in s, we have $s \upharpoonright m \in T$.

We now take $S \cdot T = \{s \upharpoonright B \mid s \in Int(S,T)\}$. We leave it to the reader to check that this is indeed a strategy for B.

We have thus defined an applicative structure G over $\mathbb{N}_\perp$. Whilst this is already a model of interest in its own right, we need to impose two further conditions on strategies to capture the essence of PCF-style computation:

Definition 7.4.4. Suppose S is a strategy for a game A.

(i) S is *well-bracketed* if whenever $smn \in S$ and n is an answer, the justifier for n in smn is the most recent unanswered question in $\lceil sm \rceil$.

(ii) A strategy S is *innocent* if the response of S to an odd-length position sm depends only on the P-view $\lceil sm \rceil$: more precisely, if

$$smn \in S \wedge t \in S \wedge tm \in P_A \wedge \lceil sm \rceil = \lceil tm \rceil \quad \text{implies} \quad tmn \in S\,,$$

where the justifier of n in tmn corresponds to the same move in $\lceil tm \rceil = \lceil sm \rceil$ as the justifier of n in smn.

The idea behind well-bracketing is clear: it ensures that subcomputations are properly nested. For innocence, the idea is that in a 'functional' style of computation, if Player invokes a subcomputation and later receives a result, Player's later behaviour cannot depend on internal details of the intervening subcomputation, but only on the result received.

It can be shown that both of the above classes of strategies are closed under application as defined above. For innocence in particular, the proof of this is non-trivial (see [124, 191]). We write G_b, G_i for the well-bracketed and innocent substructures of G respectively, and G_{ib} for their intersection.

It can also be shown directly that G_{ib} is a model of PCF: in fact, $\mathsf{G}_{ib} \cong \mathsf{SP}^0$ (see below). What is more, the other three structures turn out to provide good models for interesting non-functional extensions of PCF. (By a 'good model', we here mean one in which all finite strategies, at least, are definable in the language in question.)

- G_i models a language with non-local control flow operators (in fact, the language PCF + *catch* of Section 12.1).
- G_b models a language with imperative-style local variables of ground type (in fact, a version of Reynolds's *Idealized Algol* [243]).
- G_{ib} models a language including both control flow operators and local variables.

This illustrates how certain mathematical constraints on strategies correlate closely with the absence of certain language-level phenomena, so that by selectively imposing and relaxing these constraints, we are able to obtain models for a range of languages involving different combinations of language features. This general principle has been carried further in subsequent research on game semantics.

Next, we examine more closely the relationship between G_{ib} and SP^0. We first note that an innocent strategy $S \in \mathsf{G}_{ib}(\sigma)$ is completely determined by the set $\overline{S} = \{s \in S \mid \lceil s \rceil = s\}$. Moreover, if S is also well-bracketed, such a set $\overline{S}$ may be represented by an NSP of type σ via the following correspondence between move types and symbols within an NSP:

OQ : occurrence of 'λ'
OA : case branch (i.e. occurrence of some $j \in \mathbb{N}$ in a branch $j \Rightarrow e$)
PQ : head occurrence of a variable x (in an application $xq_0 \ldots q_{r-1}$)
PA : leaf node n

To make this more precise, we first show how to decorate an NSP $p \in \mathsf{SP}^0(\sigma)$ with certain type information. Let us write U_σ for the set of all leaves (i.e. occurrences of $\mathtt{N}$) within the syntax tree of σ, and T_σ for the set of all nodes (i.e. occurrences of all subtrees) within this tree. We shall let ζ, ξ range over U_σ and ρ, τ over T_σ. Given any $p \in \mathsf{SP}^0(\sigma)$, we shall associate an element of T_σ to every procedure subterm occurrence and every binding occurrence of a variable within p, and an element of U_σ to every expression subterm, occurrence of λ, case branch and head variable occurrence within p. We do this in a top-down way as follows:

- Associate the whole of p with the whole of σ.
- If $\lambda x_0 \ldots x_{t-1}.e$ is associated with $\tau = \tau_0, \ldots, \tau_{t-1} \to \zeta$, associate each occurrence x_i here with τ_i, the occurrence of λ with ζ, and e with ζ.
- If $e = \mathtt{case}\, xq_0 \ldots q_{r-1}.d \;\mathtt{of}\; (j \Rightarrow e_j)$ is associated with ζ of $\mathtt{N}$, and the binding occurrence of x was associated with $\rho_0, \ldots, \rho_{r-1} \to \xi$, then associate this head occurrence of x with ξ, each q_i with ρ_i, d with ξ, each case branch j with ξ, and each e_j with ζ.

It is easy to see that the set of moves in $\mathscr{G}(\sigma)$ is essentially $U_\sigma \times \{?, 0, 1, \ldots\}$. We can now transform a path through p into a sequence of such moves in the following way. Let $C[-]$ range over procedures with a single hole occurrence for an expression subterm. To any such $C[-]$ let us associate an odd-length sequence of moves $C[-]^\dagger$ by induction on the depth of the hole in $C[-]$ as follows (we use superscripts for associated elements of U_σ).

- $[\lambda^\zeta \vec{x}.-]^\dagger \;=\; (\zeta, ?)$.
- $C[\mathtt{case}\, x^\zeta \vec{q} \;\mathtt{of}\; (\cdots \mid j \Rightarrow - \mid \cdots)]^\dagger \;=\; C[-]^\dagger (\zeta, ?)(\zeta, j)$.
- $C[\mathtt{case}\, x^\zeta \cdots (\lambda^\xi \vec{y}.-) \cdots \;\mathtt{of}\; (\cdots)]^\dagger \;=\; C[-]^\dagger (\zeta, ?)(\xi, ?)$.

An occurrence of a leaf node n^ζ within p at position $C[-]$ may now be represented by the even-length sequence $C[-]^\dagger(\zeta, n)$. Likewise, an occurrence of a head variable x^ζ in $C[\mathtt{case}\, x^\zeta \vec{q} \;\mathtt{of}\; (\cdots)]$ may be represented by $C[-]^\dagger(\zeta, ?)$. We shall write $S(p)$ for the set of all sequences of both these kinds arising from p.

Furthermore, there is an evident way to endow all sequences in $S(p)$ with backward pointers. In summary, each leaf node points to the λ of the innermost enclosing procedure, each case branch points to the head variable of the corresponding application, each head variable points to the preceding λ, and each λ except the topmost one points to the head variable of which it is an argument.

It is now tedious but routine to check the following:

Proposition 7.4.5. *For any $p \in \mathsf{SP}^0(\sigma)$, the set $S(p)$ arises as $\overline{S}$ for some unique strategy $S \in \mathsf{G}_{ib}(\sigma)$; moreover, this constitutes a bijection $\mathsf{SP}^0(\sigma) \cong \mathsf{G}_{ib}(\sigma)$.* □

However, it is somewhat non-trivial to prove that *application* in G_{ib} as defined above agrees with application in SP^0. In essence, one gives an interpretation of general *meta-terms* as strategies, using application in G_{ib} to interpret meta-procedure application, and then shows by a careful analysis of game positions that NSP reduction (and hence evaluation) respects this semantics. We leave the details as an ambitious exercise.

We may now consider the relative merits of SP^0 and G_{ib}. The major attraction of G_{ib} is that it places PCF-style computation within a broader framework alongside numerous other models embodying interesting computability notions, and exhibits the particular constraints on strategies that characterize the PCF model. By contrast, NSPs as we have studied them appear to be somewhat special to PCF-style computation, and it is not clear whether similar ideas can be applied in other settings. On the other hand, the 'oracular' spirit of NSPs means that they can be directly interpreted in purely extensional type structures such as M and PC, whereas a strategy, per se, cannot naturally be 'applied' to anything except another strategy. This latter point has been decisive for our choice to focus mainly on NSPs in this book, since extensional models and Kleene computability occupy such a major share of our concerns. That said, it would be interesting to know whether there are significant *mathematical* advantages of either approach over the other for the purpose of proving results about PCF and sublanguages thereof.

A small selection of other game models of a different flavour (not based on justified sequences) will be considered in Section 12.1 (see especially Subsection 12.1.7). However, a comprehensive study of the various intensional notions of higher-order computation that arise naturally within game semantics, whilst falling squarely within our declared field of interest, would pose too large a task to be attempted in this volume.

7.5 Sequential Functionals

The remainder of this chapter is devoted to a deeper investigation of the extensional model SF. In this section, we focus on the question of which functionals are present in this model: we know they are those for which a sequential procedure happens to exist, but it is natural to look for a more intrinsic characterization of them in purely extensional terms. In Subsection 7.5.2, we shall see that such characterizations can indeed be given at certain low types. In Subsection 7.5.3, we suggest (following Jung and Stoughton [127]) that a 'good' characterization at all types would imply that equality was decidable for 'finite' sequential functions. For types of level ≤ 3 we are able to show that this is indeed the case, but an important theorem of Loader [172] shows that equality is in general undecidable at level 4. This indicates the point at which the finitary structure of SF becomes untameable.

7.5.1 Basic Properties

We begin by collecting some simple facts about sequential functionals, inspired by standard results about compact elements in Scott domains.

We shall write $\mathsf{SF}^{\mathrm{fin}}$ for the applicative structure of *finite* sequential functionals, which we here define to be those represented by some finite sequential procedure. These are closed under application by Theorem 6.3.4. Since every sequential procedure is an observational limit of finite ones, it is easy to see that $\mathsf{SF}^{\mathrm{fin}}$ is an extensional structure in its own right, and indeed is the extensional quotient of $\mathsf{SP}^{0,\mathrm{fin}}$.

We shall also work frequently in the Karoubi envelopes of SP^0 and SF, taking advantage of the finite types generated from the base types $\mathbb{N}_\perp^a = \{\perp, 0, \ldots, a-1\}$ for $a \in \mathbb{N}$. We consider $\mathbb{N}_\perp^a$ as a retract of $\mathbb{N}_\perp$ in SP^0 and hence in SF, via an idempotent sending all $n \geq a$ to $\perp$. These give rise to objects $\mathsf{SP}^{0,a}(\sigma) \lhd \mathsf{SP}^0(\sigma)$ and $\mathsf{SF}^a(\sigma) \lhd \mathsf{SF}(\sigma)$ for each σ, where $\mathsf{SF}^a(\sigma)$ is a quotient of $\mathsf{SP}^{0,a}(\sigma)$. For notational purposes, we shall often treat $\mathsf{SP}^{0,a}(\sigma)$ and $\mathsf{SF}^a(\sigma)$ as subsets of $\mathsf{SP}^0(\sigma)$ and $\mathsf{SF}(\sigma)$ respectively. The sets $\mathsf{SP}^{0,a}(\sigma)$ in general are infinite, and even contain infinite procedures, but clearly each $\mathsf{SF}^a(\sigma)$ is a finite set since $|\mathsf{SF}^a(\mathtt{N})| = a+1$ and $|\mathsf{SF}^a(\sigma \to \tau)| \leq |\mathsf{SF}^a(\tau)|^{|\mathsf{SF}^a(\sigma)|}$. We also have:

Proposition 7.5.1. $\mathsf{SF}^{\mathrm{fin}}(\sigma) = \bigcup_{a \in \mathbb{N}} \mathsf{SF}^a(\sigma)$.

Proof. Clearly every $x \in \mathsf{SF}^{\mathrm{fin}}(\sigma)$ belongs to some $\mathsf{SF}^a(\sigma)$: just take a larger than all numbers appearing in some finite procedure representing x. It remains to show that $\mathsf{SF}^a(\sigma) \subseteq \mathsf{SF}^{\mathrm{fin}}(\sigma)$; for notational simplicity we assume $\sigma = \tau \to \mathtt{N}$. Suppose $p \in \mathsf{SP}^{0,a}(\sigma)$ represents $f \in \mathsf{SF}^a(\sigma)$; then for every $g \in \mathsf{SF}^a(\tau)$ we may find $p_g \sqsubseteq p$ finite such that $p_g \cdot q = p \cdot q$ for some (and hence any) representative q of g. Since $\mathsf{SF}^a(\tau)$ is finite, the supremum of these p_g will be a finite $p_0 \sqsubseteq p$ such that $p_0 \cdot q = p \cdot q$ for all $q \in \mathsf{SP}^{0,a}(\tau)$, and hence for all $q \in \mathsf{SP}^0(\tau)$. Thus p_0 represents f, so $f \in \mathsf{SF}^{\mathrm{fin}}(\sigma)$. □

We shall write $\preceq$ for the pointwise ordering on each $\mathsf{SF}(\sigma)$; recall from Exercises 6.1.21 and 7.1.29 that this coincides with the observational ordering. If $\vec{f}, \vec{g}$ are finite sequences of functionals, we write $\vec{f} \preceq \vec{g}$ with the evident componentwise meaning. Note that $\mathsf{SF}^{\mathrm{fin}}$ is not downward-closed under $\preceq$: for example, there are many non-finite elements below $\lambda x.0$.

We may now obtain some basic structural information about $\mathsf{SF}^{\mathrm{fin}}$:

Lemma 7.5.2. *Let $S = \{f_0, \ldots, f_{n-1}\}$ be an inhabited set of finite sequential functionals of the same type σ.*

(i) S has a $\preceq$-greatest lower bound in SF, *and it is finite. Indeed, the n-ary infimum operator on* $\mathsf{SF}^{\mathrm{fin}}(\sigma)$ *is* PCF*-definable.*

(ii) If S is $\preceq$-bounded in SF, *then S has a least upper bound in* SF, *and it is finite.*

Proof. (i) We give an algorithm for computing the greatest lower bound. For any input $\vec{x}$, compute $f_0(\vec{x}), \ldots, f_{n-1}(\vec{x})$ until either we get divergence or two of the values differ. If all values agree, we output this value, otherwise output $\perp$. Clearly this algorithm is uniform in the f_i, and is represented by a finite NSP.

(ii) By Proposition 7.5.1, we may take $a \in \mathbb{N}$ sufficiently large that $f_0, \ldots, f_{n-1} \in \mathsf{SF}^a(\sigma)$. But now the set of all upper bounds for S within $\mathsf{SF}^a(\sigma)$ is finite and inhabited, and so has a greatest lower bound $f \in \mathsf{SF}^a(\sigma)$ by (i). It is easy to see that f will be the least upper bound of S within $\mathsf{SF}(\sigma)$. □

Note that the least upper bound obtained in (ii) need not be the *pointwise* least upper bound of S: for instance, there is a bounded pair of elements whose pointwise least upper bound would be the parallel convergence operator, which is not present in S. Note also that the construction of (ii) is not in general effective, since we have given no way of computing the set of upper bounds in S. We will return to the question of effectivity in Subsection 7.5.3.

Definition 7.5.3. Suppose $f \in \mathsf{SF}^{\mathrm{fin}}(\tau_0, \ldots, \tau_{r-1} \to \mathtt{N})$. A *basis* for f is a pairwise $\preceq$-incomparable set of vectors $\{\vec{g}_0, \ldots, \vec{g}_{n-1}\}$, where $g_{ij} \in \mathsf{SF}^{\mathrm{fin}}(\tau_j)$ for each i, j, such that for every $\vec{g}$ of appropriate type we have

$$f(\vec{g})\downarrow \quad \text{iff} \quad \exists i.\, \vec{g}_i \preceq \vec{g}\,.$$

Lemma 7.5.4. *Every finite sequential functional has a basis.*

Proof. Given $f \in \mathsf{SF}^{\mathrm{fin}}(\sigma)$ where $\sigma = \tau_0, \ldots, \tau_{r-1} \to \mathtt{N}$, take a such that $f \in \mathsf{SF}^a(\sigma)$, and let B be the set of $\preceq$-minimal elements in the set

$$C = \{\vec{x} \in \mathsf{SF}^a(\tau_0) \times \cdots \times \mathsf{SF}^a(\tau_{r-1}) \mid f(\vec{x}) \in \mathbb{N}\}\,.$$

Since C is finite, there will be an element of B below each element of C. Thus B is a basis for f. □

Again, this construction of a basis is not effective, since we have given no way of computing C. However, it is easy to see that if σ is of level 1, a basis for $f \in \mathsf{SF}^{\mathrm{fin}}(\sigma)$ can be computed from a finite procedure for f.

Exercise 7.5.5. Let $f, f' \in \mathsf{SF}^{\mathrm{fin}}(\tau_0, \ldots, \tau_{r-1} \to \mathtt{N})$, and suppose $\{\vec{g}_0, \ldots, \vec{g}_{n-1}\}$ and $\{\vec{g}'_0, \ldots, \vec{g}'_{m-1}\}$ are bases for f, f' respectively.

(i) Show that $f \preceq f'$ iff for all $i < n$ there exists $j < m$ with $\vec{g}'_j \preceq \vec{g}_i$ and $f(\vec{g}_i) = f'(\vec{g}'_j)$.

(ii) Show that a basis for a finite sequential functional is unique up to observational equivalence.

Lemma 7.5.6. *Suppose $f_i \in \mathsf{SF}^{\mathrm{fin}}(\tau_i)$ for $i < r$, and $b \in \mathbb{N}$. Then the* step function *$(f_0, \ldots, f_{r-1}) \mapsto b$ is present in $\mathsf{SF}^{\mathrm{fin}}(\tau_0, \ldots, \tau_{r-1} \to \mathtt{N})$, where*

$$((f_0, \ldots, f_{r-1}) \mapsto b)(\vec{y}) = \begin{cases} a & \text{if } y_i \succeq f_i \text{ for each } i\,, \\ \bot & \text{otherwise.} \end{cases}$$

Proof. For each i, let $\{g_{i0}, \ldots, g_{i(n_i-1)}\}$ be a basis for f_i. To compute $(\vec{f}) \mapsto b$ at $\vec{y}$, test whether $y_i(g_{ij}) = f_i(g_{ij})$ for each i, j. If so, return b, otherwise diverge. □

7.5.2 Some Partial Characterizations

For first-order types $\sigma = \mathtt{N}^{(r)} \rightarrow \mathtt{N}$, elements of $\mathsf{SF}(\sigma)$ will be certain functions $f : \mathbb{N}_\perp^r \rightarrow \mathbb{N}_\perp$. At this level, there are several ways to characterize exactly which functions these are; we discuss two of them here.

Definition 7.5.7. (i) Inductively on r, we declare a function $f : \mathbb{N}_\perp^r \rightarrow \mathbb{N}_\perp$ to be *Milner sequential* if either f is constant or there exists $i < r$ such that

1. f is strict in argument i: that is, if $x_i = \perp$ then $f(\vec{x}) = \perp$,
2. for any $a \in \mathbb{N}$, the function $\Lambda\vec{x}.\, f(x_0,\ldots,x_{i-1},a,x_{i+1},\ldots,x_{r-1}) \,:\, \mathbb{N}_\perp^{r-1} \rightarrow \mathbb{N}_\perp$ is Milner sequential (we shall denote this function by f_a^i).

(ii) We say $f : \mathbb{N}_\perp^r \rightarrow \mathbb{N}_\perp$ is *Vuillemin sequential* if either f is constant or for any $x_0,\ldots,x_{r-1} \in \mathbb{N}_\perp$ there exists $j < r$ such that for all $y_0,\ldots,y_{r-1} \in \mathbb{N}_\perp$, if $\vec{y} \sqsupseteq \vec{x}$ and $y_j = x_j$ then $f(\vec{y}) = f(\vec{x})$.

Theorem 7.5.8. *The following are equivalent for a function $f : \mathbb{N}_\perp^r \rightarrow \mathbb{N}_\perp$:*

1. *$f \in \mathsf{SF}(\mathtt{N}^{(r)} \rightarrow \mathtt{N})$.*
2. *f is Milner sequential.*
3. *f is monotone and Vuillemin sequential.*

Proof. $2 \Rightarrow 1$: By induction on r. If f is constant, clearly $f \in \mathsf{SF}$. Otherwise, let i be as in Definition 7.5.7(i), and for each $a \in \mathbb{N}$, inductively pick a procedure $p_a = \lambda x_0 \ldots x_{i-1} x_{i+1} \ldots x_{r-1}.e_a$ computing f_a^i. Then $\lambda\vec{x}.\mathtt{case}\ x_i\ \mathtt{of}\ (a \Rightarrow e_a)$ computes f.

$1 \Rightarrow 3$: Suppose $p = \lambda\vec{z}.e$ is a procedure for f. Given any $x_0,\ldots,x_{r-1} \in \mathbb{N}_\perp$, trace a rightward path through e as follows: at each subexpression $\mathtt{case}\ z_j\ \mathtt{of}\ (k \Rightarrow e_k)$ where $x_j \in \mathbb{N}$, choose the branch x_j. If this yields an infinite path, then clearly $f(\vec{y}) = f(\vec{x}) = \perp$ for all $\vec{y} \sqsupseteq \vec{x}$. If it leads us to a leaf node $b \in \mathbb{N}_\perp$, then $f(\vec{y}) = f(\vec{x}) = b$ for all $\vec{y} \sqsupseteq \vec{x}$. If it leads to a subexpression $\mathtt{case}\ z_j\ \mathtt{of}\ (\cdots)$ where $x_j = \perp$, then clearly $f(\vec{x}) = \perp$, and if $\vec{y} \sqsupseteq \vec{x}$ with $y_j = \perp$ then also $f(\vec{y}) = \perp$, since the computation of $f(\vec{y})$ will lead us to this same subexpression.

$3 \Rightarrow 2$: By induction on r. Suppose $f : \mathbb{N}_\perp^r \rightarrow \mathbb{N}_\perp$ is monotone and Vuillemin sequential. If f is constant, it is Milner sequential. Otherwise, consider the arguments $\vec{x} = \vec{\perp}$. We must have $f(\vec{\perp}) = \perp$, otherwise by monotonicity f would be constant. So we may take i such that $f(\vec{y}) = \perp$ whenever $y_i = \perp$. Moreover, it is an easy exercise to check that each f_a^i is Vuillemin sequential, whence f_a^i is Milner sequential by the induction hypothesis. □

The first-order sequential functions can also be characterized in a more algebraic spirit as the *strongly stable* functions: this will be covered in Section 11.4.

As shown in Subsection 7.3.1, the 'parallel' operators *pconv* and *por* are examples of non-sequential monotone first-order functions; further examples will be mentioned in Subsection 10.3.3.

Interestingly, the above characterizations as they stand do not work for the effective submodel $\mathsf{SF}^{\mathrm{eff}}$ (the extensional quotient of $\mathsf{SP}^{0,\mathrm{eff}}$). We may call a monotone function $f : \mathbb{N}_\perp^r \to \mathbb{N}_\perp$ *effective* if the set $\{(\vec{x}, a) \in \mathbb{N}_\perp^r \times \mathbb{N} \mid f(\vec{x}) = a\}$ is computably enumerable modulo coding; this is the same as saying that f is present in the model $\mathsf{PC}^{\mathrm{eff}}$. The following observation is due to Trakhtenbrot [286]:

Proposition 7.5.9. *There is an effective sequential function $f \in \mathsf{SF}(\mathtt{N}^{(2)} \to \mathtt{N})$ that is not effectively sequential (i.e. not present in $\mathsf{SF}^{\mathrm{eff}}$).*

Proof. Let $K, \overline{K} \subseteq \mathbb{N}$ denote the halting set and its complement, and define

$$f(x,y) \;=\; \begin{cases} 0 & \text{if } x \in \mathbb{N} \text{ and } (y \in \mathbb{N} \text{ or } x \in K)\,, \\ \perp & \text{otherwise.} \end{cases}$$

A non-effective sequential procedure for f is given by

$$\lambda xy.\,\mathtt{case}\; x\; \mathtt{of}\; (i \in K \Rightarrow 0 \mid i \in \overline{K} \Rightarrow \mathtt{case}\; y\; \mathtt{of}\; (j \Rightarrow 0))\,.$$

However, it is easy to show that no effective procedure for f is possible. □

One can carry the ideas of Milner and Vuillemin sequentiality just a little further. To put our finger on what these notions are really capturing, let us say a procedure $p = \lambda \vec{f}.e$ is *flat* if all applications $f\vec{q}$ occurring within e are such that none of the f_i appear free within the arguments $\vec{q}$. Thus, a flat procedure is simply a decision tree based on applications of the f_i to closed terms, with no nesting.[2]

The following exercises show how Milner and Vuillemin style characterizations can be extended a short distance into the realm of second-order types.

Exercise 7.5.10. (i) Generalize the concepts of Milner and Vuillemin sequentiality to functions $f : \mathbb{N}_\perp^{\mathbb{N}} \to \mathbb{N}_\perp$. (Note that the generalization of Milner sequentiality will have a 'coinductive' flavour.) Show that for such functions f, the following are equivalent:

1. f is represented by a flat procedure $p \in \mathsf{SP}^0(2)$.
2. f is Milner sequential and continuous.
3. f is Vuillemin sequential, monotone and continuous.

(ii) Show that every $f \in \mathsf{SF}((\mathtt{N} \to_s \mathtt{N}) \to \mathtt{N})$ is represented by a flat procedure.

(iii) Show also that every *finite* $f : \mathsf{SF}((\mathtt{N} \to \mathtt{N}) \to \mathtt{N})$ is represented by a flat procedure, but that the function $\Lambda g.g(min\ g)$ at this type is not.

(iv) Show that the functional B of Subsection 7.3.2, considered as an element of $\mathsf{SF}((\mathtt{U}^{(2)} \to \mathtt{U}) \to \mathtt{U})$, is finite but not represented by any flat procedure.

(v) Show that the functional $\Lambda fg.f(g(\top)) \in \mathsf{SF}((\mathtt{U} \to \mathtt{U})^{(2)} \to \mathtt{U})$ is not representable by a flat procedure.

[2] Flat procedures are called *normal form* procedures in [221]. As a historical aside, we note that an early hypothesis of Kleene conjectured, in effect, that every Kleene computable total functional $\Phi \in \mathsf{S}(3)$ was computable by a flat NSP. This was refuted by Kleene's student Kierstead, who gave as a simple counterexample the functional $\Lambda F.F(\Lambda x.F(\Lambda y.x))$. A more extreme counterexample is given in the proof of our Theorem 6.3.27.

We shall see in Section 11.4 that the notion of strong stability also fails to characterize SF at second order, and even at the type $(\mathtt{U} \to \mathtt{U}) \to \mathtt{U}$.

To make further progress in characterizing the second-order sequential functionals, a new idea is needed. We next present a remarkable characterization due to Sieber [271] using logical relations; this was originally presented in the context of the Scott model PC, but we adapt it here to the setting of SF. Sieber's condition provides a useful tool for ruling out specific non-sequential functionals such as those mentioned in Subsection 7.3.1 (see Exercise 7.5.15).

We shall limit our ambitions to characterizing the *finite* elements of $\mathsf{SF}(\sigma)$ where $\mathrm{lv}(\sigma) = 2$, assuming the sets $\mathsf{SF}(\tau)$ for $\mathrm{lv}(\tau) = 1$ have already been constructed. Since every such element appears in some $\mathsf{SF}^a(\sigma)$, it will be enough to characterize $\mathsf{SF}^a(\sigma)$ for an arbitrary fixed a.

The key concept in Sieber's characterization is the following:

Definition 7.5.11. (i) For any $n \in \mathbb{N}$ and any pair of sets $A \subseteq B \subseteq \{0,\dots,n-1\}$, define a relation $S^n_{A,B} \subseteq \mathbb{N}^n_\perp$ by

$$S^n_{A,B}(\vec{x}) \;\text{ iff }\; (\exists i \in A.\, x_i = \perp) \vee (\forall i,j \in B.\, x_i = x_j)\,.$$

An n-ary *ground type sequentiality relation* is an intersection of such relations $S^n_{A,B}$.

(ii) Given any relation $R_{\mathtt{N}} \subseteq \mathbb{N}^n_\perp$, we define its *lifting* R to consist of $R_{\mathtt{N}}$ along with the following relations R_τ, R_σ at first- and second-order types. (We write $\vec{x}_i$ for the vector $x_{i0},\dots,x_{i(n-1)}$, and likewise for $\vec{g}_i$.)

1. If $\tau = \mathtt{N}^{(r)} \to \mathtt{N}$, define $R_\tau \subseteq \mathsf{SF}(\tau)^n$ by

$$\begin{aligned} R_\tau(\vec{g}) \;\text{ iff }\; \forall \vec{x}_0,\dots,\vec{x}_{r-1}.\; & R_{\mathtt{N}}(\vec{x}_0) \wedge \cdots \wedge R_{\mathtt{N}}(\vec{x}_{r-1}) \Rightarrow \\ & R_{\mathtt{N}}(g_0\, x_{00} \dots x_{(r-1)0}, \dots, g_{n-1}\, x_{0(n-1)} \cdots x_{(r-1)(n-1)})\,. \end{aligned}$$

2. If $\sigma = \tau_0,\dots,\tau_{t-1} \to \mathtt{N}$ where the τ_i are of level 1, define an n-ary relation R_σ on functions $F_i : \mathsf{SF}(\tau_0) \times \cdots \times \mathsf{SF}(\tau_{t-1}) \to \mathbb{N}_\perp$ by

$$\begin{aligned} R_\sigma(\vec{F}) \;\text{ iff }\; \forall \vec{g}_0,\dots,\vec{g}_{t-1}.\; & R_{\tau_0}(\vec{g}_0) \wedge \cdots \wedge R_{\tau_{t-1}}(\vec{g}_{t-1}) \Rightarrow \\ & R_{\mathtt{N}}(F_0\, g_{00} \dots g_{(t-1)0}, \dots, F_{n-1}\, g_{0(n-1)} \cdots g_{(t-1)(n-1)})\,. \end{aligned}$$

By a *sequentiality relation* R, we shall mean the lifting of a ground type sequentiality relation $R_{\mathtt{N}}$.

(iii) An element $g \in \mathsf{SF}(\tau)$ is *invariant* under a lifting R if $R_\tau(g,\dots,g)$, and likewise for second-order functions $F : \mathsf{SF}(\tau_0) \times \cdots \times \mathsf{SF}(\tau_{t-1}) \to \mathbb{N}_\perp$.

It is admittedly hard to give much direct motivation for part (i) of this definition, beyond the fact that it serves our purpose. Part (ii) embodies the first two stages in the construction of a logical relation; however, since we are here supposing that only the first-order part of SF is provided, we are not given (and do not need) an ambient model within which the construction can be continued to higher types.

Lemma 7.5.12. *If* $\mathrm{lv}(\tau) = 1$ *and* $\mathrm{lv}(\sigma) = 2$*, every* $g \in \mathsf{SF}^a(\tau)$ *and* $F \in \mathsf{SF}^a(\sigma)$ *is invariant under every sequentiality relation.*

Proof. A straightforward induction on the size of finite NSPs representing g or F. The key idea is as follows: if R is an n-ary sequentiality relation and we have $R_{\mathbb{N}}(\vec{c})$ and $R_{\mathbb{N}}(\vec{v}_i)$ for each $i < a$, then taking d_j to be the value of $\mathtt{case}\ c_j\ \mathtt{of}\ (i \Rightarrow v_{ij})$ for each j, we also have $R_{\mathbb{N}}(\vec{d})$. This is because if $S^n_{A,B}$ is any one of the relations whose intersection defines $R_{\mathbb{N}}$, we have $\vec{c} \in S^n_{A,B}$. So either $c_j = \bot$ for some $j \in A$, in which case also $d_j = \bot$ whence $\vec{d} \in S^n_{A,B}$, or else $c_j = c \in \mathbb{N}$ (say) for all $j \in B$, in which case $d_j = v_{cj}$ for all $j \in B$. Since $\vec{v}_c \in S^n_{A,B}$, this again implies $\vec{d} \in S^n_{A,B}$ (either because $v_{cj} = \bot$ for some $j \in A$, or because v_{cj} has the same value for all $j \in B$). We leave the remaining administrative details of the induction to the reader. □

The next lemma provides a kind of converse to the above, in that it shows that the sequentiality relations are the *only* logical relations under which all first-order sequential functions are invariant. The proof of this forms the most demanding part of Sieber's argument.

Lemma 7.5.13. *If R is the lifting of some relation $R_{\mathbb{N}}$, and $0, suc, pre, ifzero \in \mathsf{SF}$ are invariant under R, then R is a sequentiality relation.*

Proof. Suppose R lifts an n-ary relation $R_{\mathbb{N}}$ and $0, suc, pre, ifzero \in \mathsf{SF}$ are invariant under R. To show that R is a sequentiality relation, it will be enough to show that $R_{\mathbb{N}}$ is the intersection of all relations $S^n_{A,B}$ that contain it. So suppose $\vec{x} \notin R_{\mathbb{N}}$; we must find some $S^n_{A,B}$ containing $R_{\mathbb{N}}$ that excludes $\vec{x}$.

Let $C \subseteq \{0, \ldots, n-1\}$ be some set of smallest size such that there is no $\vec{y} \in R_{\mathbb{N}}$ with $y_i = x_i$ for all $i \in C$. Clearly $|C| > 1$, since if $C = \{i\}$ then $(x_i, \ldots, x_i) \in R_{\mathbb{N}}$ (because 0 and *suc* are R-invariant). We now reason by cases; in each case we will either obtain some $S^n_{A,B}$ containing $R_{\mathbb{N}}$ and excluding $\vec{x}$, or else reach a contradiction. For any $i \in C$, we set $C_i = C - \{i\}$.

Case 1: $x_i = \bot$ for some $i \in C$.

Subcase 1.1: $R_{\mathbb{N}} \not\subseteq S^n_{C_i,C}$. Then there is some $\vec{u} \in R_{\mathbb{N}}$ such that $u_j \neq \bot$ for all $j \in C_i$, and $u_k \neq u_i$ for some $k \in C_i$ (so that $u_k \neq \bot$). By the minimality of C, there are vectors $\vec{v}, \vec{w} \in R_{\mathbb{N}}$ such that $\vec{v}$ agrees with $\vec{x}$ on C_i and $\vec{w}$ agrees with $\vec{x}$ on C_k. We may now piece these vectors together by defining

$$y_j \;=\; \mathit{if}\ u_j = u_k\ \mathit{then}\ v_j\ \mathit{else}\ w_j \quad (j \in C)\,.$$

Let us check that $\vec{y}$ agrees with $\vec{x}$ on the whole of C. If $i \neq j \neq k$ then $u_j, u_k \neq \bot$ so $y_j = v_j$ or $y_j = w_j$, and in either case $y_j = x_j$. For the coordinate k we have $y_k = v_k$, so again $y_k = x_k$. For the coordinate i we either have $u_i \neq \bot$, in which case $y_i = w_i = x_i$, or $u_i = \bot$, in which case $y_i = \bot = x_i$ by choice of i.

Furthermore, $\vec{v}, \vec{w} \in R_{\mathbb{N}}$ is easily seen to imply $\vec{y} \in R_{\mathbb{N}}$ using the invariance of $0, suc, pre, ifzero$. This contradicts the minimality of C.

Subcase 1.2: $R_{\mathbb{N}} \subseteq S^n_{C_i,C}$. Then we claim that $S^n_{C_i,C}$ excludes $\vec{x}$. For if $\vec{x} \in S^n_{C_i,C}$, then since $x_i = \bot$, we must have $x_k = \bot$ for some $k \in C_i$. Now choose $\vec{v}, \vec{w} \in R_{\mathbb{N}}$ so that $\vec{v}$ agrees with $\vec{x}$ on C_i and $\vec{w}$ agrees with $\vec{x}$ on C_k. If $v_i = \bot$, then $\vec{v}$ itself agrees with $\vec{x}$ on the whole of C, contradicting the minimality of C. If $v_i \in \mathbb{N}$, then take

$$y_j = \textit{if } v_j = v_i \textit{ then } w_j \textit{ else } v_j \quad (j \in C).$$

We now check that $\vec{y}$ agrees with $\vec{x}$ on C. If $i \neq j \neq k$, then either $y_j = w_j$ or $y_j = v_j$ (note that if $v_j = \bot$ then the latter will be true), and in either case $y_j = x_j$. For the coordinate k we have $v_k = x_k = \bot$, so also $y_k = \bot$. For the coordinate i we have $v_j, v_i \neq \bot$, so $y_i = w_i = x_i$.

Once again, it is easy to show that $\vec{y} \in R_{\mathbb{N}}$ by the invariance of $0, suc, pre, ifzero$, so again the minimality of C is contradicted.

Case 2: $x_i \in \mathbb{N}$ for all $i \in C$. Let T be the set of tuples $\vec{v} \in R_{\mathbb{N}}$ such that $v_i \in \mathbb{N}$ for all $i \in C$; note that $T \neq \emptyset$ since $(0, \ldots, 0) \in T$. Each tuple $\vec{v} \in T$ induces an equivalence relation on C via $i \sim j$ iff $v_i = v_j$. Choose $\vec{u} \in T$ so as to make this equivalence relation as fine as possible, i.e. such that there is no $\vec{v} \in T$ for which $\sim_v$ is strictly contained within $\sim_u$, and let B range over the equivalence classes of $\sim_u$. We will show that $R_{\mathbb{N}} \subseteq S^n_{B,B}$ for each such B, but that at least one of the $S^n_{B,B}$ excludes $\vec{x}$.

For the first claim, suppose B is the $\sim_u$-class of $i \in C$, and suppose for contradiction that $\vec{v} \in R_{\mathbb{N}} - S^n_{B,B}$. Then $v_j \in \mathbb{N}$ for all $j \in B$, and $v_k \neq v_l$ for some $k, l \in B$. This allows us to find $\vec{w} \in T$ such that $\sim_w$ is finer than $\sim_u$: namely, take

$$w_j = \textit{if } u_j = u_i \textit{ then } (\textit{if } v_j = v_k \textit{ then } u' \textit{ else } u_j) \textit{ else } u_j \quad (j \in C),$$

where $u' \in \mathbb{N}$ is chosen to be different from all u_j for $j \in C$. Then it is easy to check that $\vec{w} \in T$ and that $\sim_w$ is strictly finer than $\sim_u$, since $\sim_w$ splits B itself into two classes. This contradicts the choice of u.

For the second claim, suppose $\vec{x} \in S^n_{B,B}$ for all $\sim_u$-classes B. Since $x_i \in \mathbb{N}$ for all $i \in C$, this implies $\sim_u \subseteq \sim_x$. So there is a finite partial function $\alpha : \mathbb{N} \rightharpoonup \mathbb{N}$ such that $x_i = \alpha(u_i)$ for all $i \in C$. Using this, it is easy to construct an applicative expression $P[z^{\mathtt{N}}] : \mathtt{N}$ over $0, suc, pre, ifzero$ such that $P[u_i]$ has value x_i for all $i \in C$. Since $\vec{u} \in R_{\mathbb{N}}$, it follows that $\vec{x} \in R_{\mathbb{N}}$, contradicting the choice of $\vec{x}$. □

We may now use sequentiality relations to characterize exactly the second-order elements of SF^a:

Theorem 7.5.14. *Let* $\sigma = \tau_0, \ldots, \tau_{t-1} \to \mathtt{N}$ *where the* τ_i *are first-order. A function* $F : \mathsf{SF}^a(\tau_0) \times \cdots \times \mathsf{SF}^a(\tau_{t-1}) \to \mathbb{N}_\bot$ *is present in* $\mathsf{SF}^a(\sigma)$ *iff, regarded as a function* $\mathsf{SF}(\tau_0) \times \cdots \times \mathsf{SF}(\tau_{t-1}) \to \mathbb{N}_\bot$ *via the projections* $\mathsf{SF}(\tau_i) \to \mathsf{SF}^a(\tau_i)$, F *is invariant under all sequentiality relations.*

Proof. The left-to-right implication is given by Lemma 7.5.12. For the converse, suppose F is invariant under all sequentiality relations. Let n be the total number of tuples in $\mathsf{SF}^a(\tau_0) \times \cdots \times \mathsf{SF}^a(\tau_{t-1})$; enumerate these tuples as $\vec{g}_0, \ldots, \vec{g}_{n-1}$ where $\vec{g}_j = (g_{j0}, \ldots, g_{j(t-1)})$, and let $z_j = F(\vec{g}_j)$ for each $j < n$. Now define an n-ary relation $R_{\mathbb{N}}$ by

$$R_{\mathbb{N}}(\vec{x}) \quad \text{iff} \quad \exists G \in \mathsf{SF}(\sigma). \forall j < n. x_j = G(\vec{g}_j),$$

and let R be its lifting. To show that $F \in \mathsf{SF}^a(\sigma)$, it will suffice to show that $R_{\mathbb{N}}(\vec{z})$, since if $G \in \mathsf{SF}(\sigma)$ satisfies $z_j = G(\vec{g}_j)$ for all $j < n$ then so does the image of G in $\mathsf{SF}^a(\sigma)$, which must then coincide with F.

It is an easy exercise to verify that $0, suc, pre, ifzero$ are invariant under R: for instance, given $G_x, G_y, G_z \in \mathsf{SF}(\sigma)$ witnessing that $\vec{x}, \vec{y}, \vec{z} \in R_{\mathtt{N}}$, it is easy to concoct G witnessing that $R_{\mathtt{N}}(\mathit{ifzero}\, x_0\, y_0\, z_0, \ldots, \mathit{ifzero}\, x_{n-1}\, y_{n-1}\, z_{n-1})$. Hence by Lemma 7.5.13, R is a sequentiality relation, so $R_\sigma(F, \ldots, F)$ by hypothesis. Moreover, it is not hard to verify that $R_{\tau_i}(g_{0i}, \ldots, g_{(n-1)i})$ for each $i < t$: for instance, if τ_i has arity 2, then given sequential functionals G_x, G_y witnessing $\vec{x}, \vec{y} \in R_{\mathtt{N}}$, it is easy to concoct G witnessing $R_{\mathtt{N}}(g_{0i} x_0 y_0, \ldots, g_{(n-1)i} x_{n-1} y_{n-1})$. Since $z_j = F(g_{0j}, \ldots, g_{(t-1)j})$ for each j, we thus conclude that $R_{\mathtt{N}}(\vec{z})$. □

Exercise 7.5.15. (Challenging.) By means of a suitable sequentiality relation of arity 4, show that Curien's functional C from Subsection 7.3.1 is not present in SF. (This is done as Example 3.5 of Sieber [271]).

Another interesting application of sequentiality relations will be given as Exercise 10.3.9.

The idea of exploiting invariance under logical relations was carried much further by O'Hearn and Riecke [223], who used Sieber's ideas to give a fully abstract model of the whole of PCF using *Kripke logical relations*. However, their method uses only quite general properties of λ-definability, and does not seem to reveal much about sequentiality beyond what is already contained in the Sieber result.

We conclude this subsection by stating, without proof, a surprising theorem due to Laird [163], which precisely characterizes the sequential functionals over the *unit type* at all type levels—in our present terminology, the elements of $\mathsf{SF}^1(\sigma)$ for all σ. Syntactically, these are the functionals definable in *unary* PCF, which can be taken to be the simply typed λ-calculus with just the base type $\mathtt{U}$, the constants $\top, \bot : \mathtt{U}$, and the sequencing operator $-\,\mathit{then}\,- : \mathtt{U}, \mathtt{U} \to \mathtt{U}$ as in Section 7.2. (It is easy to show that this language defines all elements of SF^1.) Laird's result is that these are exactly the functionals over the unit type in Berry's category of *bidomains* [32]. Since all the sets in question are finite, we can in fact phrase the result in the simpler setting of *biorders*, ignoring issues of continuity.

Definition 7.5.16. (i) A *biorder* consists of a set B endowed with two partial orders $\sqsubseteq, \leq$ such that

1. $(B, \sqsubseteq)$ has a least element $\bot$ and a binary infimum operator $\sqcap$,
2. $(B, \leq)$ has the same least element $\bot$, and $x \leq y$ implies $x \sqsubseteq y$,
3. the operator $\sqcap$ is monotone with respect to $\leq$, whence if $x \leq z$ and $y \leq z$ then $x \sqcap y$ is the infimum of x, y with respect to $\leq$.

(ii) Given biorders $(B, \sqsubseteq_B, \leq_B)$ and $(C, \sqsubseteq_C, \leq_C)$, define the biorder $B \Rightarrow C$ as follows:

1. Elements of $B \Rightarrow C$ are functions $f : B \to C$ that are monotone with respect to both $\sqsubseteq$ and $\leq$, and *stable* with respect to $\leq$: that is, if $x \leq z$ and $y \leq z$ then $f(x \sqcap y) = f(x) \sqcap f(y)$.
2. $f \sqsubseteq_{B \Rightarrow C} g$ iff $f(x) \sqsubseteq_C g(x)$ for all $x \in B$.
3. $f \leq_{B \Rightarrow C} g$ iff $x \leq_B y$ implies $f(x)$ is the $\leq$-infimum of $f(y)$ and $g(x)$ for all $x, y \in B$.

We call $\sqsubseteq$ the *extensional* order, and $\leq$ the *stable* order. It can be checked that biorders form a cartesian closed category with exponentiation defined as above. We therefore obtain a type structure BD^1 over $\mathsf{T}^{\to}(\mathtt{U})$ by taking $\mathsf{BD}^1(\mathtt{U}) = \{\top, \bot\}$ (with $\bot \sqsubseteq \top$ and $\bot \leq \top$) and $\mathsf{BD}^1(\sigma \to \tau) = \mathsf{BD}^1(\sigma) \Rightarrow \mathsf{BD}^1(\tau)$. This model is related to, though different from, the *bistable* model to be discussed in Subsection 12.1.6 (for the latter, we in effect add the dual of the above stability condition on functions).

Theorem 7.5.17 (Laird). $\mathsf{BD}^1 \cong \mathsf{SF}^1$ *as applicative structures.*

However, this turns out to be something of an isolated phenomenon. Even for binary PCF (that is, PCF over the booleans), the situation is radically different, as we shall see in the next subsection.

7.5.3 Effective Presentability and Loader's Theorem

The characterizations described above work only for restricted classes of types. It is therefore natural to ask whether some similar characterization of the finite elements of SF at *all* types is possible. Our first step will be to show that this is intimately related to the problem of finding good descriptions of fully abstract models for PCF.

If $\mathbf{P}$ is any extensional model of PCF in the sense of Definition 7.1.3, with $\mathbf{P}(\mathtt{N}) = \mathbb{N}_\bot$, then as before we let $\mathbb{N}_\bot^a = \{\bot, 0, \ldots, a-1\}$ for each a, and consider $\mathbb{N}_\bot^a$ as a PCF-definable retract $\mathbf{P}^a(\mathtt{N}) \lhd \mathbf{P}(\mathtt{N})$ in $\mathbf{K}(\mathbf{P})$. We then write $\mathbf{P}^a$ for the applicative structure over $\mathbf{P}^a(\mathtt{N})$ generated by exponentiation in $\mathbf{K}(\mathbf{P})$; we shall here say an element of $\mathbf{P}$ is *finite* if it belongs to $\mathbf{P}^a$ for some $a \in \mathbb{N}$. (By Proposition 7.5.1, this accords with our earlier usage in the case of SF.) Clearly the finite elements defined in this way form an applicative substructure $\mathbf{P}^{\mathrm{fin}}$ of $\mathbf{P}$.

The following important result was obtained by Milner [193] in the context of order-extensional models consisting of directed-complete partial orders (DCPOs). However, the same proof works in our more general setting of models of PCF.

Theorem 7.5.18. *Suppose* $\mathbf{P}$ *is an extensional model of* PCF *with* $\mathbf{P}(\mathtt{N}) = \mathbb{N}_\bot$*. The following are equivalent:*

1. $\mathbf{P}$ *is fully abstract for* PCF.
2. *All finite elements of* $\mathbf{P}$ *are* PCF*-definable.*

Proof. $2 \Rightarrow 1$: Suppose all finite elements are definable, and suppose $M, M' : \sigma$ are closed PCF terms with $M \sim_{\mathrm{obs}} M'$. For notational simplicity assume $\sigma = \tau \to \mathtt{N}$; the general case is similar. For any finite element $x \in \mathbf{P}(\tau)$, we have some $N : \tau$ denoting x, so by observational equivalence we have $[\![M]\!](x) = [\![MN]\!] = [\![M'N]\!] = [\![M']\!](x)$. Now an arbitrary element $y \in \mathbf{P}(\tau)$ is an observational limit of finite ones, namely the images of y in $\mathbf{P}^a(\tau)$ for each $a \in \mathbb{N}$ (cf. Exercise 7.1.11), and it follows easily that $[\![M]\!](y) = [\![M']\!](y)$. So by extensionality we have $[\![M]\!] = [\![M']\!]$; thus $\mathbf{P}$ is fully abstract.

$1 \Rightarrow 2$: Suppose $\mathbf{P}$ is fully abstract, and assume for contradiction that σ is a type of minimal level such that not all elements of $\mathbf{P}^{\mathrm{fin}}(\sigma)$ are PCF-definable. Note that $\sigma \neq \mathtt{N}$; again, for notational simplicity we shall suppose $\sigma = \tau \to \mathtt{N}$. Let $z \in \mathbf{P}^a(\sigma)$ be some non-definable finite element, and let $\sqsubseteq$ denote the pointwise ordering on $\mathbf{P}^a(\sigma)$ induced by the usual partial order on $\mathbb{N}_\perp$. It is easy to see that there is a PCF-definable binary infimum operation $\sqcap$ on $\mathbf{P}^a(\sigma)$ with respect to $\sqsubseteq$.

Since $\mathbf{P}^a(\sigma)$ is finite, we may let $d_0,\ldots,d_{n-1}$ be the PCF-definable elements $\sqsupseteq z$ in $\mathbf{P}^a(\sigma)$, and $e_0,\ldots,e_{m-1}$ the remaining definable elements.

Case 1: $n \neq 0$. By taking the infimum of the d_i, we obtain a smallest definable element d above z. Since all elements of $\mathbf{P}^a(\tau)$ are definable by choice of σ, we may now take $u \in \mathbf{P}^a(\tau)$ definable such that $d(u) \in \mathbb{N}$ but $z(u) = \perp$. Also, since $e_i \not\sqsupseteq z$ for each i, we may take $v_0,\ldots,v_{m-1}$ such that $e_i(v_i) = \perp$ but $z(v_i) \in \mathbb{N}$. Now define $F, F' : \mathbf{P}^a(\sigma) \to \mathbb{N}_\perp$ by:

$$F(x) = \begin{cases} 0 & \text{if } \forall i < m.\, x(v_i) = z(v_i), \\ \perp & \text{otherwise,} \end{cases}$$

$$F'(x) = \begin{cases} 0 & \text{if } \forall i < m.\, x(v_i) = z(v_i) \text{ and } x(u) = d(u), \\ \perp & \text{otherwise.} \end{cases}$$

Clearly $F, F' \in \mathbf{P}^a(\sigma) \hookrightarrow \mathbf{P}(\sigma)$ are PCF-definable, say by the terms $P, P' : \sigma \to \mathtt{N}$, and it is easy to check that F, F' agree on all the definable elements d_i, e_j, but differ on z. So by the context lemma (Theorem 7.1.26) we have $P \sim_{\mathrm{obs}} P'$, but $[\![P]\!] \neq [\![P']\!]$. Thus $\mathbf{P}$ is not fully abstract.

Case 2: $n = 0$. Then there are no definable $d \sqsupseteq z$ in $\mathbf{P}^a(\sigma)$. In this case, we may define F as above but take $F'(x) = \perp$ for all x, and the argument is then similar. □

In particular, all extensional fully abstract models must share the same substructure of finite elements as SF. Milner showed that within the framework of order-extensional DCPO models, there is in fact a *unique* fully abstract model of PCF up to isomorphism—the so-called *Milner model*—and in the literature of the 1970s and 1980s, the full abstraction problem for PCF was often articulated as the challenge of finding a good mathematical characterization for this particular model. However, the nub of the problem is to find a good description of the finite elements, so we shall concentrate on that aspect here. We will return to the Milner model and its relationship to SF in the next subsection.

We have so far been vague about what we mean by a 'good' description. After all, Milner's construction of his model, and our construction of SF as a quotient of SP^0, yield mathematically sound definitions of fully abstract models—so what is lacking? At an informal level, various answers may be given. For instance, Milner's construction is 'syntactic' in the sense that the finite elements are constructed as equivalence classes of terms of PCF itself—it therefore does not offer us any independent mathematical handle on what these finite elements are, beyond what we already know from the operational definition of PCF. The construction via SP^0 arguably improves on this somewhat by abstracting away from PCF programs to the 'algorithms' they embody—and as the results of this chapter show, it allows us to obtain a good deal of structural information about SF. All the same, one can feel that it would be preferable to characterize the sequential functionals in purely 'extensional' terms rather than via the existence of suitable intensional objects.

All of these are rather imprecise criteria. However, even in the absence of a precise formulation, the so-called full abstraction problem for PCF generated a good deal of interesting and important research, yielding new approaches to modelling PCF (as in game semantics), as well as models embodying other computability notions that turned out to be of interest in their own right (see Chapters 11 and 12).

A more objective condition for a satisfactory solution to the full abstraction problem was suggested by Jung and Stoughton [127]. Namely, they suggested that such a solution should, at least, yield an *effective presentation* of the minimal model for *binary* PCF. In our present terminology, this amounts to an effective presentation of the structure SF^2.

To spell this out in a little more detail, for any $a \in \mathbb{N}$ we may define the language PCF^a to be the simply typed λ-calculus with a single base type $\mathtt{A}$ and constants

$$c_0, \ldots, c_{a-1}, \bot \,:\, \mathtt{A}\,, \qquad \mathit{branch} \,:\, \mathtt{A}^{(a+1)} \to \mathtt{A}\,,$$

along with the reduction rules

$$\mathit{branch}\; c_i\, x_0 \cdots x_{a-1} \rightsquigarrow x_i\,, \qquad \mathit{branch} \perp x_0 \cdots x_{a-1} \rightsquigarrow \perp\,,$$

plus ordinary β-reduction, applicable in the evaluation contexts generated by $[-]N$ and $\mathit{branch}[-]N_0 \ldots N_{a-1}$. In the case $a = 2$, it is natural to write $\mathtt{A}$ as $\mathtt{B}$ and $c_0, c_1, \mathit{branch}$ as $\mathit{tt}, \mathit{ff}, \mathit{if}$ respectively. In each language PCF^a, all reductions are terminating by an easy adaptation of the proof of Lemma 7.1.14; in particular it is decidable whether a given term evaluates to $\bot$. Moreover, it is not hard to see that every *finite* $p \in \mathsf{SP}^{0,a}$ is definable in PCF^a, and hence that every $f \in \mathsf{SF}^a$ is so definable.

By an effective presentation of $\mathsf{SF}^a(\mathtt{A})$, we shall mean just a listing of its $a+1$ elements. By an effective presentation of $\mathsf{SF}^a(\sigma)$ where $\sigma = \sigma_0, \ldots, \sigma_{r-1} \to \mathtt{A}$, we shall mean a complete listing of the (finitely many) elements of $\mathsf{SF}^a(\sigma)$, each specified by their graphs—this implicitly means that we must first construct effective presentations for each of the σ_i. In terms of such presentations, it is clear that the application operations in SF^a are computable and that the extensional ordering $\sqsubseteq$ is decidable. By an effective presentation of SF^a for some class T of types, we mean a computable function mapping each $\sigma \in T$ to an effective presentation of $\mathsf{SF}^a(\sigma)$. A description of SF^2 satisfying the Jung–Stoughton criterion would be one that yielded an effective presentation for all types.

Exercise 7.5.19. Let $\sigma = \sigma_0, \ldots, \sigma_{r-1} \to \mathtt{N}$. Show that any of the following pieces of information is computable from any of the others, uniformly in a and σ, and assuming that all these pieces of information are available for each of the σ_i.

1. An effective presentation of $\mathsf{SF}^a(\sigma)$.
2. The number of elements in $\mathsf{SF}^a(\sigma)$.
3. A decision procedure for the solvability of finite systems of equations of the form

$$XF_{0j}\cdots F_{(r-1)j} = B_j \quad (j < N)$$

simultaneously by a closed PCF^a term X of type σ, where the F_{ij} and B_j are closed PCF^a terms of types σ_i and $\mathtt{A}$ respectively.

4. An effective procedure for constructing a basis for a functional represented by a given $p \in \mathsf{SP}^{0,a}(\sigma \to \mathtt{A})$.
5. A decision procedure for extensional equivalence $\sim$ on $\mathsf{SP}^{0,a}(\sigma \to \mathtt{A})$.
6. A decision procedure for observational equivalence of closed PCF^a terms of type $\sigma \to \mathtt{A}$.

(Hint: for $6 \Rightarrow 1$, adapt the proof of Theorem 7.5.18.)

Notice that it is only uniformity over some class of types that makes such questions of effectivity interesting: for any *individual* type σ, the existence of an effective presentation is trivial (at least under classical reasoning), whether or not we know how to give one.

It is easy to see that Laird's characterization of SF^1 (see Theorem 7.5.17) yields an effective presentation of SF^1 at all types. It is also easy to see that the Milner and Vuillemin characterizations of first-order sequential functionals readily yield an effective presentation of any SF^a at all first-order types, so that observational equivalence at second-order types is decidable. Furthermore, we have a simple proof of the following result:

Theorem 7.5.20. *Each* SF^a *is effectively presentable at all second-order types. Equivalently, observational equivalence in* PCF^a *is decidable uniformly for all third-order types.*

Proof. To give an effective presentation of $\mathsf{SF}^a(\sigma)$ uniformly in σ where $\mathrm{lv}(\sigma) = 2$, it suffices to find a depth bound d_σ, computable from σ, such that every $F \in \mathsf{SF}^a(\sigma)$ is representable by some $p \in \mathsf{SP}^{0,a}(\sigma)$ of depth $\leq d_\sigma$. Once such a d_σ is given, we may simply enumerate all of the finitely many such p (noting that there is an obvious bound on their branching arity) and tabulate their graphs.

The key observation is that if $p = \lambda \vec{g}.e \in \mathsf{SP}^{0,a}(\sigma)$ where $\mathrm{lv}(\sigma) = 2$, then any procedure subterms within e have level 0, so no other variables than $\vec{g}$ appear anywhere in the tree for e. For any expression subterm e' within e, we thus have that $\lambda \vec{g}.e'$ represents some $F' \in \mathsf{SF}^a(\sigma)$. Now let d_σ be any upper bound on the size of $\mathsf{SF}^a(\sigma)$. If p has depth $> d_\sigma$, then there are distinct subexpressions e', e'' on the same path through p that yield the same function in $\mathsf{SF}^a(\sigma)$. It is now easy to see how p can be simplified to a smaller procedure that computes the same sequential function: assuming e'' is a proper subexpression of e', we may simply replace e' by e'' in p. We may therefore start from an arbitrary finite procedure p for F (which exists by Proposition 7.5.1), and repeatedly perform such simplifications until we arrive at a procedure of depth $\leq d_\sigma$; thus every $F \in \mathsf{SF}^a(\sigma)$ is representable by a procedure of depth $\leq d_\sigma$ as required. □

Notice that this argument breaks down at the next type level up: even at pure type $\overline{3}$, a descending path through a procedure p may enter the scope of arbitrarily many abstractions λx^0, so there is no guarantee that the same sequential function will be 'repeated' along any sufficiently long path.

Exercise 7.5.21. Use Sieber's characterization of second-order sequential functionals (Theorem 7.5.14) to give an alternative proof of Theorem 7.5.20.

For many years it seemed plausible that an effective presentation for the whole of SF^2 might be forthcoming. However, this hope was dashed when, in 1995, Loader [172] obtained the following negative result:

Theorem 7.5.22 (Loader). SF^2 *is not effectively presentable for the class of types of level 3. Equivalently,* PCF^2 *observational equivalence for types* σ *of level 4 is not decidable uniformly in* σ*.*

In essence, Loader reduced a famous undecidable problem—the *word problem* for *semi-Thue systems*—to the problem of the solvability of certain systems of *inequations* for types σ of level 3. This latter problem would be readily decidable if the solvability of equations were decidable as in item 3 of Exercise 7.5.19.

Whilst the theorem itself is of fundamental importance, the proof is long and technical, and consists of intricate syntactic arguments which in themselves shed little light on the nature of sequential functionals. We will therefore not pursue the proof further here, but refer the interested reader to [172].

There is still scope for further clarification of the threshold of undecidability for PCF^2. Whilst Loader's theorem shows that effective presentability fails for the class of *all* third-order types, it remains possible that some interesting (infinite) class of third-order types of some restricted form is effectively presentable. However, it appears that any result of this kind would require a significant advance over our present understanding.

7.6 Order-Theoretic Structure of the Sequential Functionals

The sequential procedure model SP^0 meets the demand for a mathematically satisfactory model of PCF-style computation at the intensional level. The model SF arises as the one and only extensional quotient of this, and the associated ordering $\preceq$ is certainly the canonical one, arising as both the observational and the pointwise ordering. The structure $(\mathsf{SF}, \preceq)$ and its finitary substructure $\mathsf{SF}^{\mathrm{fin}}$ are thus natural mathematical objects, even though their construction does not meet the Jung–Stoughton criterion.

We now consider how these structures relate to the *Milner model*, which, as noted in Subsection 7.5.3, may be characterized as the unique fully abstract model for PCF (up to isomorphism) within the framework of order-extensional DCPO models. Within our present setup, Milner's model is most easily defined as the *ideal completion* of $\mathsf{SF}^{\mathrm{fin}}$:

Definition 7.6.1. (i) An *ideal* within $\mathsf{SF}^{\mathrm{fin}}(\sigma)$ is an inhabited subset $I \subseteq \mathsf{SF}^{\mathrm{fin}}(\sigma)$ such that

1. I is down-closed: if $y \in \mathsf{SF}^{\mathrm{fin}}(\sigma)$ and $y \preceq x \in I$ then $y \in I$,
2. I is directed: any $x, y \in I$ have an upper bound within I.

We define $\mathsf{MM}(\sigma)$ to be the set of ideals within $\mathsf{SF}^{\mathrm{fin}}(\sigma)$, ordered by inclusion.

(ii) Given $I \in \mathsf{MM}(\sigma \to \tau)$ and $J \in \mathsf{MM}(\sigma)$, define

$$I \cdot J \;=\; \{y \in \mathsf{SF}^{\mathrm{fin}}(\tau) \mid \exists f \in I, x \in J.\, y \preceq f(x)\} \;\in\; \mathsf{MM}(\tau)\,.$$

The resulting applicative structure MM is the *Milner model.*

It is not too hard to prove that MM is an extensional type structure over $\mathbb{N}_\bot$, that each $(\mathsf{MM}(\sigma), \subseteq)$ is a directed-complete partial order, that the application operations are monotone and continuous, and that $f \subseteq g$ in $\mathsf{MM}(\sigma \to \tau)$ iff $x \subseteq y$ implies $f \cdot x \subseteq g \cdot y$.

We may readily embed SF in MM as follows: map any $f \in \mathsf{SF}(\sigma)$ to the set $I_f = \{x \in \mathsf{SF}^{\mathrm{fin}}(\sigma) \mid x \preceq f\}$. Clearly I_f is always down-closed, and it is directed by Lemma 7.5.2(ii). It is also routine to check that the mapping $f \mapsto I_f$ is injective, monotone, and respects application. We may thus identify SF with a submodel of MM, and hence obtain an interpretation of NSPs (or indeed PCF) in MM.

The natural question now is whether SF and MM coincide, or equivalently whether $(\mathsf{SF}, \preceq)$ is directed-complete in such a way that suprema of directed sets agree with those in MM. This was answered negatively by Normann [214], who constructed a $\preceq$-increasing sequence within $\mathsf{SF}^{\mathrm{fin}}(3)$ that is not bounded by any element of $\mathsf{SF}(3)$. Since this sequence will have a least upper bound in Milner's model, this shows that Milner's model will contain elements that are not present in SF, and hence not the interpretation of any sequential procedure. Later, Normann and Sazonov [221] gave an example of the same phenomenon at a second-order type (see Theorem 7.6.2 below).

This suggests that traditional styles of domain theory based on complete partial orders are not appropriate for studying the structure of SF, and motivates the search for alternative kinds of domain theory. We have already seen in Subsection 7.1.2 how the general notion of observational limit (available in all models of PCF) suffices for the more basic parts of the theory. We mention here a couple of other approaches, more specifically geared to the study of SF. In both of these approaches, the aim is to restrict attention to some class of increasing chains for which one can indeed find well-behaved least upper bounds, and with respect to which sequential functionals will appear continuous.

Escardó and Ho [84] consider *rational chains*, i.e. sequences $x_0, x_1, \ldots$ in $\mathsf{SF}(\sigma)$ such that $x_n = g(f^n(\bot))$ for some $f \in \mathsf{SF}(\tau \to \tau)$ and $g \in \mathsf{SF}(\tau \to \sigma)$. From the results of Subsection 7.1.2, it is clear that rational chains have observational limits and hence least upper bounds, and that all sequential functionals preserve these; it can moreover be shown that sequential functionals map rational chains to rational chains. The authors also discuss analogies to concepts from topology and domain theory, thereby presenting properties of the posets $(\mathsf{SF}(\sigma), \preceq)$ in a language similar to that of traditional domain theory.

Sazonov [253] gives a direct construction of the type structure SF, and also defines the concept of a *natural least upper bound*. This is a least upper bound within some space of functionals that is at the same time a pointwise least upper bound. He shows that all sequential functionals are continuous with respect to natural least upper bounds. A more general version of this analysis appears in [254].

A different line of inquiry was pursued by Normann and Sazonov [221], who asked to what extent sequential functionals might arise as unconventional least upper bounds of increasing chains, and to what extent higher-order sequential functionals are continuous with respect to such chains. They found out that except at very low type levels, many unconventional least upper bounds exist, and hardly any sequential functionals are continuous with respect to all of them.

We will here prove two theorems from [221]: the failure of chain completeness at a second-order type, and the (related) existence of non-standard least upper bounds which are not always preserved.

Theorem 7.6.2. *There is a sequence $F_0, F_1, \ldots$ of finite sequential functionals of type*

$$\sigma = (\mathtt{B}, \mathtt{N} \to \mathtt{B}) \to \mathtt{B}$$

that is $\preceq$-increasing, but that has no upper bound in $\mathsf{SF}(\sigma)$.

Proof. We will give an explicit construction of a sequence $f_0, f_1, \ldots$ of finite sequential functions of type $\mathtt{B}, \mathtt{N} \to \mathtt{B}$, and we will show that for each n there is a sequential functional F_n with the following properties:

1. $F_n(f_0) = tt$.
2. $F_n(f_m) = ff$ if $0 < m \leq n$.
3. $F_n(f_m) = \bot$ if $m > n$.

It follows easily by Lemma 7.5.2(ii) that for each n there is a $\preceq$*-smallest* F_n satisfying these properties, and these minimal F_n will clearly form an increasing chain. Moreover, we will show that there is no $F \in \mathsf{SF}(\sigma)$ sending f_0 to tt and all other f_m to ff.

The functions f_m are defined as follows:

- Let $f_0(a,b) = tt$ if $a = tt$, and $f_0(a,b) = \bot$ otherwise.
- If $m > 0$, set $f_m(a,b) = ff$ if $b = m$ or if $b < m$ and $a = ff$. Set $f_m(a,b) = \bot$ otherwise.

We now define the functionals F_n:

- Let $F_0(f) = tt$ if $f(tt, \bot) = tt$, and $F_0(f) = \bot$ otherwise.
- If $n > 0$, set

$$F_n(f) = f(f(\cdots f(f(F_{n-1}(f), n), n-1), \ldots), 0) .$$

It is easy to see that all the f_m, F_n are indeed sequential functionals.

We prove by induction on n that the above conditions 1 and 2 hold and that $F_n(f_m) \in \mathbb{B}$ iff $m \leq n$. For $n = 0$ this is easy, so assume the claim holds for $n-1$; we will show that it holds for n.

- By direct calculation and the induction hypothesis, we have $F_n(f_0) = tt$.

- Let $0 < m \leq n$. Define $G_n(f,n) = f(F_{n-1}(f),n)$ and $G_n(f,i) = f(G_n(f,i+1),i)$ for $0 \leq i < n$, so that $F_n(f) = G_n(f,0)$. Then $G_n(f_m,m) = \mathit{ff}$, whence $G_n(f_m,i) = \mathit{ff}$ for all $i \leq m$ by reverse induction on i, so $F_n(f_m) = \mathit{ff}$.
- Let $m > n$. Then $F_{n-1}(f_m) = \bot$ by the induction hypothesis, and $f_m(\bot,i) = \bot$ for all $i \leq m$, whence $F_n(f_m) = \bot$.

One can actually show that $\{f_m \mid m \leq n\}$ will be a basis for F_n, but this is not required for our proof.

We shall conclude by showing:

- If $F \in \mathsf{SF}(\mathtt{B},\mathtt{N} \to \mathtt{B}) \to \mathtt{B}$ and F terminates on f_0 and on infinitely many f_m with $m > 0$, then F is a constant functional $\lambda f.a$.

We show this by working at the level of sequential procedures. Suppose $\lambda f.e$ is a procedure representing some F satisfying the hypothesis of the above claim, and suppose p_m is some procedure representing f_m for each m. Then the evaluation of $e[f \mapsto p_m]$ terminates for $m = 0$ and for infinitely many other m.

We argue by induction on the length of the evaluation of $e[f \mapsto p_0]$. If the evaluation has length 0, then e is a constant a as required. Otherwise, e must be of the form

$$\mathtt{case}\ f(d_0,d_1)\ \mathtt{of}\ (i \Rightarrow e_i)\ ,$$

and using Lemma 6.1.13 we must have that $d_0[f \mapsto p_0]$ evaluates to tt via some shorter evaluation.

Now consider any $m > 0$ such that the evaluation of $e[f \mapsto p_m]$ terminates. Then we must have that $d_1[f \mapsto p_m]$ terminates. Now we observe that if $m > 0$ then the only proper value in the range of f_m is ff, and it clearly follows that if $m,m' > 0$ then $d_1[f \mapsto p_m]$, $d_1[f \mapsto p_{m'}]$ have the same value if both are defined. Thus, there is at most one $m > 0$ with $\ll d_1[f \mapsto p_m] \gg = m$, and for any other $m > 0$ with $F(f_m) \in \mathbb{B}$, we must have that $\ll d_0[f \mapsto p_m] \gg = \mathit{ff}$. Note that there are infinitely many such m.

We have thus shown that $\ll d_0[f \mapsto p_0] \gg = \mathit{tt}$ and that $\ll d_0[f \mapsto p_m] \gg = \mathit{ff}$ for infinitely many $m > 0$. But this contradicts our induction hypothesis, so the claim is proved. □

This theorem refutes a statement that has sometimes been regarded as a folklore result, namely that $(\mathsf{SF}(\sigma), \preceq)$ is a complete partial order for all types σ of level 2. This does indeed hold for the *pure* type $\overline{2}$, though even here a rather careful argument is necessary (see [221]). The folklore argument (which we will give as the proof of Proposition 11.4.4) implicitly assumes our first-order arguments f are always *strict* functions, so that we do not have $f(\bot) = 7$, for example. We also note in passing that the second author has recently shown that $\mathsf{SF}((\mathtt{N} \to \mathtt{N})^{(r)} \to \mathtt{N})$ is chain complete for all $r \in \mathbb{N}$; this result has not yet been published.

We now obtain an easy consequence of the above proof:

Corollary 7.6.3. *Let F be a constant functional of type $(\mathtt{B},\mathtt{N} \to \mathtt{B}) \to \mathtt{B}$. Then F is the least upper bound in SF of a strictly increasing sequence of finite sequential functionals.*

Proof. Let the constant value be a. Let f_m be as in the proof of Theorem 7.6.2, and let F_n be the least sequential functional such that $F_n(f_m) = a$ for all $m \leq n$ (this exists by virtue of Lemma 7.5.2(ii)). Then $F_n \preceq F$. Moreover, the sequence $F_0, F_1, \ldots$ is strictly increasing, since $F_n(f_m) = \bot$ when $m > n$. Finally, the proof of Theorem 7.6.2 shows that F is the least upper bound of this sequence. □

Now suppose we define $\Phi \in \mathsf{SF}(((\mathtt{B},\mathtt{N} \to \mathtt{B}) \to \mathtt{B}) \to \mathtt{U})$ by

$$\Phi(G) = \top \text{ iff } G(\bot_{\mathtt{N},\mathtt{N}\to\mathtt{N}}) = a\,.$$

Then Φ is a finite sequential functional, but is discontinuous with respect to the sequences with least upper bounds as considered in the corollary. This reinforces the message that standard domain-theoretic concepts are often inappropriate in the setting of SF. In order to gain a better understanding of this important structure, approaches such as those initiated in [84] and [253, 254] should be pursued further.

Remark 7.6.4. In unpublished work, Plotkin proved that the space of sequential functions from one concrete domain to another forms a directed-complete partial ordering. (Concrete domains were introduced in [128].) This can be seen as the most general (correct) version of what we referred to above as a folklore result.

In fact, not even $\mathsf{SF}(\mathtt{N} \to \mathtt{U})$ will be a concrete domain, since we allow non-strict functions f in our model. When f is a constant function $\lambda x.a$, f will be a compact element with infinitely many compacts below it, which is impossible in a concrete domain. Note also that many of our key examples from Section 7.3 (such as the fan functional) make crucial use of the possibility of such non-strict inputs. Thus, there are important phenomena arising in the context of PCF that cannot be accounted for using concrete domains.

7.7 Absence of a Universal Type

We conclude this chapter by showing that the model SF possesses no *universal type* in the sense of Subsection 4.2.5: that is, there is no type σ such that every $\mathsf{SF}(\tau)$ is a retract of $\mathsf{SF}(\sigma)$. (It follows immediately that SP^0 has no universal type either.) This contrasts strikingly with the fact that many other simply typed structures that feature later in the book do possess universal types (e.g. PC, SR, SA).

The non-existence of a universal type in the game model $\mathsf{G}_{ib} \cong \mathsf{SP}^0$ (see Section 7.4) has been a folklore result since the 1990s. We give here an argument that suffices also for SF, presented within the framework of NSPs. We rely heavily here on ideas and notation from Section 6.1.

Theorem 7.7.1. SF *contains no universal type in the sense of Definition 4.2.20.*

Proof. By results of Subsection 4.2.1, it will be enough to show that there is no pure type $\overline{k}$ such that $\overline{k+1}$ is a retract of $\overline{k}$ within the category $\mathscr{C}(\mathsf{SF})$ of Subsection 4.1.3. We shall argue by induction on k, using a somewhat stronger statement than this as our induction claim. If $X,Y \in \mathbf{K}(\mathsf{SF})$ where X corresponds to some simple type $\sigma \in \mathrm{T}^{\rightarrow}(\mathtt{N})$, we shall call X a *pseudo-retract* of Y if there are morphisms $t : X \to Y$ and $r : Y \to X$ in $\mathscr{K}(\mathscr{C}(\mathsf{SF}))$ such that $r \circ t \succeq id$ in the pointwise preorder on $\sigma \to \sigma$ (by Exercise 6.1.21, this coincides with the observational preorder). Taking advantage of the presence of product types in $\mathbf{K}(\mathsf{SF})$, we shall in fact show by induction on k that $\overline{k+1}$ is not a pseudo-retract of $\overline{k}^d$ for any $d \in \mathbb{N}$; in particular, it is not a pseudo-retract of $\overline{k}$, implying that $\overline{k+1}$ is not a retract of $\overline{k}$ within $\mathscr{C}(\mathsf{SF})$.

For the case $k = 0$, suppose t,r exhibit $\overline{1}$ as a pseudo-retract of $\overline{0}^d$. Then for any *maximal* element $f \in \mathsf{SP}(1)$ (representing a total function $\mathbb{N} \to \mathbb{N}$) we must have $(r \circ t)(f) = f$; but this is impossible, since the set of such maximal elements has larger cardinality than the set of elements of $\overline{0}^d$.

Now assume the result for $k-1$, and suppose for contradiction that t,r exhibit $\overline{k+1}$ as a pseudo-retract of $\overline{k}^d$. We shall regard $\overline{k}^d$ itself as presented as a retract of $\overline{0} \to \overline{k}$ in an obvious way, and henceforth write $t \in \mathsf{SP}^0((k+1) \to (0 \to k))$ and $r \in \mathsf{SP}^0((0 \to k) \to (k+1))$ for the underlying procedures that witness the pseudo-retraction; note that $r \circ t \succeq id_{k+1}$. Let z be a free variable of type $\overline{k+1}$, and set $u = \ll tz \gg \in \mathsf{SP}(0 \to k)$ and $v = \ll ru \gg \in \mathsf{SP}(k+1)$. Then $v = \ll r(tz) \gg$ by Theorem 6.1.7, so that $\ll v[z \mapsto w] \gg \succeq w$ for any $w \in \mathsf{SP}^0(k+1)$.

We first check that any v with this latter property must have the syntactic form $\lambda f^k.\mathtt{case}\ zp\ \mathtt{of}\ (\cdots)$ for some p of type $\overline{k}$. Indeed, it is clear that v does not have the form $\lambda f.n$ or $\lambda f.\bot$, and the only other alternative form is $\lambda f.\mathtt{case}\ fp'\ \mathtt{of}\ (\cdots)$. In that case, however, we would have $\ll v[z \mapsto \lambda x^k.0] \gg \cdot (\lambda y^{k-1}.\bot) = \bot$, contradicting $\ll v[z \mapsto \lambda x^k.0] \gg \cdot (\lambda y^{k-1}.\bot) \succeq (\lambda x.0)(\lambda y.\bot) = 0$.

We now focus on the subterm p in $v = \lambda f.\mathtt{case}\ zp\ \mathtt{of}\ (\cdots)$. The general direction of our argument will now be to show that $\lambda f^k.p$ represents a function of type $\overline{k} \to \overline{k}$ that dominates the identity, and that moreover our two-stage construction of v via u can be used to split this into functions $\overline{k} \to \overline{k-1}^c$ and $\overline{k-1}^c \to \overline{k}$ for some c, contradicting the induction hypothesis. An apparent obstacle to this plan is that z as well as f may appear free in p; however, it turns out that we still obtain all the properties we need if we specialize z (somewhat arbitrarily) to $\lambda x.0$.

Specifically, we claim that $\ll p[f \mapsto q, z \mapsto \lambda x.0] \gg \succeq q$ for any $q \in \mathsf{SP}^0(k)$. For suppose that $q \cdot s = n \in \mathbb{N}$ whereas $\ll p[f \mapsto q, z \mapsto \lambda x.0] \gg \cdot s \neq n$ for some $s \in \mathsf{SP}^0(k-1)$. Take $w = \lambda g.\mathtt{case}\ gs\ \mathtt{of}\ (n \Rightarrow 0)$, so $w \cdot q' = \bot$ whenever $q' \cdot s \neq n$. Then $w \preceq \lambda x.0$ pointwise, so we have $\ll p[f \mapsto q, z \mapsto w] \gg \cdot s \neq n$ by Exercise 6.1.21. By the definition of w, it follows that $\ll (zp)[f \mapsto q, z \mapsto w] \gg = \bot$, and hence that $\ll v[z \mapsto w] \gg \cdot q = \bot$, whereas $w \cdot q = 0$, contradicting $\ll v[z \mapsto w] \gg \succeq w$. We have thus shown that $\lambda f. \ll p[z \mapsto \lambda x.0] \gg \succeq id_k$.

We next show how to split the function represented by this procedure so as to go through some $\overline{k-1}^c$. Since $\ll ru \gg = \lambda f.\mathtt{case}\ zp\ \mathtt{of}\ (\cdots)$, we have that ru reduces in finitely many steps to a head normal form $\lambda f.\mathtt{case}\ zP\ \mathtt{of}\ (\cdots)$ where $\ll P \gg = p$. By working backward through this reduction sequence, we may locate the ancestor within ru of this head occurrence of z. Since r is closed, this occurs within u, and clearly it must appear as the head of some subterm $\mathtt{case}\ zP'\ \mathtt{of}\ (\cdots)$ where P is a substitution instance of P'. Now since u has type $\overline{0} \to \overline{k}$ and $z : \overline{k+1}$ is its only free variable, it is easy to see that all *bound* variables within u have pure types of level less than k. Let $h_0,\ldots,h_{c-1}$ denote the bound variables that are in scope at the relevant occurrence of zP', and suppose each h_i has type $\overline{k_i}$ where $k_i < k$. By considering the form of the head reduction sequence $ru \rightsquigarrow_h^* \lambda f.\mathtt{case}\ zP\ \mathtt{of}\ (\cdots)$, we see that $P = P'[h_0 \mapsto H_0,\ldots,h_{c-1} \mapsto H_{c-1}]$ where each $H_i : \overline{k_i}$ contains at most f and z free.

Again taking advantage of product types in $\mathbf{K}(\mathsf{SP}^0)$, and writing * for the substitution $[z \mapsto \lambda x.0]$, define morphisms

$$t' = \lambda f^k.\, \langle H_0^*,\ldots,H_{c-1}^*\rangle : \overline{k} \to (\overline{k_0} \times \cdots \times \overline{k_{c-1}}) \,,$$
$$r' = \lambda \langle h_0,\ldots,h_{c-1}\rangle .\, P'^* \ : (\overline{k_0} \times \cdots \times \overline{k_{c-1}}) \to \overline{k} \,.$$

Then $r' \circ t'$ is represented by the term $\ll \lambda f.P^* \gg = \lambda f. \ll p^* \gg$, which dominates the identity as shown above. Thus $\overline{k}$ is a pseudo-retract of $\overline{k_0} \times \cdots \times \overline{k_{c-1}}$, and hence of $\overline{k-1}^c$, which contradicts the induction hypothesis. So $\overline{k+1}$ is not a pseudo-retract of $\overline{k}^d$ after all, and the proof is complete. $\square$

It follows immediately that $\mathscr{C}(\mathsf{SP}^0)$ has no universal type. One may also prove this fact directly by means of a simpler version of the above argument, taking as our induction claim the statement that $\overline{k}$ is not a retract of $\overline{k+1}$ in the ordinary sense. The outworking of the details forms an interesting exercise.

We conclude by mentioning a recent result of Longley [185], closely related to the above:

Theorem 7.7.2. *For $k \in \mathbb{N}$, let* PCF_k *denote the sublanguage of* PCF *in which the fixed point operators Y_σ are admitted only for types σ of level $\leq k$. Then the expressive power of* PCF_k *(for denoting elements of* SF*) increases strictly with k.*

This answers a question raised by Berger in [22]. The proof, which is too long to include here, involves the identification of a substructure of SP^0 (in the manner of Section 6.3) which contains a representative of Y_σ for every σ of level $\leq k$, but not for every σ of level $k+1$. Theorem 7.7.1 also serves as a key ingredient in the proof.

An easy but interesting consequence of Theorem 7.7.2 is that there can be no finite set $B \subset \mathsf{SF}^{\mathrm{eff}}$ such that all elements of $\mathsf{SF}^{\mathrm{eff}}$ are definable in simply typed λ-calculus relative to B. If there were, then for large enough k, the language PCF_k would suffice for defining all elements of B and hence of $\mathsf{SF}^{\mathrm{eff}}$. Again, this contrasts with the situation for those models in which some low type is universal, so that some finite set B suffices for full definability.

Chapter 8
The Total Continuous Functionals

In this chapter and the next, we turn our attention again to type structures of *total* functionals over $\mathbb{N}$. From the point of view of computability theory, by far the most important such structure is the model Ct of *total continuous functionals*, which forms the subject of the present chapter.

We have already seen in Subsection 3.2.1 how the notion of continuity captures important aspects of computability, and the model Ct may be broadly described as the canonical class of 'continuous' higher-order total functionals over $\mathbb{N}$. Of course, this needs to be turned into a precise definition, but it turns out that all reasonable ways of doing so yield the same type structure, indicating that this is indeed a robust and natural class of functionals to consider.

As mentioned in Subsection 2.3.1, the total continuous functionals were first introduced in 1959 by Kleene [140] and Kreisel [152], using different but essentially equivalent constructions.[1] Kleene's approach amounts to constructing Ct as an extensional collapse of his model K_2, which can be seen as a prototypical model encapsulating the continuous character of finitary computations (see Subsection 3.2.1). Kreisel's approach was to build in a notion of continuity via a system of *formal neighbourhoods*, which in some sense correspond to finite pieces of information about functionals, and then to construct the functionals themselves as ideal 'limits' of suitable systems of neighbourhoods. The equivalence between these approaches, though not trivial to prove, was recognized at the time of discovery and was noted in both the original papers; as a result, the functionals in question are often known as the *Kleene–Kreisel functionals*. Research on this model flourished in the 1970s, with work by Bergstra, Ershov, Gandy, Grilliot, Hyland, Normann, Wainer and others yielding a variety of topological and other characterizations, as well as investigating the rich computability theory that these functionals call for. The state of the field in 1980 was described in detail in Normann's book [207]. Interest in Ct was renewed in the 1990s with further work by Berger, Normann, Plotkin and others; more recent contributions include those of Longley [179] and Escardó [80].

[1] Kleene's model was originally presented as a substructure of the full set-theoretic model S, but his approach equally well yields a definition of the standalone model Ct, which with hindsight appears more natural.

It is helpful to note at the outset that the known characterizations of Ct fall broadly into two classes. On the one hand, we may directly define each $\mathsf{Ct}(\sigma \to \tau)$ as the set of functions $\mathsf{Ct}(\sigma) \to \mathsf{Ct}(\tau)$ that are 'continuous' with respect to some topological or quasi-topological structure with which $\mathsf{Ct}(\sigma)$ and $\mathsf{Ct}(\tau)$ may be endowed. Typically, this amounts to defining Ct as the type structure generated from $\mathbb{N}$ by exponentiation in some category of 'topological spaces' or similar, e.g. the category of *compactly generated Hausdorff spaces*, or that of *limit spaces*. (Note that the usual category of topological spaces is not itself cartesian closed.) Characterizations of this sort often have the virtue of being mathematically natural, though they do not usually manifest the sense in which these functionals may be regarded as abstractly 'computable'. An alternative approach, therefore, is to start with some other structure **A** which we already accept as a model of abstractly 'computable' operations, and then arrive at Ct indirectly as the class of total functionals representable or realizable in **A**. (It is only natural that such a two-stage process is called for if we are seeking a class of *total* 'computable' functionals, since general models of computation unavoidably generate partial as well as total operations.) Kleene's construction of Ct as an extensional collapse of K_2 is a typical example of this approach. Whilst such constructions typically capture the computational character of the functionals well, they are prima facie somewhat uncanonical insofar as they refer to a particular choice of representing structure **A**. What we shall discover as we proceed, however, is that the resulting total type structure is to a remarkable extent unaffected by this choice.

Of course, definitions based on continuity offer only a rough approximation to realistic notions of effective computability, so it is natural also to ask how more genuinely effective notions might be defined in the context of Ct. Here again, one may distinguish broadly between two approaches. On the one hand, we can investigate ways of computing directly with the elements of Ct regarded abstractly as functions, without reference to any particular way of constructing them. We have seen in Chapter 5 that this is precisely what *Kleene (S1–S9) computability* offers us, so we are naturally led to consider this notion as it applies to Ct. Once again, the resulting class of computable functionals is in some sense 'canonical' in that it is defined purely in terms of the intrinsic structure of Ct. On the other hand, if we consider Ct in relation to some representing structure **A**, any reasonable notion of 'computable element' in **A** will induce one for Ct. Here too, there is a prima facie concern that the resulting substructure of Ct might turn out to be sensitive to the choice of **A** or to the notion of computability for **A** that is adopted—although once again, what we shall find is that the substructure we obtain is remarkably stable under such variations.

From the perspective of the present book, approaches to Ct via representing structures appear very natural as instances of a generic situation: we are dealing with simulations $\mathsf{Ct} \longrightarrow\!\!\!\!\triangleright\, \mathbf{A}$, and a computable substructure $\mathbf{A}^\sharp$ of **A** will naturally induce a substructure of Ct. In the 1970s, however, the focus was largely on the specific case $\mathbf{A} = K_2$, $\mathbf{A}^\sharp = K_2^{\mathrm{eff}}$, and much of the driving force for work in the area was the desire to understand the relationship between Kleene computability over Ct and the computability notion induced by K_2^{eff}. These two styles of computability were respectively dubbed 'internal' and 'external' by Normann in [210]; later, Tucker and Zucker used the terms 'abstract' and 'concrete' to express a similar distinction [290, 291]. Not surprisingly, every internally computable element of Ct turns out to be externally computable: indeed, any abstract computation with functionals can be 'simulated' at the concrete level. However, an early result of Tait [284] shows that the *fan functional* is externally computable (say in the K_2 sense) but is not Kleene computable (cf. Subsection 6.4.3). We thus have two candidates for a notion of computable element in Ct, both of which turn out to be worthy of study.

The present chapter is structured as follows. In Section 8.1, we define the model Ct as the extensional collapse of the Scott–Ershov model PC from Subsection 3.2.2. This is closely related to Kreisel's original definition via neighbourhoods, and with the benefit of hindsight, it is the choice of construction that gives the smoothest introduction to Ct. We develop some of the fundamental theory, including several versions of the crucial *density theorem* and some basic topological results, and introduce both internal and external notions of computability on Ct.

In Section 8.2, we consider a variety of alternative characterizations of Ct and establish their equivalence. These include topological or quasi-topological characterizations such as the one via limit spaces, as well as characterizations via representing structures such as K_2. We also mention a number of other characterizations without giving full proofs of the equivalences. In Section 8.3, we mention some interesting and important examples of total continuous functionals, most of which have already been considered from the perspective of PCF in Chapter 7.

The remaining sections are devoted to more advanced material. In Section 8.4, we investigate some technical consequences of the existence of *dense enumerations* within Ct as provided by the density theorem: these include a classical result of Kreisel on the relationship between Ct and the analytic hierarchy, as well as some more recent results of Escardó on the existence of search operators for certain subsets of Ct. In Section 8.5, we study the 'internal' computability theory of Ct, including both Kleene computability and the weaker notion of μ-computability. Here we give Tait's proof that the fan functional is not Kleene computable, and more generally investigate questions of relative computability within Ct.

Finally, in Section 8.6, we consider various 'external' definitions of computability on Ct and the relationships between them. Among other results, we prove a theorem of Normann [212] showing that, with reference to the construction of Ct from PC, the induced computable substructure of Ct is unaffected if we weaken the notion of computability on PC from $\mathsf{PC}^{\mathrm{eff}}$ to the class of PCF-definable elements. This, along with other results from Section 8.2, provides cumulative evidence that $\mathsf{Ct}^{\mathrm{eff}}$ is indeed a robust class of 'computable' functionals. However, there is a sense in which we can do even better than this: some general results of Longley [179] show that *any* representing model **A** satisfying certain mild conditions yields Ct as its extensional collapse, and any reasonable computable substructure $\mathbf{A}^\sharp$ of **A** then induces $\mathsf{Ct}^{\mathrm{eff}}$. We state these theorems in Subsection 8.6.2, the proofs being too long for inclusion in this book. The fact that Ct arises as the natural class of 'total' functionals within so many other models of interest is in itself a powerful motivation for studying this type structure in depth.

8.1 Definition and Basic Theory

We start with a definition of Ct using the Scott–Ershov model PC as our representing model; this approach is essentially due to Ershov [72, 74]. We also develop some of the basic structural theory, and introduce the concepts of computability with which we will be working.

We will work mostly over the type world $\mathsf{T}^{\rightarrow}(\mathtt{N})$. Following the definitions of Subsection 3.2.2, but dispensing with product and unit types, we shall regard PC and its effective submodel $\mathsf{PC}^{\mathrm{eff}}$ as models over $\mathsf{T}^{\rightarrow}(\mathtt{N} = \mathbb{N}_{\bot})$. On the few occasions when we wish to refer to product types, we shall write $\mathsf{PC}^{\times}, \mathsf{PC}^{\mathrm{eff}\times}$ for the corresponding models over $\mathsf{T}^{\rightarrow\times}(\mathtt{N} = \mathbb{N}_{\bot})$. As in earlier chapters, we will often write simply k for the pure type $\overline{k}$.

Recall that $\mathsf{PC}(\sigma \rightarrow \tau)$ is the set of monotone and continuous functions $\mathsf{PC}(\sigma) \rightarrow \mathsf{PC}(\tau)$ equipped with the pointwise ordering. Most of what we need as regards PC is contained in Subsection 3.2.2. In particular, we recall the following grammar of formal expressions for denoting compact elements of PC:

$$e ::= \bot \mid n \mid (e_0 \Rightarrow e'_0) \sqcup \cdots \sqcup (e_{r-1} \Rightarrow e'_{r-1}) \,.$$

Here $\bot$ and n denote elements of $\mathsf{PC}(\mathtt{N})$; $\sqcup$ is interpreted by joins when these exist; and $\Rightarrow$ is interpreted by the *step function* construction on compact elements

$$(a \Rightarrow b)(x) \;=\; \begin{cases} b & \text{if } x \sqsupseteq a\,, \\ \bot & \text{otherwise.} \end{cases}$$

We will also need to be a little more explicit about what we consider to be a 'legal' expression. By simultaneous induction on σ, we define a set of *valid expressions* of type σ and a binary *consistency* relation between such expressions:

- At type N, the expressions $\bot$ and n are valid, and e, e' are consistent iff $e = \bot$ or $e' = \bot$ or $e = e'$.
- If $e_0, \dots, e_{r-1}$ are valid at type σ and $e'_0, \dots, e'_{r-1}$ are valid at type τ, the expression $(e_0 \Rightarrow e'_0) \sqcup \cdots \sqcup (e_{r-1} \Rightarrow e'_{r-1})$ is valid at type $\sigma \rightarrow \tau$ iff for all $i, j < r$, either e'_i, e'_j are consistent or e_i, e_j are inconsistent. Moreover, d and d' are consistent at type $\sigma \rightarrow \tau$ iff the evident join $d \sqcup d'$ is valid at type $\sigma \rightarrow \tau$.

It is easy to verify that e denotes a compact element $[\![e]\!] \in \mathsf{PC}(\sigma)$ iff e is valid at type σ. It is also easy to check that validity and consistency are decidable properties of formal expressions (cf. Subsection 10.1.1).

We will write $\ulcorner e \urcorner$ for the Gödel number of e with respect to some fixed coding.

8.1.1 Construction via PC

Using the extensional collapse construction from Definition 3.6.12, and noting that $\mathbb{N} \subseteq \mathbb{N}_{\bot}$, we may define Ct simply as $\mathsf{EC}(\mathsf{PC}; \mathbb{N})$. Explicitly, we have a partial equivalence relation $\approx_\sigma$ on each $\mathsf{PC}(\sigma)$ given by

$$x \approx_{\mathtt{N}} y \ \text{ iff } \ x = y \in \mathbb{N}\,,$$
$$f \approx_{\sigma \to \tau} g \ \text{ iff } \ \forall x, y \in \mathsf{PC}(\sigma).\, x \approx_\sigma y \Rightarrow f(x) \approx_\tau g(y)\,.$$

We then let $\mathsf{Ct}(\sigma)$ be the set of equivalence classes for $\approx_\sigma$, and endow Ct with the application operations inherited from PC. For the general reasons explained in Chapter 3, it is immediate that Ct is an extensional TPCA over $\mathsf{T}^{\to}(\mathtt{N} = \mathbb{N})$. We shall refer to the elements of Ct as the *total continuous functionals*.

When we wish to refer to product types, we may define $\mathsf{Ct}^\times = \mathsf{EC}^\times(\mathsf{PC}^\times; \mathbb{N})$ as in Definition 3.6.12. Although we shall not labour the point, all of our main results for Ct also go through for $\mathsf{Ct}^\times$, either via simple adaptations of the proofs or by appealing to general results from Chapter 4.

Our construction of Ct yields a canonical type-respecting simulation $\mathsf{Ct} \longrightarrow\!\!\triangleright\, \mathsf{PC}$. If $x \in \mathsf{PC}(\sigma)$ and $X \in \mathsf{Ct}(\sigma)$, we write $x \Vdash_\sigma X$ (often omitting the subscript) to mean that x is a representative of the equivalence class X. (When elements of both PC and Ct are in play, we will generally use lowercase letters for the former and uppercase for the latter.) As in Subsection 3.6.3, we will also write $\mathsf{Ext} = \mathsf{Ext}(\mathsf{PC}; \mathbb{N})$ for the substructure of PC consisting of the elements x with $x \approx x$. Note that Ct is a quotient of Ext.

The effective submodel $\mathsf{PC}^{\mathrm{eff}} \subseteq \mathsf{PC}$ induces a submodel $\mathsf{Ct}^{\mathrm{eff}} \subseteq \mathsf{Ct}$ in the obvious way: we take $X \in \mathsf{Ct}^{\mathrm{eff}}$ iff there exists some $x \in \mathsf{PC}^{\mathrm{eff}}$ with $x \Vdash X$. Clearly $\mathsf{Ct}^{\mathrm{eff}}$ is closed under application, and it is immediate for general reasons that $(\mathsf{Ct}; \mathsf{Ct}^{\mathrm{eff}})$ is a relative TPCA. The elements of $\mathsf{Ct}^{\mathrm{eff}}$ are variously referred to in the literature as the *recursive(ly) continuous* or *recursive(ly) countable* functionals. It is not immediate that $\mathsf{Ct}^{\mathrm{eff}}$ is an extensional model in its own right, although we will see in Subsection 8.1.2 that this is indeed the case.

Some simple properties of the $\approx$-equivalence classes may be observed at this stage. Recall that each $\mathsf{PC}(\sigma)$ has a binary infimum operation $\sqcap$ which may be defined pointwise.

Lemma 8.1.1. *(i) If $x \in \mathsf{Ext}(\sigma)$ and $x \sqsubseteq y$, then $y \in \mathsf{Ext}(\sigma)$ and $y \approx_\sigma x$.*
(ii) If $x, y \in \mathsf{PC}(\sigma)$, then $x \approx_\sigma y$ iff $x \sqcap y \in \mathsf{Ext}(\sigma)$.

Proof. We leave (i) as an easy exercise. For (ii), we reason by induction on σ. The base case for $\mathtt{N}$ is trivial, so let us assume (ii) holds for σ and τ, and consider $f, g \in \mathsf{PC}(\sigma \to \tau)$.

First suppose $f \approx_{\sigma \to \tau} g$. To show that $f \sqcap g \in \mathsf{Ext}(\sigma \to \tau)$, it will suffice to show that $(f \sqcap g)(z) \approx_\tau f(z)$ for all $z \in \mathsf{Ext}(\sigma)$, since for any $z \approx_\sigma w$ we will then have $(f \sqcap g)(z) \approx_\tau f(z) \approx_\tau f(w) \approx_\tau (f \sqcap g)(w)$. But if $z \approx_\sigma z$ then $f(z) \approx_\tau g(z)$, so by the induction hypothesis for τ we have that $f(z) \sqcap g(z) \in \mathsf{Ext}(\tau)$, whence $(f \sqcap g)(z) = f(z) \sqcap g(z) \approx_\tau f(z)$ using (i).

Conversely, suppose $f \sqcap g \in \mathsf{Ext}(\sigma \to \tau)$. To show $f \approx_{\sigma \to \tau} g$, suppose $x \approx_\sigma y$. By the induction hypothesis for σ, $x \sqcap y \in \mathsf{Ext}(\sigma)$ so $(f \sqcup g)(x \sqcap y) \in \mathsf{Ext}(\tau)$. But now $f(x) \approx_\tau (f \sqcap g)(x \sqcap y) \approx_\tau g(y)$ using (i). $\square$

As a consequence, we may obtain a simpler characterization of Ext. Recall from Subsection 3.6.3 that we may define a substructure $\mathsf{Tot} = \mathsf{Tot}(\mathsf{PC};\mathbb{N})$ by

$$\begin{aligned} \mathsf{Tot}(\mathtt{N}) &= \mathbb{N}\,, \\ \mathsf{Tot}(\sigma \to \tau) &= \{f \in \mathsf{PC}(\sigma \to \tau) \mid \forall x \in \mathsf{Tot}(\sigma).\, f(x) \in \mathsf{Tot}(\tau)\}\,. \end{aligned}$$

In effect, this is the lifting of a unary logical relation on $\mathsf{PC}(\mathtt{N})$, whereas the relations $\approx_\sigma$ are the lifting of a binary one. In general there is no reason why these constructions should match up, but in the present case they do so:

Corollary 8.1.2. $\mathsf{Tot}(\mathsf{PC};\mathbb{N}) = \mathsf{Ext}(\mathsf{PC};\mathbb{N})$.

Proof. By induction on types. The case of $\mathtt{N}$ is trivial, so suppose the claim holds for σ and τ and consider $f \in \mathsf{PC}(\sigma \to \tau)$.

If $f \in \mathsf{Ext}(\sigma \to \tau)$, then for any $x \in \mathsf{Tot}(\sigma)$ we have $x \approx_\sigma x$ by the induction hypothesis for σ, whence $f(x) \approx_\tau f(x)$, so $f(x) \in \mathsf{Tot}(\tau)$ by the hypothesis for τ. Thus $f \in \mathsf{Tot}(\sigma \to \tau)$. Conversely, if $f \in \mathsf{Tot}(\sigma \to \tau)$, then for any $x \approx_\sigma y$ we have $x \sqcap y \in \mathsf{Tot}(\sigma)$ by Lemma 8.1.1(ii) and the hypothesis on σ; hence $f(x \sqcap y) \in \mathsf{Tot}(\tau)$ so $f(x \sqcap y) \in \mathsf{Ext}(\tau)$ by the hypothesis on τ. But now $f(x) \approx_\sigma f(x \sqcap y) \approx_\tau f(y)$ by Lemma 8.1.1(i) and the monotonicity of f. Thus $f \in \mathsf{Ext}(\sigma \to \tau)$. □

Corollary 8.1.2 is due originally to Ershov [72], but the above simple proof via Lemma 8.1.1 is due to Longo and Moggi [188].

In the terminology of Subsection 3.6.3, we have shown that the hereditarily total and hereditarily extensional elements of PC coincide, and it follows that the *modified extensional collapse* $\mathsf{MEC}(\mathsf{PC};\mathbb{N})$ coincides exactly with Ct. For brevity, we will henceforth refer to the elements of Tot or Ext as the *total* elements in PC.

8.1.2 The Density Theorem

To make further progress in understanding the structure of Ct and its relationship to PC, we require a fundamental tool known as the *density theorem*. One version of this says that every compact element $a \in \mathsf{PC}$ can be extended to a total element $\widehat{a} \sqsupseteq a$. We will also see that a version of the theorem can be framed purely in terms of Ct, without any reference to PC.

We shall first establish the density theorem for type structures over the *finite* base types $\{0,\ldots,n\}_\bot = \{\bot,0,\ldots,n\}$, then transfer it to the original model with the help of some standard retractions $\{0,\ldots,n\}_\bot \lhd \mathbb{N}_\bot$ (cf. Section 4.2). Specifically, for any n we may define an analogue PC_n of PC by

$$\begin{aligned} \mathsf{PC}_n(\mathtt{N}) &= \{0,\ldots,n\}_\bot\,, \\ \mathsf{PC}_n(\sigma \to \tau) &= \{f : \mathsf{PC}_n(\sigma) \to \mathsf{PC}_n(\tau) \mid f \text{ monotone}\}\,. \end{aligned}$$

Since all these sets are finite, it is automatic that each $f \in \mathsf{PC}_n(\sigma \to \tau)$ is continuous. Just as in the previous subsection, we may now define $\mathsf{Ct}_n = \mathsf{EC}(\mathsf{PC}_n;\{0,\ldots,n\})$

and $\mathsf{Ext}_n = \mathsf{Ext}(\mathsf{PC}_n;\{0,\ldots,n\})$; we also write $\approx_{n,\sigma}$ for the corresponding partial equivalence relation on $\mathsf{PC}_n(\sigma)$. The proof of Lemma 8.1.1 goes through for the $\approx_{n,\sigma}$, and since $\mathsf{PC}_n(\sigma)$ is finite, we may conclude that each equivalence class has a least element.

The next lemma establishes the density theorem for the PC_n.

Lemma 8.1.3. *The following hold for each type σ and each $n \in \mathbb{N}$:*
(i) Every $x \in \mathsf{PC}_n(\sigma)$ may be extended to some $\widehat{x} \in \mathsf{Ext}_n(\sigma)$.
(ii) If $x \approx_{n,\sigma} y$ then x,y have a least upper bound $x \sqcup y$ in $\mathsf{PC}_n(\sigma)$.
(iii) Every $\approx_{n,\sigma}$ equivalence class has a maximal element.

Proof. We prove all three statements simultaneously by induction on σ. In the course of doing so, we will define a particular choice of operation $x \mapsto \widehat{x}$ to which we shall refer later on.

At type $\mathtt{N}$, all three claims are trivial; here, for definiteness, we set $\widehat{\bot} = 0$. It is also trivial at each type σ that (iii) is a consequence of (ii). So let us assume the lemma holds for σ and τ, and establish (i) and (ii) for $\sigma \to \tau$.

For (i), suppose $f \in \mathsf{PC}_n(\sigma \to \tau)$, let $y_0,\ldots,y_{t-1}$ be the maximal elements in $\mathsf{Ext}_n(\sigma)$, and let $z_i \in \mathsf{Ext}_n(\sigma)$ be an extension of $f(y_i)$ for each i. We may now define an extension $\widehat{f} \sqsupseteq f$ as follows:

- If $y \in \mathsf{Ext}_n(\sigma)$, let $\widehat{f}(y) = z_i$, where i is the unique index such that $y \approx y_i$.
- Otherwise, let $\widehat{f}(y) = f(y)$.

Clearly $\widehat{f}$ is monotone and $\widehat{f} \in \mathsf{Ext}_n(\sigma \to \tau)$.

For (ii), suppose $f \approx_{n,\sigma\to\tau} g$. For any $y \in \mathsf{PC}_n(\sigma)$, construct $\widehat{y}$ using claim (i) for σ; then $f(\widehat{y}) \approx_{n,\tau} g(\widehat{y})$, so by claim (ii) for τ we see that $f(\widehat{y}), g(\widehat{y})$ have a least upper bound in $\mathsf{PC}_n(\tau)$. So $f(y), g(y)$ have an upper bound, and hence a least upper bound in $\mathsf{PC}_n(\tau)$. But now $\Lambda y.\, f(y) \sqcup g(y)$ is a least upper bound for f,g. $\square$

Next, we will show how each PC_n may be embedded in PC. In fact, we shall consider two closely related embeddings. Specifically, for each type σ we construct two different retractions $\mathsf{PC}_n(\sigma) \lhd \mathsf{PC}(\sigma)$ in the category of posets and monotone functions: that is, we define monotone functions

$$\nu_{n,\sigma}, \nu'_{n,\sigma} \;:\; \mathsf{PC}_n(\sigma) \to \mathsf{PC}(\sigma)\,, \qquad \pi_{n,\sigma}, \pi'_{n,\sigma} \;:\; \mathsf{PC}(\sigma) \to \mathsf{PC}_n(\sigma)$$

such that $\pi_{n,\sigma} \circ \nu_{n,\sigma} = \pi'_{n,\sigma} \circ \nu'_{n,\sigma} = id_{\mathsf{PC}_n(\sigma)}$.

For $\sigma = \mathtt{N}$, we define

$$\nu_{n,\mathtt{N}}(x) = x\,, \qquad \pi_{n,\mathtt{N}}(y) = \begin{cases} y & \text{if } y \leq n \in \mathbb{N}\,, \\ \bot & \text{otherwise}, \end{cases}$$

$$\nu'_{n,\mathtt{N}}(x) = x\,, \qquad \pi'_{n,\mathtt{N}}(y) = \begin{cases} y & \text{if } y \leq n \in \mathbb{N}\,, \\ n & \text{otherwise}. \end{cases}$$

Clearly these have the required properties. We now lift these retractions to higher types in the usual way:

$$\nu_{n,\sigma\to\tau}(f) = \nu_{n,\tau}\circ f\circ \pi_{n,\sigma}\,, \qquad \pi_{n,\sigma\to\tau}(g) = \pi_{n,\tau}\circ g\circ \nu_{n,\sigma}\,,$$
$$\nu'_{n,\sigma\to\tau}(f) = \nu'_{n,\tau}\circ f\circ \pi'_{n,\sigma}\,, \qquad \pi'_{n,\sigma\to\tau}(g) = \pi'_{n,\tau}\circ g\circ \nu'_{n,\sigma}\,.$$

By standard reasoning (see Proposition 4.2.1), this yields retractions $(\nu_{n,\sigma},\pi_{n,\sigma})$ and $(\nu'_{n,\sigma},\pi'_{n,\sigma}) : \mathsf{PC}_n(\sigma) \lhd \mathsf{PC}(\sigma)$ as required. We will also need the following easy facts:

Lemma 8.1.4. *(i) For each σ we have $\nu_{n,\sigma} \sqsubseteq \nu'_{n,\sigma}$ and $\pi_{n,\sigma} \sqsubseteq \pi'_{n,\sigma}$ pointwise.*

(ii) If $x \approx_{n,\sigma} x'$ then $\nu'_{n,\sigma}(x) \approx_\sigma \nu'_{n,\sigma}(x')$, and if $y \approx_\sigma y'$ then $\pi'_{n,\sigma}(y) \approx_{n,\sigma} \pi'_{n,\sigma}(y')$. In particular, $\nu'_{n,\sigma}$ maps $\mathsf{Ext}_n(\sigma)$ to $\mathsf{Ext}(\sigma)$.

Proof. For both parts the claim is immediately evident for $\sigma = \mathtt{N}$, and the general case follows by an easy induction on types (we recommend this as an exercise). □

As an easy consequence, we obtain our first version of the density theorem:

Theorem 8.1.5 (Density theorem, domain-theoretic version). *Every compact $a \in \mathsf{PC}(\sigma)$ may be extended to some $\widehat{a} \in \mathsf{Ext}(\sigma)$.*

Proof. Suppose $a \in \mathsf{PC}(\sigma)$ is compact. Let e be a valid formal expression denoting a, and let n be larger than all numbers appearing in e. Then it is clear that a is in the image of ν_n (omitting type subscripts), so that $\nu_n(\pi_n(a)) = a$. Using Lemmas 8.1.3 and 8.1.4, we now have

$$a \;=\; \nu_n(\pi_n(a)) \;\sqsubseteq\; \nu'_n(\pi_n(a)) \;\sqsubseteq\; \nu'_n(\widehat{\pi_n(a)})\,,$$

where $\nu'_n(\widehat{\pi_n(a)}) \in \mathsf{Ext}(\sigma)$. We may thus take $\widehat{a} = \nu'_n(\widehat{\pi_n(a)})$ for any suitable n. □

Remark 8.1.6. It is a familiar idea in domain theory that PC can be analysed as a 'limit' of the approximating structures PC_n, in the sense that for each σ, the idempotents $\nu_\sigma \circ \pi_\sigma$ form an increasing chain whose limit is the identity on $\mathsf{PC}(\sigma)$. It is less obvious whether there is a sense in which Ct is a 'limit' of the Ct_n, although we shall see in Section 8.2 how to give substance to this idea. In the meantime, it is worth observing that Ct_n is actually the full set-theoretic type structure S_n over $\{0,\ldots,n\}$: that is, $\mathsf{Ct}_n(\sigma \to \tau)$ consists of all total functions $\mathsf{Ct}_n(\sigma) \to \mathsf{Ct}_n(\tau)$. We leave the proof of this as an easy exercise.

We may also extract some useful effectivity information from the above proof:

Corollary 8.1.7 (Effective density theorem). *For any compact $a \in \mathsf{PC}(\sigma)$, we have $\widehat{a} \in \mathsf{PC}^{\mathrm{eff}}(\sigma)$, where $\widehat{a}$ is given as in the proof of Theorem 8.1.5. Moreover, for each σ, there is an element $\zeta_\sigma \in \mathsf{PC}^{\mathrm{eff}}(\mathtt{N} \to \sigma)$ such that $\zeta_\sigma(\lceil e \rceil) = \widehat{[\![e]\!]}$ for all valid expressions e at type σ.*

Proof. By inspection of the proof of Theorem 8.1.5, from an expression e at type σ we may effectively obtain a complete description of the finite function $\widehat{\pi_n(x)}$ for some suitable n; and uniformly in this, we may effectively decide whether $[\![d]\!] \sqsubseteq \nu'_n(\widehat{\pi_n(x)})$ for any given d of type σ. We thus obtain an effective enumeration of all compacts below the required function ζ_σ. □

Effective versions of the density theorem in more general domain-theoretic settings are given in Berger [19, 21].

With the help of Theorem 8.1.5, we may now flesh out our picture of the $\approx_\sigma$ equivalence classes:

Corollary 8.1.8. *Within* $\mathsf{PC}(\sigma)$, *each* $\approx_\sigma$ *equivalence class is upward closed, has binary meets and joins, and has a top element.*

Proof. Let E be any $\approx_\sigma$ equivalence class. Upward closure and the existence of binary meets were given by Lemma 8.1.1. For joins, suppose $f,g \in E$, and write σ as $\sigma_0,\ldots,\sigma_{r-1} \to \mathtt{N}$. If f,g have no join in $\mathsf{PC}(\sigma)$, there must be elements $y_i \in \mathsf{PC}(\sigma_i)$ such that $f(\vec{y}), g(\vec{y}) \in \mathbb{N}$ but $f(\vec{y}) \neq g(\vec{y})$. Take $a_i \sqsubseteq y_i$ compact such that $f(\vec{a}) = f(\vec{y})$ and $g(\vec{a}) = g(\vec{y})$, and for each i, let z_i be a total extension of a_i. Then $f(\vec{z}) \neq g(\vec{z})$, contradicting $f \approx_\sigma g$. Thus f,g have a join $f \sqcup g$ in $\mathsf{PC}(\sigma)$, and $f \sqcup g \in E$ since E is upward closed.

This shows that E is directed; hence E has a supremum in $\mathsf{PC}(\sigma)$, and by upward closure this is again an element of E. $\square$

Exercise 8.1.9. (i) Show that the *minimization* operator of Definition 3.3.18 yields an example of a maximal element of $\mathsf{PC}(2)$ that is not total.

(ii) Show that in $\mathsf{PC}(2)$, the equivalence class consisting of all realizers for $\Lambda f.0 \in \mathsf{Ct}(2)$ has no least element.

(iii) Show that in any $\approx$ equivalence class, the *effective* elements are closed under $\sqcap$ and $\sqcup$, but that there is an element of $\mathsf{Ct}^{\mathrm{eff}}(\mathtt{N},\mathtt{N} \to \mathtt{N})$ with no maximal effective realizer.

Note that every compact $a \in \mathsf{PC}(\sigma)$ gives rise to a set $V_a \subseteq \mathsf{Ct}(\sigma)$ consisting of all elements with a realizer extending a. Theorem 8.1.5 can then be phrased as saying simply that each V_a is inhabited. As a first example of the use of these sets, we have the following:

Corollary 8.1.10. *(i) Suppose that for all compact* $a \in \mathsf{PC}(\sigma)$ *we have some choice of element* $\overline{a} \in V_a$. *If* $F,F' \in \mathsf{Ct}(\sigma \to \tau)$ *agree on all* $\overline{a}$, *then* $F = F'$.

(ii) The applicative structure $\mathsf{Ct}^{\mathrm{eff}}$ *is extensional.*

Proof. (i) We treat the case $\tau = \mathtt{N}$, from which the general case follows easily. Suppose for contradiction that $F(X) \neq F'(X)$. Take f,f',x realizing F,F',X respectively; then by continuity there exists a compact $a \sqsubseteq x$ such that $f(a) = f(x) = F(X)$ and $f'(a) = f'(x) = F'(X)$. Hence for all $X' \in V_a$ we have $F(X') = F(X) \neq F'(X) = F'(X')$; in particular, $F(\overline{a}) \neq F'(\overline{a})$.

(ii) Taking each $\overline{a} \in \mathsf{Ct}(\sigma)$ to be the element represented by $\widehat{a}$ as in Theorem 8.1.5, we have by Corollary 8.1.7 that $\overline{a} \in \mathsf{Ct}^{\mathrm{eff}}$. So if $F,F' \in \mathsf{Ct}^{\mathrm{eff}}(\sigma \to \tau)$ agree on all of $\mathsf{Ct}^{\mathrm{eff}}(\sigma)$, then they agree on all elements $\overline{a}$, so by part (i) we have $F = F'$. $\square$

Next, we show that we can give an alternative description of the sets V_a that makes no reference to PC; from this, we will obtain a version of the density theorem framed in terms of the intrinsic structure of Ct. For this purpose, we will use our language of formal expressions for compacts in PC, but will also put these formal expressions to another use to define subsets of Ct directly:

Definition 8.1.11. Suppose e is a valid expression of type σ.

(i) The *Ershov neighbourhood* $V_e \subseteq \mathsf{Ct}(\sigma)$ consists of all $X \in \mathsf{Ct}(\sigma)$ for which there is some $x \Vdash X$ with $x \sqsupseteq [\![e]\!]$. We will sometimes write V_a for V_e where $a = [\![e]\!]$.

(ii) The *Kreisel neighbourhood* $W_e \subseteq \mathsf{Ct}(\sigma)$ is defined by recursion on the structure of e:

- $W_\perp = \mathbb{N}$ and $W_n = \{n\}$.
- If $d = (e_0 \Rightarrow e'_0) \sqcup \cdots \sqcup (e_{r-1} \Rightarrow e'_{r-1})$, then

$$W_d = (W_{e_0} \Rightarrow W_{e'_0}) \cap \cdots \cap (W_{e_{r-1}} \Rightarrow W_{e'_{r-1}}) ,$$

where

$$W \Rightarrow W' \;=\; \{F \in \mathsf{Ct} \mid \forall X \in W.\, F(X) \in W'\} .$$

Lemma 8.1.12. $V_e = W_e$ *for any valid* e.

Proof. By induction on types. At type $\mathtt{N}$ this is trivial, so suppose the claim holds for σ, τ, and consider a valid expression $d = (e_0 \Rightarrow e'_0) \sqcup \cdots \sqcup (e_{t-1} \Rightarrow e'_{r-1})$ at type $\sigma \to \tau$. If $F \in V_d$, take $f \sqsupseteq [\![d]\!]$ realizing F. Then for each $i < r$ and any $X \in W_{e_i} = V_{e_i}$ we may take $x \sqsupseteq [\![e_i]\!]$ realizing X; but then $f(x) \sqsupseteq [\![e'_i]\!]$ realizes $F(X)$, so that $F(X) \in V_{e'_i} = W_{e'_i}$. Thus $F \in W_d$. For the converse, if $F \in W_d$, take $f \Vdash F$. To see that $F \in V_d$, it will suffice to show that f and $[\![d]\!]$ are consistent in $\mathsf{PC}(\sigma \to \tau)$, since then $f \sqcup [\![d]\!]$ will also be a realizer for F. For this, we argue as in the proof of Corollary 8.1.8, using the density theorem for σ. □

Corollary 8.1.13 (Density theorem, intrinsic version). *Every Kreisel neighbourhood is inhabited.*

Proof. It is immediate from Theorem 8.1.5 that every Ershov neighbourhood is inhabited, and the corollary follows by Lemma 8.1.12. □

Since the definition of Kreisel neighbourhoods refers only to the structure of Ct, and we have a purely syntactic characterization of valid expressions, this version of the density theorem can be understood without any reference to PC. Kreisel's original approach to defining Ct was in effect to construct the meet-semilattice of Kreisel neighbourhoods purely syntactically via formal expressions, and then to obtain the functionals themselves as suitable formal 'intersections' of such neighbourhoods. In a similar vein, in our present setup we may give an intrinsic characterization of $\mathsf{Ct}^{\mathrm{eff}}$ as consisting of all $X \in \mathsf{Ct}$ such that $\{e \mid X \in W_e\}$ is computably enumerable.

Exercise 8.1.14. (i) Show that if the definition of Kreisel neighbourhoods is applied to an invalid expression e, the resulting set W_e is empty.

(ii) Show that we may have $x \Vdash_\sigma X \in W_e$ but $x \not\sqsupseteq [\![e]\!]$.

If e is a valid expression at type σ and n exceeds all numbers appearing in e, let us write $\ll e \gg_n$ for the inhabitant of W_e represented by $\nu'_n(\widehat{\pi_n([\![e]\!])}) \in \mathsf{PC}(\sigma)$ as in the proof of Theorem 8.1.5. We may now see that these inhabitants themselves admit a more intrinsic characterization. For each n, define a retraction $(\overline{\nu}_{n,\mathtt{N}}, \overline{\pi}_{n,\mathtt{N}}) : \mathsf{Ct}_n(\mathtt{N}) \lhd \mathsf{Ct}(\mathtt{N})$ by

$$\overline{\nu}_{n,\mathtt{N}}(x) = x\,, \qquad \overline{\pi}_{n,\mathtt{N}}(y) = \begin{cases} y \text{ if } y \leq n, \\ n \text{ otherwise,} \end{cases}$$

and lift this as usual to a retraction $(\overline{\nu}_{n,\sigma}, \overline{\pi}_{n,\sigma}) : \mathsf{Ct}_n(\sigma) \lhd \mathsf{Ct}(\sigma)$ for each σ. Clearly $\overline{\nu}_{n,\sigma}, \overline{\pi}_{n,\sigma}$ are respectively tracked by $\nu'_{n,\sigma}, \pi'_{n,\sigma}$ with respect to the realization of Ct by PC and Ct_n by PC_n; thus each $\ll e \gg_n$ is in the image of $\overline{\nu}_{n,\sigma}$. Conversely, if $X = \overline{\nu}_{n,\sigma}(Y)$, $y \in \mathsf{PC}_n(\sigma)$ is a maximal realizer for Y and e is a formal expression for y, then $\widehat{\pi_n([\![e]\!])} = y$ and $\nu'_n(y) \Vdash X$, whence $X = \ll e \gg_n$. So for each n, the set of elements $\ll e \gg_n$ coincides precisely with the image of $\overline{\nu}_{n,\sigma}$.

The approximation of each $\mathsf{Ct}(\sigma)$ by the spaces $\mathsf{Ct}_n(\sigma)$ will play an important role in this chapter. Importing a notation from Section 5.3, we will write $(X)_n^\sigma$ or just $(X)_n$ for $\overline{\nu}_n(\overline{\pi}_n(X))$.

Exercise 8.1.15. Suppose that $d = (e_0 \Rightarrow e'_0) \sqcup \cdots \sqcup (e_{r-1} \Rightarrow e'_{r-1})$ is valid at type $\sigma \to \tau$, and n exceeds all numbers appearing in d. By examining the proofs of Lemma 8.1.3 and Theorem 8.1.5, show that $\ll d \gg_n$ may be defined intrinsically by

$$\ll d \gg_n (X) = \begin{cases} \ll e'_i \gg_n \text{ if } \overline{\pi}_n(X) \in W_{e_i}, \\ \Lambda \vec{z}.0 \quad\;\; \text{if } \overline{\pi}_n(X) \notin W_{e_i} \text{ for all } i. \end{cases}$$

8.1.3 Topological Aspects

Next, we investigate how Ct relates to the topological notion of continuity. As we shall see, the situation is somewhat subtle: the most obvious attempt to characterize Ct topologically fails to yield any reasonable type structure at all, whereas the correct approach yields a topology with rather bizarre properties.

One's first instinct might be to suppose that the functionals in $\mathsf{Ct}(\sigma \to \tau)$ should be those that are continuous with respect to the *Kreisel topology* on $\mathsf{Ct}(\sigma)$ and $\mathsf{Ct}(\tau)$, that is, the one generated by the Kreisel neighbourhoods. Perhaps surprisingly, it turns out that neither half of this idea is correct:

Exercise 8.1.16. (i) Verify that in $\mathsf{Ct}(1)$, the constant function $\Lambda n.0$ is an *isolated point* in the sense that both $\{\Lambda n.0\}$ and its complement in $\mathbb{N}^{\mathbb{N}}$ are Kreisel open. Now define $H : \mathsf{Ct}(1) \to \mathsf{Ct}(0)$ by

$$H(g) = \begin{cases} 0 \text{ if } g = \Lambda n.0\,, \\ 1 \text{ otherwise.} \end{cases}$$

Check that H is Kreisel continuous, but $H \notin \mathsf{Ct}(2)$.

(ii) Consider the functional $G = \Lambda fn.f(2n) \in \mathsf{Ct}(1 \to 1)$. Show that G is not Kreisel continuous, since $G^{-1}\{\Lambda n.0\}$ contains no open neighbourhood of $f = \Lambda n.n \bmod 2$.

(iii) (Harder.) Show that analogous counterexamples exist even if we modify our definitions so as to exclude 'trivial' Kreisel neighbourhoods. Specifically, suppose that we omit the symbol $\perp$ from our language of formal expressions for compacts, and moreover allow formal joins $(e_0 \Rightarrow e'_0) \sqcup \cdots \sqcup (e_{r-1} \Rightarrow e'_{r-1})$ only for $r \geq 1$. Let the *modified Kreisel topology* be the one generated by the sets W_e for such restricted expressions e. Construct third-order functionals that play the roles of H and G with respect to the modified Kreisel topology.

Exercise 8.1.17. Suppose we try to construct a type structure C over $\mathsf{T}^{\to\times 1}(\mathtt{N} = \mathbb{N})$ by direct analogy with our construction of PC in Subsection 3.2.2. That is, for each σ we construct the set $\mathsf{C}(\sigma)$ along with a family $\mathscr{F}_\sigma$ of neighbourhoods on $\mathsf{C}(\sigma)$, where $\mathsf{C}(\sigma \to \tau)$ consists of the continuous functions $\mathsf{C}(\sigma) \to \mathsf{C}(\tau)$, and $\mathscr{F}_{\sigma\to\tau}$ consists of the 'Kreisel neighbourhoods' arising from $\mathscr{F}_\sigma$ and $\mathscr{F}_\tau$.

Using the counterexample H from the previous exercise, show that the resulting structure is not even cartesian closed.

These observations show that the Kreisel topology is not the appropriate one for characterizing the functionals of Ct. The right topology to consider is the one induced by the *Scott topology*. Recall from Subsection 3.2.2 that this is generated by the basic opens $U_a = \{x \mid x \sqsupseteq a\}$, where a ranges over compact elements.

Definition 8.1.18. For each type σ, the *canonical topology* on $\mathsf{Ct}(\sigma)$ is the quotient topology induced by the Scott topology on $\mathsf{Ext}(\sigma)$ (as a subspace of $\mathsf{PC}(\sigma)$). Explicitly, $V \subseteq \mathsf{Ct}(\sigma)$ is open in the canonical topology if there is some Scott open $U \subseteq \mathsf{PC}(\sigma)$ such that

$$U \cap \mathsf{Ext}(\sigma) \;=\; \{x \in \mathsf{PC}(\sigma) \mid x \Vdash X \text{ for some } X \in V\}\,.$$

Exercise 8.1.19. Show that every open set in the canonical topology is Kreisel open, but that the converse fails at type $\overline{1}$.

The following trivial fact is worth noting, as it will provide our main tool for showing sets to be open:

Proposition 8.1.20. *A set $V \subseteq Ct(\sigma)$ is open in the canonical topology if for all $X \in V$ and $x \Vdash X$, there is a compact $a \sqsubseteq x$ such that $V_a \subseteq V$.* $\square$

Remark 8.1.21. In topological terms, our first version of the density theorem (Theorem 8.1.5) clearly gives us a countable dense set in each $\mathsf{Ct}(\sigma)$ with the canonical topology, as well as a countable dense set in the space $\mathsf{Ext}(\sigma)$ of realizers.

The above definition of the canonical topology is of course tied to the presentation of Ct via PC. However, we will see later that the canonical topology also admits intrinsic characterizations as the *sequential topology* on Ct (Subsection 8.2.3), and also as the one compactly generated from the compact-open topology (Subsection 8.2.4). From now on, whenever we refer to some $\mathsf{Ct}(\sigma)$ as a topological space, we will have in mind the canonical topology unless otherwise stated.

Theorem 8.1.22. *The elements of* $\mathsf{Ct}(\sigma \to \tau)$ *are exactly the functions* $\mathsf{Ct}(\sigma) \to \mathsf{Ct}(\tau)$ *that are continuous with respect to the canonical topology.*

Proof. Suppose $F \in \mathsf{Ct}(\sigma \to \tau)$, and take $f \Vdash_{\sigma\to\tau} F$. To see that F is continuous for the canonical topology, note that f is Scott continuous, so its restriction $f' : \mathsf{Ext}(\sigma) \to \mathsf{Ext}(\tau)$ is continuous in the subspace topology. Since f' tracks $F : \mathsf{Ext}(\sigma)/{\approx_\sigma} \to \mathsf{Ext}(\tau)/{\approx_\tau}$, it follows by simple topological reasoning that F is continuous in the quotient topology.

For the converse, we shall temporarily migrate to the structures $\mathsf{Ct}^\times$ and $\mathsf{PC}^\times$ with product types. (We will not need the generalizations to $\mathsf{T}^{\to\times}(\mathtt{N})$ of any of the results proved so far.) By the canonical topology on $\mathsf{Ct}(\rho_0 \times \cdots \times \rho_{r-1})$, we shall mean the quotient topology induced by the Scott topology on $\mathsf{PC}(\rho_0 \times \cdots \times \rho_{r-1})$. Since application in PC is Scott continuous, it is clear that the application operations $\mathsf{Ct}((\rho \to \rho') \times \rho) \to \mathsf{Ct}(\rho')$ are continuous for this topology.

Now suppose $F : \mathsf{Ct}(\sigma) \to \mathsf{Ct}(\tau)$ is continuous in the canonical topology, where $\tau = \tau_0, \ldots, \tau_{r-1} \to \mathtt{N}$. From the continuity of application, it follows that the transpose $\widetilde{F} : \mathsf{Ct}(\sigma \times \tau_0 \times \cdots \times \tau_{r-1}) \to \mathsf{Ct}(\mathtt{N})$ is continuous for the canonical topology. We may now define $\widetilde{f} : \mathsf{PC}(\sigma \times \tau_0 \times \cdots \times \tau_{r-1}) \to \mathsf{PC}(\mathtt{N})$ tracking $\widetilde{F}$ as follows:

$$\widetilde{f}(x, y_0, \ldots, y_{r-1}) = n \quad \text{iff} \quad \text{there are compacts } [\![d]\!] \sqsubseteq x,\ [\![e_i]\!] \sqsubseteq y_i \text{ such that } \widetilde{F}(W_d \times W_{e_0} \times \cdots \times W_{e_{r-1}}) = \{n\}\,.$$

This is well-defined, since any two candidates for the family $[\![d]\!], [\![e_0]\!], \ldots, [\![e_{r-1}]\!]$ will have a total common extension. Clearly $\widetilde{f}$ is monotone and continuous, and $\widetilde{f}(x, \vec{y}) = \widetilde{F}(X, \vec{Y})$ whenever $x \Vdash X$ and $y_i \Vdash Y_i$ for each i. Transposing back, we obtain an element $f \in \mathsf{PC}(\sigma \to \tau)$ witnessing that $F \in \mathsf{Ct}(\sigma \to \tau)$. $\square$

The reader is warned here of a subtlety associated with product types: the canonical topology on $\mathsf{Ct}(\rho_0 \times \rho_1)$ (as defined in the above proof) is in general strictly finer than the ordinary product topology on $\mathsf{Ct}(\rho_0) \times \mathsf{Ct}(\rho_1)$. This is because a typical open neighbourhood U of a point (X_0, X_1) in $\mathsf{Ct}(\rho_0 \times \rho_1)$ will arise from some union of basic Scott open sets in $\mathsf{Ext}(\rho_0 \times \rho_1)$ which need not individually be $\approx$-closed, and such a set U need not include any sub-neighbourhood $U_0 \times U_1$ of (X_0, X_1) where U_0, U_1 are themselves open. Indeed, the following is a strong indication that the standard product topology is the wrong one to consider in this context:

Proposition 8.1.23. *Application* $\mathsf{Ct}(2) \times \mathsf{Ct}(1) \to \mathsf{Ct}(0)$ *is not continuous with respect to the product topology on* $\mathsf{Ct}(2) \times \mathsf{Ct}(1)$.

Proof. Let $F = \Lambda z.0 \in \mathsf{Ct}(2)$ and let $X = \Lambda n.0 \in \mathsf{Ct}(1)$. Let O_2 be any open set in $\mathsf{Ct}(2)$ containing F, and O_1 any open set in $\mathsf{Ct}(1)$ containing X. We will show that there exist $F' \in O_2$ and $X' \in O_1$ such that $F'(X') \neq 0$.

Take n such that $\{Y \in \mathbb{N}^{\mathbb{N}} \mid \forall i < n. Y(i) = X(i)\} \subseteq O_1$, and consider the functional $f \in \mathsf{PC}(2)$ defined by

$$f(y) = \begin{cases} 0 & \text{if } y(n) \in \mathbb{N}, \\ \bot & \text{if } y(n) = \bot. \end{cases}$$

Then $f \in \mathsf{Ext}(2)$ and $f \Vdash F$. There will therefore be a compact approximation $a \sqsubseteq f$ such that $V_a \subseteq O_2$. Since a can only cater for finitely many possible values of $y(n)$, we may find $X' \in O_1$ such that $a(x')$ is undefined for any $x' \Vdash X'$, and we can then find a total extension f' of a realizing some $F' \in O_2$ with $F'(X') = 1$. □

We now briefly survey some further properties of the canonical topology. On the positive side, we note the following:

Proposition 8.1.24. *(i) Each* $\mathsf{Ct}(\sigma)$ *is* separable, *i.e. has a countable dense subset.*

(ii) Each $\mathsf{Ct}(\sigma)$ *is* strongly disconnected*: that is, for any distinct* $F, F' \in \mathsf{Ct}(\sigma)$*, there is a clopen (i.e. closed and open) subset of* $\mathsf{Ct}(\sigma)$ *separating* F *and* F'.

(iii) Each Kreisel neighbourhood $W_e \subseteq \mathsf{Ct}(\sigma)$ *is closed in the canonical topology.*

Proof. (i) We have already noted that every canonical open set is Kreisel open; hence the elements $\ll e \gg$ from Subsection 8.1.2 form a countable dense set.

(ii) Suppose $\sigma = \sigma_0, \ldots, \sigma_{r-1} \to \mathtt{N}$, and take elements $X_i \in \mathsf{Ct}(\sigma_i)$ such that $F(\vec{X}) \neq F'(\vec{X})$. Then the inverse image of the clopen set $\{F(\vec{X})\}$ under the continuous map $\Lambda z.z(\vec{X})$ is a clopen separating F and F'.

(iii) Let $Z \subseteq \mathsf{Ext}(\sigma)$ consist of all realizers for elements of W_e, and let $\overline{Z} = \mathsf{Ext}(\sigma) - Z$. For any $x \in \overline{Z}$, we may clearly find some compact $a \sqsubseteq x$ where a is incompatible with $[\![e]\!]$; hence $U_a \cap \mathsf{Ext}(\sigma) \subseteq \overline{Z}$. This shows that $\overline{Z}$ is open in $\mathsf{Ext}(\sigma)$; hence Z is closed and W_e is closed in $\mathsf{Ct}(\sigma)$. □

It is also worth recording that the canonical topology on $\mathsf{Ct}(1)$ coincides with the usual *Baire topology* on $\mathbb{N}^{\mathbb{N}}$.

However, this is about as far as the pleasant properties go: even at type $\overline{2}$, the canonical topology is somewhat pathological. As our discussion of product types has already suggested, the canonical topology is rather hard to grasp concretely: a $\approx$-closed open set in $\mathsf{Ext}(2)$ will in general arise as a union of basic opens that are not $\approx$-closed, so the standard basis for the Scott topology in $\mathsf{PC}(2)$ sheds little direct light on the structure of open sets in $\mathsf{Ct}(2)$. The difficulty is brought into sharper focus by the following result, which has been known since the early years:

Lemma 8.1.25. *(i) No element* $F \in \mathsf{Ct}(2)$ *has a countable basis of neighbourhoods: that is, for any* F *and any countable family* $\{O_n \mid n \in \mathbb{N}\}$ *of open neighbourhoods for* n*, there is an open set* O *containing* F *but not including any* O_n *as a subset.*

(ii) $\mathsf{Ct}(2)$ *is not* metrizable, *i.e. it is not the topology associated with any metric on* $\mathsf{Ct}(2)$.

Proof. (i) We prove this for the case $F = \Lambda n.0$, the extra complications for the general case being purely notational. We may assume without loss of generality that $O_{n+1} \subseteq O_n$ for each n.

Let $f_n \in \mathsf{PC}(2)$ be the realizer for F defined by

$$f_n(y) = \begin{cases} 0 & \text{if } \forall m \le n.\, y(m) \in \mathbb{N}, \\ \bot & \text{otherwise.} \end{cases}$$

Since O_n is open, there is a compact $a_n \sqsubseteq f_n$ such that $V_{a_n} \subseteq O_n$. Here $a_n(b) \in \mathbb{N}$ only if $b(n) \in \mathbb{N}$, and a_n can cater for only finitely many values of $b(n)$, so it is easy to construct some $G_n \in V_{a_n}$ that is not constant on any of the sets V_b where b has domain $\{0, \dots, n-1\}$.

Now let $O = \mathsf{Ct}(2) - \{G_n \mid n \in \mathbb{N}\}$. It will suffice to show that O is open, since $F \in O$ and $G_n \in O_n$ but $G_n \notin O$ for each n.

Let $G \in O$ and let $g \Vdash G$. Then there will be some $n \in \mathbb{N}$ and $b \in \mathsf{PC}(1)$ with domain $\{0, \dots, n-1\}$ such that $g(b) \in \mathbb{N}$. Since $G \neq G_m$ for each $m < n$, we may find a compact approximation $a \sqsubseteq g$ such that $a(b) \in \mathbb{N}$ and a is inconsistent with each such G_m: that is, $G_m \notin V_a$ for $m < n$. But since $a(b) \in \mathbb{N}$, we also see that $G_m \notin V_a$ for $m \geq n$, as such G_m are not constant on V_b. Hence $V_a \subseteq O$. Thus the set of realizers for elements in O is open in $\mathsf{Ext}(2)$, so O is open in $\mathsf{Ct}(2)$.

(ii) This is a standard piece of topology: if $\mathsf{Ct}(2)$ were metrizable, the open balls of radius 2^{-n} around F would form a countable neighbourhood base for F. □

Recent results show that the situation is even worse than this. Schröder [265] showed that the canonical topology on $\mathsf{Ct}(2)$ is not *zero-dimensional* (that is, it has no basis of clopen sets), and is not *regular* (we cannot separate disjoint closed sets by disjoint open sets in general). There are also other results in this vein in [265].

Whilst $\mathsf{Ct}(2)$ has no countable basis, we shall see in Corollary 8.2.23 that in any $\mathsf{Ct}(\sigma)$, the closed sets V_e do form a countable *pseudobase*.

8.1.4 Computability in Ct

We are now ready to introduce the concepts of computability in which we are interested. We give here just the basic definitions and results; a much deeper study of these computability concepts will be undertaken in Sections 8.5 and 8.6.

We have already introduced one important notion of computability for Ct in the form of $\mathsf{Ct}^{\mathrm{eff}}$, the substructure induced by the effective substructure $\mathsf{PC}^{\mathrm{eff}}$ of PC. In the terminology of the introduction to this chapter, this is an example of an 'external' definition of computability. Other such external definitions are certainly possible—for instance, we may replace the realizing model PC by some other model such as K_2, or we may work with some other choice of computable substructure within PC. We will see later, however, that the resulting substructure of Ct is remarkably robust under such variations (see especially Subsection 8.6.2).

Our main purpose here is to introduce an 'internal' concept of computability for Ct—the one given by *Kleene computability*—and to begin the investigation of its relationship to the external one. We will also use this internal notion to give an alternative 'computable' version of the density theorem.

Since Ct is a total type structure over $\mathbb{N}$, we may apply the definition of Kleene computation via S1–S9 (see Section 5.1), which yields an inductively defined relation

$$\{e\}(F_0,\ldots,F_{r-1}) \;=\; n\,,$$

where e,n range over $\mathbb{N}$ and the F_i range over objects of pure type in Ct. (For the sake of continuity with the literature on Ct, we will work here with Kleene's original definition for objects of pure type, relying on the general theory from Chapter 5 to transfer our results to arbitrary types σ.) As explained in Section 5.1, the generation of any instance of this relation takes the form of a well-founded computation tree which may be of infinite (ordinal) height. It is easy to give examples of such trees whose height is some infinite countable ordinal; we will see later that there are also computation trees over Ct of uncountable height (Remark 8.3.3).

To establish that Kleene computability over Ct is a well-behaved notion, we must show that Ct is *Kleene closed*: that is, the function $\Lambda G^k.\{e\}(G,\vec{F})$ is present in $\mathsf{Ct}(k+1)$ whenever it is a total function and the F_i are in Ct (see Definition 5.1.3). This amounts to showing that Kleene computable functions are automatically continuous with respect to the canonical topology. We here prove this directly, using only the definition of S1–S9 and the knowledge of Ct that we have acquired so far.

Lemma 8.1.26. *Suppose e is a numerical index, $F_i \in \mathsf{Ct}(k_i)$ for each $i < r$, and $\{e\}(\vec{F}) = n$. Suppose too that $f_i \Vdash_{k_i} F_i$ for each i. Then there are compact elements $a_i \sqsubseteq f_i$ in $\mathsf{PC}(k_i)$ such that for all $F'_0 \in V_{a_0},\ldots,F'_{r-1} \in V_{a_{r-1}}$ we have $\{e\}(\vec{F'}) \preceq n$ (where $\preceq$ denotes Kleene inequality).*

Proof. By induction on the computation tree for $\{e\}(\vec{F}) = n$, with an induction case for each of the Kleene schemes S1–S9 (see Definition 5.1.1). All cases are trivial except those for S4, S5 and S8.

For S4 (first-order composition), suppose $e = \langle 4,e_0,e_1\rangle$, so that for some m we have $\{e_1\}(\vec{F}) = m$ and $\{e_0\}(m,\vec{F}) = n$. By the induction hypothesis we may take compacts $a'_0,\ldots,a'_{r-1}$ securing that $\{e_1\}(\vec{F'}) \preceq m$ for all relevant $\vec{F'}$, and $a''_0,\ldots,a''_{r-1}$ securing that $\{e_0\}(m,\vec{F''}) \preceq n$ for all relevant $\vec{F''}$. But a'_i and a''_i are bounded by F_i, so by setting $a_i = a'_i \sqcup a''_i$ for each i, we secure that $\{e\}(\vec{F'}) \preceq n$ whenever $F'_i \in V_{a_i}$ for each i.

A similar argument also works for S5 (primitive recursion): if the first argument is $x \in \mathbb{N}$, there will be x subcomputations whose results have to be secured by suitable finite approximations, and the componentwise join of these approximations then secures the overall result.

For S8 (higher-order function application), suppose $e = \langle 8,d\rangle$, so that we have

$$\{e\}(\vec{F}) \;=\; F_0(\Lambda H.\{d\}(H,\vec{F})) \;=\; n$$

where $F_0 \in \mathsf{Ct}(k_0)$ and H ranges over $\mathsf{Ct}(k_0-2)$. Then $\{d\}(H,\vec{F})$ is defined for every H, say with value $\Phi(H)$. So by the induction hypothesis, for each H and each $h \Vdash H$ there are compacts $b^h \sqsubseteq h$ and $a^h_i \sqsubseteq f_i$ for each i such that $\{d\}(H',\vec{F'}) \preceq \Phi(H)$ whenever $H' \in V_{b^h}$ and $F'_i \in V_{a^h_i}$ for each i, with equality holding if $\vec{F'} = \vec{F}$. We thus obtain a consistent family

$$\mathscr{G} = \{ ((b^h, a_0^h, \ldots, a_{r-1}^h) \Rightarrow \Phi(H)) \mid h \Vdash H \}$$

of compact elements in $\mathsf{PC}^\times(k_0 \times \cdots \times k_{r-1} \to \mathtt{N})$. Take $g = \bigsqcup \mathscr{G}$; then $\Lambda h.g(h, \vec{f})$ is a realizer for the total function $\Phi = \Lambda H.\{d\}(H, \vec{F})$. Thus $f_0(g) = n$, so by continuity in PC, we may take finite approximations $a_i' \sqsubseteq f_i$ and $c \sqsubseteq g$ such that $a_0'(\Lambda h.c(h, \vec{a}')) = n$, and we may now find some finite set $\mathscr{G}_0 \subseteq \mathscr{G}$ whose supremum is at least c. By amalgamating information from $\mathscr{G}_0$ with a_0', we obtain $a_0 \sqsubseteq f_0, \ldots, a_{r-1} \sqsubseteq f_{r-1}$ such that for any elements $F_i' \in V_{a_i}$ we have $\{d\}(H, \vec{F}') \preceq \Phi(H)$ for all H.

We are now ready to verify the induction claim. Suppose $F_i' \in V_{a_i}$ and $\{e\}(\vec{F}')$ is defined. Then $\Phi' = \Lambda H.\{d\}(H, \vec{F}')$ is total, so by the above we have $\Phi' = \Phi$. Moreover, this function is realized by $\Lambda h.g(h, \vec{f})$, and since $F_0' \in V_{a_0}$ where $a_0(\Lambda h.g(h, \vec{f})) = n$, it follows that $\{e\}(\vec{F}') = F_0'(\Phi') = n$. □

Theorem 8.1.27. Ct *is Kleene closed.*

Proof. Suppose $\Phi = \Lambda G^k.\{e\}(G, \vec{F})$ is total, where $F_0, \ldots, F_{r-1} \in \mathsf{Ct}$. Then for any G with $\Phi(G) = n$, the above lemma gives us a neighbourhood V_a such that $\Phi(G') = n$ for all $G' \in V_a$. Hence the set of all realizers for such G is Scott open in $\mathsf{Ext}(k)$, so that $\Phi^{-1}(n)$ itself is open in $\mathsf{Ct}(k)$. Thus Φ is continuous, and so is present in $\mathsf{Ct}(k+1)$. □

By the general considerations of Subsection 5.1.4, we therefore obtain a sub-TPCA $\mathsf{Ct}^{\mathrm{Kl}}$ consisting of the (absolutely) Kleene computable elements of Ct. We also, a fortiori, have the weaker notions of Kleene primitive recursion and μ-computability available for Ct, yielding substructures $\mathsf{Ct}^{\mathrm{prim}} \subseteq \mathsf{Ct}^{\mathrm{min}} \subseteq \mathsf{Ct}^{\mathrm{Kl}}$.

To see how $\mathsf{Ct}^{\mathrm{Kl}}$ is related to $\mathsf{Ct}^{\mathrm{eff}}$, we must extract some further information from the proof of Lemma 8.1.26. The essential idea behind this proof is that although the computation tree for $\{e\}(\vec{F}) = n$ may be infinite, only finitely much information from this tree is in practice required to certify the value of $\{e\}(\vec{F})$, since everything is continuous. More precisely, we may consider finitely branching (and hence finite) well-founded *derivation trees* whose nodes are formal assertions $\vdash \{e\}(a_0, \ldots, a_{r-1}) \preceq n$ (in which we represent the a_i syntactically via formal expressions). A close analysis of the above proof yields a decidable class of 'correct' derivation trees for such assertions (we leave the detailed formulation to the reader). The key properties of such derivations are as follows:

1. Whenever $\vdash \{e\}(a_0, \ldots, a_{r-1}) \preceq n$ is derivable, it is the case that $\{e\}(\vec{F}) \preceq n$ whenever $F_i \in V_{a_i}$ for each i.
2. Whenever $\{e\}(\vec{F}) = n$ and $f_i \Vdash F_i$ for each i, some assertion $\vdash \{e\}(a_0, \ldots, a_{r-1}) \preceq n$ is derivable, where $a_i \sqsubseteq f_i$ for each i.

Using this, we may show:

Theorem 8.1.28. $\mathsf{Ct}^{\mathrm{eff}}$ *is Kleene closed. Hence* $\mathsf{Ct}^{\mathrm{Kl}} \subseteq \mathsf{Ct}^{\mathrm{eff}}$.

Proof. Suppose $\Phi = \Lambda G^k.\{e\}(G,\vec{F})$ is total, where $F_0,\ldots,F_{r-1} \in \mathsf{Ct}^{\mathrm{eff}}$. By Theorem 8.1.27, we already know that $\Phi \in \mathsf{Ct}(k+1)$. For each i, take $f_i \in \mathsf{PC}^{\mathrm{eff}}$ realizing F_i. Now effectively enumerate all derivable assertions $\vdash \{e\}(b,\vec{a}) = n$, looking for those for which $a_i \sqsubseteq f_i$ for all i. This gives rise to a c.e. set of compact elements $(b \Rightarrow n)$, which are consistent because their Ershov neighbourhoods all contain Φ. Let g be the supremum of this set; then $g \in \mathsf{PC}^{\mathrm{eff}}$, and by property 2 above we see that g is a realizer for Φ. Thus $\Phi \in \mathsf{Ct}^{\mathrm{eff}}$.

In particular, the case $r = 0$ yields that every absolutely Kleene computable function is in $\mathsf{Ct}^{\mathrm{eff}}$. $\square$

Exercise 8.1.29. (i) Using these ideas, show that if $\Phi : \mathsf{Ct}(k_0) \times \cdots \times \mathsf{Ct}(k_{r-1}) \rightharpoonup \mathbb{N}$ is a Kleene computable partial function, there exists $g \in \mathsf{PC}^{\mathrm{eff}}(k_0,\ldots,k_{r-1} \to \mathtt{N})$ such that if $f_i \Vdash_{k_i} F_i$ for each $i < r$ and $\Phi(\vec{F}) = m$, then $g(\vec{f}) = m$.

(ii) More specifically, suppose we interpret Kleene's schemes S1–S9 in PC using the 'liberal interpretation' as described in Subsection 6.4.1. Show that if $f_i \Vdash_{k_i} F_i$ for each $i < r$ and $\{e\}(\vec{F}) = m$ in Ct, then $\{e\}(\vec{f}) = m$ in PC. (Cf. Proposition 6.4.18.)

The converse to Theorem 8.1.28 does not hold: we will see later (Theorem 8.5.3) that the *fan functional* is in $\mathsf{Ct}^{\mathrm{eff}}$ but not in $\mathsf{Ct}^{\mathrm{Kl}}$.

To conclude this section, we show how the concept of Kleene primitive recursion may be used to give a strong 'computable' version of the density theorem. This contrasts with Corollary 8.1.7, which refers to computability in the sense of $\mathsf{Ct}^{\mathrm{eff}}$. We recall here the retractions $(\overline{\nu}_n, \overline{\pi}_n) : \mathsf{Ct}_n(\sigma) \lhd \mathsf{Ct}(\sigma)$ defined in Subsection 8.1.2, along with the notation $(X)_n = (X)_n^\sigma = \overline{\nu}_n(\overline{\pi}_n(X))$.

Lemma 8.1.30. *The following are Kleene primitive recursive uniformly in $r \in \mathbb{N}$:*

1. *A functional $eq_{n,\sigma} \in \mathsf{Ct}(\sigma,\sigma \to \mathtt{N})$ such that $eq_{n,\sigma}(X,X') = 0$ iff $(X)_n = (X')_n$.*
2. *A complete listing of the elements in the range of $(-)_n^\sigma$: that is, a number $d_{n,\sigma}$ and a function $enum_{n,\sigma} \in \mathsf{Ct}(\mathtt{N} \to \sigma)$ such that*

$$\{(X)_n^\sigma \mid X \in \mathsf{Ct}(\sigma)\} \;=\; \{enum_{n,\sigma}(0),\ldots,enum_{n,\sigma}(d_{n,\sigma}-1)\}\,.$$

Proof. By induction on σ. We leave the details as an easy exercise. $\square$

Theorem 8.1.31 (Primitive recursive density theorem). *For each type σ there is an operation $\xi_\sigma \in \mathsf{Ct}^{\mathrm{prim}}(\mathtt{N} \to \sigma)$ such that whenever e is a valid expression of type σ we have $\xi_\sigma(\lceil e \rceil) \in W_e$.*

Proof. From a code $\lceil e \rceil$ it is easy to compute a suitable value of r and an index for some $(X)_n$ (relative to $enum_{n,\sigma}$) such that $(X)_n \in W_e$. The theorem now follows by the computability of $enum_{n,\sigma}$. $\square$

Let us write $\langle e \rangle$ for $\xi_\sigma(\lceil e \rceil)$ whenever e is valid at type σ, so that $\langle e \rangle \in W_e$. With a little care, one can arrange that $\langle e \rangle$ is precisely the element $\ll e \gg$ as defined in Subsection 8.1.2. Even so, Theorem 8.1.31 still does not completely supersede the effective density theorem of Corollary 8.1.7: if the primitive recursive algorithm for $\langle e \rangle$ is interpreted in PC rather than Ct, the result will be a realizer for $\langle e \rangle$, though not typically one extending $[\![e]\!]$. Indeed, a primitive recursive or even PCF-definable element extending $[\![e]\!]$ cannot be hoped for if e represents a compact element such as 'parallel or' (see Subsection 7.3.1).

The following application of Theorem 8.1.31 is folklore. Here we assume familiarity with the denotational definition of *full abstraction* from Subsection 7.1.4.

Corollary 8.1.32. Ct *is a fully abstract model for the language* $\mathsf{Klex}^{\mathrm{prim}}$ *(see Subsection 5.1.5) and for System T (see Subsection 3.2.3; also cf. Exercise 4.4.8).*

Proof. Suppose $M, M' : \sigma_0, \ldots, \sigma_{r-1} \to \mathbb{N}$ are observationally equivalent terms in $\mathsf{Klex}^{\mathrm{prim}}$; we may assume without loss of generality that M, M' are closed. Then in particular, for any valid expressions $e_0, \ldots, e_{r-1}$ of types $\sigma_0, \ldots, \sigma_{r-1}$ respectively, we have that $[\![M(\langle e_0\rangle, \ldots, \langle e_{r-1}\rangle)]\!] = [\![M'(\langle e_0\rangle, \ldots, \langle e_{r-1}\rangle)]\!]$ in Ct, since each $\langle e_i\rangle$ is $\mathsf{Klex}^{\mathrm{prim}}$ definable. By Corollary 8.1.10, it follows that $[\![M]\!] = [\![M']\!]$ in Ct.

The same argument also works for System T, since every $\mathsf{Klex}^{\mathrm{prim}}$ definable element of Ct is T-definable. □

By contrast, the full set-theoretic model S is not fully abstract for these languages. In Exercise 5.3.3 we gave, in effect, a $\mathsf{Klex}^{\mathrm{prim}}$ term $D : \overline{3}$ whose denotation in Ct coincides with that of $\lambda F.0$, so that $D \approx_{\mathrm{obs}} \lambda F.0$ in $\mathsf{Klex}^{\mathrm{prim}}$ or System T, but such that $[\![D]\!] \neq [\![\Lambda F.0]\!]$ in S(3).

8.2 Alternative Characterizations

In this section we survey a range of alternative characterizations of Ct. We begin in Subsection 8.2.1 with Kleene's original characterization as an extensional collapse of K_2, establishing the equivalence with our definition via PC, and showing that the isomorphism is itself 'computable within K_2'. In Subsection 8.2.2, we present a theorem of Bezem to the effect that the *modified* extensional collapse of K_2 again yields the same type structure. In Subsection 8.2.3, we give an intrinsic characterization of the total continuous functionals in terms of *limit spaces*—this characterization turns out to play a pivotal role, and offers some distinctive insights into the structure of Ct. From Subsection 8.2.4 onwards, we glance briefly at some further characterizations via *compactly generated spaces*, *filter spaces* and a realizability-style *reduced product* construction, without giving full proofs of the equivalences.

Yet more characterizations of Ct will emerge in Section 8.6, where we will state a theorem showing that a large class of realizability-style characterizations are bound to result in this type structure.

8.2.1 Kleene's Construction via Associates

Recall from Subsection 3.2.1 that Kleene's second model K_2 is an untyped partial combinatory algebra with underlying set $\mathbb{N}^{\mathbb{N}}$, and with an application operation designed to capture the idea of continuous dependence of the output on the input. Specifically, for $f,g \in \mathbb{N}^{\mathbb{N}}$ and $n \in \mathbb{N}$ we may define

$$(f \odot' g)(n) \simeq f(\langle n, g(0), \ldots, g(r-1)\rangle) - 1\,, \qquad \text{where } r = \mu r.\ f(\langle n, g(0), \ldots, g(r-1)\rangle) > 0\,,$$

and then define the application $f \odot g$ to be the function $\Lambda n.(f \odot' g)(n)$ whenever this is total. We shall here take as given that the structure $K_2 = (\mathbb{N}^{\mathbb{N}}, \odot)$ is a lax PCA, a fact which we will prove in Section 12.2. We say that $f \in \mathbb{N}^{\mathbb{N}}$ is an *associate* for $F : \mathbb{N}^{\mathbb{N}} \rightharpoonup \mathbb{N}^{\mathbb{N}}$ if $F(g) \simeq f \odot g$ for all $g \in \mathbb{N}^{\mathbb{N}}$.

We may represent the natural number n in K_2 via the element $\widehat{n}$, where $\widehat{n}(0) = n$ and $\widehat{n}(i+1) = 0$ for all i, and write $N = \{\widehat{n} \mid n \in \mathbb{N}\} \subseteq K_2$. We may now define a total type structure Ct^K over $\mathbb{N}$ simply by $\mathsf{Ct}^K = \mathsf{EC}(K_2;N)$, and write α for the canonical simulation $\mathsf{Ct}^K \longrightarrow\!\!\triangleright K_2$. We say $f \in K_2$ is an *associate* for $F \in \mathsf{Ct}^K(\sigma)$ if $f \Vdash^{\alpha}_{\sigma} F$; we also write $\mathsf{As}(\sigma)$ for the set of all associates for elements of $\mathsf{Ct}^K(\sigma)$. In the spirit of Subsection 3.4.2, the simulation α enables us to regard each $\mathsf{Ct}^K(\sigma)$ as an *assembly* over K_2, that is, an object of $\mathscr{A}sm(K_2)$.

Since both Ct and Ct^K are defined as extensional collapses of a representing model, it is natural to ask how these representations are related. We have already seen in Example 3.5.7 that there is an applicative simulation $\psi : \mathsf{PC} \longrightarrow\!\!\triangleright K_2$, where $f \Vdash^{\psi} x \in \mathsf{PC}(\sigma)$ iff f is (modulo coding) a possibly repetitive listing of some set of compacts in $\mathsf{PC}(\sigma)$ with supremum x. This suggests that we regard K_2 as 'more concrete' or 'more intensional' than PC: most elements of PC will have infinitely many realizers in K_2, and whilst every $h \in \mathsf{PC}(\sigma \to \tau)$ is tracked by some K_2-computable function $\mathsf{As}(\sigma) \to \mathsf{As}(\tau)$, there will be K_2-computable functions of this type that do not track any element of $\mathsf{PC}(\sigma \to \tau)$. The precise relationship between PC and K_2 will be further explored in Section 10.4.

By composing ψ with the canonical simulation $\varepsilon : \mathsf{Ct} = \mathsf{EC}(\mathsf{PC};\mathbb{N}) \longrightarrow\!\!\triangleright \mathsf{PC}$, we obtain an applicative simulation $\mathsf{Ct} \longrightarrow\!\!\triangleright K_2$ which we shall denote by β. We will say f is a *weak associate* for F at type σ if $f \Vdash^{\beta}_{\sigma} F$. This enables us to regard each $\mathsf{Ct}(\sigma)$ also as an object of $\mathscr{A}sm(K_2)$.

Our main goal here is to show that $\mathsf{Ct}^K \cong \mathsf{Ct}$. It turns out to be quite difficult to prove directly by induction on σ that $\mathsf{Ct}^K(\sigma) \cong \mathsf{Ct}(\sigma)$ as sets, but the task is made easier by adopting a stronger induction claim: namely, that $\mathsf{Ct}^K(\sigma) \cong \mathsf{Ct}(\sigma)$ *within* $\mathscr{A}sm(K_2)$. In other words, we show that both halves of the isomorphism are K_2-computable: any associate for $F \in \mathsf{Ct}^K(\sigma)$ may be transformed within K_2 to a weak associate for $F \in \mathsf{Ct}(\sigma)$, and vice versa. In the terminology of Definition 3.4.2, this will amount to showing that the simulations α and β are $\sim$-equivalent modulo the isomorphism $\mathsf{Ct}^K \cong \mathsf{Ct}$, an additional fact of interest in its own right.

Theorem 8.2.1. *For each type σ we have $\mathsf{Ct}^K(\sigma) \cong \mathsf{Ct}(\sigma)$ within $\mathscr{A}sm(K_2)$, where the isomorphisms commute with the application operations in Ct^K and* Ct.

Proof. We show this for the pure types $\overline{k}$ by induction. The general case can be proved similarly by argument induction with some notational overhead, or else deduced from the case of pure types via the results of Section 4.2. For each type $\overline{k}$, we suppose $\lceil - \rceil$ is a bijective coding of compacts in $\mathsf{PC}(k)$ as natural numbers. We will use λ^*-expressions as in Subsection 3.3.1 to denote elements of K_2, exploiting the fact that K_2 is a lax PCA.

For the type $\mathtt{N}$, the respective assembly structures on $\mathbb{N}$ are given by

$$f \Vdash^{\alpha}_{\mathtt{N}} n \quad \text{iff} \quad f(0) = n \wedge \forall i.\, f(i+1) = 0\,,$$
$$f \Vdash^{\beta}_{\mathtt{N}} n \quad \text{iff} \quad (\forall i.\, f(i) = \lceil \bot \rceil \vee f(i) = \lceil n \rceil) \wedge (\exists i.\, f(i) = \lceil n \rceil)\,.$$

It is thus easy to see that a weak associate for n may be obtained continuously from an associate for n and vice versa, and it is routine to construct elements of K_2 that effect these transformations. We thus obtain mutually inverse morphisms $U_0 : \mathsf{Ct}(\mathtt{N}) \to \mathsf{Ct}^K(\mathtt{N})$ and $V_0 : \mathsf{Ct}^K(\mathtt{N}) \to \mathsf{Ct}(\mathtt{N})$ in $\mathscr{A}sm(K_2)$, tracked say by $u_0, v_0 \in K_2$.

For the induction step, let us suppose we have mutually inverse morphisms $U_k : \mathsf{Ct}(k) \to \mathsf{Ct}^K(k)$ and $V_k : \mathsf{Ct}^K(k) \to \mathsf{Ct}(k)$ in $\mathscr{A}sm(K_2)$, tracked by u_k, v_k. We wish to establish the corresponding situation at type $k+1$.

To construct U_{k+1}, consider an arbitrary $F \in \mathsf{Ct}(k+1)$ with weak associate $f \Vdash^{\beta}_{k+1} F$. Then the corresponding function $\widehat{F} = U_0 \circ F \circ V_k$ is present in $\mathsf{Ct}^K(k+1)$ since it has the associate $\lambda^* z.u_0(h_{k0} f(v_k z))$, where h_{k0} is a realizer for ψ at types $k,0$. Moreover, the passage from F to $\widehat{F}$ is itself tracked by $\lambda^* fz.u_0(h_{k0} f(v_k z))$. This yields a morphism $U_{k+1} : \mathsf{Ct}(k+1) \to \mathsf{Ct}^K(k+1)$ in $\mathscr{A}sm(K_2)$ commuting with application.

To construct V_{k+1}, consider an arbitrary $G \in \mathsf{Ct}^K(k+1)$ with associate $g \Vdash^{\alpha}_{k+1} G$. Then the corresponding function $\check{G} = V_0 \circ G \circ U_k$ is present in $\mathscr{A}sm(K_2)$ as a morphism $\mathsf{Ct}(k) \to \mathsf{Ct}(0)$ with realizer $\check{g} = \lambda^* z.v_0(g(u_k z))$. Our main task is to construct an element $q \in \mathsf{PC}(\sigma \to \tau)$ tracking $\check{G}$ as a morphism in $\mathscr{A}sm(\mathsf{PC})$, and to show that some ψ-realizer for q is K_2-computable from $\check{g}$. This will show that $\check{G}$ is present in $\mathsf{Ct}(k+1)$, and that a weak associate for $\check{G}$ is K_2-computable from $\check{g}$ and so from g.

Let us write $\check{g}[i, m_0, \ldots, m_{t-1}] = n$ if $\check{g}(\langle i, m_0, \ldots, m_{t-1} \rangle) = n+1$ but for all $s < t$ we have $\check{g}(\langle i, m_0, \ldots, m_{s-1} \rangle) = 0$; clearly this implies that $(\check{g} \odot' h)(i) = n$ whenever $h(0) = m_0, \ldots, h(t-1) = m_{t-1}$. We may now read off from $\check{g}$ a set $\mathscr{G}$ of compact elements in $\mathsf{PC}(k+1)$ as follows:

$$\mathscr{G} \;=\; \{(a_0 \sqcup \cdots \sqcup a_{t-1} \Rightarrow n) \mid \check{g}[0, \lceil a_0 \rceil, \ldots, \lceil a_{t-1} \rceil] \;=\; n\}\,.$$

We first show that $\mathscr{G}$ is consistent. Suppose $(a \Rightarrow n), (a' \Rightarrow n') \in \mathscr{G}$ where $a = a_0 \sqcup \cdots \sqcup a_{t-1}$ and $a' = a'_0 \sqcup \cdots \sqcup a'_{t'-1}$, and suppose a, a' are consistent. By Theorem 8.1.5, we may find some total element $X \in V_{a \sqcup a'} \subseteq \mathsf{Ct}(k)$. Moreover, by Lemma 8.1.12 we see that $X \in V_a$ and $X \in V_{a'}$, so we may take realizers $x, x' \Vdash X$ extending a, a' respectively. Clearly, we may now take $h \Vdash^{\psi} a$ and $h' \Vdash^{\psi} a'$ such that $h(j) = \lceil a_j \rceil$ for all $j < t$ and $h'(j) = \lceil a'_j \rceil$ for all $j < t'$. Thus $(\check{g} \odot' h)(0) = n$ and $(\check{g} \odot' h')(0) = n'$. But since $\check{g}$ tracks $\check{G}$ in $\mathscr{A}sm(K_2)$ and $h, h' \Vdash^{\beta} X \in \mathsf{Ct}(k)$, we have that both $\check{g} \odot h$ and $\check{g} \odot h'$ are weak associates for $\check{G}(X) \in \mathsf{Ct}(0)$, hence $n = \check{G}(X) = n'$. Thus $\mathscr{G}$ is consistent.

Now define $q = \bigsqcup \mathscr{G}$. The above reasoning already makes clear that q is a realizer for $\check{G}$: given $X \in \mathsf{Ct}(k)$ and $x \Vdash X$, take any $h \Vdash^{\psi} x$ so that h is a weak associate for

$\check{G}$; then $\check{g} \odot h$ is a weak associate for $n = \check{G}(X)$ so $(\check{g} \odot' h)(0) = n$, whence there is some $(a \Rightarrow n) \in \mathscr{G}$ with $a \sqsubseteq x$, so that $q(x) = n$. This establishes that $\check{G}$ is present in $\mathsf{Ct}(k+1)$.

Clearly, any coded enumeration of $\mathscr{G}$ will serve as a weak associate for $\check{G}$. Moreover, it is easy to see that some such enumeration is K_2-computable from $\check{g}$: for example, the enumeration $\widetilde{g} \in K_2$ given by

$$\widetilde{g}(\lceil b \rceil) = \begin{cases} \lceil b \rceil & \text{if } b = (a_0 \sqcup \cdots \sqcup a_{t-1} \Rightarrow n) \text{ where } \check{g}[0, \lceil a_0 \rceil, \ldots, \lceil a_{t-1} \rceil] = n\,, \\ \lceil \bot \rceil & \text{otherwise.} \end{cases}$$

Since $\check{g}$ is in turn K_2-computable from g, we have that the passage from G to $\check{G}$ is itself tracked by an element of K_2, so as required we have a morphism $V_{k+1} : \mathsf{Ct}^K(k+1) \to \mathsf{Ct}(k+1)$ commuting with application. $\square$

Exercise 8.2.2. Verify that the realizers u_k, v_k given by the above proof may be taken to be elements of K_2^{eff}, assuming the codings used are effective. Deduce that $\mathsf{Ct}^{K,\mathrm{eff}} \simeq \mathsf{Ct}^{\mathrm{eff}}$ realizably in K_2^{eff}, where $\mathsf{Ct}^{K,\mathrm{eff}}$ is the evident substructure of Ct^K consisting of elements with an effective associate.

We now present a density theorem for Ct^K due to Kleene [140]. This gives a decidable criterion for when some finite portion u of a function $\mathbb{N} \to \mathbb{N}$ can be extended to an associate for some $F \in \mathsf{Ct}^K(\sigma)$, and shows how such an F may be computed primitively recursively from u. (This is closely related to Theorem 8.1.31, although we shall leave it to the interested reader to spell out the connections.) The machinery used in the proof will be applied in the next subsection.

Let P denote the set of finite partial functions $u : \mathbb{N} \rightharpoonup_{\mathrm{fin}} \mathbb{N}$, represented intensionally as finite lists of pairs so that the relation $u(i) = j$ is deemed decidable. Let $Q \subseteq P$ consist of those functions whose domain is some initial segment $\{0, \ldots, t-1\}$ of $\mathbb{N}$. We will henceforth let u, v range over Q, and will furthermore assume that our coding $\langle \cdots \rangle : \mathbb{N}^* \to \mathbb{N}$ has the property that $\langle \vec{x} \rangle \leq \langle \vec{y} \rangle$ whenever $\vec{x}$ is a prefix of $\vec{y}$.

Given $u, v \in Q$, define $u \boxdot v \in P$ by

$$\begin{aligned}(u \boxdot v)(n) = m \quad \text{iff } \exists r.\ & u(\langle n, v(0), \ldots, v(r-1) \rangle) = m+1 \ \wedge \\ & \forall s < r.\ u(\langle n, v(0), \ldots, v(s-1) \rangle) = 0\,.\end{aligned}$$

Clearly $\boxdot : Q \times Q \to P$ is computable, because for any u, v and n, the appropriate value of r (if it exists) will be bounded by the size of dom u. Furthermore, if $f, g \in K_2$, $u \subseteq f$, $v \subseteq g$ and $(u \boxdot v)(n) = m$, then clearly $(f \odot' g)(n) = m$.

Restricting attention to the pure types, we shall define, for each k, a binary relation $Con_k(u_0, u_1)$ on Q, expressing the idea that u_0, u_1 embody *consistent* information about some notional element $F \in \mathsf{Ct}(k)$:

1. $Con_0(u_0, u_1)$ iff there exists n such that $u_0, u_1 \subseteq \widehat{n}$.
2. $Con_{k+1}(u_0, u_1)$ iff whenever $Con_k(v_0, v_1)$ we have $Con_0(u_i \boxdot v_0, u_j \boxdot v_1)$ for all $i, j \in \{0, 1\}$.

An easy induction on k shows that the relations Con_k are decidable: in testing whether $Con_{k+1}(u_0, u_1)$ for a given u_0, u_1, there is a computable bound on the size of the v_0, v_1 that need be considered. It is also immediate from the definition that each Con_k is symmetric, and that $Con_k(u_0, u_1)$ implies $Con_k(u_0, u_0)$. We will write $Q_k = \{u \in Q \mid Con_k(u,u)\}$; thus $Con_{k+1}(u_0, u_1)$ iff $u_0, u_1 \in Q_{k+1}$ and whenever $Con_k(v_0, v_1)$ we have $Con_k(u_0 \boxdot v_0, u_1 \boxdot v_1)$. Note too that each Q_k is closed under prefixes.

Theorem 8.2.3 (Kleene density theorem). *For $k \in \mathbb{N}$ and $u_0, u_1 \in Q$, the following are equivalent.*

1. *$Con_k(u_0, u_1)$.*
2. *There is some $F \in \mathsf{Ct}(k)$ with associates $f_0 \supseteq u_0$ and $f_1 \supseteq u_1$.*

Moreover, when these conditions hold, suitable values of F, f_0, f_1 may be computed by Kleene primitive recursion, uniformly in k, u_0, u_1.

Proof. By induction on k, we shall construct Kleene primitive recursive functions $E_k : P \times P \to \mathsf{Ct}^K(k)$ and $\varepsilon_k^0, \varepsilon_k^1 : Q \times Q \to K_2$ such that whenever $Con_k(u_0, u_1)$ we have $u_0 \subseteq \varepsilon_k^0(u_0, u_1) \Vdash_k^\alpha E_k(u_0, u_1)$ and $u_1 \subseteq \varepsilon_k^1(u_0, u_1) \Vdash_k^\alpha E_k(u_0, u_1)$. The primitive recursiveness in k will then be clear from the construction.

For $k = 0$ and $k = 1$ the construction is easy and is left to the reader. For $k \geq 2$, the implication $2 \Rightarrow 1$ is also easy, so we concentrate on the algorithm for computing $E_k(u_0, u_1)$, assuming $Con_k(u_0, u_1)$. First, since there are only finitely many v and n for which $(u_0 \boxdot v)(n)$ or $(u_1 \boxdot v)(n)$ is defined, it is clear that there is a finite set $T \subseteq Q_k$, computable from u_0, u_1, such that for any $v \in Q_k$ there exists $v' \in T$ with $v' \subseteq v$, $u_0 \boxdot v' = u_0 \boxdot v$ and $u_1 \boxdot v' = u_1 \boxdot v$. Next, whenever $v_0, v_1 \in T$ and $\neg Con_0(u_0 \boxdot v_0, u_1 \boxdot v_1)$, we have $\neg Con_{k-1}(v_0, v_1)$, so there is a pair w_0, w_1 such that $Con_{k-2}(w_0, w_1)$ and $\neg Con_0(v_0 \boxdot w_0, v_1 \boxdot w_1)$; moreover, from T we may compute some finite symmetric set W of consistent pairs (w_0, w_1) that suffice for detecting all inconsistent $v_0, v_1 \in T$ in this way. We may now compute $E_k(u_0, u_1)(G)$ as follows for an arbitrary $G \in \mathsf{Ct}(k-1)$; we here write $u \cdot v$ for $(u \boxdot v)(0)$.

- See whether there exist $v_1 \in T$ and $i \in \{0, 1\}$ such that $u_i \cdot v_1 \downarrow$ and for all $(w_0, w_1) \in W$ we have $G(E_{k-2}(w_0, w_1)) \succeq v_1 \cdot w_1$.
- If so, return $u_i \cdot v_1$, otherwise return 0.

Notice that if both (v_1, i) and (v_1', i') satisfy the above criteria then v_1, v_1' are consistent, so that $Con_0(u_i \boxdot v_1, u_{i'} \boxdot v_1')$ and hence $u_i \cdot v_1 = u_{i'} \cdot v_1'$.

Finally, we show how to define a suitable associate $\varepsilon_k^0(u_0, u_1)$ for $E_k(u_0, u_1)$. Extend u_0 to an element $f_0 \in K_2$ by defining $f_0(y)$ as follows for $y \notin \mathrm{dom}\, u_0$:

- If $y = \langle n+1, \vec{z} \rangle$, set $f_0(y) = 1$.
- If $y = \langle 0, \vec{v_0} \rangle$ where for all $(w_0, w_1) \in W$ we have $v_0 \cdot \varepsilon_{k-2}^0(w_0, w_1) \downarrow$ (that is, $v_0 \cdot w \downarrow$ for some $w \subseteq \varepsilon_{k-2}^0(w_0, w_1)$), then there are two subcases:
  - If there exist $v_1 \in T$, $i \in \{0, 1\}$ such that $u_i \cdot v_1 \downarrow$ and for all $(w_0, w_1) \in W$ we have $v_0 \cdot \varepsilon_{k-2}^0(w_0, w_1) \succeq v_1 \cdot w_1$, then set $f_0(y) = u_i \cdot v_1 + 1$.
  - If not, set $f_0(y) = 1$.
- Otherwise, take $f_0(y) = 0$.

Then $u_0 \subseteq f_0$, and it is tedious but not hard to verify that $f_0 \Vdash^\alpha E_k(u_0,u_1)$; we may therefore set $\varepsilon_k^0(u_0,u_1) = f_0$. The case of ε_k^1 is treated symmetrically. □

Note that our proof shows that the elements $\varepsilon_k^0(u,u)$ for $u \in Q_k$ are dense in the set $\mathsf{Ext}(K_2;N)(k)$ of all associates for type $\bar{k}$. It is also immediate from the above proof that $E_k(u_0,u_1) = E_k(u_1,u_0)$ whenever $Con_k(u_0,u_1)$.

8.2.2 The Modified Associate Construction

As a variation on the construction of Ct^K as $\mathsf{EC}(K_2;N)$, we may consider the *modified extensional collapse* $\mathsf{MEC}(K_2;N)$. Here we first define the non-extensional model $\mathsf{ICF} = \mathsf{Tot}(K_2;N)$ (known in the literature as the model of *intensional continuous functionals*), and then define a model MCt^K as $\mathsf{EC}(\mathsf{ICF})$. This was suggested by Troelstra [288] as an alternative way of building a type structure of continuous functionals.

As we saw in Subsection 8.1.1, in the case of PC it is clear that the standard and modified extensional collapses agree, since the structures $\mathsf{Tot}(\mathsf{PC};\mathbb{N})$ and $\mathsf{Ext}(\mathsf{PC};\mathbb{N})$ coincide. For K_2, however, the situation is far less trivial: there are elements $f \in \mathsf{ICF}(2)$ that are not in $\mathsf{Ext}(K_2;N)(2)$, and at the next level, there will be elements $g \in \mathsf{Ext}(K_2;N)(3)$ with $g \odot f$ undefined for such f, so that $g \notin \mathsf{ICF}(3)$. Nonetheless, it was shown by Bezem [35] that the type structures MCt^K and Ct^K are isomorphic (computably within K_2). We prove this here using the tools developed in the proof of Theorem 8.2.3.

The following two lemmas show that the realizers $\varepsilon_k^i(u_0,u_1)$ constructed above are in fact common to both $\mathsf{Ext}(K_2;N)$ and $\mathsf{Ext}(\mathsf{ICF})$. We shall write $\approx_k'$ for the partial equivalence relation on $\mathsf{ICF}(k)$ associated with $\mathsf{MCt}^K(k)$.

Lemma 8.2.4. *If $Con_k(u_0,u_1)$, then $\varepsilon_k^i(u_0,u_1) \in \mathsf{ICF}(k)$ for $i = 0,1$.*

Proof. We argue by induction on k, taking $i = 0$ without loss of generality. The lemma is trivial for $k = 0,1$, so suppose $k \geq 2$ and $Con_k(u_0,u_1)$, To see that $f_0 = \varepsilon_k^0(u_0,u_1) \in \mathsf{ICF}(k)$, consider an arbitrary $g \in \mathsf{ICF}(k-1)$. From the construction of f_0 we see that the value of $f_0 \odot g$, if any, depends on the value of g at finitely many $w \in Q_{k-2}$. More precisely, let W be as in the proof of Theorem 8.2.3; then for all $(w_0,w_1) \in W$ we have that $\varepsilon_{k-2}^0(w_0,w_1) \in \mathsf{ICF}(k-2)$ by the induction hypothesis, so that $g \odot \varepsilon_{k-2}^0(w_0,w_1) \in N$. Let v be some segment of g such that $v \cdot \varepsilon_{k-2}^0(w_0,w_1)$ for all such w_0,w_1; then the construction of f_0 shows that $f_0(\langle 0,\vec{v}\rangle) > 0$, which suffices to show that $(f_0 \odot' g)(0)\downarrow$. Since clearly also $(f_0 \odot' g)(n+1) = 0$ for each n, we have $f_0 \odot g \in N$; and since g was arbitrary, we have shown $f_0 \in \mathsf{ICF}(k-1)$. □

For $f_0, f_1 \in K_2$, let us write $Con_k(f_0,f_1)$ if we have $Con_k(u_0,u_1)$ for all $u_0,u_1 \in P$ with $u_0 \subseteq f_0$, $u_1 \subseteq f_1$.

Lemma 8.2.5. *Suppose $f_0, f_1 \in \mathsf{ICF}(k)$. Then $f_0 \approx_k' f_1$ if and only if $Con_k(f_0,f_1)$.*

Proof. By induction on k. The case $k = 0$ is trivial, so suppose $k \geq 1$ and consider $f_0, f_1 \in \mathsf{ICF}(k)$.

First assume $f_0 \approx'_k f_1$. To show $Con_k(f_0, f_1)$, suppose $u_0 \subseteq f_0$ and $u_1 \subseteq f_1$. Given any v_0, v_1 with $Con_{k-1}(v_0, v_1)$, we have by Lemma 8.2.4 that $\varepsilon^i = \varepsilon^i_{k-1}(v_0, v_1) \in \mathsf{ICF}(k-1)$ for $i = 0, 1$. Moreover, it is easy to see that $\varepsilon^0, \varepsilon^1$ are consistent, so by the induction hypothesis we have $\varepsilon_0 \approx'_{k-1} \varepsilon_1$. Hence $f_i \odot \varepsilon_0 = f_j \odot \varepsilon_1 \in N$ for all $i, j \in \{0, 1\}$, and this shows that $u_i \boxdot v_0$ and $u_j \boxdot v_1$ are consistent for all i, j. Thus $Con_k(f_0, f_1)$.

For the converse, suppose $Con_k(f_0, f_1)$. To show $f_0 \approx'_k f_1$, suppose $g_0 \approx'_{k-1} g_1$. Then $g_0, g_1 \in \mathsf{ICF}(k-1)$ so that $f_0 \odot g_0, f_1 \odot g_1 \in N$; but also $Con_k(g_0, g_1)$ by the induction hypothesis, whence $f_0 \odot g_0 = f_1 \odot g_1$. Thus $f_0 \approx'_k f_1$. □

In particular, the above lemma shows that $\varepsilon^0(u_0, u_1) \in \mathsf{Ext}(\mathsf{ICF})$ for any consistent u_0, u_1. We may now use this knowledge to map each element of ICF to an extensional one:

Lemma 8.2.6. *For each k there is an element $\theta_k \in K_2$ such that:*

1. $\theta_k \odot f \in \mathsf{Ext}(\mathsf{ICF})(k)$ *whenever* $f \in \mathsf{ICF}(k)$,
2. $\theta_k \odot f = f$ *whenever* $f \in \mathsf{Ext}(\mathsf{ICF})(k)$.

Proof. For any $f \in K_2$, define $\Theta_k(f) \in K_2$ as follows, writing f_n for the restriction of f to $\{0, \ldots, n\}$:

- If $Con_k(f_n, f_n)$, take $\Theta_k(f)(n) = f(n)$.
- Otherwise, let $m < n$ be the largest number such that $Con_k(f_m, f_m)$, and take $\Theta_k(f)(n) = \varepsilon^0_k(f_m, f_m)(n)$.

Clearly $\Theta_k : \mathbb{N}^{\mathbb{N}} \to \mathbb{N}^{\mathbb{N}}$ is continuous, so is represented by some $\theta_k \in K_2$. Using Lemma 8.2.5, it is easy to verify that θ_k has the required properties. □

We may now obtain Bezem's theorem from [35]:

Theorem 8.2.7. $\mathsf{MCt}^K \cong \mathsf{Ct}^K$ *realizably over K_2.*

Proof. Regarding each set $\mathsf{Ct}^K(k)$ and $\mathsf{MCt}^K(k)$ as an assembly over K_2 in the obvious way, we shall show by induction on k that $\mathsf{MCt}^K(k) \cong \mathsf{Ct}^K(k)$ within $\mathscr{A}sm(K_2)$, where the isomorphisms respect application. For $k = 0$ the identity morphisms suffice; and given mutually inverse isomorphisms $I_k : \mathsf{MCt}^K(k) \to \mathsf{Ct}^K(k)$ and $J_k : \mathsf{Ct}^K(k) \to \mathsf{MCt}^K(k)$ tracked respectively by $i_k, j_k \in K_2$, we may define the corresponding isomorphisms I_{k+1}, J_{k+1} at type $k+1$ to be those tracked by

$$i_{k+1} = \lambda^* f.\lambda^* x.f(j_k x), \qquad j_{k+1} = \lambda^* g.\lambda^* y.g(i_k(\theta_k y)),$$

with θ_k as in Lemma 8.2.6.

A further application of θ_k shows that the functions present in $\mathsf{MCt}^K(k \to l)$ coincide with those in $\mathsf{Ct}^K(k \to l)$, and with the morphisms $\mathsf{Ct}^K(k) \to \mathsf{Ct}^K(l)$ in $\mathscr{A}sm(K_2)$. By the results of Section 4.2, it now follows that MCt^K and Ct^K are isomorphic at all types, with the isomorphisms realizable in K_2. □

Exercise 8.2.8. Check that the above proof effectivizes, so that MCt^K and Ct^K are isomorphic realizably over K_2^{eff}, as are the evident substructures $\mathsf{MCt}^{K,\mathrm{eff}}$ and $\mathsf{Ct}^{K,\mathrm{eff}}$.

8.2.3 Limit Spaces

We now turn to a direct characterization of the Kleene–Kreisel functionals as those satisfying a certain continuity property, without reference to any realizing objects. Among such characterizations, there is one that has so far proved to be more useful than the others, namely the characterization via *limit spaces*. The construction of this model is due to Scarpellini [255], and its equivalence to Ct is due to Hyland [120]. As we shall see, much of the computability theory of Ct can be carried out within the framework of limit spaces, leading in some cases to sharper results than other approaches; moreover, the limit space approach generalizes well to type structures over other base types such as $\mathbb{R}$.

In this paragraph, we will write ω for the set of natural numbers considered as an ordinal, and $\omega+1$ for its successor $\omega \cup \{\omega\}$. By a *sequence with limit* in a set X, we shall mean simply a function $f : (\omega+1) \to X$. We will often use notations such as $\{f(i)\}_{i\leq\omega}$ or $\{x_i\}_{i\leq\omega}$ for sequences with limit. The idea is to think of $\omega+1$ as a topological space, with $O \subseteq \omega+1$ open iff $\omega \in O \Rightarrow \exists n \in \omega. \forall i > n. i \in O$, so that $\omega = \lim_{i<\omega} i$.

If $(X, \mathscr{T})$ is a topological space, the continuous functions $\omega+1 \to X$ are precisely the *convergent sequences* f with $f(\omega) = \lim_{i<\omega} f(i)$. The set of all such sequences will be called the *limit structure* of X, and will evidently carry information about the topology $\mathscr{T}$ itself. This motivates the following definition, taken from Kuratowski [160], which axiomatizes the essential properties of such limit structures without reference to open sets.

Definition 8.2.9. A *limit space* is a set X equipped with a set L of sequences with limits $f : (\omega+1) \to X$ (called the *convergent sequences* of the limit space) satisfying the following axioms:

1. If $f : (\omega+1) \to X$ is such that $f(i) = f(\omega)$ for all but finitely many i, then $f \in L$.
2. If $f \in L$ and $g : (\omega+1) \to (\omega+1)$ is strictly increasing, then $f \circ g \in L$.
3. If $f \notin L$, there is a strictly increasing $g : (\omega+1) \to (\omega+1)$ such that for no strictly increasing $h : (\omega+1) \to (\omega+1)$ do we have that $f \circ g \circ h \in L$.

We shall write $x_i \to x$ (as $i \to \omega$) to mean that there is a convergent sequence f with $f(i) = x_i$ for all $i < \omega$ and $f(\omega) = x$. In general, a given sequence $\{x_i\}_{i<\omega}$ may have more than one limit point, although in the particular spaces we shall be interested in, limits are unique when they exist.

In the above definition, condition 1 says that any almost constant sequence converges; condition 2 says that any subsequence of a convergent sequence is convergent with at least the same limit points; and condition 3 says that if a sequence does not have x as a limit, then there is a subsequence that avoids x in the sense that no further subsequences will have x as a limit.

It is easy to check that the set of convergent sequences in any topological space $(X, \mathscr{T})$ will satisfy these axioms; we thus have an operation Lim mapping topological spaces to limit spaces. In the other direction, if (X,L) is a limit space, we may define a topology on X by letting $O \subseteq X$ be open iff for all $f \in L$ we have

$f(\omega) \in O \Rightarrow \exists n. \forall i > n. f(i) \in O$. It is routine to check that this is indeed a topology, so we have an operation Top mapping limit spaces to topological spaces.

Definition 8.2.10. (i) A limit space (X,L) is *topological* if $(X,L) = \mathsf{Lim}(\mathsf{Top}(X,L))$.
(ii) A topological space $(X,\mathscr{T})$ is *sequential* if $(X,\mathscr{T}) = \mathsf{Top}(\mathsf{Lim}(X,\mathscr{T}))$.

Exercise 8.2.11. (i) Show that if $(X,L') = \mathsf{Lim}(\mathsf{Top}(X,L))$ then $L \subseteq L'$, and if $(X,\mathscr{T}') = \mathsf{Top}(\mathsf{Lim}(X,\mathscr{T}))$ then $\mathscr{T} \subseteq \mathscr{T}'$.
(ii) Deduce that a limit space is topological iff it is in the image of Top, and a topological space is sequential iff it is in the image of Lim.

All metric spaces are clearly sequential, but there are examples of topological spaces that are not (see Remark 8.2.22 below). There are also sequential spaces that are not topological, though these will be of no concern to us.

We now turn to consider the appropriate notion of continuous function between limit spaces.

Definition 8.2.12. Let (X,L) and (Y,M) be limit spaces.
(i) A function $F : X \to Y$ is *continuous* if $F \circ f \in M$ whenever $f \in L$.
(ii) Let $(X,L) \Rightarrow (Y,M)$ be the limit space defined as follows: the underlying set is the set Z of all continuous functions $F : X \to Y$; and the limit structure consists of all $\Phi : (\omega+1) \to Z$ such that for all $f \in L$ we have $\Lambda i.\, \Phi(i)(f(i)) \in M$.

In this definition, condition (i) just says that F maps convergent sequences with limits to convergent sequences with limits, while condition (ii) says that the diagonal application of a sequence of functions to a convergent sequence of points yields a convergent sequence. In the case of metric spaces, the above definition of continuity is the standard one, and a sequence of continuous functions is convergent in the function space if it is pointwise convergent and equicontinuous (see Definition 8.4.1).

Exercise 8.2.13. (i) Check that $(X,L) \Rightarrow (Y,M)$ is indeed a limit space. Verify that limit spaces and continuous functions form a cartesian closed category.
(ii) Show that we may have $F_i(x) \to F(x)$ for every $x \in X$ but $F_i \not\to F$.

We may now readily obtain a type structure over $\mathbb{N}$:

Definition 8.2.14. (i) Let $\mathsf{Ct}^L(\mathtt{N})$ denote the limit space $(\mathbb{N}, L_{\mathbb{N}})$, where $f \in L_{\mathbb{N}}$ iff $\exists n. \forall i > n. f(i) = f(\omega)$. Given limit spaces $\mathsf{Ct}^L(\sigma)$ and $\mathsf{Ct}^L(\tau)$, let $\mathsf{Ct}^L(\sigma \to \tau)$ denote the limit space $\mathsf{Ct}^L(\sigma) \Rightarrow \mathsf{Ct}^L(\tau)$ as defined above.
(ii) Let Ct^L be the extensional applicative structure over $\mathbb{N}$ consisting of the sets $\mathsf{Ct}^L(\sigma)$ with the inherent application operations.

Since the category of limit spaces is cartesian closed, it is automatic that Ct^L is a TPCA and hence a total type structure over $\mathbb{N}$. We shall prove the equivalence $\mathsf{Ct}^L \cong \mathsf{Ct}$ in due course (Theorem 8.2.21), but before doing so, we shall pause to develop some computability theory for Ct^L purely on the basis of the above construction, in order to illustrate the power and utility of the limit space definition.

One of the most striking aspects of the limit space approach is that it leads to some non-trivial information on moduli of convergence. We here import the concept of a *modulus function* which also played a role in Section 5.3:

Definition 8.2.15. Let $\sigma = \sigma_0, \ldots, \sigma_{r-1} \to \mathtt{N}$ and let $\{F_i\}_{i \leq \omega}$ be any sequence with limit in $\mathsf{Ct}^L(\sigma)$. We say that $G : \mathsf{Ct}^L(\sigma_0) \times \cdots \times \mathsf{Ct}^L(\sigma_{r-1}) \to \mathbb{N}$ is a *modulus function* for this sequence if

$$\forall \vec{x} \in \mathsf{Ct}^L(\sigma_0) \times \cdots \times \mathsf{Ct}^L(\sigma_{r-1}).\ \forall i \geq G(\vec{x}).\ F_i(\vec{x}) = F_\omega(\vec{x}) .$$

The following pleasing theorem, due to Hyland [120], shows that functions in Ct^L always have modulus functions within Ct^L. We give here a more recent proof due to the second author. The argument is non-constructive (cf. Remark 9.5.7).

Theorem 8.2.16. *Let $\{F_i\}_{i \leq \omega}$ be any sequence with limit in $\mathsf{Ct}^L(\sigma)$ where $\sigma = \sigma_0, \ldots, \sigma_{r-1} \to \mathtt{N}$. Then the following are equivalent:*

1. *F is convergent, that is, $F_i \to F_\omega$.*
2. *There is a modulus function G for F within $\mathsf{Ct}^L(\sigma)$.*

Proof. For notational simplicity we treat just the case $r = 1$ and write τ for σ_0; the same argument works in the general case.

For the implication $2 \Rightarrow 1$, assume $G \in \mathsf{Ct}^L(\sigma)$ is a modulus for F, and suppose $x_i \to x_\omega$ is a convergent sequence in $\mathsf{Ct}^L(\tau)$. Let $m = G(x_\omega)$, let n be a modulus of convergence for $G(x_i) \to G(x_\omega)$, and let p be a modulus of convergence for $F_\omega(x_i) \to F_\omega(x_\omega)$. Then for all $i > m, n, p$ we have that $F_i(x_i) = F_\omega(x_i) = F_\omega(x_\omega)$. Thus F is itself convergent according to Definition 8.2.12(ii).

For $1 \Rightarrow 2$, assume F is convergent. Define $G : \mathsf{Ct}^L(\tau) \to \mathbb{N}$ by

$$G(x) \ = \ \mu m.\ \forall i \geq m.\ F_i(x) = F_\omega(x) .$$

Then G is everywhere defined and is clearly a modulus function for F, so it suffices to show that $G \in \mathsf{Ct}^L(\sigma)$. If not, there will be a convergent sequence $x_i \to x_\omega$ in $\mathsf{Ct}^L(\tau)$ such that $G(x_i) \not\to G(x_\omega)$ in $\mathbb{N}$. By restricting to a suitable subsequence, we may in fact assume that $G(x_i) \neq G(x_\omega)$ for all $i < \omega$.

Case 1: $\{G(x_i) \mid i \in \omega\}$ is bounded. Then $G(x_i)$ takes some value t infinitely often, so by restricting again to a suitable subsequence, we may as well assume that $G(x_i) = t$ for all i. Thus F has modulus t at each i, that is, for all $i < \omega$ and all $j \geq t$ we have $F_j(x_i) = F_\omega(x_i)$. But F_ω and each F_j are continuous, so by taking limits as $i \to \omega$, we have $F_j(x_\omega) = F_\omega(x_\omega)$ for all $j \geq t$. This shows that $G(x_\omega) \leq t$; but since $G(x_\omega) \neq G(x_0) = t$, we must have $G(x_\omega) \leq t-1$. In particular, $F_{t-1}(x_\omega) = F_\omega(x_\omega)$, whence by continuity of F_{t-1} and F_ω we have $F_{t-1}(x_i) = F_\omega(x_i)$ for some $i < \omega$. But we already know that $F_j(x_i) = F_\omega(x_i)$ for all $j \geq t$, so the modulus of F at x_i is at most $t-1$, contradicting $G(x_i) = t$.

Case 2: $\{G(x_i) \mid i \in \omega\}$ is unbounded. Then again, by taking a suitable subsequence, we may assume $G(x_0), G(x_1), \ldots$ is strictly increasing and $G(x_0) \neq 0$. For each i, let $j_i = G(x_i) - 1$; then since $G(x_i)$ is the minimal modulus for F at x_i, we must have $F_{j_i}(x_i) \neq F_\omega(x_i)$. Thus $F_{j_i}(x_i) \neq F_\omega(x_\omega)$ for all sufficiently large i. But this is a contradiction, since $F_{j_0}, F_{j_1}, \ldots$ is a subsequence of F itself and so ought to be convergent. This completes the proof. □

Whilst a modulus function for a convergent sequence F is not itself uniformly computable from F_ω and the F_i, it turns out that we may uniformly compute a modulus for a convergent sequence $F_i(f_i) \to F_\omega(f_\omega)$ if we already have moduli for $F_i \to F_\omega$ and for $f_i \to f_\omega$. The construction here uses ideas of Grilliot [111] which also featured in Section 5.3.

Theorem 8.2.17. *Suppose $f_i \to f_\omega$ in $\mathsf{Ct}^L(\sigma)$ with modulus $g \in \mathsf{Ct}^L(\sigma)$.*
(i) If $F \in \mathsf{Ct}^L(\sigma)$, then a modulus function for $F(f_i) \to F(f_\omega)$ is μ-computable uniformly in F, f_ω, $\Lambda i.f_i$ and g.
(ii) If $F_i \to F_\omega$ in $\mathsf{Ct}^L(\sigma \to \tau)$ with modulus G, then a modulus function for $F_i(f_i) \to F_\omega(f_\omega)$ is μ-computable uniformly in F_ω, $\Lambda i.F_i$, f_ω, $\Lambda i.f_i$, G and g.

Proof. (i) We shall treat the case $\tau = \mathtt{N}$, the extension to arbitrary τ being an easy exercise.

For each $n \in \mathbb{N}$, define $h_n \in \mathsf{Ct}^L(\sigma)$ by

$$h_n(\vec{x}) = \begin{cases} f_\omega(\vec{x}) \text{ if } F(f_i) = F(f_\omega) \text{ whenever } n \leq i \leq g(\vec{x})\,, \\ f_i(\vec{x}) \text{ if } i \text{ is minimal such that } n \leq i \leq g(\vec{x}) \text{ and } F(f_i) \neq F(f_\omega)\,. \end{cases}$$

Then $h_n = f_\omega$ if $F(f_i) = F(f_\omega)$ for all $i \geq n$, otherwise $h_n = f_i$ for the smallest $i \geq n$ such that $F(f_i) \neq F(f_\omega)$. But now $\mu n.F(h_n) = F(f_\omega)$ is a modulus for $F_i(f_i) \to F_\omega(f_\omega)$, μ-computable in the data.

(ii) The proof of (i) clearly adapts to functionals of two arguments. By specializing to the application operation itself, we obtain (ii). □

We may apply this theorem to sharpen our earlier idea of approximating an arbitrary element x by elements from some fixed countable dense set. In Subsection 8.1.4, we considered a family of idempotents $(-)^\sigma_n \in \mathsf{Ct}(\sigma \to \sigma)$ for $n \in \mathbb{N}$, each with finite range, and clearly the definition of such idempotents makes equally good sense in Ct^L:

$$(m)^{\mathtt{N}}_n = \begin{cases} m \text{ if } m \leq n\,, \\ n \text{ if } m > n\,, \end{cases} \qquad (F)^{\sigma \to \tau}_n = (-)^\tau_n \circ F \circ (-)^\sigma_n\,.$$

It is easy to show directly that the range of each $(-)^\sigma_n$ is a finite subset of $\mathsf{Ct}^L(\sigma)$ and corresponds to $\mathsf{S}_n(\sigma)$, where S_n is the full set-theoretic type structure over $\{0, \ldots, n\}$.

In the setting of limit spaces, what we naturally obtain from this is not just a countable dense set in each $\mathsf{Ct}^L(\sigma)$, but also a construction of a particular sequence from this dense set converging to any given element x, with a modulus of convergence computable from x alone (cf. Theorem 5.3.5):

Lemma 8.2.18. *Let $x \in \mathsf{Ct}^L(\sigma)$. Then the sequence $(x)_0, (x)_1, \ldots$ is Kleene primitive recursive uniformly in x, and converges to x. Moreover, there is a modulus $g \in \mathsf{Ct}^L(\sigma)$ for $(x)_i \to x$ which is μ-computable uniformly in x.*

Proof. The primitive recursiveness is clear from the definition of the $(x)_i$, and the convergence is shown by an easy induction on σ. As regards the modulus g, it is routine to construct a primitive recursive functional $\Phi \in \mathsf{Ct}^L(1 \to \sigma)$ such that $\Phi(id) = x$ and $\Phi(\theta_i) = (x)_i$ for each i, where $\theta_i = (-)^{\mathrm{N}}_i$. But the θ_i clearly converge to id with a readily computable modulus of convergence, so by Theorem 8.2.17 we obtain a modulus for $(x)_i \to x$, μ-computable in x alone. □

Remark 8.2.19. In a similar vein, by working with Ct^L, the fact that our type structure is Kleene closed (cf. Theorem 8.1.27) may be proved in a somewhat sharpened form: not only is the value of $\{e\}(F_0, \ldots, F_{r-1})$ continuous in the arguments F_i, but suitable 'moduli of continuity' are Kleene computable uniformly in the relevant data. More precisely, if for each j we have $F_{ji} \to F_j$ with modulus G_j, and we have terminating computations for $\{e\}(\vec{F}) = n$ and $\{e\}(\vec{F_j}) = n_j$ for each j, then a modulus for $n_j \to n$ is Kleene computable in e and the F_j, F_{ji}, G_i.

To prove this, we write a self-referential algorithm for computing the modulus, apply the second recursion theorem, and then argue by induction on the computation tree for $\{e\}(\vec{F})$ much as in the proof of Theorem 8.1.27. (Cf. Theorem 5.3.7.)

The above results indicate that the definition of Ct^L is well suited to the development of a rich computability theory. In the sequel, we shall sometimes return to $\mathsf{Ct}^L(\sigma)$ when we wish to stress that what we are defining or proving makes no reference to representing objects of any kind.

It is now time to prove that $\mathsf{Ct}^L \cong \mathsf{Ct}$. In the course of doing so, we will also show that the canonical topology on each $\mathsf{Ct}(\sigma)$ is the one generated by the limit structure on $\mathsf{Ct}^L(\sigma)$, and conversely, that this limit structure is the one given by convergence in the canonical topology (cf. Definition 8.2.10). We begin with some simple topological observations:

Lemma 8.2.20. *(i) Each* $\mathsf{Tot}(\sigma)$ *(with the Scott topology) is a sequential space.*
(ii) Each $\mathsf{Ct}(\sigma)$ *(with the canonical topology) is a sequential space.*

Proof. (i) It will suffice to show that any countably based space $(X, \mathscr{T})$ is sequential. As noted in Exercise 8.2.11, if $U \in \mathscr{T}$ then U is *sequentially open*: that is, any sequence (x_i) converging under $\mathscr{T}$ to $x \in U$ is eventually within U. Conversely, suppose $U \notin \mathscr{T}$. Take $x \in U$ with no open neighbourhood inside U, and take $V_0 \supseteq V_1 \supseteq \cdots$ a neighbourhood base for x in $\mathscr{T}$. For each i, we may pick $x_i \in V_i - U$; then (x_i) converges to x and witnesses that U is not sequentially open. Thus the sequential topology on X coincides with $\mathscr{T}$.

(ii) It is routine to check that any quotient of a sequential space is sequential. □

The key to proving the equivalence $\mathsf{Ct}^L \cong \mathsf{Ct}$ will be to characterize the topologically convergent sequences in each $\mathsf{Ct}(\sigma)$, and show that they coincide with those in the limit structure. This information is made explicit in the following theorem:

Theorem 8.2.21. *The following hold for each type* σ.
(i) $\mathsf{Ct}^L(\sigma) = \mathsf{Ct}(\sigma)$.
(ii) For any sequence with limit $(F_i)_{i \leq \omega}$ *in* $\mathsf{Ct}(\sigma)$*, the following are equivalent:*

1. $F_i \to F_\omega$ *in the topology on* $\mathsf{Ct}(\sigma)$.
2. $F_i \to F_\omega$ *in the limit structure on* $\mathsf{Ct}^L(\sigma)$.
3. *There is a convergent sequence* $f_i \to f_\omega$ *in* $\mathsf{Ext}(\sigma)$ *such that* $f_i \Vdash_\sigma F_i$ *for all* i.

Proof. We prove both parts simultaneously by induction on σ. Both (i) and (ii) are trivial at type $\mathtt{N}$, so consider a type $\sigma \to \tau$ where (i) and (ii) hold for σ, τ and all structurally smaller types.

To establish (i) at type $\sigma \to \tau$, first note that any $F \in \mathsf{Ct}(\sigma)$ will map convergent sequences to convergent sequences, since any representative $f \Vdash_\sigma F$ in $\mathsf{Ext}(\sigma)$ will do so. Thus $\mathsf{Ct}(\sigma \to \tau)$ will correspond to a subset of $\mathsf{Ct}^L(\sigma \to \tau)$.

For the other inclusion, suppose $F \in \mathsf{Ct}^L(\sigma \to \tau)$. Write τ as $\tau_0, \ldots, \tau_{r-1} \to \mathtt{N}$, and recalling that limit spaces form a cartesian closed category, regard F as a continuous function $\mathsf{Ct}(\sigma) \times \mathsf{Ct}(\tau_0) \times \cdots \times \mathsf{Ct}(\tau_{r-1}) \to \mathbb{N}$ in the sense of limit spaces. Note that part (ii) of the induction hypothesis says that all relevant notions of convergence are equivalent in $\mathsf{Ct}(\sigma)$ and each $\mathsf{Ct}(\tau_j)$, and the limit structure on products is defined componentwise in the obvious way. We will write ρ for the product type $\sigma \times \tau_0 \times \cdots \times \tau_{r-1}$.

To show that $F \in \mathsf{Ct}(\rho \to \mathtt{N})$, we shall construct a domain representation f for F. Let a range over compacts in $\mathsf{PC}(\sigma)$, and b_j over compacts in $\mathsf{PC}(\tau_j)$ for each $j < r$. Using the notation of Definition 8.1.11, let $\mathscr{F}$ be the set of all step functions $((a, \vec{b}) \Rightarrow n)$ in $\mathsf{PC}(\rho \to \mathtt{N})$ such that for all $X \in V_a$ and $Y_j \in V_{b_j}$ we have $F(X, \vec{Y}) = n \in \mathbb{N}$. As in earlier proofs, an easy application of the density theorem shows that $\mathscr{F}$ is consistent, so we may take $f = \bigsqcup \mathscr{F}$ in $\mathsf{PC}(\rho \to \mathtt{N})$.

If we can show that $f \in \mathsf{Tot}(\rho \to \mathtt{N})$, it will follow that $f \Vdash F$ (recall that Tot = Ext within PC). So suppose f is not total, and take $X, \vec{Y} \in \mathsf{Ct}$ represented by $x, \vec{y} \in \mathsf{PC}$ such that $f(x, \vec{y}) = \bot$. Take $x_0, x_1, \ldots$ an increasing sequence of compacts with supremum x, and $y_{j0}, y_{j1}, \ldots$ an increasing sequence of compacts with supremum y_j for each j. Also set $n = F(X, \vec{Y})$.

For each i, since $f(x, \vec{y}) = \bot$ we have $((x_i, y_{0i}, \ldots, y_{(r-1)i}) \Rightarrow n) \notin \mathscr{F}$, so by the density theorem we may pick $X_i \in V_{x_i}$, $Y_{ji} \in V_{y_{ji}}$ such that $F(X_i, Y_{0i}, \ldots, Y_{(r-1)i}) \neq n$. But now we have $X_i \to X$ and $Y_{ji} \to Y_i$ in the topologies on $\mathsf{Ct}(\sigma)$, $\mathsf{Ct}(\tau_i)$, and hence in the corresponding limit structures. But this contradicts the assumption that F is limit-continuous; hence f is total after all and $f \Vdash F$. This establishes that F is present in $\mathsf{Ct}(\rho \to \mathtt{N})$, completing the proof of (i) for type $\sigma \to \tau$.

For (ii), we wish to show that statements 1–3 are equivalent for $\sigma \to \tau$. The implication $3 \Rightarrow 1$ is trivial. For $2 \Rightarrow 3$, we may reformulate 2 as saying that F is a limit-continuous function $(\omega + 1) \times \mathsf{Ct}^L(\sigma) \to \mathsf{Ct}^L(\tau)$ with respect to the evident limit structure on $\omega + 1$. Since $\omega + 1$ may be identified with the image of a certain continuous idempotent on $\mathsf{Ct}(1)$, this amounts to saying that the associated function $\overline{F} : \mathsf{Ct}(1) \times \mathsf{Ct}^L(\sigma) \to \mathsf{Ct}^L(\tau)$ is limit-continuous. We may now apply the argument from (i) to show that $\overline{F}$ is represented by a domain element $\overline{f} \in \mathsf{PC}(1 \times \sigma \to \tau)$, which in turn yields a convergent sequence of realizers $f_i \to f_\omega$ tracking $F_i \to F_\omega$. Thus $2 \Rightarrow 3$.

To show $1 \Rightarrow 2$, suppose $F_i \to F_\omega$ in the topology. As above, we will consider the F_i, F_ω as elements of $\mathsf{Ct}(\rho \to \mathtt{N})$, which by (i) we know to coincide with $\mathsf{Ct}^L(\rho \to \mathtt{N})$.

To show that $F_i \to F_\omega$ in the limit structure, suppose $X_i \to X$ in $\mathsf{Ct}(\sigma)$ and $Y_{ji} \to Y_j$ in each $\mathsf{Ct}(\tau_j)$. By part (ii) of the induction hypothesis, these may be tracked by convergent sequences of realizers $x_i \to x$ in $\mathsf{PC}(\sigma)$ and $y_{ji} \to y_j$ in $\mathsf{PC}(\tau_j)$.

Let $n = F_\omega(a,\vec{b})$. Since $F \in \mathsf{Ct}^L(\rho \to \mathtt{N})$, we may assume by discarding finitely many initial terms that $F_\omega(X_i,\vec{Y}_i) = n$ for all i. We wish to show that $F_i(X_i,\vec{Y}_i) = n$ for all sufficiently large i; and since $F_i \to F$ topologically, it will be enough to show that

$$O \;=\; \{G \in \mathsf{Ct}(\rho \to \mathtt{N}) \mid \forall i.\, G(X_i,\vec{Y}_i) = n\}$$

is open in $\mathsf{Ct}(\sigma)$.

So suppose $G \in O$ and $g \Vdash G$. Then $g(x,\vec{y}) = n$, so there are compacts $c \sqsubseteq g$ and $(a,\vec{b}) \sqsubseteq (x,\vec{y})$ such that $c(a,\vec{b}) = n$. We then have $(a,\vec{b}) \sqsubseteq (x_i,\vec{y}_i)$ for almost all i, since $(x_i,\vec{y}_i) \to (x,\vec{y})$ in the Scott topology, so for these i we have $c(x_i,\vec{y}_i) = n$. But for the finitely many remaining i, we have that F_i is continuous, so there will be compact approximations $c_i \sqsubseteq g$ securing that the value on $(x_i,\vec{y}_i)$ is n. But now the join c' of c with these c_i will still be a compact approximation to g, and any $G' \in V_{c'}$ will be constantly n on the sequence in question. Thus O is open in the canonical topology, and this completes the proof. □

Remark 8.2.22. In Lemma 8.1.23 we showed that application is not continuous with respect to the product topology on $\mathsf{Ct}(2) \times \mathsf{Ct}(1)$, although it is continuous for the canonical topology on $\mathsf{Ct}(2 \times 1)$. Since these two topologies give rise to the same set of convergent sequences, and the latter topology is sequential, we conclude that the product topology on $\mathsf{Ct}(2) \times \mathsf{Ct}(1)$ is not sequential.

We now note a small corollary to the above theorem. Recall the definition of V_p in Definition 8.1.11. Following Schröder [264], let us define a *pseudobase* in a topological space X to be a collection $\mathscr{B}$ of sets V such that whenever $x \in O$ where O is open in X and $x_n \to x$ in the topology, there is a $V \in \mathscr{B}$ such that $x \in V$, $V \subseteq O$ and $x_n \in V$ for all but finitely many n.

Corollary 8.2.23. *The set of neighbourhoods $V_{p,\sigma}$ is a pseudobase for* $\mathsf{Ct}(\sigma)$.

Proof. An easy consequence of the implication $1 \Rightarrow 3$ in Theorem 8.2.21(ii). □

This is in fact an instance of a much more general result. Each topological space $\mathsf{Ct}(\sigma)$ is clearly an object in the category $\mathbf{QCB}_0$ of T_0 spaces that are quotients of countably based spaces, a category with strong closure properties introduced by Battenfeld, Schröder and Simpson [11]. Schröder characterized QCB_0 spaces as the sequential spaces with a countable pseudobase.

We conclude this subsection with a further characterization of Ct due to Bergstra [31] which follows readily from the limit space characterization, and which gives some insight into the place of Ct within the poset of all total type structures over $\mathbb{N}$. Recall from Section 5.3 that the existential quantifier ${}^2\exists$ is the prototypical example of a discontinuous functional in $\mathsf{S}(2)$.

We shall here let $\mathbf{A}^{\leq k}$ denote a pure type structure over $\mathbb{N}$ defined up to and including the type $\mathbf{A}(k)$. Since we may interpret Kleene's S1–S9 in such a structure, it is clear what is meant by saying such a structure is Kleene closed.

Theorem 8.2.24. Ct *is hereditarily the maximal Kleene closed type structure not containing* ${}^2\exists$. *More precisely, at each level k, if* $\mathbf{A}^{\leq k}$ *is Kleene closed where* $\mathbf{A}(l) = \mathsf{Ct}(l)$ *for all* $l < k$ *and* $\mathbf{A}(k)$ *properly extends* $\mathsf{Ct}(k)$, *then* ${}^2\exists \in \mathbf{A}(2)$.

Proof. We have seen that Ct is Kleene closed, and clearly ${}^2\exists \notin \mathsf{Ct}(2)$. Now suppose $\mathbf{A}^{\leq k}$ satisfies the above hypotheses for some k; then clearly $k \geq 2$. Take $F \in \mathbf{A}(k) - \mathsf{Ct}(k)$; then by Theorem 8.2.21, F is not limit-continuous, so there is a convergent sequence $x_i \to x$ in $\mathbf{A}(k-1) = \mathsf{Ct}(k-1)$ such that $F(x_i) \not\to F(x)$. By selecting a suitable subsequence, we may assume that $F(x_i) \neq F(x)$ for all i.

As in the proof of Theorem 8.2.21, we can regard $x_i \to x$ as given by a limit-continuous function $(\omega+1) \to \mathsf{Ct}(k-1)$, where $(\omega+1)$ may be represented as a retract of $\mathsf{Ct}(1)$ associated with some standard convergent sequence $g_i \to g$ in $\mathsf{Ct}(1)$. There is therefore a function $X \in \mathsf{Ct}(1 \to (k-1))$ such that $X(g_i) = x_i$ for each i and $X(g) = x$. Encode X in a standard way as an element $X' \in \mathsf{Ct}(k)$, so that $X' \in \mathbf{A}(k)$. Since there is then a Kleene computable functional mapping (g', X', F) to $F(X(g'))$ for any $g' \in \mathsf{Ct}(1)$, and $\mathbf{A}$ is Kleene closed, we see that $\mathbf{A}(2)$ contains an element E mapping each g_i to $F(x_i) \neq F(x)$ and g to $F(x)$. If $g_i \to g$ is suitably chosen, the presence of ${}^2\exists \in \mathbf{A}(2)$ follows immediately. □

8.2.4 Compactly Generated Spaces

To conclude this section, we will look briefly at three further characterizations of Ct that will not play a role in the rest of the book. Here we shall content ourselves with stating the relevant results, referring to the literature for details of the proofs.

The first of these is a topological characterization due to Hyland [120], also discussed in detail in Normann [207]. In essence, this says that Ct arises as the type structure over $\mathbb{N}$ within a cartesian closed category well known from classical topology: the category of *compactly generated Hausdorff spaces*. Explicitly, we may define a type structure as follows:

Definition 8.2.25. (i) Let $(X, \mathscr{T})$ be any topological space. We say a set $U \subseteq X$ is *CG-open* if for every compact $K \subseteq X$, the set $U \cap K$ is open in the subspace topology on K. The set of CG-open sets constitutes a new topology $\mathscr{T}^{CG}$ on X, known as the topology *compactly generated from* $\mathscr{T}$.

(ii) We say $(X, \mathscr{T})$ is *compactly generated* if $\mathscr{T}^{CG} = \mathscr{T}$.

(iii) Let X, Y be topological spaces and $X \Rightarrow Y$ the set of continuous functions $X \to Y$. The *compact-open* topology on $X \Rightarrow Y$ is the one generated by the sub-basis consisting of sets

$$B_{K,O} \;=\; \{f \in \mathrm{Ct}(X,Y) \mid K \subseteq f^{-1}(O)\} \qquad (K \subseteq X \text{ compact}, O \subseteq Y \text{ open}).$$

(iv) We now define topological spaces $\mathsf{Ct}^{CG}(\sigma)$ by induction on σ:

- $\mathsf{Ct}^{CG}(\mathtt{N}) = \mathtt{N}$ with the discrete topology.

- $\mathsf{Ct}^{CG}(\sigma \to \tau) = \mathsf{Ct}^{CG}(\sigma) \Rightarrow \mathsf{Ct}^{CG}(\tau)$, endowed with the topology compactly generated from the compact-open topology.

This defines a type structure Ct^{CG} over $\mathbb{N}$. Since the spaces $\mathsf{Ct}^{CG}(\sigma)$ are clearly Hausdorff, and the category of compactly generated Hausdorff spaces is cartesian closed with exponentials $X \Rightarrow Y$ given as in (iv) above, we see that Ct^{CG} arises naturally as the type structure over $\mathbb{N}$ within this category.

In the spaces in question, a set will be compact if and only if it is sequentially compact, and a convergent sequence with a limit point is also a compact set. Using methods similar to the proof of Theorem 8.2.21, one may establish the following:

Theorem 8.2.26. $\mathsf{Ct}^{CG} \cong \mathsf{Ct}$ *as type structures, and for each* σ, $\mathsf{Ct}^{CG}(\sigma) \cong \mathsf{Ct}(\sigma)$ *as topological spaces.*

The proof makes use of various topological characterizations of compact sets which we shall present in Theorem 8.4.2 and Exercise 8.4.3, along with the following lemma, again due to Hyland.

Lemma 8.2.27. *The compact sets in* $\mathsf{Ct}(\sigma \to \tau)$ *(with the canonical topology) are exactly the compact sets in the compact-open topology on* $\mathsf{Ct}(\sigma) \Rightarrow \mathsf{Ct}(\tau)$.

8.2.5 Filter Spaces

Our next characterization, also due to Hyland, is via *filter spaces*. We start with some very general definitions.

Definition 8.2.28. (i) A *filter* on a set X is a family F of inhabited subsets of X such that

1. if $U \in F$ and $U \subseteq V \subseteq X$ then $V \in F$,
2. F is closed under finite intersections.

(ii) A filter F on X is an *ultrafilter* if for all $U \subseteq X$, either $U \in F$ or $X - U \in F$. An ultrafilter F is *principal* if there exists $x \in X$ such that $F = F_x = \{U \subseteq X \mid x \in U\}$.

(iii) If G is any set of subsets of X such that every finite collection of sets from G has an inhabited intersection, then G *generates* the filter

$$< G > \;=\; \{U \subseteq X \mid \bigcap_{i<n} V_i \subseteq U \text{ for some } V_0, \ldots, V_{n-1} \in G \text{ where } n > 0\}.$$

Intuitively, filters on X may be viewed as ways approximating elements that may or may not actually exist within X. If the intersection of the elements of the filter is inhabited, we may regard the filter as approximating the elements of this intersection. If the filter has an empty intersection, we may view it as representing some kind of ideal element of X. For instance, the Fréchet filter on $\mathbb{N}$, consisting of all sets with finite complement, can be seen as representing some ideal infinite number. This is essentially the idea employed by Kreisel in his original construction of Ct.

Definition 8.2.29. (i) A *filter space* $(X, \mathscr{F})$ consists of a non-empty set X and an assignment to each $x \in X$ of a family $\mathscr{F}_x$ of filters on X, such that

1. the principal ultrafilter F_x is in $\mathscr{F}_x$,
2. if $F \in \mathscr{F}_x$ and $F \subseteq G$ where G is a filter, then $G \in \mathscr{F}_x$.

(ii) Let $(X, \mathscr{F})$, $(Y, \mathscr{G})$ be filter spaces. A function $f : X \to Y$ is *(filter space) continuous* if for all $x \in X$ and $F \in \mathscr{F}_x$, the filter generated by $\{f(U) \mid U \in F\}$ is in $\mathscr{G}_{f(x)}$ (where $f(U)$ denotes the image of U under f).

The set $\mathbb{N}$ may be given the filter space structure $\mathscr{N}$, where $\mathscr{N}_n$ consists of just the principal ultrafilter F_n. Moreover, if $(X, \mathscr{F})$, $(Y, \mathscr{G})$ are filter spaces, the set Z of filter space continuous maps $X \to Y$ may itself be given a filter space structure $\mathscr{H}$ defined as follows. If $f \in Z$ and H is a filter on Z, we take $H \in \mathscr{H}_f$ iff for all $x \in X$, $F \in \mathscr{F}_x$ and $V \in H$, the set

$$G = \{\bigcup_{g \in V} g(U) \mid U \in F\}$$

generates a filter in $\mathscr{G}_{f(x)}$. We leave it to the reader to check that $(Z, \mathscr{H})$ is indeed a filter space.

We thus have the means to associate a filter space $\mathsf{Ct}^F(\sigma)$ with each type σ, yielding a type structure Ct^F over $\mathbb{N}$. The following fact is due to Hyland [120, 121]; see also Normann [207]:

Theorem 8.2.30. $\mathsf{Ct}^F \cong \mathsf{Ct}$.

The proof makes crucial use of filters generated from the Kreisel neighbourhoods W_e (see Definition 8.1.11).

8.2.6 Ct *as a Reduced Product*

So far, we have emphasized the broad distinction between 'direct' characterizations of Ct via sets with structure and 'indirect' characterizations via realizability over some representing model. Our final characterization in some sense brings these worlds together, by showing how a definition somewhat akin to that of Ct^L can be presented as a realizability construction over a TPCA, albeit one of a very different character from PC or K_2.

Recall that any element $x \in \mathsf{Ct}(\sigma)$ may be approximated by a convergent sequence $(x)_0, (x)_1, \ldots$, where each $(x)_n$ belongs to a subspace of $\mathsf{Ct}(\sigma)$ isomorphic to $\mathsf{S}_n(\sigma)$. This suggests the idea that convergent sequences of this kind can be viewed as representations or 'realizers' for elements of Ct. For each type σ, let us therefore consider the set

$$\mathsf{R}(\sigma) = \prod_{n<\omega} \mathsf{S}_n(\sigma) ,$$

where we think of each $z \in \mathsf{R}(\sigma)$ as a sequence $z_0, z_1, \ldots$ with $z_n \in \mathsf{S}_n(\sigma)$. Since each S_n is a TPCA, we may regard the sets $\mathsf{R}(\sigma)$ as forming a TPCA R with application defined componentwise.

In order to single out the *convergent* sequences, let us start with the following partial equivalence relation $\approx$ on $\mathsf{R}(\mathtt{N})$:

$$z \approx w \text{ iff } \exists n, m. \forall i \geq n. z_i = w_i = m .$$

Here $z \approx w$ says that z, w converge to the same value m. We now define a type structure Ct^R over $\mathbb{N}$ simply as the extensional collapse $\mathsf{EC}(\mathsf{R}; \approx)$. Notice that the lifting of $\approx$ to higher types is somewhat reminiscent of the definition of the convergence relation in a limit space $(X, L) \Rightarrow (Y, M)$ (see Definition 8.2.12), although here we are restricting attention to convergent sequences of a certain special form.

The following result is proved in Normann, Palmgren and Stoltenberg-Hansen [222]. Here we leave the proof to the reader as an advanced exercise.

Theorem 8.2.31. $\mathsf{Ct}^R \cong \mathsf{Ct}$.

Note that this is an example of a situation where we have a natural topology on the set of realizers and a natural topology on the set of realized objects, but the map from the former to the latter is not continuous:

Exercise 8.2.32. Suppose that each $\mathsf{S}_n(\sigma)$ is endowed with the discrete topology, and $\mathsf{R}(\sigma)$ with the product topology. Show that the only topology on $\mathsf{Ct}(\sigma)$ such that the quotient map $\mathsf{Ext}(\mathsf{R}; \approx)(\sigma) \to \mathsf{Ct}(\sigma)$ is continuous is the indiscrete topology.

8.3 Examples of Continuous Functionals

We now look at some important and illuminating examples of total continuous functionals. In fact, all three of the examples we consider are closely related to PCF-definable functionals already discussed in Section 7.3, but we shall reconsider them here from the perspective of Ct. These examples played an important role in the history of the subject, and will also feature in our deeper study of computability in Section 8.5.

8.3.1 Bar Recursion

Bar recursion, as introduced by Spector [276], is in essence a means of defining functions by transfinite recursion on well-founded trees. We refer the reader to Subsection 7.3.3 for an intuitive explanation. There, Spector's version of bar recursion was presented as a third-order functional

$$\mathsf{BR}_\tau \,:\, \overline{2} \to (\mathtt{N}^* \to \tau) \to (\mathtt{N}^* \to (\mathtt{N} \to \tau) \to \tau) \to (\mathtt{N}^* \to \tau) ,$$

where for convenience we allow ourselves the use of a type $\mathtt{N}^*$ of finite lists of natural numbers, which serve as addresses for nodes in an infinite tree.

The defining clauses for BR_τ are as follows. We let $\vec{x}$ range over $\mathtt{N}^*$, write $|\vec{x}|$ for the length of $\vec{x}$, and write $\vec{x}0^\omega$ for the function $\mathbb{N} \to \mathbb{N}$ mapping i to x_i if $i < |\vec{x}|$ and 0 otherwise.

1. $\mathsf{BR}(F,L,G)(\vec{x}) = L(\vec{x})$ if $F(\vec{x}0^\omega) < |\vec{x}|$.
2. $\mathsf{BR}(F,L,G)(\vec{x}) = G(\vec{x}, \Lambda z \in \mathbb{N}. \mathsf{BR}(F,L,G)(\vec{x};z))$ if $F(\vec{x}0^\omega) \geq |\vec{x}|$.

In Subsection 7.3.3, these clauses were understood as recursively defining a functional BR_τ within a model of PCF^Ω such as PC. In Theorem 7.3.1 it was shown that the resulting functional is hereditarily total and extensional, and thus represents an element in $\mathsf{Ct} = \mathsf{EC}(\mathsf{PC};\mathtt{N})$ (we shall denote this element also by BR_τ).

However, it is also possible to interpret the above clauses as a definition in Ct directly, since they refer only to total functionals. Indeed, this shows how the functional in question may be computed internally within Ct without any reference to realizers:

Theorem 8.3.1. *The functional* $\mathsf{BR}_\tau \in \mathsf{Ct}$ *as defined above is the unique functional satisfying the above two clauses (interpreted within* Ct*), and is Kleene computable within* Ct.

Proof. It is clear that $\mathsf{BR}_\tau \in \mathsf{Ct}$, as defined via PC, satisfies clauses 1 and 2 interpreted within Ct, so it remains to show the uniqueness and Kleene computability.

For uniqueness, the argument is essentially the same as for Theorem 7.3.1. Fix $F \in \mathsf{Ct}(2)$. Then for any $g \in \mathsf{Ct}(1)$, there is clearly some $n > F(g)$ such that whenever $\vec{x}$ is an initial segment of g of length at least n, we have $F(\vec{x}0^\omega) = F(g) < |\vec{x}|$. Thus we have a well-founded tree T_F consisting of those sequences $\vec{x}$ such that $F(\vec{y}0^\omega) \geq |\vec{y}|$ for some $\vec{y}$ extending $\vec{x}$ (allowing $\vec{y} = \vec{x}$). For any $\vec{x} \notin T_F$, we have $F(\vec{x}0^\omega) < |\vec{x}|$ so the value of $\mathsf{BR}_\tau(F,L,G)(\vec{x})$ is uniquely determined by clause 1 above. For $\vec{x} \in T_F$, an easy induction on the ordinal height of $\vec{x}$ within T shows that $\mathsf{BR}_\tau(F,L,G)(\vec{x})$ is again uniquely determined, using clause 1 or 2 as appropriate.

In the terminology of Section 5.1, with the help of the recursion theorem it is easy to construct a Kleene expression M such that

$$[\![M]\!](F,L,G)(\vec{x}) \;\simeq\; \begin{cases} L(\vec{x}) & \text{if } F(\vec{x}0^\omega) < |\vec{x}|\,, \\ G(\vec{x}, \Lambda z \in \mathbb{N}.\, [\![M]\!](F,L,G)(\vec{x};z)) & \text{otherwise} \end{cases}$$

for all $F,L,G,\vec{x}$ of appropriate type. The above induction now clearly shows that $[\![M]\!]$ is total and coincides with BR_τ. Thus BR_τ is Kleene computable over Ct. □

It follows by Theorem 8.1.28 that $\mathsf{BR}_\tau \in \mathsf{Ct}^{\mathrm{eff}}$.

On the other hand, we noted in Corollary 6.3.33 that bar recursion is not $\mathsf{Klex}^{\mathrm{min}}$ or even $\mathrm{T}+min$ definable in a model such as Ct. Thus, BR_τ shows that the power of Kleene computability over Ct surpasses that of $\mathrm{T}+min$.

As an aside, we also recall from Corollary 6.3.32 that even the System T recursor rec_1 in Ct is not definable in the weaker language $\mathsf{Klex}^{\mathrm{min}}$. An alternative proof of this, not relying on the machinery of nested sequential procedures, will be given in Subsection 8.5.2.

Remark 8.3.2. One may think of bar recursion as a way to extend the power of (System T) primitive recursion whilst remaining within the realm of total functionals. Historically, bar recursion in Ct has played a major role in proof-theoretic studies, most notably in Spector's computational interpretation for full classical analysis (see the discussion in Subsection 7.3.3).

Remark 8.3.3. Since any countable ordinal may be represented by some continuous $F \in \mathsf{Ct}(2)$ as explained in Subsection 7.3.3, we may use bar recursion to find a Kleene index e such that the height of the computation of $\{e\}(F)$ is unbounded in the set of countable ordinals as F ranges over $\mathsf{Ct}(2)$. Thus for any $\Phi \in \mathsf{Ct}(4)$, the computation of

$$\{d\}(\Phi) = \Phi(\Lambda F \in \mathsf{Ct}(2).\{e\}(F))$$

has uncountable ordinal height.

By contrast, we will see in Section 8.5 that the computation of any $\{e\}(\phi)$ where $\phi \in \mathsf{Ct}(3)$ has countable ordinal height.

8.3.2 The Fan Functional

Our next example is also one that featured in Section 7.3. Suppose we are given some $G \in \mathsf{Ct}((\mathtt{N} \rightarrow \mathtt{B}) \rightarrow \mathtt{N})$, where the type $\mathtt{B}$ denotes the set $\mathbb{B}$ of booleans. By applying König's lemma to the tree of finite sequences $\vec{x} \in \mathbb{B}^*$ such that G is not constant on the set of functions $\mathbb{N} \rightarrow \mathbb{B}$ extending $\vec{x}$, we see that any such G will be *uniformly continuous*. The fan functional, by definition, returns the minimal modulus of uniform continuity for such a G:

$$\mathsf{FAN}(G) \;=\; \mu n. \forall f, f' \in \mathbb{B}^{\mathbb{N}}.\ (\forall i < n. f(i) = f'(i)) \;\Rightarrow\; G(f) = G(f')\,.$$

It can be easily read off from our results in Subsection 7.3.4 that FAN is representable by an element of $\mathsf{PC}^{\mathrm{eff}}$, and so is present in $\mathsf{Ct}^{\mathrm{eff}}(((\mathtt{N} \rightarrow \mathtt{B}) \rightarrow \mathtt{N}) \rightarrow \mathtt{N})$. However, one may also give a simple direct proof of this that does not rely on Berger's PCF definition of the fan functional:

Theorem 8.3.4. *The functional* $\mathsf{FAN} : \mathsf{Ct}((\mathtt{N} \rightarrow \mathtt{B}) \rightarrow \mathtt{N}) \rightarrow \mathbb{N}$ *as defined above is tracked by an element of* $\mathsf{PC}^{\mathrm{eff}}$*, and hence present in* $\mathsf{Ct}^{\mathrm{eff}}$.

Proof. Let us say a compact element $a \in \mathsf{PC}((\mathtt{N} \rightarrow \mathtt{B}) \rightarrow \mathtt{N})$ *covers* $\mathbb{B}^{\mathbb{N}}$ if it is itself a total element. If $g \in \mathsf{PC}((\mathtt{N} \rightarrow \mathtt{B}) \rightarrow \mathtt{N})$ is an arbitrary total element, then applying König's lemma to the tree of finite sequences $\vec{x}$ such that g is not constant on the set of $f \in \mathsf{PC}(\mathtt{N} \rightarrow \mathtt{B})$ extending $\vec{x}$, we can construct a compact element $a \sqsubseteq g$ covering $\mathbb{B}^{\mathbb{N}}$ as a finite join of step functions $(b \Rightarrow m)$.

Clearly it is decidable whether a given compact element covers $\mathbb{B}^{\mathbb{N}}$, so we may obtain an effective enumeration of all covering compacts a. Moreover, from a syntactic description of such an a, it is easy to extract an n such that $a(f) = a(f')$ whenever $f(i) = f(i')$ for all $i < n$; and by testing the value of a on elements corresponding to the 2^n possible sequences $\vec{x}$ of length n, we may easily find the least n' such that $a(f) = a(f')$ whenever $f(i) = f(i')$ for all $i < n'$. It is then the case that if G is realized by any $g \sqsubseteq p$ then $\mathsf{FAN}(G) = n$. Thus by taking the supremum of the elements $(a \Rightarrow n')$ obtained in this way, we obtain an effective realizer for FAN. □

Whilst FAN may at first sight appear similar in character to BR, there is a deep difference: bar recursion is Kleene computable over Ct, whereas the fan functional is not. We will prove the latter fact in Section 8.5, where the fan functional will play a major role in our investigations of relative Kleene computability.

We will refer to the space $\mathbb{B}^{\mathbb{N}}$ as the *Cantor space*; it is the prototypical compact subset of $\mathbb{N}^{\mathbb{N}}$. Of course, there are many other compact subsets: indeed, a set $C \subset \mathbb{N}^{\mathbb{N}}$ is compact if and only if it is closed and pointwise bounded, and each such set will have its own modulus of uniform continuity operator. In fact, we may generalize the fan functional to one that computes moduli of uniform continuity over compact sets $C_g = \{f \in \mathbb{N}^{\mathbb{N}} \mid \forall n \in \mathbb{N}.\, f(n) \leq g(n)\}$ uniformly in $g \in \mathbb{N}^{\mathbb{N}}$:

$$\mathsf{FAN}^*(G,g) \;=\; \mu n.\forall f, f' \in \mathbb{N}^{\mathbb{N}}.\ (\forall i < n.\, f(i) = f'(i) \leq g(i)) \;\Rightarrow\; G(f) = G(f')\,.$$

Exercise 8.3.5. Show that FAN^* is present in $\mathsf{Ct}^{\mathrm{eff}}(2, 1 \to 0)$, and is Kleene primitive recursive relative to FAN.

The following exercise gives another interesting application of the fan functional. This introduces an idea which we will explore in more depth in Subsection 8.4.1.

Exercise 8.3.6. By a *selection function* for a set $S \subseteq \mathsf{Ct}(\sigma)$, we mean a functional $\varepsilon_S \in \mathsf{Ct}((\sigma \to \mathtt{B}) \to \sigma)$ such that $\varepsilon_S(P) \in S$ for any $P \in \mathsf{Ct}(\sigma \to \mathtt{B})$, and if there exists $x \in S$ with $P(x) = tt$ then $P(\varepsilon_S(P)) = tt$.

(i) Show that a selection function for $C_g \subseteq \mathsf{Ct}(1)$ may be obtained uniformly in g via a functional $\varepsilon \in \mathsf{Ct}(1, (1 \to \mathtt{B}) \to 1)$ that is primitive recursive relative to FAN.

(ii) Show that any inhabited subset $S \subseteq C_g$, even if not closed, has a selection function $\varepsilon_S \in \mathsf{Ct}((1 \to \mathtt{B}) \to 1)$, although there need not be such a function in $\mathsf{Ct}^{\mathrm{eff}}$. (The definition of ε_S must in this case be more specifically tailored to the set S.)

The proof of the totality of the fan functional is non-constructive in that it appeals to König's lemma. Indeed, it can be seen that the non-constructivity is essential. In Definition 9.1.10 we shall use the so-called *Kleene tree* to define a total computable functional $K : (\mathbb{B}^{\mathbb{N}})^{\mathrm{eff}} \to \mathbb{N}$ that is not uniformly continuous (writing $(\mathbb{B}^{\mathbb{N}})^{\mathrm{eff}}$ for the set of computable binary sequences). On the other hand, if the totality of the fan functional were constructively provable, the proof would yield a modulus of uniform continuity even for K.

In fact, the totality of FAN requires only what is known as the *weak König's lemma* or WKL—that is, the statement that any infinite binary tree has an infinite branch. Experience has shown that WKL is a very natural principle to consider as a first step beyond 'constructive' mathematics: for instance, many important theorems in real analysis turn out to be equivalent to WKL relative to a weak version of second-order arithmetic (see Simpson [273]).[2] This in turn suggests that the fan functional may be just what we need in order to realize the computational content of such theorems. Here we mention two examples of this: the computation of *suprema* and *integrals* for continuous functions $[-1,1] \to \mathbb{R}$. We have already described these applications in Subsection 7.3.4, so only a small addendum is needed here to show how they fit within the framework of Ct.

Recall from Subsection 7.3.4 that we may give *signed binary digit* representations of real numbers $x \in [-1,1]$ as functions $r : \mathbb{N} \to \mathbb{D}$, where $\mathbb{D}$ is the set of digits $\{-1,0,1\}$. Using a base type $\mathtt{D}$ for $\mathbb{D}$, we may thus use the type $\mathtt{rep} = \mathtt{N} \to \mathtt{D}$ for real number representations. The story we told for PC in Subsection 7.3.4 can now be recast (slightly more abstractly) in the setting of Ct: any continuous function $[-1,1] \to [-1,1]$ may be represented by an element of $\mathsf{Ct}(\mathtt{rep} \to \mathtt{rep})$, and one may construct functionals

$$\mathsf{Sup}, \mathsf{Int} \in \mathsf{Ct}^{\mathrm{eff}}((\mathtt{rep} \to \mathtt{rep}) \to \mathtt{rep})$$

that respectively compute suprema and integrals for such functions. These functionals are readily shown to be Kleene computable in Ct relative to FAN, and indeed the functionals $\mathsf{Sup}, \mathsf{Int} \in \mathsf{PC}$ described in Subsection 7.3.4 are realizers for them. We leave it to the interested reader to fill in the details.

8.3.3 The Γ Functional

Although the fan functional is not itself Kleene computable, for some years it seemed a plausible conjecture that every element of Ct should be Kleene computable relative to FAN along with some function $\mathbb{N} \to \mathbb{N}$. This was refuted in the 1970s, when Gandy discovered the so-called Γ *functional*, and Hyland [120] showed that it is not Kleene computable in FAN and any type 1 function. With hindsight, Γ can be seen as an operator equivalent to *modified bar recursion* (see Subsection 7.3.3 and Remark 8.3.8 below), although the latter concept was not articulated until much later on. We present Γ here both as a further example of the computational power admitted by $\mathsf{Ct}^{\mathrm{eff}}$, and as a useful reference point for our more general investigation of relative Kleene computability in Section 8.5.

A little notation will be useful. If $f : \mathbb{N} \to \mathbb{N}$ and $a \in \mathbb{N}$, we shall write af for the function given by $(af)(0) = a$, $(af)(n+1) = f(n)$. If $F \in \mathsf{Ct}(2)$ and $a \in \mathbb{N}$, we define $F_a \in \mathsf{Ct}(2)$ by $F_a(g) = F(ag)$.

Theorem 8.3.7. *There exists a unique object $\Gamma \in \mathsf{Ct}(3)$ such that the equation*

---

[2] The second-order system in question, known as $\mathbf{RCA}_0$, employs classical logic, but can be regarded as an essentially 'constructive' system in that it only allows us to prove the existence of subsets of $\mathbb{N}$ for which a decision procedure can be given.

$$\Gamma(F) \;=\; F_0(\Lambda n \in \mathbb{N}.\Gamma(F_{n+1}))$$

holds for every $F \in \mathsf{Ct}(2)$, and this Γ has a computable realizer.

Proof. First we prove uniqueness. Given F, let T_F be the tree consisting of all finite sequences $\vec{x} \in \mathbb{N}^*$ such that F is not constant on the set of all $f : \mathbb{N} \to \mathbb{N}$ extending $\vec{x}$. We show by induction on the ordinal height of T_F that there is only one possible value for $\Gamma(F)$ consistent with the above equation. If T_F has height 0 (i.e. T_F is empty) then F is a constant function $\Lambda f.t$ and $\Gamma(F) = t$ is the only possibility. Otherwise, we have that each $T_{F_{n+1}}$ is isomorphic to a proper subtree of T_F with smaller height than T_F, so by the induction hypothesis there is a unique possible value for each $\Gamma(F_{n+1})$. Thus the function $\Lambda n.\Gamma(F_{n+1})$ and hence the value of $\Gamma(F)$ itself is uniquely determined.

This argument in fact shows that there exists a unique set-theoretic function $\Gamma : \mathsf{Ct}(2) \to \mathbb{N}$ satisfying the defining equation. To show that $\Gamma \in \mathsf{Ct}(3)$, it remains to show that Γ is continuous. So suppose $F \in \mathsf{Ct}(2)$ and $f \Vdash F$; we will show that there is a compact $a \sqsubseteq f$ such that Γ is constant on V_a.

Let U_f be the tree consisting of all $x_0, \ldots, x_{r-1} \in \mathbb{N}^*$ such that f is undefined on the element $g_{\vec{x}} = \bigsqcup_{i<r}(i \Rightarrow x_i) \in \mathsf{PC}(1)$. Then U_f is well-founded because f is total and continuous; we will argue by induction on the ordinal height of U_f. If the height is zero, then $f(\bot) = t \in \mathbb{N}$ say; thus F is constantly t and the compact element $(\bot \Rightarrow t)$ serves our purpose. Otherwise, we have by the induction hypothesis that the value of each $\Gamma(F_{n+1})$ is determined by a compact $a_{n+1} \sqsubseteq f$. Since computing the value m of $F_0(\Lambda n.\Gamma(F_{n+1}))$ using f requires only a finite segment b of $\Lambda n.\Gamma(F_{n+1})$, we see that a join of $(b \Rightarrow m)$ with finitely many of the a_{n+1} will suffice for a.

Since it is clearly decidable whether a given compact in $\mathsf{PC}(2)$ contains enough information to determine the value of Γ, we obtain a computably enumerable set of compacts in $\mathsf{PC}(3)$ whose supremum is a realizer for Γ. Thus $\Gamma \in \mathsf{Ct}^{\mathrm{eff}}(3)$. □

In Section 8.5 we will prove Hyland's theorem that Γ is not Kleene computable from FAN and any $f : \mathbb{N} \to \mathbb{N}$.

Remark 8.3.8. It is shown by Berger and Oliva [26] that Γ is equivalent to the modified bar recursion operator $\mathsf{MBR}_{\mathbb{N}}$ of Subsection 7.3.3 (in the sense that each is Kleene primitive recursive in the other), and also that FAN is Kleene computable in $\mathsf{MBR}_{\mathbb{N}}$ within Ct. The use of modified bar recursion to represent the computational content of certain classical proof principles is explored in [25].

Remark 8.3.9. The recursive definition of Γ can also be understood as a PCF program involving Y_3; we thus obtain a PCF-definable realizer in $\mathsf{PC}(3)$ for $\Gamma \in \mathsf{Ct}(3)$. This means that we in some sense have a sequential procedure for computing Γ, as indeed we do for FAN. It is by no means obvious whether all elements of $\mathsf{Ct}(3)$ are 'sequentially computable' in this way, although we shall see in Section 8.6 that this is indeed the case.

Exercise 8.3.10. Define $\Delta_\Gamma : \mathsf{Ct}(3) \to \mathsf{Ct}(3)$ by

$$\Delta_\Gamma(\Psi)(F) = F_0(\Lambda n.\Psi(F_{n+1}))\,.$$

Then it is clear from Theorem 8.3.7 that Γ is the unique fixed point of Δ_Γ. Show that for any $\Psi \in \mathsf{Ct}(3)$ we have that $\Delta_\Gamma^n(\Psi) \to \Gamma$ as $n \to \omega$, and that there is a modulus of convergence functional that is common to all these sequences.

8.4 Fine Structure of Ct

We now proceed to a more specialized study of some of the rich mathematical structure that Ct has to offer. In this section, we consider two topics: firstly, the properties of *compact* sets within Ct, including a variety of topological and computational characterizations of such sets; and secondly, the close relationship between Ct and the *analytic hierarchy* for second-order logical predicates (cf. Subsection 5.2.1). A common thread uniting these topics is that they both make non-trivial use of the operations $\xi_\sigma \in \mathsf{Ct}^{\mathrm{prim}}(\mathbb{N} \to \sigma)$ enumerating a countable dense subset of each $\mathsf{Ct}(\sigma)$ as in Theorem 8.1.31.

8.4.1 Characterizations of Compact Sets

An important structural role in the theory of Ct is played by the topologically compact subsets of each $\mathsf{Ct}(\sigma)$. Here we will give several possible characterizations of these sets and mention some of their ramifications. We will start with some characterizations of a topological flavour due to Hyland [120]; building on these, we will then proceed to some more 'computational' characterizations using Escardó's theory of *exhaustible* and *searchable* sets [80], one version of which we have already touched on in Subsection 7.3.5.

For notational simplicity, we will content ourselves with studying the compact subsets of $\mathsf{Ct}(\sigma)$ where $\sigma = \tau \to \mathtt{N}$, although the arguments extend to the general case without difficulty. We write q for the quotient map $\mathsf{Ext}(\sigma) \to \mathsf{Ct}(\sigma)$; recall that this is continuous.

Theorem 8.4.2 and Exercise 8.4.3 below give a selection of topological characterizations of the compact sets. For this purpose, two specialized definitions will be required:

Definition 8.4.1. (i) A set $S \subseteq \mathsf{Ct}(\tau \to \mathtt{N})$ is *bounded* if there is a function $g : \mathsf{Ct}(\tau) \to \mathbb{N}$ (not necessarily continuous) such that $f(x) \leq g(x)$ for all $f \in S$ and $x \in \mathsf{Ct}(\tau)$.

(ii) A set $S \subseteq \mathsf{Ct}(\tau \to \mathtt{N})$ is *equicontinuous* (in the sense of limit spaces) if for any convergent sequence $x_i \to x$ in $\mathsf{Ct}(\tau)$, there exists a single modulus m such that for all $f \in S$ and all $i \geq m$ we have $f(x_i) = f(x)$.

Theorem 8.4.2. *The following are equivalent for an inhabited set* $S \subseteq \mathsf{Ct}(\tau \to \mathtt{N})$*:*

1. S is compact.
2. S is closed, bounded and equicontinuous.
3. S is homeomorphic to a compact subset of $\mathbb{N}^{\mathbb{N}}$.
4. S is a continuous image of $\mathbb{B}^{\mathbb{N}}$ (Cantor space).

Proof. $1 \Rightarrow 2$: Any compact subspace of a Hausdorff space is closed. For boundedness, note that if S is compact, then for any $x \in \mathsf{Ct}(\tau)$ the image of S under the continuous map $\Lambda f.f(x)$ is a compact subset of $\mathbb{N}$, and therefore bounded. For equicontinuity, suppose $x_i \to x$. From Subsection 8.2.3 we know that the functional Δ mapping f to a modulus for $\Lambda i.f(x_i)$ will be continuous. So $\Delta(S)$ is a compact subset of $\mathbb{N}$, and hence finite. This is exactly what equicontinuity means.

$2 \Rightarrow 3$: Assume that S is closed, bounded and equicontinuous. We first show that for any $x \in \mathsf{Ct}(\tau)$ and $\widehat{x} \Vdash x$, there is a (domain-theoretically) compact $a \sqsubseteq \widehat{x}$ such that each $f \in S$ is constant on V_a. If not, take an increasing chain $a_0 \sqsubseteq a_1 \sqsubseteq \cdots$ with supremum $\widehat{x}$, and for each i take $f_i \in S$ and $x_i \in V_a$ such that $f_i(x_i) \neq f_i(x)$. Then $x_i \to x$, but no modulus of convergence for $f(x_i) \to f(x)$ works uniformly for all $f \in S$, contradicting the equicontinuity of S.

Let A be the set of all compacts a such that each $f \in S$ is constant on V_a. Since S is bounded, we may associate with each $a \in A$ a number $g(a)$ exceeding the values of all $f \in S$ on V_a. Let P be the product $\prod_{a \in A}[g(a)]$ (where $[n]$ denotes $\{0,\ldots,n-1\}$) endowed with the product topology. Since A is countable and each $[g(a)]$ is finite, and thus compact, P is homeomorphic to a compact subset of $\mathbb{N}^{\mathbb{N}}$. Let $Q \subseteq P$ consist of all A-tuples p such that $p_a = p_b$ whenever a, b are consistent in $\mathsf{PC}(\tau)$. Then Q will be a closed subset of P, and thus also homeomorphic to a compact subset of $\mathbb{N}^{\mathbb{N}}$. Any $f \in S$ will now give rise to an element $\theta(f) \in Q$, where $\theta(f)_a$ is the constant value of f on V_a. Moreover, any $p \in Q$ gives rise to an element $\gamma(p) = \bigsqcup_{a \in A}(a \Rightarrow p_a) \in \mathsf{PC}(\tau \to \mathtt{N})$; and since every element in $\mathsf{Ext}(\tau)$ extends some $a \in A$, this is a realizer for an element $q(\gamma(p)) \in \mathsf{Ct}(\tau \to \mathtt{N})$. It is also easy to see that $q(\gamma(\theta(f))) = f$ for any $f \in S$; thus $q(\gamma(\theta(S))) = S$.

Now q is continuous, and it is easy to check that γ is continuous, so $q \circ \gamma$ is continuous. Moreover, $S \subseteq (q \circ \gamma)(Q)$, so, since S is closed, $Q_0 = (q \circ \gamma)^{-1}(S)$ will be closed and hence compact, and again, homeomorphic to a compact subset of $\mathbb{N}^{\mathbb{N}}$.

Since $q \circ \gamma$ is clearly injective, it restricts to a bijection $Q_0 \to S$, where $\theta = (q \circ \gamma)^{-1} : S \to Q_0$. To establish the desired homeomorphism (which of course also shows $2 \Rightarrow 1$), it remains to show that Θ is continuous. This follows by an elementary topological argument: since $\theta^{-1} = q \circ \gamma$ on Q_0, it suffices to show that $q \circ \gamma$ maps closed sets in Q_0 to closed sets in S. But if $C \subseteq Q_0$ is closed, then C is compact, so $(q \circ \gamma)(C)$ is compact, hence closed in $\mathsf{Ct}(\tau \to \mathtt{N})$ since $\mathsf{Ct}(\tau \to \mathtt{N})$ is Hausdorff.

$3 \Rightarrow 4$: The compact subsets of $\mathbb{N}^{\mathbb{N}}$ are the closed and bounded ones, so we must show that any such set is a continuous image of $\mathbb{B}^{\mathbb{N}}$. This is well known, but for later use we give an explicit construction.

First, for any bounding function $g : \mathbb{N} \to \mathbb{N}$, it is an easy exercise to construct a continuous surjection from $\mathbb{B}^{\mathbb{N}}$ to the set $C_g = \{f \in \mathbb{N}^{\mathbb{N}} \mid \forall n \in \mathbb{N}.\, f(n) \leq g(n)\}$. Next, for any inhabited closed set $K \subseteq C_g$, we may construct an idempotent $d \in \mathsf{Ct}(1 \to 1)$ whose image is exactly K. Specifically, for $f \in \mathsf{Ct}(1)$ and $n \in \mathbb{N}$, we recursively define

$$d(f,n) = \begin{cases} f(n) & \text{if for some } f' \in K \text{ we have } \forall i \leq n.\, f'(i) = f(i)\,, \\ m & \text{if there is no such } f', \text{ but } m \text{ is the least number such that} \\ & \exists f' \in K.\, (\forall i < n.\, f'(i) = d(f,i)) \wedge f'(n) = m\,. \end{cases}$$

Clearly d is continuous, because the value of $d(f,n)$ depends on values of $f(i)$ only for $i \leq n$. It is also clear that $d(f) \in K$ for all $f \in \mathsf{Ct}(1)$, and that $d(f) = f$ if $f \in K$. Thus d is a continuous surjection $C_g \to K$, so K is a continuous image of $\mathbb{B}^{\mathbb{N}}$.

$4 \Rightarrow 1$: Since $\mathbb{B}^{\mathbb{N}}$ is compact, so is any continuous image of $\mathbb{B}^{\mathbb{N}}$. □

Exercise 8.4.3. (i) Using information from the proof of $2 \Rightarrow 1$ above, show that $S \subseteq \mathsf{Ct}(\sigma)$ is compact iff there is a compact set $R \subseteq \mathsf{Ext}(\mathsf{PC})(\sigma)$ with $q(R) = S$.

(ii) Adapt the argument to show the corresponding property for realizers in K_2 rather than PC: a set S is compact iff it has some compact set of associates.

The foregoing results are topological in character. However, an important insight due to Escardó [79, 80] is that the notion of compactness can also be understood as having a more computational significance. We shall discuss a selection of Escardó's results here, slightly adapting the treatment of [80] so as to present the material as part of the intrinsic theory of Ct.

The following concepts are variations on those of Definition 7.3.4:

Definition 8.4.4. Suppose $S \subseteq \mathsf{Ct}(\sigma)$.

(i) S is *continuously searchable* if there is an element $\varepsilon_S \in \mathsf{Ct}((\sigma \to \mathtt{B}) \to \sigma)$ such that for all $P \in \mathsf{Ct}(\sigma \to \mathtt{B})$ we have

1. $\varepsilon_S(P) \in S$,
2. if $P(x) = tt$ for some $x \in S$, then $P(\varepsilon_S(P)) = tt$.

Such an element ε_S is called a *selection function* for S.

(ii) S is *continuously exhaustible* if there is an element $\exists_S \in \mathsf{Ct}((\sigma \to \mathtt{B}) \to \mathtt{B})$ such that for all $P \in \mathsf{Ct}(\sigma \to \mathtt{B})$ we have

1. $\exists_S(P) = tt$ if there exists $x \in K$ with $P(x) = tt$,
2. $\exists_S(P) = ff$ if $P(x) = ff$ for all $x \in S$.

Such an element $\exists_S$ is called an *existential quantifier* for S.

These notions are weaker than those of Definition 7.3.4 in two respects. Firstly, we do not require the operators $\varepsilon_S, \exists_S$ to be effectively computable in any sense, although one may clearly strengthen the above notions, if desired, by requiring $\varepsilon_S, \exists_S$ to be elements of $\mathsf{Ct}^{\mathrm{eff}}$, $\mathsf{Ct}^{\mathrm{Kl}}$, $\mathsf{Ct}^{\mathrm{prim}}$ etc. Secondly, we here only consider predicates P that are defined on the whole of $\mathsf{Ct}(\sigma)$, whereas in Definition 7.3.4 we allowed predicates that were undefined outside the set S in question. We will discuss the relationship between these variants at the end of this subsection.

Exercise 8.4.5. (i) Show that the continuously searchable [resp. exhaustible] subsets of $\mathbb{N}$ are exactly the finite ones.

(ii) Show that the image of a continuously searchable [resp. exhaustible] set $S \subseteq \mathsf{Ct}(\sigma)$ via a continuous function $f : \mathsf{Ct}(\sigma) \to \mathsf{Ct}(\tau)$ is continuously searchable [resp. exhaustible]; cf. Proposition 7.3.5(ii).

We have already seen in Exercise 8.3.6 that the sets $C_g \subseteq \mathsf{Ct}(1)$ are continuously searchable. We now show how the search problem at higher types can be reduced (for suitable sets) to the search problem for compact subsets of $\mathsf{Ct}(1)$. To this end, we introduce the following key concept, using the enumeration $\xi_\tau \in \mathsf{Ct}^{\mathrm{prim}}(\mathtt{N} \to \tau)$ from Theorem 8.1.31.

Definition 8.4.6. Suppose $\sigma = \tau \to \mathtt{N}$. For any $x \in \mathsf{Ct}(\sigma)$, define its *trace* h_x to be the function $\Lambda i.x(\xi_\tau(i)) \in \mathsf{Ct}(1)$. The mapping $h = \Lambda x.h_x \in \mathsf{Ct}(\sigma \to 1)$ will be called the *trace map*, and if $S \subseteq \mathsf{Ct}(\sigma)$, the image $h(S)$ will be called the *trace of* S.

For more general types $\sigma = \tau_0, \ldots, \tau_{r-1} \to \mathtt{N}$, if desired, one may define the trace of x to be

$$h_x = \Lambda i.\, x(\xi_{\tau_0}(\pi_0(i)), \ldots, \xi_{\tau_{r-1}}(\pi_{r-1}(i))) ,$$

where $\pi_0, \ldots, \pi_{r-1}$ are the projections associated with some fixed primitive recursive bijection $\mathbb{N} \cong \mathbb{N}^r$. Note that the trace map h is itself Kleene primitive recursive, and provides an injection $\mathsf{Ct}(\sigma) \to \mathsf{Ct}(1)$.

What is perhaps surprising is that a one-sided inverse to h is continuous and even computable in many non-trivial cases—indeed, this idea provides us with the link between searchability and compactness. The topological aspects of this are dealt with by the following theorem.

Theorem 8.4.7. *The following are equivalent for an inhabited set $S \subseteq \mathsf{Ct}(\sigma)$:*

1. *S is compact.*
2. *The trace $h(S) \subseteq \mathsf{Ct}(1)$ is closed and bounded, and there exists $e \in \mathsf{Ct}(1 \to \sigma)$ such that $e(h_x) = x$ for all $x \in S$.*
3. *S is closed and continuously searchable.*
4. *S is closed and continuously exhaustible.*

Proof. $1 \Rightarrow 2$: Suppose S is compact. Since h is continuous, the image $h(S)$ is compact, hence closed and bounded. To construct e, we will first consider h as a bijection $S \to h(S)$, and show that its inverse is limit-continuous on $h(S)$. For this, it suffices to show that h itself maps open sets to open sets, or equivalently maps closed sets to closed sets. But if $A \subseteq S$ is closed, then A is compact, so $h(A)$ is compact, and since $\mathsf{Ct}(\sigma)$ is Hausdorff, this means that $h(A)$ is closed.

To extend h^{-1} to a total continuous function on the whole of $\mathsf{Ct}(1)$, we note that since $h(S)$ is compact, we may construct an idempotent $d \in \mathsf{Ct}(1 \to 1)$ with image $h(S)$ as in the proof of $3 \Rightarrow 4$ in Theorem 8.4.2. The composition $e = h^{-1} \circ d : \mathsf{Ct}(1) \to \mathsf{Ct}(\sigma)$ is then total and limit-continuous, and so is an element of $\mathsf{Ct}(1 \to \sigma)$ with the required property.

$2 \Rightarrow 3$: S is closed because $h(S)$ is closed and S is its inverse image under h. Moreover, if g is a bound for $h(S)$ and e is as in condition 2, then taking an idempotent d as above we have that S is the image of C_g under $e \circ d$. But C_g is searchable by Exercise 8.3.6(i), whence S is searchable by Exercise 8.4.5(ii).

$3 \Rightarrow 4$: As in Proposition 7.3.5(i), if ε_S is a selection operator for S, then $\exists_S = \lambda p.p(\varepsilon_S p)$ is an existential quantifier for S.

4 $\Rightarrow$ 1: It is easy to show that S is closed and bounded, so in the light of Theorem 8.4.2, it is enough to show that S is equicontinuous. Suppose $x_i \to x$. For all $f \in S$, define $\Psi(f) = \mu n.\forall i \geq n. f(x_i) = f(x)$. Applying the proof of Theorem 8.2.16 to the convergent sequence $(\Lambda f. f(x_i)) \to (\Lambda f. f(x))$, we see that $\Psi(f)$ is continuous in f. So $\Psi(S)$ is exhaustible, hence finite. So its maximum gives a modulus of convergence for $f(x_i) \to f(x)$ that works uniformly for all $f \in S$. □

As a corollary to this analysis, we may obtain a complete characterization of searchable and exhaustible sets:

Corollary 8.4.8. *The following are equivalent for an inhabited set $S \subseteq \mathsf{Ct}(\sigma)$:*

1. *The closure of S is compact.*
2. *S is continuously searchable.*
3. *S is continuously exhaustible.*

Proof. It is easy to see that an existential quantifier for S also serves as one for its closure $\overline{S}$ and vice versa; in view of Theorem 8.4.7, this establishes 1 $\Leftrightarrow$ 3. The proof of 2 $\Rightarrow$ 3 is as for 3 $\Rightarrow$ 4 in Theorem 8.4.7. Finally, if S is continuously exhaustible, then a bound g for $h(S)$ may be readily constructed, so by Exercise 8.3.6(ii) there is a continuous (but possibly non-effective) selection function for $h(S)$. Moreover, using 1 $\Rightarrow$ 2 from Theorem 8.4.7, there exists $e \in \mathsf{Ct}(1 \to \sigma)$ such that $e(h_x) = x$ for all $x \in \overline{S}$. Using e, we may now obtain a continuous selection function for S itself as in Exercise 8.4.5(ii). □

One may also ask to what extent the equivalences of Theorem 8.4.7 can be effectivized. Whilst the proof of 3 $\Rightarrow$ 4 in this theorem shows that an existential quantifier for S is λ-definable uniformly in any selection function for S, we have not so far provided any effectivity information in the other direction, in view of the non-constructive nature of our proof of 1 $\Rightarrow$ 2. This deficiency is remedied by the following theorem, which furnishes effective content for the implications 4 $\Rightarrow$ 2 $\Rightarrow$ 3.

Theorem 8.4.9. *Suppose $S \subseteq \mathsf{Ct}(\sigma)$ is inhabited, closed and continuously exhaustible. Then the following data are $\mathsf{Klex}^{\min}$ computable uniformly in $\exists_S$:*

1. *A function $g \in \mathsf{Ct}(1)$ bounding the trace $h(S)$.*
2. *An idempotent $d \in \mathsf{Ct}(1 \to 1)$ whose image is exactly $h(S)$.*
3. *A function $e \in \mathsf{Ct}(1 \to \sigma)$ such that $e \circ h = id_\sigma$.*

Moreover, a selection function for S is uniformly $\mathsf{Klex}^{\min}$ computable relative to g, d, e and the fan functional, and hence relative to $\exists_S$ and FAN.

Proof. Since a universal quantifier for S is computable uniformly in an existential quantifier, we may compute a suitable bounding function g via

$$g(i) \;=\; \mu m.\, \forall x \in S.\, h_x(i) \leq m\,.$$

The recursive definition of the idempotent d in the proof of Theorem 8.4.2 can also be readily translated into a definition in $\mathsf{Klex}^{\min}$ relative to $\forall_S$. As regards e, suppose we are given $h_x \in \mathsf{Ct}(1)$ and $y \in \mathsf{Ct}(\tau)$; we will show how to compute $x(y) \in \mathbb{N}$ with the help of $\forall_S$. First note that we have an open set $U = \{x' \in \mathsf{Ct}(\sigma) \mid x'(y) = x(y)\}$ containing x itself. Next, for each m, define a closed set

$$F_m = \{x' \in S \mid \forall i < m.\, x'(\xi_\tau(i)) = h_x(i)\}\,.$$

Then $S \supseteq F_0 \supseteq F_1 \supseteq \cdots$, and $\bigcap_m F_m = \{x\}$ since x is uniquely determined by its trace. Thus the complements of the F_m together with U form an open cover of the compact space S, so there exists some m such that $F_m \subseteq U$, and consequently $x'(y) = x''(y)$ for all $x', x'' \in F_m$. We may therefore compute $x(y)$ as

$$\mu n.\, \exists x' \in F_m.\, x'(y) = n\,, \quad \text{where } m = \mu m.\, \forall x', x'' \in F_m.\, x'(y) = x''(y)\,.$$

Finally, we note that a selection function $\varepsilon_{h(S)}$ for $h(S)$ is $\mathsf{Klex}^{\min}$ computable from g, d and FAN by Exercise 8.3.6(i), and a selection function ε_S for S itself is λ-definable from $\varepsilon_{h(S)}$ and e by Exercise 8.4.5(ii). Thus ε_S is $\mathsf{Klex}^{\min}$ computable from $\exists_S$ and FAN. □

Note that we cannot hope for a corresponding effectivization of Corollary 8.4.8, in view of the potential non-effectiveness of the selection functions here.

To conclude this subsection, we comment on the relationship to Escardó's treatment in [80]. Escardó considers searchability and exhaustibility as properties of sets $K \subseteq \mathsf{PC}(\sigma)$, the main interest being in sets of the form $q^{-1}(S)$ where $q : \mathsf{Ext} \to \mathsf{Ct}$ is the quotient map. For instance, one may say that K is (PC-)exhaustible if there is $\exists_K \in \mathsf{PC}((\sigma \to \mathsf{B}) \to \mathsf{B})$ such that whenever $P \in \mathsf{PC}(\sigma \to \mathsf{B})$ and $P(x) \in \mathbb{B}$ for all $x \in K$, we have $\exists_K(P) \in \mathbb{B}$ and $\exists_K(P) = tt$ iff $\exists x \in K.\, P(x) = tt$. (Cf. Definition 7.3.4.) In terms of such notions, Escardó obtains results very similar to Theorems 8.4.7 and 8.4.9 above, using PCF-definability in PC as the notion of effective computability.

Despite the apparent differences, it turns out that Escardó's definitions and ours match up well:

Proposition 8.4.10. *A set $q^{-1}(S)$ is PC-exhaustible iff S is closed and continuously exhaustible.*

Proof. (Outline.) In view of Theorem 8.4.7, it suffices to show that $q^{-1}(S)$ is PC-exhaustible iff S is compact. One direction is easy: if S is compact then by Exercise 8.4.3 it has a compact set R of realizers in $q^{-1}(S)$, and if P is a total predicate on S then clearly $\forall x \in S.\, P(x) = ff$ iff $\forall x \in R.\, P(x) = ff$, since each $\approx$-class is closed under $\sqcup$. Using this, it is an easy exercise to construct a continuous existential quantifier for $q^{-1}(S)$. For the converse, a more elaborate topological argument is needed (see [80, Lemma 5.5]); we saw a special case of this in Lemma 7.3.8. □

Thus, the results we have presented are essentially the same as Escardó's, although our treatment makes slightly more explicit that all relevant constructions work extensionally on realizers, and thus give rise to functions at the level of Ct. We refer the reader to [80] for more detailed information and further results.

8.4.2 The Kreisel Representation Theorem

Next, we establish a connection between Ct and the *analytic hierarchy* for predicates on $\mathbb{N}^{\mathbb{N}}$ as introduced in Definition 5.2.5. We assume familiarity here with the basic theory of logical complexity as summarized in Subsection 5.2.1.

Recall that a Π^1_1 *predicate* in general has the form $\forall f \in \mathbb{N}^{\mathbb{N}}.\phi$, where ϕ is an *arithmetical* predicate which may involve arbitrary nesting and alternation of quantifiers over $\mathbb{N}$. However, we saw in Theorem 5.2.6(iii) that any such predicate is equivalent to one in the simpler form $\forall f \in \mathbb{N}^{\mathbb{N}}.\exists z \in \mathbb{N}.\psi$ where ψ is primitive recursive. Here we will show how this normal form result may be extended to higher logical complexity classes Π^1_k, simply by allowing f to range over continuous functionals of type level k. The ideas are in essence due to Kreisel [152], though the result as we present it here was first formulated explicitly in Normann [209]. Taking a small historical licence, we will refer to this result as the *Kreisel representation theorem.*

We shall make use of the dense enumerations ξ_σ from Theorem 8.1.31, along with the corresponding trace operation $x \mapsto h_x$ as in Definition 8.4.6.

Theorem 8.4.11 (Kreisel representation theorem). *Suppose $A \subseteq \mathbb{N}^{\mathbb{N}}$ is a Π^1_k predicate, where $k \geq 1$. Then there is a primitive recursive relation $R \subseteq \mathbb{N}^{\mathbb{N}} \times \mathbb{N}^{\mathbb{N}} \times \mathbb{N}$ such that for all $f \in \mathbb{N}^{\mathbb{N}}$ we have*

$$f \in A \;\Leftrightarrow\; \forall z \in \mathsf{Ct}(k).\, \exists n \in \mathbb{N}.\, R(f, h_z, n)\,.$$

Proof. For $k = 1$, this is just Theorem 5.2.6(iii). So assume $k > 1$ where the theorem holds for $k-1$.

Since A is Π^1_k, there is a Σ^1_{k-1} set B such that

$$f \in A \;\Leftrightarrow\; \forall g \in \mathbb{N}^{\mathbb{N}}.\, \langle f, g\rangle \in B\,,$$

where $\langle -, - \rangle$ is a standard primitive recursive pairing operation on $\mathbb{N}^{\mathbb{N}}$. Applying the induction hypothesis to the complement of B, we have a primitive recursive relation S such that

$$p \notin B \;\Leftrightarrow\; \forall z \in \mathsf{Ct}(k-1).\, \exists n \in \mathbb{N}.\, S(p, h_z, n)\,.$$

By considering the function $F(z) = \mu n.\, S(p, h_z, n)$, we see that

$$\forall z \in \mathsf{Ct}(k-1).\, \exists n \in \mathbb{N}.\, S(p, h_z, n) \;\Leftrightarrow\; \exists F \in \mathsf{Ct}(k).\, \forall z \in \mathsf{Ct}(k-1).\, S(p, h_z, F(z))\,.$$

We thus have

$$f \in A \;\Leftrightarrow\; \forall g \in \mathbb{N}^{\mathbb{N}}.\, \forall F \in \mathsf{Ct}(k).\, \exists z \in \mathsf{Ct}(k-1).\, \neg S(\langle f, g\rangle, h_z, F(z))\,.$$

If for some given f, g, F there exists z such that $\neg S(\langle f, g\rangle, h_z, F(z))$, then by continuity there will be some such z of the form $\xi_{k-1}(n)$. Thus

$$f \in A \;\Leftrightarrow\; \forall g \in \mathbb{N}^{\mathbb{N}}.\, \forall F \in \mathsf{Ct}(k).\, \exists n \in \mathbb{N}.\, \neg S(\langle f,g\rangle, h_{\xi(n)}, F(\xi(n)))\,.$$

But since $\mathbb{N}^{\mathbb{N}} \times \mathsf{Ct}(k)$ is primitively recursively a retract of $\mathsf{Ct}(k)$ (see Section 4.2), we may collapse the first two quantifiers to one over $\mathsf{Ct}(k)$. Moreover, if $F' \in \mathsf{Ct}(k)$ codes the pair (g,F), it is easy to see that both g and h_F are primitive recursively computable from $h_{F'}$. Since $F(\xi(n)) = h_F(n)$, and $h_{\xi(n)}$ is primitive recursive in n, we may now recast the above formula as

$$f \in A \;\Leftrightarrow\; \forall F' \in \mathsf{Ct}(k).\, \exists n \in \mathbb{N}.\, R(f, h_{F'}, n)$$

for some primitive recursive R, and this is in the form we require. □

The same ideas also yield a normal form for Σ^1_k sets that is not simply the dual of the above. The proof of this is already implicit in the proof of the above theorem, but the corollary is so useful that it is worth spelling out the argument in detail.

Corollary 8.4.12. *Suppose $B \subseteq \mathbb{N}^{\mathbb{N}}$ is a Σ^1_k predicate, where $k \geq 1$. Then there is a primitive recursive relation $S \subseteq \mathbb{N}^{\mathbb{N}} \times \mathbb{N}^{\mathbb{N}} \times \mathbb{N}$ such that*

$$f \in B \;\Leftrightarrow\; \forall F \in \mathsf{Ct}(k+1).\, \exists n \in \mathbb{N}.\, S(f, h_F, n)\,,$$

where uniformly in any $f \notin B$, we may μ-computably construct some $F \in \mathsf{Ct}(k+1)$ such that $\forall n \in \mathbb{N}.\, \neg S(f, h_F, n)$.

Proof. By Theorem 8.4.11, there is a primitive recursive relation R such that

$$f \notin B \;\Leftrightarrow\; \forall z \in \mathsf{Ct}(k).\, \exists m \in \mathbb{N}.\, R(f, h_z, m)\,.$$

We shall show that

$$f \in B \;\Leftrightarrow\; \forall F \in \mathsf{Ct}(k+1).\, \exists n \in \mathbb{N}.\, \neg R(f, h_{\xi(n)}, F(\xi(n)))\,.$$

This is clearly of the required form, since $h_{\xi(n)}$ and $F(\xi(n))$ are primitive recursive in n and h_F.

First suppose $f \notin B$. Define $F_f(z) = \mu m.R(f, h_z, m)$; then F_f is total and present in $\mathsf{Ct}(k+1)$, and by considering its behaviour on arguments $z = \xi(n)$ we see that

$$\forall n \in \mathbb{N}.\, R(f, h_{\xi(n)}, F_f(\xi(n)))\,.$$

Thus F_f serves as a suitable counterexample, and F_f is clearly μ-computable uniformly in f as required.

On the other hand, suppose $f \in B$, and consider an arbitrary $F \in \mathsf{Ct}(k+1)$. Then there is some $z \in \mathsf{Ct}(k)$ such that $\forall m \in \mathbb{N}. \neg R(f, h_z, m)$, and in particular $\neg R(f, h_z, F(z))$. Moreover, by the continuity of R, h and F, there will be some such z of the form $\xi(n)$. Thus $\exists n \in \mathbb{N}.\, \neg R(f, h_{\xi(n)}, F(\xi(n)))$, completing the verification of the above formula for $f \in B$. □

One important application of Theorem 8.4.11 is to ascertain the logical complexity of our various representations of Ct. Recall from Subsection 8.2.1 that we have two ways of representing elements of Ct in $\mathbb{N}^{\mathbb{N}}$: via *associates* (corresponding to the standard simulation $\mathsf{EC}(K_2) \longrightarrow\!\!\triangleright K_2$) or via *weak associates* (corresponding to the composite simulation $\mathsf{EC}(\mathsf{PC}) \longrightarrow\!\!\triangleright \mathsf{PC} \longrightarrow\!\!\triangleright K_2$). A third way to represent an element $z \in \mathsf{Ct}$ in $\mathbb{N}^{\mathbb{N}}$ is by its trace h_z.

The following shows that for each of these representations, the complexity of the set of representing functions increases with the type level. Again, for notational simplicity we concentrate on the pure types.

Theorem 8.4.13. *For any $k \geq 1$, the following sets are Π^1_k complete:*

1. *The set A_{k+1} of associates for elements in* $\mathsf{Ct}(k+1)$.
2. *The set B_{k+1} of weak associates for elements in* $\mathsf{Ct}(k+1)$.
3. *The set H_{k+1} of traces h_x for $x \in \mathsf{Ct}(k+1)$.*

Proof. That A_{k+1} and B_{k+1} are Π^1_k sets is shown by an easy induction on k. As for H_{k+1}, given any $h \in \mathbb{N}^{\mathbb{N}}$ we may define a function $b_h \in \mathbb{N}^{\mathbb{N}}$ that enumerates the set of compacts $(c \Rightarrow m) \in \mathsf{PC}(k+1)$ such that $h(n) = m$ whenever c and $\xi(n)$ are compatible; then b_h is a weak associate for $f \in \mathsf{Ct}(k+1)$ iff h is the trace of f. Moreover, the formula $b_h(i) = j$ (with i, j as free variables) is equivalent to a purely arithmetical formula in terms of h, so from a Π^1_k formula for B_{k+1} we obtain a Π^1_k formula for the set $H_{k+1} = \{h \mid b_h \in B_{k+1}\}$.

To show completeness, suppose $C \subseteq \mathbb{N}^{\mathbb{N}}$ is an arbitrary Π^1_k set; we will reduce the membership problem for C to that for A_{k+1}, B_{k+1} or H_{k+1}. Using the Kreisel representation theorem, take R primitive recursive such that

$$f \in C \;\Leftrightarrow\; \forall z \in \mathsf{Ct}(k).\, \exists n \in \mathbb{N}.\, R(f, h_z, n)\,,$$

and define $F_f(z) = \mu n.R(f, h_z, n)$. Then $f \in C$ iff $F_f \in \mathsf{Ct}(k+1)$. Moreover, a 'quasi-associate' for F_f may be constructed computably in f, and this will be a true associate precisely when $f \in C$. We thus obtain a computable function $\theta : \mathbb{N}^{\mathbb{N}} \to \mathbb{N}^{\mathbb{N}}$ such that $f \in C$ iff $\theta(f) \in A_{k+1}$. The same argument works for B_{k+1}.

For H_{k+1}, a similar argument applies, but in order to ensure that a well-defined 'quasi-trace' may be constructed for any $f \in \mathbb{N}^{\mathbb{N}}$, we require the following strengthening of Theorem 8.4.11: if A is a Π^1_k set, there is a primitive recursive R such that

$$\begin{gathered} f \in A \;\Leftrightarrow\; \forall z \in \mathsf{Ct}(k).\, \exists n \in \mathbb{N}.\, R(f, h_z, n)\,, \\ \forall f \in \mathbb{N}^{\mathbb{N}}.\, \forall i \in \mathbb{N}.\, \exists n \in \mathbb{N}.\, R(f, h_{\xi(i)}, n)\,. \end{gathered}$$

We may derive this from Theorem 8.4.11 with the help of a suitable injective mapping $\iota : \mathsf{Ct}(k) \to \mathsf{Ct}(k)$ whose image avoids all elements $\xi(i)$. For instance, exploiting the fact that every $\xi(i)$ has a finite range, we may map $z \in \mathsf{Ct}(k)$ to the unbounded function $\iota(z) = \lambda w.z(w) + w(\xi_{k-2}(0))$; note that h_z is readily recoverable from $h_{\iota(z)}$. We leave the remaining details to the reader. □

Kreisel's original motivation in [152] was to give a constructive interpretation for sentences of second-order arithmetic in the spirit of the so-called *no-counterexample interpretation* for first-order arithmetic. Specifically, if ϕ is a closed second-order sentence equivalent to $\forall z \in Ct(k).\exists n \in N.R(h_z,n)$, the constructive content of ϕ (when true) may be embodied by the computable functional $\Psi_\phi = \Lambda z.\mu n.R(h_z,n)$, and the remaining content is then expressed simply by $\forall z.R(h_z,\Psi_\phi(z))$. However, as pointed out in Subsection 2.3.1, the constructive credentials of this interpretation are severely undermined by the fact that the range of z here is just as complex as the formula P that we are trying to interpret. Nonetheless, Theorem 8.4.11 turns out to be a powerful tool for analysing the computability theory of Ct; we will see further examples of this in the next subsection.

8.5 Kleene Computability and μ-Computability over Ct

In this section we will look more closely at the degree structures arising from the notion of relative Kleene computability on Ct and from the weaker notion of relative μ-computability. We will prove many instances of relative non-computability results, showing that a rich degree structure arises even among functionals of $\mathsf{Ct}^{\mathrm{eff}}$. We will also look briefly at how the classical c.e. degrees embed naturally into a class of c.e. degrees of higher-order functionals.

8.5.1 Relative Kleene Computability

We now investigate the Kleene degree structure on $\mathsf{Ct}^{\mathrm{eff}}$. We begin with Tait's theorem [284] that the fan functional is not Kleene computable. This is the prototypical non-computability result, and Tait's argument has inspired most other proofs in this area. We then proceed to Hyland's result that Γ is not computable from FAN, and thence to more advanced material, culminating in a proof that for $k \geq 3$ there is no maximal Kleene degree for functionals in $\mathsf{Ct}(k)$ or $\mathsf{Ct}^{\mathrm{eff}}(k)$.

The definitions we shall require are as follows (cf. Definition 5.1.2). In order to work with the traditional S1–S9 definition of Kleene computability, we will restrict our attention here to objects of pure type, though in the light of Theorem 4.2.9 and the discussion of Section 5.1, this is not an essential limitation. We will thus regard FAN as a functional of pure type $\overline{3}$, although of course the value of $\mathsf{FAN}(F)$ will depend only on the restriction of F to the Cantor space $C = \{0,1\}^{\mathbb{N}}$.

Definition 8.5.1. Let $F \in \mathsf{Ct}(k)$ where $k \geq 1$, and $G_i \in \mathsf{Ct}(l_i)$ for $i < s$.
(i) We say that F is *Kleene computable in* $G_0,\ldots,G_{s-1}$, and write $F \leq_{\mathsf{Kl}} G_0,\ldots,G_{s-1}$, if there is a Kleene index e at type $\overline{l_0},\ldots,\overline{l_{s-1}},\overline{k-1} \to \mathtt{N}$ such that $[\![M]\!](\vec{G},x) = F(x)$ for all $x \in \mathsf{Ct}(k-1)$.
(ii) If $l \geq 1$, we say that F is *l-obtainable* from $\vec{G}$ if F is Kleene computable in $\vec{G}$ and some $h \in \mathsf{Ct}(l)$. When $\vec{G}$ is empty, we simply say that F is *l-obtainable*.
(iii) The *k-section* of $\vec{G}$, written k–sc$(\vec{G})$, is the set of $F \in \mathsf{Ct}(k)$ with $F \leq_{\mathsf{Kl}} \vec{G}$.

The terminology in part (ii) is due to Gandy and Hyland [101].

The following key lemma shows, in effect, that the idea of Lemma 8.1.26 goes through under somewhat weakened hypotheses. We shall say $f \in \mathsf{PC}(2)$ *secures* the value of $F(G)$ (where $F \in \mathsf{Ct}(2)$, $G \in \mathsf{Ct}(1)$) if $f(g) = F(G)$ whenever $g \Vdash_1 G$. We shall also say that $a \in \mathsf{PC}(2)$ *secures* that $\{e\}(F,h) = m$ if for all $F' \in V_a$ we have $\{e\}(F',h) \preceq m$.[3]

Lemma 8.5.2. *Suppose e is a Kleene index, $F \in \mathsf{Ct}(2)$, $h \in \mathsf{Ct}(1)$, and $f \in \mathsf{PC}(2)$ secures the value of $F(G)$ for all $G \in 1\mathrm{-sc}(F,h)$. If $\{e\}(F,h) = m$, there is some compact $a \sqsubseteq f$ securing this fact.*

Proof. We shall prove this for all terminating computations $\{e\}(F,h,n_0,\vec{n})$ where $n_0,\ldots,n_{r-1} \in \mathbb{N}$, by induction on the ordinal height of the computation, as in the proof of Lemma 8.1.26. The crucial induction case is the one for S8:

$$\{e\}(F,h,n_0,\ldots,n_{r-1}) \;=\; F(\Lambda n.\{d\}(F,h,n,n_0,\ldots,n_{r-1}))\,.$$

Clearly $G = \Lambda n.\{d\}(F,h,n,n_0,\ldots,n_{r-1}) \in 1\mathrm{-sc}(F,h)$, so if $g \Vdash_1 G$ then $f(g) = F(G)$, so there is some compact $b \in \mathsf{PC}(1)$ with $f(b) = F(G)$.

Since g here may be chosen to be strict (i.e. $g(\bot) = \bot$), we may assume the domain of b is some finite set $\{v_0,\ldots,v_{t-1}\}$. But now for each $i < t$ the induction hypothesis gives us $a_i \sqsubseteq f$ such that for all $F' \in V_{a_i}$ we have $\{d\}(F',v_i,n_0,\ldots,n_{r-1}) = b(v_i)$ whenever the left-hand side terminates. Now let a be the join of the a_i together with $(b \Rightarrow f(b))$; then it is easy to check that a has the required property. □

Theorem 8.5.3 (Tait). *The fan functional is not 1-obtainable.*

Proof. Suppose $\mathsf{FAN} = \Lambda F.\{e\}(F,h)$ for some $h \in \mathsf{Ct}(1)$ and $e \in \mathbb{N}$, and let F_0 denote the constant zero functional.

Let $p \in \{0,1\}^{\mathbb{N}}$ be non-computable in h, and define $f \in \mathsf{PC}(2)$ by $f(q) = 0$ if q and p are inconsistent, while $f(q) = \bot$ whenever $q \sqsubseteq p$. Then $f(g) = F_0(g) = 0$ whenever $g \Vdash G \in 1\mathrm{-sc}(F_0,h)$, so by Lemma 8.5.2 there is a compact $a \sqsubseteq f$ such that $\Lambda F.\{e\}(F,h)$ is constant on V_a.

Since $f(p) = \bot$, we have $a(p) = \bot$, and because p will be defined on a clopen subset of Cantor space, it can readily be extended to the whole Cantor space in many ways, so the fan functional is not constant on V_a. This contradicts our assumption that $\mathsf{FAN} = \Lambda F.\{e\}(F,h)$. □

As a preparation for further results of this kind, let us re-examine the above proof in order to draw out some important points.

A key ingredient of Tait's proof is that any terminating computation $\{e\}(F,h)$ will only make use of the restriction of F to some countable set S, in this case the 1-section of F,h. The computation will then be secured by a finite portion a of any f that secures $F(G)$ for all $G \in S$, and we then argue that no such a can secure the value of the fan functional on F. In what follows, we shall use similar arguments, but

[3] Notice that our relation 'a secures $\{e\}(F,h) = m$' is not semidecidable, as the sets V_a are of high logical complexity. As an alternative, a semidecidable version of this relation (equally suitable for our present purposes) may be extracted from the proof of Theorem 8.1.28.

sometimes using a more delicately chosen countable set S in place of $1-\mathrm{sc}(F,h)$. The following lemma shows one way of constructing such an S.

Lemma 8.5.4. *Let $F \in \mathsf{Ct}(2)$ and suppose $\{e\}(F,\vec{h},\vec{n}) \downarrow$. Then there is a function $\delta = \delta_{e,F,\vec{h},\vec{n}} : \mathbb{N}^2 \to \mathbb{N}$, Kleene computable in $F,\vec{h}$, that enumerates all type 1 functions passed to F in the course of the computation—that is, for every invocation of S8 that calls F with some argument $G \in \mathsf{Ct}(1)$, we have that $G = \Lambda n.\,\delta(k,n)$ for some k.*

Proof. Using the recursion theorem, we may readily give a universal algorithm for computing $\delta e,F,\vec{h},\vec{n}$ from $e,F,\vec{h},\vec{n}$. In the case of a call to F via S8 with argument G, we may take $\delta(0,j) = G(j)$ and $\delta(1+\langle i,k\rangle, j) = \delta_i(k,j)$, where δ_i is the corresponding enumeration for the subcomputation of $G(i)$. □

We will now proceed to analyse Kleene computations with the fan functional itself as an additional argument, and in particular show that Γ is not 1-obtainable from FAN. Of course, the above lemma will no longer apply in the presence of FAN as a type 3 argument, but we can formulate an analogous statement that does.

The key idea is that whenever we apply FAN to a type 2 argument G, we only need to know the value of G on some *compact* set of type 1 arguments. (Note that we are here referring to topological compactness for subsets of $\mathsf{Ct}(1)$, as distinct from compactness in the sense of finiteness for elements of PC.) This suggests that by analogy with the above lemma, in any terminating computation $\{e\}(\mathsf{FAN},F)$ we only need to know F on some special union of compact sets. The following definition and lemma are tailored to making this idea precise:

Definition 8.5.5. A set $B \subseteq \mathbb{N}^{\mathbb{N}}$ is *effectively σ-compact* if there is a computable list $\varepsilon_0, \varepsilon_1, \ldots \in \mathbb{N}^{\mathbb{N}}$ such that

$$f \in B \ \text{ iff } \ \exists n.\forall i.\, f(i) \leq \varepsilon_n(i)\ .$$

In this case, we say that $\varepsilon_0, \varepsilon_1, \ldots$ *enumerates* the effectively σ-compact set B.

Lemma 8.5.6. *Let $F \in \mathsf{Ct}(2)$, and suppose $\{e\}(\mathsf{FAN},F) = m$. Then uniformly in e and F, we may compute an enumeration of an effectively σ-compact set B such that $\{e\}(\mathsf{FAN},F') = m$ whenever $F' \upharpoonright_B = F \upharpoonright_B$.*

Proof. We show this uniformly for all terminating computations $\{e\}(\mathsf{FAN},F,\vec{g},\vec{n})$, where the g_i are of type 1 and the n_i of type 0 (ignoring variations in the order of arguments to simplify notation). Again, the desired enumeration is computed uniformly in the data by means of the recursion theorem.

The proof is similar to that of Lemma 8.5.4. In the case of an application of F to some $g \in \mathsf{Ct}(1)$, we may add g itself to our enumeration so as to ensure $g \in B$. The only new case is for an application of FAN:

$$\{e\}(\mathsf{FAN},F,\vec{g},\vec{n}) \ = \ \mathsf{FAN}(\Lambda g.\{d\}(\mathsf{FAN},F,g,\vec{g},\vec{n}))\ .$$

By the induction hypothesis, for each g in $C = \{0,1\}^{\mathbb{N}}$ (Cantor space), there is a sequence $\varepsilon_0^g, \varepsilon_1^g, \ldots$, computable uniformly in F,g and the other data, enumerating an effectively σ-compact set B_g that suffices for fixing the value of $\{d\}(\mathsf{FAN},F,g,\vec{g},\vec{n})$.

Using a computable realizer for the fan functional, we may now compute the bound $\varepsilon_n(i) = \max_{g \in C} \varepsilon_n^g(i)$ uniformly in the data. It is then easy to see that $\varepsilon_0, \varepsilon_1, \ldots$ enumerates a suitable set B for the case in question. □

Our proof of the non-obtainability of Γ from FAN will proceed by noting that if e were a Kleene index for Γ relative to FAN, the value of $\{e\}$ at the constant zero functional F_0, say, would only depend on the restriction of F_0 to some effectively σ-compact set. It will then suffice to show that the restriction of a functional F to an effectively σ-compact B cannot suffice to pin down the true value of $\Gamma(F)$. This is achieved via the following technical lemma, which ensures a plentiful supply of suitable elements in $\mathsf{PC}(2)$ that do not constrain the value of $\Gamma(F)$.

We recall here the defining equation for Γ:

$$\Gamma(F) \;=\; F_0(\Lambda n \in \mathbb{N}.\Gamma(F_{n+1}))\,.$$

Lemma 8.5.7. *Let $\alpha : \mathbb{N} \to \mathbb{N}$ be such that $\alpha(n) > 0$ for all n. Let A_α be the set of sequences $p = (n_0, \ldots, n_{r-1}) \in \mathbb{N}^*$ such that*

1. $n_i < \alpha(i)$ for some $i < r$,
2. $r > i + \alpha(i)$ for the least such i.

Identify $(n_0, \ldots, n_{r-1}) \in A_\alpha$ with the element $\bigsqcup_{i<r}(i \Rightarrow n_i) \in \mathsf{PC}(1)$.

Suppose $p_0, \ldots, p_{k-1} \in A_\alpha$ are pairwise incompatible and $s_0, \ldots, s_{k-1} \in \mathbb{N}$. Then for any $t \in \mathbb{N}$, there exists $F \in \mathsf{Ct}(2)$ such that $F(V_{p_i}) = \{s_i\}$ for each $i < k$, and $\Gamma(F) = t$.

Proof. We will prove this uniformly for all α by induction on k. If $k = 0$, the only constraint on F is that $\Gamma(F) = t$, so we may take F to be constantly t.

Suppose $k > 0$ and $p_0, \ldots, p_{k-1}, s_0, \ldots, s_{k-1}$ are as above. Let $\beta(n) = \alpha(n+1)$ for each n; we will use the induction hypothesis for β. There are two cases.

Case 1: There is no $i < k$ such that $p_i(0) = 0$. We may then take F to be constantly s_i on each V_{p_i}, and at the same time constantly t on $V_{(0 \Rightarrow 0)}$; then $\Gamma(F) = t$.

Case 2: There is at least one $i < k$ with $p_i(0) = 0$. Then for all such i, the definition of A_α gives us that the length of p_i exceeds $\alpha(0)$. We may therefore set

$$u \;=\; \max\{p_i(\alpha(0)) \mid p_i(0) = 0\} + 1\,.$$

Furthermore, since $\alpha(0) > 0$, there are fewer than k indices i with $p_i(0) = \alpha(0)$; and if $p_i = (\alpha(0), n_1, \ldots, n_{r-1})$ then $(n_1, \ldots, n_{r-1}) \in A_\beta$. By the induction hypothesis, we may therefore pick $G \in \mathsf{Ct}(2)$ such that G is constantly s_i on V_{q_i} for all i with $p_i(0) = \alpha(0)$, and $\Gamma(G) = u$.

We now define $F \in \mathsf{Ct}(2)$ by

$$F(h) \;=\; \begin{cases} G(\Lambda n.h(n+1)) & \text{if } h(0) = \alpha(0) > 0\,, \\ s_i & \text{if } (h(0), \ldots, h(|p_i| - 1)) = p_i\,, \\ t & \text{if } h(0) = 0 \text{ and } h(\alpha(0)) = u\,, \\ 0 & \text{otherwise.} \end{cases}$$

To see that F is well-defined, note first that for a given h, the second clause can apply for at most one i since the p_i are pairwise incompatible. Next note that the first clause does not conflict with the second clause in view of how G is chosen. As for the third clause, there is no overlap with the first clause because of the condition $h(0)=0$, and no overlap with the second clause owing to the choice of u. Thus F is well-defined, and by the second clause, F is constantly s_i on V_{p_i} for each i.

It remains to compute $\Gamma(F)$. From the definition we see that $\Gamma(F) = F(h_0)$, where $h_0 = \Lambda n.\textit{if } n=0 \textit{ then } 0 \textit{ else } \Gamma(F_n)$. But $h_0(0)=0$ and $h_0(\alpha(0)) = \Gamma(F_{\alpha_0})$, and clearly $F_{\alpha_0} = G$ so that $h_0(\alpha(0)) = \Gamma(G) = u$ by choice of G. Thus the third clause applies, and $\Gamma(F) = t$ as required. □

Theorem 8.5.8 (Hyland). *The functional Γ is not 1-obtainable in* FAN.

Proof. We will prove that Γ is not Kleene computable in FAN; the same argument goes through relative to a type 1 oracle.

Suppose there were an index e such that $\Gamma(F) = \{e\}(\mathsf{FAN},F)$ for all $F \in \mathsf{Ct}(2)$. Let F_0 be the constant zero functional; then by Lemma 8.5.6 there is a computable enumeration $\varepsilon_0, \varepsilon_1$ of an effectively σ-compact set B such that $\{e\}(\mathsf{FAN},F_0)$ depends only on $F_0 \upharpoonright_B$. Let $\alpha(n) = \varepsilon_n(n)+1$ for each n, and define $A = A_\alpha$ as in Lemma 8.5.7. We will show that every $g \in B$ is secured by some $p \in A$, in the sense that $g \in V_p$. For if g is bounded by ε_j then $g(j) < \alpha(j)$, so taking i minimal such that $g(i) < \alpha(i)$, the initial segment of g of length $i+\alpha(i)$ will be in A.

Now define $f = \bigsqcup_{p\in A}(p \Rightarrow 0) \in \mathsf{PC}(2)$. Then f secures the value of $F(g)$ for every $g \in B$, so the argument of Lemma 8.5.2 shows that there is some compact $a \sqsubseteq f$ securing that $\{e\}(\mathsf{FAN},F') = \{e\}(\mathsf{FAN},F_0)$ for all $F' \in V_a$. Here we may assume that a is a join of finitely many elements $(p \Rightarrow 0)$ for $p \in A$. But this now contradicts Lemma 8.5.7, which shows that Γ itself cannot be constant on any such V_a. Hence Γ is not computable in FAN. □

The following exercise shows why working with an infinite family of compact sets is necessary in the above proof:

Exercise 8.5.9. Consider the *Kierstead functional* $K \in \mathsf{Ct}(3)$ defined by

$$K(F) \;=\; F(\Lambda n.F(\Lambda i.n))$$

(clearly this is Kleene computable). Show that there is no single compact set $B \subseteq \mathsf{Ct}(1)$ such that $K(F') = K(F_0)$ whenever $F' \upharpoonright_B = F_0 \upharpoonright_B$. (Hint: use a simplified version of the definition of F from the proof of Lemma 8.5.7.)

Next, we apply Tait's method again to obtain a surprising result: there are *effective* functionals $\Delta \in \mathsf{Ct}^{\mathrm{eff}}(3)$ that are not Kleene computable outright, but which become Kleene computable in the presence of some *non-computable* type 1 function. The construction uses the existence of computable trees over $\mathbb{N}$ (i.e. decidable prefix-closed subsets of $\mathbb{N}^*$) that have infinite branches but no computable infinite branches; some important examples of such trees will be presented in Chapter 9.

Theorem 8.5.10. *Let T be any computable tree over $\mathbb{N}$ that is not well-founded. Then there is a functional $\Delta_T \in \mathsf{Ct}^{\mathrm{eff}}(3)$ such that for any $g \in \mathsf{Ct}(1)$, Δ_T is Kleene computable in g iff T has an infinite path computable in g.*

Proof. The basic idea is to let finite paths through T serve as approximations to type 1 functions to which F may potentially be applied in the course of computing $\Delta_T(F)$. We shall construct Δ_T so that the value of $\Delta_T(F)$ can only be computed if we have access to an infinite path g through T. However, this is achieved in a much more subtle way than by simply defining $\Delta_T(F) = F(g)$, which would be non-effective in general.

If $F \in \mathsf{Ct}(2)$, $p \in T$ and $m \in \mathbb{N}$, we shall say that F *appears constant on V_p at stage m* if the following hold:

1. $|p| < m$.
2. The set $S_{p,m} = \{\xi(i) \mid i < m,\ \xi(i)$ *extends* $p\}$ has at least two elements.
3. F is constant on $S_{p,m}$.

We now define Δ_T by

$$\Delta_T(F) \;=\; \mu n.\, \forall m \geq n.\, \exists p \in T.\, F \text{ appears constant on } V_p \text{ at stage } m\,.$$

To show that Δ_T is well-defined and effective, we will first show how an upper bound for $\Delta_T(F)$ may be computed uniformly in any weak associate for F. If $h \in K_2$ is a weak associate, say $h \Vdash^{\Psi} f \Vdash F$ where $f \in \mathsf{PC}(2)$, then from h we may find a sequence $p \in T$ and compact $a \sqsubseteq f$ such that $a(p) \in \mathbb{N}$. If we take n so large that $n > |p|$ and for some $i, j < n$ we have that $\xi(i), \xi(j)$ are distinct extensions of p, then for any $m \geq n$, F will appear constant on V_p at stage m. Thus n serves as an upper bound for $\Delta_T(F)$, computable from h.

To compute $\Delta_T(F)$ itself from h, we first compute this upper bound n, then observe that for each $m < n$ we can decide whether there exists $p \in T$ such that F appears constant on V_p at stage m. This is because the only candidates for p that need to be tested are the initial segments of length $< m$ of functions $\xi(i)$ with $i < m$. We may therefore compute the minimum possible value for n, which is the required value of $\Delta_T(F)$. Since this is computable from any weak associate $h \Vdash^{\beta} f$, but the value of $\Delta_T(F)$ does not depend on the choice of h, we have by Theorem 8.2.1 that $\Delta_T \in \mathsf{Ct}^{\mathrm{eff}}(3)$.

Next, we use the Tait argument to show that if there is no infinite path through T computable from $g \in \mathsf{Ct}(1)$, then Δ_T is not Kleene computable in g. Specifically, any terminating computation $\{e\}(F_0, g)$ (where F_0 is constantly zero) will be secured by the element $f = \bigsqcup_{p \notin T}(p \Rightarrow 0) \in \mathsf{PC}(2)$, since this secures everything in the 1-section of g. So by Lemma 8.5.2 there will be a compact $a \sqsubseteq f$ securing the value of $\{e\}(F_0, g)$. To show that $\{e\}(-, g) \neq \Delta_T$, it will suffice to show that Δ_T is not constant on V_a.

At this point, we shall for simplicity suppose (as can readily be ensured) that our computable dense enumeration $\xi : \mathbb{N} \to \mathsf{Ct}(1)$ is injective. We may then consider the following function F' that is 'compatible' with a, but defined only on objects $\xi(i) \in \mathsf{Ct}(1)$:

$$F'(\xi(i)) = \begin{cases} 0 & \text{if } a \text{ secures the value 0 for } F_0(\xi(i)), \\ i+1 & \text{if } F_0(\xi(i)) \text{ is not secured by } a. \end{cases}$$

Since a is finite, we may take m large enough that for all $p \in T$, if p has an extension $\xi(i)$ secured by a, then p also has an extension $\xi(j)$ not secured by a where $j < m$. Then there is no $p \in T$ such that F' appears constant on V_p at stage m, because if p satisfied conditions 1–3 above in relation to F' and m, the constant value of F' on $S_{p,m}$ would have to be 0 by the definition of F', and then for any $\xi(i) \in S_{p,m}$ with $F'(\xi(i)) = 0$ we could find $\xi(j) \in S_{p,m}$ where $F'(\xi(j)) \neq 0$, giving a contradiction. This means that for any F'' agreeing with F' on $\xi(0), \ldots, \xi(m-1)$, we have $\Delta_T(F'') > m$. Clearly some such F'' may be found within V_a; moreover, this construction may be repeated for any sufficiently large m, so Δ_T is unbounded on V_a. But a was chosen to secure the value of $\{e\}(F_0, g)$, so e cannot be an index for Δ_T relative to g. Since e was arbitrary, we have shown that $\Delta_T \not\leq_{\mathrm{Kl}} g$.

Finally, suppose g is an infinite path through T; we will show that Δ_T is Kleene computable (indeed μ-computable) in g. Our method will be indirect, relying on Theorem 8.2.17 on moduli of convergence, which was proved using Grilliot's technique.

For any r, let $U_r = \{g' \in \mathsf{Ct}(1) \mid \forall k \leq r.\, g'(k) = g(k)\}$, and let $F \in \mathsf{Ct}(2)$ be given. Construct a sequence of functions $g_i = \xi(\delta(i))$ uniformly primitive recursive in g and i as follows.

0. Let $\delta(i) = i$ until we find two different i such that $\xi(i) \in U_0$.
1. From there on, enumerate in increasing order the remaining i such that $\xi(i) \in U_0$, until we find two different i (not already appearing at stage 0) such that $\xi(i) \in U_1$.
2. Then enumerate the i such that $\xi(i) \in U_1$ until we find two different i (not already appearing at stage 0 or 1) such that $\xi(i) \in U_2$.
3. And so on, recursively.

Clearly $g_i \to g$, with a modulus of convergence μ-computable in g. By Theorem 8.2.17, we may now compute the modulus of convergence $u \in \mathbb{N}$ for $F(g_i) \to F(g)$, uniformly μ-computably in F and g. From this we may compute an upper bound for $\Delta_T(F)$ as follows. Take r large enough that for all $i \leq u$ we have $g_i \notin U_r$ or else $F(g_i) = F(g)$, also ensuring that $g_0, \ldots, g_u$ does not extend beyond the end of stage $(r-1)$ in the above generation procedure. Then take $j > u$ such that g_j is the second new element of U_r discovered at stage r, and set $n = \delta(j)$.

We claim that n serves as an upper bound for $\Delta_T(F)$: that is, for any $m \geq n$ there exists $p \in T$ such that F appears constant on V_p at stage m. Indeed, given $m \geq n$, consider the largest r' for which $g_0, \ldots, g_{m-1}$ covers the completion of stage r' of the generation procedure, and in particular contains at least two elements of $U_{r'}$ (note that $r' \geq r$). Let $p = g(0), \ldots, g(r')$. From the construction of δ we see that $|p| = r' + 1 < m$, and by the maximality of r' we see that $S_{p,m}$ consists of precisely the elements of U_r that appear in $g_0, \ldots, g_{m-1}$. So $S_{p,m}$ has at least two members, and for any $g_i \in S_{p,m}$ we have $F(g_i) = F(g)$ (by cases according to whether $i > u$). Thus F appears constant on V_p at stage m, and we have shown that n is an upper bound for $\Delta_T(F)$.

We have already seen how $\Delta_T(F)$ itself may be μ-computed from an upper bound for $\Delta_T(F)$, so this completes the proof. □

Corollary 8.5.11. *There is a functional $\Delta \in \mathsf{Ct}^{\text{eff}}(3)$ that is Kleene computable in some $g \in \mathsf{Ct}(1)$, though not in any hyperarithmetical g.*

Proof. Let T be the second Kleene tree as in Theorem 9.1.19. Then T is computable but has no hyperarithmetical infinite paths. So by the above theorem, Δ_T has the properties required. □

Note that our method does not allow us to replace 'hyperarithmetical' by 'Π^1_1' here, since in any computable non-well-founded tree the leftmost infinite branch is computable relative to a Π^1_1 set of integers. We leave it as an open problem whether there is a 1-obtainable functional in $\mathsf{Ct}^{\text{eff}}(3)$ that is not obtainable from any Π^1_1 set of integers. The interest in this problem is that an example would require a new way of constructing continuous functionals of type 3.

As a further application, Theorem 8.5.10 not only gives us 1-obtainable effective functionals that are 'hard' to compute, but also some interesting examples that are 'easy' to compute from other functions:

Exercise 8.5.12. Let $A_0,\ldots,A_{n-1}$ be c.e. sets that are not computable. Show that there is a functional $\Delta \in \mathsf{Ct}^{\text{eff}}(3)$ that is not Kleene computable outright, but is Kleene computable relative to any one of $A_0,\ldots,A_{n-1}$.

As our climactic result on relative Kleene computability, we will show that for any $k \geq 3$, there is no maximal functional in $\mathsf{Ct}(k)$ with respect to relative 1-obtainability. This was first proved in Normann [208]; the idea is to combine the method of proof of Theorem 8.5.10 with a diagonalization argument.

First, we need to analyse the logical complexity of termination for Kleene computations over Ct:

Lemma 8.5.13. *Suppose $k \geq 3$ and $\Phi_0,\ldots,\Phi_{r-1}$ are pure type functionals in Ct of level $\leq k$. Then the relation $\{e\}(\vec{\Phi})\downarrow$ is Π^1_{k-2} uniformly in e and the traces $h_{\Phi_0},\ldots,h_{\Phi_{r-1}}$ of $\Phi_0,\ldots,\Phi_{r-1}$.*

Proof. This can be established by a concrete analysis of bottom-up computation trees in the spirit of Subsection 5.2.2, but we prefer here a more abstract argument based on the general theory of inductive definitions.

The relation $\{e\}(\vec{\Phi}) = a$ is defined by a positive inductive definition (see Definition 5.1.1). If the types of the Φ_i are bounded by $k \geq 3$, any additional objects $\vec{F}$ appearing as arguments in subcomputations will have types $\leq k-2$.

We first note that if $\{d\}(\vec{\Psi})\downarrow$, its value is arithmetically definable from d and the traces $h_{\vec{\Psi}}$. This is because a weak associate for each Ψ_i is arithmetically definable from h_{Ψ_i} as shown in the proof of Theorem 8.4.13; and the proof of Theorem 8.1.26 shows that if $\{d\}(\vec{\Psi}) = m$ then there is a finite tree certifying this fact, whose correctness is expressible arithmetically in d and these weak associates.

Thus, the only ingredient of significant complexity in the inductive definition is the assertion that for a subcomputation involving S8, the relevant function

$\Lambda G.\{d\}(\vec{\Phi},\vec{F},G)$ is total, where the types of $\vec{F}$ and G are bounded by $k-2$. Intuitively, this can be expressed by a universal quantifier over a Π^1_{k-3} set A, namely the set of traces for objects of type $\leq k-2$ (see Theorem 8.4.13).

More formally, let us consider the inductive definition of the set of finite sequences $(e,h_{\vec{\Phi}},f,m)$, where f codes a finite sequence of traces $h_{\vec{F}}$ of functionals $\vec{F}$ of types $\leq k-2$, and e and m are integers, such that $\{e\}(\vec{\Phi},\vec{F})=m$. In this definition we need a universal quantifier over the set A, and elsewhere just arithmetical expressions.

We now appeal to the following fact (a special case of a theorem of Cenzer [50]):

Claim: Let $A\subseteq\mathbb{N}^{\mathbb{N}}$ be any set, and let Γ be an operator on subsets of $\mathbb{N}^{\mathbb{N}}$ definable by a strictly positive $\Pi^1_1(A)$ formula. Then the least fixed point Γ^∞ of Γ is $\Pi^1_1(A)$.

Proof of claim: If $B\subseteq\mathbb{N}^{\mathbb{N}}$, let Γ_B be the operator on subsets of B obtained by restricting the definition of Γ to B (using $A\cap B$ in place of A). By the monotonicity of positive definitions and of restricting universal quantifiers to subsets, we have

$$\Gamma^\infty\cap B\ \subseteq\ \Gamma_B^\infty\,.$$

Moreover, by a Löwenheim–Skolem-style argument, we can show that $f\in\Gamma^\infty$ if and only if $f\in\Gamma_B^\infty$ for all countable B containing f. Finally, for each such B (coded as an element of $\mathbb{N}^{\mathbb{N}}$), the definition of Γ_B^∞ can be expressed by a monotone $\Pi^1_1(A,B)$-inductive definition over $\mathbb{N}$, and hence has a $\Pi^1_1(A,B)$ least fixed point by Exercise 5.2.9 relativized to A,B. Quantifying over all such B, we see that this characterization of Γ^∞ is $\Pi^1_1(A)$ overall.

The lemma follows, since if A is Π^1_{k-3} then a $\Pi^1_1(A)$ predicate will be Π^1_{k-2}. $\square$

Remark 8.5.14. Implicit in the use of weak associates in the above proof is the idea that the computation of $\{e\}(\vec{\Phi})$ may be interpreted in PC with respect to suitable realizers $\phi_i\Vdash\Phi_i$ as in Exercise 8.1.29. Recall that if $\{e\}(\vec{\Phi})=m$ in Ct then $\{e\}(\vec{\phi})=m$ in PC, and hence $\{e\}(\vec{p})=m$ for some compact $\vec{p}\sqsubseteq\vec{\phi}$. These PC-realizers give a level of representation intermediate between Ct and K_2.

As an aside, a similar analysis of computations gives us some interesting information about the computation trees themselves in the case of a type 4 Kleene computation over Ct (cf. Remark 8.3.3).

Corollary 8.5.15. *Suppose $\Phi\in\mathsf{Ct}(3)$ and $\{e\}(\Phi)\downarrow$. Then*

(i) a K_2 realizer for the computation tree $\mathscr{U}(e,\Phi)$ is arithmetically definable uniformly in e and an associate for Φ;

(ii) the tree $\mathscr{U}(e,\Phi)$ is of countable ordinal height.

Proof. (Sketch.) (i) If $\{e\}(\vec{\Phi})\downarrow$, the set of immediate subcomputations is arithmetical uniformly in e and any tuple of associates for $\vec{\Phi}$. Now consider the set of all finite sequences

$$(e_0,\vec{F_0},m_0),\ldots,(e_n,\vec{F_n},m_n)$$

such that $e_0=e$, $\vec{F_0}$ is the empty sequence, and $\{e_{i+1}\}(\vec{\Phi},\vec{F}_{i+1})=m_{i+1}$ is an immediate subcomputation of $\{e_i\}(\vec{\Phi},\vec{F_i})=m_i$ for all $i<n$. This representation of the

computation tree, coded as a tree on $\mathbb{N}^{\mathbb{N}}$, is arithmetically definable. Since, in the case of application of S8, we have full branching over $\mathbb{N}^{\mathbb{N}}$, the tree will be both closed and open, and we may use this to find a K_2-realizer for $\mathscr{U}(e,\Phi)$.

(ii) It is a standard result of descriptive set theory that a well-founded tree whose definition is arithmetical (or even hyperarithmetical) relative to some $\phi \in \mathbb{N}^{\mathbb{N}}$ must be of countable height: see e.g. Moschovakis [201]. □

In the proof of our main theorem, we shall make use of Lemma 8.5.13 via the following corollary. We shall say a tree $T \subseteq \mathbb{N}^*$ is *k-well-founded* if it has no infinite paths of the form h_z where $z \in \mathsf{Ct}(k)$.

Corollary 8.5.16. *Suppose given $e \in \mathbb{N}$ and $\Phi_0, \ldots, \Phi_{r-1} \in \mathsf{Ct}$ of level $\leq k$, where $k \geq 3$. Then there is a tree $T \subseteq \mathbb{N}^*$, primitive recursive uniformly in e and associates for the Φ_i, which is $(k-2)$-well-founded iff $\{e\}(\vec{\Phi})\downarrow$.*

Proof. By Lemma 8.5.13, the relation $\{e\}(\vec{\Phi})\downarrow$ is Π^1_{k-2} uniformly in e and the traces $h_{\Phi_0}, \ldots, h_{\Phi_{r-1}}$, so by the Kreisel representation theorem relativized to the h_{Φ_i}, there is a relation R, primitive recursive in the h_{Φ_i}, such that

$$\{e\}(\vec{\Phi})\downarrow \;\Leftrightarrow\; \forall z \in \mathsf{Ct}(k-2).\ \exists n \in \mathbb{N}.\ R(e,h_z,n)\ .$$

Since the truth value of $R(e,h_z,n)$ will in each instance be determined by some finite portion of h_z, we may clearly replace $\exists n.R(e,h_z,n)$ here by $\exists n.R'(e,\overline{h_z}(n))$ for some R' primitive recursive in the h_{Φ_i}, where $\overline{h_z}$ is the course-of-values function of h_z (note that n is recoverable from $\overline{h_z}(n)$). Moreover, we may without loss of generality assume that $R'(e,\overline{h_z}(n))$ implies $R'(e,\overline{h_z}(n'))$ for all $n' \geq n$.

Now define $T_e = \{q \in \mathbb{N}^* \mid \neg R'(e,q)\}$. Then T_e is primitive recursive uniformly in e and the h_{Φ_i}, and hence in e and associates for the Φ_i. Moreover, the construction ensures that $\{e\}(\vec{\Phi})\downarrow$ iff every h_z for $z \in \mathsf{Ct}(k-2)$ is secured by T_e. □

We will also need to extend parts of the Δ_T technology of Theorem 8.5.10 to higher types. This is achieved via the following technical definition and lemmas. Recall that $\lceil - \rceil$ denotes a numerical coding of the compact elements of PC at any type. Writing $\xi^{k-3} : \mathbb{N} \to \mathsf{Ct}(k-3)$ for our standard enumeration of a dense set, we shall write $\widehat{\xi}^{k-3}(i)$ for some standard choice of PC-realizer for $\xi^{k-3}(i)$ computable from i.

Definition 8.5.17. Suppose $k \geq 3$, let T be a tree of sequences of integers, and let p range over compact elements in $\mathsf{PC}(k-2)$.

(i) We say $F \in \mathsf{Ct}(k-1)$ *appears constant on V_p at stage m* if the following hold:

1. $\lceil p \rceil < m$.
2. The set $S_{p,m} = \{\xi^{k-2}(i) \mid i < m,\ \xi^{k-2}(i) \in V_p\}$ has at least two elements.
3. F is constant on $S_{p,m}$.

(ii) Let $u(p)$ denote the least number s such that $p(\widehat{\xi}^{k-3}(s))$ is undefined, or ∞ if no such s exists. Now define $h_p : \mathbb{N} \rightharpoonup \mathbb{N}$ by

$$h_p(i) = \begin{cases} p(\widehat{\xi}^{k-3}(i)) & \text{if } i < u(p)\,, \\ \text{undefined} & \text{otherwise.} \end{cases}$$

(iii) Let the *rim* of T, written $\alpha(T)$, be the supremum in $\mathsf{PC}(k-1)$ of all compact elements $(p \Rightarrow 0)$ where either V_p is a singleton or there exists $j \leq u(p)$ such that $h_p(0),\ldots,h_p(j-1) \notin T$.

The rim $\alpha(T)$ will in some sense play the role of the similar supremum f in the proof of Theorem 8.5.10. An important difference, however, is that here we will be interested in the situation for 'well-founded' trees:

Lemma 8.5.18. *If T is $(k-2)$-well-founded, then $\alpha(T)$ is a total element, and so is a genuine realizer for $F_0 = \Lambda z.0 \in \mathsf{Ct}(k-1)$.*

Proof. Suppose $z \in \mathsf{Ct}(k-2)$, with $x \in \mathsf{PC}(k-2)$ a realizer for z. Then h_z is secured by T, so take n with $(h_z(0),\ldots,h_z(n-1)) = (z(\xi(0)),\ldots,z(\xi(n-1))) \notin T$. Now take $p \sqsubseteq x$ compact so that $p(\widehat{\xi}(0)),\ldots,p(\widehat{\xi}(n-1))$ are all defined; then $u(p) \geq n$ and $(h_p(0),\ldots,h_p(n-1)) \notin T$, so $(p \Rightarrow 0)$ is one of the compact elements contributing to $\alpha(T)$. Thus $\alpha(T)(x) \sqsupseteq \alpha(T)(p) = 0$. □

In the non-well-founded situation, the following lemma generalizes information from the proof of Theorem 8.5.10. Part (iii) of the lemma is not actually needed for the proof of Theorem 8.5.20, although it has other applications, e.g. in giving a higher type version of Theorem 8.5.10 itself.

Lemma 8.5.19. *Suppose $k \geq 3$ and $T \subseteq \mathbb{N}^*$ is $(k-2)$-non-well-founded. Suppose moreover that if $k \geq 4$, there is at least one* non-constant *function $z \in \mathsf{Ct}(k-2)$ with h_z an infinite path in T. For $F \in \mathsf{Ct}(k-1)$, define*

$$\begin{aligned} \Delta_T(F) \;=\; & \mu n.\ \forall m \geq n.\ \exists p \in \mathsf{PC}(k-2) \text{ compact}. \\ & (h_p(0),\ldots,h_p(u(p)-1)) \in T\ \wedge \\ & F \text{ appears constant on } V_p \text{ at stage } m\,. \end{aligned}$$

Then:

(i) $\Delta_T \in \mathsf{Ct}(k)$, and Δ_T has a realizer computable in T.

(ii) There is no compact $a \sqsubseteq \alpha(T)$ such that Δ_T is constant on V_a.

(iii) Δ_T is Kleene computable in any $z \in \mathsf{Ct}(k-2)$ with h_z an infinite path in T, and where z is non-constant if $k \geq 4$.

Proof. We may re-use arguments from the proof of Theorem 8.5.10 with only cosmetic changes. The special treatment of constant functions and singleton neighbourhoods is a technical measure designed to avoid certain minor obstacles. □

We are now ready to combine Corollary 8.5.16 and Lemma 8.5.19 with a diagonal argument to obtain our main result.

Theorem 8.5.20. *Suppose $\Phi \in \mathsf{Ct}(k)$ where $k \geq 3$. Then there is some $\Delta \in \mathsf{Ct}(k)$ that is not 1-obtainable from Φ. Moreover, if $\Phi \in \mathsf{Ct}^{\mathrm{eff}}(k)$, we may find $\Delta \in \mathsf{Ct}^{\mathrm{eff}}(k)$ with this property.*

Proof. We shall in fact construct Δ as a functional of type $\overline{0},\overline{1},\overline{k-1} \to \overline{0}$, which is a computable retract of $\overline{k}$ as shown in Section 4.2. To motivate the definition of Δ, we will first examine the role that it is required to play in the proof.

We want to be able to show that Δ is not 1-obtainable from Φ via any $e_0 \in \mathbb{N}$ and $g_0 \in \mathsf{Ct}(1)$—that is, that the hypothesis

$$\forall e,g,F.\, \Delta(e,g,F) \;=\; \{e_0\}(\Phi,g_0,e,g,F)$$

leads to a contradiction. So assume that this hypothesis holds; then in particular

$$\Delta(e_0,g_0,F_0) \;=\; \{e_0\}(\Phi,g_0,e_0,g_0,F_0)\,,$$

where $F_0 \in \mathsf{Ct}(k-1)$ is the constant zero functional. Moreover, by Corollary 8.5.16, the termination of this computation is equivalent to the $(k-2)$-well-foundedness of a certain tree $T = T_{e_0,g_0}$, whose rim $\alpha(T)$ is therefore a realizer for F_0. It then follows that a finite portion a of this rim will secure the result, so that $\Delta(e_0,g_0,-)$ will be constant on V_a. On the other hand, for any *non-well-founded* tree T^*, Lemma 8.5.19 specifically gives us a way to construct a functional Δ_{T^*} that is not constant on V_a for any finite $a \sqsubseteq \alpha(T^*)$. This suggests that if we can construct a non-well-founded tree $T^* = T^*_{e_0,g_0}$ from T such that the rims $\alpha(T)$ and $\alpha(T^*)$ share the relevant finite portion a, then by constructing Δ so that $\Delta(e_0,g_0,F) = \Delta_{T^*}(F)$ for all $F \in \mathsf{Ct}(k-1)$, we will obtain a contradiction. This is essentially what we shall do, though some care is needed to ensure that the functional Δ we define is total.

More formally, the following constructions may be performed uniformly in any $e \in \mathbb{N}$, $g \in \mathbb{N}^{\mathbb{N}}$ and any associate ϕ for Φ; however, we shall proceed as though Φ and ϕ were fixed, and will not track the dependence on ϕ in the notation. Firstly, using Corollary 8.5.16, we may construct a tree $T_{e,g}$, primitive recursive in e,g relative to ϕ, such that $T_{e,g}$ is $(k-2)$-well-founded iff $\{e\}(\Phi,g,e,g,F_0)\downarrow$. It follows that the rim $\alpha(T_{e,g})$ is c.e. uniformly in e,g,ϕ, so we can enumerate codes for an increasing sequence of compact elements $c^0_{e,g}, c^1_{e,g}, \ldots$, uniformly computable in e,g,ϕ, with supremum $\alpha(T_{e,g})$. From ϕ we may also compute codes for an increasing sequence $b^0, b^1, \ldots$ of compact elements with supremum Φ; similarly, from g we may obtain codes for a similar increasing sequence $d^0, d^1, \ldots$ with supremum g.

Next, we show how to expand $T_{e,g}$, which may be well-founded or not, to a tree $T^*_{e,g}$ that is always non-well-founded, whilst preserving the relevant finite portion of the rim as outlined above. In the case that $\{e\}(\Phi,g,e,g,F_0)\downarrow$, say with value m, we have by Lemma 8.5.18 that $\alpha(T_{e,g}) \Vdash F_0$, so by Lemma 8.1.26 there will be approximations $b^n, c^n_{e,g}, d^n$ that suffice to secure the result, i.e. such that $\{e\}(\Phi',g',e,g',F') \preceq m$ for all $\Phi' \in V_{b^n}$, $F' \in V_{c^n_{e,g}}$ and $g' \in V_{d^n}$. We shall say simply that n *secures the result* (relative to the given approximating sequences) when this condition holds. Whenever $\{e\}(\Phi,g,e,g,F_0)\downarrow$, we will be able to find an n securing the result by means of a computable enumeration (relative to e,g,ϕ) of all derivable formal assertions $\vdash \{e'\}(\vec{a}') \preceq m'$ as explained in Subsection 8.1.4. Let us now write $N_{e,g}(n)$ to mean that some $n' \leq n$ securing the result may be found within time n; then $N_{e,g}$ is decidable relative to e,g,ϕ, and $N_{e,g}(n)$ holds for some n

iff $\{e\}(\Phi, g, e, g, F_0)\downarrow$. We will use this predicate $N_{e,g}$ to ensure that in the case of termination, enough of the rim is preserved when we pass from $T_{e,g}$ to $T^*_{e,g}$.

Let $n_{e,g}$ denote the least n such that $N_{e,g}(n)$, or ∞ if there is no such n; note that $n \leq n_{e,g}$ is decidable in n, e, g, ϕ. Now let $m_{e,g}$ denote the largest element of the set

$$\{h_p(0) \mid p \in \mathsf{PC}(k-2) \text{ compact, } c^n_{e,g}(p) = 0 \text{ for some } n \leq n_{e,g}\},$$

or ∞ if this set is unbounded. If $n_{e,g}$ is finite, then also $m_{e,g}$ is finite, since from a code for $c^{n_{e,g}}_{e,g}$ we may extract the finitely many p we need to consider. On the other hand, if $n_{e,g} = \infty$, then also $m_{e,g} = \infty$ since we are here considering all p with $\alpha(T_{e,g})(p) = 0$, and $p(\xi^{k-3}(0))$ is clearly unbounded for these (see Definition 8.5.17(ii)). In either case, the relation $m > m_{e,g}$ (as a predicate on $m \in \mathbb{N}$) is decidable in e, g, ϕ: to refute this assertion, it suffices to find suitable n, p with $h_p(0) \geq m$, and to affirm it, it suffices to compute the finite value of $n_{e,g}$, then compute $m_{e,g}$ as indicated above.

We now define the tree $T^*_{e,g}$ by $q \in T^*_{e,g}$ iff $q \in T_{e,q}$ or $q(0) > m_{e,g}$; note that $T^*_{e,g}$ is computable uniformly in e, g, ϕ. Note too that $T^*_{e,g}$ is $(k-2)$-non-well-founded, either because $T_{e,g}$ is already (in the case $\{e\}(\Phi, g, e, g, F_0)\uparrow$) or because $m_{e,g}$ is finite (in the case $\{e\}(\Phi, g, e, g, F_0)\downarrow$). Moreover, in the former case, if $k \geq 4$ then an inspection of the proof of Theorem 8.4.11 confirms that we can find a *non-constant* $z \in \mathsf{Ct}(k-2)$ with h_z an infinite path in $T_{e,g}$, as required by Lemma 8.5.19.

We may now use Lemma 8.5.19 to define

$$\Delta(e, g, F) = \Delta_{T^*_{e,g}}(F).$$

Then Δ is total, and Δ has an associate computable from ϕ, whence $\Delta \in \mathsf{Ct}$. To show that Δ is not 1-obtainable from Φ via any e_0, g_0, we may now follow the argument given at the start of the proof. Note that for our finite portion $a \sqsubseteq \alpha(T_{e_0,g_0})$ we may take the element $c^{n_{e,g}}_{e,g}$, noting that $\Delta(e_0, g_0, -)$ will be constant on V_a; moreover the construction of $T^*_{e_0,g_0}$ is designed to ensure that also $a \sqsubseteq \alpha(T^*_{e_0,g_0})$, whereas no such a can yield a constant value for $\Delta_{T^*_{e_0,g_0}}$ on V_a as shown by Lemma 8.5.19(ii).

The effectivity clause in the theorem is now immediate from the fact that, in general, an associate for Δ is computable from one for Φ. $\square$

We have thus shown that within $\mathsf{Ct}(k)$ or even $\mathsf{Ct}^{\mathrm{eff}}(k)$, where $k \geq 3$, there is no 'universal' functional with respect to 1-obtainability. However, it is natural to ask whether a functional that is universal for type k may be found within type $k+1$. We will see in the next subsection that this is indeed the case, even for a more restricted reducibility notion than 1-obtainability.

8.5.2 Relative μ-Computability

Experience has shown that whenever a functional is in any sense 'non-computable' relative to certain data, it is usually the case that it is not Kleene computable in the data; we saw several examples of such non-computability results in the previous subsection. On the other hand, whenever a functional *is* computable relative to some data, it is in practice usually computable from the data in the weaker sense of *μ-computability*. In this subsection we shall present some positive results of this kind.

First, however, we should pause to note that in principle the power of Kleene computability surpasses that of μ-computability in the context of Ct. In fact, this is already established by Corollary 6.3.32, which shows that the System T recursor rec_1 is not μ-computable in Ct. Historically, however, this result was first obtained by a very different route, not relying on the technology of nested sequential procedures, but instead exploiting the classical theory of c.e. degrees. We here present this earlier proof, due to Bergstra [31]. In contrast to Kleene's proof of the corresponding result for S [138], this avoids depending on the existence of discontinuous arguments, although it still requires the presence of non-computable ones.

We let W_e denote the domain of the eth partial computable function $\mathbb{N} \rightharpoonup \mathbb{N}$ with respect to some standard enumeration. Let T denote Kleene's T-predicate, so that $n \in W_e$ iff $\exists y. T(e,n,y)$; we shall also suppose that $T(e,n,y) = T(e,n,y') \Rightarrow y = y'$.

Definition 8.5.21. A *modulus function* for W_e is a total function $f : \mathbb{N} \to \mathbb{N}$ satisfying

$$\forall n.\, n \in W_e \;\Leftrightarrow\; \exists y < f(n).\, T(e,n,y)\,.$$

We also write Mod_e for the set of finite sequences $(y_0,\ldots,y_{r-1})$ such that for all $n < r$ we have

$$(\exists y < r. T(e,n,y)) \;\Rightarrow\; (\exists y < y_n. T(e,n,y))\,.$$

Note that Mod_e is decidable for any e, and is closed under prefixes. The idea is that if $(y_0,\ldots,y_{r-1}) \in \mathsf{Mod}_e$, there are no counterexamples below r to the claim that $(y_0,\ldots,y_{r-1})$ is an initial segment of a modulus function for W_e.

Exercise 8.5.22. Check that W_e is computable in any modulus for W_e, and that there is a modulus for W_e computable in W_e.

Bergstra introduced the following 'local jump' operators $F_e^d \in \mathsf{Ct}(0,1 \to 0)$, designed to bridge the gap between W_e and W_d:

$$F_e^d(n,f) \;=\; \begin{cases} y \text{ if there exists } r \text{ such that } \overline{f}(r) \in \mathsf{Mod}_e,\, y < r, \text{ and } T(d,n,y)\,, \\ 0 \text{ if there is no such } r. \end{cases}$$

Lemma 8.5.23. *(i) F_e^d is computable in W_d, and therefore continuous.*

(ii) W_d is computable in W_e and F_e^d.

(iii) There is a partial computable $G_e^d \sqsubseteq F_e^d$ that is defined on all (n,f) such that W_e is not computable in f.

Proof. (i) Given n and f, we may test whether $n \in W_d$, and if so find y with $T(d,n,y)$. It then suffices to test whether $\overline{f}(y+1) \in \mathsf{Mod}_e$ to determine the value of $F_e^d(n,f)$.

(ii) If f is a modulus for W_e then clearly $F_e^d(-,f)$ is a modulus for W_d. The claim now follows by Exercise 8.5.22.

(iii) Given n and f, compute G_e^d by searching for the least r such that either $\overline{f}(r) \notin \mathsf{Mod}_e$ (in which case $\overline{f}(r') \notin \mathsf{Mod}_e$ for all $r' \geq r$), or $\overline{f}(r) \in \mathsf{Mod}_e$ and $T(d,n,y)$ for some $y < r$; then return 0 or y accordingly. If W_e is not computable in f, then f is not a modulus for W_e, so there will exist some r with $\overline{f}(r) \notin \mathsf{Mod}_e$, and the search will terminate. □

Theorem 8.5.24. *There exist a functional $F \in \mathsf{Ct}(2)$ and a c.e. set W such that W is Kleene computable in F, but not μ-computable in F.*

Proof. Here we must rely on a result from classical computability theory: there is a computable sequence $e_0, e_1, \ldots$ such that W_{e_0} is computable and the Turing degrees of the W_{e_i} form a strictly increasing sequence. This is in fact an easy exercise once the classical solution to Post's problem is established.

Define

$$F(i,n,f) \;=\; F_{e_i}^{e_{i+1}}(n,f)\,, \qquad W \;=\; \{\langle i,n\rangle \mid n \in W_{e_i}\}\,.$$

By the recursion theorem for Kleene computability, we see from Lemma 8.5.23(ii) that W_{e_i} is Kleene computable in F uniformly in i, and hence that W is Kleene computable in F.

On the other hand, every third-order μ-computable functional will be definable via a normal form term of the language $\mathsf{Klex}^{\min}$ (see Subsection 5.1.5), and within this term, calls to its second-order arguments F will be nested to some finite depth j. We will show that if $\lambda F\vec{z}.M : (\overline{0},\overline{0},\overline{1} \to \overline{0}),\overline{0}^t \to \overline{0}$ is a closed normal form $\mathsf{Klex}^{\min}$ term of nesting depth j and $[\![M]\!]_{F,\vec{b}} = c$, where $F \in \mathsf{Ct}(0,0,1 \to 0)$, $\vec{b} \in \mathbb{N}_{\perp}^t$, then also $[\![M]\!]_{H_j,\vec{b}} = c$, where

$$H_j(i,n,f) \;=\; \begin{cases} F(i,n,f) & \text{if } i < j\,, \\ G_{e_i}^{e_{i+1}}(n,f) & \text{if } i \geq j\,. \end{cases}$$

We show this by induction on the term structure of M, simultaneously with the claim that $[\![\lambda\vec{z}.M]\!]_F$ is computable in W_{e_j} if M has depth j. The non-trivial case is for $M = F(I,N,\lambda x.M')$, where I,N,M' have depth $j-1$. Here, by the induction hypothesis, we have $[\![I]\!]_{F,\vec{b}} = [\![I]\!]_{H_{j-1},\vec{b}} = i$ (say), $[\![N]\!]_{F,\vec{b}} = [\![N]\!]_{H_{j-1},\vec{b}} = n$, and $[\![\lambda x.M']\!]_{F,\vec{b}} = [\![\lambda x.M']\!]_{H_{j-1},\vec{b}} = g$. Since $H_{j-1} \sqsubseteq H_j$, the equality $[\![M]\!]_{F,\vec{b}} = [\![M]\!]_{H_j,\vec{b}}$ is now clear in the case $i < j$. For $i \geq j$, the induction hypothesis says that g is computable in $W_{e_{j-1}}$, so that W_{e_i} is not computable in g. Hence by Lemma 8.5.23(iii) we have $G_{e_i}^{e_{i+1}}(n,g) = F(i,n,g)$, so again $[\![M]\!]_{F,\vec{b}} = [\![M]\!]_{H_j,\vec{b}}$. The claim that $[\![\lambda\vec{z}.M]\!]_F$ is computable in W_{e_j} now follows from the easy fact that H_j is itself computable in W_{e_j} (proved by induction on j using Lemma 8.5.23(i)).

We thus see that if W were μ-computable in F, there would be some j such that W was μ-computable in H_j, and hence computable in W_{e_j}. This is a contradiction. □

Remark 8.5.25. An inspection of the above proof reveals that W is in fact μ-computable in F along with the System T recursor rec_1; we thus retrieve the result that rec_1 itself is not μ-computable in Ct.

We now offer a positive result on μ-computability to complement the negative results of the previous subsection: namely, there exists a type $k+1$ functional that is 'universal' for type k functionals with respect to relative μ-computability.

Recall from Subsection 8.2.1 that a *weak associate* for some $F \in \mathsf{Ct}(k)$ is a K_2 realizer for a PC realizer for F. For any type level k, let $\rho_k : \mathsf{Ct}(1) \rightharpoonup \mathsf{Ct}(k)$ denote the partial continuous function that maps a weak associate for F to F itself. It follows easily from Theorem 8.5.20 that if $k \geq 3$ then no $\rho'_k : \mathsf{Ct}(1) \times \mathsf{Ct}(k-1) \rightharpoonup \mathbb{N}$ extending ρ_k can be Kleene computable in any $F \in \mathsf{Ct}(k)$. On the other hand, we do have:

Theorem 8.5.26. *For any k, there is a functional $\Theta_k \in \mathsf{Ct}^{\mathrm{eff}}(k+1)$ such that some extension of ρ_k is μ-computable in Θ_k.*

Proof. The result is trivial for $k \leq 2$, since an extension of ρ_k is then computable outright. So suppose $k \geq 3$, and let B_k denote the set of weak associates for elements of $\mathsf{Ct}(k)$. By Theorem 8.4.13, B_k is a Π^1_{k-1} set, so by Corollary 8.4.12 there is a primitive recursive relation R such that

$$f \in B_k \;\Leftrightarrow\; \exists G \in \mathsf{Ct}(k).\, \forall n.\, R(f, h_G, n)\,,$$

where a suitable G may be uniformly μ-computed from f when $f \in B_k$.

We shall define Θ_k as a functional of type $\overline{1}, \overline{k}, \overline{k-1} \to \overline{0}$, which is a retract of $\overline{k+1}$. For $f \in \mathsf{Ct}(1)$, $G \in \mathsf{Ct}(k)$ and $X \in \mathsf{Ct}(k-1)$, the value of $\Theta_k(f,G,X)$ is defined extensionally as follows:

1. If $R(f,h_G,n)$ for all n, then f is a weak associate for some unique F, and we set $\Theta_k(f,G,X) = F(X)$.
2. If n is minimal such that $\neg R(f,h_G,n)$, take the maximal $m \leq n$ such that $\overline{f}(n)$ determines a unique functional $(F)_m$ in the following sense: for some $n' \leq n$ there exists $F \in \mathsf{Ct}(k)$ with a weak associate extending $\overline{f}(n')$, and all F with weak associates extending $\overline{f}(n')$ share the same approximation $(F)_m$. Then set $\Theta_k(f,G,X) = (F)_m(X)$.

Note that in the second case, 0 is always a candidate for m (taking $n' = 0$), since $(F)_0$ is always the constant zero functional. The use of $n' \leq n$ ensures that we have a well-defined maximal value for m even when $\overline{f}(n)$ does not extend to a weak associate.

We must show that Θ_k has a computable weak associate; this will imply that $\Theta_k \in \mathsf{Ct}^{\mathrm{eff}}(k+1)$. Once this is done, the theorem follows easily, since an extension of ρ_k can be μ-computed from Θ_k as follows: given $f \in B_k$ and $X \in \mathsf{Ct}(k-1)$, compute a suitable G such that $\forall n. R(f,h_G,n)$, then return the value of $\Theta_k(f,G,X)$.

Suppose then that we are given f, a weak associate g for G, and a weak associate x for X; we wish to compute $\Theta_k(f,G,X)$ from f,g,x. For each i, let us write p_i, q_i for the compacts coded by $f(i), x(i)$ respectively, and write $\overline{p}_i = p_0 \sqcup \cdots \sqcup p_{i-1}$, $\overline{q}_i = q_0 \sqcup \cdots \sqcup q_{i-1}$ when these exist. Since h_G is computable from g, by means of an ordinary effective search we may look for one of the following pieces of information:

1. An n such that $\neg R(f, h_G, n)$.
2. Compact elements $p = \overline{p}_s \in \mathsf{PC}(k)$ and $q = \overline{q}_t \in \mathsf{PC}(k-1)$ such that $p(q) = \mathbb{N}$, along with numbers m_0, n' such that the following hold:
   - $p = \pi_{m_0}(p)$, where $\pi_{m_0} \sqsubseteq id$ is the standard projection from Subsection 8.1.2.
   - $n' \geq m_0$ and $n' \geq s$, so that $p' = \overline{p}_{n'} \sqsupseteq p$.
   - There is an $F \in \mathsf{Ct}(k)$ realized by an extension of p', and all such F share the same $(F)_{m_0}$ (note that this a decidable property of f, n', m_0).

The search will terminate, since if $f \notin B_{k+1}$ there will exist n as in item 1, while if f is a weak associate for some F then there will exist p, q as in item 2, and we may then pick m_0, n' large enough that the above conditions hold.

If we find an instance of item 1, we know that $f \notin B_{k+1}$. We may then obtain the minimal such n and compute the required value as defined by case 2 of the definition of Θ_k. Note that using the ideas of Theorem 8.2.3, it is decidable whether a given segment $\overline{f}(n')$ extends to a weak associate for some F approximated by a given $(F)_m$.

If we find an instance of item 2, we check whether there is an $n \leq n'$ such that $\neg R(f, h_G, n)$; if we find one, we again proceed as above. Otherwise, we return the value of $p(q)$. It remains to see that in this last situation, $p(q)$ is the correct value to return whether or not $R(f, h_G, n)$ for all n.

If $\forall n. R(f, h_G, n)$, then f is a weak associate for some F, and it is clear from the choice of p, q that $p(q) = F(X) = \Theta_k(f, G, X)$. Otherwise, consider the minimal n such that $\neg R(f, h_G, n)$. Then $n > n'$, and we see that there exists $F \in \mathsf{Ct}(k)$ with a weak associate extending $\overline{f}(n')$, and all such F share the same $(F)_{m_0}$. Thus m_0 is a candidate for m as in clause 2 of the definition of Θ_k, so if m is the maximal such candidate then $m \geq m_0$. But now F has a PC realizer extending p', so $(F)_m$ has a realizer extending $\pi_m(p') \sqsupseteq \pi_{m_0}(p') \sqsupseteq \pi_{m_0}(p) = p$, whence $(F)_m(X) = p(q)$.

Since the above computation procedure may clearly be embodied by a computable weak associate, we have $\Theta_k \in \mathsf{Ct}^{\mathrm{eff}}(k+1)$ as required. □

Corollary 8.5.27. *(i) If $\Theta_k \in \mathsf{Ct}^{\mathrm{eff}}(k+1)$ is as in Theorem 8.5.26, then every $F \in \mathsf{Ct}(k)$ is 1-obtainable from Θ, and every $F \in \mathsf{Ct}^{\mathrm{eff}}(k)$ is Kleene computable from Θ.*

(ii) For each pure type $k \geq 2$ there is a functional $\Delta \in \mathsf{Ct}^{\mathrm{eff}}(k+1)$ that is not k-obtainable.

Proof. Part (i) is immediate from Theorem 8.5.26, since any $F \in \mathsf{Ct}(k)$ arises as $\rho_k(f)$ where f is a weak associate for F. For (ii), apply Theorem 8.5.20 to Θ_k. □

As we have seen, μ-computability is more restricted than Kleene computability, so that the Kleene computable functionals of any type ≥ 3 split into several μ-degrees. We now work towards showing that at each type the set of Kleene computable functionals has a maximum μ-degree. For this purpose, we will make use of the ith approximation $\{e\}_i$ to the Kleene computable functional $\{e\}$ as given by Definition 5.3.6. In the setting of Ct, the key properties of this notion are as follows (note the close parallel with Theorem 5.3.7).

Lemma 8.5.28. *(i) For any e, i and $\vec{\Phi}$ there is a unique a such that $\{e\}_i(\vec{\Phi}) = a$.*

(ii) For each fixed tuple of argument types $k_0, \ldots, k_{r-1}$, the function

$$\Lambda(e,i,\vec{\Phi}).\ \{e\}_i(\vec{\Phi})\ :\ \mathbb{N}\times\mathbb{N}\times \mathsf{Ct}(k_0)\times\cdots\times\mathsf{Ct}(k_{r-1})\to\mathbb{N}$$

is Kleene primitive recursive.

(iii) If $\{e\}(\vec{\Phi})\downarrow$ then $\{e\}(\vec{\Phi}) = \lim_{i\to\infty}\{e\}_i(\vec{\Phi})$. Moreover, there is a primitive recursive $h:\mathbb{N}\to\mathbb{N}$ such that if $\{e\}(\vec{\Phi})\downarrow$ then $\{h(e)\}(\vec{\Phi})$ is defined and is a modulus of convergence for the above sequence.

Proof. Using Definition 5.3.6, (i) is easy by induction on i. For (ii), we have to overcome the obstacle that in the presence of S8, a subcomputation may involve additional arguments besides $e, i, \vec{\Phi}$. To circumvent this, take $k = \max k_i$ so that $\mathsf{Ct}(k_0)\times\cdots\times\mathsf{Ct}(k_{r-1})$ is a computable retract of $\mathsf{Ct}(k)$, say with associated idempotent $\pi_{k_0,\ldots,k_{r-1}}$. We then show that the function

$$(e,i,\langle k_0,\ldots,k_{t-1}\rangle,\Phi)\ \mapsto\ \{e\}_i(\pi_{k_0,\ldots,k_{t-1}}(\Phi))$$

is primitive recursive (where $\langle k_0,\ldots,k_{t-1}\rangle$ ranges over extensions of $\langle k_0,\ldots,k_{r-1}\rangle$ bounded by k).

For (iii), we define h using the recursion theorem, proving the relevant properties by induction on the height of the computation, and using Grilliot's method in the case for S8 (cf. the proof of Theorem 5.3.7). □

The above lemma will be applied in the next subsection. For our present purposes, we require a variation on this idea of approximating a Kleene computation—here the quality of the approximation is regulated by the depth, if any, at which a given path f exits from a given tree T.

Definition 8.5.29. Let $T\subseteq\mathbb{N}^*$ be a computable prefix-closed set of sequences of integers, and suppose $f\in\mathbb{N}^{\mathbb{N}}$ and $i\in\mathbb{N}$. We define the relation $\{e\}_{T,f,i}(\vec{\Phi}) = a$ recursively as follows:

1. If $\overline{f}(i)\notin T$, we let $\{e\}_{T,f,i}(\vec{\Phi}) = \{e\}_0(\vec{\Phi})$.
2. If $\overline{f}(i)\in T$ and $\{e\}(\vec{\Phi})\simeq\Xi(\{d_0\},\ldots,\{d_{s-1}\},\vec{\Phi})$ according to one of the Kleene schemes, we let
$$\{e\}_{T,f,i}(\vec{\Phi})\ \simeq\ \Xi(\{d_0\}_{T,f,i+1},\ldots,\{d_{s-1}\}_{T,f,i+1},\vec{\Phi})\,.$$
3. If $\overline{f}(i)\in T$ and e does not match any of the schemes, we let $\{e\}_{T,f,i} = \{e\}_{T,f,i+1}$ for all arguments.

Lemma 8.5.30. *(i) For a fixed tuple of argument types $k_0,\ldots,k_{r-1}$ and a fixed tree T, the partial function*

$$(i,f,\Phi_0,\ldots,\Phi_{r-1})\ \mapsto\ \{e\}_{T,f,i}(\vec{\Phi})$$

is Kleene computable.

(ii) If f is an infinite branch in T, then $\{e\}_{T,f,i}(\vec{\Phi}) \simeq \{e\}(\vec{\Phi})$ for all i and $\vec{\Phi}$.

(iii) If for some $j \geq i$ we have $\bar{f}(j) \notin T$, then $\{e\}_{T,f,i}(\vec{\Phi}) = \{e\}_{j-i}(\vec{\Phi})$ for the least such j.

Proof. Part (i) is an application of the recursion theorem. Part (ii) is proved by induction on the computation, simultaneously for all i, and (iii) is easy by induction on $j-i$. □

We now combine this lemma with the Kreisel representation theorem to show that there is a maximum μ-degree among the Kleene computable functionals of any given type. Again we restrict attention to the pure types, and since all computable functions of type 1 or 2 are μ-computable outright, we focus on types of level ≥ 3.

Theorem 8.5.31. *Let $k \geq 3$. Then there is a Kleene computable functional $M_k \in \mathsf{Ct}(k)$ such that all Kleene computable $F \in \mathsf{Ct}(k)$ are μ-computable in M_k.*

Proof. By Theorem 8.4.13 and Lemma 8.5.13, there is a Π^1_{k-2}-relation P such that $P(e,h)$ iff h is the trace h_ϕ of some $\phi \in \mathsf{Ct}(k-1)$ where $\{e\}(\phi)\downarrow$. Hence by Corollary 8.4.12 there is a primitive recursive relation T such that for all e and ϕ we have

$$\{e\}(\phi)\downarrow \quad \Leftrightarrow \quad \exists \psi \in \mathsf{Ct}(k-1).\, \forall n.\, T(e, \overline{h_\phi}(n), \overline{h_\psi}(n)) ,$$

where a suitable ψ may be found μ-computably in h_ϕ and e when $\{e\}(\phi)\downarrow$.

Clearly we may encode triples (e,h,h') as functions $f(e,h,h') : \mathbb{N} \to \mathbb{N}$ in such a way that for some primitive recursive tree T' and for all e,h,h' we have

$$(\forall n.\, T(e, \overline{h}(n), \overline{h'}(n))) \quad \Leftrightarrow \quad (\forall n.\, T'(\overline{f(e,h,h')}(n))) .$$

Now define $M_k \in \mathsf{Ct}(0, k-1, k-1 \to 0)$ by

$$M_k(e, \phi, \psi) \;=\; \{e\}_{T', f(e,h_\phi,h_\psi), 0}(\phi) .$$

(We may encode M_k within $\mathsf{Ct}(k)$ if so desired.) Using Lemma 8.5.30, we see that M_k is Kleene computable and total (arguing by cases on whether $\{e\}(\phi)\downarrow$). Furthermore, if e is an index for some Kleene computable F and $\phi \in \mathsf{Ct}(k-1)$, then uniformly in e, ϕ we may obtain $F(\phi)$ μ-computably in M_k as follows: from h_ϕ obtain a witness $\psi \in \mathsf{Ct}(k-1)$ as above, then return $M_k(e,\phi,\psi) = \{e\}(\phi)$. □

The above proof relativizes easily to an arbitrary $\Phi \in \mathsf{Ct}(k)$, showing that every Kleene degree within $\mathsf{Ct}(k)$ contains a maximum μ-degree. Indeed, a representative of this μ-degree is even Kleene computable uniformly in Φ.

8.5.3 Computably Enumerable Degrees of Functionals

There is an interesting consequence of Lemma 8.5.28(iii): it ties the 1-section of a continuous functional to the classical c.e. degrees. We here write c.e.(f) to mean 'computably enumerable relative to f'.

Theorem 8.5.32. *Suppose $\Phi \in \mathsf{Ct}(k)$. Then there is a type 1 function f primitive recursive in Φ such that any type 1 function g is computable in Φ iff g is computable in some c.e.(f) set A that itself is computable in Φ.*

Proof. Take $f(e,a,i) = \{e\}_i(\Phi,a)$; by Lemma 8.5.28(i) this is primitive recursive in Φ. If g is computable in Φ, there is an index e such that

$$g(a) = \{e\}(\Phi,a) = \lim_{i\to\infty} f(e,a,i) .$$

By Lemma 8.5.28(iii), the modulus of convergence $m_g(a)$ of this sequence is computable in Φ and a. Now consider the c.e.(f) set

$$A = \{(a,i) \mid \exists j > i.\ f(e,a,j) \neq f(e,a,i)\} .$$

Clearly m_g and A are computable in each other relative to f, and g is computable in m_g and f. Moreover, since m_g is computable in Φ, so is A. □

Exercise 8.5.33. Show that we may specialize f to the trace h_Φ in the statement of the previous theorem. For this we need an adjusted proof: define

$$\{\{e\}\}_i(\Phi_0,\ldots,\Phi_{r-1}) = \{e\}_i((\Phi_0)_i,\ldots,(\Phi_{r-1})_i)$$

where $(\Phi)_i$ denotes the ith approximation to Φ as in Subsection 8.1.2, and re-prove Lemma 8.5.28 for this new notion of approximating computations.

There is a natural extension of the c.e. degrees to higher types. We will not delve deeply into this here, but will give the main definitions and mention a few results. We shall use a well-known characterization of the classical c.e. degrees as the basis for our extension of the concept to higher types:

Definition 8.5.34. Suppose $\Phi \in \mathsf{Ct}(\sigma)$. We say that Φ is *of c.e. degree* if there is a computable sequence $\{\Phi_n\}_{n\in\mathbb{N}}$ with limit Φ whose modulus (as in Definition 8.2.15) is computable in Φ. A Kleene degree within Ct is called a *c.e. degree* if it contains some Φ of c.e. degree.

Lemma 8.5.35. *Suppose $\Phi \in \mathsf{Ct}(\sigma)$ where $\mathsf{lv}(\sigma) \geq 2$. Then the following are equivalent:*

1. *Φ is of c.e. degree.*
2. *There exists Ψ of the same Kleene degree as Φ whose trace h_Ψ is computable.*

Proof. We give the argument only for pure types $k \geq 2$.

1 ⇒ 2: Suppose $\{\Phi_n\}_{n\in\mathbb{N}}$ is a computable sequence converging to Φ with modulus computable in Φ. Take $\xi \in \mathsf{Ct}(k-1)$ Kleene computable but not primitive recursive, and define $\widehat{\Phi} : \mathsf{Ct}(k-1) \times \mathsf{Ct}(k-1) \to \mathbb{N}$ by

- $\widehat{\Phi}(\phi,\psi) = \Phi(\phi)$ if $\psi = \xi$,
- $\widehat{\Phi}(\phi,\psi) = \Phi_n(\phi)$ if n is minimal such that $h_\psi(n) \neq h_\xi(n)$.

We claim Φ and $\widehat{\Phi}$ are Kleene computable in each other (whence $\widehat{\Phi} \in \mathsf{Ct}$). To compute $\widehat{\Phi}(\phi,\psi)$ given Φ, first compute the modulus m for $\Phi_n \to \Phi$ at ϕ, then test whether there exists $n \leq m$ with $h_\psi(n) \neq h_\xi(n)$. If n is the least such, return $\Phi_n(\phi)$; if no such n exists, return $\Phi(\phi)$. In the other direction, we may compute $\Phi(\phi)$ simply as $\widehat{\Phi}(\phi,\xi)$.

Moreover, since ξ is not primitive recursive and the sequence Φ_n is computable, it is easy to see that $h_{\widehat{\Phi}}$ is μ-computable. Thus $\widehat{\Phi}$ satisfies the requirements for Ψ.

$2 \Rightarrow 1$: Assume h_Ψ is computable. Then the standard sequence of approximations $\{(\Psi)_n\}_{n\in\mathbb{N}}$ is clearly computable, and the modulus for $(\Psi)_n \to \Phi$ is μ-computable in Φ by Lemma 8.2.18, so Ψ itself is of c.e. degree.

If Φ is computable in Ψ, then for some index e we have $\Phi(\phi) = \{e\}(\Psi,\phi)$ for all $\phi \in \mathsf{Ct}(k-1)$. Define $\Phi_n(\phi) = \{e\}_n((\Psi)_n,\phi)$ for all n and ϕ. Then Theorem 8.2.17(ii) yields that $\Phi_n \to \Phi$ with modulus computable in Ψ, in view of the following facts:

- The sequences $\{e\}_n$ and $(\Psi)_n$ are computable outright.
- $\{e\}_n \to \{e\}$ has a computable modulus by Lemma 8.5.28(iii).
- $(\Psi)_n \to \Psi$ has a modulus computable in Ψ.

So if additionally $\Psi \leq_{\mathrm{Kl}} \Phi$, then the modulus for $\Phi_n \to \Phi$ is computable in Φ. But also the sequence Φ_n is computable since the sequence $(\Psi)_n$ is; thus Φ is of c.e. degree. □

As a consequence of this characterization, all functionals in $\mathsf{Ct}^{\mathrm{eff}}$ will be of c.e. degree. Since for $k > 2$ we have functionals in $\mathsf{Ct}^{\mathrm{eff}}(k)$ that are not $(k-1)$-obtainable (see Corollary 8.5.27(ii)), we in particular have new c.e. degrees manifesting themselves at each new type level $k > 2$.

At type 2 itself, the local jump operators F_e^d are of c.e. degree, and if W_e is non-computable but of strictly lower Turing degree than W_d, it follows readily from Lemma 8.5.23 that F_e^d will not be of the same degree as any type 1 function. (Note that part (iii) of the lemma implies that any type 1 function computable in F_e^d is computable outright.)

The availability of higher type functionals makes the study of the c.e. degrees within Ct much simpler than the study of the classical c.e. degrees. Many of the key results appear in Normann [208]; here we will illustrate just one proof technique by establishing the density of the c.e. degrees of continuous functionals.

The following definition and lemma adapt ideas from the proof of Lemma 8.5.35:

Definition 8.5.36. Suppose $T \subseteq \mathbb{N}^*$ is a tree and $\Phi \in \mathsf{Ct}(k)$. Let $\langle -,- \rangle$ denote a standard bijective pairing operation on $\mathsf{Ct}(k)$, and define $(\Phi)_T \in \mathsf{Ct}(k)$ by

- $(\Phi)_T(\langle F,G \rangle) = \Phi(F)$ if $\forall m.\, \overline{h_G}(m) \in T$,
- $(\Phi)_T(\langle F,G \rangle) = (\Phi)_m(F)$ if m is minimal such that $\overline{h_G}(m) \notin T$.

Lemma 8.5.37. *(i) $(\Phi)_T$ is μ-computable uniformly in Φ and T.*

(ii) If $\forall G \in \mathsf{Ct}(\tau).\exists n.\overline{h_G}(n) \notin T$ (i.e. T is τ-well-founded) then $(\Phi)_T$ is μ-computable in h_Φ and T.

Proof. (i) Recall that by Lemma 8.2.18 the modulus for $(\Phi)_m \to \Phi$ is μ-computable uniformly in Φ, and proceed as in the proof of Lemma 8.5.35.

(ii) For any G, the least m such that $\overline{h_G}(m) \notin T$ may be found by a simple search, and $(\Phi)_m$ may then be retrieved from h_Φ. □

Theorem 8.5.38. *The poset of c.e. degrees in* Ct *is dense.*

Proof. Suppose $\phi <_{\mathrm{KI}} \psi$ where both ϕ and ψ are of c.e. degree.

We may assume that h_ϕ and h_ψ are computable, and that both ϕ and ψ are in $\mathsf{Ct}(k)$ for some pure type $k \geq 2$ (this may be readily ensured by a simple adaptation of the construction of $\widehat{\Phi}$ from the proof of Lemma 8.5.35).

Claim 1: There is a functional $\Phi \in \mathsf{Ct}^{\mathrm{eff}}(k+2)$ such that ψ is not computable in Φ and ϕ, and such that Φ is not computable in any $\Delta \in \mathsf{Ct}(k+1)$.

Proof of claim 1: Using Corollary 8.5.27, we may take:

- $\Delta_0 \in \mathsf{Ct}^{\mathrm{eff}}(k+1)$ not computable in any $\xi \in \mathsf{Ct}(k)$,
- $\Delta_1 \in \mathsf{Ct}^{\mathrm{eff}}(k+2)$ not computable in any $\Delta \in \mathsf{Ct}(k+1)$,

so that both h_{Δ_0} and h_{Δ_1} are computable. Let T be a computable tree with h_{Δ_0} as the only infinite path, and set $\Phi = (\Delta_1)_T \in \mathsf{Ct}(k+2)$.

Clearly Δ_1 is computable in Φ and Δ_0, but Δ_1 is not computable in any $\Delta \in \mathsf{Ct}(k+1)$, so Φ cannot be computable in any $\Delta \in \mathsf{Ct}(k+1)$.

To see that ψ is not computable in Φ and ϕ, let Φ' be the partial subfunction of Φ defined on all $\langle F, G\rangle$ with $G \neq \Delta_0$; then Φ' is μ-computable. We may now show by induction on the ordinal height of computations that if $\{e\}(\Phi, \vec{\xi})\downarrow$ and the types of the $\vec{\xi}$ are bounded by k, then $\{e\}(\Phi', \vec{\xi})\downarrow$ with the same computation tree. In the crucial case of a call to Φ via S8, we have by the induction hypothesis that

$$\Lambda\zeta^k.\, \{d\}(\Phi, \vec{\xi}, \zeta) \;=\; \Lambda\zeta^k.\, \{d\}(\Phi', \vec{\xi}, \zeta)\,.$$

Since Φ' is computable outright and Δ_0 is not obtainable from the $\vec{\xi}, \zeta$, the right-hand side here cannot be of the form $\langle F, \Delta_0\rangle$; thus

$$\Phi(\Lambda\zeta^k.\, \{d\}(\Phi, \vec{\xi}, \zeta)) \;=\; \Phi'(\Lambda\zeta^k.\, \{d\}(\Phi', \vec{\xi}, \zeta))\,.$$

In particular, if ψ were computable from Φ, ϕ, it would be computable from Φ', ϕ, which it is not. This completes the proof of claim 1.

An inspection of this proof also reveals that h_Φ is computable. Indeed, writing $\xi^{k+1}(n)$ as $\langle \xi_0^{k+1}(n), \xi_1^{k+1}(n)\rangle$ we have for any n that

$$h_\Phi(n) \;=\; (\Delta_1)_T(\xi(n)) \;=\; (\Delta_1)_m(\xi_0(n)) \;\text{ where }\; m = \mu m.\, \overline{h_{\xi_1(n)}}(m) \notin T\,,$$

observing that since $\xi_1(n)$ is primitive recursive, $h_{\xi_1(n)}$ never coincides with h_{Δ_0}, which is the only infinite path through T. Thus h_Φ is computable since both T and h_{Δ_1} are computable. We may therefore take a computable tree T' with h_Φ as the only infinite path. We also take $\psi' \in \mathsf{Ct}(k+3)$ of the same Kleene degree as ψ; again, by adapting the proof of Lemma 8.5.35 we may arrange that $h_{\psi'}$ is computable. Now define $\Psi = (\psi')_{T'} \in \mathsf{Ct}(k+3)$.

Claim 2: The degree of the pair (ϕ, Ψ) is strictly between that of ϕ and ψ.

Proof of claim 2: By construction, we have that $\phi \leq_{\mathsf{Kl}} (\phi, \Psi) \leq_{\mathsf{Kl}} \psi$, so we just have to prove that the inequalities are strict. The proof parallels that of claim 1.

For the first inequality, clearly ψ is computable in Ψ and Φ, whereas ψ is not computable in ϕ and Φ by claim 1. It follows that Ψ is not computable in ϕ.

For the second inequality, we have by claim 1 that Φ is not $(k+1)$-obtainable. Let Ψ' be the partial subfunction of Ψ defined on all $\langle F, G\rangle$ with $G \neq \Phi$; then Ψ' is clearly μ-computable. As in the proof of claim 1, we may now show that if $\{e\}(\Psi, \vec{\xi}) \downarrow$ and the types of the $\vec{\xi}$ are bounded by $k+1$, then $\{e\}(\Psi', \vec{\xi}) \downarrow$ with the same computation tree. In particular, ψ cannot be computable from Ψ, ϕ since it is not computable from Ψ', ϕ. Thus the second inequality is strict.

Finally, we note that the trace h_Ψ is computable, for a reason analogous to the above argument for the computability of h_Φ. Hence Ψ is of c.e. degree by Lemma 8.5.35, and the density theorem is proved. □

8.6 Computing with Realizers

In the previous section we studied 'internal' notions of computation that treat elements of Ct purely as abstract functionals. We now turn our attention to 'external' notions, considering some of the different kinds of realizers for these functionals and the different ways of computing with them.

So far, the realizing structures PC and K_2 have played key roles. We shall see in Subsection 8.6.2 that many other realizing structures are possible, and indeed that *any* full continuous model satisfying some general conditions yields Ct as its extensional collapse. However, we shall start by focussing on the realization $\mathsf{Ct} \longrightarrow\!\!\triangleright\, \mathsf{PC}$ and examining the choice of computability notions available for PC. Clearly, any applicative substructure $\mathsf{PC}^\sharp$ of 'computable' elements in PC will induce a substructure $\mathsf{Ct}^\sharp$ consisting of the elements of Ct that have at least one realizer in $\mathsf{PC}^\sharp$.

To provide some context for our results, let us summarize what we know so far. We have seen that one natural notion of computable element in PC is given by the structure $\mathsf{PC}^{\mathrm{eff}}$; this induces the substructure $\mathsf{Ct}^{\mathrm{eff}}$ of Ct, which is considerably wider than the class $\mathsf{Ct}^{\mathrm{Kl}}$ of functionals Kleene computable over Ct, as shown by the example of the fan functional and the other results of Subsection 8.5.1. This raises the question of the precise source of this difference: informally, what exactly do we need beyond the power of Kleene computability over Ct if we are to compute additional operations such as FAN? For example, what happens if we replace $\mathsf{PC}^{\mathrm{eff}}$ here by some smaller substructure of PC, such as the one consisting of 'Kleene computable' elements?

Here we pick up a thread of discussion from Subsection 6.4.3, where we considered various definitions of Kleene computability over partial models and their relationship to the total ones. As we saw there, the best version of Kleene computability to consider in PC is the one given by the so-called *liberal interpretation*, in which the recursive definition of the computation relation is in effect construed as a grand simultaneous least fixed point. We shall here write $\mathsf{PC}^{\mathrm{Kl}}$ for the class of liberally Kleene computable elements of PC; we have seen in Theorem 7.1.39 that this coincides exactly with the class of PCF-*definable* elements of PC, or equivalently those denotable by NSPs.

One might at first expect that the substructure $\mathsf{PC}^{\mathrm{Kl}}$ will induce $\mathsf{Ct}^{\mathrm{Kl}}$, since both are based on Kleene computability. However, we have already seen a counterexample to this: the fan functional has a PC-realizer definable in PCF and hence present in $\mathsf{PC}^{\mathrm{Kl}}$ (Theorem 7.3.2), whereas the fan functional itself is not in $\mathsf{Ct}^{\mathrm{Kl}}$. It is also evident that the recursive definition of the Γ functional can be construed as a PCF definition of a realizer for Γ, whereas Γ itself is not Kleene computable.

This raises the obvious question of how much of $\mathsf{Ct}^{\mathrm{eff}}$ is covered by the substructure induced by $\mathsf{PC}^{\mathrm{Kl}}$. The answer, perhaps surprisingly, is: all of it! In other words, the $\mathsf{PC}^{\mathrm{Kl}}$ realizability of FAN and Γ are not isolated examples, but instances of a general phenomenon. This result, conjectured by Cook [58] and Berger [19, 21] and proved by Normann [212], will form the substance of Subsection 8.6.1. As the proof will show, we do not even need the full power of PCF here: to realize all elements of $\mathsf{Ct}^{\mathrm{eff}}$, it is sufficient to take the elements definable in PCF_1, the sublanguage with no fixed point operators beyond $Y_1 : (\overline{1} \to \overline{1}) \to \overline{1}$.

The moral, then, is that Kleene computability takes on a quite different meaning according to whether it is interpreted in Ct or PC. It is illuminating to contrast this with the situation for the weaker notion of μ-computability: suitably interpreted, this has essentially the same meaning in Ct and in PC. More precisely, in both models we can frame the concept of μ-computability in terms of definability in the language $\mathsf{Klex}^{\mathrm{min}}$, and a functional F is $\mathsf{Klex}^{\mathrm{min}}$ definable in Ct iff it has a realizer f that is $\mathsf{Klex}^{\mathrm{min}}$ definable in PC (Theorem 6.4.20). Indeed, the proof of this is applicable even to more powerful languages, such as $\mathrm{T} + \mathit{min}$, that remain within the realm of left-well-founded computation. One may therefore say that a crucial threshold is crossed when the operator Y_1 is added to the language: at this point, all elements of $\mathsf{Ct}^{\mathrm{eff}}$ instantly become 'computable' at the level of realizers, and no further (effective) language extensions yield any other 'computable' functionals beyond these.

The results of Subsection 8.6.1 also yield that every element of Ct has a PC realizer definable in PCF^{Ω}. In Subsection 8.6.2, we build on the general idea that 'all total continuous functionals are expressible in PCF^{Ω}', but abstract away from the specific model PC. This leads to a general result of Longley [179] saying that, under mild conditions, any full continuous model $\mathbf{A}$ that 'supports PCF^{Ω}' will give rise to Ct, and any computable substructure $\mathbf{A}^{\sharp}$ of $\mathbf{A}$ that 'supports PCF' will give rise to $\mathsf{Ct}^{\mathrm{eff}}$. This further strengthens the impression, already created by the results of Sections 8.1–8.3, that Ct is the canonical class of total continuous functions over $\mathbb{N}$ on which virtually all natural constructions are bound to converge, and that $\mathsf{Ct}^{\mathrm{eff}}$ is its canonical substructure of 'externally computable' functionals.

8.6.1 Sequential Computability and Total Functionals

In this subsection, we prove the following theorem from [212]. Here, if $g \in \mathbb{N}^{\mathbb{N}}$, we write $\widehat{g}$ for its canonical extension to an element of $\mathsf{PC}(1)$ with $\widehat{g}(\bot) = \bot$.

Theorem 8.6.1. *For each σ there is a* PCF*-computable functional $\Phi_\sigma \in \mathsf{PC}(1 \to \sigma)$ such that whenever $g \in K_2$ is a realizer for some $x \in \mathsf{PC}(\sigma)$, we have $\Phi_\sigma(\widehat{g}) \sqsubseteq x$, and if $x \in \mathsf{Tot}(\sigma)$ then $\Phi_\sigma(\widehat{g}) \in \mathsf{Tot}(\sigma)$ (whence $\Phi_\sigma(\widehat{g}) \approx x$).*

It follows immediately from this that every $X \in \mathsf{Ct}(\sigma)$ has some PC-realizer definable in PCF^{Ω}, and every $X \in \mathsf{Ct}^{\mathrm{eff}}(\sigma)$ has a realizer definable in PCF. As usual, we shall concentrate for notational simplicity on the pure types, relying on Section 4.2 for the extension to arbitrary types.

Our statement of the above theorem refers implicitly to a simulation $\mathsf{PC} \longrightarrow\!\!\!\triangleright K_2$. The simulation ψ of Example 3.5.7 is acceptable for this purpose, but it will be more convenient to work with a slightly different though equivalent simulation, which we set up as follows.

Recall our language of formal expressions e denoting compact elements $[\![e]\!]$, as introduced at the start of Section 8.1. Clearly, a given compact may be denoted by many different expressions: e.g. if $[\![e]\!] \sqsupseteq [\![e']\!]$ then $[\![e]\!] = [\![e \sqcup e']\!] = [\![e' \sqcup e]\!]$. We shall here refer to valid expressions e as *compact expressions*. We may suppose that each compact expression e is coded by a unique Gödel number $\ulcorner e \urcorner$; a *code* for a compact element a is then a Gödel number for any compact expression denoting a.

Definition 8.6.2. (i) A *basic compact* in $\mathsf{PC}(k+1)$ will be simply a step function $(b \Rightarrow m)$ where b is compact in $\mathsf{PC}(k)$ and $m \in \mathbb{N}_\bot$. Likewise, a *basic expression* in our formal language will be one of the form $(b \Rightarrow m)$.

(ii) By a *basic enumeration* of $x \in \mathsf{PC}(k+1)$ we shall mean a function $g : \mathbb{N} \to \mathbb{N}$ whose image is the set of all codes for all basic compacts $a \sqsubseteq x$.

Throughout this section, by a *(K_2-)realizer* for $x \in \mathsf{PC}$ we shall mean a basic enumeration of x. It is easy to see that this yields a realization equivalent to ψ. Note that we allow basic expressions of the form $(b \Rightarrow \bot)$; this ensures that even $\bot \in \mathsf{PC}(k+1)$ has a basic enumeration.

As a warm-up to our main proof, we shall consider the case of type $\overline{2}$, which introduces some of the key ingredients of the general case. (For types $\overline{0}$ and $\overline{1}$ the theorem is trivial.) A version of this has already appeared as Proposition 6.3.26, but we repeat the argument here for convenience.

Lemma 8.6.3. *There is a* PCF*-computable $\Phi_2 \in \mathsf{PC}(1 \to 2)$ such that whenever $g \in K_2$ basically enumerates $x \in \mathsf{PC}(2)$, we have $\Phi_\sigma(\widehat{g}) \sqsubseteq x$, and if $x \in \mathsf{Tot}(\sigma)$ then $\Phi_2(\widehat{g}) \in \mathsf{Tot}(2)$ (whence $\Phi_2(\widehat{g}) \approx x$).*

Proof. Suppose first that x and y are total and that g basically enumerates x. We must show how $x(y) \in \mathbb{N}$ may be computed in PCF from g and y. Let $y' \sqsubseteq y$ denote the total element equivalent to y but with $y'(\bot) = \bot$. For each i, let $(b_i \Rightarrow m_i) \sqsubseteq x$ be the basic expression coded by $g(i)$, where $b_i = \bigsqcup_{j<r_i}(c_{ij} \Rightarrow n_{ij})$ with $n_{ij} \in \mathbb{N}$ for each

j. Then we may test whether $b_i \sqsubseteq y'$ as follows: if any of the c_{ij} are $\bot$, return false, otherwise test whether $y(c_{ij}) = n_{ij}$ for each j. We may now compute $x(y) = x(y')$ by searching for the least i such that $b_i \sqsubseteq y'$ and $m_i \in \mathbb{N}$, then returning m_i.

It is easy to see that this computation is expressible in PCF, and that it always terminates yielding the correct value of $x(y)$. We thus obtain an element $x' \approx x$ that is PCF-definable in g. It is also clear that even if x, y are not total, the above algorithm can return a result m only if $x(y) = m$; thus $x' \sqsubseteq x$. □

Note that a total element $x \in \mathsf{Tot}(2)$ need not be PCF^{Ω}-definable in general: for instance, x may represent the 'parallel-or' operation as in Subsection 7.3.1. Thus, if we wish to transform a total x into a sequentially computable but otherwise equivalent x', the best we can do is to find $x' \sqsubseteq x$.

Let us now informally outline how we extend the above argument to pure types $k \geq 3$. The overall strategy will be the same: to compute $x(y)$, we work our way through a basic enumeration g of x until we find a basic expression $(b_i \Rightarrow m_i)$ where y 'matches' b_i, at which point we return m_i. The difficulty we have to overcome concerns the tests involving $y(c_{ij})$: whereas in the type 2 case the c_{ij} were always natural numbers (except in the case $c_{ij} = \bot$, which we could discount), in the case $k \geq 3$ they will typically be non-total elements of $\mathsf{PC}(k-2)$, so that $y(c_{ij})$ is not guaranteed to terminate. We therefore cannot risk a direct evaluation of $y(c_{ij})$ for fear of causing the whole computation to diverge. However, there are two measures we can adopt to address this problem.

First, we can look for evidence that y does not match b_i, in the form of some total element $z_{ij} \sqsupseteq c_{ij}$ such that $y(z) \neq n_{ij}$. If such an element exists, we will eventually find one by searching through some standard enumeration of a dense set of total extensions of c_{ij}. Once such negative evidence is found, we may discount $(b_i \Rightarrow m_i)$ and move on to the next basic expression $(b_{i+1} \Rightarrow m_{i+1})$.

Of course, this search will be fruitless if no such z_{ij} exists (in which case m_i will indeed be the value of $x(y)$). However, we may interleave it with a second search, which looks for evidence from further along the enumeration g that $x(y) = m_i$ (a non-trivial recursion is involved here). If this is found, we may use an arbitrary total extension of c_{ij} to play the role of z_{ij}. Even if this yields a 'false positive' in the sense that $y(z_{ij}) = n_{ij}$ for all j when y does not in fact match b_i, this will not matter, since m_i will be the correct value to return anyway.

How do we ever acquire positive evidence that y 'matches' some b_i? It might seem from the above sketch that we have simply postponed the problem by appealing to the existence of some $i' > i$ where y matches some $b_{i'}$. The solution to this, and the key to establishing termination of our algorithm, is to show that there will eventually be some $(b_i \Rightarrow m_i)$ in our enumeration, with $b_i = \bigsqcup_{j<r_i}(c_{ij} \Rightarrow n_{ij})$, such that the c_{ij} are already sufficiently good approximations to total elements to ensure that $y(c_{ij}) = n_{ij}$ for each j. Thus, even if neither of the above searches delivers a total extension $z_{ij} \sqsupseteq c_{ij}$, the information present in the c_{ij} is by itself enough to ensure a positive outcome of the test for $(b_i \Rightarrow m_i)$. Exactly how this works out will be made clear by the detailed proof, to which we now proceed.

Proof. We prove Theorem 8.6.1 for pure types $\overline{k}$ by induction on k. The cases $k = 0, 1$ are trivial, and the case $k = 2$ is given by Lemma 8.6.3. We therefore suppose $k \geq 3$, and assume that a suitable PCF-definable functional $\Phi_{k-2} \in \mathsf{PC}(1 \to (k-2))$ has already been constructed to our satisfaction.

We suppose $g \in K_2$ is a basic enumeration for $x \in \mathsf{Tot}(k)$, and $y \in \mathsf{Tot}(k-1)$ is arbitrary: our task is to compute the value of $x(y)$ from g and y. (The clause in the theorem concerning arbitrary $x \in \mathsf{PC}(x)$ will be dealt with at the very end of the proof.) We shall give an informal description of our algorithm, from which it will be clear how it may be expressed by a PCF program (or by a Kleene index if the reader prefers S1–S9).

We first need to gather some information on the behaviour of Φ_{k-2} on compact approximations. We do this via a series of claims. If $e = e_0 \sqcup \cdots \sqcup e_{r-1}$ is a formal expression where the e_i are basic, we write h_e for the 'partial basic enumeration' given by $h_e(i) = \lceil e_i \rceil$ for $i < r$ and $h_e(i)$ undefined for $i \geq r$. We also write $\widehat{h_e}$, or more simply $\widehat{e}$, for the associated strict element of $\mathsf{PC}(1)$. Since h_e extends to a basic enumeration of e, we have $\Phi_{k-2}(\widehat{e}) \sqsubseteq [\![e]\!]$ by the induction hypothesis.

Claim 1: Suppose $y \in \mathsf{Tot}(k-1)$ and $z \in \mathsf{Tot}(k-2)$. Then there is a compact expression $c : \overline{k-2}$ with $[\![c]\!] \sqsubseteq z$ such that $y(\Phi_{k-2}(\widehat{c}))\downarrow$.

Proof of claim 1: Let h be any basic enumeration of z. By the induction hypothesis, $\Phi_{k-2}(\widehat{h})$ is total, so $y(\Phi_{k-2}(\widehat{h}))\downarrow$. So for some n we have $y(\Phi_{k-2}(\widehat{h}_n))\downarrow$, where $\widehat{h}_n \sqsubseteq \widehat{h}$ has domain $\{0,\ldots,n-1\}$. Taking $c = c_0 \sqcup \cdots \sqcup c_{n-1}$ where each c_i is the basic expression coded by $h(i)$, we have that $\widehat{c} = \widehat{h}_n$. This proves the claim.

Claim 2: Suppose $y \in \mathsf{Tot}(k-1)$. If $y(\Phi_{k-2}(\widehat{c})) = m$, then $y([\![c]\!]) = m$.

Proof of claim 2: We may extend $\widehat{c}$ to a basic enumeration $\widehat{h}$ of $[\![c]\!]$ simply by adding codes for $\bot$ after the basic expressions within c have been listed. Since our induction hypothesis includes the fact that $\Phi_{k-2}(\widehat{h}) \sqsubseteq [\![c]\!]$ even when c is not total, we have $y([\![c]\!]) \sqsupseteq y(\Phi_{k-2}(\widehat{h})) \sqsupseteq y(\Phi_{k-2}(\widehat{c})) = m$. This proves claim 2.

From here on, we suppose g is a basic enumeration of some $x \in \mathsf{Tot}(k)$ realizing $X \in \mathsf{Ct}(k)$. For each i, let $(b_i \Rightarrow m_i)$ be the basic expression coded by $g(i)$, where

$$b_i \;=\; \bigsqcup_{j<r_i} (c_{ij} \Rightarrow n_{ij}) \,.$$

The next claim establishes the crucial property that for any $y \in \mathsf{Tot}(k-1)$, the enumeration g contains an element $(b_i \Rightarrow m_i)$ such that y 'satisfies' b_i as it stands. This will be used to establish the termination and correctness of our algorithm.

Claim 3: Suppose $y \in \mathsf{Tot}(k-1)$. Then there exists u such that $y(\Phi_{k-2}(\widehat{c_{uj}})) = n_{uj}$ for each $j < r_u$.

Proof of claim 3: Let $S \subseteq \mathsf{PC}(k-1)$ be the set of all basic compacts $[\![(c \Rightarrow n)]\!]$ where $y(\Phi_{k-2}(\widehat{c})) = n$. Clearly S is bounded by y: if $y(\Phi_{k-2}(\widehat{c})) = n$ then $y([\![c]\!]) = n$ by claim 2, so $[\![(c \Rightarrow n)]\!] \sqsubseteq y$. Thus S is consistent, and has a supremum $y' \sqsubseteq y$. Moreover, it follows from claim 1 that $y' \in \mathsf{Tot}(k-1)$. But now $x(y')\downarrow$, so the enumeration g must include some basic expression $(b_i \Rightarrow m_i)$ where $[\![b_i]\!] \sqsubseteq y'$.

Thus $y'([\![c_i j]\!]) = n_{ij}$ for each $j < r_i$, and by the definition of y', this means that there are expressions c'_{ij} with $[\![c'_{ij}]\!] \sqsubseteq [\![c_{ij}]\!]$ and $y(\Phi_{k-2}(\widehat{c'_{ij}})) = n_{ij}$. Now set

$b'_i = \bigsqcup_{j<r_i}(c'_{ij} \Rightarrow n_{ij})$; then $[\![b_i]\!] \sqsubseteq [\![b'_i]\!]$, so $[\![(b'_i \Rightarrow m_i)]\!] \sqsubseteq [\![(b_i \Rightarrow m_i)]\!] \sqsubseteq x$. But g enumerates all codes for all basic compacts below x, so the basic expression $(b'_i \Rightarrow m_i)$ will appear as $g(u)$ for some u, and u then satisfies the condition of claim 3.

We are now ready to describe our algorithm for computing $\Phi_k(g)$ at an argument $y \in \mathsf{PC}(k-1)$. For each i and $j < r_i$, let $\xi_{ij} : \mathbb{N} \to \mathsf{Tot}(k-2)$ be a standard enumeration of a dense subset of $\{z \in \mathsf{Tot}(k-2) \mid z \sqsupseteq [\![c_{ij}]\!]\}$, and pick $\rho_{ij} \in \mathsf{PC}(0 \to 0 \to 0)$ so that $\rho_{ij}(\bot) = \widehat{c_{ij}}$ and for each $t \in \mathbb{N}$, $\rho_{ij}(t)$ is a basic enumeration of $\xi_{ij}(t)$ extending $\widehat{c_{ij}}$. It is easy to check that ρ_{ij} may be chosen to be PCF-definable uniformly in i, j.

We shall define, by simultaneous recursion, a partial predicate $T : \mathbb{N} \to \mathbb{B}_\bot$ and a search function $\Psi : \mathbb{N} \to \mathbb{N}_\bot$ (notationally suppressing the dependence on g and y). In terms of our proof sketch above, the purpose of $T(i)$ is to test whether y matches b_i (returning a false positive only when it is harmless to do so), whilst in cases where y does not match b_i, $\Psi(i)$ will yield an index t such that some total $\xi_{ij}(t)$ can be used to show that this is the case.

Specifically, we define:

$$T(i) \;\equiv\; \bigwedge_{j<r_i} y(\Phi_{k-2}(\rho_{ij}(\Psi(i)))) = c_{ij}\,.$$

Note that the evaluation of $T(i)$ may in principle diverge if $\Psi(i)$ does; otherwise, each $\Phi_{k-2}(\rho_{ij}(\Psi(i)))$ will be total and the evaluation of each of the conjuncts above will terminate. Simultaneously with this, we compute $\Psi(i)$ as follows:

1. Search for the least t such that either $y(\Phi_{k-2}(\rho_{ij}(t))) \neq c_{ij}$ for some $j < r_i$, or $t > i$, $m_t \in \mathbb{N}$, and $T(t)$ is true.
2. Whichever of the above applies, if $m_t = m_i$ then set $\Psi(i) = t$.
3. If $m_t \neq m_i$, search for the least t' such that $y(\Phi_{k-2}(\rho_{ij}(t'))) \neq c_{ij}$ for some $j < r_i$, and set $\Psi(i) = t'$.

Finally, to compute $\Phi_k(g)(y)$ itself, we test $T(0), T(1), \ldots$ in turn until we find an i such that $m_i \in \mathbb{N}$ and $T(i)$ is true, then return m_i. It is not hard to see that Φ_k is PCF-definable and thus an element of $\mathsf{PC}(1 \to k)$.

To show that this algorithm does what is required, let u be the smallest number such that $y(\Phi_{k-2}(\widehat{c_{uj}})) = n_{uj}$ for all $j < r_u$, as in claim 3. We will show that $T(u)$ and $\Psi(u)$ have the values we would like, and moreover that this good behaviour propagates back to all $i \leq u$.

Claim 4: If $i \leq u$ then $T(i)\downarrow$, and if $T(i) = tt$ then $m_i = m_u = x(y)$. Moreover, if $i < u$, then $\Psi(i)\downarrow$.

Proof of claim 4: We use downwards induction on i, with $i = u$ as the base case. If $i = u$, then $T(i) = tt$ by the choice of u (whether or not $\Psi(u)\downarrow$), so the claim holds. So suppose $i < u$, where the claim holds for all i' with $i < i' \leq u$. Let us first verify that $\Psi(i)\downarrow$.

For each individual $t \leq u$, the test given in step 1 of the algorithm for Ψ will terminate, because $\Phi_{k-2}(\rho_{ij}(t))$ will always be total, and if $t > i$ then $T(t)\downarrow$ by the induction hypothesis. Moreover, the test will succeed if $t = u$, so the search in step 1 will terminate at some $t \leq u$. If $m_t = m_i$, then immediately $\Psi(i) = t$. If $m_t \neq m_i$, there are two subcases to consider:

- If $y(\Phi_{k-2}(\rho_{ij}(t))) \neq c_{ij}$ for some $j < r_i$, then the search for t' in step 3 of the algorithm will rediscover this value of t, and so will terminate.
- If $t > i$ and $T(t)$ is true, then by the induction hypothesis, m_t is the correct value for $x(y)$, so $x(y) \neq m_i$ and it cannot be the case that y realizes an element of V_{b_i}. So again, the search for t' in step 3 is guaranteed to terminate.

Having seen that $\Psi(i)\downarrow$, it now follows immediately that $T(i)\downarrow$.

For the remainder of the claim, if $T(i)$ is true then $\Psi(i)$ must arise as some $t > i$ for which $T(t)$ is true and $m_t = m_i$, since in all other cases $\Psi(i)$ is chosen so as to make $T(i)$ false. So by the induction hypothesis, we have $m_i = m_t = m_u = x(y)$ as required. This completes the proof of claim 4.

Since, as we have observed, we have that $T(u)$ is true, it is now immediate from our definition of Φ_k that $\Phi_k(g)(y)$ will terminate and will yield the value of $x(y)$.

It only remains to check that if g is a basic enumeration for $x \in \mathsf{PC}(k)$ then $\Phi_k(\widehat{g}) \sqsubseteq x$ even if x is non-total. This can be seen from the fact that whether x is total or not, we can only obtain $\Phi_k(\widehat{g})(y) = m$ via the above algorithm if we have found u such that $y(\Phi_{k-2}(\widehat{c_{uj}})) = n_{uj}$ for all $j < r_u$ and $m_u = m$. If this is the case, then by claim 2 we have $y([\![c_{uj}]\!]) = n_{uj}$ for each j, so $[\![b_u]\!] \sqsubseteq y$ and $x(y) = m_u$. This establishes that $\Phi_k(\widehat{g}) \sqsubseteq x$. □

Remark 8.6.4. It can be seen from our description of the algorithm for Φ_k that its expression in PCF needs no fixed point operators beyond $Y_1 : (\overline{1} \rightarrow \overline{1}) \rightarrow \overline{1}$. Thus, the sublanguage PCF_1 suffices for defining Φ_k, and hence for computing representatives for all elements of $\mathsf{Ct}^{\mathrm{eff}}$; likewise for PCF_1^Ω and Ct.

The following application of Theorem 8.6.1 is essentially due to Plotkin [239].

Corollary 8.6.5. *Let* Q *denote the closed term model* $(\mathrm{PCF}^\Omega)^0/{=_{\mathrm{op}}}$ *for* PCF^Ω *(see Subsection 3.2.3), considered as a TPCA over* $\mathbb{N}_\perp$. *Then* $\mathsf{EC}(\mathsf{Q};\mathbb{N}) \cong \mathsf{Ct}$, *and the elements represented by* PCF *terms correspond exactly to those of* $\mathsf{Ct}^{\mathrm{eff}}$.

Proof. The familiar interpretation of PCF^Ω in PC gives an applicative homomorphism $[\![-]\!] : \mathsf{Q} \rightarrow \mathsf{PC}$. Let us here write $\approx_\sigma$ for the partial equivalence relation on $\mathsf{Q}(\sigma)$ associated with the collapse $\mathsf{EC}(\mathsf{Q};\mathbb{N})$, and $\approx'_\sigma$ for the PER on $\mathsf{PC}(\sigma)$ associated with $\mathsf{EC}(\mathsf{PC};\mathbb{N}) = \mathsf{Ct}$. Identifying PCF^Ω terms with their $=_{\mathrm{op}}$ classes, we will show by induction on types that for any closed PCF^Ω terms $M, M' : \sigma$ we have $M \approx_\sigma M'$ iff $[\![M]\!] \approx'_\sigma [\![M']\!]$. Since Theorem 8.6.1 shows that every $\approx'$ class is inhabited by some PCF^Ω-denotable element, this will show that $\mathsf{EC}(\mathsf{Q};\mathbb{N}) \cong \mathsf{Ct}$.

For type $\overline{0}$, the above claim is immediate. So assume $\approx_k$ and $\approx'_k$ are in agreement, and suppose M, M' are terms of type $k+1$. If $[\![M]\!] \approx'_{k+1} [\![M']\!]$, then in particular $[\![M]\!]([\![N]\!]) = [\![M']\!]([\![N']\!])$ whenever $N \approx_k N'$, so $M \approx_{k+1} M'$. For the converse, suppose $M \approx_{k+1} M'$, and consider an arbitrary pair $x \approx'_k x'$ in $\mathsf{PC}(k)$. Let g, g' be basic enumerations of x, x' respectively; then

$$\Phi_k(g) \ \approx'_k \ x \approx'_k x' \ \approx'_k \ \Phi_k(g') ,$$

where $\Phi_k(g), \Phi_k(g')$ are PCF$^\Omega$ denotable. Thus $[\![M]\!](\Phi_k(g)) = [\![M']\!](\Phi_k(g')) \in \mathbb{N}$, and because $\Phi_k(g) \sqsubseteq x$ and $\Phi_k(g') \sqsubseteq x'$, it follows that $[\![M]\!](x) = [\![M']\!](x')$. □

Clearly the above corollary also holds for any quotient of Q, such as SP^0 or SF. In Plotkin [239], the corresponding result was actually presented for the *Milner model*, a slightly larger model than SF which we discussed in Section 7.6.

8.6.2 The Ubiquity Theorem

Whilst Theorem 8.6.1 as it stands is a result about the representation of Ct via PC, it turns out that the ideas behind the proof can be generalized considerably, leading to a wealth of other realizability-style characterizations of Ct and Ct^{eff}. We conclude the chapter with a brief account of these developments, which are presented in detail in Longley [179].

Broadly speaking, Theorem 8.6.1 involves two main features of PC:

1. It supports PCF-style computation. For our present purposes, the essence of this is captured by saying that PC is a TPCA with *numerals* and *guarded recursion* in the sense of Subsections 3.3.4 and 3.3.5. (As shown in Example 3.5.10, every such model admits an 'interpretation of PCF', at least in a weak sense.)
2. It is a *full continuous* TPCA in the sense of Subsection 3.5.1: that is, it is equipped with a numeral-respecting simulation PC $\longrightarrow\!\!\!\!\triangleright$ K_2, and has well-behaved representations of arbitrary set-theoretic functions $\mathbb{N} \to \mathbb{N}$.

We now abstract away from PC and consider an arbitrary model $\mathbf{P}$ sharing these features. Loosely speaking, the intuition is that condition 1, along with the representability of all functions $\mathbb{N} \to \mathbb{N}$, should suffice to ensure that all functionals $F \in$ Ct will have representations in $\mathbf{P}$, since any such F is expressible by some PCF$^\Omega$ program $\Phi_\sigma(g)$ which we can interpret in $\mathbf{P}$; whereas the realization $\mathbf{P} \longrightarrow\!\!\!\!\triangleright K_2$ ensures that *at most* the functionals of Ct are representable in $\mathbf{P}$, since no others are realizable within K_2.

To make this more precise, we shall need some technical conditions on our model and its continuous realization:

Definition 8.6.6. Suppose that $\mathbf{P}$, endowed with a simulation $\gamma : \mathbf{P} \longrightarrow\!\!\!\!\triangleright K_2$, is a full continuous TPCA with numerals and guarded recursion over some type world T. We write $\mathtt{N}$ for the type containing the numerals.

(i) We say $(\mathbf{P}, \gamma)$ has *semidecidable evaluation to* $\widehat{0}$ if for any type $\sigma \in$ T and any $f \in \mathbf{P}(\sigma \to \mathtt{N})$, there is an open set $R \subseteq \mathbb{N}^{\mathbb{N}}$ such that whenever $g \Vdash^{\gamma}_{\sigma} x$ we have $f \cdot x = \widehat{0}$ iff $g \in R$.

(ii) We say that $\mathbf{P}$ has *normalization for functions* $\mathbb{N} \to \mathbb{N}$ if it is equipped with an element $norm \in \mathbf{P}((\mathtt{N} \to \mathtt{N}) \to (\mathtt{N} \to \mathtt{N}))$ such that if $h, h' \in \mathbf{P}(\mathtt{N} \to \mathtt{N})$ both represent a total function $f : \mathbb{N} \to \mathbb{N}$ then $norm \cdot h$ also represents f and $norm \cdot h = norm \cdot h'$.

Condition (i) in the above definition is a way of expressing that for any f, the predicate $f(x) = \widehat{0}$ is (topologically) semidecidable at the level of realizers; the slightly eccentric formulation is designed to cater for the awkward fact that the definedness of application in K_2 is not itself semidecidable. Typical models that fail to have semidecidable evaluation to $\widehat{0}$ are those in which $\widehat{0}$ is not a 'maximal' element of $\mathbf{P}(\mathtt{N})$, such as $\mathscr{P}\omega$ (see Section 10.6) and other lattice-theoretic models.

Condition (ii) above says, in effect, that the model $\mathbf{P}$ is not 'too intensional'; it will typically hold, for example, in models where elements of $\mathbf{P}(\mathtt{N} \to \mathtt{N})$ are functions $\mathbb{N}_\perp \to \mathbb{N}_\perp$ or partial functions $\mathbb{N} \rightharpoonup \mathbb{N}$, and sometimes in other cases as well, but may fail in models of a very fine-grained or syntactic nature, such as term models for programming languages with type 1 oracles. Note that condition (ii) is slightly milder than the notion of 'normalizability at type 1' in Subsection 7.1.2.

Both conditions are satisfied in a wide range of models, particularly those consisting of algebraic domains where the type $\mathtt{N}$ is interpreted by $\mathbb{N}_\perp$. Specifically, besides PC itself, they hold in the models SP^0, SF and M of Chapters 6 and 7, the models SR and St of Chapter 11, the model SA of Section 12.1, and the various other game models mentioned in Sections 7.4 and 12.1 (in both their intensional and extensional manifestations). Moreover, the conditions apply also to TPCAs of a 'call-by-value' character as discussed in Sections 4.3 and 7.2. There are also some interesting untyped models to which they apply, such as the van Oosten model B and the *Böhm tree* model of the untyped λ-calculus (see [9]).

We are now ready to give a precise statement of our theorem. Recall that the guarded recursion operator for type ρ is denoted by Z_ρ.

Theorem 8.6.7 (Ubiquity of Ct). *Suppose $(\mathbf{P}, \gamma)$ is a full continuous TPCA with numerals and guarded recursion, with semidecidable evaluation to $\widehat{0}$ and normalization for functions $\mathbb{N} \to \mathbb{N}$. Then:*

(i) $\mathsf{EC}(\mathbf{P};\mathbb{N}) \cong \mathsf{Ct}$ within $\mathscr{A}sm(K_2)$.

(ii) For each type σ there is an element $\Phi_\sigma \in \mathbf{P}((\mathtt{N} \to \mathtt{N}) \to \sigma)$ such that whenever $g \in \mathbf{P}(\mathtt{N} \to \mathtt{N})$ represents a weak associate (in the usual sense) for some $G \in \mathsf{Ct}(\sigma)$, we have that $\Phi_\sigma \cdot g$ is a realizer for G in $\mathsf{EC}(\mathbf{P};\mathbb{N})$.

(iii) Moreover, a suitable element Φ_σ is definable by an applicative expression in the following combinators (where ρ, ρ', ρ'' range over arbitrary types):

$$k_{\rho\rho'}\ ,\ s_{\rho\rho'\rho''}\ ,\ \widehat{0}\ ,\ suc\ ,\ rec_\rho\ ,\ Z_\rho\ ,\ norm\ .$$

The operators Φ_σ play a crucial role in establishing part (i) of the theorem, as suggested by our informal remarks above. Specifically, having established an isomorphism $\mathsf{EC}(\mathbf{P};N)(k) \cong \mathsf{Ct}(k)$ within $\mathscr{A}sm(K_2)$, we use Φ_{k+1} to show that at least the functionals of $\mathsf{Ct}(k+1)$ are present in $\mathsf{EC}(\mathbf{P};N)(k+1)$, and the simulation γ makes it easy to see that at most these functionals can be present.

The proof of Theorem 8.6.7 is quite long (see [179, Section 7]); most of it is devoted to showing how the various ingredients in the proof of Theorem 8.6.1 may be adapted to this more general setting, with a suitably modified construction of the operators Φ_k. The main challenge is to find a suitable notion of 'approximation' within $\mathbf{P}$, so as to be able to generalize claim 3 from the proof in the absence of any kind of complete partial order structure on $\mathbf{P}$. The key is to show that $\mathbf{P}$, simply by virtue of being a continuous TPCA, inherits enough continuity from K_2 to make a suitable notion of non-total approximation viable.

Whilst the conditions of Definition 8.6.6 are needed in the proof of Theorem 8.6.7 (in ways we shall not attempt to explain here), the scope of the phenomenon itself appears to be somewhat wider, in that the conclusions of Theorem 8.6.7 may also be verified separately for other models such as $\mathscr{P}\omega$.

We now briefly survey some variations and extensions of the ubiquity theorem. First, we may give a corresponding result for $\mathsf{Ct}^{\mathrm{eff}}$ as follows. In the situation of Theorem 8.6.7, suppose that $\mathbf{P}^\sharp$ is a sub-TPCA of $\mathbf{P}$ containing the interpretations of all the combinators listed in part (iii). Suppose too that every element of $\mathbf{P}^\sharp$ has a realizer in K_2^{eff}, and that γ satisfies various natural effectivity conditions as detailed in [179, Section 5]. (In the cases of interest, all these conditions are easy to check for any reasonable choice of 'effective submodel'.) We may then conclude that $\mathsf{EC}((\mathbf{P};\mathbf{P}^\sharp);\mathbb{N}) \cong \mathsf{Ct}^{\mathrm{eff}}$ (realizably in K_2^{eff}), and that if $g \in \mathbf{P}^\sharp$ represents a computable weak associate for $G \in \mathsf{Ct}^{\mathrm{eff}}(\sigma)$, then $\Phi_\sigma \cdot g$ is a realizer for G in $\mathsf{EC}((\mathbf{P};\mathbf{P}^\sharp);\mathbb{N})$. Since a representative for any computable function $\mathbb{N} \rightarrow \mathbb{N}$ is programmable using the above repertoire of combinators, it follows that any $G \in \mathsf{Ct}^{\mathrm{eff}}$ has a realizer in $\mathbf{P}^\sharp$ that is programmable in this language.

Secondly, there is a uniformity in the construction of the Φ_k that is worth noting. Suppose we restrict attention to models over the type world $\mathsf{T}^{\rightarrow}(\mathtt{N})$. Then in part (iii) of the theorem, the applicative expression that defines Φ_σ for any given σ may be taken to be independent of $\mathbf{P}$ and γ: the same formal program works in all models of interest. It follows, for example, that for any $G \in \mathsf{Ct}^{\mathrm{eff}}$ we may find an applicative expression e_G that robustly implements G in all models, i.e. such that e_G denotes a realizer for G in any $\mathbf{P}$ satisfying the conditions of Theorem 8.6.7. This contrasts with the fact that the set of *all* expressions that implement G with respect to $\mathbf{P}$ may vary wildly from one model to another.

Thirdly, a precisely analogous result holds for the *modified extensional collapse* construction. Indeed, Theorem 8.6.7 as formulated above remains true when EC is replaced by MEC, although in this case the construction of the operators Φ_σ (and the programs for them) needs some adaptation, and the proof is slightly more involved. The details appear in Section 9 of [179].

Finally, the proof of Theorem 8.6.7 applies equally well to many other total type structures $\mathbf{A}$ representable in $\mathbf{P}$ besides the extensional collapse, provided they are 'sufficiently rich'. Specifically, suppose $\mathbf{P}$ and γ are as in the theorem, and $\mathbf{A}$ is a total type structure over $\mathbb{N}$ equipped with a type- and numeral-respecting simulation $\delta : \mathbf{A} \longrightarrow\!\!\triangleright \mathbf{P}$. Let $\mathscr{L}^\Omega$ denote the language of combinators from Theorem 8.6.7 augmented with a constant c_f for each $f : \mathbb{N} \rightarrow \mathbb{N}$, and suppose also that whenever $g \in \mathbf{P}(k+1)$ is $\mathscr{L}^\Omega$-definable and tracks an extensional function $G : \mathbf{A}(k) \rightarrow \mathbb{N}$, we have $G \in \mathbf{A}(k+1)$ and $g \Vdash^\delta G$. These hypotheses suffice to ensure that the ubiquity proof applies, so that $\mathbf{A} \cong \mathsf{Ct}$. This hugely extends the class of representations to which the theorem is applicable, further reinforcing the impression that it is hard to arrive at any full continuous total type structure other than Ct by any reasonable realizability construction.

An analogue of Theorem 8.6.7 for HEO will be given as Theorem 9.2.15.

Chapter 9
The Hereditarily Effective Operations

In Chapter 8, we studied the total type structure Ct based on the concept of 'continuous functionals acting on continuous data', and its substructure $\mathsf{Ct}^{\mathrm{eff}}$ of 'effective functionals on continuous data'. The purpose of the present chapter is to study a somewhat similar model based on the idea of 'effective functionals on effective data'. This model was first defined in Kreisel [152] in the context of constructive interpretations of analysis. Further investigations, including equivalence proofs for various characterizations, were undertaken by Gandy [99], Troelstra [288], Hyland [120], Ershov [73, 74] and Bezem [35]. The model is generally known in the literature as HEO, the type structure of *hereditarily effective operations.*

In the case of Ct, we saw that two broad classes of characterizations were available: those of a topological or quasi-topological kind, and those based on realizability. What we will find here is that most of the realizability-style characterizations carry over naturally to HEO via straightforward effectivizations—for instance, we may construct HEO as the extensional collapse $\mathsf{EC}(\mathsf{PC}^{\mathrm{eff}};\mathbb{N})$ of the effective submodel $\mathsf{PC}^{\mathrm{eff}}$ of PC. In addition, there are realizability characterizations of HEO with no natural counterpart for Ct: in particular, we may obtain HEO simply as $\mathsf{EC}(K_1)$. (This is the construction with which the name HEO is usually associated in the literature.) On the other hand, the topological approaches do not carry over so readily, and much of the topological theory of Ct makes little sense for HEO. There are also some interesting examples and counterexamples that give the theory of HEO a flavour quite distinct from that of Ct.

To bring out the extent of the analogy with Ct, we begin in Section 9.1 by reviewing the material from Sections 8.1 and 8.2 that carries over directly to the effective setting. We start with the structure $\mathsf{EC}(\mathsf{PC}^{\mathrm{eff}};\mathbb{N})$—when constructed in this way, we shall refer to our model as HECt, the class of *hereditarily effective continuous operations*. We also discuss a crucial example, based on the well-known *Kleene tree*, that reveals the difference in character between HECt and $\mathsf{Ct}^{\mathrm{eff}}$.

In Section 9.2, we show how this model may be characterized without any explicit reference to continuity. First, we establish the isomorphism HECt $\cong$ HEO ($= \mathsf{EC}(K_1)$). At type 2, this is a classical result known as the *Kreisel–Lacombe–Shoenfield theorem* [155]—we discuss this result in some detail before showing how it extends to higher types. Next, we prove a theorem of Bezem [35] showing that HEO coincides with the *modified* extensional collapse of K_1. We also outline a 'ubiquity theorem' analogous to that of Subsection 8.6.2: *any* effective TPCA with certain properties is bound to yield HEO as its extensional collapse.

In Section 9.3 we present some negative results, in particular an example due to Gandy [101] of an element in HEO(3) that is not Kleene computable. In Section 9.4 we study questions of logical complexity, showing that HEO is related to the arithmetical hierarchy much as Ct is related to the analytical one.

In Section 9.5, we look briefly at some other type structures. Just as HECt is obtained by relativizing the definition of Ct to the set $\mathbb{N}^{\mathbb{N}}_{\mathrm{eff}}$ of computable functions, we may also relativize the construction to other sets $A \subseteq \mathbb{N}^{\mathbb{N}}$ such as the set of hyperarithmetical functions. We show how the properties of the resulting models are intimately related to the closure properties of A, and use this analysis to establish the logical strength of the assertions from Section 8.3 that the functionals FAN and Γ exist in Ct.

9.1 Constructions Based on Continuity

We begin with some constructions in which a notion of continuity is explicitly built into the definition—these will be straightforward adaptations of material from Chapter 8. For the sake of readability, we will repeat some of the definitions here, but will rely on Chapter 8 for the proofs where these effectivize routinely. In Subsection 9.1.2 we introduce the *Kleene tree* and some of its relatives, and use these to illustrate how our model differs from $\mathsf{Ct}^{\mathrm{eff}}$ at types $\overline{2}$ and $\overline{3}$.

9.1.1 Hereditarily Effective Continuous Functionals

We assume familiarity with the model PC and its effective submodel $\mathsf{PC}^{\mathrm{eff}}$ as described in Subsection 3.2.2; we shall consider these as models over $\mathsf{T}^{\to}(\mathtt{N} = \mathbb{N}_{\bot})$. In particular, we recall again our formal language for compact elements:

$$e \ ::= \ \bot \mid n \mid (e_0 \Rightarrow e'_0) \sqcup \cdots \sqcup (e_{r-1} \Rightarrow e'_{r-1}) \,.$$

A syntactic characterization of the *valid* expressions in this language (those that denote compact elements $[\![e]\!] \in \mathsf{PC}(\sigma)$) was given at the start of Section 8.1.

In the first instance, we shall define our model as

$$\mathsf{HECt} \ = \ \mathsf{EC}(\mathsf{PC}^{\mathrm{eff}};\mathbb{N}) \,.$$

The type structure obtained in this or similar ways is also known in the literature by such names as the *hereditarily recursive(ly) continuous* (or *countable*) *functionals*, reflecting its kinship with Ct and $\mathsf{Ct}^{\mathrm{eff}}$. However, the difference between HEO and

$\mathsf{Ct}^{\mathrm{eff}}$ is that an element $F \in \mathsf{Ct}^{\mathrm{eff}}(\sigma \to \tau)$ is (the restriction of) a total continuous functional defined on the whole of $\mathsf{Ct}(\sigma)$, whereas an element $F \in \mathsf{HECt}(\sigma \to \tau)$ only needs to be defined on $\mathsf{HECt}(\sigma)$. Clearly, both $\mathsf{HECt}(1)$ and $\mathsf{Ct}^{\mathrm{eff}}(1)$ consist of just the total computable functions $\mathbb{N} \to \mathbb{N}$. However, we shall see in Subsection 9.1.2 that there are functionals in $\mathsf{HECt}(2)$ that cannot be extended to total continuous functionals on the whole of $\mathbb{N}^{\mathbb{N}}$, so that $\mathsf{Ct}^{\mathrm{eff}}(2) \subseteq \mathsf{HECt}(2)$ but not vice versa.

We will write $\approx_\sigma$ for the partial equivalence relation on $\mathsf{PC}^{\mathrm{eff}}$ associated with the extensional collapse $\mathsf{EC}(\mathsf{PC}^{\mathrm{eff}};\mathbb{N})$, and $\mathsf{Ext}(\sigma)$ for the substructure $\mathsf{Ext}(\mathsf{PC}^{\mathrm{eff}};\mathbb{N})$ of $\mathsf{PC}^{\mathrm{eff}}$. (These notations were used with analogous but different meanings in Chapter 8.) Since each $\mathsf{PC}^{\mathrm{eff}}(\sigma)$ is closed under binary infima in $\mathsf{PC}(\sigma)$, we straightforwardly have the following analogues of results from Subsection 8.1.1:

Proposition 9.1.1. *(i) If $x \in \mathsf{Ext}(\sigma)$ and $x \sqsubseteq y$ then $y \in \mathsf{Ext}(\sigma)$ and $y \approx_\sigma x$.*
(ii) If $x,y \in \mathsf{PC}^{\mathrm{eff}}(\sigma)$, then $x \approx_\sigma y$ iff $x \sqcap y \in \mathsf{Ext}(\sigma)$.
(iii) $\mathsf{Tot}(\mathsf{PC}^{\mathrm{eff}};\mathbb{N}) = \mathsf{Ext}(\mathsf{PC}^{\mathrm{eff}};\mathbb{N})$.

Proof. Exactly as for Lemma 8.1.1 and Corollary 8.1.2. □

It follows immediately that the modified extensional collapse $\mathsf{MEC}(\mathsf{PC}^{\mathrm{eff}};\mathbb{N})$ coincides with HECt.

Since all compact elements of PC are present in $\mathsf{PC}^{\mathrm{eff}}$, the various versions of the density theorem for Ct also hold for HECt, with the same proofs. For example:

Theorem 9.1.2. *Every compact $a \in \mathsf{PC}(\sigma)$ extends to some $\widehat{a} \in \mathsf{Ext}(\sigma)$. Moreover, there is an element $\zeta_\sigma \in \mathsf{PC}^{\mathrm{eff}}(\mathtt{N} \to \sigma)$ such that whenever $\lceil e \rceil$ codes a formal expression e, we have that $\zeta_\sigma(\lceil e \rceil)$ is some total extension $\widehat{[\![e]\!]}$ of $[\![e]\!]$.*

Proof. As for Theorem 8.1.5 and Corollary 8.1.7; indeed, exactly the same function $\zeta_\sigma \in \mathsf{PC}^{\mathrm{eff}}(\mathtt{N} \to \sigma)$ does duty in both settings. □

As in Corollary 8.1.8, it follows that each $\approx_\sigma$ equivalence class has binary joins as well as binary meets; however, they need not contain a top element (see Exercise 8.1.9(iii)).

We also, as for Ct, write $(X)_n^\sigma$ or just $(X)_n$ for the standard nth approximation to an element $X \in \mathsf{HECt}(\sigma)$; we may define this explicitly by

$$(m)_n^{\mathtt{N}} = \min(m,n)\,, \qquad (F)_n^{\sigma\to\tau} = \Lambda X.(F((X)_n^\sigma))_n^\tau\,.$$

Again as for Ct, it is useful to construe our formal expressions e as specifying certain subsets of $\mathsf{HECt}(\sigma)$:

Definition 9.1.3. If e is a valid expression of type σ, the *Kreisel neighbourhood* $W_e \subseteq \mathsf{HECt}(\sigma)$ is defined recursively as follows:

- $W_\perp = \mathbb{N}$ and $W_n = \{n\}$.

- If $d = (e_0 \Rightarrow e'_0) \sqcup \cdots \sqcup (e_{r-1} \Rightarrow e'_{r-1})$, then

$$W_d = (W_{e_0} \Rightarrow W_{e'_0}) \cap \cdots \cap (W_{e_{r-1}} \Rightarrow W_{e'_{r-1}}) \,,$$

where

$$W \Rightarrow W' \;=\; \{F \in \mathsf{HECt} \mid \forall X \in W.\, F(X) \in W'\} \,.$$

Proposition 9.1.4. *Every Kreisel neighbourhood in* HECt *is inhabited by some element* $(X)_n$.

Proof. As for Corollary 8.1.13. □

The semilattice of Kreisel neighbourhoods can thus be seen as a common skeleton shared by both Ct and HECt—the difference between these models is only a matter of which infinite families of neighbourhoods have inhabited intersections.

The following exercise indicates another sense in which the elements $(X)_n$ may be regarded as 'dense' in HECt (cf. Lemma 8.2.18):

Exercise 9.1.5. Adapt the proof of Theorem 5.3.5 to show that whenever $F \in \mathsf{HECt}(\sigma \to \mathtt{N})$ and $X \in \mathsf{HECt}(\sigma)$, we have $F(X) = \lim_{n\to\infty}(F)_n(X)$, with a modulus of convergence uniformly μ-computable in F and X.

We will also be interested in the interpretation of Kleene's S1–S9 in HECt. With some effort, it is possible to adapt the proofs in Subsection 8.1.4 to show that HECt is Kleene closed; however, this fact will be much more easily obtained once the characterization as $\mathsf{EC}(K_1)$ is available, so we defer this to Subsection 9.2.1. In the meantime, it is immediate that HECt is a model for Kleene primitive recursion (S1–S8), and the proof of Theorem 8.1.31 carries over to yield the following version of the density theorem:

Theorem 9.1.6. *For each* σ *there is an operation* $\xi_\sigma \in \mathsf{HECt}^{\mathrm{prim}}(\mathtt{N} \to \sigma)$ *such that for all valid expressions* e *of type* σ *we have* $\xi_\sigma(\lceil e \rceil) \in W_e$. □

We turn next to some alternative characterizations of HECt, mirroring those of Section 8.2. Here we recall from Subsection 3.2.1 the definition of *Kleene's second model* $K_2 = (\mathbb{N}^{\mathbb{N}}, \odot)$ and its effective submodel K_2^{eff} consisting of the computable functions $\mathbb{N} \to \mathbb{N}$. Writing N for the set $\{\widehat{n} \mid n \in \mathbb{N}\}$ where $\widehat{n}(0) = n$ and $\widehat{n}(i+1) = 0$ for all i, we may define a total type structure over $\mathbb{N}$ by

$$\mathsf{HECt}^K \;=\; \mathsf{EC}(K_2^{\mathrm{eff}}; N) \,.$$

This comes with a simulation $\mathsf{HECt}^K \longrightarrow\!\!\!\triangleright\, K_2^{\mathrm{eff}}$, allowing us to view each $\mathsf{HECt}^K(\sigma)$ as an object in $\mathcal{A}sm(K_2^{\mathrm{eff}})$. Since we have a simulation $\psi : \mathsf{PC}^{\mathrm{eff}} \longrightarrow\!\!\!\triangleright\, K_2^{\mathrm{eff}}$ (see Example 3.5.7), we can also regard each $\mathsf{HECt}(\sigma)$ as an assembly over K_2^{eff}. The proof of Theorem 8.2.1 now effectivizes routinely to yield:

Theorem 9.1.7. $\mathsf{HECt}^K \cong \mathsf{HECt}$ *within* $\mathcal{A}sm(K_2^{\mathrm{eff}})$. □

The Kleene density theorem also carries over straightforwardly. As in Subsection 8.2.1, we write Q for the set of finite partial functions $\mathbb{N} \rightharpoonup \mathbb{N}$ whose domain is an initial segment of $\mathbb{N}$, and $Con_k(u,v)$ for the syntactic *consistency relation* on Q, capturing the idea that u,v embody consistent information about an object of type $\overline{k}$ (recall that Con_k is decidable). The proof of Theorem 8.2.3 then goes through to yield:

Theorem 9.1.8. *For $k \in \mathbb{N}$ and $u_0, u_1 \in Q$, the following are equivalent:*

1. $Con_k(u_0, u_1)$.
2. *There is some $F \in \mathsf{HECt}^K(k)$ with associates $f_0 \supseteq u_0$ and $f_1 \supseteq u_1$.*

Moreover, when these conditions hold, suitable values of F, f_0, f_1 may be computed by Kleene primitive recursion, uniformly in k, u_0, u_1. □

We may now obtain the effective analogue of Theorem 8.2.7. Here MHECt^K denotes the modified extensional collapse of K_2^{eff}, obtained as $\mathsf{EC}(\mathsf{Tot}(K_2^{\mathrm{eff}};N))$.

Theorem 9.1.9. $\mathsf{MHECt}^K \cong \mathsf{HECt}^K$ *within* $\mathscr{A}sm(K_2^{\mathrm{eff}})$. □

Not all of the remaining characterizations of Ct from Section 8.2 are readily amenable to effectivization. We will see in Section 9.3 that there are functionals in HECt that are not even limit-continuous, and also that the evident effectivization of the reduced product construction (Subsection 8.2.6) fails to deliver HECt.

9.1.2 The Kleene Tree

Considering $\mathsf{Ct}^{\mathrm{eff}}$ as a standalone extensional type structure (see Corollary 8.1.10), it is immediate from the definitions that $\mathsf{Ct}^{\mathrm{eff}}(2) \subseteq \mathsf{HECt}(2)$. We now present the archetypal example showing that the reverse does not hold. This will be based on the *Kleene tree*, a computable binary tree with an infinite branch but no computable infinite branch.

We shall write $e \bullet_n i$ for the result, if any, of running the eth Turing machine on the input i for at most n steps, so that $e \bullet_n i \preceq e \bullet i$. Clearly, the domain of the partial function $\Lambda i.e \bullet_n i$, and the function itself restricted to this domain, are primitive recursive uniformly in e and n.

It is a well-known result from classical computability theory that the sets

$$A = \{e \mid e \bullet 0 = 0\}, \qquad B = \{e \mid e \bullet 0 = 1\}$$

are computably inseparable—that is, there is no total computable $f : \mathbb{N} \to \mathbb{N}$ with $A \subseteq f^{-1}(0)$, $B \subseteq f^{-1}(1)$. This was used by Kleene [138, Theorem LII] to define a binary tree as follows; similar examples had also been given by Lacombe [161] and Zaslavskiĭ [305].

Definition 9.1.10. The *Kleene tree* K consists of all sequences $(a_0,\ldots,a_{n-1}) \in \{0,1\}^*$ such that for each $e < n$ we have

$$e \bullet_n 0 = 0 \Rightarrow a_e = 0\,, \qquad e \bullet_n 0 = 1 \Rightarrow a_e = 1\,.$$

It is clear from the definition that K is a primitive recursive set of binary sequences and is prefix-closed. The crucial property of this tree is the following:

Proposition 9.1.11. *K contains an infinite path, but no computable infinite path.*

Proof. Define sets A,B as above, and let $f : \mathbb{N} \to \{0,1\}$ be any function with $A \subseteq f^{-1}(0)$, $B \subseteq f^{-1}(1)$. It is easy to see that $(f(0),\ldots,f(n-1)) \in K$ for all n; we therefore have an infinite path in K. Conversely, suppose f defines an infinite path; then for any $e \in A$ we may take n large enough that $e < n$ and $e \bullet_n 0 = 0$, and since $(f(0),\ldots,f(n-1)) \in K$ we must have $f(e) = 0$. Likewise, if $e \in B$ then $f(e) = 1$; thus f separates A and B and so cannot be computable. $\square$

We now use this tree to define a second-order functional F_K as follows:

$$F_K(f) \;=\; \mu n.\,(f(0),\ldots,f(n-1)) \notin K\,.$$

Proposition 9.1.12. *The functional F_K is present in $\mathsf{HECt}(2)$, but is not the restriction of any total continuous function $\mathbb{N}^{\mathbb{N}} \to \mathbb{N}$, and so is not present in $\mathsf{Ct}^{\mathrm{eff}}(2)$.*

Proof. It is immediate from Proposition 9.1.11 that F_K is total on $\mathsf{HECt}(1)$, and since K is computable, it is easy to find a realizer for F_K within $\mathsf{PC}^{\mathrm{eff}}(2)$; thus $F_K \in \mathsf{HECt}(2)$. To see that F_K is not the restriction of a functional in $\mathsf{Ct}(2)$, let f be an infinite path in K as above, and consider the standard approximating sequence $(f)_n \to f$. Clearly $F_K((f)_n) > n$ for each n, so if $F : \mathbb{N}^{\mathbb{N}} \to \mathbb{N}$ extends F_K then we cannot have $F((f)_n) \to F(f)$. $\square$

The presence of F_K in $\mathsf{HECt}(2)$ has repercussions at the next type level. Recall from Subsection 8.3.2 that every $F \in \mathsf{Ct}((\mathtt{N} \to \mathtt{B}) \to \mathtt{N})$ is uniformly continuous: indeed, there is a third-order functional $\mathsf{FAN} \in \mathsf{Ct}(3)$ that returns the modulus of uniform continuity for any given G:

$$\mathsf{FAN}(G) \;=\; \mu n.\forall f,f' \in \{0,1\}^{\mathbb{N}}.\,(\forall i < n.f(i) = f'(i)) \;\Rightarrow\; G(f) = G(f')\,.$$

It is clear that no functional in $\mathsf{HECt}(3)$ can meet this specification, since F_K is manifestly not uniformly continuous. Indeed, if $\phi \in \mathsf{PC}^{\mathrm{eff}}(3)$ is the standard realizer for $\mathsf{FAN} \in \mathsf{Ct}(3)$ and $\psi \in \mathsf{PC}^{\mathrm{eff}}(2)$ is a realizer for $F_K \in \mathsf{HECt}(2)$, we will have $\phi(\psi) = \bot$. (This idea implicitly featured in the proof that FAN is not Kleene computable—see Theorem 8.5.3.) The intuition here is that F_K and FAN cannot coexist within the same world of total computable functionals. The proof of the following theorem gives formal substance to this idea:

Theorem 9.1.13. *There is no effective total type structure $\mathbf{C}$ over $\mathbb{N}$ (in the sense of Definition 3.5.12) such that $\mathsf{HECt} \preceq \mathbf{C}$ and $\mathsf{Ct}^{\mathrm{eff}} \preceq \mathbf{C}$ in the poset $\mathscr{T}(\mathbb{N})$ (see Subsection 3.6.4).*

Proof. Suppose we have $\mathsf{HECt} \preceq \mathbf{C}$ and $\mathsf{Ct}^{\mathrm{eff}} \preceq \mathbf{C}$ where $\mathbf{C}$ is effective; then $\mathbf{C}(1)$ must consist of exactly the total computable functions on $\mathbb{N}$, so every second-order functional in HECt or $\mathsf{Ct}^{\mathrm{eff}}$ is present as it stands in $\mathbf{C}$. Working first in HECt, for any $d \in \mathbb{N}$ let K_d consist of all tuples $(a_0, \ldots, a_{n-1}) \in K$ where $d \bullet 0$ does not terminate in fewer than n steps, and define $G_{K_d} \in \mathsf{HECt}(2)$ by

$$G_{K_d}(f) \;=\; \bar{f}(\mu n.\,(f(0), \ldots, f(n-1)) \notin K_d)\,,$$

where $\bar{f}(n) = \langle f(0), \ldots, f(n-1) \rangle$. For all d with $d \bullet 0 \uparrow$, this yields the same functional G_K, whereas if $d \bullet 0 \downarrow$ in exactly n steps, clearly G_{K_d} is also present in $\mathsf{Ct}^{\mathrm{eff}}$, and $\mathsf{FAN}(G_{K_d}) = n$. Moreover, the mapping $G = \Lambda d.G_{K_d}$ is evidently present in $\mathsf{HECt}(0, 1 \to 0)$.

Now consider G as an element of $\mathbf{C}$. Let $\mathsf{FAN}' \in \mathbf{C}(3)$ represent FAN, and suppose $\mathsf{FAN}'(G_K) = m$. Then $d \bullet 0 \uparrow$ iff $\mathsf{FAN}'(G(d)) = m$ and $d \bullet 0$ does not halt within m steps, which contradicts the undecidability of the halting problem. □

As an immediate corollary of Theorem 9.1.13, there is no largest effective total type structure over $\mathbb{N}$. One can see this as saying that no analogue of Church's thesis is available for total computable functionals (see also Theorem 11.1.4).

More particularly, the above shows that FAN has no counterpart within $\mathsf{HECt}(3)$ itself. Since the example F_K may also be lifted to type 3 via a standard embedding, we may say that $\mathsf{HECt}(3)$ and $\mathsf{Ct}^{\mathrm{eff}}(3)$ are incomparable, in that each contains functionals with no counterpart in the other. This offers a good example of the non-functoriality of EC (see Subsection 3.6.3): the inclusion $\mathsf{PC}^{\mathrm{eff}} \hookrightarrow \mathsf{PC}$ does not induce any sensible simulation $\mathsf{EC}(\mathsf{PC}^{\mathrm{eff}}; \mathbb{N}) \longrightarrow\!\!\!\triangleright\, \mathsf{EC}(\mathsf{PC}; \mathbb{N})$. It is worth observing, however, that we do not need an object as powerful as FAN to exemplify this:

Exercise 9.1.14. Writing h_G for the trace of G as in Definition 8.4.6 (for G in $\mathsf{Ct}^{\mathrm{eff}}(2)$ or $\mathsf{HECt}(2)$), define $\Psi : \mathsf{Ct}^{\mathrm{eff}}(2) \to \mathbb{N}$ by $\Psi(G) = \mu n.\ h_G(n) \neq h_{F_K}(n)$. Show that $\Psi \in \mathsf{Ct}^{\mathrm{eff}}(3)$ and Ψ is μ-computable, but that Ψ is incompatible with every element of $\mathsf{HECt}(3)$.

As another application of the Kleene tree, we will show that the functional $\Gamma \in \mathsf{Ct}^{\mathrm{eff}}(3)$ of Subsection 8.3.3 has no counterpart in HEO. Recall that $F_{a_0 \ldots a_{r-1}}$ denotes $\Lambda f.F(a_0 \ldots a_{r-1} f)$, where $(a_0 \ldots a_{r-1} f)(i) = a_i$ for $i < n$ and $(a_0 \ldots a_{r-1} f)(i) = f(i-n)$ for $i \geq n$.

Theorem 9.1.15. *There is no $\Gamma \in \mathsf{HEO}(3)$ such that $\Gamma(F) = F_0(\Lambda n.\Gamma(F_{n+1}))$ for all $F \in \mathsf{HEO}(2)$.*

Proof. We shall construct an $F \in \mathsf{HEO}(2)$ for which the above equation cannot be satisfied. For this purpose, we will construe elements of $\mathsf{HEO}(1)$ simply as functions $\mathbb{N} \to \{0, 1, 2\}$, and will regard the Kleene tree K as a set of sequences over $\{1, 2\}$.

Define $F(f)$ by means of the following algorithm:

- Search for the least n such that one of the following holds:

  1. $f(n) = 0$.

2. The sequence $f(0),\ldots,f(n) \in \{1,2\}^*$ is outside K.

- In case 2, return the value 1. In case 1, return $f(n+1)+f(n+2)$.

From the characteristic property of the Kleene tree, it is clear that $F(f) \in \mathbb{N}$ for all total computable f; thus $F \in \mathsf{HEO}(2)$.

Now suppose there were an element Γ with $\Gamma(G) = G_0(\Lambda n.\Gamma(G_{n+1}))$ for all $G \in \mathsf{HEO}(2)$. We shall obtain a contradiction based on the intuition that the computation of $\Gamma(F)$ attempts to count all the leaf nodes in the Kleene tree. Let K^+ consist of all the sequences in K plus their immediate successors in $\{1,2\}^*$, and for any $\vec{a} \in K$, let $K^+_{\vec{a}}$ denote the set of sequences in K^+ commencing with $\vec{a}$.

Claim 1: $\Gamma(F_{\vec{a}}) \geq 1$ for any $\vec{a} \in K^+$.

Proof of claim 1: Let j be the least number such that $\vec{a}1^j \notin K$. Then clearly $\Gamma(F_{\vec{a}1^j}) = F_{\vec{a}1^j0}(\cdots) = 1$, and for any $i < j$ we see that $\Gamma(F_{\vec{a}1^i}) \geq \Gamma(F_{\vec{a}1^{i+1}})$, so $\Gamma(F_{\vec{a}1^i}) \geq 1$ by reverse induction on i, and the case $i = 0$ yields the claim.

Claim 2: If $\vec{a} \in K^+$ and $K^+_{\vec{a}}$ has at least m leaf nodes, then $\Gamma(F_{\vec{a}}) \geq m$.

Proof of claim 2: By induction on m. The case $m = 1$ is given by claim 1. If $K^+_{\vec{a}}$ has at least $m > 1$ leaf nodes, then $\vec{a}$ is not itself a leaf in K^+, so $\vec{a} \in K$ and we have

$$\Gamma(F_{\vec{a}}) \;=\; F_{\vec{a}0}(\Lambda n.\Gamma(F_{\vec{a}(n+1)})) \;=\; \Gamma(F_{\vec{a}1}) + \Gamma(F_{\vec{a}2}) \,.$$

Moreover, since $K^+_{\vec{a}1}$ and $K^+_{\vec{a}2}$ have at least one leaf node each, there exist $m_1, m_2 < m$ with $m_1 + m_2 = m$ such that $K^+_{\vec{a}i}$ has at least m_i leaf nodes. It follows by the induction hypothesis that $\Gamma(F_{\vec{a}}) \geq m$. This establishes claim 2.

We now obtain a contradiction by specializing to the empty sequence: K^+ has infinitely many leaf nodes, so no finite value for $\Gamma(F)$ is possible. □

Exercise 9.1.16. (i) Let $\widehat{\Gamma} \in \mathsf{PC}^{\mathrm{eff}}(3)$ be the canonical realizer for $\Gamma \in \mathsf{Ct}^{\mathrm{eff}}(3)$ obtained by interpreting the defining equation for Γ as a least fixed point definition. Deduce from the above theorem that $\widehat{\Gamma}$ is not a realizer for an element of $\mathsf{HECt}(3)$.

(ii) Combine ideas from the proofs of Theorems 9.1.13 and 9.1.15 to show that $\mathsf{HEO}(3)$ contains no element that agrees with the usual Γ functional on the subset $\mathsf{Ct}^{\mathrm{eff}}(2) \subseteq \mathsf{HEO}(2)$.

To conclude this section, we mention three interesting variations on the Kleene tree phenomenon. The second and third of these will play roles later in this chapter; Theorem 9.1.19 also featured in the proof of Corollary 8.5.11.

Theorem 9.1.17 (Lacombe). *There is a primitive recursive binary tree L that contains no computable infinite path, but in which the set of infinite paths has measure $> \frac{1}{2}$ in the standard Lebesgue measure on Cantor space.*

Proof. Let us call a sequence $(a_0,\ldots,a_{n-1}) \in \{0,1\}^*$ an *exit point* if there exists $e < n-1$ such that for all $i \leq e+1$ we have $e \bullet_n i = a_i$. We then let L be the set of sequences s such that neither s nor any prefix of s is an exit point. Clearly L is a primitive recursive binary tree.

To show that L has no computable infinite paths, suppose e indexes a total computable function $f : \mathbb{N} \to \{0,1\}$. Take $n > e+1$ so large that $e \bullet_n 0,\ldots,e \bullet_n (e+1)$

are all defined; then $(e \bullet 0, \ldots, e \bullet (n-1))$ is an exit point as witnessed by e itself. Thus f is not an infinite path in L.

Now let P be the set of infinite paths in L, with $\overline{P}$ its complement in Cantor space. Clearly $\overline{P}$ is the union of all the basic open sets associated with exit points. Moreover, for any $e \in \mathbb{N}$, the elements of $\overline{P}$ arising from exit points witnessed by e are precisely the infinite sequences commencing with $e \bullet 0, \ldots, e \bullet (e+1)$, and the set of such sequences has measure $2^{-(e+2)}$. Summing over all possible e (and noting that not every $e \in \mathbb{N}$ indexes a total function), we see that the measure of $\overline{P}$ is less than $\sum_{e \in \mathbb{N}} 2^{-(e+2)} = \frac{1}{2}$, so P itself has measure $> \frac{1}{2}$. □

Our next example is a countably branching tree that illustrates two phenomena not possible for computable binary trees. First, it contains just a single infinite path—it is an easy exercise to show that this is not possible in the binary case. Secondly, this unique path has Turing degree $\mathbf{0}'$, the degree of the usual halting set—a property which we will exploit in the proof of Theorem 9.5.4. By contrast, it can be shown that any computable binary tree with an infinite path must contain a path of *low degree* (i.e. of some degree d such that the Turing jump of d coincides with $\mathbf{0}'$).

Theorem 9.1.18. *There is a primitive recursive tree $J \subseteq \mathbb{N}^*$ that contains just a single infinite path h, and this path is of degree $\mathbf{0}'$.*

Proof. For later convenience, we will actually define J as a tree over the set $\mathbb{N}_1$ of non-zero natural numbers. Let H be the usual halting set $\{e \mid e \bullet 0 \downarrow\}$, with H_n denoting the usual approximation $\{e \mid e \bullet_n 0 \downarrow\}$. We may suppose that $H_0 = \emptyset$.

Define $h : \mathbb{N} \to \mathbb{N}_1$ by setting $h(n) = 1$ if $n \notin H$, and $h(n) = j+1$ if j is minimal such that $n \in H_j$. Now let J consist of all sequences $(a_0, \ldots, a_{n-1}) \in \mathbb{N}_1^*$ such that for all $i < n$ we have

1. if $a_i = 1$ then $i \notin H_n$,
2. if $a_i > 1$ then $a_i = h(i)$.

It is easy to see that J is primitive recursive, that h is the only infinite path in J, and that H is computable in h. □

Our final example shows that we may also find computable trees in which all infinite paths are of far higher degree:

Theorem 9.1.19. *There is a primitive recursive tree $T \subseteq \mathbb{N}^*$ that contains an infinite path, but no hyperarithmetical (i.e. Δ^1_1) infinite path.*

Proof. It is a classical result from descriptive set theory that the set HYP of Δ^1_1 functions is itself Π^1_1. One way to see this is that the hyperarithmetical functions are exactly the total functions of the form $\Lambda n.\{e\}({}^2\exists, n)$ (Theorem 5.4.1), and the relation $\{e\}({}^2\exists, n_0, \ldots, n_{r-1}) = m$ is given by a Π^1_1 inductive definition and so is itself Π^1_1 (Exercise 5.2.9). Standard quantifier manipulations then show that the predicate $\mathsf{HYP}(f) \equiv \exists e. \forall n. \{e\}({}^2\exists, n) = f(n)$ is Π^1_1.

It follows by Proposition 5.2.6(iii) that there is a primitive recursive relation R such that

$$f \in \mathsf{HYP} \text{ iff } \forall g \in \mathbb{N}^{\mathbb{N}}.\ \exists n \in \mathbb{N}.\ R(\bar{f}(n), \bar{g}(n), n)\ ,$$

where $\bar{h}(n) = \langle h(0), \ldots, h(n-1) \rangle$. Take $(a_0, \ldots, a_{k-1}) \in T$ iff

$$\forall n.\ 2n+1 < k \Rightarrow \neg R(\langle a_0, a_2, \ldots, a_{2n} \rangle, \langle a_1, a_3, \ldots, a_{2n+1} \rangle, n)\ .$$

If h is any infinite branch in T, let $f(m) = h(2m)$ and $g(m) = h(2m+1)$ for all m. Then $\forall n. \neg R(\bar{f}(n), \bar{g}(n), n)$, so f cannot be in HYP, whence $h \notin \mathsf{HYP}$.

On the other hand, for any $f \notin \mathsf{HYP}$, there is a g such that $\forall n. \neg R(\bar{f}(n), \bar{g}(n), n)$, and then f and g may be merged to yield an infinite branch in T. $\square$

9.2 Alternative Characterizations

The isomorphisms obtained in Section 9.1 are in one sense unsurprising, in that the characterizations in question are all in some way based on an idea of effective continuity. What is more remarkable is that many other 'effective' constructions with no explicit notion of continuity built in also give rise to the same structure—in other words, continuity often arises as a *consequence* of effectivity in a non-trivial way. The prototype of this phenomenon is the equivalence $\mathsf{HECt} \cong \mathsf{HEO}(= \mathsf{EC}(K_1))$, a higher type version of the *Kreisel–Lacombe–Shoenfield theorem* which we establish in Subsection 9.2.1. In Subsection 9.2.2 we show that also $\mathsf{MEC}(K_1) \cong \mathsf{HEO}$, while in Subsection 9.2.3 we state a ubiquity theorem: the extensional collapse of *any* effective TPCA with certain properties is bound to yield HEO. This indicates that HEO is indeed the canonical total type structure arising from the idea of 'effective functionals acting on effective data'.

9.2.1 The Kreisel–Lacombe–Shoenfield Theorem

We define HEO simply as $\mathsf{EC}(K_1)$; our main goal in this subsection will be to show that $\mathsf{HEO} \cong \mathsf{HECt}$. Indeed, we will show that the isomorphism is itself 'effective' in the following sense. Each type $\mathsf{HEO}(\sigma)$ can readily be viewed as an assembly over K_1; and since we have an evident simulation $\mathsf{PC}^{\mathrm{eff}} \longrightarrow\!\!\!\triangleright\, K_1$, whereby each $x \in \mathsf{PC}^{\mathrm{eff}}$ is realized by any Kleene index for an enumeration of the compacts bounded by x, the same is true for each type $\mathsf{HECt}(\sigma)$. We will show that the type structures HEO and HECt are actually isomorphic within $\mathscr{A}sm(K_1)$. Throughout this section, the symbol $\Vdash$ (with no superscript) will refer to the canonical realizability relation associated with $\mathsf{EC}(K_1)$.

The key step here is to show that every type 2 functional in HEO is effectively continuous—a classical result due to Kreisel, Lacombe and Shoenfield [155]. We shall present two proofs of this important result: the original one from [155], and an alternative due to Gandy [99]. These proofs are genuinely different, and it is instructive to compare them.

Let $\mathbb{N}_{\mathrm{eff}}^{\mathbb{N}}$ denote the set of total computable functions $\mathbb{N} \to \mathbb{N}$. As usual, we shall say a function $F : \mathbb{N}_{\mathrm{eff}}^{\mathbb{N}} \to \mathbb{N}$ is *continuous* if

$$\forall g \in \mathbb{N}_{\mathrm{eff}}^{\mathbb{N}}.\ \exists j \in \mathbb{N}.\ \forall g' \in \mathbb{N}_{\mathrm{eff}}^{\mathbb{N}}.\ (\forall i < j.\, g'(i) = g(i)) \Rightarrow F(g') = F(g)\,.$$

Any j satisfying this condition with respect to a given g will here be called a *modulus* for F at g.[1] In general, we shall write

$$B(g,j) \ = \ \{g' \in \mathbb{N}_{\mathrm{eff}}^{\mathbb{N}} \mid \forall i < j.\, g'(i) = g(i)\}\,,$$

and let $\xi(g,j,0), \xi(g,j,1), \ldots$ be a fixed primitive recursive enumeration of the eventually zero functions within $B(g,j)$, which form a dense subset of $B(g,j)$. If $e \Vdash_1 g$, we also let $\widehat{\xi}(e,j,t)$ denote a standard choice of Kleene index for $\xi(g,j,t)$, primitively recursively computable from e, j, t.

The first of our two proofs is probably the easier to grasp, and has close affinities with other proofs in the book, such as those of Grilliot (Theorem 5.3.5), Normann (Theorem 8.6.1) and Plotkin (Lemma 10.3.10).

Theorem 9.2.1. *Every $F \in \mathsf{HEO}(2)$ is continuous. Moreover, a modulus of continuity for F at any $g \in \mathbb{N}_{\mathrm{eff}}^{\mathbb{N}}$ is computable uniformly in g and a realizer $c \Vdash_2 F$ (in the ordinary sense of Turing machines with type 1 oracles).*

Proof. Suppose F, c, g are as above, and let H be the halting set $\{d \in \mathbb{N} \mid d \bullet 0 \downarrow\}$. Our basic strategy will be as in the proof of Theorem 9.1.13: parametrically in $d \in \mathbb{N}$, we shall define an approximation g_d to g in such a way that whenever $d \notin H$ we will have $g_d = g$ and hence $F(g_d) = F(g)$; whereas if $d \in H$ then g_d is, whenever possible, chosen so that $F(g_d) \neq F(g)$. However, the undecidability of the halting problem means that there must be some $d \in H$ with $F(g_d) = F(g)$; and for this d, the impossibility of forcing $F(g_d) \neq F(g)$ will give us what we need to obtain a modulus for F at g.

The quality of the approximation g_d to g will be regulated by the time taken to evaluate $d \bullet 0$. Specifically, for any $g : \mathbb{N} \to \mathbb{N}$ and $d \in \mathbb{N}$, we define a *partial* function $g_d : \mathbb{N} \rightharpoonup \mathbb{N}$ as follows.

1. If $d \bullet 0$ does not terminate within i steps, set $g_d(i) = g(i)$.
2. If $d \bullet 0$ terminates in exactly j steps where $j \leq i$, search for the least t such that $F(\xi(g,j,t)) \neq F(g)$, and then set $g_d(i) = \xi(g,j,t)(i)$. (If no such t exists, $g_d(i)$ will be undefined.)

Clearly, if $e \Vdash_1 g$, a realizer e_d for g_d is computable uniformly in c, d, e. In particular, to implement step 2 above, we search for the smallest t such that $c \bullet \widehat{\xi}(e,j,t) \neq c \bullet e$.

Since $g_d = g$ whenever $d \notin H$, we have $c \bullet e_d = c \bullet e$ for all $d \notin H$; and since $\{d \mid c \bullet e_d = c \bullet e\}$ is semidecidable but $\overline{H}$ is not, we must also have $c \bullet e_d = c \bullet e$ for some $d \in H$. Moreover, we may find some such d by a simple search, computably

[1] In other chapters, by contrast, the term 'modulus' sometimes refers specifically to the least such j. This variation in usage is relatively innocuous, however, since in all situations that feature in this book, the least modulus is readily computable from an arbitrary one and the other relevant data.

in c,e. Let d be obtained in this way, and suppose that $d \bullet 0$ terminates in exactly j steps. We shall write $J : \mathbb{N}^2 \rightharpoonup \mathbb{N}$ for the partial computable function that computes this j from c,e by the method indicated here.

We now show that $j = J(c,e)$ is a modulus for F at g. First, we note that for all *eventually zero* functions $g' \in B(g,j)$ we must have $F(g') = F(g)$: otherwise, the search in item 2 above would find some t with $F(\xi(g,j,t)) \neq F(g)$, and we would then have $g_d = \xi(g,j,t)$ so that $F(g_d) \neq F(g)$, contrary to our choice of d. Now consider an arbitrary $g' \in B(g,j)$. Applying the above argument to g', we see that there is some j' such that $F(g'') = F(g')$ for all eventually zero $g'' \in B(g',j')$. So taking g'' eventually zero in $B(g,j) \cap B(g',j')$, we have $F(g') = F(g'') = F(g)$. Thus j is a modulus for F at g, computable from c,e. In particular, F is continuous.

A little more work is needed to show how a modulus is computable purely from c and g, without recourse to the realizer e. We do this as follows. Given $c \Vdash_2 F$ and $g \in \mathbb{N}^{\mathbb{N}}_{\mathrm{eff}}$, search for a pair (e,j) such that

- $J(c,e) = j$,
- $c \bullet J(c,e) = c \bullet e$,
- $e \bullet i = g(i)$ for all $i < j$.

Once such a pair is found, return j as the modulus.

This search is bound to terminate, since if e is a genuine realizer for g then the pair $(e, J(c,e))$ satisfies all the above conditions. Moreover, if (e,j) satisfies these conditions, then whether or not e indexes a total function, we see from our algorithm for J and the above argument that there can be no t with $c \bullet \widehat{\xi}(e,j,t) \neq c \bullet e$, whence again j is a modulus for F at g. This completes the proof. □

The second proof, due to Gandy, involves an ingenious use of self-reference that is not present in the first proof. This more subtle approach allows us to strengthen our result slightly: the procedure for computing the modulus will terminate even for certain c that do not realize any $F \in \mathsf{HEO}(2)$. The usefulness of this additional information will become clear in Subsection 9.2.2 below.

We shall refer here to the non-extensional type structure $\mathsf{HRO} = \mathsf{Tot}(K_1;\mathbb{N})$, known as the model of *hereditarily recursive operations*. The elements of $\mathsf{HRO}(1)$ are just the indices for total computable functions on $\mathbb{N}$, and all realizers for elements of $\mathsf{HEO}(2)$ are present in $\mathsf{HRO}(2)$, but $\mathsf{HRO}(2)$ also includes indices for total operations on $\mathsf{HRO}(1)$ that are not extensional.

As a slight variation on our earlier notation, we shall let $\xi(e,j,0), \xi(e,j,1), \ldots$ (where $e \in \mathbb{N}$) be a standard primitive recursive enumeration of all eventually zero functions g such that $g(i) = e \bullet i$ for all $i < j$. Such an enumeration makes sense whenever $e \bullet 0, \ldots, e \bullet (j-1)$ are all defined, whether or not e indexes a total function. As before, we may write $\widehat{\xi}(e,j,t)$ for a standard index for $\xi(e,j,t)$ primitively recursively computable from e,j,t.

We now restate our theorem incorporating the extra information we shall obtain.

Theorem 9.2.2. *Every $F \in \mathsf{HEO}(2)$ is continuous. Moreover, there exists a computable function $M : \mathbb{N} \times \mathbb{N}^{\mathbb{N}}_{\mathrm{eff}} \rightharpoonup \mathbb{N}$ such that $M(c,g)\downarrow$ whenever $c \in \mathsf{HRO}(2)$ and $g \in \mathbb{N}^{\mathbb{N}}_{\mathrm{eff}}$, and furthermore if $c \Vdash_2 F$ then $M(c,g)$ is a modulus for F at g.*

Proof. Suppose given $c \in \mathsf{HRO}(2)$ and $e \in \mathbb{N}$, where we think of e as a potential realizer for some $g \in \mathbb{N}^{\mathbb{N}}_{\mathrm{eff}}$. Using Kleene's second recursion theorem (as embodied by the guarded recursion operator of Subsection 3.3.5), we may construct an index z_{ce}, computable from c, e, satisfying the following self-referential property for all $i \in \mathbb{N}$. Informally, we think of z_{ce} as an index for some 'critical approximation' to the function realized by e, playing a role similar to e_d in the previous proof. As before, $c \bullet_i n$ denotes the result (if any) of computing $c \bullet n$ for at most i steps.

$$z_{ce} \bullet i \simeq \begin{cases} \text{undefined} & \text{if } c \bullet e \uparrow , \\ e \bullet i & \text{if } c \bullet_i z_{ce} \uparrow \text{ or } c \bullet_i z_{ce} \neq c \bullet e , \\ \xi(e,j,t)(i) & \text{if } c \bullet_i z_{ce} = c \bullet e,\ j \leq i \text{ is minimal such that } c \bullet_j z_{ce} \downarrow, \\ & \text{and } t \text{ is minimal such that } c \bullet \widehat{\xi}(e,j,t) \neq c \bullet e . \end{cases}$$

In connection with the third clause, note that $z_{ce} \bullet i$ will be undefined if not all of $e \bullet 0, \ldots, e \bullet (j-1)$ are defined, or if there is no t such that $c \bullet \widehat{\xi}(e,j,t) \neq c \bullet e$.

Suppose $c \in \mathsf{HRO}(2)$ and $e \in \mathsf{HRO}(1)$, so that $c \bullet e \downarrow$. We will show first that $c \bullet z_{ce} \downarrow$. If not, the above specification for z_{ce} tells us that $z_{ce} \bullet i \simeq e \bullet i$ for all i, so $z_{ce} \in \mathsf{HRO}(1)$. But $c \in \mathsf{HRO}(2)$ so $c \bullet z_{ce} \downarrow$, a contradiction. Thus $c \bullet z_{ce} \downarrow$ after all.

We may therefore find the least j such that $c \bullet_j z_{ce} \downarrow$, where $j = J(c,e)$ for some partial computable $J : \mathbb{N}^2 \rightharpoonup \mathbb{N}$. We now adopt the hypotheses that $c \Vdash_2 F$ and $e \Vdash_1 g$, and will show that, in this case, j is a modulus for F at g.

It is now easy to see that $c \bullet z_{ce} = c \bullet e$: if not, the specification of z_{ce} would tell us that $z_{ce} \bullet i = e \bullet i$ for all i, so $z_{ce} \Vdash_1 g$ and $c \bullet z_{ce} = F(g) = c \bullet e$ after all. Moreover, there cannot exist t such that $c \bullet \widehat{\xi}(e,j,t) \neq c \bullet e$: if there were, we would have $z_{ce} \bullet i = \xi(e,j,t)(i)$ for all i, and the extensionality of c would then yield $c \bullet z_{ce} = c \bullet \widehat{\xi}(e,j,t)$, contradicting $c \bullet z_{ce} = c \bullet e$.

We thus have $F(\xi(g,j,t)) = F(g)$ for all t. Just as in the first proof, we can now extend this to show that $F(g') = F(g)$ for all $g' \in B(g,j)$: indeed, for a suitable j' we will have $F(\xi(g',j',t)) = F(g')$ for all t, and we may then easily choose t, t' so that $\xi(g,j,t) = \xi(g',j',t')$.

This shows that a modulus for F at g is computable as $J(c,e)$, where $J(c,e) \downarrow$ for any $c \in \mathsf{HRO}(2)$ and $e \in \mathsf{HRO}(1)$. To show that a modulus $M(c,g)$ is computable just from c and g, we argue as in the previous proof. Given any $c \in \mathsf{HRO}(2)$ and $g \in \mathbb{N}^{\mathbb{N}}_{\mathrm{eff}}$, we may search for a pair (e,j) such that $J(c,e) = j$, $c \bullet z_{ce} = c \bullet e$, and $e \bullet i = g(i)$ for all $i < j$; when such a pair is found, return j as the value of $M(c,g)$. As in the first proof, an inspection of the algorithm for J confirms that M does what is required. □

Exercise 9.2.3. (i) Show that the procedure for computing moduli given in the proof of Theorem 9.2.1 will not terminate for arbitrary $c \in \mathsf{HRO}(2)$.

(ii) Show that no *extensional* modulus operator is computable: that is, there is no operation N within $\mathsf{HEO}(2,1 \to 0)$ itself such that $N(F,g)$ is a modulus for F at g for every $F \in \mathsf{HEO}(2)$, $g \in \mathsf{HEO}(1)$.

Either of the above proofs readily yields the following information:

Corollary 9.2.4. $\mathsf{HEO}(2) \cong \mathsf{HECt}(2)$ *within* $\mathcal{A}sm(K_1)$.

Proof. Suppose given $F = \mathsf{HECt}(2) = \mathsf{EC}(\mathsf{PC}^{\mathrm{eff}};\mathbb{N})(2)$ realized by $\widehat{F} \in \mathsf{PC}^{\mathrm{eff}}(2)$, and c an index for an enumeration of the compacts bounded by $\widehat{F}$. For any $g \in \mathbb{N}^{\mathbb{N}}_{\mathrm{eff}}$, we may compute $F(g)$ from c and any index e for g as follows: search the enumeration for some compact

$$(((b_0 \Rightarrow m_0) \sqcup \cdots \sqcup (b_{r-1} \Rightarrow m_{r-1})) \Rightarrow n)$$

where $e \bullet b_i = m_i$ for each i; when this is found, return n. This establishes a morphism $\mathsf{HECt}(2) \to \mathsf{HEO}(2)$ in $\mathcal{A}sm(K_1)$ commuting with application.

Conversely, suppose $c \Vdash_2 F \in \mathsf{HEO}(2)$. Then we may computably enumerate a set of compacts in $\mathsf{PC}^{\mathrm{eff}}(2)$ as follows: search for all pairs (e,j) satisfying the conditions listed in the proof of Theorem 9.2.1 (or Theorem 9.2.2 if preferred); and whenever such a pair is found, output the compact element

$$(((0 \Rightarrow e \bullet 0) \sqcup \cdots \sqcup ((j-1) \Rightarrow e \bullet (j-1))) \;\Rightarrow\; (c \bullet e)) \,.$$

It is an easy exercise to check that the supremum of these compacts exists and is a realizer for F in $\mathsf{PC}^{\mathrm{eff}}$. It is then clear that the construction yields the required morphism $\mathsf{HEO}(2) \to \mathsf{HECt}(2)$ in $\mathcal{A}sm(K_1)$. □

Before leaving the topic of type 2 effective operations, it is worth noting how the above theorems relate to other results. As we have seen, the Kreisel–Lacombe–Shoenfield theorem says that continuity is automatic for total effective operations acting on total type 1 functions. In Section 10.4 we will present a somewhat similar theorem due to Myhill and Shepherdson, saying that continuity is likewise automatic for partial effective operations acting on partial type 1 functions. Perhaps surprisingly, however, the same is not true for *partial* operations acting on *total* functions. The following intriguing counterexample is due to Friedberg [97].

Example 9.2.5. Define a partial functional $F : \mathbb{N}^{\mathbb{N}}_{\mathrm{eff}} \rightharpoonup \mathbb{N}$ as follows:

$$F(g) \;=\; \begin{cases} 0 & \text{if } g = \Lambda i.0\,, \\ 0 & \text{if } z = \mu i.\, g(i) \neq 0 \wedge \exists e < z. \forall i \leq z.\, e \bullet i = g(i)\,, \\ \text{undefined otherwise.} \end{cases}$$

We claim that the domain of F is not open, since it contains no open neighbourhood of $\Lambda i.0$. Indeed, for any n, we may pick g_n so that $g_n(i) = 0$ for all $i < n$, and $g_n(n)$ is larger than $e \bullet n$ for all $e < n$; we then have $g_n \in B(\Lambda i.0, n)$ but $F(g_n)\uparrow$. Thus, whilst F is of course continuous on its domain, it is not continuous if regarded as a function $\mathbb{N}^{\mathbb{N}}_{\mathrm{eff}} \to \mathbb{N}_\perp$.

Nonetheless, F is computable at the level of indices: there is a partial computable $f : \mathbb{N} \rightharpoonup \mathbb{N}$ such that whenever $d \Vdash_1 g$ we have $f(d) \simeq F(g)$. Specifically, take

$$f(d) \;=\; \begin{cases} 0 & \text{if } \forall i \leq d.\, d \bullet i = 0\,, \\ 0 & \text{if } z = \mu i.\, d \bullet i \neq 0 \wedge \exists e < z. \forall i \leq z.\, e \bullet i = d \bullet i\,, \\ \text{undefined otherwise.} \end{cases}$$

Clearly f is computable, and it is an enjoyable exercise to verify that f correctly represents F.

Next, we proceed to extend Corollary 9.2.4 to higher types. For this purpose, we shall need a mild generalization of Theorem 9.2.1, in order to cover effective operations F that are defined only on certain sets $A \subseteq \mathbb{N}^{\mathbb{N}}_{\mathrm{eff}}$. All we really require here is the existence of a computable dense subset of A to play the role of the eventually zero functions. Specifically:

Corollary 9.2.6. *Suppose $A \subseteq \mathbb{N}^{\mathbb{N}}_{\mathrm{eff}}$, and let B be the set of (graphs of) all finite subfunctions of functions in A. Suppose moreover that B is decidable, and that uniformly in any $b \in B$, we may compute a function $\varepsilon(b) \in A$ extending b.*

Let $F : A \to \mathbb{N}$ be realized by c, in the sense that whenever $e \Vdash_1 g \in A$, we have $c \bullet e = F(g)$. Then some $F' : \mathbb{N}^{\mathbb{N}} \to \mathbb{N}$ extending F is computable in the sense of Turing machines with oracles, uniformly in c.

Proof. It is routine to check that the proofs of Theorem 9.2.1 and Corollary 9.2.4 go through under these hypotheses, replacing $\xi(g,j,t)$ by $\varepsilon(\zeta(g,j,t))$ where $\zeta(g,j,0)$, $\zeta(g,j,1),\ldots$ enumerates all finite extensions of $\overline{g}(j)$. This yields an element of $\mathsf{PC}^{\mathrm{eff}}(2)$ realizing F, computable uniformly from c, and it is then easy to see how an extension of F itself is computable, treating the argument g as an oracle. □

For the purpose of our main theorem, it will be convenient to work with the structure $\mathsf{HECt}^K = \mathsf{EC}(K_2^{\mathrm{eff}}; N)$ from Subsection 9.1.1 (recall that $N = \{\widehat{n} \mid n \in \mathbb{N}\}$ where $\widehat{n}(0) = n, \widehat{n}(i+1) = 0$). This comes with a simulation $\mathsf{HECt}^K \longrightarrow\!\!\!\!\triangleright\, K_2^{\mathrm{eff}}$, which may be composed with a standard simulation $K_2^{\mathrm{eff}} \longrightarrow\!\!\!\!\triangleright\, K_1$.

Theorem 9.2.7. $\mathsf{HEO} \cong \mathsf{HECt}^K$ *within $\mathscr{A}sm(K_1)$.*

Proof. For notational simplicity we will concentrate on the pure types. By induction on k, we will construct morphisms $\alpha_k : \mathsf{HEO}(k) \to \mathsf{HECt}^K(k)$ and $\beta_k : \mathsf{HECt}^K(k) \to \mathsf{HEO}(k)$ in $\mathscr{A}sm(K_1)$ commuting with application.

The case $k = 0$ is easy: effectively within K_1, we may translate between a number n and an index for the $\widehat{n} \in K_2$. So assume we have mutually inverse morphisms α_k, β_k as required, and consider the situation for $k+1$. The morphism β_{k+1} is easy to construct: given $\Phi \in \mathsf{HECt}^K(k+1)$ realized by $\phi \in K_2^{\mathrm{eff}}$ with index $e \in K_1$, we may obtain $\beta_{k+1}(\Phi) \in \mathsf{HEO}(k+1)$ as $\Lambda G.\Phi(\alpha_k(G))$, and a K_1 realizer for this is provided by $\lambda^* y.v_0(re(u_k y))$, where r realizes the simulation $K_2^{\mathrm{eff}} \longrightarrow\!\!\!\!\triangleright\, K_1$, u_k tracks α_k, and v_0 tracks β_0. Thus β_{k+1} itself is tracked by $\lambda^* ey.v_0(re(u_k y))$.

For α_{k+1}, we use Corollary 9.2.6. Let A be the set of all elements of $K_2^{\mathrm{eff}} = \mathbb{N}^{\mathbb{N}}_{\mathrm{eff}}$ that realize elements of $\mathsf{HECt}^K(k)$, and let B be the set of finite subfunctions of elements of A. Both the decidability of B and the uniform computability of some $\varepsilon(b) \in A$ extending $b \in B$ are provided by the Kleene density theorem for HECt^K (Theorem 9.1.8).

Now suppose we are given $\Psi \in \mathsf{HEO}(k+1)$ realized by $d \in \mathbb{N}$. If v_k tracks β_k and u_0 tracks α_0, then $c = \lambda^* x.u_0(d(v_k x))$ tracks the corresponding morphism $\mathsf{HECt}^K(k) \to N$ in $\mathscr{A}sm(K_1)$. But now $c \bullet e$ is defined whenever e is an index for an element of A; moreover, $c \bullet e = c \bullet e'$ whenever e, e' index the same element of A, since e, e' then represent the same element of $\mathsf{HECt}^K(k)$. Thus c realizes some $F : A \to \mathbb{N}$, and Corollary 9.2.6 applies. We thus conclude that some extension of F is computable uniformly in c and hence in d, treating the argument $g \in A$ as an oracle. It follows easily that there is an element $h \in K_2^{\mathrm{eff}}$, computable uniformly in d, such that $h \odot g = F(g)$ for all $g \in A$, and it is now clear that h is a K_2 realizer for Ψ as an element of $\mathsf{HECt}^K(k+1)$. Since a K_1 realizer for h may be uniformly computed from d, we have constructed the morphism $\alpha_{k+1} : \mathsf{HEO}(k+1) \to \mathsf{HECt}^K(k+1)$ in $\mathcal{A}sm(K_1)$ as required. $\square$

We have seen that $\mathsf{HECt}^K \cong \mathsf{HECt}$ within $\mathcal{A}sm(K_2^{\mathrm{eff}})$ (Theorem 9.1.7), so these are certainly isomorphic within $\mathcal{A}sm(K_1)$, and we conclude that $\mathsf{HEO} \cong \mathsf{HECt}$ within K_1. We will henceforth use HEO as our preferred notation for this structure.

As an alternative route, a domain-theoretic version of the Kreisel–Lacombe–Shoenfield theorem has been given by Berger [19]. This allows one to obtain the equivalence $\mathsf{HEO} \cong \mathsf{HECt}$ without going via HECt^K.

We may now immediately read off a version of the computability of 'moduli of continuity' at higher types. We recall here the definition of Kreisel neighbourhoods W_e from Definition 9.1.3.

Corollary 9.2.8. *Suppose $F \in \mathsf{HEO}(k+1)$ and $g \in \mathsf{HEO}(k)$. Uniformly in K_1 indices for F and g, we may compute a code a for a compact element of type $\overline{k}$ such that $g \in W_a$ and F is constant on W_a.*

Proof. For HECt this is immediate: search the enumerations of ($\mathsf{PC}^{\mathrm{eff}}$ realizers for) F and g respectively looking for compacts $(a \Rightarrow n)$ and a' with $a \sqsubseteq a'$; when such a pair is found, return a. Since every $g' \in W_a$ has a $\mathsf{PC}^{\mathrm{eff}}$ realizer extending a, this has the required property.

The corresponding property for HEO now follows by Theorem 9.2.7. $\square$

This is also an appropriate point at which to note:

Theorem 9.2.9. HEO *is Kleene closed.*

Proof. A routine exercise in effective transfinite recursion (see Section 5.1) yields a computable function θ such that whenever $\{e\}(\Phi_0, \ldots, \Phi_{r-1}) = n$ and $d_i \Vdash_{k_i} \Phi_i$ for each $i < r$, we have that $\theta(e, \langle k_0, \ldots, k_{r-1}\rangle, \langle d_0, \ldots, d_{r-1}\rangle) = n$. It follows that if Ψ is total and Kleene computable relative to $\Phi_0, \ldots, \Phi_{r-1} \in \mathsf{HEO}$, then $\Psi \in \mathsf{HEO}$. $\square$

9.2.2 The Modified Extensional Collapse of K_1

We now prove another non-trivial equivalence result, due to Bezem [35]: HEO coincides with the modified extensional collapse $\mathsf{MHEO} = \mathsf{MEC}(K_1)$ (in other words, the structure $\mathsf{EC}(\mathsf{HRO})$). In this model, each type σ is represented by a partial equivalence relation on a set of 'actual' realizers for functionals of type σ, contained within a possibly larger set of 'potential' realizers (the set $\mathsf{HRO}(\sigma)$). It is easy to see that at types 0 and 1, the sets of actual and potential realizers coincide. At type 2, however, there are elements e in HRO that are not actual realizers, since they do not act extensionally on the type 1 realizers. This means that there will be realizers d for elements of $\mathsf{HEO}(3)$ that are not actual realizers within $\mathsf{MHEO}(3)$, since we may have $d \bullet e \uparrow$ for some potential type 2 realizers e, and it is far from obvious whether $\mathsf{HEO}(3)$ and $\mathsf{MHEO}(3)$ even contain the same functionals. The relationship between the two models becomes even less apparent as one passes to higher types.

Whilst the types of MHEO are in principle best thought of in terms of both actual and potential realizers, we will here treat each $\mathsf{MHEO}(\sigma)$ simply as an assembly over K_1 embodying just the actual realizers, and will treat the potential realizers in a more ad hoc way. We will show, in fact, that the type structure HEO and MHEO are isomorphic within $\mathcal{A}sm(K_1)$: that is, we may effectively translate a realizer for Φ in $\mathsf{HEO}(\sigma)$ to a realizer for the same Φ in $\mathsf{MHEO}(\sigma)$, and vice versa.

The proof we give here is a somewhat streamlined version of Bezem's argument, and in its basic outline runs parallel to the proof of the isomorphism $\mathsf{MEC}(K_2;N) \cong \mathsf{EC}(K_2;N)$ in Subsection 8.2.2 (also from Bezem [35]). We shall make serious use of the isomorphisms established in the preceding subsection, and the flexibility of presentation that these afford.

Fix $k \geq 1$. Let us here write P_k for the assembly representing $\mathsf{EC}(\mathsf{PC}^{\mathrm{eff}};\mathbb{N})(k)$: that is, P_k has underlying set $\mathsf{HEO}(k)$, and realizers for $G \in P_k$ are indices for enumerations of compacts in $\mathsf{PC}^{\mathrm{eff}}(k)$ approximating a realizer $\widehat{G}$ for G. As we have already seen, P_k is isomorphic within $\mathcal{A}sm(K_1)$ to any of our other presentations of $\mathsf{HEO}(k)$. We will write E_k for the set of all realizers for elements of P_k, and $\approx_k$ for the equivalence relation on E_k induced by P_k.

Let us also write C_{k+1} for the set of all $c \in K_1$ such that $c \bullet e \downarrow$ for all $e \in E_k$, and $D_{k+1} \subseteq C_{k+1}$ for the set of such c that act extensionally on E_k (i.e. such that $e \approx_k e'$ implies $c \bullet e = c \bullet e'$). We shall write $\approx_{k+1}$ for the induced equivalence relation on D_k (taking $c \approx_{k+1} c'$ iff for all e, e', $e \approx_k e'$ implies $c \bullet e = c' \bullet e$), and Q_{k+1} for the corresponding assembly. Since, by construction, Q_{k+1} is the exponential $P_k \Rightarrow N$ within $\mathcal{A}sm(K_1)$, it is isomorphic within $\mathcal{A}sm(K_1)$ to any of our presentations of $\mathsf{HEO}(k+1)$.

The idea is to think of D_{k+1} as the set of 'actual realizers' for functionals $\Phi \in \mathsf{HEO}(k+1)$, and C_{k+1} as some wider set of 'potential realizers'. We shall show that it is possible to retract down from C_{k+1} to D_{k+1} in the following sense: there is a computable total function $\theta_{k+1} : C_{k+1} \to D_{k+1}$ such that whenever $c \in D_{k+1}$ we have $\theta_{k+1}(c) \approx_{k+1} c$. This will provide the key to showing (inductively) that $\mathsf{MHEO} \cong \mathsf{HEO}$ at all types.

To associate an element of D_{k+1} with an arbitrary $c \in C_{k+1}$, we shall use a variant of Gandy's proof of the Kreisel–Lacombe–Shoenfield theorem, which associates not only a modulus at each $e \in E_k$ to every element of D_{k+1}, but also a 'pseudo-modulus' at each e to every element of C_{k+1}. From these pseudo-moduli for c one can attempt to build a 'graph' for an element of $\mathsf{PC}^{\mathrm{eff}}(k+1)$, and this will lead us to a total functional $\Phi_c \in \mathsf{HEO}(k+1)$ realized by some $\theta(c) \in D_{k+1}$.

Lemma 9.2.10. *There is a computable function $L_k : \mathbb{N} \times \mathbb{N} \rightharpoonup \mathbb{N}$ such that*

1. *whenever $c \in C_{k+1}$, $e \in E_k$ and e realizes $G \in P_k$ via some $\widehat{G} \in \mathsf{PC}^{\mathrm{eff}}(k)$, $L_k(c,e)$ is defined and is a code a for some compact $[\![a]\!] \in \mathsf{PC}^{\mathrm{eff}}(k)$ such that $[\![a]\!] \sqsubseteq \widehat{G}$,*
2. *if moreover $c \in D_{k+1}$ and c realizes $\Phi \in Q_{k+1}$, then Φ is constant on W_a.*

Proof. A straightforward higher type generalization of the proof of Theorem 9.2.2. We use the fact that for each code a for a compact at type k, we have an enumeration $\{\xi(a,t)\}_{t\in\mathbb{N}}$ of a dense subset of W_a, Kleene primitive recursive uniformly in a,t, and an associated choice of realizers $\widehat{\xi}(a,t)$, again primitive recursive uniformly in a,t. The existence of such enumerations is an easy consequence of Theorem 9.1.6 and its proof.

Suppose given $c \in C_{k+1}$ and $e \in E_k$, where e realizes $G \in P_k$ via $\widehat{G} \in \mathsf{PC}^{\mathrm{eff}}(k)$. Using the second recursion theorem, computably in c,e we may construct an index z_{ce} such that the following holds for all $i \in \mathbb{N}$:

$$z_{ce} \bullet i \;\simeq\; \begin{cases} \text{undefined if } c \bullet e\uparrow\,, \\ e \bullet i \qquad\quad \text{if } c \bullet_i z_{ce}\uparrow \text{ or } c \bullet_i z_{ce} \neq c \bullet e\,, \\ \xi(a,t)(i) \;\; \text{if } c \bullet_i z_{ce} = c \bullet e,\; j \leq i \text{ is minimal such that } c \bullet_j z_{ce}\downarrow, \\ \qquad\qquad\quad [\![a]\!] = [\![e \bullet 0]\!] \sqcup \cdots \sqcup [\![e \bullet (j-1)]\!], \\ \qquad\qquad\quad \text{and } t \text{ is minimal such that } c \bullet \widehat{\xi}(a,t) \neq c \bullet e\,. \end{cases}$$

Exactly as in the proof of Theorem 9.2.2, we may show that $c \bullet z_{ce}\downarrow$. We may therefore, computably in c,e, find the least j such that $c \bullet_j z_{ce}\downarrow$, then let $L_k(c,e)$ be a code for the compact element $[\![e \bullet 0]\!] \sqcup \cdots \sqcup [\![e \bullet (j-1)]\!]$. This establishes item 1 of the lemma.

For item 2, we argue as in the proof of Theorem 9.2.2 that $c \bullet z_{ce} = c \bullet e$, and also that there cannot exist t with $c \bullet \widehat{\xi}(a,t) \neq c \bullet e$. This shows that $\Phi(\xi(a,t)) = \Phi(G)$ for all t, and for any $G' \in W_a$ we likewise have some a' such that $\Phi(\xi(a',t)) = \Phi(G')$ for all t, whence $\Phi(G') = \Phi(G)$ under innocuous assumptions on ξ. □

Lemma 9.2.11. *There is a computable function $\theta_{k+1} : C_{k+1} \to D_{k+1}$ such that $\theta_{k+1}(c) \approx_{k+1} c$ whenever $c \in D_{k+1}$.*

Proof. Given $c \in C_{k+1}$, we may compute a list γ_c of compacts in $\mathsf{PC}^{\mathrm{eff}}(k+1)$ as follows. Using the function L from the previous lemma, enumerate all pairs (e,a) such that $c \bullet e\downarrow$, $L(c,e) = a$, and a codes a compact of type k (note that there will be some such pairs with $e \notin E_k$). For each such pair, output the compact $(a \Rightarrow c \bullet e)$, as long as this is consistent with all compacts output previously. If an inconsistency is found, stop the enumeration. Let b denote the supremum of the (consistent) compacts already output; let $\widehat{\xi}(b)$ be any total element of $\mathsf{PC}^{\mathrm{eff}}(k+1)$ extending b, and continue by outputting some standard sequence of compacts with supremum $\widehat{\xi}(b)$ (this can be done computably using the effective version of the density theorem).

We first show that γ_c enumerates a total element of $\mathsf{PC}^{\mathrm{eff}}(k+1)$ for any $c \in C_{k+1}$. In the case where an inconsistency arises, this is clear as γ_c will enumerate the relevant $\widehat{\xi}(b)$. If no inconsistency is ever found, we must show that for any total $\widehat{G} \in \mathsf{PC}^{\mathrm{eff}}(k)$ representing any $G \in \mathsf{HEO}(k)$, there is some $(a \Rightarrow n)$ in γ_c with $[\![a]\!] \sqsubseteq \widehat{G}$. To see this, take e any index for an enumeration of $\widehat{G}$, so that $e \in E_k$; then by Lemma 9.2.10, $a = L(c,e)$ is a code such that $[\![a]\!] \sqsubseteq \widehat{G}$, and the pair (e,a) will eventually appear in the enumeration that generates γ_c, so that the compact $(a \Rightarrow c \bullet e)$ is output. Thus, in either case, the supremum of the compacts listed in γ_c yields a total element $\widehat{\Phi}_c \in \mathsf{PC}^{\mathrm{eff}}(k+1)$ realizing some $\Phi_c \in \mathsf{HEO}(k+1)$.

Since the mapping $c \mapsto \gamma_c$ is computable, it is clear that $\Phi_c(G)$ is computable from c and any P_k-realizer e for G; hence from c alone we may compute a realizer $\theta(c)$ for Φ_c as an element of Q_{k+1}. We thus have our total computable function $\theta : C_{k+1} \to D_{k+1}$ as desired.

It remains to see that if $c \in D_{k+1}$ then $\theta(c) \approx_{k+1} c$. But any $c \in D_{k+1}$ realizes some $\Phi \in Q_{k+1}$, and in this case, by the proof of Lemma 9.2.10, all $(a \Rightarrow n)$ listed in γ_c will represent properties of Φ, and hence will be consistent with each other. (Even if $(a \Rightarrow n)$ is generated from some $e \notin E_k$, the same argument shows that $\Phi \in W_{(a \Rightarrow n)}$.) Since we have already seen that $\widehat{\Phi}_c$ is a total element, it must actually be a realizer for Φ, whence $\theta(c) \approx_{k+1} c$. $\square$

Theorem 9.2.12. $\mathsf{MHEO} \cong \mathsf{HEO}$ *within* $\mathscr{A}sm(K_1)$.

Proof. Let $|\mathsf{MHEO}(k)|$ denote the set of realizers for elements of $\mathsf{MHEO}(k)$, so that $|\mathsf{MHEO}(k)| \subseteq \mathsf{HRO}(k)$. We shall show by induction on k that

1. $\mathsf{MHEO}(k) \cong \mathsf{HEO}(k)$ as assemblies over K_1, where for $k > 0$ the isomorphism respects application,
2. there is a total computable function $\phi_k : \mathsf{HRO}(k) \to |\mathsf{MHEO}(k)|$ such that if c is already a realizer for some $G \in \mathsf{MHEO}(k)$ then $\phi_k(c)$ also realizes G.

For $k = 0, 1$, both claims are trivial, since here $\mathsf{HEO}(k)$ and $\mathsf{MHEO}(k)$ are exactly the same assembly and $|\mathsf{MHEO}(k)| = \mathsf{HRO}(k)$. So assume both claims hold for $k \geq 1$, and consider the situation for $k+1$.

For claim 1, any realizer e for some $\Phi \in \mathsf{MHEO}(k+1)$ tracks a morphism $\mathsf{MHEO}(k) \to N$ within $\mathscr{A}sm(K_1)$, so using the assembly isomorphism $\mathsf{MHEO}(k) \cong \mathsf{HEO}(k)$, we effectively obtain a realizer for the corresponding morphism $\mathsf{HEO}(k) \to N$, i.e. a realizer for $\Phi \in \mathsf{HEO}(k+1)$. Conversely, if d realizes $\Phi \in \mathsf{HEO}(k+1)$, v_k tracks the isomorphism $\mathsf{MHEO}(k) \to \mathsf{HEO}(k)$, and p is a Kleene index for ϕ_k, then $\lambda^* x.d(v(px))$ acts totally on $\mathsf{HRO}(k)$ and represents Φ on realizers in $|\mathsf{MHEO}(k)|$, and hence realizes $\Phi \in \mathsf{MHEO}(k+1)$. Thus $\lambda^* dx.d(v(px))$ tracks the required morphism $\mathsf{HEO}(k+1) \to \mathsf{MHEO}(k+1)$.

For claim 2, suppose $c \in \mathsf{HRO}(k+1)$; then $c \bullet e\downarrow$ for all $e \in \mathsf{HRO}(k)$, and in particular for all $e \in |\mathsf{MHEO}(k)|$. Let w track the assembly isomorphism $P_k \to \mathsf{MHEO}(k)$, and let $c' = \lambda^* y.c(wy)$. Then $c' \bullet e\downarrow$ for all $e \in E_k$, so $c' \in C_{k+1}$; and if c realizes $\Phi \in \mathsf{MHEO}(k+1)$ then $c' \in D_{k+1}$ and c' realizes $\Phi \in Q_{k+1}$. We are therefore in the situation of Lemma 9.2.11, and may map c' to a suitable realizer in D_{k+1} using a realizer for θ. We thus have a realizer for some $\Phi_c \in Q_{k+1}$, where $\Phi_c = \Phi$ if c realizes Φ. Finally, since $Q_{k+1} \cong \mathsf{HEO}(k+1)$ within $\mathscr{A}sm(k+1)$, and we have already shown that $\mathsf{HEO}(k+1) \cong \mathsf{MHEO}(k+1)$, we may transport Φ_c to an element of $\mathsf{MHEO}(k+1)$. Since all of this is effective uniformly in c, we have constructed the required mapping $\phi_{k+1} : \mathsf{HRO}(k+1) \to |\mathsf{MHEO}(k+1)|$. □

9.2.3 Ubiquity of HEO

We now briefly review some further characterizations of HEO. First, we note that Theorem 8.6.1 has a precise counterpart in the hereditarily effective setting:

Theorem 9.2.13. *For each σ there is a* PCF*-computable $\Phi_\sigma \in \mathsf{PC}^{\mathrm{eff}}(1 \to \sigma)$ such that whenever $g \in K_2^{\mathrm{eff}}$ is a realizer for some $x \in \mathsf{PC}^{\mathrm{eff}}(\sigma)$, we have $\Phi_\sigma(\widehat{g}) \sqsubseteq x$, and if $x \Vdash_\sigma X \in \mathsf{HEO}(\sigma)$ then also $\Phi_\sigma(\widehat{g}) \Vdash_\sigma X$.*

Proof. Exactly as for Theorem 8.6.1. □

As a consequence, we have that if $\mathsf{Q}^{\mathrm{eff}}$ is the closed term model $\mathrm{PCF}^0/{=_{\mathrm{op}}}$, then $\mathsf{EC}(\mathsf{Q}^{\mathrm{eff}};\mathbb{N}) \cong \mathsf{HEO}$ (cf. Corollary 8.6.5); likewise for $\mathsf{EC}(\mathsf{SP}^{0,\mathrm{eff}};\mathbb{N})$ or $\mathsf{EC}(\mathsf{SF}^{\mathrm{eff}};\mathbb{N})$.

Building on this, it turns out that we may obtain a result analogous to Theorem 8.6.7, stating that the extensional collapse of *any* effective TPCA with certain properties is bound to yield HEO. Again, we give here only a bare statement of the result, referring the reader to Longley [179] for the lengthy proof.

The notion of an effective TPCA was introduced in Definition 3.5.12. As in Subsection 8.6.2, some further technical conditions are required.

Definition 9.2.14. Suppose that $\mathbf{P}$, endowed with a simulation $\gamma : \mathbf{P} \longrightarrow\!\!\triangleright K_1$, is an effective TPCA with numerals and guarded recursion over some type world T. We write $\mathtt{N}$ for the type containing the numerals.

We say $(\mathbf{P},\gamma)$ has *semidecidable evaluation to* $\widehat{0}$ if

1. the set $\{e \mid e \Vdash^{\gamma}_{\mathtt{N}} \widehat{0}\}$ is c.e.,
2. for some choice of $a \in K_1$ tracking $\cdot_{\sigma\mathtt{N}}$ as in Definition 3.4.7, we have that if $d \Vdash^{\gamma}_{\sigma\to\mathtt{N}} f$, $e \Vdash^{\gamma}_{\sigma} x$ and $a \bullet d \bullet e \Vdash^{\gamma}_{\mathtt{N}} y$, then $f \cdot x = y$.

Item 2 here is a very mild condition saying that the realizers a can be chosen so as not to give misleading results. Note that we have no need here for a notion of normalization at type 1 as in Subsection 8.6.2—this means that even extremely intensional models, including the term models for languages such as PCF + *timeout* (see Subsection 12.2.1), will fall within the scope of our theorem.

Theorem 9.2.15 (Ubiquity of HEO). *Suppose $(\mathbf{P},\gamma)$ is an effective TPCA with numerals and guarded recursion, with semidecidable evaluation to $\widehat{0}$.*

(i) $\mathsf{EC}(\mathbf{P};\mathbb{N}) \cong \mathsf{HEO}$ within $\mathcal{A}sm(K_1)$.

(ii) For each type σ there is an element $\Phi_\sigma \in \mathbf{P}((\mathtt{N}\to\mathtt{N})\to\sigma)$ such that whenever $g \in \mathbf{P}(\mathtt{N}\to\mathtt{N})$ represents an associate for $G \in \mathsf{HEO}(\sigma)$, we have that $\Phi_\sigma \cdot g$ is a realizer for G in $\mathsf{EC}(\mathbf{P};\mathbb{N})$.

(iii) Moreover, a suitable element Φ_σ is definable by an applicative expression in the following combinators:

$$k_{\rho\rho'}\ ,\ s_{\rho\rho'\rho''}\ ,\ \widehat{0}\ ,\ suc\ ,\ rec_\rho\ ,\ Z_\rho\ .$$

As in the full continuous case, the proof is essentially an elaboration of that of Theorem 8.6.1. The main additional task is to show that **P**, by virtue of being an effective TPCA, inherits enough 'continuity' from K_1 (thanks to the Kreisel–Lacombe–Shoenfield phenomenon) to support the required use of approximations. There are also some further subtleties, not present in the continuous case, that are designed to avoid the need for a type 1 normalizer.

Many of the variants and extensions mentioned in Subsection 8.6.2 are also applicable here, though again the situation in the effective case is typically more subtle (see [179] for details). Here we single out just one point for mention. Under slightly stronger hypotheses, an analogue of Theorem 9.2.15 for the modified extensional collapse construction can be obtained (Theorem 9.7 of [179]). What is of interest is that Gandy-style arguments using the second recursion theorem turn out to be essential here, in contrast to the proof of Theorem 9.2.15 itself which requires only the original Kreisel–Lacombe–Shoenfield method. This is, of course, entirely consistent with our experience in Subsection 9.2.2.

9.3 Some Negative Results

Theorem 9.2.15 shows that it is somewhat difficult to find a realizability-style construction of a type structure over $\mathbb{N}$, involving only effective data, that does not result in HEO. Even so, it is not the case that *every* 'effective' construction of a type structure yields this model. We discuss here two ways in which this can fail—the first rather trivial, the second involving a sophisticated counterexample within HEO itself.

Firstly, we note that one cannot obtain HEO by straightforwardly restricting the *reduced product* construction of Ct to effective sequences. As in Subsection 8.2.6, let S_n denote the full set-theoretic type structure over $\{0,\ldots,n\}$, and let $\mathsf{R} = \prod_{n<\omega} \mathsf{S}_n$. Let $\mathsf{R}^{\mathrm{eff}} \subseteq \mathsf{R}$ be the substructure consisting of all computable sequences $(z_0, z_1, \ldots) \in \prod_{n<\omega} \mathsf{S}_n(\sigma)$ (with respect to any reasonable enumeration of the $\mathsf{S}_n(\sigma)$), and let $\approx$ be the partial equivalence relation on $\mathsf{R}^{\mathrm{eff}}(\mathtt{N})$ that relates sequences converging to the same value in $\mathbb{N}$. The extensional collapse $\mathsf{EC}(\mathsf{R}^{\mathrm{eff}};\approx)$ then represents the natural effectivization of the construction of Ct in Subsection 8.2.6.

To see that even $\mathsf{EC}(\mathsf{R}^{\mathrm{eff}};\approx)(1)$ contains non-computable functions, let $C \subseteq \mathbb{N}$ be c.e. but not computable, and let e be a Kleene index for a partial computable function with domain C. For each n, define $z_n \in \mathsf{S}_n(1)$ by $z_n(i) = 0$ if $e \bullet_n i \downarrow$, and $z_n(i) = 1$ otherwise. Then the z_n clearly form a computable sequence $z \in \mathsf{R}^{\mathrm{eff}}(1)$, and z is a realizer for the characteristic function of C. Thus $\mathsf{EC}(\mathsf{R}^{\mathrm{eff}};\approx)(1) \not\cong \mathsf{HEO}(1)$.

Secondly, let us consider the model HK as the class of *hereditarily Kleene computable* functionals: take $\mathsf{HK}(\mathtt{N}) = \mathbb{N}$, and having defined $\mathsf{HK}(\sigma_0), \ldots, \mathsf{HK}(\sigma_{r-1})$, let $\mathsf{HK}(\sigma_0, \ldots, \sigma_{r-1} \to \mathtt{N})$ be the set of Kleene (S1–S9) computable total functionals

$\mathsf{HK}(\sigma_0) \times \cdots \times \mathsf{HK}(\sigma_{r-1}) \to \mathbb{N}$. This model has received very little attention to date; here we will satisfy ourselves with showing that $\mathsf{HK} \not\cong \mathsf{HEO}$.

It is easy to see that $\mathsf{HK}(\sigma) \cong \mathsf{HEO}(\sigma)$ when $\mathrm{lv}(\sigma) \leq 2$, and hence that $\mathsf{HK}(3) \subseteq \mathsf{HEO}(3)$. However, the following theorem shows that the latter inclusion is proper. This result was obtained by Gandy around 1965; a published proof appeared in Gandy and Hyland [101].

Theorem 9.3.1. *There is a functional $\Delta \in \mathsf{HEO}(3)$ that is not Kleene computable over* HEO.

Proof. We first set up some notation. For each finite sequence $a = a_0 \ldots a_{n-1} \in \mathbb{N}^*$, let $a^\star$ denote the sequence

$$0^{a_0}1 \cdots 10^{a_{n-1}}1 \, ,$$

and for each function $f : \mathbb{N} \to \mathbb{N}$, let $f^\star : \mathbb{N} \to \{0,1\}$ denote the function given by the sequence

$$0^{f(0)}10^{f(1)}1 \cdots .$$

Here 0^k denotes a sequence of k zeros, seen as a word. We also write $a^\star 0^\omega$ for the function given by the sequence

$$0^{a_0}1 \cdots 10^{a_{n-1}}1000 \cdots .$$

Note that if t is empty, then $t^\star$ is empty and $t^\star 0^\omega$ is the constant zero function.

Any finite binary sequence s can now be written uniquely in the form $a^\star 0^k$. Moreover, every function $\alpha : \mathbb{N} \to \{0,1\}$ is either of the form $f^\star$ for some $f : \mathbb{N} \to \mathbb{N}$ or of the form $a^\star 0^\omega$ for some $a \in \mathbb{N}^*$.

Let $T \subseteq \mathbb{N}^*$ be the tree of Theorem 9.1.19; recall that this has an infinite path but no arithmetical or even hyperarithmetical one. Now define $\Delta : \mathsf{HEO}(2) \to \mathbb{N}$ by

$$\Delta(F) \;=\; \max\,\{F(a^\star 0^\omega) \mid a \in T\} \, .$$

Claim 1: $\Delta(F)$ is well defined for every $F \in \mathsf{HEO}(2)$.

Proof of claim 1: Suppose for contradiction that $S = \{F(a^\star 0^\omega) \mid a \in T\}$ is unbounded. Then we may inductively construct an infinite sequence $z = z_0 z_1 \ldots \in \{0,1\}^{\mathbb{N}}$ such that $S_n = \{F(a^\star 0^\omega) \mid a \in T,\ z_0 \ldots z_{n-1} \sqsubseteq a^\star\}$ is unbounded for each n, and then select $a^0, a^1, \ldots \in T$ such that $z_0 \ldots z_{n-1} \sqsubseteq a^{n\star}$ (whence $z = \lim_{n\to\infty} a^{n\star}0^\omega$) and $F(a^{0\star}0^\omega) < F(a^{1\star}0^\omega) < \cdots$. Clearly this may be done in such a way that z is arithmetical. There are now two cases to consider:

1. $z = a^\star 0^\omega$ for some a. Then $z \in \mathrm{dom}\, F$, and by continuity of F at z we have $F(a^{n\star}0^\omega) = F(z)$ for all sufficiently large n, contradicting the assertion that $\{F(a^{n\star}0^\omega)\}_{n\in\mathbb{N}}$ is strictly increasing.
2. $z = f^\star$ for some f. Then for any initial segment a of f, we will have $a^\star \sqsubseteq a^{n\star}$ for sufficiently large n, whence $a \sqsubseteq a^n$ for such n, so $a \in T$. Thus f is an infinite path in T; but this is impossible since f is arithmetical.

Claim 2: $\Delta \in \mathsf{HEO}(3)$.

Proof of claim 2: Let $F \in \mathsf{HEO}(2)$, and let $\phi \in \mathsf{PC}^{\mathrm{eff}}(2)$ be a realizer for F considered as an element of $\mathsf{HECt}(2)$. For all $z \in \{0,1\}^{\mathbb{N}}$, we claim that at least one of the following will hold:

1. ϕ secures z: that is, $\phi(z) \in \mathbb{N}$ if we consider z as an element of $\mathsf{PC}(1)$.
2. There exists $a \notin T$ such that $a^\star$ is an initial segment of z.

To see this, recall that z is of the form $a^\star 0^\omega$ or $f^\star$. In the first case, h is computable so ϕ secures h. In the second case, either f has an initial segment $a \notin T$, in which case assertion 2 holds, or f is an infinite path in T. But in this latter case, ϕ must secure h, since otherwise we may readily construct an *arithmetical* $h' = f'^\star$ where f' is an infinite path in T, much as in the proof of claim 1. But then f' itself is arithmetical, a contradiction.

We now show how to compute $\Delta(F)$ from a K_1-realizer e for ϕ. Let U_F be the binary tree of all sequences s such that ϕ does not secure s and whenever $b^\star \sqsubseteq s$ we have $b \in T$. Then the above shows that U_F has no infinite paths, and hence is finite by König's lemma. Moreover, it is easy to see that a complete listing of the nodes of U_F may be computed uniformly in e. Writing V_F for the set of sequences sz where s is a leaf of U_F and $z \in \{0,1\}$, we may now compute $\Delta(F)$ as

$$\max\ (\{F(a^\star 0^\omega) \mid a \in T,\ a^\star \in U_F\} \cup \{\phi(s) \mid s \in V_F,\ \forall b.\ b^\star \sqsubseteq s \Rightarrow b \in T\})\ .$$

(Regarding the second of these sets, note that any s satisfying the given condition must be secured by ϕ.) To see that this indeed yields $\Delta(F)$, observe that for any $a \in T$, either $a^\star \in U_F$ or $F(a^\star 0^\omega) = \phi(a^\star 0^\omega) = \phi(s)$ for some s satisfying the comprehension condition of the second set. This establishes claim 2.

Claim 3: Δ is not Kleene computable.

Proof of claim 3: An application of the 'Tait argument' as discussed in Subsection 8.5.1. Suppose for contradiction that e is a Kleene index such that $\{e\}(F) = \Delta(F)$ for all $F \in \mathsf{Ct}_{\mathrm{eff}}(2)$. Let $F_0 = \Lambda g.0$; then $\{e\}(F_0) = 0$. Let f be an infinite path in T, and take $\phi \in \mathsf{PC}(2)$ so that $\phi(g) = 0$ if g is incompatible with $f^\star$, $\phi(g) = \bot$ otherwise, where g ranges over $\mathsf{PC}(1)$. Then ϕ secures all total computable functions, so as in the proof of Lemma 8.5.2 there will be a compact $p \sqsubseteq \phi$ securing the value of $\{e\}(F_0)$.

Since p is finite and does not secure $f^\star$, there must be an n such that p does not secure $\overline{f}(n)^\star 0^\omega$. There will thus be a computable extension ψ of p realizing a G such that $G(\overline{f}(n)^\star 0^\omega) \neq 0$, so that $\Delta(G) \neq \Delta(F_0)$. But $\{e\}(G) = \{e\}(F_0)$ by the choice of p, so e is not a Kleene index for Δ after all. □

The following extracts a little more mileage from this counterexample:

Exercise 9.3.2. (i) Show that $\mathsf{HEO}(1)$ and $\mathsf{HEO}(2)$ carry an evident *limit space* structure (in the sense of Subsection 8.2.3) such that application is continuous.

(ii) Taking T and Δ as in the proof of Theorem 9.3.1, let c^n be some node of depth n in T computable from n, and define $G_n \in \mathsf{HEO}(2)$ by $G_n(g) = 1$ if $c^{n\star}$ is an initial segment of g, and $G_n(g) = 0$ otherwise. Show that $G_n \to \Lambda g.0$ in $\mathsf{HEO}(2)$, but that $\Delta(G_n) \not\to \Delta(\Lambda g.0)$, so that Δ is not limit-continuous.

9.4 Logical Complexity

In this section we will see that much of the theory of logical complexity from Subsection 8.4.2 transfers by analogy to HEO. The main difference is that Ct captures the full complexity of the analytic hierarchy, whereas the analogous results for HEO concern only the arithmetic hierarchy, since the definition of each set $\mathsf{HEO}(k)$ can easily be formalized within first-order number theory. Only when we consider Kleene computations over HEO do higher complexity classes enter the picture.

We will focus particularly on the presentation of HEO as $\mathsf{EC}(K_1)$, and will use $\equiv_\sigma$ to denote the partial equivalence relation on $\mathbb{N}$ that yields $\mathsf{EC}(K_1)(\sigma)$. We first ascertain the logical complexity of these relations as manifested by their direct definition. Note that the relation $\equiv_0$ is simply equality on $\mathbb{N}$, whose complexity is as low as anyone could wish for.

Lemma 9.4.1. *If $k \geq 1$, then $\equiv_k$ is a Π^0_{k+1} relation.*

Proof. For $k = 1$ this is well known, but we include the proof here as part of an argument by induction on k. Let $T(e,n,y)$ be Kleene's T-predicate and U the associated output function, so that $T(e,n,y)$ means 'y codes a complete course of computation for $e \bullet n$' and $U(y)$ is the final result of the course of computation coded by y. For any $k \geq 1$, we then have

$$e \equiv_k e' \ \text{ iff } \ \forall n,n'.\,(n \equiv_{k-1} n' \Rightarrow \exists y,y'.\, T(e,n,y) \wedge T(e',n',y') \wedge U(y) = U(y'))\,.$$

When $k = 1$, $\equiv_{k-1}$ is simply the equality relation, so the formula for $\equiv_1$ is obviously equivalent to a Π^0_2 formula. For $k > 1$, the induction hypothesis says that $\equiv_{k-1}$ is a Π^0_k relation, and it is then easy to transform the formula for $\equiv_k$ into an equivalent Π^0_{k+1} formula as in Proposition 5.2.2. $\square$

In Theorem 8.4.13, we saw that the familiar representations of $\mathsf{Ct}(k)$ (for $k \geq 1$) have complexity Π^1_{k-1}. The above proposition can be seen as the effective analogue of this, in that replacing $\mathbb{N}^\mathbb{N}$ by $\mathbb{N}^\mathbb{N}_{\mathrm{eff}}$ in the definition of Π^1_{k-1} relations yields exactly the Π^0_{k+1} relations. We now show that the proof of the Kreisel representation theorem also relativizes to HEO, yielding a normal form for Π^0_{k+1} relations on $\mathbb{N}$.

We here define the *trace* h_n of a number n to be the constant function $\Lambda i.n$, and the trace h_f of a first-order function f to be f itself. For $F \in \mathsf{HEO}(k)$ where $k \geq 2$, we define $h_F(i) = F(\xi_\sigma(i))$ as usual, with ξ_σ as in Theorem 9.1.6.

Theorem 9.4.2. *Suppose $A \subseteq \mathbb{N}$ is Π^0_{k+1} where $k \geq 1$. Then there is a primitive recursive relation $R \subseteq \mathbb{N} \times \mathbb{N}^\mathbb{N} \times \mathbb{N}$ such that for all $a \in \mathbb{N}$ we have*

$$a \in A \quad \Leftrightarrow \quad \forall z \in \mathsf{HEO}(k-1).\, \exists n \in \mathbb{N}.\, R(a,h_z,n)\,.$$

Proof. For $k = 1$, this is essentially just the normal form for Π^0_2-statements, so assume that $k > 1$ where the lemma holds for $k-1$. Note that the lemma at type $k-1$ trivially extends to subsets of $\mathbb{N}^n$ for any n.

Let B be Σ^0_k such that $a \in A$ iff $\forall b \in \mathbb{N}.\,(a,b) \in B$. Then the complement of B is Π^0_k, so by the induction hypothesis, there is a primitive recursive relation S such that

$$(a,b) \notin B \;\Leftrightarrow\; \forall z \in \mathsf{HEO}(k-2).\ \exists n.\ S(a,b,h_z,n)\,.$$

By considering the function $F(z) = \mu n.\,S(a,b,h_z,n)$, we see that

$$(a,b) \in B \;\Leftrightarrow\; \forall F \in \mathsf{HEO}(k-1).\ \exists z \in \mathsf{HEO}(k-2).\ \neg S(a,b,h_z,F(z))\,.$$

By the continuity of S, this is equivalent to

$$(a,b) \in B \;\Leftrightarrow\; \forall F \in \mathsf{HEO}(k-1).\ \exists i \in \mathbb{N}.\ \neg S(a,b,h_{\xi(i)},F(\xi(i)))\,,$$

which again can be written in the form

$$(a,b) \in B \;\Leftrightarrow\; \forall F \in \mathsf{HEO}(k-1).\ \exists n \in \mathbb{N}.\ R'(a,b,h_F,n)$$

for some primitive recursive relation R'. Since $\mathbb{N} \times \mathsf{HEO}(k-1)$ is a computable retract of $\mathsf{HEO}(k-1)$, we may now write a definition of A of the required form. □

Exercise 9.4.3. Formulate a corresponding characterization of Σ^0_k relations by analogy with Corollary 8.4.12.

The above theorem has an obvious consequence:

Corollary 9.4.4. *The set $E_k = \{e \mid e \equiv_k e\}$ is Π^0_{k+1} complete for $k \geq 1$.*

Proof. Let A be any Π^0_{k+1} set. By Theorem 9.4.2, there is a primitive recursive relation R such that

$$a \in A \;\Leftrightarrow\; \forall z \in \mathsf{HEO}(k-1).\ \exists n.\ R(a,h_z,n)\,.$$

Now for any $a \in N$, the partial functional $F_a : \mathsf{HEO}(k-1) \rightharpoonup \mathbb{N}$ defined by $F_a(z) \simeq \mu n.\ R(a,h_z,n)$ clearly has a K_1 realizer r_a that is primitive recursive in a, and we now have $a \in A$ iff F_a is total iff $r_a \equiv_k r_a$. We have thus reduced the membership problem of A to that of E_k. □

This establishes the complexity of the canonical K_1 realization of HEO. However, the same idea can also be used to show that there can be no well-behaved K_1 realization of HEO with lower complexity than this, suggesting that one should regard Π^0_{k+1} as the 'intrinsic complexity' of $\mathsf{HEO}(k)$:

Exercise 9.4.5. Suppose $\gamma : \mathsf{HEO} \longrightarrow\!\!\triangleright K_1$ is any numeral-respecting simulation such that the (partial) interpretation of $\mathsf{Klex}^{\min}$ terms in HEO is computable with respect to γ. That is, for each type σ there is a partial computable operation ι_σ on closed $\mathsf{Klex}^{\min}$ terms of type σ such that for any closed $M : \sigma$ and $F \in \mathsf{HEO}(\sigma)$ we have

$$\iota_\sigma(M){\downarrow} \;\wedge\; \iota_\sigma(M) \Vdash^{\gamma}_{\sigma} F \quad \text{iff} \quad [\![M]\!]_{\mathsf{HEO}}{\downarrow} \;\wedge\; [\![M]\!]_{\mathsf{HEO}} = F\,.$$

Show that for $k \geq 1$, the set of γ-realizers for elements of $\mathsf{HEO}(k)$ has complexity at least Π^0_{k+1}.

In the case of Ct, we used the Kreisel representation theorem extensively to investigate the concept of (relative) Kleene computability. We do not have any parallel applications in the case of HEO, one reason being that here the complexity of termination for Kleene computations far exceeds the complexity of the model:

Theorem 9.4.6. *Suppose $F \in \mathsf{HEO}(k)$ for some $k \geq 2$. Then a set $A \subseteq \mathbb{N}$ is Π^1_1 iff A is Kleene semicomputable over* HEO *relative to F (that is, there is a Kleene index e such that $a \in A$ iff $\{e\}(F,a)\downarrow$).*

Proof. Suppose first that A is Kleene semicomputable in F. If $x_0,\dots,x_{r-1} \in \mathbb{N}$, let us write $V(e,\langle x_0,\dots,x_{r-1}\rangle,\langle k_0,\dots,k_{r-1}\rangle,n)$ to mean that each x_i is a K_1 realizer for some $X_i \in \mathsf{HEO}(k_i)$, where $k_i < k$, and $\{e\}(F,X_0,\dots,X_{r-1}) = n$. Mirroring the inductive definition of the relation $\{e\}(F,X_0,\dots,X_{r-1}) = n$ itself, we obtain an inductive definition of V, which we may use to ascertain its logical complexity.

The key case that gives rise to the complexity of V is the one for S8, and in particular for an application of the type k functional F:

$$\{e\}(F,\vec{X}) \;=\; F(\Lambda Y \in \mathsf{HEO}(k-2).\,\{d\}(F,Y,\vec{X}))\;.$$

Here we require that $\Lambda Y.\,\{d\}(F,Y,\vec{X})$ is total, and to express this at the level of realizers, we need a universal quantification over the domain of $\equiv_{k-2}$, which is Π^0_{k-1}. Thus, the set V is defined via a strictly positive Π^0_k inductive definition, and hence is a Π^1_1 set by Exercise 5.2.9.

We shall establish the converse first for the case $k = 2$. Given any computable tree $T \subseteq \mathbb{N}^*$ with index t, we may use the recursion theorem for S1–S9 (Theorem 5.1.5) to construct an index e_t, primitive recursive in t, such that for any $s \in \mathbb{N}^*$ we have

$$\{e_t\}(F,\langle s\rangle) \;\simeq\; \begin{cases} 0 & \text{if } s \notin T\,, \\ F(\Lambda x \in \mathbb{N}.\,\{e_t\}(F,\langle sx\rangle)) & \text{if } s \in T\,. \end{cases}$$

Writing ε for the empty sequence, we then have that $\{e_t\}(F,\langle\varepsilon\rangle)\downarrow$ iff T is well-founded. It follows that any Π^1_1 set arises as the domain of some function $\{e\}(F,-)$.

For $k > 2$, the same proof can be made to work by considering $\mathbb{N}$ as a computable retract of $\mathsf{HEO}(k-2)$. □

One can argue that for Ct, μ-computability is just as important as Kleene computability if we are interested in internal computability notions. For HEO it may be even more important, in view of the huge gap in complexity highlighted above. We conclude this section by showing that no such gap arises for μ-computability.

Theorem 9.4.7. *Suppose $F \in \mathsf{HEO}(k)$ for some $k \geq 2$. Then $A \subseteq \mathbb{N}$ is μ-semicomputable relative to F iff A is Π^0_k.*

Proof. We consider the case $k = 2$ first. Suppose A is Π^0_2; then there is a Σ^0_1 set B such that $a \in A$ iff $\forall b.\,(a,b) \in B$. But every such B is the domain of a classically computable, and hence μ-computable, partial function $g : \mathbb{N}^2 \rightharpoonup \mathbb{N}$, and now $a \in A$ iff $F(\Lambda b.g(a,b))\downarrow$, so A is μ-semicomputable in F.

For the converse, let $\widehat{F} \in \mathsf{PC}^{\mathrm{eff}}(2)$ be a realizer for F. Recall that $\mathsf{Klex}^{\mathrm{min}}$ is a formal language for μ-computations, and let $[\![-]\!]^l$ denote the liberal (or equivalently the intermediate) interpretation of $\mathsf{Klex}^{\mathrm{min}}$ in $\mathsf{PC}^{\mathrm{eff}}$ as described in Subsection 6.4.1. Note that this is more generous than the usual interpretation $[\![-]\!]$ of $\mathsf{Klex}^{\mathrm{min}}$ in HEO: e.g. we may have $\widehat{F}(\Lambda a.[\![M]\!]^l(\widehat{F},a)) \in \mathbb{N}$ even when $\Lambda a.[\![M]\!]^l(\widehat{F},a)$ is not total. Moreover, it is easy to see that the relation

$$[\![M]\!]^l(\widehat{F},a_0,\ldots,a_{r-1}) = m$$

is Σ^0_1 in $\lceil M \rceil, \vec{a}, m$ and a K_1-realizer for $\widehat{F}$, since all true statements of this kind are certified by a finite tree (cf. Subsection 8.1.4).

Recall that any $\mathsf{Klex}^{\mathrm{min}}$ term M has finite nesting depth. By induction on this depth, it is easy to show that the following are equivalent (where $a_0,\ldots,a_{r-1} \in \mathbb{N}$):

1. $[\![M]\!](F,a_0,\ldots,a_{r-1})\downarrow$.
2. For every subcomputation $[\![N]\!](F,\vec{b})$ that appears in the top-down computation tree for $[\![M]\!](F,\vec{a})$, we have $[\![N]\!]^l(\widehat{F},\vec{b}) \in \mathbb{N}$.

Since the appearance of a subcomputation $[\![N]\!](F,\vec{b})$ in the top-down tree for $[\![M]\!](F,\vec{a})$ may be witnessed by finitary information, we see that condition 2 is Π^0_2 uniformly in the data. This establishes our theorem for $k = 2$.

We now proceed to the case $k > 2$. Suppose $A \subseteq \mathbb{N}$ is Π^0_k. By Theorem 9.4.2, there is a primitive recursive relation R such that

$$a \in A \;\Leftrightarrow\; \forall z \in \mathsf{HEO}(k-2).\ \exists n.\ R(a,h_z,n)\ .$$

Define $\Phi(F,a) \simeq F(\Lambda z.\,\mu n.R(a,h_z,n))$; then $\Phi(F,a)\downarrow$ iff $a \in A$, so A is μ-semicomputable in F.

For the converse, we proceed much as for $k = 2$. Consider the relation

$$[\![M]\!]^l(\widehat{F},\widehat{X}_0,\ldots,\widehat{X}_{r-1}) \;=\; m\,,$$

where $\widehat{F} \in \mathsf{PC}^{\mathrm{eff}}(k)$ is a realizer for F and the $\widehat{X}_i$ are realizers for objects X_i of types $\leq k-2$. This relation is Σ^0_1 uniformly in $\lceil M \rceil, m$ and K_1-realizers for the $\widehat{X}_i$, whilst the set of K_1-realizers for objects of type $\leq k-2$ is itself Π^0_{k-1} by Corollary 9.4.4. The formula stating that $[\![M]\!](F,X_0,\ldots,X_{r-1})\downarrow$ will involve a universal quantification over this Π^0_{k-1} set, all other ingredients being of low complexity, so the domain of $[\![M]\!]$ is itself Π^0_k. $\square$

We leave our discussion of logical complexity at this point. There is scope for future investigation of whether there are other applications of the representation theorem to μ-computations over HEO, parallel to the applications in Chapter 8.

9.5 Other Relativizations of the Continuous Functionals

We began this chapter by pointing out, in effect, that the construction of Ct can be 'relativized' to yield the model HECt $\cong$ HEO, simply by replacing the set $\mathbb{N}^{\mathbb{N}}$ throughout the definition by the set $\mathbb{N}^{\mathbb{N}}_{\mathrm{eff}}$ of computable functions. For the construction via K_2 this is immediately clear, but it is also true for the construction of PC, suitably formulated: having defined the poset of compact elements, the elements of $\mathsf{PC}(\sigma)$ itself can be constructed as formal suprema of compacts enumerated by functions $\mathbb{N} \to \mathbb{N}$, and relativizing to $\mathbb{N}^{\mathbb{N}}_{\mathrm{eff}}$ then yields $\mathsf{PC}^{\mathrm{eff}}$.

In this section, we start with the simple observation that such relativized constructions of Ct make sense for *any* set $A \subseteq \mathbb{N}^{\mathbb{N}}$ with modest closure properties, yielding a model Ct^A. We then present some results showing how the contents of Ct^A can depend on what further closure properties A possesses. We also look briefly at one special case: the relativization of Ct to the set HYP of hyperarithmetical functions. Much of this territory is unexplored, and we do no more than offer some initial observations.

The preliminary stage of this development can be presented as an exercise:

Exercise 9.5.1. Suppose $A \subseteq \mathbb{N}^{\mathbb{N}}$ is *computationally closed*: that is, if $f_0, \ldots, f_{n-1} \in A$ and $g : \mathbb{N} \to \mathbb{N}$ is Turing computable in $f_0, \ldots, f_{n-1}$, then $g \in A$.

(i) Give the natural definitions of the relativized type structure Ct^A using relativizations of PC and K_2 respectively, and show that these are equivalent.

(ii) Show that Ct^A is Kleene closed.

Naturally, the contents of Ct^A will depend on the contents of A. The idea now is that we can investigate the conditions under which certain important functionals such as FAN and Γ are present in Ct^A. Interestingly, it turns out that these conditions have more to do with the closure properties of A than with its 'size'. In particular, we shall consider the effect of the following two closure conditions:

Definition 9.5.2. Let $A \subseteq \mathbb{N}^{\mathbb{N}}$ be computationally closed.

(i) *A satisfies the weak König's lemma* (WKL) if every infinite binary tree present in A (modulo standard coding) contains an infinite path in A.

(ii) A is *closed under jump* if for every $g \in A$, there is a function $g' \in A$ that codes the halting set for Turing computability relative to g.

These properties have a well-established significance in the arena of formal systems of second-order arithmetic (see Simpson [273]). The usual language for such systems is a two-sorted predicate logic involving both number variables and (type 1) function variables. By an *ω-model* for this language, we mean simply a set $A \subseteq \mathbb{N}^{\mathbb{N}}$ to be used as the range of the function variables, the number variables being taken to range over the standard set $\mathbb{N}$. If A is computationally closed, then A is already an ω-model for a weak second-order system known as $\mathbf{RCA}_0$, often used as a baseline for the study of second-order systems. Saying that such a set A satisfies WKL is tantamount to saying that it models the stronger system $\mathbf{WKL}_0$ obtained by adding a formalization of WKL to $\mathbf{RCA}_0$, while saying that A is closed under jump amounts to saying that it models the system $\mathbf{ACA}_0$ obtained by adding the *arithmetical comprehension* axiom. It is easy to show that if A is closed under jump then A satisfies WKL, mirroring the fact that WKL is provable in $\mathbf{ACA}_0$. However, one can construct sets A that satisfy WKL but are not closed under jump,[2] and this provides one way of seeing that arithmetical comprehension is not derivable in $\mathbf{WKL}_0$.

Let PC^A denote the relativization of PC to some computationally closed A, and let us say an element $\Phi \in \mathsf{PC}^A$ is *A-total* if it realizes some element of $\mathsf{EC}(\mathsf{PC}^A;\mathbb{N})$. Let us also write $\widehat{\mathsf{FAN}} \in \mathsf{PC}^{\mathrm{eff}}(3)$ for the standard realizer for $\mathsf{FAN} \in \mathsf{Ct}^{\mathrm{eff}}(3)$, as constructed in the proof of Theorem 8.3.4. From results already established in Subsections 8.3.2 and 9.1.2, we may now obtain the following:

Theorem 9.5.3. *Suppose A is computationally closed. Then* $\widehat{\mathsf{FAN}}$ *is A-total iff A satisfies WKL.*

Proof. Suppose $G \in \mathsf{Ct}^A((\mathtt{N} \rightarrow \mathtt{B}) \rightarrow \mathtt{N})$ where A satisfies WKL, and suppose $\widehat{G} \in \mathsf{PC}^A$ is a realizer for G—say the supremum of the compacts enumerated by $g \in A$. Let $\widehat{G}_0 \sqsubseteq \widehat{G}_1 \sqsubseteq \cdots$ be the evident sequence of compact approximations to $\widehat{G}$ arising from g, and define a binary tree T_G by

$$(a_0,\ldots,a_{n-1}) \in T_G \;\text{ iff }\; \widehat{G}_n(\vec{a}) \;=\; \bot$$

(viewing $\vec{a}$ as an element of $\mathsf{PC}(1)$ in the obvious way). Then T_G is present in A. It follows that T_G is finite, otherwise T_G would have an infinite path $f \in A$ and we would have $\widehat{G}(f) = \bot$. From the construction of $\widehat{\mathsf{FAN}}$, it is now easy to see that $\widehat{\mathsf{FAN}}(G) \in \mathbb{N}$.

Conversely, if A does not satisfy WKL, there is some infinite tree T in A with no infinite path in A. Define a functional G_T by

$$G_T(f) \;\simeq\; \mu n.\, (f(0),\ldots,f(n-1)) \notin T\,.$$

Clearly $\mathsf{Ct}^A(1) = A$, so $G_T(f)\downarrow$ for all $f \in \mathsf{Ct}^A(1)$; moreover, G_T has a realizer $\widehat{G}_T \in \mathsf{PC}^A(2)$, so $G_T \in \mathsf{Ct}^A(2)$. On the other hand, since T has infinite depth, it is easy to show that $\widehat{\mathsf{FAN}}(\widehat{G}_T) = \bot$; thus $\widehat{\mathsf{FAN}}$ is not A-total. □

This is a model-theoretic result, but the same argument also yields a proof-theoretic counterpart: relative to RCA_0, the totality of $\widehat{\mathsf{FAN}}$ is provably equivalent to WKL. This is an example of a result of *Reverse Mathematics*, a research programme that seeks to investigate the logical strength of mathematical theorems by showing what axioms they imply (see Simpson [273]).

A more sophisticated result in this vein is the following. Let $\Gamma \in \mathsf{Ct}(3)$ be the functional of Subsection 8.3.3. As in Exercise 9.1.16, we write $\widehat{\Gamma}$ for the element of $\mathsf{PC}^{\mathrm{eff}}(3)$ obtained by interpreting the defining equation $\Gamma(G) = G_0(\Lambda n.\Gamma(G_{n+1}))$ as a least fixed point definition.

[2] Such a set A can be constructed, for example, using functions of *low* c.e. degree (see the remark preceding Theorem 9.1.18). For this and other constructions of such sets, see Simpson [273].

Theorem 9.5.4. *Suppose A is computationally closed. Then $\widehat{\Gamma}$ is A-total iff A is closed under jump.*

Proof. Suppose A is closed under jump and $F \in \mathsf{PC}^A(2)$. We may construct a top-down computation tree for $\widehat{\Gamma}(F)$ by repeatedly expanding each node $\widehat{\Gamma}(F')$ to $F'_0(\widehat{\Gamma}(F'_1), \widehat{\Gamma}(F'_2), \ldots)$. It is easy to show that if $\widehat{\Gamma}(F) = \bot$ then one of the following must hold:

1. For some such node, we have $\widehat{\Gamma}(F'_i) \in \mathbb{N}$ for each i, so that $\Lambda i.\widehat{\Gamma}(F'_i) \in \mathsf{PC}^A(1)$ is A-total, but $F'_0(\Lambda i.\widehat{\Gamma}(F'_i)) = \bot$. Thus F'_0, and hence F itself, is not A-total.
2. There is an infinite sequence $a_0, a_1, \ldots$, computable in the jump of any realizer for F, such that $\widehat{\Gamma}(F_{a_0,\ldots,a_{n-1}}) = \bot$ for each n. We then see that $F(a_0, a_1, \ldots) = \bot$, and that $(a_0, a_1, \ldots) \in A$ since A is closed under jump. So again, F is not A-total.

We have thus shown that if F is A-total then $\widehat{\Gamma}(F) \in \mathbb{N}$. In other words, $\widehat{\Gamma}$ is total.

For the converse, suppose that $g \in A$ but no element of A solves the halting problem relative to g. We will show that $\widehat{\Gamma}$ is not A-total by an elaboration of the proof of Theorem 9.1.15, using the tree J of Theorem 9.1.18 (relativized to g) in place of the usual Kleene tree. Specifically, we shall construct $F \in \mathsf{PC}^A(2)$ such that $\widehat{\Gamma}(F) = \bot$, but $F(f) \in \mathbb{N}$ for all total arguments f apart from h, the unique path through J. Since h solves the halting problem relative to g, we have $h \notin A$, so $F(f) \in \mathbb{N}$ for all $f \in \mathsf{Ct}^A(1)$, establishing that $F \in \mathsf{Ct}^A(2)$. Since F is constructed so that $\Gamma(F) = \bot$, it will follow that $\widehat{\Gamma}$ is not A-total.

We here interpret the definition of the halting set H and its approximations H_n with reference to Turing computability relative to g. As in Theorem 9.1.18, we suppose $H_0 = \emptyset$, and define

$$h(n) \;=\; 1 \;\text{ if } n \notin H\,, \qquad h(n) \;=\; (\mu j.n \in H_j) + 1 \;\text{ otherwise},$$

and let J consist of all sequences $(a_0, \ldots, a_{n-1}) \in \mathbb{N}_1^*$ such that

$$\forall i < n.\ (a_i = 1 \Rightarrow i \notin H_n) \wedge (a_i > 1 \Rightarrow a_i = h(i))\,.$$

We now define $F \in \mathsf{PC}^A(2)$. To compute $F(f)$, search for the least n such that either $f(n) = 0$ or $f(n) > 0$ and $(f(0), \ldots, f(n-1)) \notin J$. In the second case, we let $F(f) = n$, while in the first case, we let

$$F(f) \;=\; \max\,\{f(n+1), f(n+2), \ldots, f(n+1+f(n+1))\}\,.$$

It is easy to see that if f is total and $f \neq h$ then $F(f)\downarrow$; hence $F \in \mathsf{Ct}^A(2)$ by the above discussion.

To show that $\widehat{\Gamma}(F) = \bot$, we will establish the following.

Claim: If $\vec{a} = (a_0, \ldots, a_{n-1}) \in \mathbb{N}_1^*$ and $\widehat{\Gamma}(F_{\vec{a}}) = m \in \mathbb{N}$, then there is no sequence in J of length m comparable with $\vec{a}$.

Proof of claim: By induction on the length of the evaluation of $\widehat{\Gamma}(F_{\vec{a}})$. (We present the proof here in an operational style that can be routinely translated into a strictly denotational argument.)

If the evaluation is immediate in that $F_{\vec{a}0}$ is constantly m, this is because $\vec{a}$ contains a prefix $(a_0,\ldots,a_{m-1}) \notin J$, so the claim holds. If the evaluation is not immediate, we have

$$\widehat{\Gamma}(F_{\vec{a}}) \;=\; F(\vec{a},0,\widehat{\Gamma}(F_{\vec{a}1}),\widehat{\Gamma}(F_{\vec{a}2}),\ldots,\widehat{\Gamma}(F_{\vec{a}(1+\widehat{\Gamma}(F_{\vec{a}1}))}),\ldots)\,,$$

and the search in the procedure for F terminates at the displayed value of 0. The computation therefore triggers subcomputations of $\widehat{\Gamma}(F_{\vec{a}1}),\ldots,\widehat{\Gamma}(F_{\vec{a}(1+\widehat{\Gamma}(F_{\vec{a}1}))})$, which must all be shorter than that of $\widehat{\Gamma}(F_{\vec{a}})$. In particular, we must evaluate $\widehat{\Gamma}(F_{\vec{a}1}) = k$.

Case 1: $n \notin H$. Then every sequence in J comparable with $\vec{a}$ is also comparable with $\vec{a}1$. By the induction hypothesis, there is no such sequence of length k, and since $m \geq k$, the conclusion follows.

Case 2: $n \in H$. By an easy reverse induction, we see that the value $m = \widehat{\Gamma}(F_{\vec{a}})$ must have arisen from some $F_{\vec{a}\vec{c}}$ whose evaluation is immediate, whence $m \geq n$. So suppose $\vec{a}\vec{b}$ is an extension of $\vec{a}$ of length m in J. By the argument from case 1, this cannot be an extension of $\vec{a}1$, so $b_0 > 1$, and by the definition of J we have $b_0 = (\mu j.n \in H_j)+1$, so $n \in H_{b_0}$. If $b_0 \geq k$, then the first k elements of $\vec{a}1b_1b_2\ldots$ form a sequence in J of length k comparable with $\vec{a}1$, contradicting the induction hypothesis for $\vec{a}1$. However, if $b_0 < k$, then $\widehat{\Gamma}(F_{\vec{a}b_0})$ is one of the values whose maximum yields m, and the prefix of $\vec{a}\vec{b}$ of length $\widehat{\Gamma}(F_{\vec{a}b_0})$ then refutes the induction hypothesis for $\vec{a}b_0$.

This establishes the claim. As a consequence, taking $\vec{a}$ to be empty, we see that $\widehat{\Gamma}(F)$ must be undefined, since J contains sequences of arbitrary length. □

For $k > 3$, the relationship between A and $\mathsf{Ct}^A(k)$ is virtually unexplored. The only thing we know here is the following, which can be proved using the evident relativization of the Kreisel representation theorem. We say a formula ϕ is *absolute* for $(A,\mathbb{N}^{\mathbb{N}})$ if for any valuation of its free variables in A, ϕ has the same truth-value whether the bound variables are construed as ranging over A or $\mathbb{N}^{\mathbb{N}}$.

Exercise 9.5.5. Let $A \subseteq \mathbb{N}^{\mathbb{N}}$ be computationally closed.

(i) Show that if, for all n, all Π^1_n formulae are absolute for $(A,\mathbb{N}^{\mathbb{N}})$, then Ct^A is isomorphic (over $\mathbb{N}$) to an applicative substructure of Ct.

(ii) Conversely, show that if Ct^A arises as a substructure of Ct, then for all n, all Π^1_n formulae are absolute for $(A,\mathbb{N}^{\mathbb{N}})$.

To conclude, we will look briefly at the case where A is the set HYP of hyperarithmetical functions. Since HYP satisfies both WKL and closure under jump, both FAN and Γ naturally live in $\mathsf{Ct}^{\mathsf{HYP}}$. However, the totality of a Kleene computable or even μ-computable operation can have a different meaning in $\mathsf{Ct}^{\mathsf{HYP}}$ and in Ct: if T is the tree of Theorem 9.1.19 and $F_T(f) \simeq \mu n.\bar{f}(n) \notin T$ for all f, then $F_T \in \mathsf{Ct}_{\mathsf{HYP}}(2)$, but F_T is not total in the real world.

A possible reason for interest in $\mathsf{Ct}^{\mathsf{HYP}}$ is that it helps us to see which properties of $\mathbb{N}^{\mathbb{N}}$ are true for quite general reasons, and which depend essentially on the completeness of $\mathbb{N}^{\mathbb{N}}$. As an example, we will outline how the limit space characterization of Ct can be adapted to HYP. At the end, we will observe that the limit space approach requires arithmetical comprehension in order to work.

For each pure type k, we shall inductively define the following data: a set $\mathsf{Ct}^{\mathsf{HYP},L}(k)$ of functions $\mathsf{Ct}^{\mathsf{HYP},L}(k-1) \to \mathbb{N}$ (or natural numbers when $k = 0$), a set of *hyperarithmetical sequences* in $\mathsf{Ct}^{\mathsf{HYP},L}$, a set of *convergent* such sequences, and an interpretation of $\xi_k(i)$ in $\mathsf{Ct}^{\mathsf{HYP},L}$ for each i (cf. Theorem 9.1.6).

For $k = 0$, we take $\mathsf{Ct}^{\mathsf{HYP},L}(0) = \mathbb{N}$; the hyperarithmetical sequences are just the hyperarithmetical sequences in the ordinary sense, and the convergent sequences are just the convergent sequences in the ordinary sense. In this case, all convergent sequences are almost constant, and in particular hyperarithmetical.

Now assume that all ingredients are defined for k. A function $F : \mathsf{Ct}^{\mathsf{HYP},L} \to \mathbb{N}$ is called hyperarithmetical if its trace $h_F^{k+1}(i) = F(\xi_k(i))$ is hyperarithmetical in the ordinary sense. We then take $F \in \mathsf{Ct}^{\mathsf{HYP},L}(k+1)$ iff F is hyperarithmetical and continuous with respect to all convergent hyperarithmetical sequences in $\mathsf{Ct}^{\mathsf{HYP},L}(k)$.

A sequence $\{F_i\}_{i\in\mathbb{N}}$ in $\mathsf{Ct}^{\mathsf{HYP},L}(k+1)$ is called hyperarithmetical if $\{h_{F_i}^{k+1}\}_{i\in\mathbb{N}}$ is hyperarithmetical in the usual sense. Such a sequence $\{F_i\}_{i\in\mathbb{N}}$ has F as a limit if $F(x) = \lim_{i\to\infty} F_i(x_i)$ whenever $x = \lim_{i\to\infty} x_i$ where $\{x_i\}_{i\in\mathbb{N}}$ is hyperarithmetical.

The definition of ξ_i^{k+1} is now given exactly as for $\mathsf{Ct}(k+1)$.

Exercise 9.5.6. (i) Show that ξ_i^k is hyperarithmetical for all k and i.

(ii) Show that if $\{x_i\}_{i\in\mathbb{N}}$ is a hyperarithmetical sequence with x as a pointwise limit, then x is hyperarithmetical.

(iii) Suppose $k > 0$ and $\{x_i\}_{i\in\mathbb{N}}$ is a hyperarithmetical sequence in $\mathsf{Ct}^{\mathsf{HYP},L}(k)$. Show that $\{x_i\}_{i\in\mathbb{N}}$ has a limit x in the sense of $\mathsf{Ct}^{\mathsf{HYP},L}(k)$ if and only if this sequence has a modulus function within $\mathsf{Ct}^{\mathsf{HYP},L}(k)$.

(iv) Show by induction on k that $\mathsf{Ct}^{\mathsf{HYP}}(k) = \mathsf{Ct}^{\mathsf{HYP},L}(k)$, using the characterization of $\mathsf{Ct}^{\mathsf{HYP}}$ as $\mathsf{EC}(\mathsf{PC}^{\mathsf{HYP}};\mathbb{N})$. As an additional part of the induction hypothesis, show that $x = \lim_{i\to\infty} x_i$ in the sense of $\mathsf{Ct}^{\mathsf{HYP},L}$ iff there is a hyperarithmetical sequence $y = \lim_{i\to\infty} y_i$ of realizers in PC.

Remark 9.5.7. Such an approach will not work in the case of HECt, because it is false that if a computable sequence of functions $\mathbb{N} \to \mathbb{N}$ converges pointwise to a computable function then there exists a computable modulus. As a counterexample, let h be as in the proof of Theorem 9.1.18, and for $n > 1$, define $f_n(m) = 1$ if $h(m) = n$, otherwise $f_n(m) = 0$. Then the f_n are computable and have $\Lambda m.0$ as their pointwise limit, but we can compute h itself from a modulus of convergence.

This shows that the non-constructivity in the proof of Theorem 8.2.16 is essential. In fact, the claim that every pointwise convergent sequence of functions has a modulus can be shown classically to imply arithmetical comprehension.

Chapter 10
The Partial Continuous Functionals

In this chapter we will take a closer look at the type structure PC of partial continuous functionals and its effective analogue $\mathsf{PC}^{\mathrm{eff}}$, as introduced in Subsection 3.2.2. These models have already played key roles in Chapters 8 and 9 in the construction of the total type structures Ct and HEO; they also featured in Chapter 7 as leading examples of models for Plotkin's PCF. Our main purpose in this chapter is to survey a range of different characterizations of PC and $\mathsf{PC}^{\mathrm{eff}}$, offering cumulative evidence that these are indeed natural mathematical objects. We also develop more systematically certain concepts that appeared in Chapter 8, such as the Scott topology.

We start with some purely mathematical descriptions independent of any programming language or dynamic model of computation. In Section 10.1, we begin with a purely order-theoretic definition of PC, then show that it may also be characterized topologically using the *compact-open* topology on function spaces, or alternatively in the category of *limit spaces* using its natural cartesian closed structure. This consolidates and extends the material of Subsection 3.2.2. In Section 10.2, we introduce Plotkin's domain $\mathbb{T}^{\omega}$, a universal object in the category of *coherent domains*, which yields an untyped model equivalent to PC, so that PC arises naturally within its Karoubi envelope.

In Section 10.3, we characterize $\mathsf{PC}^{\mathrm{eff}}$ in terms of programming languages. In particular, following Plotkin [236] and Sazonov [251], we consider extensions of PCF with a 'parallel-or' operator (as in Subsection 7.3.1) and an infinitary 'parallel exists' operator. The extension with both of these operators is a language with precisely the power of $\mathsf{PC}^{\mathrm{eff}}$. In Section 10.4, we give some further characterizations of a computational nature, showing that PC and $\mathsf{PC}^{\mathrm{eff}}$ arise from partial equivalence relations on (respectively) Kleene's second model K_2 and his first model K_1. The latter result can be seen as a higher-order generalization of the classical *Myhill–Shepherdson theorem* (Theorem 10.4.4).

In Section 10.5, we present a characterization due to Longo and Moggi [188], from which it follows that PC and $\mathsf{PC}^{\mathrm{eff}}$ are in some sense *maximal* among continuous and effective type structures respectively. Finally, in Section 10.6, we briefly touch on Scott's *graph model* $\mathscr{P}\omega$, another model of interest in its own right but with a close relationship to PC.

The partial continuous functionals are also well treated in other books which the reader may consult for further background and intuition: e.g. Amadio and Curien [6], Stoltenberg-Hansen, Lindström and Griffor [279], and Streicher [283]. For a survey of domain theory, we also recommend the paper of Abramsky and Jung [2].

10.1 Order-Theoretic and Topological Characterizations

Our initial definition of PC in Subsection 3.2.2 was topological in spirit, and was chosen to emphasize the intuitive idea of 'continuity with respect to finitary information'. From this, we then derived some facts about the order-theoretic structure of each $\mathsf{PC}(\sigma)$. Here, by contrast, we will start with a purely order-theoretic definition, and develop its theory to the point where its equivalence to the topological one becomes apparent. We then show how the topological description can itself be recast more systematically in terms of the compact-open topology, and also mention a quasi-topological characterization of PC in terms of limit spaces.

10.1.1 Definition and Order-Theoretic Structure

The fundamental order-theoretic notions we shall use are the following:

Definition 10.1.1. (i) In any poset $(X,\sqsubseteq)$, we say a subset $D\subseteq X$ is *directed* if it is non-empty and for all $x,y\in D$ there exists $z\in D$ with $x\sqsubseteq z$ and $y\sqsubseteq z$. We say $(X,\sqsubseteq)$ is a *directed-complete partial order*, or *DCPO*, if every directed $D\subseteq X$ has a least upper bound or *limit* within X, written as $\bigsqcup D$.

(ii) Suppose X,Y are DCPOs. A function $f:X\to Y$ is *monotone* if $f(x)\sqsubseteq_Y f(x')$ whenever $x\sqsubseteq_X x'$. A monotone function f is *continuous* if $f(\bigsqcup D)=\bigsqcup f(D)$ whenever $D\subseteq X$ is directed.

(iii) We write $X\Rightarrow Y$, or Y^X, for the set of monotone and continuous functions $X\to Y$ endowed with the pointwise ordering.

For the rest of this chapter, we will use 'continuous' to mean 'monotone and continuous' when speaking of DCPOs. It is an easy exercise to check that if X,Y are DCPOs then so is Y^X, and indeed that DCPOs and continuous functions form a cartesian closed category with exponentials given as above. We are therefore ready for our definition of PC. Here we shall not bother to include product types (by contrast with Subsection 3.2.2 where we specifically wanted a strongly cartesian closed model).

Definition 10.1.2. For each type $\sigma\in\mathsf{T}^{\to}(\mathtt{N})$, we define a DCPO $\mathsf{PC}(\sigma)$ as follows: $\mathsf{PC}(\mathtt{N})=\mathbb{N}_\bot$ (with the usual partial ordering), and $\mathsf{PC}(\sigma\to\tau)=\mathsf{PC}(\sigma)\Rightarrow\mathsf{PC}(\tau)$. Endowing these sets with the canonical application operations, we obtain the type structure PC.

The DCPOs $\mathsf{PC}(\sigma)$ enjoy some special properties that give us a good handle on their structure. A key role is played here by the notion of *compactness*. If $(X, \sqsubseteq)$ is any DCPO, we say $x \in X$ is compact if whenever $D \subseteq X$ is directed and $x \sqsubseteq \bigsqcup D$, we have $x \sqsubseteq y$ for some $y \in D$. We write X^{comp} for the set of compact elements of X.

Theorem 10.1.3. *Each DCPO* $\mathsf{PC}(\sigma)$ *is a* Scott domain. *That is to say,*

1. $\mathsf{PC}(\sigma)$ *is* bounded-complete*: every bounded subset* $D \subseteq \mathsf{PC}(\sigma)$ *has a least upper bound or* join *(again written* $\bigsqcup D$*),*
2. $\mathsf{PC}(\sigma)$ *is* algebraic*: for any* $x \in \mathsf{PC}(\sigma)$*, the set* K_x *of compact elements below* x *is directed and* $\bigsqcup K_x = x$.

Proof. It is trivial that $\mathbb{N}_\perp$ is a Scott domain. For function types, we shall prove more generally that if X, Y are Scott domains then so is Y^X (so that Scott domains form a full cartesian closed subcategory of the category of DCPOs).

To see that Y^X is bounded-complete, suppose $D \subseteq Y^X$ is bounded by g. Then for each $x \in X$, the set $D_x = \{f(x) \mid f \in D\}$ is bounded by $g(x)$, and so has a least upper bound since Y is bounded-complete. Set $h(x) = \bigsqcup D_x$ for each x. It is easy to check that h is monotone and continuous, and is thus the least upper bound of D in Y^X.

To show that Y^X is algebraic, we shall give an explicit description of its compact elements. As in Subsection 3.2.2, if $a \in X^{\mathrm{comp}}$ and $b \in Y^{\mathrm{comp}}$, we write $(a \Rightarrow b)$ for the *step function* $X \to Y$ given by

$$(a \Rightarrow b)(x) \;=\; \begin{cases} b & \text{if } a \sqsubseteq x\,, \\ \perp & \text{otherwise.} \end{cases}$$

It is easy to check that $(a \Rightarrow b)$ is continuous. Moreover, $(a \Rightarrow b)$ is itself compact in Y^X: if $(a \Rightarrow b) \sqsubseteq \bigsqcup D$ with D directed, then $b \sqsubseteq \bigsqcup D_a$, whence $b \sqsubseteq f(a)$ for some $f \in D$ and so $(a \Rightarrow b) \sqsubseteq f$.

Now let F be the set of all finite joins of such step functions $(a \Rightarrow b)$ that exist within Y^X. It is easy to show that each $c \in F$ is compact. Moreover, for any $f \in Y^X$, the set $F_f = \{c \in F \mid c \sqsubseteq f\}$ is clearly directed since Y^X is bounded-complete. To complete the proof that Y^X is algebraic, it will suffice to check that $f = \bigsqcup F_f$.

Clearly $\bigsqcup F_f \sqsubseteq f$. We shall show that $f(x) \sqsubseteq \bigsqcup F_f(x)$ for an arbitrary $x \in X$. Since Y is algebraic, it will suffice to show $b \sqsubseteq \bigsqcup F_f(x)$ for an arbitrary $b \in K_{f(x)}$. But since X is also algebraic, we have $b \sqsubseteq f(x) = \bigsqcup f(K_x)$, whence $b \sqsubseteq f(a)$ for some $a \in K_x$, so $(a \Rightarrow b) \in F_f$. We thus have $b \sqsubseteq \bigsqcup F_f(a) \sqsubseteq \bigsqcup F_f(x)$ as required. □

From the above proof, we may also read off the fact that the finite joins of step functions are the *only* compacts in Y^X: if f is compact, then since $f = \bigsqcup F_f$ we must have $f = g$ for some $g \in F_f$. Note too that as a special case of bounded-completeness, each $\mathsf{PC}(\sigma)$ has a least element $\perp_\sigma$.

We shall write $\mathcal{SD}om$ for the category of Scott domains and continuous functions. Historically, the $\mathsf{PC}(\sigma)$ were the first 'domains' to be considered: see Scott [256]. The crucial importance of the compact elements within Scott domains is indicated by the following standard domain-theoretic construction:

Exercise 10.1.4. In any poset $(X,\sqsubseteq)$, an *ideal* is a directed subset $I \subseteq X$ such that $x \sqsubseteq y \in I$ implies $x \in I$. Show that the set of ideals, ordered by inclusion, forms an algebraic DCPO whose compacts are the *prime ideals* $I_x = \{y \in X \mid y \sqsubseteq x\}$ for $x \in X$. (We call this DCPO the *ideal completion* of X.) Show too that any algebraic DCPO is isomorphic to the ideal completion of its compact elements.

We shall say that two elements x,y in a Scott domain X are *consistent*, and write $x \uparrow y$, if x,y have an upper bound (and hence a least upper bound) in X. We write $x \sharp y$ to mean x,y have no upper bound.

The above analysis suggests that a suitable formal language for denoting compact elements may be given by the grammar from Subsection 3.2.2:

$$e ::= \bot \mid n \mid (e_0 \Rightarrow e'_0) \sqcup \cdots \sqcup (e_{r-1} \Rightarrow e'_{r-1}) \, .$$

Of course, even a 'well-typed' expression e need not denote an element of PC, since the relevant joins may not exist. Nevertheless, it is easy to show, by simultaneous induction on the type, that the following two properties of formal expressions e,e' are (primitively recursively) decidable for each type σ.

1. e denotes an element of $\mathsf{PC}(\sigma)$.
2. e,e' denote consistent elements of $\mathsf{PC}(\sigma)$.

For type $\mathtt{N}$, both properties are trivially decidable. For the step case $\sigma \to \tau$, the decidability of property 1 comes from the following readily verified fact:

Proposition 10.1.5. *If $a_0,\ldots,a_{r-1} \in \mathsf{PC}(\sigma)^{\mathrm{comp}}$ and $b_0,\ldots,b_{r-1} \in \mathsf{PC}(\tau)^{\mathrm{comp}}$, then*

$$(a_0 \Rightarrow b_0) \sqcup \cdots \sqcup (a_{r-1} \Rightarrow b_{r-1})$$

exists in Y^X iff for all $i,j<r$, either $a_i \sharp a_j$ or $b_i \uparrow b_j$. □

The decidability of property 2 follows, since $c_0 \sqcup \cdots \sqcup c_{r-1}$ and $c'_0 \sqcup \cdots \sqcup c'_{r'-1}$ denote consistent elements iff $c_0 \sqcup \cdots \sqcup c_{r-1} \sqcup c'_0 \sqcup \cdots \sqcup c'_{r'-1}$ denotes an element.

Recalling that every element in PC is the supremum of the finite elements below it, we may now call an element $x \in \mathsf{PC}(\sigma)$ *effective* if the set K_x is computably enumerable (that is, if there is a c.e. set of expressions e whose denotations yield precisely K_x). We write $\mathsf{PC}^{\mathrm{eff}}(\sigma)$ for the set of effective elements in $\mathsf{PC}(\sigma)$. The following exercise shows that these sets constitute a substructure $\mathsf{PC}^{\mathrm{eff}}$:

Exercise 10.1.6. (i) If $\sigma = \sigma_0,\ldots,\sigma_{t-1} \to \mathtt{N}$, show that any compact in $\mathsf{PC}(\sigma)$ is a least upper bound of compacts of the form $(a_0 \Rightarrow (\cdots(a_{t-1} \Rightarrow n)\cdots))$.

(ii) Give an explicit characterization of the order relation on compact elements of type $\sigma = \sigma_0,\ldots,\sigma_{t-1} \to \mathtt{N}$ in terms of the order relations on compacts at each σ_i. Deduce that at the level of formal expressions, the order and equality relations on compacts are decidable, and hence application of compacts is computable.

(iii) Show that if $f \in \mathsf{PC}(\sigma \to \tau)$ and $x \in \mathsf{PC}(\sigma)$ then

$$f(x) \;=\; \bigsqcup \{c(a) \mid c \in K_f, a \in K_x\} \, .$$

Using (ii), deduce that if both f and x are effective then so is $f(x)$.

10.1.2 Topological and Limit Characterizations

We now turn to consider PC from a topological perspective. The most frequently used topology on the sets $\mathsf{PC}(\sigma)$ is the *Scott topology*—this may be motivated by the idea that open subsets of $\mathsf{PC}(\sigma)$ correspond to 'semidecidable properties'. To formalize this, let Σ be the DCPO $\{\bot, \top\}$ with $\bot \sqsubseteq \top$; we shall regard continuous maps $X \to \Sigma$ as semidecidable properties of elements $x \in X$.

Definition 10.1.7. The *Scott topology* on a DCPO X consists of all sets $h^{-1}(\top)$ where $h : X \to \Sigma$ is continuous in the sense of Definition 10.1.1.

It is easy to see that this is indeed a topology: open sets are closed under unions because the pointwise maximum of any family of continuous functions $X \to \Sigma$ is itself continuous, and under binary intersections because the pointwise minimum of two such functions is continuous. If X is a Scott domain, the Scott topology on X is T_0 (i.e. if x, y belong to exactly the same open sets then $x = y$), but it is not Hausdorff (indeed, the only open set containing $\bot$ is the whole of X). This means that compact sets do not need to be closed.

It is easy to see that $U \subseteq X$ is Scott-open iff U is both upward-closed ($y \sqsupseteq x \in U$ implies $y \in U$) and inaccessible via directed suprema (if D is directed and $\bigsqcup D \in U$ then U contains some $x \in D$). In particular, $V_x = \{y \in X \mid y \not\sqsubseteq x\}$ is Scott-open for any $x \in X$.

We may now correlate order-theoretic concepts with topological ones. If $x \in X$, we write U_x for its *upper set* $\{y \in X \mid x \sqsubseteq y\}$. The following facts are easy exercises.

Proposition 10.1.8. *Let X, Y be algebraic DCPOs.*

(i) A function $f : X \to Y$ is continuous in the sense of Definition 10.1.1 iff it is continuous with respect to the Scott topologies on X, Y.

(ii) A set U_x is Scott-open iff x is (order-theoretically) compact.

(iii) The sets U_a for $a \in X^{\mathrm{comp}}$ form a basis *for the Scott topology, i.e. any Scott-open U is a union of such sets.*

(iv) Every set U_x is compact in the topological sense. □

We are now in a position to see that our present definition of PC agrees with the one in Subsection 3.2.2: the sets $\mathscr{F}_\sigma$ are precisely the bases $\mathscr{B}(\mathsf{PC}(\sigma))$, and so a function $\mathsf{PC}(\sigma) \to \mathsf{PC}(\tau)$ is continuous in the sense of Subsection 3.2.2 iff it is Scott continuous. However, the definition of $\mathscr{F}_{\sigma\to\tau}$ given in Subsection 3.2.2 was slightly ad hoc. We can now improve on this, and so obtain a characterization of PC using only elementary concepts from topology.

Recall that if X, Y are topological spaces and $\mathrm{Ct}(X,Y)$ is the set of continuous functions $X \to Y$, the *compact-open* topology on $\mathrm{Ct}(X,Y)$ is the one generated by the sub-basis consisting of sets

$$B_{C,O} = \{f \in \mathrm{Ct}(X,Y) \mid f(C) \subseteq O\} \qquad (C \subseteq X \text{ compact}, O \subseteq Y \text{ open}) .$$

If $\mathsf{PC}(\sigma)$ and $\mathsf{PC}(\tau)$ are equipped with the Scott topology, we know already from Proposition 10.1.8(i) that $\mathrm{Ct}(\mathsf{PC}(\sigma), \mathsf{PC}(\tau))$ is the set $\mathsf{PC}(\sigma \to \tau)$.

Theorem 10.1.9. *Let* $X = \mathsf{PC}(\sigma)$ *and* $Y = \mathsf{PC}(\tau)$ *be endowed with the Scott topology. Then the compact-open topology on* $\mathrm{Ct}(\mathsf{PC}(\sigma), \mathsf{PC}(\tau))$ *coincides with the Scott topology on* $\mathsf{PC}(\sigma \to \tau)$.

Proof. If $a \in X^{\mathrm{comp}}$ and $b \in Y^{\mathrm{comp}}$, we have that $U_{a \Rightarrow b} = B_{U_a, U_b}$ where U_a, U_b are both compact and open. Since every Scott-open $U \in \mathsf{PC}(\sigma \to \tau)$ is a union of finite intersections of sets $U_{a \Rightarrow b}$, every Scott-open set is in the compact-open topology.

Conversely, suppose $f \in B_{C,O}$ where $C \subseteq X$ is compact and $O \subseteq Y$ is open. For every $x \in C$ there is an order-theoretically compact element $a \sqsubseteq x$ such that $f(a) \in O$, and hence a compact element $b \sqsubseteq f(a)$ with $b \in O$. Since C is topologically compact, there will be a finite set of such pairs (a_i, b_i) with $C \subseteq \bigcup_i U_{a_i}$. But now the basic Scott-open neighbourhood $U_{\bigsqcup_i (a_i \Rightarrow b_i)}$ contains f and is contained in $B_{C,O}$. Thus every set in the compact-open topology is Scott-open. $\square$

It follows immediately that PC may be constructed from the topological space $\mathbb{N}_\perp$ by interpreting arrow types as compact-open function spaces.

Although the category of topological spaces is not cartesian closed, the compact-open topology is a natural one to consider for certain kinds of spaces. In typical well-behaved cartesian closed categories of topological spaces such as **QCB** (see [89]), exponentials coincide with compact-open function spaces for the finite types over $\mathbb{N}_\perp$, though not for all objects in the category. For instance, we saw in Subsection 8.2.4 that for the finite types over $\mathbb{N}$, one must use the topology *compactly generated from* the compact-open one to obtain the appropriate function spaces.

We conclude this section with two points of peripheral interest. First, we offer a characterization of PC using the cartesian closed category of *limit spaces* (this was also used in Chapter 8 to characterize Ct). Whilst the result is of minor importance, it provides a useful exercise in the manipulation of higher type functionals. We refer to Subsection 8.2.3 for the definition of limit spaces and exponentials thereof.

Exercise 10.1.10. We regard $\mathbb{N}_\perp$ as a limit space in which all sequences have $\perp$ as a limit, while only sequences that are almost constantly a will have $a \in \mathbb{N}$ as a limit. Now define limit spaces $L_\perp(\sigma)$ by $L_\perp(\mathtt{N}) = \mathbb{N}_\perp$ and $L_\perp(\sigma \to \tau) = L_\perp(\sigma) \Rightarrow L_\perp(\tau)$.

By induction on the type σ, prove that

1. $L_\perp(\sigma) = \mathsf{PC}(\sigma)$ as sets,
2. the limit structure on $L_\perp(\sigma)$ agrees with that induced by the Scott topology,
3. the topology induced by this limit structure on $\mathsf{PC}(\sigma)$ is just the Scott topology.

Secondly, we touch briefly on an alternative interpretation for the base type $\mathtt{N}$, namely the domain of so-called *lazy natural numbers*. Let $\mathbb{N}^\infty$ be the poset consisting of the set $(\{0,1\} \times \mathbb{N}) \cup \{\infty\}$, with the partial order $\sqsubseteq$ generated by the following:

- $(0,m) \sqsubseteq (i,n)$ if $m \leq n$.
- $(0,n) \sqsubseteq \infty$ for all n.

Note that $(1,m) \sqsubseteq (i,n)$ iff $(i,n) = (1,m)$. It is easy to see that $\mathbb{N}^\infty$ is a Scott domain, with all elements except ∞ being compact. The idea is that this domain allows for partial information regarding the value of a number: we think of $(0,n)$ as representing the information 'at least n', while $(1,n)$ represents 'exactly n'. The element ∞ will typically represent a process that produces partial information forever without arriving at a definite number n. (Note that the elements $(0,n)$ will not be representable by terms in PCF, for example.)

Let us now write PC^∞ for the type structure over $\mathbb{N}^\infty$ given by exponentiation in $\mathscr{SD}om$. The following exercise investigates aspects of this structure:

Exercise 10.1.11. (i) Show that for each σ, $\mathsf{PC}(\sigma)$ is a retract of $\mathsf{PC}^\infty(\sigma)$ in $\mathscr{SD}om$.

(ii) (More substantial.) Discuss the truth or otherwise of the following claims:

1. We may construct PC^∞ using the compact-open topology on function spaces.
2. We may construct PC^∞ by exponentiation in the category of limit spaces.

10.2 A Universal Domain

In this section we present a result of Plotkin [237] to the effect that a certain domain $\mathbb{T}^\omega$ is a reflexive object, and indeed is universal for an important subcategory of $\mathscr{SD}om$ (see Subsection 4.2.5). Using the general machinery of Chapter 4, we deduce that the model PC arises naturally within the Karoubi envelope of $\mathbb{T}^\omega$ (i.e. its category of retracts)—thus, the whole structure of PC at all type levels is in some sense already present within the set of continuous endofunctions on $\mathbb{T}^\omega$.

The definition of $\mathbb{T}^\omega$ is very simple:

Definition 10.2.1. Let $\mathbb{T}$ be the poset $\{tt, ff, \bot\}$, with the partial order generated by $\bot \sqsubseteq tt$, $\bot \sqsubseteq ff$. Let $\mathbb{T}^\omega$ be the poset of functions $\mathbb{N} \to \mathbb{T}$, ordered pointwise.

Intuitively, we may regard elements of $\mathbb{T}^\omega$ as *partial predicates* on $\mathbb{N}$ giving both positive and negative information, although we shall treat them formally as functions. It is easy to see that $\mathbb{T}^\omega$ is a Scott domain: the compact elements $z \in \mathbb{T}^\omega$ are those with $z(n) = \bot$ for all but finitely many n. Extending our earlier notation slightly, we shall write $(m \Rightarrow i)$ for the element sending $m \in \mathbb{N}$ to $i \in \mathbb{B} = \{tt, ff\}$ and everything else to $\bot$.

A Scott domain is *countably based* if it has only countably many compacts. We will consider countably based Scott domains with the following important property:

Definition 10.2.2. A Scott domain X is *coherent* if every set $D \subseteq X$ that is pairwise consistent (i.e. such that $x \uparrow y$ for any x,y) is also globally consistent (i.e. D has an upper bound in X, and hence a least upper bound).

It is not hard to check directly that each $\mathsf{PC}(\sigma)$ is coherent. However, all we shall formally require at this stage is the following:

Proposition 10.2.3. $\mathbb{N}_\bot$, $\mathbb{T}^\omega$ *and* $\mathbb{T}^\omega \Rightarrow \mathbb{T}^\omega$ *are coherent.*

Proof. We show this for $\mathbb{T}^\omega \Rightarrow \mathbb{T}^\omega$, the case of $\mathbb{T}^\omega$ being simpler and that of $\mathbb{N}_\bot$ being trivial. Suppose $D \subseteq \mathbb{T}^\omega \Rightarrow \mathbb{T}^\omega$ is pairwise consistent. Define $f : \mathbb{T}^\omega \Rightarrow \mathbb{T}^\omega$

by setting $f(z)(n) = i \in \mathbb{B}$ iff there exists $d \in D$ with $d(z)(n) = i$. This is a sound definition since if $d(z)(n) = tt$ and $d'(z)(n) = \mathit{ff}$ then $d \sharp d'$. Clearly f is continuous and is an upper bound for D. □

Proposition 10.2.4. *Any retract of $\mathbb{T}^\omega$ in $\mathcal{SDom}$ is coherent and countably based.*

Proof. Suppose $(t,r) : X \lhd \mathbb{T}^\omega$ is a retraction pair in $\mathcal{SDom}$. For coherence, note that if $D \subseteq X$ is pairwise consistent then so is $t(D)$; hence $t(D)$ has an upper bound z, and $r(z)$ is then an upper bound for D. For a countable basis, we show that every $a \in X^{\mathrm{comp}}$ arises as $r(b)$ for some $b \in \mathbb{T}^{\omega\mathrm{comp}}$. Given a, let D be the set of compacts $c \sqsubseteq t(a)$ in $\mathbb{T}^\omega$; then $t(a) = \bigsqcup D$ so $a = \bigsqcup r(D)$, whence there exists $r(b) \in r(D)$ with $r(b) \sqsubseteq a$. But also $r(b) \sqsubseteq a$, so $a = r(b)$. □

We now give Plotkin's proof that, conversely, every countably based coherent Scott domain X is a retract of $\mathbb{T}^\omega$ in $\mathcal{SDom}$.

Theorem 10.2.5. *If X is a countably based coherent Scott domain, there exist $t : X \to \mathbb{T}^\omega$ and $r : \mathbb{T}^\omega \to X$ continuous with $r \circ t = id_X$.*

Proof. Let $e_0, e_1, \ldots$ be an enumeration of some set of compacts in X sufficient to generate the whole of X under least upper bounds. For instance, in the case of $\mathbb{T}^\omega \Rightarrow \mathbb{T}^\omega$, we may suppose $e_n = (p_n \Rightarrow (m_n \Rightarrow i_n))$ where $p_n \in \mathbb{T}^{\omega\mathrm{comp}}$, $m_n \in \mathbb{N}$ and $i_n \in \mathbb{B}$. The reader wishing to keep an eye on this special case should note that for $x \in \mathbb{T}^\omega \Rightarrow \mathbb{T}^\omega$ we have

$$\begin{aligned} e_n \sqsubseteq x \quad &\text{iff} \quad x(p_n)(m_n) = i_n\,, \\ e_n \sharp x \quad &\text{iff} \quad \exists p \sqsupseteq p_n.\, x(p)(m_n) = \neg i_n\,, \\ e_k \sharp e_n \quad &\text{iff} \quad p_k \uparrow p_n \wedge m_k = m_n \wedge i_k = \neg i_n\,. \end{aligned}$$

In general, given any $x \in X$, define $t(x) \in \mathbb{T}^\omega$ by

$$t(x)(n) = \begin{cases} tt & \text{iff } e_n \sqsubseteq x\,, \\ \mathit{ff} & \text{iff } e_n \sharp x\,, \\ \bot & \text{otherwise.} \end{cases}$$

It is easy to see that the resulting function $t : X \to \mathbb{T}^\omega$ is monotone. For continuity, suppose $D \subseteq X$ is directed. If $t(\bigsqcup D)(n) = tt$, the compactness of e_n yields that $t(x)(n) = tt$ for some $x \in D$. On the other hand, if $t(\bigsqcup D)(n) = \mathit{ff}$, we must have $e_n \sharp x$ for some $x \in D$, otherwise $D \cup \{e_n\}$ is pairwise consistent so that e_n and $\bigsqcup D$ have an upper bound. Thus t is continuous.

In the other direction, we define $r : \mathbb{T}^\omega \to X$ by

$$r(z) = \bigsqcup E_z\,, \text{ where } E_z = \{e_n \mid z(n) = tt \wedge \forall k < n.\, e_k \sharp e_n \Rightarrow z(k) = \mathit{ff}\}\,.$$

Here E_z is pairwise consistent by construction, so that the required least upper bound exists. Again, it is clear that r is monotone. For continuity, if D is directed and $e_n \in E_{\bigsqcup D}$, it is easy to see that $e_n \in E_z$ for some $z \in D$, so that $E_{\bigsqcup D} = \bigcup_{z \in D} E_z$, whence $r(\bigsqcup D) = \bigsqcup_{z \in D} r(z)$.

It remains to show that $r \circ t = id_X$. Suppose $x \in X$. Whenever $e_n \sqsubseteq x$, we have $t(x)(n) = tt$, whereas if $k < n$ with $e_k \sharp e_n$ then $e_k \sharp x$ so $t(x)(k) = ff$. Thus $e_n \in E_{t(x)}$. Since x is the supremum of such e_n, this shows that $x \sqsubseteq r(t(x))$. Conversely, if $e \in E_{t(x)}$, then $e = e_n$ for some n where $t(x)(n) = tt$, and this requires $e_n \sqsubseteq x$. Since $r(t(x))$ is the supremum of such e, this shows that $r(t(x)) \sqsubseteq x$. Since x was arbitrary, we have $r \circ t = id$ as required. $\square$

In particular, we have $(\mathbb{T}^\omega \Rightarrow \mathbb{T}^\omega) \lhd \mathbb{T}^\omega$. We may therefore regard $\mathbb{T}^\omega$ as an untyped λ-algebra as in Theorem 4.1.29 (using the concretion $\mathrm{Hom}(1,-)$).

We shall write $\mathcal{CSD}om$ for the category of countably based coherent Scott domains, or equivalently retracts of $\mathbb{T}^\omega$. By standard abstract reasoning, this is a cartesian closed subcategory of $\mathcal{SD}om$: if $X, Y \lhd \mathbb{T}^\omega$ then $X \Rightarrow Y \lhd (\mathbb{T}^\omega \Rightarrow \mathbb{T}^\omega) \lhd \mathbb{T}^\omega$ (see Proposition 4.2.1). What is more, since $\mathbb{T}^\omega$ is a universal object in $\mathcal{CSD}om$, by Proposition 4.2.21 we have an inclusion $\mathcal{CSD}om \hookrightarrow \mathcal{K}(\mathbb{T}^\omega)$ that preserves products and exponentials. Since also $\mathbb{N}_\bot \lhd \mathbb{T}^\omega$, all the $\mathsf{PC}(\sigma)$ are retracts of $\mathbb{T}^\omega$, so that PC may be obtained from $\mathbb{N}_\bot$ by exponentiation in $\mathbf{K}(\mathbb{T}^\omega)$.

Remark 10.2.6. Although $\mathbb{T}^\omega$ is not itself one of the $\mathsf{PC}(\sigma)$, it is easy to see that $\mathbb{T}^\omega \lhd \mathsf{PC}(1)$. Thus $\mathbb{T}^\omega$ appears in the Karoubi envelope of PC (considered as a λ-algebra), and it follows that $\mathbb{T}^\omega$ and PC are equivalent λ-algebras in the sense of Subsection 4.1.4.

Exercise 10.2.7. Show how the type structure PC^∞ over $\mathbb{N}^\infty$ (see Subsection 10.1.2) may likewise be constructed within $\mathbf{K}(\mathbb{T}^\omega)$.

Exercise 10.2.8. By slightly extending the definition of effective elements given in Subsection 10.1.1, show that the computable elements of $\mathbb{T}^\omega$ form a sub-λ-algebra $\mathbb{T}^{\omega\mathrm{eff}}$, and that $\mathsf{PC}^{\mathrm{eff}}$ arises as the type structure over $\mathbb{N}_\bot$ within $\mathbf{K}(\mathbb{T}^{\omega\mathrm{eff}})$.

Having constructed the model PC by the above route, its ordering, compact elements and Scott topology may in principle be recovered from its bare applicative structure. It is more satisfying, however, to see that these may be inherited directly from $\mathbb{T}^\omega$, as the following result shows. If f is an idempotent on $\mathbb{T}^\omega$, we write $Fix(f)$ for the set $\{x \in \mathbb{T}^\omega \mid f(x) = x\}$ with the partial order inherited from $\mathbb{T}^\omega$.

Theorem 10.2.9. *Suppose $(t,r) : X \to \mathbb{T}^\omega$ is a retraction pair in $\mathcal{SD}om$, and let $f = t \circ r$. Then:*

(i) $Fix(f)$ and X are isomorphic as posets.

(ii) If $D \subseteq Fix(f)$ is directed, its supremum in $Fix(f)$ coincides with that in $\mathbb{T}^\omega$.

(iii) An element $a \in Fix(f)$ is compact within $Fix(f)$ iff there is a compact b in $\mathbb{T}^\omega$ with $b \sqsubseteq a$ and $f(b) = a$.

(iv) A set $V \subseteq Fix(f)$ is Scott-open in $Fix(f)$ iff there exists U Scott-open in $\mathbb{T}^\omega$ such that $V = U \cap Fix(f)$.

Proof. (i) Trivial, since t and $r \upharpoonright_{Fix(f)}$ establish a bijection $X \cong Fix(f)$ and both preserve the ordering.

(ii) Again trivial, since t is continuous. (Note, however, that t need not preserve all joins that exist in X.)

(iii) The left-to-right implication is given by the proof of Proposition 10.2.4. For the converse, suppose $b \in \mathbb{T}^{\omega\mathrm{comp}}$ with $b \sqsubseteq a = f(b)$, and suppose $D \subseteq Fix(f)$ is directed with $a \sqsubseteq \bigsqcup D$. Since b is compact, we have $b \sqsubseteq x$ for some $x \in D$, whence $a \sqsubseteq f(b) \sqsubseteq f(x) = x$. Thus a is compact in $Fix(f)$.

(iv) Let us here write U_x for the upper set of x within $\mathbb{T}^{\omega}$, and V_y for the upper set of y within $Fix(f)$, so that $V_y = U_y \cap Fix(f)$. Suppose $V \subseteq Fix(f)$ is Scott-open in $Fix(f)$. Given any $y \in Fix(f)$, take $a \sqsubseteq y$ compact in $Fix(f)$ so that $y \in V_a \subseteq V$, and using (iii) above, take b compact in $\mathbb{T}^{\omega}$ with $b \sqsubseteq a = f(b)$. Now if $b \sqsubseteq x \in Fix(f)$ then $a = f(b) \sqsubseteq f(x) = x$, so U_b is Scott-open in $\mathbb{T}^{\omega}$ with $U_b \cap Fix(f) = V_a$, which contains y and is contained within V. Since y was arbitrary, V is a union of such sets $U_b \cap Fix(f)$, so that $V = U \cap Fix(f)$ for some U Scott-open in $\mathbb{T}^{\omega}$.

Conversely, let U be Scott-open in $\mathbb{T}^{\omega}$ and suppose $y \in U \cap Fix(f)$. Let D be a directed set of compacts in $Fix(f)$ such that $y = \bigsqcup D$ (in $Fix(f)$, and hence in $\mathbb{T}^{\omega}$). Since U is open in $\mathbb{T}^{\omega}$, we may take $a \in D$ such that $a \in U$; but now $V_a \sqsubseteq U \cap Fix(f)$ with $y \in V_a$. Since y was arbitrary, $U \cap Fix(f)$ is a union of such sets V_a, and so is Scott-open in $Fix(f)$. □

We thus see that the construction via $\mathbf{K}(\mathbb{T}^{\omega})$ naturally yields the sets $\mathsf{PC}(\sigma)$ together with their Scott topology.

Remark 10.2.10. We have shown that a (countably based) Scott domain is a retract of $\mathbb{T}^{\omega}$ iff it is coherent; thus the category of coherent Scott domains embeds fully in the Karoubi envelope of $\mathbb{T}^{\omega}$. However, the model $\mathbf{K}(\mathbb{T}^{\omega})$ turns out to be much wider than this—in fact, it corresponds to the category of *coherent* ω*-continuous domains* (see [237, Theorem 11]). On the other hand, we may precisely characterize the idempotents f that arise from embeddings of Scott domains in $\mathbb{T}^{\omega}$: they are those such that

$$f(x) \;=\; \bigsqcup \{f(b) \mid b \in \mathbb{T}^{\omega\mathrm{comp}}, b \sqsubseteq f(x), b \sqsubseteq f(b)\} \;\text{ for all } x \in \mathbb{T}^{\omega}\,.$$

This is shown as Theorem 16 of [237], where such idempotents are called *partial closures*. In particular, if $f \sqsubseteq id_{\mathbb{T}^{\omega}}$ then f is a partial closure; however, the idempotents with which we have been chiefly concerned above do not satisfy $f \sqsubseteq id$.

10.3 PCF with Parallel Operators

We now turn our attention to programming languages that correspond in computational power to PC and $\mathsf{PC}^{\mathrm{eff}}$. The material here is due mainly to Plotkin [236] and Sazonov [251].

We have already seen in Example 7.1.4 that these type structures are models of PCF^{Ω} and PCF respectively, and so admit adequate interpretations of these languages (Theorem 7.1.16). However, we have also seen in Subsection 7.3.1 that there are simple functions present in $\mathsf{PC}^{\mathrm{eff}}$, such as the parallel convergence tester *pconv* and the parallel-or operator *por*, that are not denotable by any nested sequential procedure and hence are not definable in PCF. This raises the question of what needs to be added to PCF in order to obtain the full strength of $\mathsf{PC}^{\mathrm{eff}}$.

Following [236], we will extend PCF to a language PCF^+ by adding a 'parallel-or' operator, and will prove that all *compact* elements of PC are PCF^+-definable. We will show, however, that there remain other elements of $\mathsf{PC}^{\mathrm{eff}}$ that are not definable in PCF^+, such as the continuous existential quantifier $exists \in \mathsf{PC}^{\mathrm{eff}}(2)$ (to be defined below). Finally, we will introduce the language PCF^{++}, including both 'parallel-or' and 'exists' operators, and show that the PCF^{++}-definable elements of PC are exactly those of $\mathsf{PC}^{\mathrm{eff}}$. Likewise, all elements of PC are PCF^{++}-definable relative to a type 1 oracle. This means that $\mathsf{PC}^{\mathrm{eff}}$ and PC may be constructed up to isomorphism as the extensional term models for PCF^{++} and $\mathrm{PCF}^{++\Omega}$ respectively.

Throughout this section, we will work with a version of PCF enriched with a ground type $\mathtt{B}$ of booleans as described at the start of Section 7.3. Semantically we interpret this by the Scott domain $\mathbb{T} = \mathbb{B}_\perp$ as defined in Section 10.2. Strictly speaking, then, PC and $\mathsf{PC}^{\mathrm{eff}}$ will now denote models over $\mathsf{T}^{\rightarrow}(\mathtt{N} = \mathbb{N}_\perp, \mathtt{B} = \mathbb{B}_\perp)$. However, since $\mathbb{B}_\perp \lhd \mathbb{N}_\perp$, this is not an essential difference, and all the foregoing theory applies to the new versions of PC and $\mathsf{PC}^{\mathrm{eff}}$.

10.3.1 Finitary Parallel Operators

We recall from Chapter 7 the semantic definition of the parallel-or operator $por \in \mathsf{PC}^{\mathrm{eff}}(\mathtt{B},\mathtt{B} \rightarrow \mathtt{B})$:

$$\begin{aligned} por\, x\, y &= tt \text{ if } x = tt \text{ or } y = tt\,, \\ por\, x\, y &= f\!f \text{ if } x = f\!f \text{ and } y = f\!f\,, \\ por\, x\, y &= \perp \text{ otherwise.} \end{aligned}$$

In Subsection 7.3.1 we showed that *por*, like the simpler operator *pconv*, is not the denotation of any PCF term. The original proof in [236] used the operational semantics of PCF. Our treatment slightly simplified this by the use of nested sequential procedures, a tool which we shall exploit again in Theorem 10.3.8.

Our first task is to extend PCF with a new constant to denote *por*; we shall write this constant also as $por : \mathtt{B},\mathtt{B} \rightarrow \mathtt{B}$. The operational semantics of PCF as given in Chapter 7 may now be augmented with the reduction rules

$$por\; tt\; x \rightsquigarrow tt\,, \qquad por\; x\; tt \rightsquigarrow tt\,, \qquad por\; f\!f\; f\!f \rightsquigarrow f\!f\,,$$

and the new primitive evaluation contexts

$$por\,[-]\,M\,, \qquad por\,M\,[-]\,.$$

We denote the resulting language PCF + *por* by PCF$^+$. In contrast to PCF, the reduction relation of PCF$^+$ is *non-deterministic*: e.g. the term $por\,(suc\,\widehat{0})(suc\,\widehat{0})$ may be reduced in one step to either $por\,\widehat{1}\,(suc\,\widehat{0})$ or $por\,(suc\,\widehat{0})\,\widehat{1}$.

We may now check that our interpretation of PCF$^+$ in PC is *adequate*: for any closed ground type term M and ground type value v, we have $M \rightsquigarrow^* v$ iff $[\![M]\!] = [\![v]\!]$ in PC. The proof is in fact a trivial extension of the adequacy proof for PCF given in Subsection 7.1.3. First, the proof of soundness clearly goes through since the above reduction rules are sound for our semantic operator *por*. Next, we note that the logical relation R in the proof of Lemma 7.1.14 is also satisfied by *por*, so that the lemma applies when the language PCF$^-$ is augmented by *por*; the proof of Theorem 7.1.16 then goes through unchanged.

It follows immediately from this that although $\rightsquigarrow$ in PCF$^+$ is non-deterministic, *evaluation* of ground type terms is deterministic: any closed PCF$^+$ term of ground type may be reduced to at most one value.

Exercise 10.3.1. If preferred, a deterministic reduction relation for PCF$^+$ can be specified using inductive clauses such as the following:

$$\text{if} \quad M \rightsquigarrow M' \quad \text{then} \quad por\,M\,N \rightsquigarrow por\,N\,M' \,.$$

In effect this implements a 'fair' reduction strategy for $por\,M\,N$ ensuring that the evaluation of both arguments will make progress.

Work out the details of an operational semantics of PCF$^+$ along these lines, and verify that it is adequate with respect to the denotational interpretation.

There are also other possible choices of language primitive yielding the same expressive power as PCF$^+$. For instance, Plotkin [236] originally used a *parallel conditional* operator $pif_{\mathtt{N}} : \mathtt{B},\mathtt{N},\mathtt{N} \rightarrow \mathtt{N}$ defined denotationally by

$$\begin{array}{ll} pif_{\mathtt{N}}\; tt\;\; x\,y = x\,, & pif_{\mathtt{N}}\, z\,x\,x = x\,, \\ pif_{\mathtt{N}}\, ff\;\; x\,y = y\,, & pif_{\mathtt{N}}\, z\,x\,y = \bot \text{ in all other cases.} \end{array}$$

In PCF + $pif_{\mathtt{N}}$, it is easy to define a similar operator $pif_{\mathtt{B}} : \mathtt{B},\mathtt{B},\mathtt{B} \rightarrow \mathtt{B}$ satisfying the same clauses but with x,y ranging over $\mathsf{PC}(\mathtt{B})$.

The following interdefinability result is due to Stoughton [281]:

Lemma 10.3.2. *The elements por and $pif_{\mathtt{N}}$ are interdefinable modulo* PCF*: that is, there are* PCF *terms M,N such that $[\![M]\!](por) = pif_{\mathtt{N}}$ and $[\![N]\!](pif_{\mathtt{N}}) = por$ in* PC.

Proof. We may define *por* from $pif_{\mathtt{N}}$ via $pif_{\mathtt{B}}$ as follows:

$$por \;=\; \lambda xy.\, pif_{\mathtt{B}}\, x\; tt\; y\,.$$

It is easy to check that this yields a denotationaly correct PCF definition of *por* relative to $pif_{\mathtt{N}}$. For the other direction, we note that $pif_{\mathtt{N}}\, z\,x\,y$, when defined, is the least $w \in N$ such that

$$(z = tt \wedge w = x) \vee (z = ff \wedge w = y) \vee (w = x \wedge w = y)\,,$$

and that this expression also has a defined truth-value for all smaller w. We may capture this in PCF$+por$ as follows, using some obvious syntactic sugar:

$$
\begin{aligned}
pand\,x\,y &= \neg(por\,(\neg x)(\neg y))\,,\\
f_0\,z\,x\,y\,w &= pand\,(z = tt)(w = x)\,,\\
f_1\,z\,x\,y\,w &= pand\,(z = ff)(w = y)\,,\\
f_2\,z\,x\,y\,w &= pand\,(w = x)(w = y)\,,\\
g\,z\,x\,y\,w &= por\,(f_0\,z\,x\,y\,w)\,(por\,(f_1\,z\,x\,y\,w)\,(f_2\,z\,x\,y\,w))\,,\\
pif_{\mathtt{N}}\,z\,x\,y &= \mu w.\,g\,z\,x\,y\,w\,.
\end{aligned}
$$

It is routine to check that this gives a correct definition of $pif_{\mathtt{N}}$. $\square$

The main significance of PCF$^+$ lies in the following denotational result. For each type σ, we suppose given a standard effective coding $\lceil - \rceil : \mathsf{PC}(\sigma)^{\mathrm{comp}} \to \mathbb{N}$.

Theorem 10.3.3. *(i) Every compact in* PC *is definable in* PCF$^+$.
(ii) For each type σ*, the map* $\lceil a \rceil \mapsto a : \mathbb{N} \to \mathsf{PC}(\sigma)$ *is* PCF$^+$*-definable.*

Proof. (i) We broadly follow the argument from Plotkin [236], showing by induction on the structure of σ that every $a \in \mathsf{PC}(\sigma)^{\mathrm{comp}}$ is PCF$^+$-definable. For ground types this is trivial, so suppose $\sigma = \sigma_0, \ldots, \sigma_{r-1} \to \gamma$ where $r > 0$ and γ is $\mathtt{N}$ or $\mathtt{B}$.

In the light of Exercise 10.1.6, we see that any $a \in \mathsf{PC}(\sigma)^{\mathrm{comp}}$ has the form

$$
a \;=\; a_0 \sqcup \cdots \sqcup a_{n-1}\,, \quad \text{where } a_i = (\vec{b}_i \Rightarrow t_i),\; b_{ij} \in \mathsf{PC}(\sigma_j)^{\mathrm{comp}},\; t_i \in \mathsf{PC}(\gamma)\,.
$$

We first show that each a_i is PCF$^+$-definable. Let $c_i = (b_{i1} \Rightarrow \cdots \Rightarrow b_{i(r-1)} \Rightarrow t_i)$ and suppose $b_{i0} = \bigsqcup_j (\vec{d}_j \Rightarrow l_j)$ (where the $\vec{d}_j$ may be empty). By the induction hypothesis, c_i and each d_{jk} are PCF$^+$-definable, so we may define a_i via

$$
a_i(x_0, \ldots, x_{r-1}) \;=\; \begin{cases} c_i(x_1, \ldots, x_{r-1}) & \text{if } x_0(\vec{d}_j) = l_j \text{ for each } j\,,\\ \bot & \text{otherwise.} \end{cases}
$$

We now show by induction on n that a is PCF$^+$-definable. The case $n = 0$ is trivial, and we have just done the case $n = 1$, so suppose $n > 1$. There are two cases.

Case 1: b_{ij} and $b_{i'j}$ are consistent for all i, i', j. Then all the t_i are the same element t. We may then define a as follows by iterated use of por (here written as $\vee$):

$$
a(\vec{x}) \;=\; \mathit{if}\ (a_0(\vec{x}) = t \vee \cdots \vee a_{n-1}(\vec{x}) = t)\ \mathit{then}\ t\ \mathit{else}\ \bot\,.
$$

Case 2: There are i, i', j such that $b_{ij} \sharp b_{i'j}$. Let $a' = \bigsqcup_{k \neq i} a_k$ and $a'' = \bigsqcup_{k \neq i'} a_k$. By the inner induction hypothesis, both a' and a'' are PCF$^+$-definable. Moreover, by the outer induction hypothesis, there are definable arguments $\vec{e}$ such that $u = b_{ij}(\vec{e})$ and $u' = b_{i'j}(\vec{e})$ are inconsistent at ground type.

Using Lemma 10.3.2, we may now define the join $a = a' \sqcup a''$ by

$$
a(\vec{x}) \;=\; pif\ (x_j(\vec{e}) = u)\ (a'(\vec{x}))\ (a''(\vec{x}))\,.
$$

(ii) For each σ, the above construction by recursion on the size of a can itself be embodied by a PCF$^+$ program acting on codes for compact elements. $\square$

One consequence of Theorem 10.3.3 is the following. We say an interpretation $[\![-]\!]$ is *inequationally fully abstract* if $[\![M]\!] \sqsubseteq [\![M']\!] \Leftrightarrow \forall C[-].C[M] \sqsubseteq C[M']$, where M, M' range over closed terms and $C[-]$ ranges over ground type contexts.

Corollary 10.3.4. PC *is, up to isomorphism, the unique type structure* $\mathbf{T}$ *over* $\mathbb{N}_\perp, \mathbb{B}_\perp$ *consisting of algebraic DCPOs (with continuous application) and admitting a compositional, adequate, inequationally fully abstract interpretation of* PCF$^+$.

Proof. Inequational full abstraction for PC follows easily from the definability of all finite elements (cf. Subsection 7.1.4). So suppose $\mathbf{T}$ is a type structure satisfying the above conditions. We show by induction on σ that $\mathbf{T}(\sigma) \cong \mathsf{PC}(\sigma)$. For base types this is given by assumption, so suppose $\sigma = \sigma_0, \ldots, \sigma_{r-1} \to \mathtt{N}$.

First, we may embed $\mathsf{PC}(\sigma)^{\mathrm{comp}}$ in $\mathbf{T}(\sigma)$: if a is compact in $\mathsf{PC}(\sigma)$, take a PCF$^+$ term M_a denoting a, and let $\widehat{a}$ be the interpretation of M_a in $\mathbf{T}$. Since both PC and $\mathbf{T}$ are inequationally fully abstract, we have $a \sqsubseteq b$ iff $\widehat{a} \sqsubseteq \widehat{b}$. Thus directed sets of compacts in $\mathsf{PC}(\sigma)$ map to directed sets in $\mathbf{T}(\sigma)$, so we may extend our mapping to the whole of $\mathsf{PC}(\sigma)$ by taking $\widehat{x} = \bigsqcup\{\widehat{a} \mid a \sqsubseteq x \text{ compact}\}$. Clearly $\widehat{x}$ agrees with x in its action on arguments of types $\sigma_0, \ldots, \sigma_{r-1}$.

To see that this mapping is surjective, suppose $y \in \mathbf{T}(\sigma)$, and consider the set $A_y = \{a \in \mathsf{PC}(\sigma)^{\mathrm{comp}} \mid \widehat{a} \sqsubseteq y\}$. Clearly if $a, a' \in A_y$ then $a \uparrow a'$ and $a \sqcup a' \in A_y$. So take $x = \bigsqcup A_y$. By comparing the action of $\widehat{x}$ and y on an arbitrary tuple of compact arguments $b_0, \ldots, b_{r-1}$, it is easy to check that $y = \widehat{x}$. $\square$

A further indication of the expressive power of PCF$^+$ is that it suffices for defining all elements of $\mathsf{PC}^{\mathrm{eff}}$ at certain low types. To see the extent of this, let us work in the Karoubi envelope of PC, and take $\mathsf{PC}(1_v) \lhd \mathsf{PC}(1)$ to be the type of *strict* continuous functions $\mathbb{N}_\perp \to \mathbb{N}_\perp$, or isomorphically the type of all functions $\mathbb{N} \to \mathbb{N}_\perp$. We may then define $\mathsf{PC}(2_v) = \mathsf{PC}(1_v) \Rightarrow \mathbb{N}_\perp$, and note that $\mathsf{PC}(2_v) \lhd \mathsf{PC}(2)$. (The subscript v here means 'call-by-value'.) We shall regard $\mathsf{PC}(2_v)$ as a subset of $\mathsf{PC}(2)$, and write $\mathsf{PC}^{\mathrm{eff}}(2_v)$ for the induced subset of $\mathsf{PC}^{\mathrm{eff}}(2)$.

Theorem 10.3.5. *Every element of* $\mathsf{PC}^{\mathrm{eff}}(2_v)$ *is* PCF$^+$*-definable.*

Proof. Suppose $f \in \mathsf{PC}^{\mathrm{eff}}(2_v)$, and write f as $\bigsqcup_{i \in \mathbb{N}}(q_i \Rightarrow t_i)$, where $i \mapsto \lceil (q_i \Rightarrow t_i) \rceil$ is computable. Here the t_i are natural numbers and the q_i are in effect finite partial functions $\mathbb{N} \rightharpoonup_{\mathrm{fin}} \mathbb{N}$.

We first define an operation $Cons \in \mathsf{PC}(1_v \to \mathbb{N} \to \mathbb{B})$ which tests, as far as possible, whether a given $x \in \mathsf{PC}(1_v)$ is consistent with a given $q : \mathbb{N} \rightharpoonup_{\mathrm{fin}} \mathbb{N}$. Here '&' denotes 'parallel-and'.

$$Cons(x, \lceil q \rceil) = (x(u_0) = v_0)\ \&\ \cdots\ \&\ (x(u_{n-1}) = v_{n-1}))\,,$$
$$\text{where } q = \textstyle\bigsqcup_{j<n}(u_j \Rightarrow v_j)\,.$$

Clearly this is expressible in PCF$^+$. Note that if $x \in \mathsf{PC}(1_v)$ then $Cons(x, \lceil q \rceil)$ yields tt if $x \sqsupseteq q$, ff if $x \sharp q$, and $\perp$ otherwise.

Now let $\widehat{f} \in \mathsf{PC}(0 \to 2_v)$ be the minimal solution to the equation

$$\widehat{f}(i,x) \;=\; pif_{\mathbb{N}}\ (Cons(x,\lceil q_i \rceil))\ t_i\ (\widehat{f}(i+1,x))\,.$$

Clearly this is PCF$^+$-definable. We shall show that $\widehat{f}(0) = f$.

Suppose first that $f(x)\downarrow$. Then there exists i such that $q_i \sqsubseteq x$, and then $\widehat{f}(x,i) = t_i = f(x)$. From this we see by downwards induction that $\widehat{f}(x,j) = t_i$ for all $j \leq i$, since for every such j we have either $t_j = t_i$ or $Cons(x,\lceil q_j \rceil) = \mathit{ff}$. In particular, $\widehat{f}(0,x) = f(x)$.

Conversely, let $\widehat{f_k}$ denote the kth approximation to $\widehat{f}$. By induction on k we see that if $\widehat{f_{k+1}}(x,j) = t$ then there must exist $i \leq j+k$ such that $q_i \sqsubseteq x$ and $t_i = t$. We conclude that $\widehat{f}(0,x) = f(x)$ for all x. □

The following special case will be used in the proof of Theorem 10.3.11.

Corollary 10.3.6. *Let σ be any first-order type. Then there is a* PCF$^+$*-definable element $\varepsilon_\sigma \in \mathsf{PC}(\mathtt{N},\mathtt{N} \to \sigma)$ such that for each compact a,*

1. $\varepsilon_\sigma(\lceil a \rceil, \bot) = a$,
2. $\{\varepsilon_\sigma(\lceil a \rceil, m)\}_{m \in \mathbb{N}}$ *enumerates all compact extensions of a.*

Proof. The required operation ε_σ is clearly an element of $\mathsf{PC}^{\mathrm{eff}}(\mathtt{N},\mathtt{N} \to \sigma)$, which is a PCF-definable retract of $\mathsf{PC}^{\mathrm{eff}}(2_v)$. Hence ε_σ is PCF$^+$-definable. □

10.3.2 Infinitary Parallel Operators

It is now natural to ask whether PCF$^+$ is powerful enough to define *all* effective elements in PC. This was answered negatively by Plotkin [236], who introduced a continuous existential quantifier *exists*, and showed that it is not definable in PCF$^+$ or its oracle version PCF$^{+\Omega}$ (cf. Section 7.1). He showed, however, that all effective elements *are* definable in the language PCF^{++} = PCF$^+$ + *exists*, so that $\mathsf{PC}^{\mathrm{eff}}$ may be characterized as the type structure of PCF^{++}-definable functionals.

Denotationally, the operator $exists \in \mathsf{PC}^{\mathrm{eff}}((\mathtt{N} \to \mathtt{B}) \to \mathtt{B})$ is specified by

$$exists(f) \;=\; \begin{cases} \mathit{ff} & \text{if } f(\bot) = \mathit{ff}\,, \\ \mathit{tt} & \text{if } \exists n \in \mathbb{N}.\, f(n) = \mathit{tt}\,, \\ \bot & \text{otherwise.} \end{cases}$$

On the operational side, we may extend PCF$^+$ with a constant $exists : (\mathtt{N} \to \mathtt{B}) \to \mathtt{B}$, and augment the definition of the reduction relation with the following clauses (note that we now define $\rightsquigarrow$ by simultaneous recursion with its reflexive-transitive closure $\rightsquigarrow^*$). Here $\Omega : \mathtt{N}$ is some standard term denoting $\bot$, such as $Y_{\mathtt{N}}\ suc$.

- If $M\Omega \rightsquigarrow^* \mathit{ff}$ then $exists\ M \rightsquigarrow \mathit{ff}$.
- If $M\widehat{n} \rightsquigarrow^* \mathit{tt}$ then $exists\ M \rightsquigarrow \mathit{tt}$.

Exercise 10.3.7. Extend the adequacy result for PCF$^+$ to PCF^{++}. The adaptation of the proof in Subsection 7.1.3 requires a little work, since we must work with finite approximants to $\exists$ as well as to the Y_σ, but the basic method is otherwise the same.

We now show that *exists* is not definable in PCF$^+$. Our proof is essentially that of Plotkin [236], but presented here in a more denotational style using nested sequential procedures.

Theorem 10.3.8. *The element exists* $\in \mathsf{PC}$ *is not definable in* PCF$^{+\Omega}$.

Proof. Suppose $\lambda f.M$ were a PCF$^{+\Omega}$ term defining *exists*. Treating occurrences of *por* as variables, we obtain a PCF$^\Omega$ term $\lambda por.\lambda f.M$. Recalling from Subsection 7.1.7 that the interpretation of PCF$^\Omega$ in PC factors through nested sequential procedures, we obtain an NSP p such that $[\![p]\!](por) = exists$. We will show however that there is no such p.

For each n, let $s_n \in \mathsf{PC}(\mathtt{N} \to \mathtt{B})$ map n to tt and everything else to $\bot$, and let $u = \Lambda z.ff \in \mathsf{PC}(\mathtt{N} \to \mathtt{B})$. Specifically, we will show that if $[\![p]\!](por)(u)\downarrow$ and $[\![p]\!](por)(s_n)\downarrow$ for all n in some infinite set $A \subseteq \mathbb{N}$, then there is an infinite set $B \subseteq A$ such that $[\![p]\!](por)(s_n) = [\![p]\!](por)(u)$ for all $n \in B$. It follows from this that $[\![p]\!](por) \neq exists$.

Suppose $p = \lambda por.\lambda f.e$ satisfies the hypotheses. By continuity of the interpretation of NSPs, there is some finite $p_0 \sqsubseteq p$ such that $[\![p_0]\!](por)(u)\downarrow$. We argue by induction on the size of the smallest such p_0, and reason by cases on the form of e. If t is an NSP term and $g \in \mathsf{PC}(\mathtt{N} \to \mathtt{B})$, we will write $[\![t]\!]\langle g\rangle$ for $[\![t]\!]_{por \mapsto por,\, f \mapsto g}$.

Case 1: e is a constant. Then the conclusion holds trivially with $B = A$.

Case 2: $e = \mathtt{case}\ por\ q_0\ q_1\ \mathtt{of}\ (tt \Rightarrow e_0 \mid ff \Rightarrow e_1)$. Let $A = \{n \mid [\![p]\!](por)(s_n)\downarrow\}$. There are three subcases, which by symmetry we can reduce to two.

Subcase 2.1: $[\![q_0]\!]\langle u\rangle = [\![q_1]\!]\langle u\rangle = ff$. For each $n \in A$, at least one of $[\![q_0]\!]\langle s_n\rangle$, $[\![q_1]\!]\langle s_n\rangle$ is defined. By symmetry, we assume $[\![q_0]\!]\langle s_n\rangle$ is defined for infinitely many $n \in A$, and let B be the set of such n.

By the induction hypothesis, there is an infinite set $C \subseteq B$ such that $[\![q_0]\!]\langle s_n\rangle = [\![q_0]\!]\langle u\rangle = ff$ for all $n \in C$. Since $[\![por\ q_0\ q_1]\!]\langle s_n\rangle\downarrow$ for all $n \in C$, we must have $[\![q_1]\!]\langle s_n\rangle\downarrow$ for all $n \in C$. But now again by the induction hypothesis, there is an infinite set $D \subseteq C$ such that $[\![q_1]\!]\langle s_n\rangle = ff$ for all $n \in D$, so for $n \in D$ we have $[\![e]\!]\langle s_n\rangle = [\![e_1]\!]\langle s_n\rangle$. Applying the induction hypothesis to e_1 and D now yields the desired conclusion.

Subcase 2.2: $[\![q_0]\!]\langle u\rangle = tt$. If for infinitely many $n \in A$ we have $[\![q_0]\!]\langle s_n\rangle\downarrow$, by the induction hypothesis we can find an infinite set $B \subseteq A$ such that $[\![q_0]\!]\langle s_n\rangle = tt$ for all $n \in B$, so that $[\![e]\!]\langle s_n\rangle = [\![e_0]\!]\langle s_n\rangle$ for all $n \in B$. Applying the induction hypothesis to e_0 and B then yields the desired conclusion.

If, however, $[\![q_0]\!]\langle s_n\rangle\downarrow$ for only finitely many $n \in A$, let B consist of all the others. We then must have $[\![q_1]\!]\langle s_n\rangle = tt$ for all $n \in B$, and can continue as above.

Case 3: $e = \mathtt{case}\ f\ d\ \mathtt{of}\ (n \Rightarrow e_n)$. Then for all $n \in A$ we have $[\![f\ d]\!]\langle s_n\rangle = s_n([\![d]\!]\langle s_n\rangle)\downarrow$, whence $[\![d]\!]\langle s_n\rangle = n$. But for all m,n we have that $s_m \uparrow s_n$, whence $[\![d]\!]\langle s_m\rangle \uparrow [\![d]\!]\langle s_n\rangle$. So this case is impossible. $\square$

One consequence of Theorem 10.3.8 is that the set of $\mathrm{PCF}^{+\Omega}$-definable functionals is not closed under least upper bounds of chains. Indeed, there is a chain of compact (and hence PCF^{+}-definable) elements approximating *exists* which has no $\mathrm{PCF}^{+\Omega}$-definable upper bound.

As an aside, the following exercise uses a very different proof technique to show that *por* is not PCF-definable from *exists* (this was stated in Sazonov [251]).

Exercise 10.3.9. On ground types $\mathsf{PC}(\gamma)$, let R_γ be the ternary relation given by

$$R_\gamma(x,y,z) \;\Leftrightarrow\; x=\bot \vee y=\bot \vee z=\bot \vee x=y=z\,,$$

and let R be the lifting of the R_γ to a logical relation. Note that R_γ is a *sequentiality relation* in the sense of Definition 7.5.11, so R is satisfied by all the PCF^{Ω} constants. Show too that R is satisfied by *exists*, but not by *por*. Deduce that *por* is not definable in PCF^{Ω} + *exists*.

Next, we work towards showing that PCF^{++} suffices to define all elements of $\mathsf{PC}^{\mathrm{eff}}$, and likewise, $\mathrm{PCF}^{++\Omega}$ suffices to define all elements of PC. The following lemma shows, in effect, that for second-order types σ, the mapping $t : \mathsf{PC}(\sigma) \to \mathbb{T}^\omega$ constructed in the proof of Theorem 10.2.5 is PCF^{++}-definable.

Lemma 10.3.10. *For any second-order type σ, there is a PCF^{++}-definable element $Cons_\sigma \in \mathsf{PC}(\sigma \to (\mathtt{N} \to \mathtt{B}))$ such that for all $x \in \mathsf{PC}(\sigma)$ and $a \in \mathsf{PC}(\sigma)^{\mathrm{comp}}$ we have*

$$Cons_\sigma(x, \lceil a \rceil) \;=\; \begin{cases} tt & \textit{iff } a \sqsubseteq x\,, \\ ff & \textit{iff } a \,\sharp\, x\,, \\ \bot & \textit{otherwise.} \end{cases}$$

Proof. As a representative example, we shall treat the case $\sigma = \overline{1}, \overline{0} \to \overline{0}$ (which is all we shall require for the proof of the following theorem). If $\lceil a \rceil = n$, let us write

$$a \;=\; p_n \;=\; p_{n0} \sqcup \cdots \sqcup p_{n(k_n-1)}\,, \quad \text{where } p_{ni} = (q_{ni} \Rightarrow r_{ni} \Rightarrow t_{ni}) \text{ for each } i < k_n.$$

Using Corollary 10.3.6, we may readily construct PCF^{+}-definable operations $\delta \in \mathsf{PC}(\mathtt{N}^{(3)} \to 1)$ and $\gamma \in \mathsf{PC}(\mathtt{N}^{(3)} \to 0)$ such that:

1. $\delta(n,i,\bot) = q_{ni}$ and $\gamma(n,i,\bot) = r_{ni}$ for each $n \in \mathbb{N}$ and $i < k_n$,
2. $\{(\delta(n,i,m), \gamma(n,i,m))\}_{m \in \mathbb{N}}$ enumerates all compact extensions of (q_{ni}, r_{ni}), uniformly in n and $i < k_n$.

Now define

$$Cons'_\sigma(x,n,i) \;=\; \neg\, exists\; (\lambda z.\, x(\delta(n,i,z))\,(\gamma(n,i,z)) \neq t_{ni})\,.$$

It is easy to check that

$$Cons'_\sigma(x,n,i) \;=\; \begin{cases} tt & \text{iff } p_{ni} \sqsubseteq x\,, \\ ff & \text{iff } p_{ni} \,\sharp\, x\,, \\ \bot & \text{otherwise.} \end{cases}$$

Finally, since k_n is computable from n, we may use 'parallel-and' (here denoted by &) to define a suitable element $Cons_\sigma$:

$$Cons_\sigma(x,n) = Cons'_\sigma(x,n,0) \;\&\; \cdots \;\&\; Cons'_\sigma(x,n,k_n-1)\,. \qquad \square$$

We now come to the main completeness theorem of Plotkin [236] and Sazonov [251] (see also Feferman [93]).

Theorem 10.3.11. *Every element of* $\mathsf{PC}^{\mathrm{eff}}$ *is definable in* PCF^{++}.

Proof. We will show that each $\mathsf{PC}(\sigma)$ is a PCF^{++}-definable retract of the universal domain $\mathbb{T}^\omega$. To make sense of this, let us represent $\mathbb{T}^\omega$ as a PCF-definable retract of $\mathsf{PC}(1)$ in an obvious way; we will henceforth proceed as if $\mathbb{T}^\omega$ were actually a type in PCF.

The key step is to establish that both halves of the retraction $(\mathbb{T}^\omega \Rightarrow \mathbb{T}^\omega) \lhd \mathbb{T}^\omega$ are PCF^{++}-definable. First note that $\mathbb{T}^\omega \Rightarrow \mathbb{T}^\omega$ is a definable retract of $\mathsf{PC}(1 \to 1)$, and that every compact in $\mathbb{T}^\omega \Rightarrow \mathbb{T}^\omega$ is also one in $\mathsf{PC}(1 \Rightarrow 1)$. By taking $Cons_\sigma$ as in Lemma 10.3.10 (for $\sigma = \overline{1} \to \overline{1}$) and restricting to arguments $\lceil a \rceil$ such that $a \in (\mathbb{T}^\omega \Rightarrow \mathbb{T}^\omega)$, we recover exactly the map $t : (\mathbb{T}^\omega \Rightarrow \mathbb{T}^\omega) \to \mathbb{T}^\omega$ from the proof of Theorem 10.2.5; thus this map is PCF^{++}-definable. Moreover, this is one half of a retraction, the other half being an effective element of $\mathbb{T}^\omega \Rightarrow (\mathbb{T}^\omega \Rightarrow \mathbb{T}^\omega)$. However, we see by Theorem 10.3.5 that *all* effective elements of this type are PCF^{+}-definable, since $\mathbb{T}^\omega \Rightarrow (\mathbb{T}^\omega \Rightarrow \mathbb{T}^\omega)$ is evidently a PCF-definable retract of $\mathsf{PC}((\mathtt{N} \to_s \mathtt{N}) \to \mathtt{N})$.

Since $\mathbb{N}_\perp$ is clearly a PCF-retract of $\mathbb{T}^\omega$, an easy induction on types now shows that every $\mathsf{PC}(\sigma)$ is a PCF^{++}-retract of $\mathbb{T}^\omega$. As every effective element of $\mathbb{T}^\omega$ itself is PCF-definable, the theorem follows. $\square$

Corollary 10.3.12. PC, $\mathsf{PC}^{\mathrm{eff}}$ *are respectively isomorphic to* PCF$^{++\Omega 0}/=_{\mathrm{obs}}$ *and* PCF$^{++0}/=_{\mathrm{obs}}$, *the type structures of closed terms of* PCF$^{++\Omega}$ *and* PCF^{++} *modulo observational equivalence. Under these isomorphisms, the ordering* $\sqsubseteq$ *on* PC, $\mathsf{PC}^{\mathrm{eff}}$ *corresponds to the observational preorder on terms.* $\square$

Exercise 10.3.13. (i) Show that every $x \in \mathsf{PC}(\sigma)$ is PCF^{++}-definable relative to any enumeration of (code for) the compacts bounded by x.

(ii) Show that for each type σ there is a PCF^{++} program $E_\sigma : \mathtt{N} \to \sigma$ such that for any closed PCF^{++} term $M : \sigma$ we have that $[\![E_\sigma \lceil M \rceil]\!] = [\![M]\!]$ in PC, where $\lceil M \rceil$ is a Gödel number for M.

10.3.3 Degrees of Parallelism

A few papers have been devoted to the degree theory induced by the notion of relative PCF-computability. If $x, y \in \mathsf{PC}^{\mathrm{eff}}$, we say x is PCF-*computable in* y if there is some PCF-definable $f \in \mathsf{PC}^{\mathrm{eff}}$ such that $f(y) = x$. This defines a preorder on $\mathsf{PC}^{\mathrm{eff}}$, and the induced equivalence classes are known as *degrees of parallelism.*

The poset of degrees of parallelism was introduced by Sazonov [251]. Like other posets of degrees, its structure appears to be extremely complicated. Sazonov mentioned several examples of distinct degrees and some relationships between them. For example, we have seen that *por* and *exists* represent two incomparable degrees (Theorem 10.3.8 and Exercise 10.3.9). To mention one other example, the parallel convergence tester *pconv* defined in Subsection 7.3.1 is easily seen to be computable in both *por* and *exists* separately; it follows that *pconv* represents a strictly lower degree than *por*.

A few other results in a similar spirit appeared in Trakhtenbrot [287]. Bucciarelli [44] undertook a somewhat more systematic study of degrees of parallelism for first-order functions, a line of investigation pursued further by Panangaden [228]. In principle, these degrees give rise to many intermediate notions of computability between PCF and PCF^{++}; however, so far none of these intermediate notions have established themselves as being of independent mathematical interest.[1]

10.4 Realizability over K_2 and K_1

We now consider some other ways of characterizing PC and $\mathsf{PC}^{\mathrm{eff}}$ computationally, quite different in flavour from those of the previous subsection. Specifically, in contrast to the relatively 'extensional' style of computation offered by PCF^{++}, we consider low-level 'intensional' representations of functions given by Kleene's models K_2 and K_1 respectively, and ask which higher-order functionals are computable if one is given access to these intensional representations. The answer, as it turns out, is that we can compute no more or less than the functionals of PC [resp. $\mathsf{PC}^{\mathrm{eff}}$], so that we obtain new characterizations of these type structures as extensional collapses of K_2 and K_1 (given a suitable presentation of $\mathbb{N}_\perp$). We revert at this point to considering PC, $\mathsf{PC}^{\mathrm{eff}}$ as models with just the single ground type $\mathtt{N}$. We assume familiarity here with the definitions of K_1 and K_2 (Example 3.1.9 and Subsection 3.2.1), and with the material on modest assemblies and partial equivalence relations in Subsections 3.3.6 and 3.6.3.

The characterization of PC via K_2 is straightforward, so we give just a brief treatment here. First, we recall from Example 3.5.7 that there is an obvious applicative simulation $\psi : \mathsf{PC} \longrightarrow\!\!\triangleright K_2$: we take $g \Vdash^{\psi}_{\sigma} x$ iff the range of $g : \mathbb{N} \to \mathbb{N}$ is some set of (codes for) compacts with supremum x. This allows us to regard each $\mathsf{PC}(\sigma)$ as a modest assembly over K_2. The key observation is the following:

Lemma 10.4.1. *Within $\mathcal{M}od(K_2)$ we have $\mathsf{PC}(\sigma \to \tau) \cong \mathsf{PC}(\sigma) \Rightarrow \mathsf{PC}(\tau)$, where $\Rightarrow$ denotes exponentiation of modest assemblies.*

Proof. (Sketch.) We have seen in Example 3.5.7 that application in PC is realized in K_2, so the passage from a realizer for $f \in \mathsf{PC}(\sigma \to \tau)$ to one for $f : \mathsf{PC}(\sigma) \to \mathsf{PC}(\tau)$

[1] A possible exception comes from the work of Sazonov [253], who has shown that PCF^{+}-definability corresponds to a notion of *wittingly consistent* strategy.

in $\mathscr{M}od(K_2)$ is itself K_2-realizable. Conversely, any $f : \mathsf{PC}(\sigma) \to \mathsf{PC}(\tau)$ is tracked by some $h \in K_2$, so it is clear that if $g \Vdash^{\psi}_{\sigma} x$ and $b \sqsubseteq f(x)$ is among the compacts listed by $h \odot g$ then there is some finite $a \sqsubseteq x$ with $b \sqsubseteq f(a)$. This implies that f is continuous. Moreover, a listing of enough elements $(a \Rightarrow b)$ to generate f may readily be extracted from h, again within K_2 itself. □

Furthermore, it is easy to check that $\mathsf{PC}(\mathtt{N})$, viewed as a partial equivalence relation, is isomorphic in $\mathscr{P}er(K_2)$ to the following equivalence relation $\approx$ representing $\mathbb{N}_\perp$. Here $|$ is the 'pre-application' operation from Subsection 3.2.1.

$$f \approx g \;\text{ iff }\; f \,|\, (\Lambda i.0) \simeq g \,|\, (\Lambda i.0)\,.$$

We may therefore construct PC from $\approx$ by repeated exponentiation in $\mathscr{P}er(K_2)$. In other words:

Theorem 10.4.2. $\mathsf{PC} \cong \mathsf{EC}(K_2;\approx)$, *and the simulations* $\psi : \mathsf{PC} \longrightarrow\!\!\!\triangleright K_2$ *and* $\varepsilon : \mathsf{EC}(K_2;\approx) \longrightarrow\!\!\!\triangleright K_2$ *are equivalent modulo this identification.* □

Exercise 10.4.3. Show that under the above isomorphism, the elements of $\mathsf{PC}^{\mathrm{eff}}$ correspond to the equivalence classes represented by an element of K_2^{eff}. Show also that $\mathsf{PC}^{\mathrm{eff}} \cong \mathsf{EC}(K_2^{\mathrm{eff}};\approx)$.

The above results are in a sense unsurprising in that both PC and K_2 have the concept of continuity built into their definition. A deeper result, due essentially to Ershov [71], is that one can similarly characterize $\mathsf{PC}^{\mathrm{eff}}$ using K_1—a model whose definition makes no reference to continuity. The key here is a classical theorem of Myhill and Shepherdson [202], which establishes a link between effectivity and continuity at type 2. We start by proving this theorem in its standard form.

Recall that $\simeq$ means that if either side is defined, so is the other and they are equal, while $\preceq$ means that if the left-hand side is defined, so is the right and they are equal. For present purposes we consider the value $\perp$ to be 'undefined'. We write ϕ_e for the partial computable function $\mathbb{N} \rightharpoonup \mathbb{N}$ with index e in some standard enumeration, so that $\phi_e(n) \simeq e \bullet n$.

Theorem 10.4.4 (Myhill–Shepherdson). *(i) Let* $f : \mathbb{N} \rightharpoonup \mathbb{N}$ *be partial computable such that* $\phi_d = \phi_e$ *implies* $f(d) \simeq f(e)$ *for all* $d, e \in \mathbb{N}$. *Then*

1. $\phi_d \subseteq \phi_e$ *implies* $f(d) \preceq f(e)$,
2. *if* $f(e)\downarrow$, *there is an index* d *such that* ϕ_d *is finite,* $\phi_d \subseteq \phi_e$, *and* $f(d) = f(e)$,
3. *there exists* $F \in \mathsf{PC}^{\mathrm{eff}}((\mathtt{N} \to \mathtt{N}) \to \mathtt{N})$ *such that* $F(\phi_e) \simeq f(e)$ *for all* $e \in \mathbb{N}$.

(ii) If $F \in \mathsf{PC}^{\mathrm{eff}}((\mathtt{N} \to \mathtt{N}) \to \mathtt{N})$, *there is a partial computable* $f : \mathbb{N} \rightharpoonup \mathbb{N}$ *such that* $f(e) \simeq F(\phi_e)$ *for all* $e \in \mathbb{N}$.

Proof. We give somewhat informal descriptions of the required algorithms; more formal details may be found in standard texts such as [245, 65].

(i) For claim 1, suppose $\phi_d \subseteq \phi_e$. Consider the partial function $\psi(c,n)$ computed as follows: compute $d \bullet n$ and $c \bullet 0$ in parallel until one terminates. If $c \bullet 0$ terminates first, let $\psi(c,n) \simeq e \bullet n$; otherwise let $\psi(c,n) \simeq d \bullet n$. Thus $\psi(c,-)$ is ϕ_e if $c \bullet 0 \downarrow$, and ϕ_d otherwise.

Now suppose for contradiction that $f(d) = m \in \mathbb{N}$ while $\neg f(e) = m$. Let $g(c)$ denote an index for $\psi(c,-)$ computed from c, so that $\phi_{g(c)}$ is ϕ_e or ϕ_d according to whether $c \bullet 0 \downarrow$. Then $f(g(c)) = m$ iff $c \bullet 0 \uparrow$, contradicting the undecidability of the halting problem.

For claim 2, suppose $f(e) = m \in \mathbb{N}$. Let ϕ_e^k denote the approximation to ϕ_e which runs the algorithm for ϕ_e only on inputs $\leq k$ and only for up to k steps. Now define $\chi(c,n) = \phi_e^k(n)$ if $c \bullet 0$ terminates in k steps, $\chi(c,n) \simeq \phi_e(n)$ otherwise; it is easy to see that χ is computable. Thus $\chi(c,-) = \phi_e$ iff $c \bullet 0 \uparrow$.

So if $h(c)$ is an index for $\chi(c,-)$ computed from c, we have $f(h(c)) = m$ whenever $c \bullet 0 \uparrow$. But since the halting problem is undecidable, we must also have $f(h(c)) = m$ for some c with $c \bullet 0 \downarrow$, say in k steps. Setting $d = h(c)$, we have $\phi_d = \phi_e^k \subseteq \phi_e$ and $f(d) = m = f(e)$ as required.

Claims 1 and 2 together show that there exists $F \in \mathsf{PC}(2)$ with $F(\phi_e) \simeq f(e)$ for all e. For claim 3, it remains to show that F is effective. But if $p \in \mathsf{PC}(1)^{\mathrm{comp}}$ and $m \in \mathbb{N}$, we have $(p \Rightarrow m) \sqsubseteq F$ iff $f(e_p) = m$, where e_p is an index for p computable from p. Thus the relation $(p \Rightarrow m) \sqsubseteq F$ is semidecidable in p and m.

(ii) Given $F \in \mathsf{PC}^{\mathrm{eff}}(2)$ and $e \in \mathbb{N}$, let $f(e)$ be computed as follows. Search among all finite $p : \mathbb{N} \rightharpoonup \mathbb{N}$ for one such that $p \sqsubseteq \phi_e$ and $F(p) \downarrow$, and then set $f(e) = F(p)$. Since $p \sqsubseteq \phi_e \wedge F(p) \downarrow$ is semidecidable in p uniformly in e, we see that f is partial computable, and clearly does the trick. □

Remark 10.4.5. An inspection of the above proof shows that the passage from f to F may be effected by a functional in $\mathsf{PC}^{\mathrm{eff}}(1 \to 2)$, and that from F to f by one in $\mathsf{PC}^{\mathrm{eff}}(2 \to 1)$.

For our present purpose, some mild generalizations of this theorem are required. We say a set X of partial computable functions $\mathbb{N} \rightharpoonup \mathbb{N}$ is *down-closed* if $\phi_e \sqsubseteq \phi_{e'} \in X$ implies $\phi_e \in X$. The following corollary relativizes much of Theorem 10.4.4(i) to such sets X, and at the same time generalizes to operations on r arguments. If Y is any poset, we say a function $F : Y \to \mathbb{N}_\perp$ is *continuous* if it is monotone and preserves whatever directed suprema exist within Y.

Corollary 10.4.6. *Let $X_0, \ldots, X_{r-1}$ be down-closed sets of partial computable functions $\mathbb{N} \rightharpoonup \mathbb{N}$, considered as posets ordered by $\subseteq$. Suppose $f : \mathbb{N}^r \rightharpoonup \mathbb{N}$ is partial computable such that $f(d_0, \ldots, d_{r-1}) \simeq f(e_0, \ldots, e_{r-1})$ whenever $\phi_{d_i} = \phi_{e_i} \in X_i$ for each i. Then there is a continuous function $F : X_0 \times \cdots \times X_{r-1} \to \mathbb{N}_\perp$ such that $F(\phi_{e_0}, \ldots, \phi_{e_{r-1}}) \simeq f(e_0, \ldots, e_{r-1})$ whenever $\phi_{e_i} \in X_i$ for each i.* □

Proof. A trivial adaptation of the proof of Theorem 10.4.4(i). □

To characterize $\mathsf{PC}^{\mathrm{eff}}$ in terms of K_1, we shall work with the applicative morphism $\gamma : \mathsf{PC}^{\mathrm{eff}} \longrightarrow\!\!\!\triangleright K_1$ defined as follows: $e \Vdash_\sigma^\gamma x$ iff the range of ϕ_e is a set of (codes for) compacts in $\mathsf{PC}(\sigma)$ with supremum x. (Note that ϕ_e need not be total.) To see that application in $\mathsf{PC}^{\mathrm{eff}}$ is realized in K_1, suppose $d \Vdash_{\sigma \to \tau}^\gamma f$ and $e \Vdash_\sigma^\gamma x$. From these it is easy to construct c such that

$$\text{ran } \phi_c = \{p(q_0 \sqcup \cdots \sqcup q_{r-1}) \mid p \in \text{ran } \phi_d, \, q_0, \ldots, q_{r-1} \in \text{ran } \phi_e\} .$$

Clearly $c \Vdash^{\gamma}_{\tau} f(x)$. Since such a c is computable from d and e, the conclusion follows.

We may now identify each $\mathsf{PC}^{\text{eff}}(\sigma)$ with a modest assembly over K_1 via its γ-realization. Before proceeding further, it is an easy exercise to check that $\mathsf{PC}^{\text{eff}}(\mathtt{N})$ is isomorphic within $\mathcal{M}od(K_1)$ to the assembly $N_\perp$ defined as follows:

$$|N_\perp| = \mathbb{N}_\perp , \qquad e \Vdash_{N_\perp} n \text{ iff } e \bullet 0 = n , \qquad e \Vdash_{N_\perp} \perp \text{ iff } e \bullet 0 = \uparrow .$$

Lemma 10.4.7. $\mathsf{PC}^{\text{eff}}(\sigma_0, \ldots, \sigma_{r-1} \to \mathtt{N}) \cong \mathsf{PC}^{\text{eff}}(\sigma_0) \times \cdots \times \mathsf{PC}^{\text{eff}}(\sigma_{r-1}) \Rightarrow N_\perp$ *in* $\mathcal{M}od(K_1)$.

Proof. We treat the case $r = 1$, the extension to the general case being purely bureaucratic in the light of Corollary 10.4.6.

We have just seen that application in PC^{eff} is K_1-realizable, so we may pass from a realizer for $f \in \mathsf{PC}^{\text{eff}}(\sigma \to \mathtt{N})$ to one for $f : \mathsf{PC}^{\text{eff}}(\sigma) \to \mathsf{PC}^{\text{eff}}(\mathtt{N}) \cong N_\perp$, computably within K_1. For the converse, we note that

$$X_\sigma = \{\phi_e : \mathbb{N} \rightharpoonup \mathbb{N} \mid e \Vdash^{\gamma}_{\sigma} x \text{ for some } x \in \mathsf{PC}^{\text{eff}}(\sigma)\}$$

is a down-closed set of partial computable functions. So given any morphism $f : \mathsf{PC}^{\text{eff}}(\sigma) \to N_\perp$ tracked by $d \in K_1$ say, we have that ϕ_d gives rise to a partial computable function $g : \| X_\sigma \| \rightharpoonup \mathbb{N}$ such that $g(e) \simeq g(e')$ whenever $\phi_e = \phi_{e'} \in X_\sigma$. Now by Corollary 10.4.6, g lifts to a continuous function $G : X_\sigma \to \mathsf{PC}^{\text{eff}}(\mathtt{N})$ such that $G(\phi_e) = f(x)$ whenever $e \Vdash^{\gamma}_{\sigma} x$. But this readily implies that f itself is monotone and continuous, in view of the following easy observations:

- If $x \sqsubseteq x' \in \mathsf{PC}^{\text{eff}}(\sigma)$, there exist $e \Vdash^{\gamma}_{\sigma} x$ and $e' \Vdash^{\gamma}_{\sigma} x'$ such that $\phi_e \sqsubseteq \phi_{e'}$.
- If $x = \bigsqcup_{i \in I} x_i \in \mathsf{PC}^{\text{eff}}(\sigma)$, there exist $e \Vdash^{\gamma}_{\sigma} x$ and $e_i \Vdash^{\gamma}_{\sigma} x_i$ for each i such that $\phi_e = \bigsqcup \phi_{e_i}$.

Furthermore, the set of all step functions $(a \Rightarrow b) \sqsubseteq f$ is semidecidable uniformly in d: given a, b we compute a realizer for $f(a)$ and search the corresponding enumeration for an occurrence of $b \in \mathbb{N}$. Thus an enumeration of all such step functions may be computed uniformly in d, and this serves as a realizer for $f \in \mathsf{PC}^{\text{eff}}(\sigma \to \mathtt{N})$, computable from d within K_1. □

Recasting this in terms of partial equivalence relations, we see that $\mathsf{PC}^{\text{eff}}(\mathtt{N})$ corresponds up to isomorphism to the PER $\approx$ defined by

$$e \approx e' \quad \text{iff} \quad e \bullet 0 \simeq e' \bullet 0 ,$$

and that the whole of PC^{eff} may be constructed from this by exponentiation in $\mathcal{P}er(K_1)$. In other words:

Theorem 10.4.8. $\mathsf{PC}^{\text{eff}} \cong \mathsf{EC}(K_1; \approx)$, *and the simulations* $\gamma : \mathsf{PC}^{\text{eff}} \longrightarrow\!\!\!\!\triangleright\, K_1$ *and* $\varepsilon : \mathsf{EC}(K_1; \approx) \longrightarrow\!\!\!\!\triangleright\, K_1$ *are equivalent modulo this identification.* □

Exercise 10.4.9. Show that the simulation $\mathrm{PCF}^{++0}/{=_{\mathrm{obs}}} \longrightarrow\!\!\triangleright K_1$ induced by Gödel numbering of closed PCF^{++} terms is also equivalent to γ modulo the identification of Corollary 10.3.12.

The coincidence of $\mathsf{EC}(K_1;\approx)$ and $\mathrm{PCF}^{++0}/{=_{\mathrm{obs}}}$ is perhaps especially striking, since both of these models arise from purely 'computational' definitions, with no extrinsic requirement of continuity. Informally, one may say that PCF^{++} represents a purely 'extensional' style of computation somewhat in the spirit of Kleene computability (but with parallelism), whereas the functionals of $\mathsf{EC}(K_1;\approx)$ are computable at an intensional level by realizers that 'just happen' to behave extensionally. That these two quite different definitions of computability should agree at all finite types is on the face of it a remarkable and surprising fact.

Exercise 10.4.10. (i) Define a 'call-by-value' analogue of the structure $\mathsf{EC}(K_1;\approx)$, using ordinary equality on $\mathbb{N}$ as the PER representing $\mathtt{N}$. Formulate and prove a call-by-value analogue of Theorem 10.4.8 using the technology of Section 4.3.

(ii) Define a call-by-value version of PCF^{++} for which the analogue of Corollary 10.3.12 holds (cf. Subsection 7.2.2).

10.5 Enumeration and Its Consequences

In general, one way of trying to characterize computable operations $A \to B$ is to say that they map computable sequences over A to computable sequences over B. Indeed, an early attempt at defining the computable functions $f : \mathbb{R} \to \mathbb{R}$ was first to define the computable sequences of reals, then require that f respected such sequences. If, in addition, we require that f has a computable modulus of continuity, we obtain one of the characterizations of Grzegorczyk [112], which is equivalent to all other reasonable definitions of computability for functions $\mathbb{R} \to \mathbb{R}$.

We here present a result of Longo and Moggi [188] which uses this idea to give a characterization of $\mathsf{PC}^{\mathrm{eff}}$. The idea is that by recursion on σ we simultaneously construct the set $\mathsf{PC}^{\mathrm{eff}}(\sigma)$ and the set of computable sequences $\mathbb{N} \to \mathsf{PC}^{\mathrm{eff}}(\sigma)$. The beauty of this construction is that there is no explicit reference to any kind of computability beyond the familiar first-order notion. Using similar ideas, we will then show that $\mathsf{PC}^{\mathrm{eff}}$ has the status of a *maximal* effective type structure over $\mathbb{N}_\bot$. We also outline similar results for PC, using $\mathbb{N}^{\mathbb{N}}$ in place of $\mathbb{N}$.

As we saw in Section 10.4, we can regard each $\mathsf{PC}^{\mathrm{eff}}(\sigma)$ as an object in $\mathcal{M}od(K_1)$ via the simulation γ, and each $\mathsf{PC}(\sigma)$ as an object in $\mathcal{M}od(K_2)$ via the simulation ψ. The key to our results will be the following useful property of these assemblies. Here N denotes the object in $\mathcal{M}od(K_1)$ corresponding to the PER $(\mathbb{N},=)$, and N^N denotes the object in $\mathcal{M}od(K_2)$ corresponding to $(\mathbb{N}^{\mathbb{N}},=)$.

Lemma 10.5.1 (Enumeration lemma). *(i) For each σ there is a morphism ε_σ : $N \to \mathsf{PC}^{\mathrm{eff}}(\sigma)$ in $\mathcal{M}od(K_1)$ such that $\varepsilon_\sigma(e) = x$ whenever $e \Vdash^{\gamma}_{\sigma} x$.*

(ii) For each σ there is a morphism $\varepsilon_\sigma : N^N \to \mathsf{PC}(\sigma)$ in $\mathcal{M}od(K_2)$ such that $\varepsilon_\sigma(p) = x$ whenever $p \Vdash^{\psi}_{\sigma} x$.

Proof. (i) Define finite approximations $\phi_e^k \subseteq \phi_e$ as in the proof of Theorem 10.4.4, and a set X_σ of partial enumerations as in the proof of Lemma 10.4.7. Now define $E_\sigma : \mathbb{N} \to X_\sigma$ by

$$E_\sigma(e) \;=\; \begin{cases} \phi_e & \text{if } \phi_e \in X_\sigma \,, \\ \phi_e^k & \text{if } k \text{ is maximal such that } \phi_e^k \in X_\sigma \,. \end{cases}$$

Clearly E_σ is computable, and it is easy to see that a K_1 realizer for E_σ also realizes a morphism $\varepsilon_\sigma : N \to \mathsf{PC}^{\mathrm{eff}}(\sigma)$ with the required property.

(ii) Define $E_\sigma : K_2 \to K_2$ so that $E_\sigma(p)(n) = p(n)$ if the compacts $p(0),\dots,p(n)$ are consistent, and $E_\sigma(p)(n) = \lceil \bot \rceil$ otherwise. Clearly E_σ is continuous, and any K_2 realizer for it also realizes ε_σ as required. □

We give our main results for $\mathsf{PC}^{\mathrm{eff}}$ first. The following definition is a mild reformulation of that in [188]. Here *fst*, *snd* denote computable projections $\mathbb{N} \to \mathbb{N}$ corresponding to some computable pairing $\mathbb{N}^2 \to \mathbb{N}$.

Definition 10.5.2. For each type σ, we define a set $\mathsf{Q}^{\mathrm{eff}}(\sigma)$ along with a set $\mathsf{Q}^{\mathrm{eff}*}(\sigma)$ of total functions $\mathbb{N} \to \mathsf{Q}^{\mathrm{eff}}(\sigma)$ as follows:

- $\mathsf{Q}^{\mathrm{eff}}(\mathtt{N}) = \mathbb{N}_\bot$, and $\mathsf{Q}^{\mathrm{eff}*}(\mathtt{N})$ consists of the partial computable functions $\mathbb{N} \to \mathbb{N}_\bot$.
- $\mathsf{Q}^{\mathrm{eff}}(\sigma \to \tau)$ consists of all $f : \mathsf{Q}^{\mathrm{eff}}(\sigma) \to \mathsf{Q}^{\mathrm{eff}}(\tau)$ such that $h \in \mathsf{Q}^{\mathrm{eff}*}(\sigma)$ implies $f \circ h \in \mathsf{Q}^{\mathrm{eff}*}(\tau)$.
- $\mathsf{Q}^{\mathrm{eff}*}(\sigma \to \tau)$ consists of all $g : \mathbb{N} \to \mathsf{Q}^{\mathrm{eff}}(\sigma \to \tau)$ such that

$$h \in \mathsf{Q}^{\mathrm{eff}*}(\sigma) \;\Rightarrow\; \Lambda n.\, g(\mathit{fst}\, n)(h(\mathit{snd}\, n)) \in \mathsf{Q}^{\mathrm{eff}*}(\tau)\,.$$

This yields a type structure $\mathsf{Q}^{\mathrm{eff}}$ over $\mathbb{N}_\bot$.

Theorem 10.5.3 (Longo–Moggi). $\mathsf{Q}^{\mathrm{eff}} \cong \mathsf{PC}^{\mathrm{eff}}$.

Proof. We use the characterization of Theorem 10.4.8. We show by simultaneous induction on σ that $\mathsf{Q}^{\mathrm{eff}}(\sigma) = \mathsf{PC}^{\mathrm{eff}}(\sigma)$ and that $\mathsf{Q}^{\mathrm{eff}*}(\sigma)$ coincides with the set of morphisms $N \to \mathsf{PC}^{\mathrm{eff}}(\sigma)$ in $\mathcal{M}od(K_1)$. For the type N, both claims are trivial. For $\sigma \to \tau$, we first note that any morphism $f : \mathsf{PC}^{\mathrm{eff}}(\sigma) \to \mathsf{PC}^{\mathrm{eff}}(\tau)$ in $\mathcal{M}od(K_1)$ maps morphisms $h : N \to \mathsf{PC}^{\mathrm{eff}}(\sigma)$ to morphisms $f \circ h : N \to \mathsf{PC}^{\mathrm{eff}}(\sigma)$, so that $f \in \mathsf{Q}^{\mathrm{eff}}(\sigma \to \tau)$ using the induction hypothesis. Conversely, we have the morphism $\varepsilon_\sigma : N \to \mathsf{PC}^{\mathrm{eff}}(\sigma)$ from Lemma 10.5.1(i), so if $f \in \mathsf{Q}^{\mathrm{eff}}(\sigma \to \tau)$ then $f \circ \varepsilon_\sigma$ is a morphism $N \to \mathsf{PC}^{\mathrm{eff}}(\tau)$ by the induction hypotheses. But any K_1-realizer for this morphism is clearly also a realizer for $f : \mathsf{PC}^{\mathrm{eff}}(\sigma) \to \mathsf{PC}^{\mathrm{eff}}(\tau)$ in $\mathcal{M}od(K_1)$. This establishes $\mathsf{Q}^{\mathrm{eff}}(\sigma \to \tau) = \mathsf{PC}^{\mathrm{eff}}(\sigma \to \tau)$.

For $\mathsf{Q}^{\mathrm{eff}*}(\sigma \to \tau)$, the argument is similar. Any morphism $N \to \mathsf{PC}^{\mathrm{eff}}(\sigma \to \tau)$ induces a mapping from morphisms $N \to \mathsf{PC}^{\mathrm{eff}}(\sigma)$ to morphisms $N \to \mathsf{PC}^{\mathrm{eff}}(\tau)$ as specified. Conversely, if $g(0), g(1), \dots$ is any sequence over $\mathsf{PC}^{\mathrm{eff}}(\sigma \to \tau)$ with the specified property, then in particular $\Lambda n.g(\mathit{fst} n)(\varepsilon_\sigma(\mathit{snd} n))$ is a morphism $N \to \mathsf{PC}^{\mathrm{eff}}(\tau)$, and now a K_1 realizer for $\Lambda n_0.\Lambda n_1.g(n_0)(\varepsilon_\sigma(n_1))$ yields a K_1 realizer for $g(n_0)$ computably in n_0. □

Next, we use Lemma 10.5.1 to say something about the place of $\mathsf{PC}^{\mathrm{eff}}$ within the poset $\mathscr{T}(N_\perp)$ of type structures over $\mathbb{N}_\perp$ as introduced in Subsection 3.6.4. Recall from Definition 3.5.12 that such a type structure $\mathbf{T}$ is *effective* if it admits a numeral-respecting applicative simulation $\mathbf{T} \longrightarrow\!\!\!\triangleright K_1$, allowing us to view each $\mathbf{T}(\sigma)$ as an object in $\mathscr{M}od(K_1)$. We say the effective type structure $\mathbf{T}$ is *standard* if $\mathbf{T}(\mathtt{N})$ is canonically isomorphic to $N_\perp$ within $\mathscr{M}od(K_1)$.

Theorem 10.5.4. $\mathsf{PC}^{\mathrm{eff}}$ *is a maximal standard effective type structure over* $\mathbb{N}_\perp$. *Moreover, all standard realizations* $\varepsilon : \mathsf{PC} \longrightarrow\!\!\!\triangleright K_1$ *are* $\sim$*-equivalent in the sense of Definition 3.4.2.*

Proof. Suppose ε is a standard effective realization of $\mathbf{T}$ and $\mathsf{PC}^{\mathrm{eff}} \preceq \mathbf{T}$ in $\mathscr{T}(N_\perp)$. We shall show by induction on σ that $\mathbf{T}(\sigma) \cong \mathsf{PC}^{\mathrm{eff}}(\sigma)$ within $\mathscr{M}od(K_1)$. For type $\mathtt{N}$ this is given by the hypothesis, since $\mathsf{PC}^{\mathrm{eff}}(\mathtt{N}) \cong N_\perp$. For type $\sigma \to \tau$, assume we have $\mathbf{T}(\sigma) \cong \mathsf{PC}^{\mathrm{eff}}(\sigma)$ and $\mathbf{T}(\tau) \cong \mathsf{PC}^{\mathrm{eff}}(\tau)$ in $\mathscr{M}od(K_1)$. It is immediate that in $\mathscr{M}od(K_1)$ we have a canonical morphism

$$\beta_{\sigma\tau} : \mathbf{T}(\sigma \to \tau) \;\to\; (\mathbf{T}(\sigma) \Rightarrow \mathbf{T}(\tau)) \cong (\mathsf{PC}^{\mathrm{eff}}(\sigma) \Rightarrow \mathsf{PC}^{\mathrm{eff}}(\tau)) \cong \mathsf{PC}^{\mathrm{eff}}(\sigma \to \tau)\,.$$

To construct an inverse to this, note that the mapping $\varepsilon_{\sigma\to\tau}$ of Lemma 10.5.1 is clearly representable within $\mathsf{PC}^{\mathrm{eff}}(\mathtt{N} \to \sigma \to \tau)$, and hence within $\mathbf{T}(\mathtt{N} \to \sigma \to \tau)$ since $\mathsf{PC}^{\mathrm{eff}} \preceq \mathbf{T}$. We thus obtain the composite morphism

$$N \;\to\; \mathbf{T}(\mathtt{N}) \;\to\; \mathbf{T}(\sigma \to \tau)\,,$$

and any K_1 realizer for this clearly tracks $\beta_{\sigma\tau}^{-1}$. □

Very similar results also hold for PC and K_2; the only real difference is that we must use $\mathbb{N}^{\mathbb{N}}$ (the underlying set of K_2) in place of $\mathbb{N}$. For $p \in \mathbb{N}^{\mathbb{N}}$, we may set $\mathit{fst}(p)(n) = p(2n)$ and $\mathit{snd}(p)(n) = p(2n+1)$.

Definition 10.5.5. For each type σ, we define a set $\mathsf{Q}(\sigma)$ along with a set $\mathsf{Q}^*(\sigma)$ of total functions $\mathbb{N}^{\mathbb{N}} \to \mathsf{Q}(\sigma)$ as follows:

- $\mathsf{Q}(\mathtt{N}) = \mathbb{N}_\perp$, and $\mathsf{Q}^*(\mathtt{N})$ consists of the continuous functions $\mathbb{N}^{\mathbb{N}} \to \mathbb{N}_\perp$.
- $\mathsf{Q}(\sigma \to \tau)$ consists of all $f : \mathsf{Q}(\sigma) \to \mathsf{Q}(\tau)$ such that $h \in \mathsf{Q}^*(\sigma) \Rightarrow f \circ h \in \mathsf{Q}^*(\tau)$.
- $\mathsf{Q}^*(\sigma \to \tau)$ consists of all $g : \mathbb{N}^{\mathbb{N}} \to \mathsf{Q}(\sigma \to \tau)$ such that

$$h \in \mathsf{Q}^*(\sigma) \;\Rightarrow\; \Lambda p.\, g(\mathit{fst}\ p)(h(\mathit{snd}\ p)) \in \mathsf{Q}^*(\tau)\,.$$

Theorem 10.5.6. $\mathsf{Q} \cong \mathsf{PC}$.

Proof. Analogous to the proof of Theorem 10.5.3, using Lemma 10.5.1(ii). □

Recall from Definition 3.5.14 that a continuous type structure $\mathbf{T}$ is one equipped with a numeral-respecting simulation $\mathbf{T} \longrightarrow\!\!\!\triangleright K_2$. We say $\mathbf{T}$ is *standard* if $\mathbf{T}(\mathtt{N})$ is isomorphic within K_2 to the modest assembly $\mathsf{PC}(\mathtt{N})$ as given in Section 10.4. Exactly by analogy with Theorem 10.5.4, we now have:

Theorem 10.5.7. PC *is a maximal standard continuous type structure over* $\mathbb{N}_\perp$, *and all standard realizations* $\varepsilon : \mathsf{PC} \longrightarrow\!\!\!\triangleright K_2$ *are* $\sim$*-equivalent.* □

10.6 Scott's Graph Model $\mathscr{P}\omega$

Much of Scott's early work in denotational semantics focussed not on the domains $\mathsf{PC}(\sigma)$ as such, but on *algebraic lattices*, and in particular on the universal algebraic lattice $\mathscr{P}\omega$ (see Scott [259]). We conclude the chapter with a brief glance at this model, both because of its close relationship to PC and because of its interest as a computability model in its own right. For more on $\mathscr{P}\omega$, see Barendregt [9].

An *algebraic lattice* is simply a Scott domain with a greatest element $\top$; such structures have the pleasing property that every subset has a least upper bound. Clearly, algebraic lattices are closed under exponentiation as given by the proof of Theorem 10.1.3. A very simple example of an algebraic lattice is $\mathscr{P}\omega$, the powerset of $\mathbb{N}$ ordered by subset inclusion. The key fact we shall need is the following:

Proposition 10.6.1. *Any algebraic lattice X with a countable set of compacts is a retract of $\mathscr{P}\omega$ via continuous maps.*

Proof. Let $c : \mathbb{N} \to X$ enumerate the compacts of X. Then we may define a map $X \to \mathscr{P}\omega$ by $x \mapsto \{n \mid c(n) \sqsubseteq x\}$, and a map $\mathscr{P}\omega \to X$ by $A \mapsto \bigsqcup\{c(n) \mid n \in A\}$. Clearly these are both continuous and compose to give id_X. □

In particular, we have $\mathscr{P}\omega \Rightarrow \mathscr{P}\omega \lhd \mathscr{P}\omega$, so $\mathscr{P}\omega$ has the structure of an untyped λ-algebra. For a certain way of enumerating the compacts of $\mathscr{P}\omega \Rightarrow \mathscr{P}\omega$, the resulting application operation on $\mathscr{P}\omega$ is as follows:

$$A \cdot B \;=\; \{n \mid \exists m_0, \ldots, m_{r-1} \in B.\, \langle\langle m_0, \ldots, m_{r-1}\rangle, n\rangle \in A\}\,.$$

We also obtain the sub-λ-algebra $\mathscr{P}\omega^{\mathrm{eff}}$ consisting of the computably enumerable subsets of $\mathbb{N}$; clearly these are closed under application.

The domains $\mathsf{PC}(\sigma)$ are not algebraic lattices as they lack a top element. However, the following result shows how we may easily retrieve PC from $\mathscr{P}\omega$. We here let $N_\perp$ denote the set of elements in $\mathscr{P}\omega$ with at most one member.

Theorem 10.6.2. $\mathsf{PC} \cong \mathsf{EC}(\mathscr{P}\omega; N_\perp)$.

We leave the proof as an exercise. Note that we can actually identify each $\mathsf{PC}(\sigma)$ with a subset of $\mathscr{P}\omega$, representing $x \in \mathsf{PC}(\sigma)$ by the set of compacts $\sqsubseteq x$.

Despite their great mathematical elegance, algebraic lattices are widely felt to be somewhat unsatisfactory as a way of modelling languages like PCF, mainly because the non-denotable top element has no obvious computational meaning, and its presence means that the important notion of *inconsistency* between two elements is not captured by the model. However, there is another possible way to look at $\mathscr{P}\omega$, namely as a model for a *non-deterministic* flavour of computation. Here we think of an element $A \in \mathscr{P}\omega$ as the set of possible values for a non-deterministic process of type `N`. (Indeed, in [177] it is indicated how the Karoubi envelope of $\mathscr{P}\omega^{\mathrm{eff}}$ yields a fully abstract and universal model for an extension of PCF with non-deterministic choice.) From this perspective, Theorem 10.6.2 can be seen as saying (broadly) that the functionals computable by non-deterministic means that happen to yield deterministic results are exactly those of PC.

Chapter 11
The Sequentially Realizable Functionals

In the previous chapter, we saw how the model PC offers a 'maximal' class of partial computable functionals strictly extending SF (in the sense of the poset $\mathscr{T}(\mathbb{N}_\perp)$ of Subsection 3.6.4). In the present chapter, we show that SF can also be extended in a very different direction to yield another class SR of 'computable' functionals which is in some sense incompatible with PC. This class was first identified by Bucciarelli and Ehrhard [45] as the class of *strongly stable* functionals; later work by Ehrhard [69], van Oosten [294] and Longley [176] established the computational significance of these functionals, investigated their theory in some detail, and provided a range of alternative characterizations.

Much of the interest of SR lies in the fact that, by contrast with PC, its elements are all 'sequentially computable', albeit in a sense more generous than that of PCF-sequentiality. Indeed, the functionals in question were termed *sequentially realizable (SR)* in [176] in order to emphasize this aspect. Our primary definition of SR will reflect this: the elements of SR will be the functionals representable by elements of the van Oosten model B for sequential computation (see Subsection 3.2.4), in much the same way that the elements of Ct or PC may be defined as the functionals representable by associates in K_2.

We will give a somewhat brief introduction to this class of functionals here, referring the reader to [176] for further information. In Section 11.1 we define the model SR, establish its basic properties, and present a simple example to illustrate its incompatibility with PC. In Section 11.2 we obtain the central structural result, namely that the type $\overline{2}$ is universal within SR. Many other results flow easily once this key fact is in place. In Section 11.3 we give an alternative characterization of SR, showing that it is in some sense a 'maximal' type structure of sequential functions; while in Section 11.4 we show that it is classically equivalent to the strongly stable model. In Section 11.5 we investigate the order-theoretic structure of SR, which is quite different in flavour from that of either PC or SF. Finally, in Section 11.6 we look briefly at some different but related type structures, to give some idea of the geography of the poset $\mathscr{T}(\mathbb{N}_\perp)$ as a whole and the place of SR within it.

11.1 Basic Structure and Examples

We assume familiarity with the van Oosten model $\mathsf{B} = (\mathbb{N}_\bot^\mathbb{N}, \cdot)$ and its effective submodel $\mathsf{B}^{\mathrm{eff}}$ as defined in Subsection 3.2.4. This will be studied as an intensional model of interest in its own right in Section 12.1. The main fact we need to import here is that B is a total combinatory algebra (with $(\mathsf{B};\mathsf{B}^{\mathrm{eff}})$ a relative combinatory algebra)—see Theorem 12.1.14. We will also use the obvious fact that application in B is continuous with respect to the pointwise order on $\mathbb{N}_\bot^\mathbb{N}$.

We shall choose to represent each element $a \in \mathbb{N}_\bot$ by the element $p_a \in \mathsf{B}$, where $p_a(0) = a$ and $p_a(n+1) = \bot$ (many equivalent representations are possible). We construct our models of interest as follows.

Definition 11.1.1. (i) Let $N_\bot$ be the set $\{p_a \mid a \in \mathbb{N}_\bot\}$, and define $\mathsf{SR} = \mathsf{EC}(\mathsf{B};N_\bot)$.

(ii) Let $\mathsf{SR}^{\mathrm{eff}}$ denote the substructure of SR consisting of the elements of SR represented by some element of $\mathsf{B}^{\mathrm{eff}}$.

As usual, an element of B representing some $f \in \mathsf{SR}$ will be called a *realizer* for f.

It is now immediate for general reasons that SR is an extensional type structure over $\mathbb{N}_\bot$, and hence a $\lambda\eta$-algebra, with $(\mathsf{SR};\mathsf{SR}^{\mathrm{eff}})$ a relative $\lambda\eta$-algebra (see Proposition 4.1.3). For notational purposes, we will henceforth treat SR as a *canonical* type structure, writing application in SR as ordinary function application. As usual, we will freely extend these structures in the canonical way with base types $\mathtt{U}$ and $\mathtt{B}$ when convenient, as explained in Section 4.2.

Proposition 11.1.2. SR *and* $\mathsf{SR}^{\mathrm{eff}}$ *are models of* PCF$^\Omega$ *and* PCF *respectively, in the sense of Definition 7.1.3.*

Proof. It is an easy exercise to construct realizers for suitable functions *suc*, *pre*, *ifzero*, and for the functions C_f in the case of SR. For Y_σ, we use that the definition of EC yields a logical relation R between SR and B. For any $k \in \mathbb{N}$, let Y_σ^k denote the interpretation of the λ-term $\lambda f^{\sigma\to\sigma}.f^k(\bot_\sigma)$ in SR, and let y^k denote the interpretation of the corresponding untyped term in B (interpreting $\bot$ as $\Lambda n.\bot$). By Lemma 4.4.3, we have $R(Y_\sigma^k, y^k)$ for each k. Moreover, the y^k form an increasing chain with respect to the pointwise order on B, since we have $y^k = g^k(\bot)$ for a certain λ-definable g. Define $y = \bigsqcup y^k$; then it is easy to check that y represents an element $Y_\sigma \in \mathsf{SR}((\sigma\to\sigma)\to\sigma)$, so that $R(Y_\sigma, y)$. Indeed, we have $Y_\sigma \in \mathsf{SR}^{\mathrm{eff}}$, since y itself is effective. Moreover, by continuity in B we have $y\cdot f = f\cdot(y\cdot f)$ for any $f \in \mathsf{B}$, and it follows that $Y_\sigma(F) = F(Y_\sigma(F))$ for any $F \in \mathsf{SR}(\sigma\to\sigma)$. Finally, it is clear by continuity that y is an observational limit of the y_k in B, whence also Y_σ is an observational limit of the Y_σ^k as required by Definition 7.1.3. □

By the general results of Section 7.1, it follows that we have adequate interpretations of PCF$^\Omega$ and PCF in SR and $\mathsf{SR}^{\mathrm{eff}}$ respectively, and hence that $\mathsf{SF} \preceq \mathsf{SR}$ and $\mathsf{SF}^{\mathrm{eff}} \preceq \mathsf{SR}^{\mathrm{eff}}$ in the poset $\mathscr{T}(\mathbb{N}_\bot)$ of all type structures over $\mathbb{N}_\bot$. On the other hand, the parallel convergence tester of Subsection 7.3.1 is clearly not realizable in B and hence not present in SR, so we do not have $\mathsf{PC} \preceq \mathsf{SR}$.

The following example illustrates what is distinctive about SR.

Example 11.1.3. $\mathsf{SR}(\mathtt{U} \to \mathtt{U})$ contains exactly the three elements $\Lambda x.\top, \Lambda x.x, \Lambda x.\bot$. Moreover, there is a functional *strict?* $\in \mathsf{SR}^{\mathrm{eff}}((\mathtt{U} \to \mathtt{U}) \to \mathtt{B})$ such that

$$strict?(\Lambda x.\top) = f\!f\,, \qquad strict?(\Lambda x.x) = tt\,, \qquad strict?(\Lambda x.\bot) = \bot\,.$$

To show this, one must construct $r \in \mathsf{B}^{\mathrm{eff}}$ for this functional. Informally, given any $g \in \mathsf{SR}(\mathtt{U} \to \mathtt{U})$ realized by $t \in \mathsf{B}$, r will check by examination of t that $g(\top) = \top$ (otherwise r diverges), and will also see whether t ever 'looks at' its argument of type $\mathtt{U}$. We leave the detailed construction of r as an exercise.

We thus see that *strict?* is a perfectly extensional function and is in a sense 'sequentially computable', at least at the intensional level of realizers. Indeed, *strict?* can easily be implemented in many 'sequential' programming languages, e.g. using exceptions or state to detect whether the given $f : \mathtt{U} \to \mathtt{U}$ evaluates its argument: see Longley [175]. Loosely speaking, one may think of SR as consisting of PCF-sequential functionals together with the function *strict?* and 'all things like it'.

It is clear that *strict?* itself is not PCF-definable, since its behaviour is not monotone with respect to the pointwise order: we have $\Lambda x.x \preceq \Lambda x.\top$ but $tt \not\preceq f\!f$. This non-monotonicity also shows that *strict?* has no counterpart in PC, so that we do not have $\mathsf{SR} \preceq \mathsf{PC}$ or $\mathsf{SR}^{\mathrm{eff}} \preceq \mathsf{PC}^{\mathrm{eff}}$ in $\mathscr{T}(\mathbb{N}_\bot)$. Indeed, combining this with Theorems 10.5.7 and 10.5.4, we can conclude the following:

Theorem 11.1.4. *There can be no continuous type structure* $\mathbf{P} \in \mathscr{T}(\mathbb{N}_\bot)$ *subsuming both* PC *and* SR. *Likewise, there can be no effective* $\mathbf{P} \in \mathscr{T}(\mathbb{N}_\bot)$ *subsuming both* $\mathsf{PC}^{\mathrm{eff}}$ *and* $\mathsf{SR}^{\mathrm{eff}}$. □

The second half of this theorem was presented in [176] as a kind of 'anti-Church's thesis' for partial computable functionals of higher type.

As a general setting for our investigation of SR, we will find it convenient to work within the category $\mathscr{P}er(\mathsf{B})$, or equivalently $\mathscr{M}od(\mathsf{B})$ (see Subsection 3.6.3). By definition of $\mathsf{EC}(\mathsf{B};P)$, the types $\mathsf{SR}(\sigma)$ can be viewed as the objects generated via exponentiation from an evident PER representing $N_\bot$. Since also $\mathscr{P}er(\mathsf{B})$ has idempotent splittings, we have that the Karoubi envelope $\mathscr{K}(\mathsf{SR})$ embeds fully and faithfully in $\mathscr{P}er(\mathsf{B})$ (see Proposition 4.1.17 and the remark following it). A large repertoire of datatypes of interest is therefore naturally available within this category. In fact, everything of interest also works within the subcategory $\mathscr{P}er(\mathsf{B};\mathsf{B}^{\mathrm{eff}})$, consisting of morphisms with an effective realizer.

We draw particular attention to the *call-by-value* interpretation of the simple types—in effect, their interpretation in the partial λ-algebra corresponding to SR. Since a natural number object N is available in $\mathscr{P}er(\mathsf{B})$, and $\mathscr{P}er(\mathsf{B})$ carries an evident partiality structure yielding a lift functor $L : \mathscr{P}er(\mathsf{B}) \to \mathscr{P}er(\mathsf{B})$ (see Subsection 4.2.4), we can define the relevant objects $\mathsf{SR}(\sigma_v)$ by

$$\mathsf{SR}(\mathtt{N}_v) = N\,, \qquad \mathsf{SR}((\sigma \to \tau)_v) = \mathsf{SR}(\sigma) \Rightarrow L(\mathsf{SR}(\tau))\,.$$

Some concrete presentations of the objects $\mathsf{SR}(1_v)$ and $\mathsf{SR}(2_v)$ will help to clarify this definition. We first note that $L(N) = N_\bot$: in fact, since we will be concentrating on the pure types, this is all we really need to know about L. We therefore have that $\mathsf{SR}(1_v) = N_\perp^N$. It is easy to see that this object, viewed as a partial equivalence relation, can be taken to be simply the equality relation on B; thinking of it as a modest assembly (see Subsection 3.3.6), every element of B is simply a realizer for itself. This is the presentation we will always have in mind when dealing with the object $\mathsf{SR}(1_v)$.

As regards $\mathsf{SR}(2_v)$, it is clear that one possible presentation is as the (total) equivalence relation $\sim$ on B, where $f \sim f'$ iff $f \mid g = f' \mid g$ for all $g \in \mathsf{B}$. Here $\mid : \mathbb{N}_\perp^\mathbb{N} \times \mathbb{N}_\perp^\mathbb{N} \to \mathbb{N}_\perp$ denotes a version of application that treats its first argument as coding a single decision tree rather than a forest of them (cf. Subsection 3.2.4):

$$\begin{aligned} f \mid g \;=\; a \text{ iff } & \exists n_0, m_0, \ldots n_{r-1}, m_{r-1}. \\ & \forall i < r.\, f(\langle m_0, \ldots, m_{i-1} \rangle) = 2n_i \;\wedge \\ & \forall i < r.\, g(n_i) = m_i \;\wedge \\ & f(\langle m_0, \ldots, m_{r-1} \rangle) = 2a+1\,. \end{aligned}$$

Presenting this as a modest assembly, if $f \in \mathsf{B}$ and $F \in \overline{2}_v$, we shall write $f \Vdash F$ to mean $F = \Lambda g. f \mid g$.

Another more restricted presentation of $\mathsf{SR}(2_v)$ will also play an important role:

Definition 11.1.5. Consider $f \in \mathsf{B}$ as a decision tree for $\Lambda g. f \mid g$ as above. We say f is *irredundant* if the following hold for all $m_0, \ldots, m_{r-1} \in \mathbb{N}$ and all $t < r$:

1. If $f(\langle m_0, \ldots, m_{r-1} \rangle) \downarrow$ then $f(\langle m_0, \ldots, m_{t-1} \rangle)$ is defined and is a question (i.e. an even number).
2. If $f(\langle m_0, \ldots, m_{r-1} \rangle) \downarrow$ then $f(\langle m_0, \ldots, m_{r-1} \rangle) \neq f(\langle m_0, \ldots, m_{t-1} \rangle)$.

We write $f \Vdash_{\mathrm{irr}} F$ to mean that f is irredundant and $F = \Lambda g. f \mid g$.

Informally, an irredundant decision tree has no inaccessible nodes, and never asks the same question twice along any path. The fact that $\Vdash$ and $\Vdash_{\mathrm{irr}}$ define isomorphic objects within $\mathscr{P}er(\mathsf{B})$ is given by the following:

Proposition 11.1.6. *There exists $Irr \in \mathsf{B}^{\mathrm{eff}}$ such that if $f \Vdash F$ then $Irr \cdot f \Vdash_{\mathrm{irr}} F$.* □

We leave the proof as an interesting exercise (or see Proposition 2.10 of [176]).

Remark 11.1.7. Either of the above presentations makes it clear that the functionals in $\mathsf{SR}(2_v)$ are precisely those of the form $\Lambda g. f \cdot g$ for some $f \in \mathsf{B}$. Since the operation $\cdot$ is itself PCF-definable (see the proof of Proposition 7.1.30), we immediately have every $F \in \mathsf{SR}(2_v)$ is PCF^Ω-definable. Thus SR coincides with SF at the type $\overline{2}_v$, and indeed at all second-order call-by-value types, so we have just a single notion of 'sequential functional' at such types. It is at second-order *call-by-name* types that a difference arises, as illustrated by *strict?*.

This is a convenient place to mention another illuminating example of an SR functional, one that lives naturally at a third-order type.

Example 11.1.8. Consider the application of an arbitrary $F \in \mathsf{SR}(2_\nu)$ (realized by $f \in \mathsf{B}$) to an arbitrary $g \in \mathsf{SR}(1_\nu)$ (realized by itself). Assuming $F(g) = a \in \mathbb{N}$, the computation of $f \mid g$ will interrogate g at finitely many arguments i; there is therefore some smallest finite subfunction $g_F \sqsubseteq g$ such that $F(g_F) = a$. We may call g_F the *modulus* of F at g; it may be canonically represented by a list of pairs $(i_0, g(i_0)), \ldots (i_{r-1}, g(i_{r-1}))$ where $i_0 < i_1 < \cdots < i_{r-1}$, which in turn may be represented by a number $M(F,g) \in \mathbb{N}$. If $F(g) = \bot$, we may define $M(F,g) = \bot$. It is easy to check that this modulus functional M is present as an element of $\mathsf{SR}((2,1 \to 0)_\nu)$.

Notice that the requirement $i_0 < \cdots < i_{r-1}$ means that the returned value of $M(F,g)$ reveals no information about the *order* in which the values of g were interrogated. If this were not the case, M would fail to be extensional with respect to different choices of realizer for the same F.

It was observed by Alex Simpson that the functionals *strict*? and M are interdefinable relative to PCF: see [176, Proposition 9.11].

We conclude this section with some general machinery relating to type $\overline{2}_\nu$ functionals and their realizers.

Definition 11.1.9. (i) Given $F \in \mathsf{SR}(2_\nu)$, we define its *trace*, *tr* F, to be the set of all pairs $(g, F(g))$ where $g : \mathbb{N} \rightharpoonup \mathbb{N}$ is some minimal finite partial function such that $F(g)\downarrow$. We say F is *finite* if *tr* F is finite.

(ii) We write $F' \sqsubseteq_{\mathrm{tr}} F$ if *tr* $F' \subseteq$ *tr* F,

Clearly, F is determined by *tr* F, since $F(h) = n$ iff $(g,n) \in$ *tr* F for some $g \sqsubseteq h$. As in the example above, if $F(h) = n$, we will write h_F for the unique smallest subfunction of h such that $F(h_F) = n$.

If $f \Vdash_{\mathrm{irr}} F$, we say a node α in f is *accessible* if there is some $g \in \mathsf{B}$ such that the computation of $f \mid g$ requires the value of $f(\langle\alpha\rangle)$ (in the obvious sense). We also say α is an *answer node* if $f(\langle\alpha\rangle)$ is an odd number. We now see that there is a bijection between elements of *tr* F and accessible answer nodes of f. Explicitly, $(g, F(g)) \in$ *tr* F corresponds to α iff $f(\langle\alpha\rangle) = 2F(g)+1$ and whenever $\beta;m$ is a prefix of α and $f(\langle\beta\rangle) = 2n$, we have $g(n) = m$.

Lemma 11.1.10. *If $F' \sqsubseteq_{\mathrm{tr}} F$ and $g \Vdash_{\mathrm{irr}} F$, there exists $g' \sqsubseteq g$ with $g' \Vdash_{\mathrm{irr}} F'$.*

Proof. We may obtain g' by restricting the domain of g to prefixes of answer nodes corresponding to elements of *tr* F'. □

Exercise 11.1.11. Show that the dual of Lemma 11.1.10 fails: if $g' \Vdash_{\mathrm{irr}} F' \sqsubseteq_{\mathrm{tr}} F$, there need not exist g with $g' \sqsubseteq g \Vdash_{\mathrm{irr}} F$.

11.2 A Universal Type

We now prove the main structural result: the type $\overline{3}_\nu$ is a retract of $\overline{2}_\nu$ (e.g. within the category $\mathscr{P}er(\mathsf{B})$). From this it follows easily that all types $\overline{k}_\nu$ are retracts of $\overline{2}_\nu$, and hence that all $\mathsf{SR}(\sigma)$ are retracts of $\mathsf{SR}(2_v)$ or indeed $\mathsf{SR}(2)$. Our presentation here somewhat streamlines and simplifies that of [176].

One half of the retraction is easy to describe. If $\mathsf{SR}(1_v)$ is presented as the identity relation on B and $\mathsf{SR}(2_v)$ as the total equivalence relation $\sim$ mentioned in Section 11.1, then the identity mapping on B realizes a canonical surjection $j : \overline{1}_v \to \overline{2}_v$ within $\mathscr{P}er(\mathsf{B})$, where $j(f) = \Lambda g.f \mid g$ for any f. Obviously we cannot expect j to have a one-sided inverse, since one cannot continuously extract a canonical choice of decision tree for every $F \in \mathsf{SR}(2_v)$. Suppose, however, that we 'lift' j to the next type level by defining

$$J \;=\; (\Lambda\Phi : \mathsf{SR}(2_v \to 0).\, \Phi \circ j) \;:\; \mathsf{SR}(3_v) \to \mathsf{SR}(2_v)\,.$$

Then it turns out that J does have a one-sided inverse $H : \mathsf{SR}(2_v) \to \mathsf{SR}(3_v)$, exhibiting $\overline{3}_v$ as a retract of $\overline{2}_v$.

The intuition behind H is as follows. An arbitrary element $F \in \mathsf{SR}(2_v)$—that is, a morphism $F : \mathsf{SR}(1_v) \to \mathsf{SR}(0)$ within $\mathscr{P}er(\mathsf{B})$—may be naturally regarded as representing an element of $\mathsf{SR}(3_v)$ iff F is extensional with respect to the equivalence relation $\sim$ on $\mathsf{SR}(1_v) \cong \mathsf{B}$. Clearly, not all F are extensional in this way, but the idea of H is, roughly speaking, to map each F to the best approximation to F that is extensional. A first attempt at capturing this idea might be to define

$$H(F)(G) \;=\; n \in \mathbb{N} \;\;\text{iff}\;\; \forall g \Vdash G.\, F(g) = n\,.$$

This particular definition will not work, because there is no morphism H in $\mathscr{P}er(\mathsf{B})$ with exactly this property, but it conveys something of the flavour of the construction. For the correct definition of H, two modifications are needed.

First, we need to work with $\Vdash_{\mathrm{irr}}$ rather than $\Vdash$. Intuitively, since there are fewer irredundant realizers g than arbitrary ones for a given G, it will be more feasible to check effectively whether $F(g) = n$ for all of them.

Secondly, our condition for $H(F)(G) = n$ will require not just that F yields the same numerical result on all realizers g of G, but also that it consumes the same amount of information about G in each case. This leads us to focus on the 'minimal subfunction' G' of G that embodies this information, rather than on G itself. In essence, we want to say that F yields the same result on all realizers g' for G', and moreover uses up all information about G' in each case. The correct specification of H captures this idea:

Definition 11.2.1. An *H-functional* is a morphism $H : \mathsf{SR}(2_v) \to \mathsf{SR}(3_v)$ in $\mathscr{P}er(\mathsf{B})$ such that

$$H(F)(G) \;=\; n \in \mathbb{N} \;\;\text{iff}\;\; \exists G' \sqsubseteq_{\mathrm{tr}} G.\, \forall g' \Vdash_{\mathrm{irr}} G'.\, F(g') = n \,\wedge\, j(g'_F) = G'\,.$$

We will show that an H-functional exists in $\mathscr{P}er(\mathsf{B})$ and has an effective realizer. We will then show that any H-functional is necessarily a one-sided inverse to J.

As regards the existence of an H-functional, the following shows that Definition 11.2.1 at least yields a well-defined set-theoretic function $H(F) : \mathsf{SR}(2_v) \to \mathbb{N}_\perp$ for any F.

Proposition 11.2.2. *For any $F, G \in \mathsf{SR}(2_\nu)$, if there exists some $G' \sqsubseteq_{\mathrm{tr}} G$ such that $\forall g \Vdash_{\mathrm{irr}} G'.\ F(g) = n \wedge j(g_F) = G'$, then G' is unique and is finite. Hence there is at most one n for which the condition for $H(F)(G) = n$ is satisfied.*

Proof. Given $G', G'' \sqsubseteq_{\mathrm{tr}} G$ both satisfying the above condition with respect to some n, take any $g \Vdash_{\mathrm{irr}} G$, and restrict g to $g' \Vdash_{\mathrm{irr}} G'$, $g'' \Vdash_{\mathrm{irr}} G''$ as in Lemma 11.1.10. Then $F(g') = n$, so $g_F = g'_F$, whence $j(g_F) = G'$; but likewise $j(g_F) = G''$, so $G' = G''$. Note that g'_F is finite and $j(g_F) = G'$, so that *tr* G' is finite. Finally, since $F(g') = n$ and $g' \sqsubseteq g$, we have $F(g) = n$, so n itself is uniquely determined. □

Note that this shows that if $H(F)(G) = n$ then $\forall g \Vdash_{\mathrm{irr}} G.\ F(g) = n$.

The key to the 'computability' of H lies in the following crucial fact:

Lemma 11.2.3. *If $F \in \mathsf{SR}(2_\nu)$ is finite, there is a finite set of irredundant realizers $h_0, \ldots, h_{r-1}$ for F, each with finite domain, such that any $h \Vdash_{\mathrm{irr}} F$ extends some h_i. Moreover, the set $\{h_0, \ldots, h_{r-1}\}$ can be effectively computed from tr F (where we represent finite partial functions $N \rightharpoonup N$ by their graphs).*

Proof. Clearly, a minimal irredundant realizer for F is one whose domain consists only of prefixes of answer nodes, and any $h \Vdash_{\mathrm{irr}} F$ extends some such minimal realizer. If h' is a minimal irredundant realizer, the answer nodes of h' correspond to the finitely many elements of *tr* F. Moreover, if $\langle \alpha \rangle \in \mathrm{dom}\, h'$ corresponds to $(g, F(g)) \in tr\, F$ (with $h'(\langle \alpha \rangle) = 2F(g) + 1$), then the restriction of h to prefixes of α is in effect determined by some total ordering of the graph of g: if dom g consists of the distinct elements $n_0, \ldots, n_{r-1}$, then $h(\langle g(n_0), \ldots, g(n_{i-1}) \rangle) = 2n_i$ for each $i < r$, and $\alpha = g(n_0), \ldots, g(n_{r-1})$. Since dom h' is the union of such restrictions, it is now clear that only finitely many such h' can exist, and that a list of them is effectively computable from *tr* F. □

Lemma 11.2.4. *For any $F \in \mathsf{SR}(2_\nu)$, there exists an element $H(F) \in \mathsf{SR}(3_\nu)$ such that the condition of Definition 11.2.1 is satisfied for every $G \in \mathsf{SR}(2_\nu)$.*

Proof. Suppose $g \Vdash_{\mathrm{irr}} G \in \mathsf{SR}(2_\nu)$. We will describe a process for computing $H(F)(G)$ given g; we leave it to the reader to check that this computation can be carried out within B itself—that is, there is some $r \in \mathsf{B}$ (dependent on F) such that $r \mid g = H(F)(G)$.

First compute $F(g)$. If this yields a numeral n, then compute also the modulus g_F as a finite graph. Then g_F realizes some $G' \sqsubseteq_{\mathrm{tr}} G$, where *tr* G' is finite and may be computed from g_F. Using Lemma 11.2.3, we may now enumerate all minimal irredundant realizers of G'. For each such realizer h in turn, we may check whether $F(h) = n$ and also $h_F = h$ (so that $j(h_F) = G'$). If the answers are all positive, return n as the value of $H(F)(G)$, otherwise diverge.

To show that $H(F)(G)$ is well-defined, we must check that this computation yields the same result on all possible realizers $g \Vdash_{\mathrm{irr}} G$. So suppose $g_0, g_1 \Vdash_{\mathrm{irr}} G$ where the above computation yields n on g_0, and let G'_i be the functional realized by g_{iF} for $i = 0, 1$. Since $G'_0 \sqsubseteq_{\mathrm{tr}} G$ and $g_1 \Vdash_{\mathrm{irr}} G$, we may find $g_0^* \sqsubseteq g_1$ such that $g_0^* \Vdash_{\mathrm{irr}} G'_0$; but then g_0^* is one of the realizers h for which we have checked that

$F(h) = n$ and $h_F = h$. It follows that $g^*_{0F} = g^*_0$ whence $g_{1F} = g^*_0$, and hence that $G'_1 = G'_0$. We now see that the above procedure will check exactly the same set of realizers h whether applied to g_0 or to g_1, so that the result will be n in both cases.

Next we check that $H(F)(G)$, thus defined, satisfies the specification of Definition 11.2.1. First suppose $H(F)(G) = n$; we wish to show that

$$\exists G' \sqsubseteq_{\mathrm{tr}} G.\, \forall g' \Vdash_{\mathrm{irr}} G'.\, F(g') = n \wedge j(g'_F) = G' .$$

Take any $g \Vdash_{\mathrm{irr}} G$ and set $G' = j(g_F)$ as above, so that $G' \sqsubseteq_{\mathrm{tr}} G$. For any $g' \Vdash_{\mathrm{irr}} G'$, there is a minimal irredundant $h' \Vdash_{\mathrm{irr}} G'$ with $h' \sqsubseteq g'$, and this will be among the h for which our computation has checked that $F(h) = n$ and $h_F = h$. It follows that $F(g') = n$ and $g'_F = h'$, whence $j(g'_F) = G'$.

Conversely, suppose there exists $G' \sqsubseteq_{\mathrm{tr}} G$ as above, and consider the computation of $H(F)(G)$ using an arbitrary $g \Vdash_{\mathrm{irr}} G$. As before, we may restrict g to some $g' \Vdash_{\mathrm{irr}} G'$; we then have $F(g') = n$ and $j(g'_F) = G'$, whence $g_F = g'_F$ and $j(g_F) = G'$; thus G' coincides with the functional G' featuring in the computation of $H(F)(G)$. But now our condition on G' ensures that $F(h) = n$ and $h_F = h$ for all minimal $h \Vdash_{\mathrm{irr}} G'$, so that the computation of $H(F)(G)$ will indeed return n. □

We have thus shown that H is well-defined as a set-theoretic function from $\mathsf{SR}(2_\nu)$ to $\mathsf{SR}(3_\nu)$. It remains to check that the passage from F to $H(F)$ can itself be effected within $\mathsf{B}^{\mathrm{eff}}$:

Theorem 11.2.5. *An H-functional exists in $\mathscr{P}er(\mathsf{B})$ and has a realizer in $\mathsf{B}^{\mathrm{eff}}$.*

Proof. This amounts to showing that a realizer for the algorithm described in the proof of Lemma 11.2.4 can itself be computed from a realizer $f \Vdash_{\mathrm{irr}} F$ via some element of $\mathsf{B}^{\mathrm{eff}}$. Again we leave the details to the reader; for an outline see Proposition 7.7 of [176]. □

We now show that any H-functional is indeed a one-sided inverse to J.

Lemma 11.2.6. *If $g \Vdash G$ and $\Phi(G) = n$, then $j(g_{\Phi\circ j})$ is the unique $\sqsubseteq_{\mathrm{tr}}$-smallest subfunction $G_\Phi \sqsubseteq_{\mathrm{tr}} G$ such that $\Phi(G_\Phi) = n$.*

Proof. Suppose $G' \sqsubseteq_{\mathrm{tr}} G$ and $\Phi(G') = n$. Then we may take $g' \sqsubseteq g$ such that the accessible answer nodes of g' correspond to the elements of $tr\, G'$; we then have that $G' = j(g')$, so that $(\Phi \circ j)(g') = n$. But since $g' \sqsubseteq g$, we now have $g_{\Phi\circ j} \sqsubseteq g'$. So every accessible answer node in $g_{\Phi\circ j}$ is an accessible answer node in g', whence $j(g_{\Phi\circ j}) \sqsubseteq_{\mathrm{tr}} j(g') = G'$. Thus $j(g_{\Phi\circ j})$ has the stated minimality property. □

Proposition 11.2.7. *If $H : \mathsf{SR}(2_\nu) \to \mathsf{SR}(3_\nu)$ is an H-functional, then $H \circ J = id_{3_\nu}$.*

Proof. Suppose H is an H-functional and $\Phi \in \mathsf{SR}(3_\nu)$; we wish to show that $H(J(\Phi))(G) = \Phi(G)$ for any $G \in \mathsf{SR}(2_\nu)$. But for any $n \in \mathbb{N}$, we have by Definition 11.2.1 that $H(J(\Phi))(G) = H(\Phi \circ j)(G) = n$ iff

$$\exists G' \sqsubseteq_{\mathrm{tr}} G.\, \forall g' \Vdash_{\mathrm{irr}} G'.\, \Phi(j(g')) = n \wedge j(g'_{\Phi\circ j}) = G' .$$

But if $g' \Vdash_{\mathrm{irr}} G'$ then $j(g') = G'$, so the first conjunct here is equivalent to $\Phi(G') = n$, which implies $\Phi(G) = n$. So if $H(J(\Phi))(G) = n$ then $\Phi(G) = n$. Conversely, if $\Phi(G) = n$, pick $g \Vdash_{\mathrm{irr}} G$ and let $G_\Phi = j(g_{\Phi\circ j})$ as in Lemma 11.2.6. Then for all $g' \Vdash_{\mathrm{irr}} G_\Phi$ we have $j(g') = G_\Phi$ so $\Phi(j(g')) = n$; also setting $G'' = j(g'_{\Phi\circ j})$ yields $G'' \sqsubseteq_{\mathrm{tr}} G_\Phi \sqsubseteq_{\mathrm{tr}} G$ with $\Phi(G'') = n$, whence $G'' = G_\Phi$ by the minimality of G_Φ. □

The following theorem pulls all this together:

Theorem 11.2.8. *Within $\mathcal{P}er(\mathsf{B})$ or even $\mathcal{P}er(\mathsf{B};\mathsf{B}^{\mathrm{eff}})$, every call-by-value type $\mathsf{SR}(\sigma_v)$, and every call-by-name type $\mathsf{SR}(\sigma)$, is a retract of $\mathsf{SR}(2_v)$ and hence of $\mathsf{SR}(2)$.*

Proof. Working in $\mathcal{P}er(\mathsf{B};\mathsf{B}^{\mathrm{eff}})$, the above morphisms J and H constitute a retraction $\overline{3}_v \lhd \overline{2}_v$. This lifts inductively to $\overline{k+1}_v \lhd \overline{k}_v$ for all $k \geq 2$, whence $\overline{k}_v \lhd \overline{2}$ for all k. Since $\overline{0} \lhd \overline{1}_v$, we inductively have $\overline{k} \lhd \overline{k+1}_v$ for any k, so also $\overline{k} \lhd \overline{2}_v$ for any k. By Proposition 4.2.3 and Remark 7.2.4 it follows that $\overline{\sigma} \lhd \overline{2}_v$ and $\overline{\sigma_v} \lhd \overline{2}_v$ for all σ. Finally, $\overline{2}_v \lhd \overline{2}$, so all the above holds with $\overline{2}$ in place of $\overline{2}_v$. □

We now derive some easy consequences of this theorem. The first of these can be seen as an analogue of Theorem 7.1.32:

Corollary 11.2.9. *For any object $X \lhd \mathsf{SR}(2_v)$ in $\mathcal{M}od(\mathsf{B};\mathsf{B}^{\mathrm{eff}})$, there is a morphism $I_X : \mathsf{SR}(1) \to X$ such that $I_X(f) = F$ whenever $f \Vdash_X F$. Hence for any type σ there is an element $I_\sigma \in \mathsf{SR}^{\mathrm{eff}}((1 \to \sigma)_v)$ such that $I_\sigma(f) = F$ whenever $f \Vdash F \in \mathsf{SR}(\sigma_v)$, and similarly for the call-by-name types.*

Proof. If $(X, \Vdash)$ is any modest set at all endowed with a retraction $(T,R) : \mathsf{SR}(2_v)$ in $\mathcal{M}od(\mathsf{B};\mathsf{B}^{\mathrm{eff}})$, take $r \in \mathsf{B}^{\mathrm{eff}}$ tracking R; then for any $f \Vdash F \in X$ we have $r \cdot f \Vdash R(F) \in \mathsf{SR}(2_v)$, so $j(f) = R(F)$ with j as above, whence $T(j(f)) = T(R(F)) = F$. So let $I_\sigma = \Lambda f.T(j(r \cdot f)) : \mathsf{SR}(1_v) \to X$. If $X = \mathsf{SR}(\sigma_v)$ then I_σ can be presented as an element of $\mathsf{SR}((1 \to \sigma)_v)$, while if $X = \mathsf{SR}(\sigma)$ then I_σ can be packaged as an element of $\mathsf{SR}(1 \to \sigma)$ using the retraction $\overline{1}_v \lhd \overline{1}$. □

Another corollary addresses the question of a programming language corresponding to SR. We write PCF_v for the call-by-value version of PCF introduced in Subsection 7.2.2, and PCF_v^Ω for its obvious oracle counterpart. The content of Subsection 7.2.2 shows, in effect, how any closed PCF_v term $M : \sigma$ may be interpreted as an element of $\mathsf{SR}(\sigma_v)_\perp$.

Corollary 11.2.10. *(i) Every element of each $\mathsf{SR}(\sigma_v)$ [resp. $\mathsf{SR}^{\mathrm{eff}}(\sigma_v)$] is definable in PCF_v^Ω [resp. in PCF_v] relative to a certain element $H \in \mathsf{SR}((2 \to 3)_v)$.*

(ii) Likewise, every element of each $\mathsf{SR}(\sigma)$ [resp. $\mathsf{SR}^{\mathrm{eff}}(\sigma)$] is definable in PCF^Ω [resp. in PCF] relative to an element $H' \in \mathsf{SR}(2 \to 3)$.

Proof. (i) By adapting the proof of Proposition 7.1.30, it is easy to see that the morphism $j : \overline{1}_v \to \overline{2}_v$ is PCF_v-definable, whence so is J. By Remark 7.2.4, It follows readily that all the retraction pairs $\sigma_v \lhd \overline{2}_v$ are PCF_v-definable relative to H; in particular, for any σ there is a $\mathrm{PCF}_v + H$-definable surjection $\mathsf{SR}(2_v) \to \mathsf{SR}(\sigma_v)$. But again by the adapted proof of Proposition 7.1.30, clearly every element of $\mathsf{SR}(2_v)$ [resp. $\mathsf{SR}^{\mathrm{eff}}(2_v)$] is definable in PCF_v^{Ω} [resp. PCF_v], and the result follows.

(ii) Using the obvious retractions $\overline{2}_v \lhd \overline{2}$ and $\overline{3}_v \lhd \overline{3}$, we may represent H by an element $H' \in \mathsf{SR}(2 \to 3)$; moreover, the element $H'' = t_{2\to 3}(H')$ (with t as in Subsection 7.2.2) is easily shown to have the same power as H relative to PCF_v. The result now follows by Theorem 7.2.5. □

Since there is an evident retraction $(\overline{2} \to \overline{3})_v \lhd \overline{3}_v$, one may in principle replace H here by a functional of type $\overline{3}_v$. However, it is shown in [176, Section 9.4] that there is no element $K \in \mathsf{SR}(\sigma)$ where σ is a level 2 (call-by-name) type such that all elements of SR are definable in $\mathrm{PCF}^{\Omega} + K$.

11.3 SR as a Maximal Sequential Model

Having established the universality of type $\overline{2}_v$ in SR, many other results follow easily. Our first example will be an alternative presentation of SR due to Colson and Ehrhard [57], leading to a characterization of SR as the *maximal* model for which the type $\overline{2}_v$ functionals are the familiar sequential ones (see Subsection 7.5.2 and Remark 11.1.7). This, in turn, yields a simple 'ubiquity' result: *any* sequential model with sufficient computational power to include H must coincide with SR.

Our treatment here will be more direct than that in [176]. The following can be seen as a close analogue of Theorems 10.5.3 and 10.5.6, and the proof is essentially the same. We work here within the category $\mathcal{M}od(\mathsf{B})$; for convenience we write U for the object $\mathsf{SR}(2_v) \cong N_\perp^N$. We define $fst, snd : U \to U$ by $fst(f)(n) = f(2n)$ and $snd(f)(n) = f(2n+1)$.

Theorem 11.3.1. *Suppose $X, Y \lhd U$ in $\mathcal{M}od(\mathsf{B})$.*

(i) A set-theoretic function $F : |X| \to |Y|$ is present in $(X \Rightarrow Y)$ iff every morphism $h : U \to X$ yields a morphism $F \circ h : U \to Y$.

(ii) A function $G : |U| \times |X| \to |Y|$ is present as a morphism $U \to (X \Rightarrow Y)$ iff every morphism $h : U \to X$ yields a morphism $\Lambda r. G(fst\ r, h(snd\ r)) : U \to Y$.

Proof. (i) The forwards implication is trivial. For the converse, consider the morphism $I_X : U \to X$ of Corollary 11.2.9, so that $I_X(g) = x$ whenever $g \Vdash_X x$. Take f realizing $F \circ I_X : U \to Y$; then clearly f realizes F itself.

(ii) Again the forwards implication is trivial. For the converse, consider the object $U \times X$ as presented by $r \Vdash (u,x) \Leftrightarrow fst\ r = u \wedge snd\ r \Vdash x$, and take g realizing $\Lambda r. G(fst\ r, I_X(snd\ r))$. Then g realizes $G : U \times X \to Y$, since if $r \Vdash (u,x)$ then $G(fst\ r, I_X(snd\ r)) = G(u,x)$. Hence G is present as a morphism $U \to (X \Rightarrow Y)$ by transposing. □

This theorem can be read as defining the set $\mathsf{SR}(\sigma)$ together with the set of morphisms $U \to \mathsf{SR}(\sigma)$ by induction on σ, starting from the fact that $\mathsf{SR}(\mathtt{N}) \cong \mathbb{N}_\perp$ and the morphisms $U \to \mathsf{SR}(\mathtt{N})$ are the familiar sequential functionals as in $\mathsf{SF}(2_v)$ (i.e.

those computable in B). This is the characterization of SR given in [57]. Moreover, by interpreting the above theorem in the category $\mathscr{M}od(\mathsf{B};\mathsf{B}^{\mathrm{eff}})$, we obtain an analogous characterization of the substructure $\mathsf{SR}^{\mathrm{eff}}$, starting with the effectively sequential functionals as in $\mathsf{SF}^{\mathrm{eff}}(2_v)$ (i.e. those computable in $\mathsf{B}^{\mathrm{eff}}$).

Remark 11.3.2. In [176, Section 6], a more categorical formulation of this presentation was discussed. Let $\mathscr{M}$ denote the monoid of sequential endofunctions on $\mathbb{N}_\perp^{\mathbb{N}}$ (i.e. those representable in B), and consider the functor category $[\mathscr{M}^{op}, \mathscr{S}et]$, consisting of sets endowed with a right $\mathscr{M}$-action and $\mathscr{M}$-equivariant functions between them. It is well known that such functor categories are cartesian closed. Now consider the $\mathscr{M}$-set $S_{\mathbb{N}}$ consisting of sequential functions $\mathbb{N}_\perp^{\mathbb{N}} \to \mathbb{N}_\perp$, with the $\mathscr{M}$-action given by precomposition, and inductively define $S_{\sigma\to\tau} = (S_\sigma \Rightarrow S_\tau)$. It is shown in [176] that the applicative structure obtained by applying the functor $\mathrm{Hom}(1,-)$ to the S_σ is canonically isomorphic to SR, essentially via Theorem 11.3.1.

Just as in the case of PC, this characterization of SR yields an interesting maximality property:

Theorem 11.3.3. *(i) Within the poset $\mathscr{T}(\mathbb{N}_\perp)$ of Subsection 3.6.4, SR [resp. $\mathsf{SR}^{\mathrm{eff}}$] is maximal among type structures $\mathbf{P}$ with $\mathbf{P}(2_v) \cong \mathsf{SF}(2_v)$ [resp. $\mathsf{SF}^{\mathrm{eff}}(2_v)$].*

(ii) Suppose $\mathbf{P}$ is any type structure over $\mathbb{N}_\perp$ such that $\mathbf{P}(2_v) \cong \mathsf{SF}(2_v)$ [resp. $\mathsf{SF}^{\mathrm{eff}}(2_v)$] and H is present within $\mathbf{P}((2,2\to 0)_v)$. Then $\mathbf{P} \cong \mathsf{SR}$ [resp. $\mathsf{SR}^{\mathrm{eff}}$].

Proof. (i) Suppose for contradiction that $\mathbf{P}$ is strictly above SR in $\mathscr{T}(\mathbb{N}_\perp)$, where $\mathbf{P}(2_v) \cong \mathsf{SF}(2_v)$, and let $\sigma \to \tau$ be a type of minimal level such that there is some G present in $\mathbf{P}(1_v \to (\sigma \to \tau))$ but not in $\mathsf{SR}(1_v \to (\sigma \to \tau))$. Then by Theorem 11.3.1(ii), there exists $h \in \mathbf{P}(1_v \to \sigma)$ with $\Lambda r.G(\mathit{fst}\,r, h(\mathit{snd}\,r)) \notin \mathbf{P}(1_v \to \tau)$, which contradicts elementary closure properties of $\mathbf{P}$.

(ii) Suppose $\mathbf{P}$ satisfies the hypotheses with respect to $\mathsf{SF}(2_v)$. Then since $H : \mathsf{SR}(2_v) \to \mathsf{SR}(3_v)$ is surjective, all functionals in $\mathsf{SR}(3_v)$ are present in $\mathbf{P}(3_v)$, and by the argument of part (i), no others can be. Moreover, we clearly have a retraction $\overline{3}_v \lhd \overline{2}_v$ in $\mathbf{P}$ agreeing with that in SR, and it follows easily that the same is true with any σ in place of $\overline{3}_v$; hence $\mathbf{P} \cong \mathsf{SR}$. The effective case is similar. □

This can be seen as a modest kind of 'ubiquity' result for the model SR. We anticipate that Theorem 11.3.3(ii) can be used to verify that many other 'sequential' models, such as the game models $\mathsf{G}, \mathsf{G}_i, \mathsf{G}_b$ from Section 7.4, yield SR as their extensional collapse, though in some cases the details have yet to be worked out.

It is also possible that a deeper or more far-reaching ubiquity theorem for SR can be established. For instance, we do not know whether SR can be characterized as the *unique* maximal type structure (satisfying mild conditions) with $\mathsf{SR}(2_v) \cong \mathsf{SF}(2_v)$.

Exercise 11.3.4. Use Theorem 11.3.3(ii) to verify that $\mathsf{SR}^{\mathrm{eff}} \cong \mathsf{EC}(\mathsf{B}^{\mathrm{eff}}; N_\perp)$. (This was the definition of $\mathsf{SR}^{\mathrm{eff}}$ employed in [176].)

11.4 The Strongly Stable Model

We now turn to the original construction of SR as the *strongly stable* model of Bucciarelli and Ehrhard [45]. This can be seen as a counterpart to the classical domain-theoretic construction of PC, showing that the SR functionals can be characterized in purely extensional terms by means of a preservation property.

We here present the model within the framework of *hypercoherences* as in [68]. In this approach, elements of a type σ are sets of 'information atoms' subject to some consistency properties, and a somewhat subtle notion of *coherence* is used to determine which mappings $\sigma \to \tau$ are present in the model. For further intuition we refer to [68]. We write $A \subseteq^*_{\mathrm{fin}} B$ to mean that A is a finite non-empty subset of B.

Definition 11.4.1. (i) A *hypercoherence* X is a countable set $|X|$ together with a set Γ_X of non-empty finite subsets of $|X|$ containing all singletons $\{a\}$ for $a \in |X|$.

(ii) The *domain* $D(X)$ generated by a hypercoherence X is the set of all $x \subseteq |X|$ such that $u \in \Gamma_X$ for all $u \subseteq^*_{\mathrm{fin}} x$. We refer to elements $x \in D(X)$ as *states* of X, and regard them as forming a complete partial order under inclusion.

(iii) A set $A \subseteq^*_{\mathrm{fin}} D(X)$ is *coherent* if for all $u \subseteq^*_{\mathrm{fin}} |X|$, $u \lhd A$ implies $u \in \Gamma_X$, where $u \lhd A$ means

$$(\forall a \in u.\ \exists x \in A.\ a \in x) \ \wedge\ (\forall x \in A.\ \exists a \in u.\ a \in x)\ .$$

We write $\mathscr{C}(X)$ for the set of such coherent subsets.

(iv) If X, Y are hypercoherences, a function $f : D(X) \to D(Y)$ is *strongly stable* if f is monotone and continuous and for any $A \in \mathscr{C}(X)$ we have $f(A) \in \mathscr{C}(Y)$ and $f(\bigcap A) = \bigcap f(A)$. We write HC for the category of hypercoherences, where morphisms $X \to Y$ are strongly stable functions $D(X) \to D(Y)$.

For example, we have a hypercoherence N given by $|N| = \mathbb{N}$ with Γ_N the set of singletons; note that $D(N) \cong \mathbb{N}_\perp$. Moreover:

Proposition 11.4.2. HC *is cartesian closed.*

Proof. The product $X \times Y$ is defined by $|X \times Y| = |X| \sqcup |Y|$, with $\Gamma_{X\times Y}$ consisting of all $u \subseteq^*_{\mathrm{fin}} |X| \sqcup |Y|$ such that $u \subseteq |X| \Rightarrow u \in \Gamma_X$ and $u \subseteq |Y| \Rightarrow u \in \Gamma_Y$. For the exponential Y^X, we let $|Y^X|$ be the set of pairs (x, b) where $x \in D(X)$ is finite and $b \in |Y|$, and let Γ_{Y^X} consist of all $w \subseteq^*_{\mathrm{fin}} |Y^X|$ such that

$$\pi_0(w) \in \mathscr{C}(X) \ \Rightarrow\ (\pi_1(w) \in \Gamma_Y \ \wedge\ (\pi_1(w) \text{ a singleton} \ \Rightarrow\ \pi_0(w) \text{ a singleton}))\ .$$

We leave the remaining details to the reader (see also [68]). □

It may seem puzzling that $\Gamma_{X\times Y}$ and Γ_{Y^X} are not usually closed under subsets. Some insight into the reason for this may be gained by verifying that the *Gustave* function $D(N^3) \to D(N)$ (to be mentioned in Section 11.6) is not strongly stable.

We now have an interpretation $[\![\sigma]\!]_{\mathsf{HC}}$ for the simple types $\sigma \in \mathscr{T}^{\to}(\mathtt{N})$, and this gives rise to a type structure SSt, where $\mathsf{SSt}(\sigma) = D([\![\sigma]\!]_{\mathsf{HC}})$. We call SSt the *strongly stable model*; we will now work towards showing that it coincides with SR.

We start with the morphisms $N^r \to N$ in HC. Note that $D(N^r) \cong \mathbb{N}_\bot^r$, and $\mathscr{C}(N^r)$ consists of those $A \subseteq_{\mathrm{fin}}^* \mathbb{N}_\bot^r$ such that for each $i < r$, the projection $\pi_i(A) \subseteq \mathbb{N}_\bot$ either contains $\bot$ or is a singleton (in other words, $\pi_i(A)$ is closed under meets).

Proposition 11.4.3. *The morphisms $N^r \to N$ in* HC *are exactly the Milner sequential functions $\mathbb{N}_\bot^r \to \mathbb{N}_\bot$ as in Definition 7.5.7.*

Proof. By induction on r. The case $r = 1$ is trivial since the Milner sequential functions $\mathbb{N}_\bot \to \mathbb{N}_\bot$ are exactly the monotone ones, so suppose $r > 1$. First suppose $f : \mathbb{N}_\bot^r \to \mathbb{N}_\bot$ is Milner sequential; then it is certainly monotone and continuous. Suppose now $A \in \mathscr{C}(N^r)$. If f is constant, clearly $f(A) \in \mathscr{C}(N)$ and $f(\bigcap A) = \bigcap f(A)$, so suppose $i < r$ is a sequentiality index as in Definition 7.5.7. If $\bot \in \pi_i(A)$, then $\bot \in f(A)$, so clearly $f(A) \in \mathscr{C}(N)$ and $f(\bigcap A) = \bot = \bigcap f(A)$. Otherwise, $\pi_i(A)$ is some singleton $\{a\}$, and the corresponding function $f_a^i : \mathbb{N}_\bot^{r-1} \to \mathbb{N}_\bot$ is Milner sequential. So f_a^i is strongly stable by the induction hypothesis, whence it follows easily that $f(\bigcap A) = \bigcap f(A)$.

Conversely, suppose $f : \mathbb{N}_\bot^r \to \mathbb{N}_\bot$ is strongly stable. Then all functions f_a^i (where $i < r$, $a \in \mathbb{N}$) are clearly strongly stable, so are Milner sequential by the induction hypothesis. So if f is not Milner sequential, we must have that $f(\bot, \ldots, \bot) = \bot$ (since f is not constant) and that for each $i < r$ there exists $\vec{x}^i \in \mathbb{N}_\bot^r$ with $x_i^i = \bot$ but $f(\vec{x}^i) \neq \bot$. Let $A = \{\vec{x}^0, \ldots, \vec{x}^{r-1}\}$; then A is coherent since $\bot \in \pi_i(A)$ for each A. But now either $f(\vec{x}^i) \neq f(\vec{x}^j)$ for some i, j, in which case $f(A) \subseteq \mathbb{N}_\bot$ is not coherent, or all $f(\vec{x}^i)$ have the same value $m \in \mathbb{N}$, in which case $f(\bigcap A) = f(\bot, \ldots, \bot) = \bot$ but $\bigcap f(A) = m$. Either way, the strong stability of f is contradicted. □

Next we consider functions $\mathbb{N}_\bot^{\mathbb{N}} \to \mathbb{N}_\bot$. We shall represent $\mathbb{N}_\bot^{\mathbb{N}}$ using the hypercoherence N^ω, where $|N^\omega| = \mathbb{N} \times \mathbb{N}$, and Γ_{N^ω} consists of sets $u \subseteq_{\mathrm{fin}}^* \mathbb{N} \times \mathbb{N}$ such that if $\pi_0(u)$ is a singleton then so is $\pi_1(u)$. Then $D(N^\omega) \cong \mathbb{N}_\bot^{\mathbb{N}}$; in fact, N^ω is the product of ω copies of N in HC.

For each r, there is an evident retraction $N^r \lhd N^\omega$ within HC, where the corresponding idempotent $e_r : N^\omega \to N^\omega$ simply discards all pairs $(m,n) \in |N^\omega|$ with $m \geq i$. This gives rise to an idempotent E_r on $(N^\omega \Rightarrow N)$ in HC; moreover, any $F \in D(N^\omega \Rightarrow N)$ is clearly the supremum of the $E_r(F)$ for $r \in \mathbb{N}$. But by Proposition 11.4.3, the image of E_r coincides with the set of Milner sequential functions $\mathbb{N}_\bot^r \to \mathbb{N}_\bot$; thus every element of $D(N^\omega \Rightarrow N)$ is a supremum of representatives of such functions. To show that $D(N^\omega \to N) \cong \mathsf{SR}(2_v)$, it now suffices to show:

Proposition 11.4.4. $\mathsf{SR}(2_v)$, *ordered pointwise, is a CPO with pointwise suprema of chains.*

Proof. Since the identity on $\mathsf{SR}(2_v)$ arises as the limit of some standard sequence of idempotents whose images consist of finite elements as in Definition 11.1.9(i), it is clearly enough to consider chains $F_0 \sqsubseteq F_1 \sqsubseteq \cdots$ where each F_i is finite. For any r, we may find a sequence $f_0 \sqsubseteq \cdots \sqsubseteq f_{r-1}$ in B such that each f_i is a minimal irredundant realizer for F_i: for instance, take some minimal $f_{r-1} \Vdash_{\mathrm{irr}} F_{r-1}$, and restrict f_{r-1} to $f_i \Vdash_{\mathrm{irr}} F_i$ as in Lemma 11.1.10. But by Lemma 11.2.3, each F_i has only finitely many minimal irredundant realizers altogether; hence by König's lemma there is an infinite chain $f_0 \sqsubseteq f_1 \sqsubseteq \cdots$ with $f_i \Vdash_{\mathrm{irr}} F_i$ for all i. Take $f = \bigsqcup f_i$ in B; then clearly f realizes some F which is the pointwise supremum of the F_i. □

Following Colson and Ehrhard [57], we may now show that SSt agrees with SR at all type levels using the idea of Theorem 11.3.1, with N^ω playing the role of U. The key induction step is given by the following.

Lemma 11.4.5. *Suppose X,Y are hypercoherences.*

(i) A function $F : D(X) \to D(Y)$ is strongly stable iff every morphism $h : N^\omega \to X$ in HC *yields a morphism $F \circ h : N^\omega \to Y$.*

(ii) A function $G : D(N^\omega \times X) \to D(Y)$ is strongly stable iff every morphism $h : N^\omega \to X$ yields a morphism $\Lambda r. G(fst\ r, h(snd\ r)) : N^\omega \to Y$.

Proof. (Sketch.) (i) The forwards implication is easy. So suppose every morphism $h : N^\omega \to X$ yields a morphism $F \circ h$ in HC. To show that F is monotone and continuous, we shall show that every ω-chain in $D(X)$ is the image of a certain 'generic chain' in $D(N^\omega)$. Specifically, define $c_i = \{(j,0) \mid j < i\} \in D(N^\omega)$ for each i; then clearly $c_0 \sqsubseteq c_1 \sqsubseteq \cdots$. Now given any chain $x_0 \sqsubseteq x_1 \sqsubseteq \cdots$ in $D(X)$, we can define $h : D(N^\omega) \to D(X)$ by $h(z) = x_0 \cup \bigcup_{(j,0)\in z}(x_{j+1} - x_j)$. It is routine to check that h is strongly stable. Hence $F \circ h$ is strongly stable by hypothesis, so that $F(x_0 \sqsubseteq x_1 \sqsubseteq \cdots) = (F \circ h)(c_0 \sqsubseteq c_1 \sqsubseteq \cdots)$ is an increasing chain in $D(Y)$, and $F(\bigsqcup x_i) = (F \circ h)(\bigsqcup c_i) = \bigsqcup (F \circ h)(c_i) = \bigsqcup F(x_i)$.

It remains to show that F preserves coherent sets and their meets. For this, we show that every $A \in \mathscr{C}(X)$ is the image of some $C \in \mathscr{C}(N^\omega)$ via some strongly stable $h' : N^\omega \to X$. If A is a singleton, this is easy. If $A = x_0, \ldots, x_{k-1}$ where $k \geq 2$, let $C = \{g_0^k, \ldots, g_{k-1}^k\}$, where

$$g_j^k = \{(j+1,0), \ldots, (k-1,k-j-2), (0,k-j-1), \ldots, (j-1,k-2)\} \in D(N^\omega) .$$

It is easy to check that C is coherent, but that no proper subset of C of size > 1 is coherent. We now take $x' = \bigcap x_i$, and define $h'(z) = x' \cup \bigcup_{g_j^k \subseteq z}(x_j - x')$. It is routine to check that h' is strongly stable, and that $h'(C) = A$. But now since $F \circ h'$ is strongly stable, we may deduce that $F(A)$ is coherent and that $F(\bigcap A) = \bigcap F(A)$.

The proof of (ii) is analogous. □

Theorem 11.4.6. $\mathsf{SSt} \cong \mathsf{SR}$.

Proof. By induction on σ, we show both that $\mathsf{SSt}(\sigma) \cong \mathsf{SR}(\sigma)$ and that the morphisms $N^\omega \to [\![\sigma]\!]_{\mathsf{HC}}$ correspond to elements of $\mathsf{SR}(1 \to \sigma)$. The base case holds since we have shown that $D(N^\omega \to N) \cong \mathsf{SR}(2_\nu)$. The induction step follows from Theorem 11.3.1 and Lemma 11.4.5. □

Finally, we note that the proof of Theorem 11.4.6 is non-constructive, because of the appeal to König's lemma in the proof of Proposition 11.4.4. Indeed, it is shown in [176, Section 5.3] that the non-constructivity is essential, in that the equivalence fails in the *effective* setting: there is a c.e. element of $\mathsf{SSt}(\mathtt{N}^{(3)} \to \mathtt{N})$ with no counterpart in $\mathsf{SR}^{\mathrm{eff}}$. (Cf. Proposition 7.5.9 for SF.) The constructive inequivalence between strong stability and sequential realizability is one reason for maintaining a terminological distinction between these notions.

11.5 Order-Theoretic Structure

In Section 11.1 we noted that there are SR functionals that are not monotone with respect to the pointwise order. However, it turns out that there is another partial order structure on SR that yields good properties, markedly different in character from the order theory of either PC or SF.

The relevant ordering on SR may be described in a number of equivalent ways:

Theorem 11.5.1. *The following are equivalent for $x, y \in \mathsf{SR}(\sigma)$.*

(i) $x \preceq_{\mathrm{obs}} y$ in the sense of Definition 7.1.7: that is, for all $G \in \mathsf{SR}(\sigma \to \mathtt{N})$ we have $G(x) = n \Rightarrow G(y) = n$.

(ii) Regarding $\mathsf{SR}(\sigma)$ as a modest set, there exist $r \Vdash x$ and $s \Vdash y$ with $r \sqsubseteq s$.

(iii) There exists $P \in \mathsf{SR}(\mathtt{U} \to \sigma)$ with $P(\bot) = x$ and $P(\top) = y$.

(iv) $x \subseteq y$ where x, y are viewed as elements of $\mathsf{SSt}(\sigma) = D([\![\sigma]\!]_{\mathsf{HC}})$.

(v) (For types $\sigma = \tau \to \tau'$.) $x \leq y$ in the stable *order: that is, if $z \subseteq w \in \mathsf{SSt}(\tau)$ then $x(z) = y(z) \cap x(w)$.*

Proof. (ii) $\Rightarrow$ (i): If $r \Vdash x$, $s \Vdash y$ and g tracks G as a morphism $\mathsf{SR}(\sigma) \to \mathsf{SR}(\mathtt{N})$, then $g \cdot r \Vdash G(x)$ and $g \cdot s \Vdash G(y)$. But if $r \sqsubseteq s$ then $g \cdot r \sqsubseteq g \cdot s$ so $G(x) \sqsubseteq G(y) \in \mathbb{N}_\bot$.

(iii) $\Rightarrow$ (ii): Take $r' \Vdash \bot$ and $s' \Vdash \top$ with $r' \sqsubseteq s'$. If p tracks P as a morphism $\mathsf{SR}(\mathtt{U}) \to \mathsf{SR}(\sigma)$, then taking $r = p \cdot r'$ and $s = p \cdot s'$ gives us realizers as required.

(iv) $\Rightarrow$ (iii): Let 1 denote the hypercoherence with $|1| = \{*\}$. If $x \subseteq y$ within $\mathsf{SSt}(\sigma)$, we may define a function $P : D(1) \to D([\![\sigma]\!])$ by $P(\emptyset) = x$, $P(\{*\}) = y$, and it is clear that P is strongly stable. Thus P exists as an element of $\mathsf{SR}(\mathtt{U} \to \sigma)$ with the required properties.

(i) $\Rightarrow$ (iv): Suppose $x \not\subseteq y \in \mathsf{SSt}(\sigma)$, and take $a \in x - y$. Define $G : \mathsf{SSt}(\sigma) \to \mathsf{SSt}(\mathtt{N})$ by $G(z) = 0$ if $a \in z$, $G(z) = \bot$ otherwise. Clearly G is strongly stable and hence exists in $\mathsf{SR}(\sigma \to \mathtt{N})$, and $G(x) = 0$ but $G(y) = \bot$. Thus $x \not\preceq_{\mathrm{obs}} y$.

(iv) $\Rightarrow$ (v): Suppose $x \subseteq y$ and $z \subseteq w$. Clearly $x(z) \subseteq y(z) \cap x(w)$. So suppose $a \in y(z) \cap x(w)$. Then there exists $(u, a) \in y$ such that $u \subseteq z \subseteq w$; also there exists $(v, a) \in x \subseteq y$ such that $v \subseteq w$. Hence $u \cup v \subseteq w$, so $u \cup v \in \Gamma_{[\![\tau]\!]}$, which implies $\{u, v\}$ is coherent. But also $\{(u, a), (v, a)\} \subseteq y$ so $\{(u, a), (v, a)\} \in \Gamma_{[\![\sigma]\!]}$, whence $u = v$ by the definition of Γ_{YX} (see the proof of Proposition 11.4.2). It follows that $a \in x(z)$.

(v) $\Rightarrow$ (iv): Suppose $x \not\subseteq y$, and take $(u, a) \in x - y$. If there is no $v \subseteq u$ with $(v, a) \in y$, then $a \in x(u) - y(u)$ so $x(u) \neq x(u) \cap y(u)$; thus $x \not\leq y$ in the stable order. Otherwise, if $v \subseteq u$ and $(v, a) \in y$, then we have $a \in y(v) \cap x(u)$, but not $a \in x(v)$, since if $(v', a) \in x$ with $v' \subseteq v$, then $v' = u$ as above, whereas $v \neq u$. $\square$

From description (iv) above (for example), it is clear that the relation in question is a partial order on each $\mathsf{SR}(\sigma)$, and that application is monotone. It is also clear from description (iv) that each $\mathsf{SR}(\sigma)$ is a CPO, with suprema of chains given pointwise. Using Lemma 11.1.10, it follows easily that any chain $x_0 \preceq x_1 \preceq \cdots$ in $\mathsf{SR}(\sigma)$ can be mirrored by a chain of realizers $f_0 \sqsubseteq f_1 \sqsubseteq \cdots$ in B, with $f_i \Vdash x_i$. Thus, suprema of chains are always *observational limits* in the sense of Subsection 7.1.2. This contrasts with the more pathological situation for SF as described in Section 7.6.

11.6 The Stable Model and Its Generalizations

The existence of two interesting but incomparable type structures over $\mathbb{N}_\perp$, namely PC and SR, naturally raises the question of whether there are other such type structures not comparable with either of these. We give here a brief survey of what is known in this area.

Perhaps the most prominent alternative to PC and SR is Berry's *stable model* [32], which we define here within the framework of *coherence spaces* [104]. Note that Definition 11.4.1 can be naturally seen as an elaboration of this definition.

Definition 11.6.1. (i) A *coherence space* X is a countable set $|X|$ together with a reflexive, symmetric binary relation $\frown_X$ on $|X|$, called the *coherence relation* of X.

(ii) The *domain* $C(X)$ generated by a coherence space X is the set of all $x \subseteq |X|$ such that if $a,b \in X$ then $a \frown_X b$. The coherence relation is extended to $C(X)$ via $x \frown_X y$ iff $x \cup y \in C(X)$.

(iii) A function $f : C(X) \to C(Y)$ is *stable* if f is monotone and continuous and for any $x \frown y \in C(X)$ we have $f(x \cap y) = f(x) \cap f(y)$.

(iv) If X,Y are coherence spaces, we define a coherence space $X \Rightarrow Y$ as follows: $|X \Rightarrow Y| = \{(x,b) \mid x \text{ finite} \in C(X),\ b \in |Y|\}$, and $(x,b) \frown_{X \Rightarrow Y} (x',b')$ iff either $x \cup x' \notin C(X)$ or $(x,b) = (x',b')$ or ($b \frown_Y b'$ but $b \neq b'$).

It can be checked that the category of coherence spaces and stable functions is cartesian closed, with exponentials given by $\Rightarrow$, and there is also an obvious coherence space N with $|N| = \mathtt{N}$ and $m \frown n$ iff $m = n$. We thus obtain a type structure St over $\mathbb{N}_\perp$, which we call the stable model.

Note that St excludes the parallel convergence function *pconv* of Subsection 7.3.1. Intuitively, St lies somewhere between PC and SSt, but is incomparable with either in the sense of $\mathscr{T}(\mathbb{N}_\perp)$. In particular, the function *strict?* defined in Section 11.1 is present in St and SSt but not in PC, while a certain function $\mathit{Gustave} : \mathbb{N}_\perp^3 \to \mathbb{N}_\perp$ (due to Berry) is present in St and PC but not in SSt. We may define *Gustave* as the smallest monotone function $\mathbb{N}_\perp^3 \to \mathbb{N}_\perp$ such that

$$\mathit{Gustave}(0,1,\perp) = 0\,, \qquad \mathit{Gustave}(\perp,0,1) = 1\,, \qquad \mathit{Gustave}(1,\perp,0) = 2\,.$$

In fact, Paolini [229] has shown that every finite element of St is definable in PCF + *strict?* + *Gustave*, so that St is fully abstract for this language.

An intriguing computational interpretation of the notion of stability is developed by Asperti [7]. As an interesting avenue for further work, it seems likely that this will lead to a characterization of St analogous to our construction of SR via B.

The model St is not the only model incomparable with PC and SSt. In [176, Section 9] it is shown that there is an infinite sequence of type structures St^r over $\mathbb{N}_\perp$ for $r \geq 2$, with $\mathsf{St}^2 = \mathsf{St}$, such that none of $\mathsf{PC}, \mathsf{St}^2, \mathsf{St}^3, \ldots, \mathsf{SR}$ are comparable in $\mathscr{T}(\mathbb{N}_\perp)$. In general, the model St^r coincides with SSt at type $\mathtt{N}^{(r)} \to \mathtt{N}$ and below, though not at $\mathtt{N}^{(r+1)} \to \mathtt{N}$; thus, in a loose informal sense, the St^r 'converge' towards SSt. An interesting open question is whether there are any (reasonable) continuous type structures over $\mathbb{N}_\perp$ that are not comparable with any of those listed above.

Chapter 12
Some Intensional Models

In the preceding chapters, broadly speaking, we have been working our way gradually from 'weaker' to 'stronger' notions of computability (see the diagram on page 33). This has culminated in Chapters 10 and 11 with the study of the two type structures PC and SR that represent *maximal* notions of computable functional subject to certain constraints (Theorems 10.5.7 and 11.3.3). If we wish to consider higher-order computability notions even more powerful than these, we are therefore led to abandon the framework of computable *functionals*—that is, of extensional models (over $\mathbb{N}$ or $\mathbb{N}_\perp$)—and instead to explore the world of *intensional* models, in which operations of type $\sigma \rightarrow \tau$ are not simply functions, but finer-grained objects such as algorithms, strategies or programs. Our purpose in this chapter is to give a taste of the study of such intensional models.

Such a study may seem initially disorienting, because in place of ordinary functions we are led to seek out other less familiar kinds of mathematical objects—and as we shall see, even such basic notions as composition for such objects can prove to be quite subtle. Nevertheless, we shall see that at least some intensional models exhibit a rich and beautiful mathematical theory worthy to stand alongside that of the extensional models we have studied.

An exhaustive treatment of intensional models is out of the question here: much of the landscape has yet to be explored, and even what is known would be enough to fill another book. (Some of the territory we have not covered will be briefly surveyed in Chapter 13.) In the present chapter, we shall confine our attention to those intensional notions that have already played major roles in earlier parts of the book. The bulk of the chapter (Section 12.1) is devoted to an in-depth study of the *sequential algorithms* model—an example *par excellence* of an intensional λ-algebra with a satisfying theory and a wealth of non-trivially equivalent characterizations. We consider this in conjunction with its untyped counterpart, the van Oosten model B, which has already featured in Chapters 6, 7 and 11. In Section 12.2 we turn our attention to Kleene's second model K_2, which has played a major role in Chapters 8–10, and has served as our benchmark for the concept of *continuous* (finitary) operations on countable data (Definition 3.5.14). This model is more intensionally 'fine-grained' than B, even to the extent that it is no longer a (partial) λ-algebra, and offers interesting possibilities for modelling 'time-sensitive' flavours of computation. Finally, in Section 12.3, we reach the natural terminus of our subject: the 'maximally intensional' model K_1, in which operations are in effect presented as fully explicit programs. A consideration of the status of this model provides an opportunity to draw together some threads from the book as a whole.

12.1 Sequential Algorithms and van Oosten's B

In this section, we study an intensional computability notion represented by two equivalent models: a certain (simply typed) λ-algebra SA of *sequential algorithms*, and van Oosten's (untyped) model B as introduced in Subsection 3.2.4. We begin with an intuitive description of the kinds of intensional objects that constitute SA.

A good starting point is the notion of a *decision tree* for a sequential functional $F : \mathbb{N}_\bot^\mathbb{N} \to \mathbb{N}_\bot$ as described in Subsection 3.2.4. It is clear that such trees are intensional objects, insofar as many trees may represent the same extensional functional: for instance, the functional $F = \Lambda g.g(0) + g(1)$ may be represented by a tree which (along any path) features the query ?0 followed by ?1, or the other way round. Whereas elements of B are themselves partial functions $\mathbb{N} \rightharpoonup \mathbb{N}$ which may be construed as *representing* such decision trees, an element of type $\overline{2}$ in SA will be something more like the decision tree itself.

It is convenient to formalize computations with decision trees as interactions between two players (much in the spirit of Section 7.4). Labelled nodes in the tree correspond to Player (P) moves, whilst labelled edges correspond to Opponent (O) moves, so that a path through the tree will correspond to an alternating sequence of O- and P-moves. (In fact, our plays will always start with an O-move, although this is not shown in the diagram of Subsection 3.2.4.) Each simple type σ will be interpreted by a *game*, specifying a set of O- and P-moves, along with the set of move sequences that represent meaningful interactions for operations of that type. A particular operation of type σ will then be represented by a *strategy* for Player: essentially a partial function telling Player what move to make given the sequence of moves so far. Such a strategy may be visualized as a decision tree, though it will be technically more convenient to represent it as a set S of even-length plays such that for any odd-length play s there is at most one P-move x with $sx \in S$.

Since game plays impose a definite order on the events that occur in a computation, it is not surprising that strategies (or decision trees) typically contain information relating to evaluation order: witness the two strategies for $\Lambda f.f(0) + f(1)$ mentioned above. However, a further point should also be made at this stage. In contrast to other related game models (see Subsection 12.1.7), the model SA is not sensitive to *repetitions* of a given evaluation: no distinction is made at the level of sequential algorithms between the programs $\lambda f.f(0) + f(0)$ and $\lambda f.f(0) * 2$.[1] A

[1] On the face of it, the decision trees of Subsection 3.2.4 do seem to allow one to make this distinction. However, this turns out to be a small piece of noise that disappears when we pass to the corresponding categorical λ-algebra as described in Subsection 4.1.3.

possible motivation for this is that we are trying to model a 'state-free' style of computation in which the second evaluation of $f(0)$ must necessarily behave in exactly the same way as the first; there is therefore no way (in such a setting) even to detect whether such a repeated evaluation has occurred. Thus it is natural in our semantics to elide such 'repeated moves' from our game plays. Even so, we shall see that sequential algorithms allow for some quite complex interactions at higher types, going well beyond the power of PCF-style computation.

The model we will call SA is actually a full subcategory of the category of *concrete data structures* and sequential algorithms as originally introduced by Berry and Curien [33]. We here present SA more in the style of game models, drawing on ideas from Abramsky and Jagadeesan [1], Lamarche [166] and Curien [61]. We begin in Subsection 12.1.1 by defining a very simple game model, the category of *sequential data structures*. Whilst this already captures many aspects of the kind of sequential interaction described above, a limitation of this category is that it cannot model operations that use their argument more than once (as in the two invocations of f in $\lambda f.f(0)+f(1)$). This is remedied in Subsection 12.1.2 where we move to a richer model $\mathscr{SA}$, which turns out to be cartesian closed. From this we can extract a simply typed model SA in a standard way (Subsection 12.1.3). In Subsection 12.1.4, we 'discover' B as a reflexive object within $\mathscr{SA}$, and use this construction to establish that B is indeed a λ-algebra.

In Subsection 12.1.5, we show that the (effective) sequential algorithms model is precisely matched in computational power by a certain programming language PCF + *catch*, giving further insight into the kinds of computation the model supports. In Subsection 12.1.6, we present a surprising 'extensional' characterization of SA due to Laird, much in the spirit of the domain-theoretic characterizations of PC and SR in earlier chapters. Finally, in Subsection 12.1.7, we briefly situate the sequential algorithms model in relation to other game models, including those described in Section 7.4, along with a few others not treated in detail in this book.

12.1.1 Sequential Data Structures

In order to present our construction of the sequential algorithms model $\mathscr{SA}$, it will be convenient to start with a very simple category of games, the category $\mathscr{SD}$ of *sequential data structures*, then obtain $\mathscr{SA}$ as a somewhat more elaborate category with the same objects. For a more leisurely introduction to the basic intuitions of game semantics, we recommend Abramsky and McCusker [3] or Curien [62].

A little specialized notation will be useful. For any sets X,Y, we write $\mathrm{Alt}(X,Y)$ for the set of finite sequences $z_0,\ldots,z_{n-1}$ where $z_i \in X$ for i even, $z_i \in Y$ for i odd. If $L \subseteq \mathrm{Alt}(X,Y)$, we write L^{odd}, L^{even} for the sets of odd- and even-length sequences in L respectively. As usual we write Z^* for the set of all finite sequences over Z, and use $\sqsubseteq$ for the prefix relation on sequences.

For the purpose of this section, we concretely define the disjoint union $X+Y$ to be the set $\{(x,0) \mid x \in X\} \cup \{(y,1) \mid y \in Y\}$. We shall sometimes denote such elements (z,i) by z^i in order to emphasize that z is the main value of interest and i is just administrative tagging; we may also write $((z,i),j)$ as $z^{i,j}$. Given any sequence s whose elements are pairs, we write s_i for the inverse image under $z \mapsto z^i$ of the subsequence of s consisting of elements z^i (thus, superscripts represent tagging while subscripts represent untagging). On the other hand, we shall sometimes ignore such coding niceties and treat X,Y simply as subsets of $X+Y$ when the intention is clear.

Definition 12.1.1. (i) A *(sequential) data structure* A consists of disjoint countable sets O_A, P_A of *opponent* and *player* moves respectively, together with a non-empty prefix-closed set $L_A \subseteq \mathrm{Alt}(O_A,P_A)$ of *legal positions*. We let $M_A = O_A \cup P_A$.

(ii) A *(player) strategy* for A is a non-empty set $S \subseteq L_A^{\mathrm{even}}$ such that

1. if $sxy \in S$ then $s \in S$,
2. if $sxy, sxy' \in S$ then $y = y'$.

We write $\mathrm{Str}(A)$ for the set of strategies for A, and $\mathrm{Str}^{\mathrm{fin}}(A)$ for the set of finite strategies.

(iii) Given data structures A,B, we may form data structures $A\otimes B$, $A\times B$ and $A \multimap B$ as follows:

$$\begin{aligned}
O_{A\otimes B} &= O_A + O_B\,, \quad P_{A\otimes B} = P_A + P_B\,,\\
L_{A\otimes B} &= \{s \in \mathrm{Alt}(O_{A\otimes B},P_{A\otimes B}) \mid s_0 \in L_A,\ s_1 \in L_B\}\,,\\
O_{A\times B} &= O_A + O_B\,, \quad P_{A\times B} = P_A + P_B\,,\\
L_{A\times B} &= \{s \in \mathrm{Alt}(O_{A\times B},P_{A\times B}) \mid (s_0 \in L_A \wedge s_1 = \varepsilon) \vee (s_1 \in L_B \wedge s_0 = \varepsilon)\}\,,\\
O_{A\multimap B} &= P_A + O_B\,, \quad P_{A\multimap B} = O_A + P_B\,,\\
L_{A\multimap B} &= \{s \in \mathrm{Alt}(O_{A\multimap B},P_{A\multimap B}) \mid s_0 \in L_A,\ s_1 \in L_B\}\,.
\end{aligned}$$

By a *play* of A we shall mean a finite or infinite sequence of moves all of whose finite prefixes belong to L_A. Note in particular that a play in $A \multimap B$ must begin (if non-empty) with a move from O_B; this may then be followed by zero or more pairs consisting of an O_A-move followed by a P_A-move before the next move from B, which will be a P_B-move.

We now have what we need to construct a category.

Definition 12.1.2. The category $\mathscr{SD}$ of sequential data structures is defined as follows:

- Objects are sequential data structures.
- Morphisms $S : A \to B$ are strategies for $A \multimap B$.
- The identity $A \to A$ is the *copycat* strategy

$$\begin{aligned}
id_A &= \{s \in L_{A\multimap A}^{\mathrm{even}} \mid \forall t \sqsubseteq^{\mathrm{even}} s.\ t_0 = t_1\}\\
&= \{x_0{}^1 x_0{}^0 x_1{}^0 x_1{}^1 x_2{}^1 x_2{}^0 x_3{}^0 x_3{}^1 \cdots \mid x_0x_1x_2x_3 \ldots \in L_A\}\,.
\end{aligned}$$

- The composition of $S : A \to B$ and $T : B \to C$ is given by

$$T \circ S = \{s \restriction_{A,C} \mid s \in (M_A + M_B + M_C)^*,\ s \restriction_{A,B} \in S,\ s \restriction_{B,C} \in T\}$$

where $s \upharpoonright_{A,C}$, $s \upharpoonright_{A,B}$, $s \upharpoonright_{B,C}$ have the obvious meanings.

The idea behind the identity strategy is that Player simply copies the moves played by Opponent from the right side to the left or vice versa. To gain some experience with the basic definitions, the reader should check that such plays s are indeed exactly characterized by the property that $t_0 = t_1$ for all even initial subsequences t of s. For composition, the idea is that in response to a given O_C or P_A move, an internal dialogue between S and T is possible while both play moves in B; this ends when either S plays an O_A move or T plays a P_C move, and this is treated as the next 'externally visible' move played by the composite strategy $T \circ S$.

It is now routine to check that $\mathscr{SD}$ is a category. For instance, to show associativity for the composition of morphisms $S : A \to B$, $T : B \to C$ and $U : C \to D$, one checks that both $(U \circ T) \circ S$ and $U \circ (T \circ S)$ coincide with the set

$$\{s \upharpoonright_{A,D} \mid s \in (M_A + M_B + M_C + M_D)^*,\ s \upharpoonright_{A,B} \in S,\ s \upharpoonright_{B,C} \in T,\ s \upharpoonright_{C,D} \in U\}\,.$$

It is also easy to see that $\times$ is the cartesian product operation in $\mathscr{SD}$; we leave the definitions of the projections and pairing operation as an easy exercise. To understand why $\otimes$ is not a cartesian product, the reader should verify that there is in general no diagonal morphism $A \to A \otimes A$.

We write I for the sequential data structure with no moves; clearly I is the terminal object in $\mathscr{SD}$, and $I \otimes A \cong I \times A \cong A$ for any A. In fact, $(\mathscr{SD}, I, \otimes, \multimap)$ is a *symmetric monoidal closed category*, although our treatment here will not require familiarity with this concept.

12.1.2 A Category of Sequential Algorithms

The category $\mathscr{SD}$ as it stands is not cartesian closed. However, we may now build a category $\mathscr{SA}$ with the same objects as $\mathscr{SD}$ but with different morphisms. We opt here for a relatively concrete and self-contained presentation; the general categorical story behind this will be indicated in Remark 12.1.10.

The essential point is that a play of $A \multimap B$ allows us to pursue only a single path through the game tree for A, whereas in order to model operations that use their argument more than once (such as $\lambda f.f(0) + f(1)$), some more general kind of exploration of this tree is needed. To capture this, we shall define a new game $!A$ such that paths through A correspond to sequential 'explorations' of A; a morphism $A \to B$ in $\mathscr{SA}$ will then be a strategy for $!A \multimap B$. Our notion of exploration will allow a play to backtrack to any previously visited even position in A, and then play an O-move different from any played before.

Definition 12.1.3. For any data structure A, we define a data structure $!A$ as follows. Moves are given by

$$O_{!A} = \{(t,x) \mid tx \in L_A^{\mathrm{odd}}\}\,, \qquad P_{!A} = P_A\,.$$

Legal positions in $L_{!A}$ are sequences $s \in \mathrm{Alt}(O_{!A}, P_{!A})$ satisfying the following:

1. If an O-move (t,x) is immediately followed in s by a P-move y, then $txy \in L_A^{\mathrm{even}}$.
2. If an O-move (txy,z) appears in s, then the adjacent OP move pair $(t,x)y$ must appear at some earlier point in s.
3. No O-move (t,x) appears more than once in s.

A *sequential algorithm* from A to B is now a strategy for $!A \multimap B$.

This definition is partly motivated by the fact that if $s \in L_{!A}$, then $\| s \|$ is a finite strategy for A, where

$$\| s \| \;=\; \{\varepsilon\} \cup \{txy \mid (t,x)y \text{ is an OP move pair within } s\} \,.$$

For any data structure B and any $s \in L_B$, we shall also write $|s|$ for the set of even prefixes of s; note that $|s|$ is a strategy for B of a special 'linear' kind.

Composition of sequential algorithms, and the associativity thereof, are cumbersome to handle directly in terms of the above definition. We therefore resort at this point to a slightly more abstract characterization of sequential algorithms due to Curien [60], which also provides further insight into the nature of the concept.

We begin with an informal explanation. The idea is to characterize a sequential algorithm R from A to B by what it does when pitted against an Opponent who is himself following some finite 'counter-strategy'. Recall that the Opponent will be responsible for P-moves in A and for O-moves in B. For the purpose of generating a play in $!A \multimap B$, which may involve a non-linear exploration of A but can only play linearly in B, we may therefore consider Opponent to be armed with a finite strategy S for A and an odd-length position t in B (although strictly speaking only the odd moves of the latter are supplied by Opponent). A sequential algorithm $R : A \to B$ (as defined above) will engage with this Opponent data, supplying the O_A and P_B moves and thus generating a play in $!A \multimap B$. This will proceed until either R goes undefined (if it has no response to some Opponent move), or the data in either S or t are exhausted—that is, either R plays an O_A-move to which S has no answer, or R plays a P_B-move that responds to the last O_B-move in t. The idea now is to represent R abstractly by a partial function f which maps such Opponent data (of type $\mathrm{Str}^{\mathrm{fin}}(A) \times L_B^{\mathrm{odd}}$) to the next move (if any) that R will play after these data have run out. (For bureaucratic reasons, it is actually convenient for f to return the complete play in A or B that ends in this move, i.e. a play in L_A^{odd} or L_B^{even}, rather than just the move itself.) By means of various conditions on f, we can axiomatize those functions that do indeed arise from sequential algorithms R as given above, thus ensuring a perfect correspondence between abstract and concrete algorithms.

To formalize this idea, some additional terminology and notation will be useful. For any finite sequence s, we write s^{even} for the longest even prefix of s, and (if s is non-empty) s^{odd} for its longest odd prefix and s_{last} for its last element. If $S \in \mathrm{Str}(A)$ and $t \in L_A^{\mathrm{odd}}$, we say

- t is *filled* in S if $tx \in S$ for some $x \in P_A$,
- t is *accessible* from S if t is not filled in S but $t^{\mathrm{even}} \in S$.

We now present the formal definition of abstract sequential algorithms, followed by some informal explanation of the various technical conditions.

Definition 12.1.4. Let A,B be sequential data structures. An *abstract sequential algorithm* from A to B is a partial function $f : \mathrm{Str}^{\mathrm{fin}}(A) \times L_B^{\mathrm{odd}} \rightharpoonup L_A^{\mathrm{odd}} + L_B^{\mathrm{even}}$ satisfying the following conditions:

1. If $f(S,t) = q \in L_A^{\mathrm{odd}}$ then q is accessible from S.
2. If $f(S,t) = r \in L_B^{\mathrm{even}}$ then $r = tx$ for some $x \in P_B$.
3. If $f(S,t) = q \in L_A^{\mathrm{odd}}$, $S \subseteq S'$, $t \sqsubseteq t'$ and q is not filled in S', then $f(S',t') = q$.
4. If $f(S,t) = r \in L_B^{\mathrm{even}}$ and $S \subseteq S'$, then $f(S',t) = r$.
5. If $f(S',t')$ is defined, $S \subseteq S'$ and $t \sqsubseteq t'$, then $f(S,t)$ is defined.
6. If $f(S,txy) \in L_B^{\mathrm{even}}$ then $f(S,t) = tx$.

An abstract sequential algorithm f is *affine* if additionally the following holds:

7. If $f(S,t)$ is defined, then for some $s \in S$ we have $f(|s|,t) = f(S,t)$.

Condition 1 says that if f plays an O_A-move, this follows on from some even position in S, as one would expect if this were the first move in A outside the tree represented by S. Likewise, condition 2 says that if f plays a P_B-move, this follows on from the given position t. Conditions 3 and 4 are monotonicity conditions. Condition 3 captures the idea that if the play reaches a position q in A to which S has no response, then the same point will be reached when playing against any longer t' and any larger S' that contains no response to q. Likewise, condition 4, says that if the play reaches a position r in A to which t contains no response, the same point will be reached against any larger S' (note that in this case any t' strictly longer than t would necessarily contain a response to r). Condition 5 says that if the algorithm f has not gone undefined by the time S' or t' is exhausted, it will certainly not go undefined before some smaller S,t are exhausted (note, however, that $f(S,t)$ may well represent a move to which S' or t' contains a response). Condition 6 is a hygiene condition saying that f is only defined on positions $t \in L_B^{\mathrm{odd}}$ whose P-moves are actually those that would be played by f. Finally, the special condition 7 is a way of characterizing those algorithms that only explore a single path through A, and thus correspond to strategies for $A \multimap B$.

Many of these points are further clarified by the proof of the following correspondence result:

Theorem 12.1.5. *There is a canonical bijection between strategies for $!A \multimap B$ and abstract sequential algorithms from A to B.*

Proof. If f is an abstract sequential algorithm $A \to B$, we define a set $R_f \subseteq L_{!A \multimap B}^{\mathrm{even}}$ inductively as follows. Intuitively, at each stage, the response of R_f to an odd position sx will be simply the response of f to the A strategy and B position extracted from sx.

- $\varepsilon \in R_f$.
- If $s \in R_f$ where both $s \restriction_{!A}$ and $s \restriction_B$ are even (so that an O_B move is expected next), and $f(\| s \restriction_{!A} \|, (sx) \restriction_B) = p$ where $x \in O_B$ and $sx \in L_{!A \multimap B}^{\mathrm{odd}}$, then $sx(p_{\mathrm{end}}) \in R_f$.

- If $s \in R_f$ where both $s \upharpoonright_{!A}$ and $s \upharpoonright_B$ are odd (so that a P_A move is expected), and $f(\| (sx) \upharpoonright_{!A} \|, s \upharpoonright_B) = p$ where $x \in P_A$ and $sx \in L^{\mathrm{odd}}_{!A \multimap B}$, then $sx(p_{\mathrm{end}}) \in R_f$.

Here we take $p_{\mathrm{end}} = p_{\mathrm{last}} \in P_B$ if $p \in L^{\mathrm{even}}_B$, and $p_{\mathrm{end}} = (p^{\mathrm{even}}, p_{\mathrm{last}}) \in O_{!A}$ if $p \in L^{\mathrm{odd}}_A$. It is immediate by construction that R_f is a strategy for $!A \multimap B$, even without reference to the conditions of Definition 12.1.4.

Conversely, if $R \in \mathrm{Str}(!A \multimap B)$, $S \in \mathrm{Str}^{\mathrm{fin}}(A)$ and $t \in L^{\mathrm{odd}}_B$, we define $f_R(S,t)$ by playing out R against S and t until one of S,t is exhausted, as suggested by the informal explanation above. Formally, let $play(R,S,t)$ be the unique maximal sequence $s \in L_{!A \multimap B}$ such that $s^{\mathrm{even}} \in R$, $\| s \upharpoonright_{!A} \| \subseteq S$ and $s^{\mathrm{odd}} \upharpoonright_B \sqsubseteq t$. Now define

$$f_R(S,t) \;\simeq\; \begin{cases} uz & \text{if } play(R,S,t) \text{ is even and ends with a move } (u,z) \in O_{!A}\,, \\ v \upharpoonright_B & \text{if } v = play(R,S,t) \text{ is even and } v \upharpoonright_B = tx \text{ for some } x \in P_B\,. \end{cases}$$

It is an easy exercise to verify that f_R satisfies each of the conditions 1–6 of Definition 12.1.4 (this makes precise the above intuitive explanation of these conditions).

It remains to show that these constructions are mutually inverse. Given $R \in \mathrm{Str}(!A \multimap B)$, an easy induction on i shows that R and R_{f_R} contain exactly the same sequences of length $2i$. Conversely, given an abstract algorithm f and any S,t as above, we may construct a play $x_0y_0x_1y_1\ldots$ for $!A \multimap B$ inductively as follows. Let x_0 be the first element of t. Given $s_i = x_0y_0\ldots x_i$, take $y_i = f(\| s_i \upharpoonright_{!A} \|, |s_i \upharpoonright_B|)$ if defined. If $y_i \in P_B$ with $(s_iy_i) \upharpoonright_B \sqsubseteq t$, let x_{i+1} be the next element of t after $(s_iy_i) \upharpoonright_B$; and if $y_i = (t,z) \in O_{!A}$ and $tzx \in S$, let $x_{i+1} = x$. Let s be the maximal play constructed in this way; it is routine to check that $f(S,t) = u$ iff s is even, $\| s \upharpoonright_{!A} \| \subseteq S$, $s^{\mathrm{odd}} \upharpoonright_B \sqsubseteq t$ and $s_{\mathrm{last}} = u_{\mathrm{end}}$. On the other hand, an easy induction shows that every even prefix of s is present in R_f, and hence that $f_{R_f}(S,t) = u$ under exactly the same conditions; thus $f = f_{R_f}$. Again, we leave the detailed checking to the reader. □

Exercise 12.1.6. Show that there is also a canonical bijection between strategies for $A \multimap B$ and abstract *affine* sequential algorithms from A to B.

We now make abstract sequential algorithms into a category. The identity algorithm on a data structure A is given by

$$id_A(S,t) \;=\; \begin{cases} tx \in L^{\mathrm{even}}_A & \text{if } tx \in S\,, \\ t' \in L^{\mathrm{odd}}_A & \text{if } t' \text{ is the shortest odd prefix of } t \text{ not filled in } S\,. \end{cases}$$

If $f : A \to B$ and $g : B \to C$ are abstract sequential algorithms, their composition $g \circ f : \mathrm{Str}^{\mathrm{fin}}(A) \times L^{\mathrm{odd}}_C \rightharpoonup L^{\mathrm{odd}}_A + L^{\mathrm{even}}_C$ is given by

$$(g \circ f)(S,u) \;\simeq\; \begin{cases} r & \text{if } g(f.S,u) = r \in L^{\mathrm{even}}_C\,, \\ p & \text{if } g(f.S,u) = q \in L^{\mathrm{odd}}_B \text{ and } f(S,q) = p \in L^{\mathrm{odd}}_A\,, \end{cases}$$

where we define

$$f.S \;=\; \{tx \in L^{\mathrm{even}}_B \mid f(S,t) = tx\}\,.$$

Proposition 12.1.7. *With the above definitions, sequential data structures and sequential algorithms form a category, which we denote by $\mathcal{SA}$.*

Proof. It is easy to check that id_A is an abstract sequential algorithm. Given f, g as above, it is trivial to check that $g \circ f$ satisfies conditions 1, 2, 4 and 6 of Definition 12.1.4. For condition 3, if $(g \circ f)(S,u) = p \in L_A^{\mathrm{odd}}$ then $f(S,q) = p$ where $q = g(f.S,u)$; hence if $S \subseteq S'$, $u \sqsubseteq u'$ and p is not filled in S' then $f(S',q) = p \in L_A^{\mathrm{odd}}$. But q cannot be filled in $f.S'$, otherwise $qx \in f.S'$ for some x so that $f(S',q) = qx \in L_B^{\mathrm{even}}$; we therefore have $q = g(f.S',u')$, whence $(g \circ f)(S',u') = p$ as required.

For condition 5, suppose $(g \circ f)(S',u')$ is defined, where $S \subseteq S'$ and $u \sqsubseteq u'$. Then $v' = g(f.S',u')$ is defined, whence also $v = g(f.S,u)$ is defined since clearly $f.S \subseteq f.S'$. If $v \in L_C^{\mathrm{even}}$ then $(g \circ f)(S,u) = v$ and we are done. So suppose $v \in L_B^{\mathrm{odd}}$, so that $(g \circ f)(S,u) \simeq f(S,v)$. If v is not filled in $f.S'$, then $v' = v$ so $f(S,v) \simeq f(S,v')$, which is defined because $f(S',v')$ is defined. If v is filled in $f.S'$, then $f(S',v)$ is defined by the definition of $f.S'$, whence also $f(S,v)$ is defined and we are done. Thus abstract sequential algorithms are closed under composition.

For the identity laws, suppose $f : A \to B$ is an abstract algorithm and $S \in \mathrm{Str}^{\mathrm{fin}}(A)$, $t \in L_B^{\mathrm{odd}}$. As regards $f \circ id_A$, clearly $id_A.S = S$, so for $r \in L_B^{\mathrm{even}}$ we have $(f \circ id_A)(S,u) = r$ iff $f(S,u) = r$. Also if $f(S,u) = q \in L_A^{\mathrm{odd}}$ then q is accessible from S so $id_A(S,q) = q$, whence $(f \circ id_A)(S,u) = q$; conversely if $(f \circ id_A)(S,u) = q$ then $f(S,u)$ is some $q' \in L_A^{\mathrm{odd}}$ and then $q' = q$ by the above.

As regards $id_B \circ f$, if $f(S,u) = r \in L_B^{\mathrm{even}}$ then $r \in f.S$ so $(id_B \circ f)(S,u) = id_B(f.S,u) = r$, and the converse also holds. Also if $f(S,u) = p \in L_A^{\mathrm{odd}}$ then u is not filled in $f.S$, so taking q to be the shortest odd prefix of u not filled in $f.S$, we have $id_B(f.S,u) = q$. Now $q \sqsubseteq u$ and $f(S,u) = p$ so $f(S,q)$ is defined by condition 5; however, $f(S,q) \notin L_B^{\mathrm{even}}$ since q is not filled in $f.S$. So $f(S,q) \in L_A^{\mathrm{odd}}$, so $f(S,q) = f(S,u) = p$ by condition 3. It follows that $(id_B \circ f)(S,u) = p$, and again the converse argument goes through.

For associativity, suppose we have abstract algorithms $f : A \to B$, $g : B \to C$ and $h : C \to D$, and suppose $S \in \mathrm{Str}^{\mathrm{fin}}(A)$, $u \in L_D^{\mathrm{odd}}$. It is immediate from the definitions that $(g \circ f).S = g.(f.S)$, and it follows readily that the values of both $(h \circ (g \circ f))(S,u)$ and $((h \circ g) \circ f)(S,u)$, when defined, coincide with

$$\begin{cases} t \ \text{ if } h(g.f.S,u) = t \in L_D^{\mathrm{even}}\,, \\ p \ \text{ if } h(g.f.S,u) = r \in L_C^{\mathrm{odd}},\ g(f.S,r) = q \in L_B^{\mathrm{odd}}, \text{ and } f(S,q) = p \in L_A^{\mathrm{odd}}\,. \end{cases}$$ □

Exercise 12.1.8. Under the correspondence established by Exercise 12.1.6, show that for abstract affine sequential algorithms, the above definition of composition in $\mathscr{SA}$ agrees with the definition of composition in $\mathscr{SD}$. Deduce that abstract affine sequential algorithms are closed under composition.

Theorem 12.1.9. *$\mathscr{SA}$ is cartesian closed.*

Proof. The operation $\times$ from Definition 12.1.1(iii) is a cartesian product operation in $\mathscr{SA}$. The definitions of projection and pairing are straightforward and are left to the reader.

For exponentiation, we take $A \Rightarrow B = \,!A \multimap B$. The application morphism is constructed as follows. First, let α_{AB} be the evident 'copycat' strategy for the game $((!A \multimap B) \otimes !A) \multimap B$, consisting of plays in which every O-move x is followed immediately by the P-move $\overline{x}$ (that is, the counterpart of x in the other copy of $!A$ or

B). Next, for any object C we have a strategy δ_C for $!C \multimap C$, consisting of all even plays s where $s \upharpoonright_{!C}$ and $s \upharpoonright_C$ agree under the evident inclusion $L_C \hookrightarrow L_{!C}$. We thus obtain a strategy $\alpha_{AB} \circ (\delta_{A \Rightarrow B} \times id_{!A})$ for $(!(A \Rightarrow B) \otimes !A) \multimap B$. But for any objects C, D, it is easy to see that $!C \otimes !D$ coincides exactly with $!(C \times D)$ modulo some evident bijections on move sets; we therefore have a strategy for $!((A \Rightarrow B) \times A) \multimap B$, yielding a morphism $app_{AB} : (A \Rightarrow B) \times A \rightarrow B$ of $\mathcal{SA}$ via the correspondence of Theorem 12.1.5.

Finally, given a sequential algorithm $f : C \times A \rightarrow B$, we may define its transpose $\widehat{f} : C \rightarrow A \Rightarrow B$. Given any $S \in \mathrm{Str}^{\mathrm{fin}}(C)$ and $t \in L^{\mathrm{odd}}_{A \Rightarrow B}$, set $u = f(\langle S, |t \upharpoonright_A| \rangle, t \upharpoonright_B)$ if this is defined, and then set

$$\widehat{f}(S,t) \;\simeq\; \begin{cases} u & \text{if } u_{\mathrm{last}} \in O_C\,, \\ t u_{\mathrm{last}} & \text{if } u_{\mathrm{last}} \in O_A \text{ or } u_{\mathrm{last}} \in P_B\,. \end{cases}$$

To show that $\widehat{f}$ is the unique morphism such that $f = app \circ (\widehat{f} \times id)$, one shows that plays of app against $\langle \widehat{f}.R, S \rangle$ and u correlate precisely with plays of f against $\langle R, S \rangle$ and u. We omit the tedious details. □

Remark 12.1.10. (i) Alternatively, for a more systematic though slightly longer route to the above result (avoiding the use of abstract sequential algorithms), one may show that the operation ! is the object part of a certain *comonad* on $\mathcal{SD}$—that is, a functor $! : \mathcal{SD}$ endowed with natural transformations $\varepsilon : ! \rightarrow id$ and $\delta : ! \rightarrow !^2$ satisfying certain properties—and that $\mathcal{SA}$ arises as its *co-Kleisli* category. One may go further and show that this comonad satisfies suitable axioms for a categorical model of intuitionistic linear logic, from which it follows for general abstract reasons that its co-Kleisli category is cartesian closed. (This approach is followed in [1].) We shall return to these abstract ideas briefly in Subsection 12.1.7.

(ii) The original category of 'sequential algorithms', as introduced by Berry and Curien [33], was a somewhat larger category than $\mathcal{SA}$ whose objects were the *concrete data structures* of Kahn and Plotkin [128]. As explained in [61], our $\mathcal{SA}$ is in effect the full subcategory of this consisting of *filiform* concrete data structures.

12.1.3 Simply Typed Models

We may now readily extract a λ-algebra over $\mathsf{T}^{\rightarrow}(\mathtt{N})$ from $\mathcal{SA}$. We start from a sequential data structure N defined by:

$$O_N = \{?\}\,, \qquad P_N = \mathbb{N}\,, \qquad L_N = \{\varepsilon, ?\} \cup \{?n \mid n \in \mathbb{N}\}\,.$$

Note that strategies for N correspond to elements of $\mathbb{N}_\perp$. As usual, we may now take $A_{\mathtt{N}} = N$, $A_{\sigma \rightarrow \tau} = A_\sigma \Rightarrow A_\tau$, and define a total applicative structure SA by $\mathsf{SA}(\sigma) = \mathrm{Str}(A_\sigma)$, with application induced by the application morphisms in $\mathcal{SA}$. Since $\mathcal{SA}$ is cartesian closed, it is automatic that SA is a λ-algebra. We also obtain a subalgebra $\mathsf{SA}^{\mathrm{eff}}$ in an obvious way: a strategy $S \in \mathrm{Str}(A_\sigma)$ is said to be effective iff it is computably enumerable as a subset of $L^{\mathrm{even}}_{A_\sigma}$ relative to any natural enumerations of O_{A_σ} and P_{A_σ}.

The following is now routine to check, following the pattern of Example 7.1.4:

Proposition 12.1.11. SA *is a model of* PCF$^\Omega$, *and* SA^{eff} *is a model of* PCF, *in the sense of Subsection 7.1.1.* □

Exercise 12.1.12. (i) Construct two different elements of $\mathsf{SA}(0^{(2)} \to 0)$ representing the addition function $\mathbb{N}_\bot^2 \to \mathbb{N}_\bot$. Construct a non-PCF-definable element of $\mathsf{SA}((0^{(2)} \to 0) \to 0)$ that maps these two algorithms to 0 and 1 respectively.

(ii) Show however that the programs $\lambda x.x+x$ and $\lambda x.x*2$ (with obvious algorithms for $+$ and $*$) denote the same element of $\overline{1}$.

A non-trivial result of Laird [162] shows that SA is actually a quotient of the innocent (non-well-bracketed) game model G_i defined in Section 7.4.

We shall also have occasion to work with the call-by-value analogue of SA: that is, the partial λ-algebra SA_v over $\mathbb{N}$, generated from SA as in Proposition 4.3.4 using the obvious partiality structure and definedness predicate. All we shall really need for our present purposes, however, is a concrete description of the pure types $\overline{1}, \overline{2}, \overline{3}$ in this model. The set $\mathsf{SA}_v(1)$ consists, in effect, of strategies for the game N^ω, where

$$O_{N^\omega} = P_{N^\omega} = \mathbb{N}\,, \qquad L_{N^\omega} = \{\varepsilon\} \cup \{n \mid n \in \mathbb{N}\} \cup \{nm \mid n,m \in \mathbb{N}\}\,.$$

(It is easy to see that N^ω is indeed the categorical product of ω copies of N.) Thus, a play of this game consists of (at most) a natural number played by Opponent, followed by a natural number played by Player; we may informally think of these moves as a 'question' and 'answer' respectively. The sets $\mathsf{SA}_v(2)$ and $\mathsf{SA}_v(3)$ consist of strategies for $N^\omega \Rightarrow N$ and $(N^\omega \Rightarrow N) \Rightarrow N$ respectively. The situation is similar for the effective call-by-value model $\mathsf{SA}_v^{\mathrm{eff}}$ likewise obtained from $\mathsf{SA}^{\mathrm{eff}}$.

By the general results of Subsection 7.2.2, both SA_v and $\mathsf{SA}_v^{\mathrm{eff}}$ admit adequate interpretations of the call-by-value language PCF$_v$. It is easy to see how the example of Exercise 12.1.12(i) may be repackaged as an element of $\mathsf{SA}_v^{\mathrm{eff}}$ not definable in PCF$_v$.

12.1.4 B *as a Reflexive Object*

We now show that N^ω is a *universal object* for $\mathscr{SA}$ in the sense of Subsection 4.2.5. From this we may retrieve the definition of van Oosten's untyped model B of Subsection 3.2.4, thus showing that B is a λ-algebra. For typographical convenience, throughout this subsection and the next we shall refer to N^ω as B.

Now let A be an arbitrary sequential data structure; we shall see how the game A may be 'embedded' in B. To motivate the construction, we first observe that a *strategy* for A may be represented by a partial function $L_A^{\mathrm{odd}} \rightharpoonup P_A$, and since both these sets are countable, this may be coded as a partial function $\mathbb{N} \rightharpoonup \mathbb{N}$, i.e. a strategy for B. Likewise, any partial function $\mathbb{N} \rightharpoonup \mathbb{N}$ can, with suitable pruning, be regarded as a strategy for A. The idea is that we can define sequential algorithms $A \to B$ and $B \to A$ that induce these mappings on strategies.

A small variation on the above is helpful, given that our intention is to end up with the λ-algebra B. For any $s \in L_A^{odd}$ we write $s^\dagger \in O_A^*$ for the subsequence of s consisting of all O-moves; we also set $L_A^\dagger = \{s^\dagger \mid s \in L_A^{odd}\}$. Clearly, a strategy for A can equally well be represented by a partial function $L_A^\dagger \rightharpoonup P_A$, which we may code as a strategy for B by fixing on injections $i : L_A^\dagger \to \mathbb{N}$ and $j : P_A \to \mathbb{N}$. This accords with the concrete definition of B: recall from Subsection 3.2.4 that the address of a node in a decision tree is specified by a list $m_0, \ldots, m_{i-1}$, corresponding in effect to the sequence of 'Opponent moves' required to reach that node.

We now proceed to define sequential algorithms $\sigma : A \to B$ and $\rho : B \to A$ such that $\rho \circ \sigma = id_A$. The idea behind these algorithms may be conveyed by some representative game plays: a typical play for σ (abusing notation slightly) will have the form

$$i(x_0 \ldots x_{r-1})^1 {x_0}^0 {y_0}^0 \ldots {x_{r-1}}^0 {y_{r-1}}^0 j(y_{r-1})^1$$

where $x_0 y_0 \ldots x_{r-1} y_{r-1} \in L_A^{even}$, whilst a typical play for ρ has the form

$${x_0}^1 i(x_0)^0 j(y_0)^0 {y_0}^1 \;\; {x_1}^1 i(x_0 x_1)^0 j(y_1)^0 {y_1}^1 \;\; {x_2}^1 i(x_0 x_1 x_2)^0 j(y_2)^0 {y_2}^1 \;\; \cdots$$

where $x_0 y_0 x_1 y_1 x_2 y_2 \ldots$ is a play for A. One may then see intuitively that if plays of these kinds are 'composed', the result will be simply a copycat play on A, since repeated move pairs $(x_k, 0)(y_k, 0)$ are elided in the game $A \multimap A$.

For the purpose of our formal treatment, it is preferable to define ρ, σ as abstract sequential algorithms $\rho : \mathrm{Str}^{fin}(B) \times L_A^{odd} \rightharpoonup L_A^{even} + L_B^{odd}$ and $\sigma : \mathrm{Str}^{fin}(A) \times L_B^{odd} \rightharpoonup L_B^{even} + L_A^{odd}$:

$$\rho(S,t) = \begin{cases} ty \in L_A^{even} & \text{if } i(u^\dagger) j(x) \in S \text{ for all } ux \sqsubseteq^{even} ty\,, \\ i(u^\dagger) \in L_B^{odd} & \text{if } u \text{ is the shortest odd prefix of } t \\ & \text{with } i(u^\dagger) \text{ not filled in } S\,, \end{cases}$$

$$\sigma(T, i(t^\dagger)) = \begin{cases} i(t^\dagger) j(y) \in L_B^{even} & \text{if } ty \in T\,, \\ ux \in L_A^{odd} & \text{if } u \in T \text{ but } ux \sqsubseteq^{odd} t \text{ is not filled in } T\,. \end{cases}$$

It is easily checked that the above definition yields the same values (if any) for $\sigma(T, i(t^\dagger))$ and $\sigma(T, i(t'^\dagger))$ if $t^\dagger = t'^\dagger$. A routine calculation now shows that $\rho \circ \sigma = id_A$, making A a retract of B. Thus B is a universal object in $\mathcal{SA}$.

Exercise 12.1.13. Show that B is not a universal object in the category $\mathcal{SD}$, but that $\mathcal{SD}$ does have a universal object U, where $O_U = P_U = \mathbb{N}$ and $L_U = \mathrm{Alt}(O_U, P_U)$.

As a special case, let us consider the object $A = B^B = !B \multimap B$. Modulo the obvious identification of $O_{!B}$ with $\mathbb{N}$, we have $O_A = P_A = \mathbb{N} + \mathbb{N}$, so that an element of $L_A^\dagger$ will in general have the form ${m_0}^1 {m_1}^0 \ldots {m_r}^0 \ldots$. We may therefore adopt the following concrete codings $i : L_A^\dagger \to \mathbb{N}$ and $j : P_A \to \mathbb{N}$; recall that $\langle \cdots \rangle$ is a standard coding $\mathbb{N}^* \to \mathbb{N}$.

$$i({m_0}^1 {m_1}^0 \ldots {m_r}^0) = \langle m_0, \ldots, m_r \rangle \,, \qquad j(n^l) = 2n + l \;\; (l = 0, 1) \,.$$

Notice that these codings match those used in the definition of B itself. Using this, we obtain a retraction $(\sigma, \rho) : B^B \lhd B$ as explained above, so by Theorem 4.1.29 we may extract an untyped λ-algebra whose elements are the morphisms $1 \to B$ and whose application operation is induced by ρ. But morphisms $1 \to B$ are precisely strategies for B itself, which correspond exactly to elements of $\mathbb{N}_\bot^\mathbb{N}$, and it is not hard to see that the application induced by ρ agrees precisely with van Oosten's application operation $\cdot$ as defined concretely in Subsection 3.2.4. Thus, the applicative structure extracted from the reflexive object B is nothing other than van Oosten's model B, and we conclude:

Theorem 12.1.14. B *is a* λ*-algebra, and hence a total combinatory algebra.* □

Since the morphisms σ, ρ are effective with respect to any standard enumeration of moves, it is easy to see that we also obtain a relative λ-algebra $(\mathsf{B}; \mathsf{B}^{\mathrm{eff}})$, where $\mathsf{B}^{\mathrm{eff}}$ is the substructure consisting of the computable partial functions $\mathbb{N} \rightharpoonup \mathbb{N}$.

Since B is a universal object within $\mathscr{SA}$, it follows that B and $\mathscr{SA}$ generate the same Karoubi envelope. Moreover, there is an evident retraction $B \lhd N \Rightarrow N$ in $\mathscr{SA}$, so that $\mathsf{SA}(1) = N \Rightarrow N$ is also a universal object in $\mathscr{SA}$. It follows that B and SA are equivalent as λ-algebras in the sense of Definition 4.1.23, and hence as computability models (see Exercise 4.1.19). Likewise, the relative λ-algebra $(\mathsf{SA}; \mathsf{SA}^{\mathrm{eff}})$ is equivalent to $(\mathsf{B}; \mathsf{B}^{\mathrm{eff}})$.

12.1.5 A Programming Language for Sequential Algorithms

We have already noted that $\mathsf{SA}^{\mathrm{eff}}$ contains non-extensional operations that are clearly not definable in PCF. We now show that if PCF is extended with a certain non-extensional operator called *catch*, all effective sequential algorithms become definable. The character of *catch* will thus shed further light on the kinds of computations that sequential algorithms can represent.

The language PCF + *catch* was introduced by Cartwright and Felleisen [48] under the name SPCF, and the relationship with sequential algorithms was established by Cartwright, Curien, Felleisen and Kanneganti in [49, 129]. We here give a simple proof of the definability theorem by exploiting the universality of type $\overline{1}$.

It will be convenient to establish our results first in the call-by-value setting, and then to transfer them to the call-by-name one. As already noted in Subsection 12.1.3, we have a call-by-value model $\mathsf{SA}_v^{\mathrm{eff}}$ in which we may interpret terms of PCF_v. (More precisely, it is the closed PCF_v terms $M : \sigma$ whose evaluation terminates that receive denotations as elements of $\mathsf{SA}_v^{\mathrm{eff}}(\sigma)$.) Our goal is now to extend PCF_v with a constant $catch_v : \overline{3}$, interpreted by a certain element of $\mathsf{SA}_v^{\mathrm{eff}}(3)$ (also called $catch_v$) in such a way that every element of $\mathsf{SA}_v^{\mathrm{eff}}$ is definable in $\mathrm{PCF}_v + catch_v$.

The intuition is that $catch_v$ will allow us to detect whether a given $F \in \mathsf{SA}_v(2)$ is a 'constant' algorithm, or whether it attempts to evaluate its (type $\overline{1}$) argument at some value $n \in \mathbb{N}$. If F is the algorithm which immediately returns the result $m \in \mathbb{N}$ without interrogating its argument, the value of $catch\ F$ will be $2m+1$; but if F, when presented with an argument $g \in \mathsf{SA}(1)$, begins by evaluating $g(n)$, the value of $catch\ F$ will be $2n$.

To formalize this idea, we first define an effective strategy $catch'$ which exemplifies the same essential behaviour at a slightly simpler type. Specifically, let $catch'$ be the strategy for $(!B \multimap N) \multimap N$ consisting of all plays of either of the following forms, and even prefixes thereof:

$$?^1 ?^{1,0} m^{1,0} (2m+1)^1 \,, \qquad ?^1 ?^{1,0} (\varepsilon, n)^{0,0} (2n)^1 \,.$$

We now lift this to a strategy in $\mathsf{SA}_v^{\mathrm{eff}}(3)$ as follows. First, we recall that $B \cong N^\omega$, so we may regard $catch'$ as a strategy for $(N^\omega \Rightarrow N) \multimap N$. Next, we may compose with the canonical strategy for $!(N^\omega \Rightarrow N) \multimap (N^\omega \Rightarrow N)$ to obtain an effective strategy $catch_v$ for $(N^\omega \Rightarrow N) \Rightarrow N$, that is, an element of $\mathsf{SA}_v^{\mathrm{eff}}(3)$.

Intuitively, one may imagine evaluating $catch_v\ F$ by applying F to a 'dummy argument' g; if g is applied to a value n, we jump out of this subcomputation and return our final result $2n$. Indeed, it is possible to give a standalone operational semantics for $\mathrm{PCF}_v + catch_v$ that formalizes this idea, and to show that our interpretation in SA is adequate for this semantics (this is done e.g. in [49]), though we shall not go into this here. In real-world programming languages, versions of $catch_v$ may typically be implemented using features such as exceptions or continuations that provide 'non-local jumping'.

Exercise 12.1.15. (i) Write a program in $\mathrm{PCF}_v + catch_v$ that implements the operation *strict?* from Example 11.1.3 (or rather some corresponding operation within the structure SA_v).

(ii) Show that $catch_v$ itself is actually PCF_v-definable from a similar strategy of the simpler type $((\mathtt{B} \to \mathtt{N}) \to \mathtt{N}) \to \mathtt{N}$ (semantically, a strategy for $(N^2 \Rightarrow N) \Rightarrow N$).

We now show that $\mathrm{PCF}_v + catch_v$ defines all elements of $\mathsf{SA}_v^{\mathrm{eff}}$. As with the proof of definability for $\mathrm{PCF} + H$ and $\mathrm{SR}^{\mathrm{eff}}$ in Section 11.2, our proof follows the general pattern indicated in Subsection 4.2.5; the key is to show that both halves of a certain retraction $\overline{2} \lhd \overline{1}$ in SA_v (that is, a retraction $N^B \lhd B$ within $\mathscr{SA}$) are definable in $\mathrm{PCF}_v + catch_v$.

To define this retraction, we adapt the definition of the retraction $B^B \lhd B$ from Subsection 12.1.4. Up to some evident bijections, we have $O_{N^B} = \mathbb{N} + \{?\}$ and $P_{N^B} = \mathbb{N} + \mathbb{N}$, so that an element of $L^\dagger_{N^B}$ has the form $?^1 {m_0}^0 \ldots {m_{r-1}}^0 \ldots$. We now define the following codings $i : L^\dagger_{N^B} \to \mathbb{N}$ and $j : P_{N^B} \to \mathbb{N}$; recall that $\langle \cdots \rangle$ is a standard coding $\mathbb{N}^* \to \mathbb{N}$.

$$i(?^1 {m_0}^0 \ldots {m_{r-1}}^0) \;=\; \langle m_0, \ldots, m_{r-1} \rangle \,, \qquad j(n^l) \;=\; 2n + l \;\; (l = 0,1) \,.$$

Using these, we may construct a retraction $(\sigma,\rho) : (B \Rightarrow N) \lhd B$ exactly as in Subsection 12.1.4. Note that the operation $\mathbb{N}_\perp^\mathbb{N} \times \mathbb{N}_\perp^\mathbb{N} \to \mathbb{N}_\perp$ induced by ρ is exactly the operation $\mid$ defined in Section 11.1.

Lemma 12.1.16. *(i) The morphism $\rho : B \to N^B$ is definable by a* PCF_v *term $R[f] : \overline{2}_v$ with free variable $f : \overline{1}_v$.*

(ii) The morphism $\sigma : N^B \to B$ is definable by a $\mathrm{PCF}_v + catch_v$ *term $S[F] : \overline{1}_v$ with free variable $F : \overline{2}_v$.*

Proof. (i) The PCF_v term in question is a minor modification of a program given in Subsection 7.1.5. Using some obvious pseudocode notation, we define:

$$\begin{aligned} play\ f\ g\ \langle m_0,\ldots,m_{r-1}\rangle &= case\ f\,\langle m_0,\ldots,m_{r-1}\rangle\ of \\ &\qquad (2n \Rightarrow let\ m = g(n)\ in\ play\ f\ g\ \langle m_0,\ldots,m_{r-1},m\rangle \\ &\qquad \mid 2n+1 \Rightarrow n)\,, \\ R[f] &= \lambda g^1.\, play\ f\ g\ \langle\rangle\,. \end{aligned}$$

A tedious calculation (which we omit) is needed to verify that $R[f]$ denotes ρ.

(ii) The construction of $S[F]$ involves a programming trick with $catch_v$. Given any finite portion of a partial function $\mathbb{N} \rightharpoonup \mathbb{N}$, represented by a graph p (i.e. a list of ordered pairs), we may use $catch_v$ to detect the first value outside the domain of p (if any) on which F tries to interrogate its argument. To determine the label at a given node $\langle m_0,\ldots,m_{r-1}\rangle$ in the tree for F, we may (recursively) compute the label n'_k at $\langle m_0,\ldots,m_{k-1}\rangle$ for each $k<r$; if $n'_k = 2n_k$ for each k, we may then use the graph $p = (n_0,m_0),\ldots(n_{r-1},m_{r-1})$ to determine the next step carried out by F.

We implement this idea by the following pseudocode, in which $\langle\langle\cdots\rangle\rangle$ denotes an effective coding operation $(\mathbb{N}\times\mathbb{N})^* \to \mathbb{N}$, and :: denotes augmentation of a graph by a new pair, so that $(n,m) :: \langle\langle (n_0,m_0),\ldots,(n_{r-1},m_{r-1})\rangle\rangle = \langle\langle (n,m),(n_0,m_0),\ldots,(n_{r-1},m_{r-1})\rangle\rangle$.

$$\begin{aligned} mask\ p^0\ g^1 &= \lambda z^0.\, case\ p\ of \\ &\qquad (\langle\rangle \Rightarrow g(z) \\ &\qquad \mid (n,m) :: p' \Rightarrow if\ z=n\ then\ m\ else\ mask\ p'\ g\ z)\,, \\ next\ \langle m_0,\ldots,m_{r-1}\rangle\ F^2\ p^0 &= case\ catch_v(\lambda g.\, F(mask\ p\ g))\ of \\ &\qquad (2m+1 \Rightarrow if\ r=0\ then\ 2m+1\ else\ diverge \\ &\qquad \mid 2n \Rightarrow if\ r=0\ then\ 2n \\ &\qquad\qquad else\ next\ \langle m_1,\ldots,m_{r-1}\rangle\ F\ ((n,m_0) :: p))\,, \\ S[F^2] &= \lambda l^0.\, next\ l\ F\ \langle\rangle\,. \end{aligned}$$

Again, we omit the tedious verification that $S[F]$ denotes σ. □

The proof of the following now proceeds along standard lines:

Theorem 12.1.17. *Every element of* $\mathsf{SA}_v^{\mathrm{eff}}$ *is definable in* $\mathrm{PCF}_v + catch_v$.

Proof. Working in $\mathsf{SA}_v^{\mathrm{eff}}$, the $\mathrm{PCF}_v + catch_v$ definable retraction $\overline{2}_v \lhd \overline{1}_v$ lifts readily to definable retractions $\overline{k+1}_v \lhd \overline{k}_v$, whence we obtain a definable retraction $\overline{k}_v \lhd \overline{1}_v$ for each $k > 0$. But by the general results of Section 4.2, every type σ is a PCF_v-definable retract of some pure type, so every σ is a retract of $\overline{1}$; in particular, there is always a $\mathrm{PCF}_v + catch_v$ definable surjection from $\mathsf{SA}_v(1)$ to $\mathsf{SA}_v(\sigma)$. But clearly all elements of $\mathsf{SA}_v(1)$ are PCF_v-definable, since these correspond simply to partial computable functions $\mathbb{N} \rightharpoonup \mathbb{N}$; thus every element of $\mathsf{SA}_v(\sigma)$ is $\mathrm{PCF}_v + catch_v$ definable. □

For a call-by-name version of this, we use the evident retraction $N^\omega \lhd N^N$ to lift $catch_v$ to an effective strategy for $(N^N \Rightarrow N) \Rightarrow N$, i.e. an element of $\mathsf{SA}^{\mathrm{eff}}(3)$, which we call *catch*. We thus have an interpretation of $\mathrm{PCF} + catch$ in $\mathsf{SA}^{\mathrm{eff}}$, where *catch* is a constant of type $\overline{3}$. As a routine application of Theorem 7.2.5, we now have:

Corollary 12.1.18. *Every element of* $\mathsf{SA}^{\mathrm{eff}}$ *is definable in* $\mathrm{PCF} + catch$. □

Exercise 12.1.19. Show that $\mathsf{SA}^{\mathrm{eff}}$, considered as a model of PCF, is *simple* in the sense of Subsection 7.1.2. From this and Corollary 12.1.18, deduce that the interpretation of $\mathrm{PCF} + catch$ in $\mathsf{SA}^{\mathrm{eff}}$ (and hence in SA) is *fully abstract* in the denotational sense (see Subsection 7.1.4).

12.1.6 An Extensional Characterization

We now present a remarkable characterization of SA due to Laird [165] (see also Curien [63]). The idea is to consider a mild extension of the definition of sequential algorithms that incorporates a non-recoverable error value $\top$, which we treat as dual to the element $\bot$ representing divergence. If this is done, we actually obtain an *extensional* model $\mathsf{SA}^\top$ over $\mathbb{N}_\bot^\top$ with good mathematical structure, from which the original model SA can easily be extracted as a certain substructure. What is more, we may abstractly characterize which functions $\mathsf{SA}^\top(\sigma) \to \mathsf{SA}^\top(\tau)$ are represented in $\mathsf{SA}^\top(\sigma \to \tau)$ by means of certain preservation properties; this yields a mathematical characterization of SA much in the spirit of traditional domain theory, complementing the more dynamic perspective offered by games and strategies.

We shall work here with the same notion of sequential data structure as in Subsection 12.1.1, but a different notion of strategy which, in response to a given odd position t, may now play a special *error* value $\top$ as an alternative to playing an ordinary P-move. These errors are non-recoverable: no further play can happen once an error has occurred.

Definition 12.1.20. Let A be a sequential data structure as in Definition 12.1.1.

(i) Define the set $L_A^{\top\mathrm{even}}$ of $\top$-*positions* of A as

$$L_A^\top = L_A \cup \{t\top \mid t \in L_A^{\mathrm{odd}}\},$$

and write $L_A^{\top\mathrm{even}}$ for the set of $\top$-positions of even length.

(ii) A $\top$-*strategy* for A is now a non-empty set $S \subseteq L_A^{\top\text{even}}$ such that S is closed under even prefixes, and if $ty, ty' \in S$ then $y = y'$. We write $\mathrm{Str}^{\top}(A)$ for the set of $\top$-strategies for A.[2]

If $S \in \mathrm{Str}^{\top}(A)$, we write $\downarrow S$ for the set of plays in S not involving $\top$. Clearly $\downarrow S \in \mathrm{Str}^{\top}(A)$, $\downarrow\downarrow S = \downarrow S$, and $\mathrm{Str}(A) = \{\downarrow S \mid S \in \mathrm{Str}^{\top}(A)\}$.

Let us now pick out some structure present in $\mathrm{Str}^{\top}(A)$. Firstly, there is an equivalence relation $\updownarrow$ on $\mathrm{Str}^{\top}(A)$ given by $S \updownarrow T$ iff $\downarrow S = \downarrow T$; that is, S and T are equivalent if they only differ as regards the raising of errors. Secondly, we may define a partial order $\sqsubseteq$ on $\mathrm{Str}^{\top}(A)$ which treats $\top$ as a maximal element (this is the key idea behind Laird's approach). To make this precise, for $s, s' \in L_A^{\top\text{even}}$, let us write $s \preceq s'$ if either s is error-free and s is a prefix of s', or $s' = t'\top$ where t' is some prefix of s. For $S, T \in \mathrm{Str}^{\top}(A)$, we now take $S \sqsubseteq T$ if for all $s \in S$ there exists $s' \in T$ with $s \preceq s'$. It is easy to check that $\preceq$ is a partial ordering, whence so is $\sqsubseteq$, and also that $\downarrow S$ is the $\sqsubseteq$-least element equivalent to S.

The following marshals some abstract properties possessed by $\updownarrow$ and $\sqsubseteq$, showing that they give $\mathrm{Str}^{\top}(A)$ the structure of a *bistable bi-CPO*. (The terminology arises from an equivalent presentation of this structure in terms of two partial orderings—see [165].) If $D, E \subseteq \mathrm{Str}^{\top}(A)$, we write $D \updownarrow E$ to mean that for all $S \in D$, $T \in E$ there exist $S' \in D$, $T' \in E$ such that $S \sqsubseteq S' \updownarrow T' \sqsupseteq T$.

Proposition 12.1.21. $(\mathrm{Str}^{\top}(A), \sqsubseteq, \updownarrow)$ *is a bistable bi-CPO. That is:*

1. *Any equivalence class K for $\updownarrow$ is a distributive lattice with respect to $\sqsubseteq$: i.e. any $S, T \in K$ have both a least upper bound $S \vee T$ and a greatest lower bound $S \wedge T$ in K, and both distributivity laws hold.*
2. *If $S \updownarrow T$, then $S \vee T$, $S \wedge T$ are respectively the least upper bound and greatest lower bound of S, T within* $\mathrm{Str}^{\top}(A)$.
3. $(\mathrm{Str}^{\top}(A), \sqsubseteq)$ *is a directed-complete partial order.*
4. *If $D, E \subseteq$* $\mathrm{Str}^{\top}(A)$ *are directed sets and $D \updownarrow E$, then $\bigsqcup D \updownarrow \bigsqcup E$ and*

$$\bigsqcup D \wedge \bigsqcup E \;=\; \bigsqcup \{S \wedge T \mid S \in D,\ T \in E,\ S \updownarrow T\}\,.$$

Proof. For property 1, we note that each equivalence class K is a Boolean algebra: if $\bot_K$ is the least element of K and $Z_K = \{sx \in L_A^{\text{odd}} \mid s \in \bot_K\,,\ \nexists y.sxy \in \bot_K\}$, then identifying an arbitrary $T \in K$ with $\{t \in Z \mid t\bot \in T\}$, we see that $K \cong \mathscr{P}(Z)$. Since $\vee$ and $\wedge$ correspond simply to union and intersection, property 2 also follows easily.

For property 3, given $D \subseteq \mathrm{Str}^{\top}(A)$ directed, let

$$S \;=\; \{s \in \bigcup D \mid \forall t \sqsubseteq s.\, t\top \in \bigcup D \Rightarrow s = t\top\}\,.$$

It is routine to check that S is a $\top$-strategy and is the supremum of D. We leave property 4 as an exercise (recall that directed sets are non-empty). □

[2] Laird [165] gives a mathematically more elegant, though perhaps less intuitive, treatment of strategies in which an error-raising play $t\top$ is represented simply by the odd position t.

Note also that $\downarrow S$ is the $\sqsubseteq$-least element equivalent to S, so we may abstractly retrieve $\mathrm{Str}(A)$ from the structure $(\mathrm{Str}^\top(A), \sqsubseteq, \updownarrow)$ as the set of strategies that are $\sqsubseteq$-minimal within their $\updownarrow$-equivalence class.

Next, we define a way of applying a $\top$-strategy $S \in \mathrm{Str}^\top(A \multimap B)$ to $T \in \mathrm{Str}^\top(A)$; notice that any errors arising from either S or T are 'propagated' into the result.

$$\begin{aligned} app(S,T) = & \{s \upharpoonright_B \mid s \in S \ \top\text{-free}, \ s \upharpoonright_A \in T\} \cup \\ & \{(s \upharpoonright_B)\top \mid s\top \in S, \ s \upharpoonright_A \in T\} \cup \\ & \{(s \upharpoonright_B)\top \mid s \in S, \ (s \upharpoonright_A)\top \in T\} . \end{aligned}$$

It is routine to check that $app(S,T) \in \mathrm{Str}^\top(B)$. We may also promote a $\top$-strategy $S \in \mathrm{Str}^\top(A)$ to $!S$ in $\mathrm{Str}^\top(!A)$ in an obvious way:

$$!S \ = \ \{(s_0,x_0)y_0 \dots (s_{r-1},x_{r-1})y_{r-1} \in L_{!A}^{\top\mathrm{even}} \mid \forall i < r. \ s_i x_i y_i \in S\} .$$

Again, it is routine to check that $!S$ in $\mathrm{Str}^\top(!A)$.

Given $S \in \mathrm{Str}^\top(A \Rightarrow B)$ and $T \in \mathrm{Str}^\top(A)$, we may therefore define the application $S \cdot T = app(S, !T) \in Str^\top(B)$; we also write Φ_S for the function $T \mapsto S \cdot T$. The crucial point is that this notion of application is *extensional*:

Lemma 12.1.22. *If $S, S' \in \mathrm{Str}^\top(A \Rightarrow B)$ and $\Phi_S = \Phi_{S'}$, then $S = S'$.*

Proof. Suppose $S \neq S'$. Let $t \in L_A^{\mathrm{odd}}$ be a minimal play to which the responses (if any) of S, S' differ; without loss of generality we may suppose $ty \in S$ but $ty \notin S'$. If y is either $\top$ or a move from P_B, let $T = \| t \upharpoonright_{!A} \| \in \mathrm{Str}(A)$ with $\| - \|$ as in Subsection 12.1.2; then $t \upharpoonright_{!A}$ is itself a play of $!T$, and it follows easily that $S \cdot T$ contains the play $(t \upharpoonright_B)y$ whereas $S' \cdot T$ does not; thus $\Phi_S \neq \Phi_{S'}$. If y is a move from $O_{!A}$, let $T = \| (t \upharpoonright_{!A})y\top \|$; then $(t \upharpoonright_{!A})y\top$ is a play of $!T$, so $S \cdot T$ contains the play $(t \upharpoonright_B)\top$ whereas $S' \cdot T$ does not, whence again $\Phi_S \neq \Phi_{S'}$. □

We now show how the functions induced by $\top$-strategies for $A \Rightarrow B$ may be precisely characterized via preservation properties. At an abstract level, if $(X, \sqsubseteq_X, \updownarrow_X)$ and $(Y, \sqsubseteq_Y, \updownarrow_Y)$ are bistable bi-CPOs as in Proposition 12.1.21, we say a function $f : X \to Y$ is *bistable* if for all $a, b \in X$ we have

$$a \updownarrow_X b \ \Rightarrow \ f(a) \updownarrow_Y f(b) \, , \ f(a \vee b) \ = \ f(a) \vee f(b) \, , \ f(a \wedge b) \ = \ f(a) \wedge f(b) \, .$$

Proposition 12.1.23. *For any $S \in \mathrm{Str}^\top(A \Rightarrow B)$, the map $\Phi_S : \mathrm{Str}^\top(A) \to \mathrm{Str}^\top(B)$ is monotone, bistable and continuous with respect to $\sqsubseteq$ and $\updownarrow$.*

Proof. For monotonicity, suppose $T \sqsubseteq T'$ in $\mathrm{Str}^\top(A)$. To show $S \cdot T \sqsubseteq S \cdot T'$, take $u \in S \cdot T$; we wish to find $u' \in S \cdot T'$ with $u \preceq u'$. Take $s \in S$ with $t = s \upharpoonright_A \in T$ and $s \upharpoonright_B = u$, and take $t' \in T'$ such that $t \preceq t'$. If t is error-free, this means also $t \in T'$ and we may take $u' = u \in S \cdot T'$. If t' ends in $\top$, then by playing t' against some prefix of s we again obtain some $u' \in S \cdot T'$ with $u \preceq u'$.

For bistability, suppose $T \updownarrow T'$ in $\mathrm{Str}^\top(A)$. Then T, T' each extend $\downarrow T$ simply by the addition of certain plays $t\top$, so from the definition of application it is clear that

$S \cdot T$, $S \cdot T'$ each extend $S \cdot (\downarrow T)$ simply by the addition of certain plays $u\top$; thus $S \cdot T \updownarrow S \cdot (\downarrow T) \updownarrow S \cdot T'$. As regards meets and joins, we reason as in the proof of Proposition 12.1.21. Let

$$Z = \{tx \in L_A^{\mathrm{odd}} \mid t \in \downarrow T\ ,\ \nexists y.txy \in \downarrow T\}\ ,$$
$$W = \{uz \in L_B^{\mathrm{odd}} \mid u \in S \cdot (\downarrow T)\ ,\ \nexists w.uzw \in \downarrow T\}\ ,$$

so that the $\updownarrow$-classes of $\downarrow T$ and $S \cdot (\downarrow T)$ may be identified with $\mathscr{P}(Z)$ and $\mathscr{P}(W)$ respectively. From the deterministic nature of S, we see that for any $uz \in W$, either uz is unfilled in all $S \cdot T''$ where $T'' \updownarrow T$, or there is a unique $tx \in Z$ such that for all $T'' \updownarrow T$ we have $uz\top \in S \cdot T''$ iff $tx\top \in T''$. From this it is easy to see that meets and joins are preserved as required—indeed, the action of S on the equivalence class is encapsulated by a lattice homomorphism $\mathscr{P}(Z) \to \mathscr{P}(W)$.

We leave continuity as an easy exercise using the description of suprema from the proof of Proposition 12.1.21. □

Theorem 12.1.24. *Any monotone, bistable and continuous function* $\Phi : \mathrm{Str}^\top(A) \to \mathrm{Str}^\top(B)$ *arises as* Φ_S *for some* $S \in \mathrm{Str}^\top(A \Rightarrow B)$.

Proof. Given Φ monotone, bistable and continuous, we shall retrieve S first as an *abstract* algorithm $f : \mathrm{Str}^{\top\mathrm{fin}}(A) \times L_B^{\mathrm{odd}} \rightharpoonup L_A^{\mathrm{odd}} + L_B^{T\mathrm{even}}$ in the spirit of Definition 12.1.4; we will then convert this to an element of $\mathrm{Str}^\top(A \Rightarrow B)$ in the manner of Theorem 12.1.5.

Given any $T \in \mathrm{Str}^{\mathrm{fin}}(A)$ and $u \in L_B^{\mathrm{odd}}$, there are three cases to consider. If $\Phi(T)$ contains an element uz for some necessarily unique z (possibly $\top$), we set $f(T,u) = uz$. If $\Phi(T)$ contains no such uz and neither does $\Phi(T')$ for any $T' \updownarrow T$, we let $f(T,u)$ be undefined. If $\Phi(T)$ contains no such uz but $uz \in \Phi(T')$ for some $T' \updownarrow T$, then necessarily $z = \top$. We will show that there is some specific $t \in L_A^{\mathrm{odd}}$ such that for all $T' \updownarrow T$ we have $u\top \in \Phi(T')$ iff $t\top \in T'$; we will then set $f(T,u) = t$.

As in previous proofs, define $Z \subseteq L_A^{\mathrm{odd}}$ so that the equivalence class of T corresponds to $\mathscr{P}(Z)$, and let $\mathscr{I} \subseteq \mathscr{P}(Z)$ correspond to the set $\{T' \updownarrow T \mid u\top \notin \Phi(T')\}$. Note that $\mathscr{I}$ is down-closed, and that $\emptyset \in \mathscr{I}$ but $\mathscr{I} \neq \mathscr{P}(Z)$, so $Z \notin \mathscr{I}$. Note too that by the preservation properties of Φ, $\mathscr{I}$ is closed under joins and so is directed, and that $I = \bigsqcup \mathscr{I} \in \mathscr{I}$. We will show that there exists $t \in Z$ such that $I = Z - \{t\}$. (In lattice-theoretic parlance, $\mathscr{I}$ is a *principal ideal* on $\mathscr{P}(Z)$.) On the one hand, we have $I \neq Z$ since $I \in \mathscr{I}$ but $Z \notin \mathscr{I}$. On the other hand, if $Z - I$ contains more than one element, we may express I as the intersection of two proper supersets J, J'. But now $J, J' \notin \mathscr{I}$ whereas $J \cap J' \in \mathscr{I}$, contradicting that Φ preserves meets. So there exists $t \in Z$ with the stated properties, and we may set $f(T,u) = t$ as proposed.

Even without formulating the general notion of an abstract $\top$-algorithm, we may convert our function f to a $\top$-strategy $S \in \mathrm{Str}^\top(A \Rightarrow B)$ using the construction from the proof of Theorem 12.1.5. This construction goes through without difficulty in the presence of $\top$, and it is immediate that it delivers a $\top$-strategy.

It remains to show that $S \cdot T = \Phi(T)$ for $T \in \mathrm{Str}^\top(A)$. We here outline the method without entering into details. To show that $u \in S \cdot T$ implies $u \in \Phi(T)$, we argue by induction on u. By considering the play $s \in S$ that gives rise to u, and examining the way S is generated, we obtain (via some case splits) conditions on s that show $u \in \Phi(T)$. For the converse, we again argue by induction on u. Here we use continuity to select some finite $T_0 \sqsubseteq T$ such that $u \in \Phi(T_0)$. To obtain a play in S showing that $u \in S \cdot T$, we inductively build a sequence $x_0 y_0 x_1 y_1 \ldots$ of moves in $!A$ that is compatible with both S and T_0. Since T_0 is finite and O-moves may not be repeated, this sequence will eventually terminate, and the conditions under which it does so give a sequence $s \in S$ yielding $u \in S \cdot T$ as required. We leave the remaining details as an exercise (the reader may consult [63] or [163] for inspiration). □

We have thus characterized the functions induced by $\top$-strategies precisely by means of a preservation property. To complete the picture, we show that exponentiation in $\mathscr{SA}$ can also be captured in terms of bistable bi-CPOs.

Theorem 12.1.25. *(i) The category $\mathscr{BB}$ of bistable bi-CPOs and monotone, bistable and continuous functions is cartesian closed.*
(ii) If $A, B \in \mathscr{SA}$ then $\mathrm{Str}^\top(A \Rightarrow B) \cong \mathrm{Str}^\top(A) \Rightarrow \mathrm{Str}^\top(B)$ within $\mathscr{BB}$.

Proof. (i) Finite products in $\mathscr{BB}$ are as expected, with $\sqsubseteq$ and $\updownarrow$ given componentwise. If X, Y are bistable bi-CPOs, we define $X \Rightarrow Y$ as follows. The underlying set consists of all morphisms $f : X \to Y$ in $\mathscr{BB}$, and the partial order $\sqsubseteq$ is given pointwise. For the equivalence relation, we take $f \updownarrow g$ iff for all x we have $f(x) \updownarrow g(x)$, and for all $x \updownarrow y$ we have

$$f(x) \wedge g(y) = f(y) \wedge g(x), \qquad f(x) \vee g(y) = f(y) \vee g(x).$$

The verification that this defines a bistable bi-CPO with the required universal property is lengthy but entirely straightforward, and we leave it to the reader.

(ii) Suppose A, B are objects in $\mathscr{SA}$. We have already seen that $\mathrm{Str}^\top(A \Rightarrow B)$, as a set, is isomorphic to the set of morphisms $\mathrm{Str}^\top(A) \to \mathrm{Str}^\top(B)$ in $\mathscr{BB}$. For the partial order, first note that if $S \sqsubseteq S'$ in $\mathrm{Str}^\top(A \Rightarrow B)$, then for any $T \in \mathrm{Str}^\top(A \Rightarrow B)$ we have $S \cdot T \sqsubseteq S' \sqsubseteq T$ by monotonicity of the action of $\lambda f. f \cdot T \in \mathrm{Str}^\top((A \Rightarrow B) \Rightarrow B)$. Conversely, if $S \not\sqsubseteq S'$, a slight refinement of the proof of Lemma 12.1.22 can be used to construct a strategy T and a play $u \in S \cdot T$ not dominated by any $u' \in S' \cdot T$; hence $S \cdot T \not\sqsubseteq S' \cdot T$.

For the equivalence relation, first suppose $S \updownarrow S'$ in $\mathrm{Str}^\top(A \Rightarrow B)$. Then for any $T \in \mathrm{Str}^\top(A)$ we have $S \cdot T \updownarrow S' \cdot T$ by the bistability of $\lambda f. f \cdot T$. Moreover, if $T \updownarrow T'$, then working in $(A \Rightarrow B) \times A$ we have $(S, T) \updownarrow (S', T')$ and $(S, T) \wedge (S', T') = (S \wedge S', T \wedge T')$, so by the bistability of application we have $S \cdot T \wedge S' \cdot T' = (S \wedge S') \cdot (T \wedge T')$, and likewise for $S \cdot T' \wedge S' \cdot T$. The equation for $\vee$ holds similarly.

Conversely, suppose $S \not\updownarrow S'$. By symmetry, we may take $sxy \in S - S'$ of minimal length such that $y \neq \top$. If $y \in P_B$, then as in Lemma 12.1.22 we may construct T such that $(sxy) \upharpoonright B \in S \cdot T - S \cdot T'$, so we have a counterexample to $\forall T. S \cdot T \updownarrow S' \cdot T$. If $y \in O_A$, let $T = \| (sx) \upharpoonright_{!A} \|$ and $T' = \| (sxy\top) \upharpoonright_{!A} \|$ so that $T \updownarrow T'$. Writing $uz = (sx) \upharpoonright_B \in L_B^{\mathrm{odd}}$, we clearly have $u \in S \cdot T$ and uz unfilled in $S \cdot T$ but $uz\top \in S \cdot T'$. Now consider the action of S' on T and T'. If $S \cdot T \not\updownarrow S' \cdot T$ or $S \cdot T' \not\updownarrow S' \cdot T'$ then we again have a counterexample to $\forall S. S \cdot T \updownarrow S' \cdot T$, so we may assume $S \cdot T \updownarrow S' \cdot T \updownarrow S \cdot T' \updownarrow S' \cdot T'$. So if uz is filled in any of these, it is filled with $\top$.

There are now three cases to consider. If uz is filled in $S' \cdot T$, it is filled in $S \cdot T' \wedge S' \cdot T$ but not in $S \cdot T$, yielding a counterexample to the '$\wedge$' property. If uz is not filled in $S' \cdot T'$, it is filled in $S \cdot T'$ but not in $S' \cdot T'$ or $S \cdot T$, yielding a counterexample to the '$\vee$' property. The remaining case, however, is impossible: if uz is filled in $S' \cdot T'$ but not in $S \cdot T$, then sx must be filled in S' with some $y' \in O_{!A}$ that is answered in T'; but the only such move y' that has not already occurred in s is y, giving $sxy \in S'$, a contradiction. □

We therefore have a construction of SA purely in terms of bistable bi-CPOs. Let N be the bistable bi-CPO $\mathbb{N}_\bot^\top = \mathbb{N} \cup \{\bot, \top\}$, with $\sqsubseteq, \updownarrow$ respectively generated by $\bot \sqsubseteq n \sqsubseteq \top$ and $\bot \updownarrow \top$. Working in $\mathscr{BB}$, let $E_\mathbb{N} = N$ and $E_{\sigma \to \tau} = E_\sigma \Rightarrow E_\tau$. Then the underlying sets of these objects yield an extensional type structure $\mathsf{SA}^\top$ over $\mathbb{N}_\bot^\top$, whilst the elements minimal within their equivalence class form the substructure SA (a non-extensional applicative structure over $\mathbb{N}_\bot$).

12.1.7 Relation to Other Game Models

We conclude this section by briefly situating our model SA in relation to other known game models. We have already remarked that SA is a quotient of the model G_i of Section 7.4 (see Laird [162]). Here we point out its relationship to another family of models, constructed from $\mathscr{SD}$ in much the same way as SA but using a different definition of the '!' operator.

We will here write $!_1$ for the operator ! as given by Definition 12.1.3. As we have seen, the role of this operator is intuitively to allow operations to invoke their arguments more than once (as in $\lambda f.f(0)+f(1)$), by allowing computations to explore more than one path through the tree for a game A. However, there are many other ways of achieving this effect. For instance, a rather simple kind of 're-usability' is provided by the following operator, due to Hyland [123]:

Definition 12.1.26. For any sequential data structure A we define a data structure $!_2A$ as follows. Moves are given by $O_{!_2A} = O_A \times \mathbb{N}$, $P_{!_2A} = P_A \times \mathbb{N}$. Legal positions in $L_{!_2A}$ are sequences $s \in \mathrm{Alt}(O_{!_2A}, P_{!_2A})$ such that

1. for each $i \in \mathbb{N}$, $s_i \in L_A$,
2. for every occurrence in s of a move $(y, i+1)$, there is an earlier occurrence in s of a move (x, i).

It is also possible to combine the ideas behind $!_1$ and $!_2$ to define a *repetitive backtracking* operator $!_3$ (see Harmer, Hyland and Melliès [113]).

All three of these operators (and some others besides) turn out to provide instances of the general abstract situation mentioned in Remark 12.1.10(i): they each form the object part of a comonad on $\mathscr{SD}$ yielding a model of intuitionistic linear logic, so that their co-Kleisli categories are cartesian closed for general reasons. All of these models embody different kinds of intensional information and support distinctive styles of computation. For example, the model based on $!_2$ distinguishes between $\lambda f.f(0)+f(0)$ and $\lambda f.f(0)*2$ (not surprisingly, given that repetitions of earlier moves are allowed for in $!_2A$), and models a language with a *coroutine* operator (see Laird [164]), which in turn allows us to simulate the use of stateful (local) variables of ground type. A selection of programming languages that define all effective strategies in each of these models is briefly presented in Longley [182]. However, there is more work to be done on these and related models, and in the general area of intensional models arising naturally from game semantics.

These different notions of re-usability also furnish an example of another interesting phenomenon, touched on briefly in Subsection 4.1.4. The associated simply typed models (call them $\mathsf{SA}_1, \mathsf{SA}_2, \mathsf{SA}_3$) are equivalent as computability models in the sense of Definition 3.4.4 (they are all equivalent to B), but not as λ-algebras since their Karoubi envelopes are different (as is witnessed by the fact that they support different kinds of operations of simple type). Since each of the corresponding CCCs contains a universal object, we also obtain examples of *untyped* models that are equivalent as PCAs but not as λ-algebras.

12.2 Kleene's Second Model K_2

We now turn more briefly to another intensional model that has played a major role in this book: Kleene's second model K_2, first introduced in Kleene and Vesley [146] and briefly described in Subsection 3.2.1. As explained there, K_2 plays the role of a basic model for continuous computation on countable data; in this capacity it has featured especially in our studies of Ct and PC in Chapters 8 and 10. Here we consider it as an intensional model of some interest in its own right.

We first establish the main result needed by earlier chapters: K_2 is a lax partial combinatory algebra in the sense of Subsection 3.1.6. We shall also see that it is sufficiently 'intensional' that the equations of λ-calculus break down. Even so, K_2 offers intriguing possibilities for modelling a certain flavour of 'time-sensitive' computation, an idea we briefly illustrate with a language PCF + *timeout*. Finally, we touch on the relationship between K_2 and models such as B and $\mathscr{P}\omega$, and on its place within the universe of all intensional models.

For the basic intuition behind K_2, we refer to Subsection 3.2.1. We here recall the definition of the operation $\odot : \mathbb{N}^{\mathbb{N}} \times \mathbb{N}^{\mathbb{N}} \rightharpoonup \mathbb{N}^{\mathbb{N}}$, making explicit that this is defined via a *total* function $\mathbb{N}^{\mathbb{N}} \times \mathbb{N}^{\mathbb{N}} \to \mathbb{N}_{\bot}^{\mathbb{N}}$, which we here call $\odot'$.

$$f \odot' g = \Lambda n.\ f(\langle n, g(0), \ldots, g(k_{fgn}-1)\rangle) - 1\,,$$
$$\text{where } k_{fgn} = \mu k.f(\langle n, g(0), \ldots, g(k-1)\rangle) > 0\,.$$

We take $f \odot g$ to be $f \odot' g$ if this is in $\mathbb{N}^{\mathbb{N}}$, undefined otherwise, and define K_2 to be the untyped applicative structure $(\mathbb{N}^{\mathbb{N}}, \odot)$. Clearly, if f, g are computable then $f \odot' g$ is partial computable, so we obtain a substructure K_2^{eff}.

We now investigate which partial functions on $\mathbb{N}^{\mathbb{N}}$ are representable within K_2 and K_2^{eff}. We here use $\vec{p}, \vec{q}$ to range over finite sequences of natural numbers, with $\sqsubseteq$

the prefix ordering on such sequences. If $\vec{p} = p_0, \ldots, p_{k-1}$, we define

$$U_{\vec{p}} \;=\; \{g \in \mathbb{N}^{\mathbb{N}} \mid \forall i < k.\, g(i) = p_i\}\,.$$

The collection of all such sets forms a countable basis for the *Baire topology* on $\mathbb{N}^{\mathbb{N}}$.

Definition 12.2.1. Let r be fixed, and let $\mathscr{F}$ be a (finite or infinite) set of tuples $(\vec{p}_0, \ldots, \vec{p}_{r-1}, n, m) \in \mathbb{N}^{*r} \times \mathbb{N} \times \mathbb{N}$. We say $\mathscr{F}$ is a *graph* for $F : (\mathbb{N}^{\mathbb{N}})^r \to \mathbb{N}_\perp^{\mathbb{N}}$ if

1. for all $(\vec{p}_0, \ldots, \vec{p}_{r-1}, n, m)$ in $\mathscr{F}$, we have $F(g_0, \ldots, g_{r-1})(n) = m$ whenever $g_i \in U_{\vec{p}_i}$ for each i,
2. whenever $F(g_0, \ldots, g_{r-1})(n) = m$, there is a unique $(\vec{p}_0, \ldots, \vec{p}_{r-1}, n, m)$ in $\mathscr{F}$ with $g_i \in U_{\vec{p}_i}$ for each i.

We say that F is *continuous* if it has a graph in this sense, and *effective* if it has a computably enumerable graph.

It is clear that the above definition of continuity coincides with the usual one with respect to the Baire topology on $\mathbb{N}^{\mathbb{N}}$ and the Scott topology on $\mathbb{N}_\perp^{\mathbb{N}}$. We may also apply these definitions to *partial* functions $F : (\mathbb{N}^{\mathbb{N}})^r \rightharpoonup \mathbb{N}^{\mathbb{N}}$, identifying F with a total function $\overline{F}$ such that $\overline{F}(\vec{g}) = \Lambda n.\perp$ whenever $F(\vec{g})$ is undefined. We may now observe that $\odot'$ is continuous (indeed effective), whereas $\odot$ is not: the definedness of $(f \odot g)(0)$ will in general depend on an infinite amount of information about f and g.

Lemma 12.2.2. *Suppose that $F : (\mathbb{N}^{\mathbb{N}})^{r+1} \to \mathbb{N}_\perp^{\mathbb{N}}$ is continuous. Then there is a continuous $\widehat{F} : (\mathbb{N}^{\mathbb{N}})^r \to \mathbb{N}^{\mathbb{N}}$ such that $\widehat{F}(\vec{g}) \odot' h = F(\vec{g}, h)$ for all $g_0, \ldots, g_{r-1}, h \in \mathbb{N}^{\mathbb{N}}$. Moreover, $\widehat{F}$ may be taken to be effective if F is.*

Proof. Suppose $\mathscr{F}$ is a graph for F. For each n, let $u_0^n, u_1^n, \ldots$ be an enumeration of the tuples in $\mathscr{F}$ with penultimate component n. Let $\widehat{F} : (\mathbb{N}^{\mathbb{N}})^r \to \mathbb{N}^{\mathbb{N}}$ be any total function such that for all $g_0, \ldots, g_{r-1} \in \mathbb{N}^{\mathbb{N}}$, $n \in \mathbb{N}$ and $\vec{q} = q_0, \ldots, q_k$ we have

$$\widehat{F}(\vec{g})(\langle n, q_0, \ldots, q_{k-1}\rangle) \;=\; \begin{cases} m+1 & \text{if } \exists j < k.\, \exists \vec{p}_0, \ldots, \vec{p}_{r-1}.\, \exists \vec{q}\,' \sqsubseteq \vec{q}. \\ & \quad u_j^n = (\vec{p}_0, \ldots, \vec{p}_{r-1}, \vec{q}\,', n, m) \;\wedge \\ & \quad g_0 \in U_{\vec{p}_0} \;\wedge \cdots \wedge\; g_{r-1} \in U_{\vec{p}_{r-1}}, \\ 0 & \text{if no such } m, j, \vec{p}_0, \ldots, \vec{p}_{r-1}, \vec{q}\,' \text{ exist.} \end{cases}$$

(Notice the diagonal flavour of this definition, with k playing a dual role.) The function $\widehat{F}$ is well-defined, since $m+1$ will be returned only when $F(\vec{g}, h)(n) = m$ for all $h \in U_{\vec{q}}$. To see that $\widehat{F}(\vec{g}) \odot' h = F(\vec{g}, h)$, first suppose $(\widehat{F}(\vec{g}) \odot' h)(n) = m$. Then there exists k such that if $\vec{q} = q_0, \ldots, q_{k-1}$ where $q_i = h(i)$ for each i then $\widehat{F}(\vec{g})(\langle n, q_0, \ldots, q_{k-1}\rangle) = m+1$; hence there exists a tuple $u_j^n \in \mathscr{F}$ witnessing that $F(\vec{g}, h)(n) = m$. Conversely, if $F(\vec{g}, h)(n) = m$, then there exists a unique tuple $u_j^n = (\vec{p}_0, \ldots, \vec{p}_{r-1}, \vec{q}, n, m')$ in $\mathscr{F}$ with $g_i \in U_{\vec{p}_i}$ for each i and $h \in U_{\vec{q}}$, and for this tuple we have $m' = m$. Let $k = \max(j+1, \mathrm{length}(\vec{q}))$; then we see that $\widehat{F}(\vec{g})(\langle n, h(0), \ldots, h(k'-1)\rangle)$ has value 0 when $k' < k$ and $m+1$ when $k' = k$. Thus $(\widehat{F}(\vec{g}) \odot' h)(n) = m$.

It is clear by construction that $\widehat{F}$ is continuous: $\widehat{F}(\vec{g})(\langle n, q_0, \ldots, q_{k-1} \rangle)$ can depend only on values of the g_i embodied in tuples u_j^n for $j < k$. Finally, if $\mathscr{F}$ is c.e., the condition for returning $m+1$ becomes decidable (where m itself is computable from $k, n, \vec{q}$), so $\widehat{F}$ may be chosen to be effective. $\square$

As a first application of this lemma, let $F : (\mathbb{N}^{\mathbb{N}})^2 \to \mathbb{N}^{\mathbb{N}}$ be the first projection, with $\mathscr{F}$ its canonical graph (which is clearly effective). By the lemma, we obtain $\widehat{F} : \mathbb{N}^{\mathbb{N}} \to \mathbb{N}^{\mathbb{N}}$ effective such that $\widehat{F}(g) \odot h = g$ for all g, h. Indeed, the above construction yields the obvious 'clean' choice of $\widehat{F}$ with $\widehat{F}(g)(\langle n \rangle) = g(n) + 1$ for each n (this was our reason for separating out graph elements by their penultimate component). By a second application of the lemma, we obtain an effective element $k \in \mathbb{N}^{\mathbb{N}}$ such that $k \odot g \odot h = g$ for all g, h.

To show that K_2 is a lax PCA, we likewise need to construct a suitable element s; this we do with the help of the following lemma.

Lemma 12.2.3. *There is an effective (hence continuous) operation $S : (\mathbb{N}^{\mathbb{N}})^3 \to \mathbb{N}_{\perp}^{\mathbb{N}}$ such that $S(f,g,h) = e$ whenever $(f \odot h) \odot (g \odot h) = e \in \mathbb{N}^{\mathbb{N}}$.*

Proof. By interpreting the definition of $\odot'$ as a sequential 'program' in the obvious way (e.g. in PCF$_v$), we can clearly extend $\odot'$ to a sequentially computable, and hence continuous and effective, operation $\odot'' : \mathbb{N}_{\perp}^{\mathbb{N}} \times \mathbb{N}_{\perp}^{\mathbb{N}} \to \mathbb{N}_{\perp}^{\mathbb{N}}$. Now define S by $S(f,g,h) = (f \odot' h) \odot'' (g \odot' h)$; then it is easy to see that S is continuous and effective. Moreover, if $(f \odot h) \odot (g \odot h) = e$ then $f \odot h \downarrow$ and $g \odot h \downarrow$, whence $f \odot' h = f \odot h$ and $g \odot' h = g \odot h$, and it follows that $(f \odot' h) \odot'' (g \odot' h) = e$. $\square$

By starting with S and applying Lemma 12.2.2 three times, we obtain an effective element s such that $s \odot f \odot g$ is always defined and $s \odot f \odot g \odot h = e$ whenever $(f \odot h) \odot (g \odot h) = e$. We have thus shown:

Theorem 12.2.4. *K_2 is a lax PCA, with K_2^{eff} a lax sub-PCA.* $\square$

Remark 12.2.5. The particular element s given by the above construction will be highly sensitive to the order of enumeration of the graph for S. By paying closer attention to the sequential nature of the procedure for computing S, one can arrive at a choice of s which is canonical relative to the choice of coding $\langle \cdots \rangle$, though we will not pursue this here.

As things stand, our element s does not make K_2 an ordinary PCA, since we will sometimes have $s \odot f \odot g \odot h \downarrow$ even when $f \odot' h$ or $g \odot' h$ is non-total, so that $f \odot h \uparrow$ or $g \odot h \uparrow$. We may rectify this (somewhat artificially) by replacing $\odot''$ above by a stricter version $\odot'''$, where

$$(f \odot''' g)(n) \;=\; m \text{ iff } f(n) \downarrow \;\wedge\; g(n) \downarrow \;\wedge\; (f \odot'' g) \;=\; m\,.$$

This will yield an element s' satisfying the ordinary PCA axiom

$$s' \odot f \odot g \odot h = e \;\simeq\; (f \odot h) \odot (g \odot h) = e\,.$$

However, the lax structure is arguably the mathematically natural one, and suffices for our purposes in earlier chapters.

It turns out, however, that K_2 fails to be a *p*λ*-algebra* in the sense of Definition 4.3.1, at least under the natural interpretation of $p\lambda$-terms. We here explain where the problem arises; whilst not rigorously excluding the possibility of some alternative $p\lambda$-algebra structure on K_2, our argument renders the existence of such a structure somewhat implausible.

We have in mind a partial interpretation $[\![-]\!]_-$ of λ-terms that interprets constants and variables in the obvious way, and λ-abstraction as suggested by Lemma 12.2.2. To interpret application, we have the choice between $\odot''$ and $\odot'''$; however, the former choice will not validate the equation $(\lambda xyz.(xz)(yz))abc \simeq (ac)(bc)$, which is derivable in $p\lambda$-calculus. To see the problem with using $\odot'''$, let a be any constant from K_2, and consider the interpretations of the $p\lambda$-equivalent terms

$$\lambda x.a\,, \qquad \lambda x.((\lambda y.a)(\lambda z.x))\,.$$

As noted above in relation to k, the element $[\![\lambda x.a]\!]$ will map $\langle n\rangle$ to $a(n)+1$; note that the evaluation of $[\![\lambda x.a]\!] \odot g$ at n does not require any values of g at all. By contrast, because we are modelling application by $\odot'''$, the evaluation of $(\lambda y.a)(\lambda z.x)$ at n (under the valuation $x \mapsto g$) will entail the evaluation of $(\lambda z.x)$ at n (again under $x \mapsto g$). Depending on the value of n, this may in turn require the value of some $g(n')$. Thus, the element $[\![\lambda x.((\lambda y.a)(\lambda z.x))]\!]$ will not be the simple element $[\![\lambda x.a]\!]$ described above, but a more complex element that demands values of its argument g. Thus, the interpretation $[\![-]\!]_-$ does not respect $=_{p\lambda}$.

It is clear that this problem is robust under numerous variations in our definitions, as long as the interpretation of $\lambda x.a$ is the 'clean' one (and it is hard to imagine that any other interpretation could make things better). Furthermore, the argument will be applicable to any plausible definition of 'partial λ-algebra' and is not sensitive to the specific treatment of undefinedness, since our example makes no use of potentially undefined subterms. We leave it as an open question how close one can come to some reasonable kind of 'partial λ-algebra' via a suitable structure on K_2.

12.2.1 Intensionality in K_2

One can understand the above discussion as showing that the natural interpretation of λ-terms in K_2 is 'too intensional': the value of $[\![M]\!]$ retains traces of the syntactic structure of M that are obliterated in its $p\lambda$-equivalence class. This provides one indication that K_2 should be thought of as 'more intensional' than B or any of the other models so far studied. To give some intuition for the particular flavour of intensionality that K_2 offers, we shall mention some examples present in simply typed structures derived from K_2.

We first consider the hereditarily total operations arising from K_2—in effect the *intensional continuous functionals* as introduced by Kreisel [153] (see also

Section 2.6 of Troelstra [288]). The version we will use here may be defined as $\mathsf{ICF}^1 = \mathsf{Tot}(K_2;\mathbb{N}^\mathbb{N})$; since this is an applicative structure over $\mathbb{N}^\mathbb{N}$ rather than $\mathbb{N}$, we shall use β rather than $\mathtt{N}$ as our base type symbol. This definition is similar in spirit, though not formally identical, to the definition of ICF adopted in Subsection 8.2.2.

The best-known example of intensionality in ICF is the following:

Exercise 12.2.6. (i) Show that ICF^1 contains a *local modulus of continuity* operation, that is, an element $M \in \mathsf{ICF}^1((\beta \to \beta) \to \beta \to \beta)$ such that for all $F \in \mathsf{ICF}^1(\beta \to \beta)$, $g \in \mathsf{ICF}^1(\beta)$ and $n \in \mathbb{N}$ we have

$$\forall g' \in \mathsf{ICF}^1(\beta).\ (\forall i < (M \odot F \odot g)(n).\ g'(i) = g(i)) \Rightarrow (F \odot g')(n) \ = \ (F \odot g)(n)\,.$$

(ii) Recall from Subsection 8.2.1 that $\mathsf{Ct} = \mathsf{EC}(K_2;\mathbb{N})$ where $\mathbb{N}$ is identified with a certain subset of $\mathbb{N}^\mathbb{N}$. Show that no such local modulus of continuity operation exists within Ct (e.g. at type $(\overline{1} \to \overline{1}) \to \overline{1} \to \overline{1}$).

For our other source of examples, we consider *partial* operations representable in K_2. To this end, we introduce the following representation of $\mathbb{N}_\perp$. For $n,t \in \mathbb{N}$, let us define $p^t_n \in \mathbb{N}^\mathbb{N}$ by $p^t_n(i) = 0$ if $i < t$ and $p^t_n(i) = n+1$ if $i \geq t$. In accordance with the spirit of K_2, we may think of p^t_n as representing a process that evaluates to n in t steps, or in 'time' t. We also set $p_\infty = \Lambda i.0$; we may think of this as a process that runs forever without reaching a result. We now set $P = \{p^t_n \mid n,t \in \mathbb{N}\} \cup \{p_\infty\}$, and define $\mathsf{PICF} = \mathsf{Tot}(K_2;P)$, writing $\mathtt{N}$ for the base type.

We can use PICF to model a 'time-sensitive' version of PCF of the sort considered by Escardó [78].[3] The idea is that in our operational semantics for PCF (as in Subsection 3.2.3), certain reduction steps are deemed to be accompanied by a 'clock tick'. Many choices are possible here; for simplicity we shall follow [78] in decreeing that only instances of the rule $Y_\sigma f \rightsquigarrow f(Y_\sigma f)$ generate a tick. Whilst this does not correspond to a very realistic notion of computation time, it is sufficient to ensure that any infinite reduction sequences will yield an infinite number of ticks, since it is only the Y rule that introduces the possibility of non-termination into the semantics. If M is a closed PCF term, we write $M \rightsquigarrow^{*t} n$ if $M \rightsquigarrow^{*} n$ with t clock ticks, and $M \rightsquigarrow^{*\infty}$ if the reduction sequence for M is infinite.

We can now give an interpretation $[\![-]\!]$ of PCF in PICF such that

$$[\![M]\!] \ = \ p^t_n \ \text{ if } M \rightsquigarrow^{*t} n\,, \qquad [\![M]\!] \ = \ p_\infty \ \text{ if } M \rightsquigarrow^{*\infty}\,.$$

Since PICF is automatically a TPCA, it will suffice to give interpretations for the PCF constants (see Example 3.5.10). For the arithmetical constants this is straightforward: e.g. for *ifzero* we require an element of $\mathsf{PICF}(\mathtt{N}^{(3)} \to \mathtt{N})$ such that

$$\mathit{ifzero} \odot p^t_0 p^u_m q \ = \ p^{t+u}_m\,, \qquad \mathit{ifzero} \odot p^t_{n+1} q p^u_m \ = \ p^{t+u}_m\,,$$

[3] In [78], such a language (including the *timeout* operator of Exercise 12.2.7) is interpreted in Escardó's *metric model*, a mathematically very natural model of an intensional nature which we do not have space to discuss here. The metric model for PCF is significantly more abstract than K_2, and is indeed realizable over it via an applicative simulation.

and armed with Lemma 12.2.2 it is easy to construct such an element. For Y_σ where $\sigma = \sigma_0, \ldots, \sigma_{r-1} \to \mathtt{N}$, consider the continuous function $\Phi : \mathbb{N}_\bot^\mathbb{N} \times (\mathbb{N}^\mathbb{N})^{r+1} \to \mathbb{N}_\bot^\mathbb{N}$ given by

$$\Phi(y, f, \vec{x}) \;=\; tick(f \odot'' (Y \odot'' f) \odot'' x_0 \cdots \odot'' x_{r-1}) \,,$$

where $tick(g)(0) = 0$ and $tick(g)(n+1) \simeq g(n)$. Noting that Lemma 12.2.2 readily goes through with an extra parameter of type $\mathbb{N}_\bot^\mathbb{N}$, by applying the lemma $r+1$ times we obtain a continuous function $\widehat{\Phi} : \mathbb{N}_\bot^\mathbb{N} \to \mathbb{N}_\bot^\mathbb{N}$, intuitively representing $y \mapsto \lambda f \vec{x}. f(yf)\vec{x}$ with a clock tick built in. We now interpret Y_σ by the least fixed point of $\widehat{\Phi}$. We leave it to the reader to check that $Y_\sigma \in \mathsf{PICF}((\sigma \to \sigma) \to \sigma)$, and that the resulting interpretation $[\![-]\!]$ satisfies the soundness property mentioned above.

The following illustrates the kind of time-sensitive intensional operation that PICF supports.

Exercise 12.2.7. (i) Construct an element $timeout \in \mathsf{PICF}(\mathtt{N}^{(2)} \to \mathtt{N})$ such that

$$\begin{aligned} timeout \odot p_m^u \odot p_n^t &= p_{n+1}^{u+t} \quad \text{if } t \le m \,, \\ timeout \odot p_m^u \odot p_n^t &= p_0^{u+m} \quad \text{if } t > m \,, \\ timeout \odot p_m^u \odot p_\infty &= p_0^{u+m} \,. \end{aligned}$$

Give reduction rules for a construct *timeout* that yield the corresponding timed evaluation behaviour.

(ii) Show how versions of the 'parallel-or' and 'exists' operations from Section 10.3 may be implemented in PCF + *timeout*.

(iii) Devise a suitable operational semantics for the language PCF + *catch* + *timeout*, with *catch* as in Subsection 12.1.5. Show that the closed term model for PCF + *catch* + *timeout* yields $\mathsf{PC}^{\mathrm{eff}}$ as its extensional collapse, whereas it follows from results of Section 12.1 that the smaller language PCF + *catch* yields the incomparable type structure $\mathsf{SR}^{\mathrm{eff}}$. (This illustrates the non-functoriality of EC as mentioned in Subsection 3.6.3.)

We conclude our investigation of K_2 by clarifying its relationship with the models B and $\mathscr{P}\omega$ (cf. the diagram on page 33). The following exercise offers further support for the (informal) idea that K_2 is 'more intensional' than either of these models. It also makes use of the possibility of representing partial operations within K_2, as already illustrated by PICF.

Exercise 12.2.8. (i) Construct applicative simulations $\gamma : \mathsf{B} \longrightarrow\!\!\triangleright K_2$, $\delta : K_2 \longrightarrow\!\!\triangleright \mathsf{B}$ such that $\delta \circ \gamma \sim id_{K_2}$ and $\gamma \circ \delta \preceq id_\mathsf{B}$ but $id_\mathsf{B} \not\preceq \gamma \circ \delta$. (Loosely speaking, B does not have access to its own elements at the deeper intensional level represented by their K_2 realization. This is analogous to the relationship between $\Lambda^0/{=_\beta}$ and K_1—see Example 3.5.3.)

(ii) Show that the simulation γ above is essentially multi-valued, i.e. there is no single-valued simulation $\gamma' \sim \gamma$. (Thus, K_2 is more fine-grained than B insofar as each element of B is represented by several elements of K_2.)

(iii) Show that (i) and (ii) above also hold with $\mathscr{P}\omega$ in place of B. Show too that the effective analogues of all these results hold (i.e. using $K_2^{\mathrm{eff}}, \mathsf{B}^{\mathrm{eff}}, \mathscr{P}\omega^{\mathrm{eff}}$).

The above results suggest an intriguing conceptual problem. Are there other models, intuitively 'more intensional' than K_2, which are related to K_2 as K_2 is to B? Or is there some sense in which K_2 is a 'maximally intensional' model? Convincing answers to such questions would significantly advance our understanding of the overall geography of the space of computability models.

Remark 12.2.9. Research involving K_2 has also flourished in several other directions not covered here. The model was first explicitly introduced by Kleene and Vesley [146] in order to provide realizability interpretations for aspects of Brouwer's intuitionism, including principles such as bar induction and the fan theorem. This has given rise to a stream of work using K_2 and ICF to prove metamathematical results on intuitionistic formal systems (see for example [288, Section 2.6]). Work on the K_2 realizability universe is still ongoing, a recent example being Normann [219].

The category of assemblies over K_2 also provides a natural setting for the theory of Type Two Effectivity as developed by Weihrauch [302] and others. This is a version of computable analysis in which real numbers (and other objects) are conceived as presented by data on an infinite tape; the connections with K_2 are investigated in Bauer [12]. In a similar vein, $\mathscr{A}sm(K_2)$ turns out to contain many other 'quasi-topological' subcategories of interest, such as the category **QCB** of quotients of countably based spaces [11].

Finally, a generalization of the construction of K_2, with some interesting applications to the general theory of PCAs, has recently been studied by van Oosten [296].

12.3 Kleene's First Model K_1

Whilst it appears plausible that K_2 may in some way have the status of a 'maximally intensional' model, this is certainly not the case for K_2^{eff}. For effectively given (countable) models generally, the ultimate intensional realization is given in terms of Kleene's first model K_1. As explained in Section 3.1, this is the untyped applicative structure $(\mathbb{N}, \bullet)$, where $e \bullet n$ is the result (if any) of running the eth Turing machine on the input n. The fact that this defines a PCA is an immediate consequence of the s-m-n theorem from basic computability theory (see Example 3.1.9).

One may see K_1 as the natural 'end of the road' for our subject: since all data values are here presented as natural numbers, higher-order computability in effect collapses to first-order computability. We complete our technical exposition with a few observations indicating how K_1 fits into our overall picture and drawing together some threads from earlier in the book.

First, we may observe that K_1 is very far from being a $p\lambda$-algebra (interpreting λ-terms as in Subsection 3.3.1 for some natural choice of k and s). Revisiting the example used in the previous subsection (now with $a \in \mathbb{N}$), we see that the term $\lambda x.a$ now denotes the index of a Turing machine T_a which simply writes a representation of a on its output tape (with no reference to the input), while $\lambda x.(\lambda y.a)(\lambda z.x)$ denotes the index of a machine which does the following: given x, construct an index

w for the machine T_x, and simulate the application of T_a to w. Indeed, with a little care in the choice of k, s, we can arrange that the interpretation $[\![-]\!]$ of λ-terms in K_1 is actually injective: in effect, $[\![M]\!]$ incorporates the syntax tree of M as an explicit 'program' for computing the desired result.

To make a connection with programming languages, the intensionality of K_1 is close to what is provided by the *quote* operator of Lisp. In slightly idealized form, if M is a closed program of any type, the program $quote(M)$ will evaluate to a Gödel number $\lceil M \rceil$ for M, giving programs full intensional access to their own syntax. The reader may enjoy working out a suitable definition for a language PCF + *quote* and showing that it has a sound interpretation in K_1 (cf. Longley [174]).

The highly intensional character of K_1 is also inherited by the structure $\mathsf{HRO} \cong \mathsf{Tot}(K_1; \mathbb{N})$, known as the model of *hereditarily recursive operations*. This model was first considered by Kreisel [151]; like K_1 itself, it has played a significant role in the metamathematical study of intuitionistic formal systems (see [288, Chapter 2]). It also featured in our study of HEO in Subsection 9.2.2. We mention here just one example to illustrate the power of HRO:

Exercise 12.3.1. (i) Using Theorem 9.2.2, show that there exists an element $M \in \mathsf{HRO}(2, 1 \to 0)$ that computes local moduli of continuity for extensional type 2 operations: i.e. for $F \in \mathsf{HRO}(2)$ and $g \in \mathsf{HRO}(1)$ with $F \in \mathsf{Ext}(\mathsf{HRO})(2)$, we have

$$\forall g' \in \mathsf{HRO}(1).\ (\forall i < (M \bullet F \bullet g).\, g'(i) = g(i)) \Rightarrow (F \bullet g')(n) = (F \bullet g)(n)\ .$$

(ii) Show that no such element exists within HEO.

We now turn to consider how K_1 is related to other models via simulations. In Subsection 3.5.1 we defined an *effective TPCA* to be a TPCA A with numerals endowed with a numeral-respecting simulation $A \longrightarrow\!\!\!\!\triangleright\, K_1$, and all the models branded in this book as 'effective' do indeed admit such a simulation. The following curiosity emphasizes the importance of respecting the numerals:

Exercise 12.3.2. Let T be any non-trivial Turing degree, and $K_1^T = (\mathbb{N}, \bullet^T)$ the evident relativization of K_1 to T. Show that there is an applicative simulation $\gamma^T : K_1^T \longrightarrow\!\!\!\!\triangleright\, K_1$ given by $\gamma^T(n, n')$ iff $n' \bullet^T 0 = n$. Show however that γ^T does not respect numerals.

On the other hand, if A is any untyped PCA A, or more generally any TPCA with numerals and minimization, we obtain a numeral-respecting simulation $K_1 \longrightarrow\!\!\!\!\triangleright\, A$ (see Exercise 3.5.2). Putting these facts together, we easily obtain:

Proposition 12.3.3. *If A is an effective TPCA with numerals and minimization, we have applicative simulations $\gamma : A \longrightarrow\!\!\!\!\triangleright\, K_1$ and $\delta : K_1 \longrightarrow\!\!\!\!\triangleright\, A$ with $\gamma \circ \delta \sim id_{K_1}$.* □

For most of the effective models with minimization considered in this book ($\mathsf{SP}^{0,\mathrm{eff}}, \mathsf{SF}^{\mathrm{eff}}, \mathsf{PC}^{\mathrm{eff}}, \mathsf{SR}^{\mathrm{eff}}, \mathsf{SA}^{\mathrm{eff}}, \mathsf{B}^{\mathrm{eff}}, K_2^{\mathrm{eff}}$), we also have $\delta \circ \gamma \preceq id_A$, although this is in some cases a non-trivial result (see for example Corollary 7.1.36 and the ensuing remark). Another instance of this abstract situation is provided by the term model $\Lambda^0/{=_\beta}$ for untyped λ-calculus; this was exactly the scenario discussed in Subsection 1.1.3 and Example 3.5.3 as a motivating example for the theory of simulations.

We conclude with an open-ended conceptual problem related to the one we posed for K_2. Are there any (constructible) effective models that can be naturally seen as intermediate between K_2 and K_1 in terms of intensionality? A convincing negative answer to this question might perhaps be loosely regarded as a kind of constructive counterpart to the Continuum Hypothesis: by analogy with the idea that there is 'nothing between' the sets $\mathbb{N}^\mathbb{N}$ and $\mathbb{N}$ in terms of cardinality, we would have an assertion that at the level of computational objects there is 'nothing between' an oracle of type $\mathbb{N} \to \mathbb{N}$ and a Kleene index for it. Again, progress in this direction would be likely to deepen our understanding of the fundamental notion of a 'constructive presentation of a mathematical object' which has been a major underlying motif of this book.

Chapter 13
Related and Future Work

For the bulk of this book, our focus has been on the theory of computability models as a field of interest in its own right, with particular emphasis on the more established parts of the subject, and with relatively little attention to applications or to other related areas of research. It is therefore fitting to end with a more outward- and forward-facing chapter, mentioning some connections with other fields and some possible avenues for future exploration. This chapter can be seen as continuing various threads of discussion from Section 1.2 and Chapter 2.

Since there are many topics here clamouring for attention, our coverage will inevitably be rather terse: our purpose is merely to provide some pointers to the literature, indicating which topics are already extensively explored, which are currently active, and which seem ripe for future development.

Roughly speaking, we shall begin with topics directly relevant to the subject matter of this book, then work our way outwards in concentric circles towards other neighbouring areas and potential applications. In Section 13.1 we discuss how our programme of mapping out the landscape of computability notions might be deepened and broadened; in Section 13.2 we consider a loose selection of research areas intellectually adjacent to ours; while in Section 13.3 we mention some further topics of a more applied character. However, our classification of topics is somewhat arbitrary, since many of them would in principle qualify for more than one of these three categories.

13.1 Deepening and Broadening the Programme

We begin with some possibilities for continuing the programme of research represented by this book. First, we mention some remaining challenges and open problems associated with the models we have studied. Next, we touch on some other classes of higher-order models not considered in detail in this book. Finally, we consider various ways in which our programme could be broadened to embrace models of computation that are not higher-order in the sense we have been considering.

13.1.1 Challenges and Open Problems

Whilst a number of open problems have been mentioned in the course of our exposition, it is perhaps an indicator of the maturity of the parts of the subject we have concentrated on that there are rather few outstanding questions of major importance concerning the models we have principally discussed. There are still some obvious gaps in our knowledge regarding sublanguages of PCF and their relative computational power (see e.g. the end of Subsection 6.3.4), but here we expect that the relatively recent technology of nested sequential procedures is likely to yield further advances in the near future.

A more general challenge is to obtain a better structural understanding of the various extensional models, especially SF. The topological approach initiated by Scott in [256] was successful in the case of PC, but as shown by Normann [214], it yields too many functionals in the sequential setting. The question, then, is how to modify concepts from topology or domain theory in order to describe the structure of models such as SF, or alternatively find completely new concepts yielding relevant classes of mathematical structures. There are some pioneering results in [84, 253, 254, 221], but there is still more to do.

A further interesting direction to pursue is to find further results in the vein of our 'ubiquity' theorems 8.6.7 and 9.2.15—that is, general results establishing that a certain model is the *only* one that can arise from a certain kind of construction. For instance, we expect that the type structure SR is considerably more ubiquitous than the results of Chapter 11 alone indicate—indeed, we conjecture that it should be the unique maximal 'sequential' type structure over $\mathbb{N}_\perp$ satisfying some reasonable conditions. (Here 'sequential' means 'realizable over van Oosten's B'.) A more ambitious goal would be to obtain a complete classification of maximal (continuous or effective) type structures over $\mathbb{N}_\perp$ subject to some conditions: at present, the only natural candidates we are aware of are PC, SR and the models St^r of Section 11.6, along with the effective counterparts of all these.

13.1.2 Other Higher-Order Models

There are also a wealth of other intensional and untyped models not studied in Chapter 12, but which fit perfectly well within our framework of higher-order models. We mention here several (somewhat overlapping) classes of examples.

- Syntactic models (i.e. term models) for lambda calculi, combinatory calculi and other 'programming languages' of interest. Whilst these often have the character of artificial rather than mathematically natural models, they have the virtue that the language or calculus in question makes very explicit the kinds of computations that the model supports. Representative examples of such calculi include:
  - Variants of typed or untyped λ-calculi, e.g. lazy λ-calculus [4], call-by-value λ-calculus [235], infinitary λ-calculus [131].

  - In particular, (variants of) Parigot's *λμ-calculus* [230, 250], an extension of λ-calculus with a continuation-style construct allowing non-local jumps.
  - Extensions of PCF with primitives for features such as local state, exceptions or fresh name generation.

- Other untyped λ-algebras of a more semantic character:

  - Graph models akin to Scott's $\mathscr{P}\omega$ [70, 240], and other relational models in a somewhat similar spirit [192, 46]. In some cases, the non-trivial computability theory of these models is already indicated by known connections with programming languages: for instance, McCusker [192] relates certain graph models to an imperative style of computation.
  - Lattice and domain-theoretic models such as Scott's D_∞ [258], and analogous models in the realm of stable or strongly stable domains [32, 45].
  - *Filter models*, in which elements are defined indirectly via the set of 'types' that can be assigned to them [10].
  - Models based on *Böhm trees* and their variants [9].

  For work prior to 1984, the canonical reference is Barendregt [9]; a more recent text is Amadio and Curien [6]. Laurent [170] discusses models of the λμ-calculus.
- *Game models*. A small sample of such models has been briefly discussed in Section 7.4 and Subsection 12.1.7, but there is a wide field of other known game models that we have not touched on (see [102] for an overview and further references). Many game models constructed to date have been specifically designed as fully abstract models for particular programming languages; however, it is also of interest to proceed from some simple and natural game model to a language that matches it, as in [182].
- Escardó's *metric model* [78], a mathematically natural model incorporating an abstract notion of 'time-sensitivity', closely related to PCF + *timeout* (see Subsection 12.2.1).

Several of these models have already received considerable attention, and we believe that our framework of higher-order models may offer a useful setting for integrating and consolidating what is known about them. There are also outstanding questions to be answered: for instance, the correspondences between semantic models and programming languages are not worked out in all cases, nor are the simulations or even equivalences that naturally exist between these models. A full treatment of all the models mentioned above would probably fill another book as long as this one.

Perhaps the biggest challenge, however, is to find some organizing principle that allows us to map out the geography of this space of models to some extent. For instance, the natural partial ordering of models which we were able to exploit in the extensional setting (cf. Subsection 3.6.4) is no longer available in the intensional world. An ambitious challenge for the long-term future would be to find some useful mathematical classification of intensional models subject to some reasonable constraints.

13.1.3 Further Generalizations

Another way to broaden our programme is to generalize the notion of higher-order model itself. We mention here two natural generalizations. As far as abstract theory is concerned, these are both subsumed by the framework of Longley [183]; however, little has been done to date as regards the marshalling of relevant examples.

First, we can replace the notion of cartesian product structure by that of a *monoidal* or *affine* one. Loosely speaking, if $\mathbf{C}$ is a computability model, we call an operation $\otimes$ a *monoidal* structure on $\mathbf{C}$ if any $f \in \mathbf{C}[A,B]$ and $g \in \mathbf{C}[C,D]$ yield an operation $f \otimes g \in \mathbf{C}[A \otimes C, B \otimes D]$ such that certain associativity and unitary axioms are satisfied—however, we do not require the existence of projections in $\mathbf{C}[A \otimes B, A]$ and $\mathbf{C}[A \otimes B, B]$, nor of diagonals in $\mathbf{C}[A, A \otimes A]$. If projections are required but not diagonals, we call $\otimes$ an *affine* structure.

Such weaker notions of product have played important roles in many areas of theoretical computer science, particularly in connection with linear logic [106]. Of particular interest from our perspective is that they can be used to give an account of *stateful* data. Suppose for example that A represents some type of objects with a persistent internal state (in an object-oriented language such as Java). In the absence of a diagonal or 'cloning' operation for A, there will typically be no way to combine two operations $f \in \mathbf{C}[A,B]$ and $g \in \mathbf{C}[A,C]$ into a single operation $f \in \mathbf{C}[A, B \otimes C]$: given $a \in A$, there will be no way to compute both $f(a)$ and $g(a)$, since for example the internal state of a may change during the computation of $f(a)$, so that the original state is no longer available for the purpose of computing $g(a)$. Much work in game semantics for example uses affine products to model stateful computation for precisely this reason.

Secondly, we can generalize to *non-deterministic* models: here, an operation $f \in \mathbf{C}[A,B]$ need not be a partial function, but might be any kind of relation from A to B, capturing the idea that many values of $f(a)$ may be possible. Our theory of higher-order models is shown in [183] to generalize to the non-deterministic setting in two different ways. The idea of non-determinism arises at many levels in computing practice; however, a systematic investigation of such models has yet to be undertaken.

Finally, as suggested in Chapter 1, we could throw off the shackles of the higher-order framework altogether and pursue a much more general investigation of computability models, including (for example) those arising from concurrency and automata theory. The prospects for such a broad programme, including some preliminary examples, are discussed in [183]. One hope is that the general theory of computability models and simulations (or something like it) might provide a framework within which a wide range of computation models from across theoretical computer science might be brought under a common roof and their relationships explored.

13.2 Neighbouring Areas

13.2.1 Computability over Non-discrete Spaces

In this book, we have essentially been concerned with issues of computability for functionals of finite type over the base type N. As we have seen in Chapter 4, the relevant models are typically capable of interpreting many other kinds of data such as finite lists or finite trees. However, if we are interested in including base types representing *non-discrete* mathematical structures, we are faced with new challenges, not only in choosing the model for this base type and for functionals of finite types over it, but in choosing the concept of computation that makes sense for the model.

We here survey some of the extensions of typed computability where the set $\mathbb{R}$ of reals is added as a base type. This has, to a limited extent, been extended further to computable metric spaces of various kinds, and even to arbitrary topological spaces, but this is mainly virgin territory, where the first problem may be to decide which questions to ask.

There are two classical ways to define the reals within set theory: as the set of Dedekind cuts in $\mathbb{Q}$, or as a set of equivalence classes of Cauchy sequences. Neither of these as they stand are suitable as a basis for computing with reals, but slight modifications of them are. There are two alternatives to consider:

1. $x \in \mathbb{R}$ is represented by the double cut $\{q \in \mathbb{Q} \mid q < x\}$ and $\{q \in \mathbb{Q} \mid x < q\}$.
2. $x \in \mathbb{R}$ is represented by any Cauchy sequence $(q_n)_{n \in \mathbb{N}}$ from $\mathbb{Q}$ with x as a limit, and with a rate of Cauchyness at least as good as $(2^{-n})_{n \in \mathbb{N}}$.

The disadvantage of 1 is that the unique representation of x does not depend continuously on x—a problem that arises for any attempt to represent each real by one single object in a disconnected space—so alternative 2 is preferred. Variants of 2, in which a real essentially is represented by a total sequence $(a_n)_{n \in \mathbb{Z}}$, are:

- Represent x by (coded) enumerations of the two sets in the double cut.
- Represent x by a so-called *signed binary digit representation*

$$x = a_0 + \sum_{n>0} a_n \cdot 2^{-n} ,$$

  where $a_0 \in \mathbb{Z}$ and $a_{n+1} \in \{-1, 0, 1\}$ for all $n \in \mathbb{N}$.

Switching from $\mathbb{Z}$ to $\mathbb{N}$, we can thus represent reals as elements of $\mathbb{N}^{\mathbb{N}}$, i.e. elements of type $\mathtt{N} \to \mathtt{N}$. Operations of simple type over $\mathbb{R}$, such as continuous functions on the reals and operators for finding the integral or maximal value of a function over $[0,1]$, can thus be represented by elements of the type structures over $\mathbb{N}$ that we have considered. Moreover, all reasonable ways of representing $\mathbb{R}$ via $\mathbb{N}^{\mathbb{N}}$ turn out to be computably equivalent in all models of interest.

Di Gianantonio [66, 67] defined an extension of PCF that models algorithms for exact computations over the reals based on this kind of representation. Simpson [272] used the PCF program for the fan functional to give programs for computing the maximum of a continuous function over $[0,1]$ or its Riemann integral (these are touched on in Subsection 7.3.4).

A different flavour of approach has been to consider the reals as the maximal elements in a domain of *partial* reals. Using so-called *continuous domains*, we let a partial real just be a closed interval $[a,b]$, and the interval is seen as an approximation to any of its elements. Escardó [77, 76] devised an extension of PCF called *Real* PCF, whose denotational semantics in the category of domains uses both $\mathbb{N}_\bot$ and the partial reals as base domains.

These two extensions of PCF are both tailor-made to deal with real number calculations and will have denotational semantics involving domains, but they are nevertheless not equivalent. As with the integers, there is no canonical way to make the concept of a *partial functional of finite type* precise. However, it may come as a greater surprise that here the concept of a hereditarily *total* continuous functional of finite type is not uniquely determined either. Indeed, the class of hereditarily total functionals of pure type 3 over the reals arising from the 'intensional' approach (as in the semantics of di Gianantonio's calculus) does not coincide with that arising from the 'extensional' approach (as in the semantics of Escardó's Real PCF). This highly non-trivial result was obtained by Schröder [266], building on earlier work of Bauer, Escardó, Simpson and Normann [15, 213].

If one is not concerned with the need for PCF-like calculi, one might wish to define a structure of total continuous functionals over $\mathbb{R}$ directly. One natural way to do so is to work within the category of limit spaces or, equivalently, within the category **QCB** of quotients of countably based spaces [11, 89]. It turns out that the resulting type structure coincides with the hereditarily total functionals arising from the extensional approach, so this total type structure is a strong candidate for the most natural one. If we accept this as the canonical choice, the challenge may be to understand what it may mean to compute over this structure—for instance, in a style analogous to Kleene's S1–S9.

One problem with Real PCF is that the evaluation strategy is non-deterministic, so the evaluation of a real number to some desired accuracy may involve a branching tree of subexecutions. (This is related to the fact that even the additive structure of the reals is inherently non-sequential in this setting, as shown by Escardó, Hofmann and Streicher [88].) This suggests the challenge of finding some more restricted type structure of 'sequential' functionals over the reals, thereby giving an alternative and more realistic model for exact computations in practice.

There have been a few attempts to study typed computability for hereditarily total functionals when we extend the class of base types further. The main achievements have been when the base type is a computable metric space, but even then we require some internal structure, as in Banach spaces, to be able to get anywhere. Interested readers may consult Normann [216, 217]. Functional programming with datatypes containing partial data of this generality has not been systematically investigated in the literature.

13.2.2 Computability in Analysis and Physics

The study of computability for non-discrete spaces leads naturally to an investigation of computability in real or complex analysis, and even in physics. An interesting line of work in this area was initiated by Pour-El and Richards [241, 242], who showed among other things that solutions to certain wave equations arising in physics (viewed as operators on certain Banach spaces) may be non-computable in a precise sense. This by itself is not the whole story, however, since the computability or otherwise of some operator will sometimes depend on the choice of *norm* for the space in question (the mathematical issues are clarified in [280]). This chimes in with one of the main ideas of this book: namely, that issues of computability are often sensitive to the precise way in which the inputs are presented.

Broadly speaking, the work of Pour-El and Richards relies only on the 'computability theory' offered by Type Two Effectivity (see Section 12.2), and not on any higher types. However, insofar as solution operators for classes of differential equations (for instance) can be regarded as higher type operations, it is likely that the line of work described in Section 13.2.1 may offer an alternative perspective on these questions. Again, the whole area awaits exploration.

13.2.3 Normal Functionals and Set Recursion

In Section 5.4, we touched briefly on the computability theory of *normal* functionals, and on set recursion. Recall that ${}^{k+2}\exists$ essentially is the quantifier $\exists x \in \mathrm{S}_k$, and that a functional Φ of type $k+2$ is normal if ${}^{k+2}\exists$ is Kleene computable in Φ.

The subject began with the successful generalization of classical computability theory to *metarecursion*: the *hyperarithmetical* sets are the subsets of $\mathbb{N}$ that we can define by extending arithmetical definitions transfinitely through all ordinals below ω_1^{CK} (the least ordinal that is not the order type of a computable well ordering of $\mathbb{N}$). As shown by Kleene, these coincide with the Δ^1_1-definable subsets of $\mathbb{N}$. It turns out that the hyperarithmetical sets also coincide with the subsets of $\mathbb{N}$ computable in ${}^2\exists$ (as noted in Theorem 5.4.1), and with those appearing in Gödel's universe of constructible sets up to ω_1^{CK}. For an exposition of the theory of hyperarithmetical sets, see Rogers [245] or Sacks [248].

To make a long story short and incomplete, ω_1^{CK} was seen as a generalization of the ordinal ω, the order type of $\mathbb{N}$, for which the μ operator made sense. This offered a clean generalization of degree theory, and led to α-recursion theory [269], the degree theory of admissible ordinals, and thence to β-recursion theory [98], where β is some inadmissible limit ordinal. However, this branch of higher computability theory falls outside even the natural extensions of the scope of the present book.

Kleene's S1–S9 represents the other strand growing out of the hyperarithmetical theory. Kleene's main motivation as presented in [138, 139] was to investigate higher analogues of the hyperarithmetical sets, and to do so by investigating the computability theory of ${}^{k+2}\exists$ for $k > 0$. In particular, one defines the *hyperanalytical sets* as the subsets of $\mathbb{N}^{\mathbb{N}}$ computable in ${}^{3}\exists$ and some $f \in \mathbb{N}^{\mathbb{N}}$.

However, the concept of a hyperanalytical set proved less fruitful than that of a hyperarithmetical set. Whilst hyperarithmetical sets and metarecursion theory have found applications, for instance in descriptive set theory, we cannot claim the same for hyperanalytical sets or for any of the higher analogues investigated in higher computability theory. The hyperanalytical sets are too complex to appear as a natural class in descriptive set theory or any other branch of mathematics, whilst the class of sets computable in some quantifier of finite type is too narrow to be of interest in axiomatic set theory. The further extension to *set recursion*—essentially the computability theory of bounded quantifiers in the universe of sets—has not found significant applications either. Research on set recursion was lively for a while, with the investigation of problems arising from generalized degree theory and of the extent to which set recursion differs from admissible set theory, but now the key properties seem to be well understood, and in order to stay alive, the subject needs a fresh area of applications.

There is a beautiful connection between 1-sections of normal functionals and models for weak set theories. Given a countable set X of type 1 functions closed under relative computability, its *companion* will be the transitive set $C(X)$ of sets with codes in X. If X is the 1-section of a normal functional, its companion will be a model of Kripke–Platek set theory; conversely, Sacks [247] showed by a forcing construction that the set of functions in any countable, transitive model of Kripke–Platek set theory will be the 1-section of a type 2 normal functional.

There is also a grey zone between the continuous functionals and the normal functionals. Corollary 5.3.8 is a dichotomy result that separates the two worlds: from the 1-section of a functional Φ we can tell whether Φ is *continuous-like* or *normal-like*. The question is whether this dichotomy goes deeper: that is, whether every 1-section is either the 1-section of a continuous functional or the 1-section of a normal functional. Sacks [247] proved that the 1-section of any normal functional is the 1-section of a normal functional of type 2, but the question is open for non-normal functionals from which ${}^{2}\exists$ is computable. Putting it another way, if X is the 1-section of some such functional, we do not yet know if the companion of X satisfies the Kripke–Platek axioms.

It is even open whether the 1-section of any *continuous* functional of type 3 is the 1-section of a functional of type 2. Solutions to these problems would be likely to shed interesting new light on the nature of algorithms for functionals.

For a more detailed recent exposition of higher generalizations of classical computability theory, see Normann [218].

13.2.4 Complexity and Subrecursion

Yet another related area concerns questions of *computational complexity* for higher-order operations. In the literature one sometimes finds ad hoc analyses of complexity issues for specific operations: most notably, Escardó [80] discusses the complexity of his algorithms for quantification over infinite sets, comparing them with those of Berger. In general, however, it is far from clear what we should mean by the complexity of a higher-order program, or what kind of 'cost measure' should be used to define complexity classes. Among the higher-order complexity classes proposed to date (see [125] for a survey), the most prominent has been the class of *basic feasible functionals*, a conservative extension of the class of polynomial-time first-order functionals. Beyond this, however, the overall shape of the higher-order complexity landscape remains very unclear, and even some of the most fundamental concepts still await clarification.

In another direction, Schwichtenberg [267] has investigated the first-order functionals definable in System T (or in PCF) with restrictions on the type levels of the recursions involved, correlating these classes with the *extended Grzegorczyk hierarchy*. Finally, in a remarkable connection between higher types and classical complexity theory, Kristiansen [156] has shown that if one restricts attention to functions definable via System T terms without *suc*, one obtains a correlation with complexity classes such as LOGSPACE, P, PSPACE, LINSPACE and EXP.

13.2.5 Transfinite Types

Our main theme in this book has been the link between computability and functionals of *finite type*. We have considered several models where each finite type is interpreted as some (structured) set. Now the question is whether there are types of computational interest that are not covered by this approach.

Consider the set of well-founded trees of integers with branching over $\mathbb{N}$. Using sequence numbering, we may of course code each such tree as a set of finite sequences of numbers, but this does not enforce well-foundedness, and as a datatype it is more natural to view the set as the least fixed point of the equation

$$\mathsf{Tree} = \mathbb{N} \oplus (\mathbb{N} \to \mathsf{Tree}) \, .$$

This datatype does not naturally fit into the framework we have used so far, and its 'type level' is essentially the first uncountable ordinal.

Martin-Löf type theory with universes allows constructions like the type of well-founded trees, and even allows definitions by transfinite recursion of indexed families of types. This type theory is based on syntax and rules of deduction, but set-theoretical models have been constructed, based on domain theory [227], on realizability and assemblies [282], or on domains with totality [298, 299]. In order to be natural, a full classical model of this type theory requires the existence of inaccessible cardinals; a natural question is therefore what the complexity will be if one manages to restrict function spaces via some kind of continuity.

Such problems led Berger [20] on the one hand, and Kristiansen and Normann [157, 158, 159] on the other, to formalize the concept of *totality* in order to make sense of transfinite versions of the continuous functionals. It turns out that the closure ordinal of ${}^3\exists$ plays the role of an inaccessible cardinal, and that the Kreisel representation theorem can be extended to the hyperanalytical hierarchy—see Normann [211].

We will not go into details here of the computability theory for functionals of transfinite types. There is unexplored territory here, e.g. in combining constructive type theory and set theory with classical approaches to functional algorithms.

13.2.6 Abstract and Generalized Computability Theories

With the increasing number of explicit generalizations of classical computability theory came a natural desire to systematize these generalizations via axiomatizations. Kreisel [154] is a seminal paper motivating the task of generalizing computability theory. Moschovakis [199] may be viewed as the first serious attempt to axiomatize *(pre)computation theories*. The approach by Moschovakis is partly motivated by the example of Kleene computability, where we in some sense find well-founded trees of computations leading to ordinal ranks on terminating generalized algorithms. Fenstad [96] is more explicit in introducing the subcomputation relation as a primitive. The latter book gives an overview of the field as of 1980; later expositions have been given by Hinman [118] and Normann [218].

Partial combinatory algebras represent the same level of abstraction as precomputation theories, with a wide range of models. The axioms of computation theories that go beyond precomputation theories are not so relevant for the topics of this book. They try to capture abstractly a situation where the class of terminating computations is defined inductively, where each computation has an ordinal rank, and where one may, effectively and uniformly in a terminating computation, search over the set of other computations of lower rank. If we are dealing with Kleene computations relative to partial functionals, the base for an oracle call need not be unique, so there is no clear concept of subcomputation. When we work with a typed structure of extensional, total functionals, the computation trees will be well defined, but they do not fit with axioms stating that one may (generalized computably) search over computation trees. Computation with nested sequential procedures does obey the axioms of generalized computability theory to a greater extent; however, the classical axioms try to capture *bottom-up* evaluations, while NSPs give rise to *top-down* ones.

The philosophy of Moschovakis is that computing essentially is recursion, which in our context means evaluations of least fixed points of computable operators. This is of course in accordance with the semantics of PCF, but the continuity requirement in most models for PCF means that only the weaker axioms from axiomatic computability theory are typically satisfied.

In conclusion, we claim that the axiom systems studied in the late 1960s and 1970s are to some extent applicable to the kinds of higher-order operations studied in this book, but that they throw little light on the topics we focus on.

13.3 Applications

13.3.1 Metamathematics and Proof Mining

Historically, as explained in Subsections 1.2.1 and 1.2.2, one of the most important applications of higher-order computability has been to give computational interpretations for logical systems, leading both to metamathematical theorems about these systems and to the possibility of extracting computational information from proofs carried out in them. Although work in this area is already highly developed, most of it to date has concentrated on just a small handful of computability notions, especially those represented by Ct, by HEO and by Kleene computability.[1] (All three of these featured already in Kreisel's seminal paper [152].) There is therefore scope for exploring whether the other models studied in this book can find good metamathematical applications. A few preliminary observations in this vein are made in [174], where it is indicated how realizability over SF, SR and B respectively yield different flavours of constructive real analysis. Likewise, it may be that recent advances such as the sequential procedure perspective on Kleene computability open up interesting new applications.

We also note that realizability interpretations over Kleene's models K_1 and K_2 have for many decades received significant attention. Broadly speaking, these depict mathematics as conceived (respectively) by the Russian school of recursive mathematics and by a version of Brouwerian intuitionism with continuity principles. The metamathematical use of both of these models is still alive and well, as witnessed by the recent papers [13, 219].

13.3.2 Design and Logic of Programming Languages

In Subsection 1.2.3 we sketched a general manifesto for starting from some mathematically natural model of computation and designing a clean and beautiful programming language around it. Whilst this idea scored an early success in terms of the impact of Scott's model PC and language LCF on the development of functional programming languages, it should be admitted that since then this ideology has had little practical impact, and speculative suggestions for semantically inspired language primitives (as in Longley [175, 182]) remain underdeveloped.[2] By contrast, a far bigger impact on practical language design has resulted from research focussing

[1] Good use has also been made of the class of *hereditarily majorizable functionals*, which provide an interesting model for bar induction—see e.g. [288]. These have not been studied in this book since they do not appear to represent a natural computability notion: for instance, they are not closed under Kleene computation.

[2] An extensive attempt at designing a programming language based on game semantics was made by the first author in the period 2005–2009 [181]. It may yet bear fruit.

not on particular models of computability, but on type systems (Pierce [232]) and on abstract categorical structures such as monads (Wadler [300]).

It may be that the primary benefits of studying particular mathematical models have less to do with language design than with logics for reasoning about programs (see for example [187]). For some sample applications of game semantics to program verification, see [226]. We also mention two lines of inquiry initiated by the first author which have not so far been extensively pursued; both of these exploit specific higher-order operators that are available within certain models.

1. The use of certain 'encapsulation' operators within game models to give a semantic treatment of *data abstraction*, as pioneered in Wolverson's Ph.D. thesis [304] (cf. [182]).
2. The use of higher-order operators to interpret computational effects such as exceptions, state, or fresh name generation within models that at face value do not support these features [180] (cf. Subsection 1.2.4).

13.3.3 Applications to Programming

In Subsection 7.3.5 we mentioned some recent work on possible programming applications for non-trivial higher-order operators, e.g. to certain very general kinds of search problems [81, 86]. It seems plausible that such ideas may lead to genuine practical applications, for instance to the implementation of certain kinds of searches in a modular way.

The abovementioned examples have the distinction that they rely only on PCF-style computation (or even less). On the other hand, there is also a prima facie case for exploring programming applications at the more intensional end of the spectrum, since it is here that one would expect both maximum expressive power and maximum efficiency for implementable operations.

Finally, aspects of the theory of higher-order models have formed the basis for implementations of tools for aiding program development—for example the Minlog system [30] which extracts programs from natural deduction proofs, and the system RZ [14] which uses the ideas of realizability over TPCAs to generate interface code from mathematical specifications of datatypes.

All in all, we see a wide range of potential extensions and applications for the theory we have covered, and it is hard to anticipate which will be the most fruitful. We hope that by rendering this theory more accessible to a wide audience, this book will contribute to the exploration of some of these application areas, as well as promoting the study of higher-order computability itself as a locus of fruitful interaction between logic and computer science.

References

1. Abramsky, S., Jagadeesan, R.: Games and full completeness for multiplicative linear logic. Journal of Symbolic Logic **59(2)**, 543–574 (1994)
2. Abramsky, S., Jung, A.: Domain theory. In: Abramsky, S., Gabbay, D., Maibaum, T.E. (eds.) Handbook of Logic in Computer Science, pp. 1-168. Oxford University Press (1994)
3. Abramsky, S., McCusker, G.: Game semantics. In: Schwichtenberg, H., Berger, U. (eds.) Computational Logic: Proceedings of the 1997 Marktoberdorf Summer School, pp. 1-56. Springer, Heidelberg (1999)
4. Abramsky, S., Ong, C.-H.L.: Full abstraction in the lazy lambda calculus. Information and Computation **105(2)**, 159–268 (1993)
5. Abramsky, S., Jagadeesan, R., Malacaria, P.: Full abstraction for PCF. Information and Computation **163(2)**, 409–470 (2000)
6. Amadio, R., Curien, P.-L.: Domains and Lambda Calculi. Cambridge University Press (1998)
7. Asperti, A.: Stability and computability in coherent domains. Information and Computation **86(2)**, 115–139 (1990)
8. Asperti, A., Longo, G.: Categories, Types and Structures. MIT Press (1990)
9. Barendregt, H.P.: The Lambda Calculus: Its Syntax and Semantics. Elsevier, Amsterdam (1985)
10. Barendregt, H.P., Coppo, M., Dezani-Ciancaglini, M.: A filter lambda model and the completeness of type assignment. Journal of Symbolic Logic **48(4)**, 931–940 (1983)
11. Battenfeld, I., Schröder, M., Simpson, A.: A convenient category of domains. Electronic Notes in Theoretical Computer Science **172**, 69–99 (2007)
12. Bauer, A.: A relationship between equilogical spaces and Type Two Effectivity. Mathematical Logic Quarterly **48(1)**, 1–15 (2002)
13. Bauer, A.: On the failure of fixed-point theorems for chain-complete lattices in the effective topos. Theoretical Computer Science **430**, 43–50 (2012)
14. Bauer, A., Stone, C.A.: RZ: a tool for bringing constructive and computable mathematics closer to programming practice. Journal of Logic and Computation **19(1)**, 17–43 (2009)
15. Bauer, A., Escardó, M.H., Simpson, A.: Comparing functional paradigms for exact real-number computation. In: Proceedings of the 29th International Colloquium on Automata, Languages and Programming, pp. 488-500. Springer, London (2002)
16. Beeson, M.: Foundations of Constructive Mathematics. Springer, Berlin Heidelberg (1985)
17. Berardi, S., Bezem, M., Coquand, T.: On the computational content of the axiom of choice. Journal of Symbolic Logic **63(2)**, 600–622 (1998)
18. Berger, U.: Totale Objekte und Mengen in der Bereichtheorie. PhD thesis, Ludwig Maximilians-Universität, München (1990)
19. Berger, U.: Total sets and objects in domain theory. Annals of Pure and Applied Logic **60(2)**, 91–117 (1993)

20. Berger, U.: Continuous functionals of dependent and transfinite types. In: Cooper, S.B., Truss, J.K. (eds.) Models and Computability, pp. 1-22. Cambridge University Press (1999)
21. Berger, U.: Computability and totality in domains. Mathematical Structures in Computer Science **12(3)**, 281–294 (2002)
22. Berger, U.: Minimisation vs. recursion on the partial continuous functionals. In: Gärdenfors, P., Woleński, J., Kijania-Placek, K. (eds.) In the Scope of Logic, Methodology and Philosophy of Science: Volume One of the 11th International Congress of Logic, Methodology and Philosophy of Science, Cracow, August 1999, pp. 57-64. Kluwer, Dordrecht (2002)
23. Berger, U.: Continuous semantics for strong normalization. Mathematical Structures in Computer Science **16(5)**, 751–762 (2006)
24. Berger, U: From coinductive proofs to exact real arithmetic. In: Grädel, E. and Kahle, R. (eds.) Computer Science Logic: 23rd International Workshop, CSL 2009, 18th Annual Conference of the EACSL, Coimbra, Portugal, September 7-11, 2009, Proceedings, pp. 132-146. Springer (2009)
25. Berger, U., Oliva, P.: Modified bar recursion and classical dependent choice. In: Baaz, M., Friedman, S.D., Krajicek, J. (eds.) Logic Colloquium '01: Proceedings of the Annual European Summer Meeting of the Association for Symbolic Logic, Vienna, 2001, pp. 89-107. ASL (2005)
26. Berger, U., Oliva, P.: Modified bar recursion. Mathematical Structures in Computer Science **16(2)**, 163–183 (2006)
27. Berger, U., Schwichtenberg, H.: An inverse of the evaluation functional for typed λ-calculus. In: Vemuri, R. (ed.) Proceedings of the Sixth Annual IEEE Symposium on Logic in Computer Science, pp. 203-211. IEEE, Los Alamitos (1991)
28. Berger, U., Schwichtenberg, H., Seisenberger, M.: The Warshall algorithm and Dickson's lemma: Two examples of realistic program extraction. Journal of Automated Reasoning **26(2)**, 205–221 (2001)
29. Berger, U., Berghofer, S., Letouzey, P., Schwichtenberg, H.: Program extraction from normalization proofs. Studia Logica **82**, 25–49 (2006)
30. Berger, U., Miyamoto, K., Schwichtenberg, H., Seisenberger, M.: Minlog: A tool for program extraction supporting algebras and coalgebras, In: Proceedings of the 4th International Conference on Algebra and Coalgebra in Computer Science, pp. 393–399. Springer, Heidelberg (2011)
31. Bergstra, J.: Continuity and Computability in Finite Types. PhD thesis, University of Utrecht (1976)
32. Berry, G.: Stable models of typed λ-calculi. In: Proceedings of the Fifth Colloquium on Automata, Languages and Programming, pp. 72-89. Springer, London (1978)
33. Berry, G., Curien, P.-L.: Sequential algorithms on concrete data structures. Theoretical Computer Science **20(3)**, 265–321 (1982)
34. Berry, G., Curien, P.-L., Lévy, J.-J.: Full abstraction for sequential languages: the state of the art. In: Nivat, M., Reynolds, J.C. (eds.) Algebraic methods in semantics, pp. 89-132. Cambridge University Press (1985)
35. Bezem, M.: Isomorphisms between HEO and $\mathrm{HRO}^{\mathrm{E}}$, ECF and $\mathrm{ICF}^{\mathrm{E}}$. Journal of Symbolic Logic **50(2)**, 359–371 (1985)
36. Birkedal, L., van Oosten, J., Relative and modified relative realizability. Annals of Pure and Applied Logic **118(1-2)**, 115–132 (2002)
37. Bishop, E.: Foundations of Constructive Analysis. Academic Press, New York (1967)
38. Bloom, B., Riecke, J.G., LCF should be lifted. In: Rus, T. (ed.) Proceedings of Conference on Algebraic Methodology and Software Technology, pp. 133-136. Department of Computer Science, University of Iowa (1989)
39. Böhm, C., Gross, W.: Introduction to the CUCH. In: Caianiello, E. (ed.) Automata Theory, pp. 35-65. Academic Press, New York (1966)
40. Brouwer, L.E.J.: Over de onbetrouwbaarheid der logische principes (On the unreliability of the principles of logic). Tijdschrift voor Wijsbegeerte **2**, 152–158 (1908), English translation in [42], pp. 107-111

41. Brouwer, L.E.J.: Über Definitionsbereiche von Funktionen (On the domains of definition of functions). Mathematische Annalen **97**, 60–75 (1927), English translation in [42], pp. 390-405
42. Brouwer, L.E.J.: Collected Works, Volume 1 (Heyting, A., ed.). North-Holland, Amsterdam (1975)
43. Bucciarelli, A.: Another approach to sequentiality: Kleene's unimonotone functions. In: Brookes, S. (ed.) Mathematical Foundations of Programming Semantics: 9th International Conference, New Orleans, USA, April 1993, Proceedings, pp. 333-358. Springer (1994)
44. Bucciarelli, A.: Degrees of parallelism in the continuous type hierarchy. Theoretical Computer Science **177(1)**, 59–71 (1997)
45. Bucciarelli, A., Ehrhard, T., Sequentiality and strong stability. In: Vemuri, R. (ed.) Proceedings of the Sixth Annual IEEE Symposium on Logic in Computer Science, pp. 138-145. IEEE, Los Alamitos (1991)
46. Bucciarelli, A., Ehrhard, T., Manzonetto, G.: Not enough points is enough. In: Duparc, J., Henzinger, T.A. (eds.) CSL'07: Proceedings of 16th Computer Science Logic, pp. 298-312. Springer (2007)
47. Cardone, F., Hindley, J.R.: Lambda-calculus and combinators in the 20th century. In: Gabbay, D.M., Woods, J. (eds.) Handbook of the History of Logic Volume 5, Logic from Russell to Church, pp. 723-817. Elsevier, Amsterdam (2009)
48. Cartwright, R., Felleisen, M.: Observable sequentiality and full abstraction. In: Proceedings of the 19th ACM SIGPLAN-SIGACT Symposium on Principles of Programming Languages, pp. 328-342. ACM Press, New York (1992)
49. Cartwright, R., Curien, P.-L., Felleisen, M.: Fully abstract semantics for observably sequential languages. Information and Computation **111(2)**, 297–401 (1994)
50. Cenzer, D: Inductively defined sets of reals. Bulletin of the American Mathematical Society **80(3)**, 485–487 (1974)
51. Church, A.: A set of postulates for the foundation of logic, parts I and II. Annals of Mathematics, Series 2 **33**, 346–366 (1932) and **34**, 839–864 (1933)
52. Church, A.: An unsolvable problem in elementary number theory. American Journal of Mathematics **58(2)**, 345–363 (1936)
53. Church, A.: A formulation of the simple theory of types. Journal of Symbolic Logic **5(2)**, 56–68 (1940)
54. Cockett, R., Hofstra, P.: Introduction to Turing categories. Annals of Pure and Applied Logic **156(2-3)**, 183–209 (2008)
55. Cockett, R., Hofstra, P.: Categorical simulations. Journal of Pure and Applied Algebra **214(10)**, 1835–1853 (2010)
56. Cockett, R., Hofstra, P.: An introduction to partial lambda algebras. Draft paper available from http://mysite.science.uottawa.ca/phofstra/lambda.pdf (2007)
57. Colson, L., Ehrhard, T.: On strong stability and higher-order sequentiality. In: Abramsky, S. (ed.) Proceedings of the Ninth Annual IEEE Symposium on Logic in Computer Science, pp. 103-108. IEEE (1994)
58. Cook, S.: Computability and complexity of higher type functions. In: Moschovakis, Y. (ed.) Logic from Computer Science: Proceedings of a Workshop held November 1989, pp. 51-72. Springer, New York (1990)
59. Cooper, S.B.: Computability Theory. Chapman & Hall/CRC (2004)
60. Curien, P.-L.: Categorical combinators, sequential algorithms and functional programming (2nd ed.) Birkhäuser, Boston (1993)
61. Curien, P.-L.: On the symmetry of sequentiality. In: Brookes, S. (ed.) Mathematical Foundations of Programming Semantics: 9th International Conference, New Orleans, USA, April 1993, Proceedings, pp. 29-71. Springer (1994)
62. Curien, P.-L.: Notes on game semantics. Course notes available from http://www.pps.univ-paris-diderot.fr/ curien/Game-semantics.pdf (2006)
63. Curien, P.-L.: Sequential algorithms as bistable maps. In: Bertot, Y., Huet, G., Lévy, J.-J., Plotkin, G.: From Semantics to Computer Science: Essays in Honour of Gilles Kahn, pp. 51-70. Cambridge University Press (2009)

64. Curry, H.B.: Grundlagen der kombinatorischen Logik. American Journal of Mathematics **52(3)**, 509–536 (1930)
65. Cutland, N.J.: Computability: An Introduction to Recursive Function Theory. Cambridge University Press (1980)
66. di Gianantonio, P.: A Functional Approach to Computability on Real Numbers. PhD thesis, University of Pisa (1993)
67. di Gianantonio, P.: Real number computability and domain theory. Information and Computation **127(1)**, 11–25 (1996)
68. Ehrhard, T.: Hypercoherences: a strongly stable model of linear logic. Mathematical Structures in Computer Science **3(4)**, 365–385 (1993)
69. Ehrhard, T.: Projecting sequential algorithms on strongly stable functions. Annals of Pure and Applied Logic **77(3)**, 201–244 (1996)
70. Engeler E.: Representation of varieties in combinatory algebras. Algebra Universalis **25(1)**, 85–95 (1988)
71. Ershov, Yu.L.: Computable functionals of finite types. Algebra and Logic **11(4)**, 203–242 (1972)
72. Ershov, Yu.L.: Maximal and everywhere defined functionals. Algebra and Logic **13(4)**, 210–225 (1974)
73. Ershov, Yu.L.: Hereditarily effective operations. Algebra and Logic **15(6)**, 400–409 (1976)
74. Ershov, Yu.L.: Model $\mathbb{C}$ of partial continuous functionals. In: Gandy, R.O., Hyland, J.M.E. (eds.) Logic Colloquium 76, pp. 455-467. North-Holland, Amsterdam (1977)
75. Ershov, Yu.L.: Theory of numberings. In: Griffor, E.R. (ed.) Handbook of Computability Theory, pp. 473-503. Elsevier (1999)
76. Escardó, M.H.: PCF extended with real numbers. Theoretical Computer Science **162(1)**, 79–115 (1996)
77. Escardó, M.H.: PCF Extended with Real Numbers: A Domain-Theoretic Approach to Higher-Order Exact Real Number Computation. PhD Thesis, Imperial College, University of London (1997)
78. Escardó, M.H.: A metric model of PCF. Presented at Workshop on Realizability Semantics and Applications, Trento, 1999. Preprint at http://www.cs.bham.ac.uk/ mhe/papers/ (1999)
79. Escardó, M.H.: Synthetic topology of data types and classical spaces. Electronic Notes in Theoretical Computer Science **87**, 21–156 (2004)
80. Escardó, M.H.: Exhaustible sets in higher-type computation. Logical Methods in Computer Science **4(3)**, 37 pages (2008)
81. Escardó, M.H.: Infinite sets that admit fast exhaustive search. In: 22nd Annual IEEE Symposium on Logic in Computer Science, 2007, Wroclaw, Poland, Proceedings, pp. 443-452. IEEE (2007)
82. Escardó, M.H.: Algorithmic solution of higher type equations. Journal of Logic and Computation **23(4)**, 839–854 (2013)
83. Escardó, M.H.: Infinite sets that satisfy the principle of omniscience in any variety of constructive mathematics. Journal of Symbolic Logic **78(3)**, 764–784 (2013)
84. Escardó, M.H., Ho, W.K.: Operational domain theory and topology of sequential programming languages. Information and Computation **207**, 411–437 (2009)
85. Escardó, M.H., Oliva, P.: Selection functions, bar recursion, and backward induction. Mathematical Structures in Computer Science **20(2)**, 127–168 (2010)
86. Escardó, M.H., Oliva, P.: Sequential games and optimal strategies. Proceedings of the Royal Society A **467(2130)**, 1519–1545 (2011)
87. Escardó, M.H., Oliva, P.: Bar recursion and products of selection functions. Journal of Symbolic Logic **80(1)**, 1–28 (2015)
88. Escardó, M.H., Hofmann, M., Streicher, T.: On the non-sequential nature of the interval-domain model of real-number computation. Mathematical Structures in Computer Science **14(6)**, 803–814 (2004)
89. Escardó, M.H., Lawson, J., Simpson, A.: Comparing cartesian closed categories of (core) compactly generated spaces. Topology and its Applications **143(1-3)**, 105–145 (2004)

90. Faber, E., van Oosten, J.: More on geometric morphisms between realizability toposes. Theory and Applications of Categories **29**, 874–895 (2014)
91. Faber, E., van Oosten, J.: Effective operations of type 2 in PCAs. Submitted for publication (2014)
92. Feferman, S.: A language and axioms for explicit mathematics. In: Crossley, J.N. (ed.) Algebra and Logic: Papers from the 1974 Summer Research Institute of the Australian Mathematical Society, pp. 87–139. Springer, Berlin (1975)
93. Feferman, S.: Inductive schemata and recursively continuous functionals. In: Gandy, R.O., Hyland, J.M.E. (eds.) Logic Colloquium 76, pp. 373-392. North-Holland, Amsterdam (1977)
94. Feferman, S.: Theories of finite type related to mathematical practice. In: Barwise, J. (ed.) Handbook of Mathematical Logic, pp. 913-971. Elsevier, Amsterdam (1977)
95. Felleisen, M.: On the expressive power of programming languages. Science of Computer Programming **17(1-3)**, 35–75 (1991)
96. Fenstad, J.E.: General Recursion Theory. Springer, Berlin (1980)
97. Friedberg, R.M.: Un contre-exemple relatif aux fonctionelles récursives, Comptes Rendus de l'Académie des Sciences, Paris **247**, 852–854 (1958)
98. Friedman, S.D.: An introduction to β-recursion theory. In: Fenstad, J.E., Gandy, R.O., Sacks, G.E. (eds.) Generalized Recursion Theory II: Proceedings of the 1977 Oslo Symposium, pp. 111-126. North-Holland, Amsterdam (1978)
99. Gandy, R.O.: Effective operations and recursive functionals (abstract). In: Löb, M.H.: Meeting of the Association for Symbolic Logic, Leeds 1962. Journal of Symbolic Logic **27(3)**, pp. 378-379 (1962)
100. Gandy, R.O.: General recursive functionals of finite types and hierarchies of functionals. Annales de la Faculté des Sciences de l'Université de Clermont-Ferrand **35**, 202–242 (1967)
101. Gandy, R.O., Hyland, J.M.E.: Computable and recursively countable functions of higher type. In: Gandy, R.O., Hyland, J.M.E. (eds.) Logic Colloquium 76, pp. 407-438. North-Holland, Amsterdam (1977)
102. Ghica, D.R.: Applications of game semantics: from program analysis to hardware synthesis. In: 24th Annual IEEE Symposium on Logic in Computer Science, Los Angeles, Proceedings, pp. 17-26. IEEE (2009)
103. Ghica, D.R., Tzevelekos, N.: A system-level game semantics. Electronic Notes on Theoretical Computer Science **286**, 191–211 (2012)
104. Girard, J.-Y.: The System *F* of variable types, fifteen years later. Theoretical Computer Science **45(2)**, 159–192 (1986)
105. Girard, J.-Y.: Proof Theory and Logical Complexity: volume I. Bibliopolis, Naples (1987)
106. Girard, J.-Y.: Linear logic. Theoretical Computer Science **50(1)**, 1–101 (1987)
107. Girard, J.-Y., Lafont, Y., Taylor, P.: Proofs and Types. Cambridge University Press (1989)
108. Gödel, K.: Über eine bisher noch nicht benützte Erweiterung des finiten Standpunktes (On a hitherto unutilized extension of the finitary standpoint). Dialectica **12**, 280–287 (1958), reprinted with English translation in [109]
109. Gödel, K.: Collected Works, Volume II: Publications 1938-1974. Oxford University Press (1990)
110. Grilliot, T.J.: Selection functions for recursive functionals. Notre Dame Journal of Formal Logic **10**, 225–234 (1969)
111. Grilliot, T.J.: On effectively discontinuous type-2 objects. Journal of Symbolic Logic **36(2)**, 245–248 (1971)
112. Grzegorczyk, A.: On the definitions of computable real continuous functions. Fundamenta Mathematicae **44**, 61–71 (1957)
113. Harmer, R., Hyland, J.M.E., Melliès, P.-A.: Categorical combinatorics for innocent strategies. In: 22nd Annual IEEE Symposium on Logic in Computer Science, 2007, Wroclaw, Poland, Proceedings, pp. 379-388. IEEE (2007)
114. Harrington, L.A., MacQueen, D.B.: Selection in abstract recursion theory. Journal of Symbolic Logic **41(1)**, 153–158 (1976)
115. Henkin, L.: Completeness in the theory of types. Journal of Symbolic Logic **15(2)**, 81–91 (1950)

116. Hinata, S., Tugué, T.: A note on continuous functionals. Annals of the Japan Association for the Philosophy of Science **3**, 138–145 (1969)
117. Hinman, P.G.: Recursion-Theoretic Hierarchies. Springer, Berlin (1978)
118. Hinman, P.G.: Recursion on abstract structures. In: Griffor, E.R. (ed.) Handbook of Computability Theory, pp. 315-359. Elsevier (1999)
119. Honsell, F., Sannella, D.: Prelogical relations. Information and Computation **178(1)**, 23–43 (2002)
120. Hyland, J.M.E.: Recursion Theory on the Countable Functionals. PhD thesis, University of Oxford (1975)
121. Hyland, J.M.E.: Filter spaces and continuous functionals. Annals of Mathematical Logic **16(2)**, 101–143 (1979)
122. Hyland, J.M.E.: First steps in synthetic domain theory. In: Carboni, A., Pedicchio, M.C., Rosolini, G. (eds.) Category Theory: Proceedings of the International Conference in Como, 1990, pp. 131-156. Springer, Berlin (1991)
123. Hyland, J.M.E.: Game semantics. In: Pitts, A., Dybjer, P. (eds.) Semantics and Logics of Computation, pp. 131-184. Cambridge University Press (1997)
124. Hyland, J.M.E., Ong, C.-H.L.: On full abstraction for PCF: I, II, and III. Information and computation **163**, 285–408 (2000)
125. Irwin, R.J., Kapron, B., Royer, J.S.: On characterizations of the basic feasible functionals I. Journal of Functional Programming **11(1)**, 117–153 (2001)
126. Johnstone, P.T., Robinson, E.: A note on inequivalence of realizability toposes. Mathematical Proceedings of the Cambridge Philosophical Society **105(1)**, 1–3 (1989)
127. Jung, A., Stoughton, A.: Studying the fully abstract model of PCF within its continuous function model. In: Bezem, M., Groote, J.F. (eds.) Proceedings of the International Conference on Typed Lambda Calculi and Applications, pp. 230-244. Springer, London (1993)
128. Kahn, G., Plotkin, G.D.: Concrete domains. Theoretical Computer Science **121**, 178–277 (1993), first appeared in French as INRIA technical report (1978)
129. Kanneganti, R., Cartwright, R., Felleisen, M.: SPCF: its model, calculus, and computational power. In: Proceedings of REX Workshop on Semantics and Concurrency, pp. 318-347. Springer, London (1993)
130. Kechris, A.S., Moschovakis, Y.N., Recursion in higher types. In: Barwise, J. (ed.) Handbook of Mathematical Logic, pp. 681-737. Elsevier, Amsterdam (1977)
131. Kennaway, J.R., Klop, J.W., Sleep, M.R., de Vries, F.J.: Infinitary lambda calculus. Theoretical Computer Science **175(1)**, 93–125 (1997)
132. Kierstead, D.P.: A semantics for Kleene's j-expressions. In Barwise, J., Keisler, H.J., Kunen, K. (eds.) The Kleene Symposium, pp. 353-366. North-Holland, Amsterdam (1980)
133. Kleene, S.C.: λ-definability and recursiveness. Duke Mathematical Journal **2**, 340–353 (1936)
134. Kleene, S.C.: Recursive predicates and quantifiers. Transactions of the American Mathematical Society **53(1)**, 41–73 (1943)
135. Kleene, S.C.: On the interpretation of intuitionistic number theory. Journal of Symbolic Logic **10(4)**, 109–124 (1945)
136. Kleene, S.C.: Introduction to metamathematics. Van Nostrand, New York (1952)
137. Kleene, S.C.: Arithmetical predicates and function quantifiers. Transactions of the American Mathematical Society **79(2)**, 312–340 (1955)
138. Kleene, S.C.: Recursive functionals and quantifiers of finite types I. Transactions of the American Mathematical Society **91(1)**, 1–52 (1959)
139. Kleene, S.C.: Recursive functionals and quantifiers of finite types II. Transactions of the American Mathematical Society **108(1)**, 106–142 (1963)
140. Kleene, S.C.: Countable functionals. In: Heyting, A. (ed.) Constructivity in Mathematics: Proceedings of the Colloquium held in Amsterdam, 1957, pp. 81-100. North-Holland, Amsterdam (1959)
141. Kleene, S.C.: Recursive functionals and quantifiers of finite types revisited I. In: Fenstad, J.E., Gandy, R.O., Sacks, G.E. (eds.) Generalized Recursion Theory II: Proceedings of the 1977 Oslo Symposium, pp. 185-222. North-Holland, Amsterdam (1978)

142. Kleene, S.C.: Recursive functionals and quantifiers of finite types revisited II. In Barwise, J., Keisler, H.J., Kunen, K. (eds.) The Kleene Symposium, pp. 1-29. North-Holland, Amsterdam (1980)
143. Kleene, S.C.: Recursive functionals and quantifiers of finite types revisited III. In: G. Metakides (ed.) Patras Logic Symposion, 1980, Proceedings, pp. 1-40. North-Holland, Amsterdam (1982)
144. Kleene, S.C.: Unimonotone functions of finite types (Recursive functionals and quantifiers of finite types revisited IV). In: A. Nerode and R.A. Shore (eds.) Proceedings of the AMS-ASL Summer Institute on Recursion Theory, pp. 119-138. American Mathematical Society, Providence (1985)
145. Kleene, S.C.: Recursive functionals and quantifiers of finite types revisited V. Transactions of the American Mathematical Society **325**, 593–630 (1991)
146. Kleene, S.C., Vesley, R.E.: The Foundations of Intuitionistic Mathematics, especially in relation to Recursive Functions. North-Holland, Amsterdam (1965)
147. Kohlenbach, U.: Theory of Majorizable and Continuous Functionals and their Use for the Extraction of Bounds from Non-Constructive Proofs: Effective Moduli of Uniqueness for Best Approximations from Ineffective Proofs of Uniqueness. PhD thesis, Frankfurt (1990)
148. Kohlenbach, U.: Foundational and mathematical uses of higher types. In: Sieg, W., Sommer, R., Talcott, C. (eds.) Reflections on the Foundations of Mathematics: Essays in Honor of Solomon Feferman, pp. 92-116. A K Peters (2002)
149. Kohlenbach, U.: Applied Proof Theory: Proof Interpretations and their Use in Mathematics. Springer (2008)
150. Koymans, K.: Models of the lambda calculus. Information and Control **52(3)**, 306–332 (1982)
151. Kreisel, G.: Constructive Mathematics. Notes of a course given at Stanford University (1958)
152. Kreisel, G.: Interpretation of analysis by means of constructive functionals of finite types. In: Heyting, A. (ed.) Constructivity in Mathematics: Proceedings of the Colloquium held in Amsterdam, 1957, pp. 101-128. North-Holland, Amsterdam (1959)
153. Kreisel, G.: On weak completeness of intuitionistic predicate logic. Journal of Symbolic Logic **27(2)**, 39–58 (1962)
154. Kreisel, G.: Some reasons for generalizing recursion theory. In: Gandy, R.O., Yates, C.M.E. (eds.) Logic Colloquium '69: Proceedings of the Summer School and Colloquium in Mathematical Logic, Manchester, pp. 139-198. North-Holland, Amsterdam (1971)
155. Kreisel, G. Lacombe, D., Shoenfield, J.R.: Partial recursive functionals and effective operations. In: Heyting, A. (ed.) Constructivity in Mathematics: Proceedings of the Colloquium held in Amsterdam, 1957, pp. 290-297. North-Holland, Amsterdam (1959)
156. Kristiansen, L.: Higher types, finite domains and resource-bounded Turing machines. Journal of Logic and Computation **22(2)**, 281–304 (2012)
157. Kristiansen, L., Normann, D.: Interpreting higher computations as types with totality. Archive for Mathematical Logic **33(4)**, 243–259 (1994)
158. Kristiansen, L., Normann, D.: Semantics for some constructors of type theory. Symposia Gaussiana, Proceedings of the 2nd Gauss Symposium, pp. 201-224. De Gruyter, Berlin (1995)
159. Kristiansen, L., Normann, D.: Total objects in inductively defined types. Archive for Mathematical Logic **36(6)**, 405–436 (1997)
160. Kuratowski, C.: Topologie, Volume I. Polska Akademia Nauk, Warszawa (1952)
161. Lacombe, D.: Remarques sur les opérateurs récursifs et sur les fonctions récursives d'une variable réelle. Comptes Rendus de l'Académie des Sciences, Paris **241**, 1250–1252 (1955)
162. Laird, J.: A Semantic Analysis of Control. PhD thesis, University of Edinburgh (1998)
163. Laird, J.: Sequentiality in bounded biorders. Fundamenta Informaticae **65(1-2)**, 173–191 (2004)
164. Laird, J.: A calculus of coroutines. Theoretical Computer Science **350(2-3)**, 275–291 (2006)
165. Laird, J.: Bistable biorders: a sequential domain theory. Logical Methods in Computer Science **3(2)**, 22 pages (2007)
166. Lamarche, F.: Sequentiality, games and linear logic. In: Proceedings of CLICS Workshop, Aarhus University (1992)

167. Lambek, J., Scott, P.J.: Introduction to higher order categorical logic. Cambridge University Press (1988)
168. Lassen, S.B., Levy, P.: Typed normal form bisimulation. In: Duparc, J., Henzinger, T.A. (eds.) CSL'07: Proceedings of 16th Computer Science Logic, pp. 283-297. Springer (2007)
169. Landin, P.J.: Correspondence between ALGOL 60 and Church's lambda notation, parts I and II. Communications of the ACM **8(2,3)**, 89–101 and 158–167 (1965)
170. Laurent, O.: On the denotational semantics of the untyped lambda-mu calculus. Preprint at http://perso.ens-lyon.fr/olivier.laurent/lmmodels.pdf (2004)
171. Loader, R.: Notes on simply typed lambda calculus. University of Edinburgh report ECS-LFCS-98-381 (1998)
172. Loader, R.: Finitary PCF is not decidable. Theoretical Computer Science **266(1-2)**, 341–364 (2001)
173. Longley, J.R.: Realizability Toposes and Language Semantics. PhD thesis ECS-LFCS-95-332, University of Edinburgh (1995)
174. Longley, J.R.: Matching typed and untyped realizability. Electronic Notes in Theoretical Computer Science **23(1)**, 74–100 (1999)
175. Longley, J.R.: When is a functional program not a functional program? In: ICFP '99: Proceedings of the Fourth ACM SIGPLAN International Conference on Functional Programming, Paris, pp. 1-7. ACM Press (1999)
176. Longley, J.R.: The sequentially realizable functionals. Annals of Pure and Applied Logic **117(1)**, 1–93 (2002)
177. Longley, J.R.: Universal types and what they are good for. In: Zhang, G.-Q., Lawson, J., Liu, Y.-M., Luo, M.-K. (eds.) Domain Theory, Logic and Computation: Proceedings of the 2nd International Symposium on Domain Theory, pp. 25-63. Kluwer, Boston (2004)
178. Longley, J.R.: Notions of computability at higher types I. In: Cori, R., Razborov, A., Todorčević, S., Wood, C.: Logic Colloquium 2000: Proceedings of the Annual European Summer Meeting of the Association for Symbolic Logic, Paris, 2000, pp. 32-142. ASL, Illinois (2005)
179. Longley, J.R.: On the ubiquity of certain total type structures. Mathematical Structures in Computer Science **17(5)**, 841–953 (2007)
180. Longley, J.R.: Interpreting localized computational effects using operators of higher type. In: A. Beckmann, C. Dimitracopoulos and B. Loewe (eds.) Logic and Theory of Algorithms: Fourth Conference on Computability in Europe, Athens, Proceedings, pp. 389–402. Springer (2008).
181. Longley, J.R.: The Eriskay project homepage. http://homepages.inf.ed.ac.uk/jrl/Eriskay/ (2009)
182. Longley, J.R.: Some programming languages suggested by game models. Electronic Notes in Theoretical Computer Science **249**, 117-134 (2009)
183. Longley, J.R.: Computability structures, simulations and realizability. Mathematical Structures in Computer Science **24(2)**, 49 pages (2014) doi: 10.1017/S0960129513000182
184. Longley, J.R.: Bar recursion is not T+min definable. Informatics Research Report EDI-INF-RR-1420, University of Edinburgh (2015)
185. Longley, J.R.: The recursion hierarchy for PCF is strict. Informatics Research Report EDI-INF-RR-1421, University of Edinburgh (2015)
186. Longley, J.R., Plotkin, G.D.: Logical full abstraction and PCF. In: Ginzburg, J. Khasidashvili, Z., Vogel, C., Levy, J.-J., Vallduví, E. (eds.) Tbilisi Symposium on Logic, Language and Computation, pp. 333-352. SiLLI/CSLI (1997).
187. Longley, J.R., Pollack, R.A.: Reasoning about CBV functional programs in Isabelle/HOL. In: Slind, K., Bunker, A., Gopalakrishnan, G. (eds.) Theorem Proving in Higher Order Logics, 17th International Conference, Utah, Proceedings, pp. 201-216. Springer (2004)
188. Longo, G. Moggi, E.: The hereditarily partial functionals and recursion theory in higher types. Journal of Symbolic Logic **49(4)**, 1319–1332 (1984)
189. Mac Lane, S.: Categories for the Working Mathematician (second edition). Springer (1978)
190. McCarthy, J.: Recursive functions of symbolic expressions and their computation by machine, Part I. Communications of the ACM **3(4)**, 184–195 (1960)

191. McCusker, G.: Games and Full Abstraction for a Functional Metalanguage with Recursive Types. Springer, London (1998)
192. McCusker, G.: A graph model for imperative computation. Logical Methods in Computer Science **6(1)**, 35 pages (2010)
193. Milner, R.: Fully abstract models of typed λ-calculi. Theoretical Computer Science **4(1)**, 1–22 (1977)
194. Mitchell, J.: Foundations for Programming Languages. MIT Press (1996)
195. Moggi, E.: The Partial Lambda Calculus. PhD thesis ECS-LFCS-88-63, University of Edinburgh (1988)
196. Moldestad, J.: Computations in Higher Types. Springer, New York (1977)
197. Moldestad, J., Normann, D.: Models of recursion theory. Journal of Symbolic Logic **41(4)**, 719–729 (1976)
198. Moschovakis, Y.N.: Hyperanalytic predicates. Transactions of the American Mathematical Society **129(2)**, 249–282 (1967)
199. Moschovakis, Y.N.: Axioms for computation theories — first draft. In: Gandy, R.O., Yates, C.M.E. (eds.) Logic Colloquium '69: Proceedings of the Summer School and Colloquium in Mathematical Logic, Manchester, pp. 199-255. North-Holland, Amsterdam (1971)
200. Moschovakis, Y.N.: On the basic notions in the theory of induction. In: Butts, R.E., Hintikka, J. (eds.) Logic, Foundations of Mathematics and Computability Theory: Proceedings of the Fifth International Congress of Logic, Methodology and Philosophy of Science, London, Ontario, pp. 207-236. Reidel, Dordrecht (1977)
201. Moschovakis, Y.N.: Descriptive Set Theory (second edition). American Mathematical Society (2009)
202. Myhill, J.R., Shepherdson, J.C.: Effective operations on partial recursive functions. Zeitschrift für mathematische Logik und Grundlagen der Mathematik **1** 310–317 (1955)
203. Nickau, H.: Hereditarily sequential functionals. In: Third International Symposium on Logical Foundations of Computer Science, St. Petersburg, 1994, pp. 253-264. Springer, Heidelberg (1995)
204. Niggl, K.-H.: M^{ω} considered as a programming language. Annals of Pure and Applied Logic **99(1-3)**, 73–92 (1999)
205. Normann, D.: Set recursion. In: Fenstad, J.E., Gandy, R.O., Sacks, G.E. (eds.) Generalized Recursion Theory II: Proceedings of the 1977 Oslo Symposium, pp. 303-320. North-Holland, Amsterdam (1978)
206. Normann, D.: A classification of higher type functionals. In: Proceedings from the 5th Scandinavian Logic Symposium, Aalborg 1979, pp. 301-308. Aalborg University Press (1979)
207. Normann, D.: Recursion on the Countable Functionals. Springer, Berlin (1980)
208. Normann, D.: The continuous functionals, computations, recursions and degrees. Annals of Mathematical Logic **21(1)**, 1–26 (1981)
209. Normann, D.: Countable functionals and the projective hierarchy. Journal of Symbolic Logic **46(2)**, 209–215 (1981)
210. Normann, D.: External and internal algorithms on the continuous functionals. In: G. Metakides (ed.) Patras Logic Symposion, 1980, Proceedings, pp. 137-144. North-Holland, Amsterdam (1982)
211. Normann, D.: Closing the gap between the continuous functionals and recursion in ${}^3E$. Archive for Mathematical Logic **36(8)**, 405–436 (1997)
212. Normann, D.: Computability over the partial continuous functionals. Journal of Symbolic Logic **65(3)**, 1133–1142 (2000)
213. Normann, D.: Comparing hierarchies of total functionals. Logical Methods in Computer Science **1(2)**, 28 pages (2005)
214. Normann, D.: Sequential functionals of type 3. Mathematical Structures in Computer Science **16(2)**, 279–289 (2006)
215. Normann, D.: Computing with functionals—computability theory or computer science? Bulletin of Symbolic Logic **12(1)**, 43–59 (2006)
216. Normann, D.: A rich hierarchy of functionals of finite types. Logical Methods in Computer Science **5(3)**, 21 pages (2009)

217. Normann, D.: Banach spaces as data types. Logical Methods in Computer Science **7(2)**, 20 pages (2011)
218. Normann, D.: Higher generalizations of the Turing model. In R. Downey (ed.) Turing's Legacy: Developments from Turing's Ideas in Logic, pp. 397-433. ASL, Cambridge, MA (2014)
219. Normann, D.: The extensional realizability model of continuous functionals and three weakly non-constructive classical theorems. Logical Methods in Computer Science **11(1)**, 27 pages (2015)
220. Normann, D., Rørdam, C.: The computational power of M^{ω}. Mathematical Logic Quarterly **48(1)**, 117–124 (2002)
221. Normann, D., Sazonov, V.Yu.: The extensional ordering of the sequential functionals. Annals of Pure and Applied Logic **163(5)**, 575–603 (2012)
222. Normann, D., Palmgren, E., Stoltenberg-Hansen, V.: Hyperfinite type structures. Journal of Symbolic Logic **64(3)**, 1216–1242 (1999)
223. O'Hearn, P., Riecke, J.: Kripke logical relations and PCF. Information and Computation **120(1)**, 107–116 (1995)
224. Okasaki, C.: Even higher-order functions for parsing, or Why would anyone ever want to use a sixth-order function? Journal of Functional Programming **8(2)**, 195–199 (1998)
225. Ong, C.-H.L.: Correspondence between operational and denotational semantics. In: Abramsky, S., Gabbay, D.M., Maibaum, T.S.E. (eds.) Handbook of Logic in Computer Science (Volume 4): Semantic Modelling, pp. 269-356. Oxford University Press (1995)
226. Ong, C.-H.L.: Verification of higher-order computation: a game-semantic approach. In: Drossopoulou, S. (ed.) 17th European Symposium on Programming, Budapest, 2008, Proceedings, pp. 299-306. Springer (2008)
227. Palmgren, E., Stoltenberg-Hansen, V.: Domain interpretations of Martin-Löf's partial type theory. Annals of Pure and Applied Logic **48(2)**, 135–196 (1990)
228. Panangaden, P.: On the expressive power of first-order boolean functions. Theoretical Computer Science **266(1-2)**, 543–567 (2001)
229. Paolini, L.: A stable programming language. Information and Computation **204(3)**, 339–375 (2006)
230. Parigot, M.: $\lambda\mu$-calculus: an algorithmic interpretation of classical natural deduction. In: Voronkov, A. (ed.) Proceedings of the International Conference on Logic Programming and Automated Reasoning, pp. 190-201. Springer, London (1992)
231. Pierce, B.C.: Basic Category Theory for Computer Scientists. MIT Press, Cambridge, MA (1991)
232. Pierce, B.C.: Types and Programming Languages. MIT Press, Cambridge, MA (2002)
233. Platek, R.A.: Foundations of Recursion Theory. PhD thesis, Stanford University (1966)
234. Plotkin, G.D.: Lambda-definability and logical relations. Memorandum SAI-RM-4, University of Edinburgh (1973)
235. Plotkin, G.D.: Call-by-name, call-by-value and the λ-calculus. Theoretical Computer Science **1(2)**, 125–159 (1975)
236. Plotkin, G.D.: LCF considered as a programming language. Theoretical Computer Science **5(3)**, 223–255 (1977)
237. Plotkin, G.D.: $\mathbb{T}^{\omega}$ as a universal domain. Journal of Computer and System Sciences **17(2)**, 209–236 (1978)
238. Plotkin, G.D.: Lambda-definability in the full type hierarchy. In: Seldin, J.P., Hindley, J.R. (eds.) To H.B. Curry: Essays on Combinatory Logic, Lambda Calculus and Formalism, pp. 363-373. Academic Press (1980)
239. Plotkin, G.D.: Full abstraction, totality and PCF. Mathematical Structures in Computer Science **9(1)**, 1–20 (1999)
240. Plotkin, G.D.: Set-theoretical and other elementary models of the λ-calculus. Theoretical Computer Science **121(1-2)**, 351-409 (1993). Subsumes unpublished memorandum: A set-theoretical definition of application, University of Edinburgh (1972)

241. Pour-El, M.B., Richards, I.: Non-computability in analysis and physics, a complete determination of the class of noncomputable linear operators. Advances in Mathematics **48(1)**, 44–74 (1983)
242. Pour-El, M.B., Richards, I.: Computability in Analysis and Physics. Springer, Berlin (1989)
243. Reynolds, J.C.: The essence of Algol. In: de Bakker, J.W., van Vliet, J.C. (eds.) Proceedings of 1981 International Symposium on Algorithmic Languages, pp. 345-372. North-Holland, Amsterdam (1981)
244. Riecke, J.: Fully abstract translations between functional languages. Mathematical Structures in Computer Science **3(4)**, 387–415 (1993)
245. Rogers, H.: Theory of Recursive Functions and Effective Computability. McGraw-Hill, New York (1967)
246. Rosolini, G.: Continuity and Effectiveness in Topoi. PhD thesis, Carnegie Mellon University (1986)
247. Sacks, G.E.: The 1-section of a type n object. In: Fenstad, J.E., Hinman, P.G. (eds.) Generalized Recursion Theory, Proceedings of the 1972 Oslo Symposium, pp. 81-96. North-Holland, Amsterdam (1974)
248. Sacks, G.E.: Higher recursion theory. Springer, Berlin (1990)
249. Sacks, G.E.: E-recursion. In: Griffor, E.R. (ed.) Handbook of Computability Theory, pp. 301-314. Elsevier (1999)
250. Saurin, A.: Böhm theorem and Böhm trees for the $\Lambda\mu$-calculus. Theoretical Computer Science **435**, 106–138 (2012)
251. Sazonov, V.Yu.: Degrees of parallelism in computations. In: Mazurkiewicz, A. (ed.) Mathematical Foundations of Computer Science 1976, Proceedings, 5th Symposium, Gdańsk, 517-523. Springer, Berlin (1976)
252. Sazonov, V.Yu.: Expressibility in D. Scott's LCF language. Algebra and Logic **15(3)**, 192–206 (1976)
253. Sazonov, V.Yu.: An inductive definition and domain theoretic properties of fully abstract models of PCF and PCF+. Logical Methods in Computer Science **3(3)**, 50 pages (2007)
254. Sazonov, V.Yu.: Natural non-DCPO domains and f-spaces. Annals of Pure and Applied Logic **159(3)**, 341–355 (2009)
255. Scarpellini, B.: A model for bar recursion of higher types. Compositio Mathematica **23**, 123–153 (1971)
256. Scott, D.S.: A type-theoretical alternative to ISWIM, CUCH, OWHY. Unpublished note (1969)
257. Scott, D.S.: Models for the λ-calculus. Manuscript (1969)
258. Scott, D.S.: Continuous lattices. In: Lawvere, F.W. (ed.) Toposes, Algebraic Geometry and Logic, pp. 97-136. Springer, Berlin (1972)
259. Scott, D.S.: Data types as lattices. SIAM Journal of Computing **5(3)**, 522–587 (1976)
260. Scott, D.S.: Relating theories of the λ-calculus. In: Seldin, J.P., Hindley, J.R. (eds.) To H.B. Curry: Essays on Combinatory Logic, Lambda Calculus and Formalism, pp. 403-450. Academic Press (1980)
261. Scott, D.S.: A type-theoretical alternative to ISWIM, CUCH, OWHY. In: A collection of contributions in honor of Corrado Böhm on the occasion of his 70th birthday. Theoretical Computer Science **121(1-2)**, 411–440 (1993), edited version of [256]
262. Schönfinkel, M.: Über die Bausteine der mathematischen Logik. Mathematische Annalen **92(3-4)**, 305–316 (1924)
263. Schröder, M.: Admissible Representations for Continuous Computations. PhD thesis, Fern-Universität Hagen (2003)
264. Schröder, M.: Extended admissibility. Theoretical Computer Science **284(2)**, 519–538 (2002)
265. Schröder, M.: The sequential topology on $\mathbb{N}^{\mathbb{N}^{\mathbb{N}}}$ is not regular. Mathematical Structures in Computer Science **19(5)**, 943–957 (2009)
266. Schröder, M.: $\mathbb{N}^{\mathbb{N}^{\mathbb{N}}}$ does not satisfy Normann's condition. ACM Computing Research Repository abs/1010.2396 (2010)
267. Schwichtenberg, H.: Classifying recursive functions. In: Griffor, E.R. (ed.) Handbook of Computability Theory, pp. 533-586. Elsevier (1999)

268. Schwichtenberg, H., Wainer, S.S.: Proofs and Computations. ASL, Cambridge University Press (2011)
269. Shore, R.A.: α-recursion theory. In: Barwise, J. (ed.) Handbook of Mathematical Logic, pp. 653-680. Elsevier, Amsterdam (1977)
270. Sieber, K.: Relating full abstraction results for different programming languages. In: Nori, K.V., Veni Madhavan, C.E., Proceedings of the Tenth Conference on Foundations of Software Technology and Theoretical Computer Science, pp. 373-387. Springer, London (1990)
271. Sieber, K.: Reasoning about sequential functions via logical relations. In: Fourman, M.P., Johnstone, P.T., Pitts, A.M. (eds.) Applications of Categories in Computer Science, pp. 258-269. Cambridge University Press (1992)
272. Simpson, A.: Lazy functional algorithms for exact real functionals. In: Brim, L., Gruska, J., Zlatuška, J. (eds.) Mathematical Foundations of Computer Science 1998: 23rd International Symposium, Brno, Proceedings, pp. 456-464. Springer, Berlin (1998)
273. Simpson, S.G.: Subsystems of Second Order Arithmetic (second edition). Cambridge University Press (2010)
274. Sipser, M.: Introduction to the Theory of Computation (third edition). Cengage Learning, Boston (2013)
275. Smith, P.: An Introduction to Gödel's Theorems (second edition). Cambridge University Press (2013)
276. Spector, C.: Provably recursive functionals of analysis: a consistency proof of analysis by an extension of principles formulated in current intuitionistic mathematics. In: Dekker, J. (ed.) Recursive Function Theory, Proceedings of Symposia in Pure Mathematics, Volume 5, pp. 1-27. AMS, Providence (1962)
277. Staples, J.: Combinator realizability of constructive finite type analysis. In: Mathias, A.R.D., Rogers, H. (eds.) Cambridge Summer School in Mathematical Logic, pp. 253–273. Springer, New York (1973)
278. Statman, R.: Logical relations and the typed λ-calculus. Information and Control **65(2-3)**, 85–97 (1985)
279. Stoltenberg-Hansen, V., Lindström, I., Griffor, E.R.: Mathematical Theory of Domains. Cambridge University Press (1994)
280. Stoltenberg-Hansen, V., Tucker, J.V.: Concrete models of computation for topological algebras. Theoretical Computer Science **219(1-2)**, 347–378 (1999)
281. Stoughton, A.: Interdefinability of parallel operators in PCF. Theoretical Computer Science **79(2)**, 357–358 (1991)
282. Streicher, T.: Semantics of Type Theory: Correctness, Completeness and Independence Results. Birkhäuser, Boston (1991)
283. Streicher, T.: Domain-theoretic Foundations of Functional Programming. World Scientific (2006)
284. Tait, W.W.: Continuity properties of partial recursive functionals of finite type. Unpublished note (1958)
285. Tait, W.W.: Intensional interpretations of functionals of finite type I. Journal of Symbolic Logic **32(2)**, 198–212 (1967)
286. Trakhtenbrot, M.B.: On representation of sequential and parallel functions. In: Bečvář, J. (ed.) Mathematical Foundations of Computer Science 1975 4th Symposium, pp. 411-417. Springer, Berlin (1975)
287. Trakhtenbrot, M.B.: Relationships between classes of monotonic functions. Theoretical Computer Science **2(2)**, 225–247 (1976)
288. Troelstra, A.S. (ed.): Metamathematical Investigation of Intuitionistic Arithmetic and Analysis. Springer, Berlin (1973)
289. Troelstra, A.S.: Realizability. In: S.R. Buss (ed.) Handbook of Proof Theory, pp. 407-473. Elsevier, Amsterdam (1998)
290. Tucker, J.V., Zucker, J.I., Computable functions and semicomputable sets on many-sorted algebras. In: Abramsky, S;, Gabbay, D., Maibaum, T. (eds.) Handbook of Logic in Computer Science, Volume V: Logical and Algebraic Methods, pp. 317-523. Oxford University Press (2000)

291. Tucker, J.V., Zucker, J.I., Abstract versus concrete computation on metric partial algebras. ACM Transactions on Computational Logic **5(4)**, 611–668 (2004)
292. Turing, A.M.: On computable numbers, with an application to the Entscheidungsproblem. Proceedings of the London Mathematical Society, Series 2, **42(1)**, 230–265 (1937)
293. Turing, A.M.: Systems of logic based on ordinals. Procedings of the London Mathematical Society, Series 2, **45(1)**, 161-228 (1939)
294. van Oosten, J.: A combinatory algebra for sequential functionals of finite type. In: Cooper, S.B., Truss, J.K. (eds.) Models and Computability, pp. 389-405. Cambridge University Press (1999)
295. van Oosten, J.: Realizability: An Introduction to its Categorical Side. Elsevier, Amsterdam (2008)
296. van Oosten, J.: Partial combinatory algebras of functions. Notre Dame Journal of Formal Logic **52(4)**, 431–448 (2011)
297. Vuillemin, J.: Proof Techniques for Recursive Programs. PhD thesis, Stanford University (1973)
298. Waagbø, G.: Domains-with-Totality Semantics for Intuitionistic Type Theory. PhD thesis, University of Oslo (1997)
299. Waagbø, G.: Denotational semantics for intuitionistic type theory using a hierarchy of domains with totality. Archive for Mathematical Logic **38(1)**, 19–60 (1999)
300. Wadler, P.: Comprehending monads. In: Proceedings of the 1990 ACM Conference on LISP and Functional Programming, pp. 61-78. ACM Press, New York (1990)
301. Wainer, S.S.: The 1-section of a non-normal type-2 object. In: Fenstad, J.E., Gandy, R.O., Sacks, G.E. (eds.) Generalized Recursion Theory II: Proceedings of the 1977 Oslo Symposium, pp. 407-417. North-Holland, Amsterdam (1978)
302. Weihrauch, K.: Computable Analysis: An Introduction. Springer, Berlin (2000)
303. Winskel, G.: Event structures. In: Reisig, W., Rozemburg, G. (eds.) Advances in Petri Nets 1986, Part II, pp. 325-392. Springer, London (1987)
304. Wolverson, N.: Game Semantics for an Object-Oriented Language. PhD thesis, University of Edinburgh, available at http://hdl.handle.net/1842/3216 (2009)
305. Zaslavskiĭ, I.D. Refutation of some theorems of classical analysis in constructive analysis (in Russian). Uspekhi Matematichekikh Nauk **10**, 209-210 (1955)

Index

For mathematical notation and terminology, only the points of definition and other occurrences of particular significance are referenced. For authors' names, all occurrences are referenced.

Printed in the United States
By Bookmasters